Water Management

GREEN CHEMISTRY AND CHEMICAL ENGINEERING

Series Editor: Sunggyu Lee
Ohio University, Athens, Ohio, USA

Hydrogen Energy and Vehicle Systems
Scott E. Grasman

Chemicals from Biomass: Integrating Bioprocesses into Chemical Production Complexes for Sustainable Development
Debalina Sengupta, Ralph W. Pike

Magneto Luminous Chemical Vapor Deposition
Hirotsugu Yasuda

Particle Technology and Applications
Sunggyu Lee, Kimberly H. Henthorn

Carbon-Neutral Fuels and Energy Carriers
Nazim Z. Muradov, T. Nejat Veziroğlu

Oxide Semiconductors for Solar Energy Conversion: Titanium Dioxide
Janusz Nowotny

Water for Energy and Fuel Production
Yatish T. Shah

Managing Biogas Plants: A Practical Guide
Mario Alejandro Rosato

The Water-Food-Energy Nexus: Processes, Technologies, and Challenges
Iqbal M. Mujtaba, Rajagopalan Srinivasan, and Nimir Elbashir

Hemicelluloses and Lignin in Biorefineries
Jean-Luc Wertz, Magali Deleu, Séverine Coppée, and Aurore Richel

Materials in Biology and Medicine
Sunggyu Lee, David Henthorn

Resource Recovery to Approach Zero Municipal Waste
Mohammad J. Taherzadeh, Tobias Richards

Hydrogen Safety
Fotis Rigas, Paul Amyotte

Nuclear Hydrogen Production Handbook
Xing L. Yan, Ryutaro Hino

Water Management: Social and Technological Perspectives
Iqbal M. Mujtaba, Thokozani Majozi, and Mutiu Kolade Amosa

For more information about this series, please visit: https://www.crcpress.com/Green-Chemistry-and-Chemical-Engineering/book-series/CRCGRECHECHE

Water Management

Social and Technological Perspectives

Edited by

Iqbal M. Mujtaba
Thokozani Majozi
Mutiu Kolade Amosa

CRC Press
Taylor & Francis Group
Boca Raton London New York

CRC Press is an imprint of the
Taylor & Francis Group, an **informa** business

CRC Press
Taylor & Francis Group
6000 Broken Sound Parkway NW, Suite 300
Boca Raton, FL 33487-2742

Printed on acid-free paper

International Standard Book Number-13: 978-1-1380-6724-0 (Hardback)
International Standard Book Number-13: 978-1-315-15877-8 (eBook)

Library of Congress Cataloging-in-Publication Data

Names: Mujtaba, I. M., editor. | Majozi, Thokozani, editor. | Amosa, Mutiu Kolade, editor.
Title: Water management : social and technological perspectives / edited by Professor Iqbal M. Mujtaba, Professor Thokozani Majozi and Dr. Mutiu Kolade Amosa.
Description: First editor. | Boca Raton : Taylor & Francis, a CRC title, part of the Taylor & Francis imprint, a member of the Taylor & Francis Group, the academic division of T&F Informa, plc, [2019] | Series: Green chemistry and chemical engineering | Includes bibliographical references and index. |
Identifiers: LCCN 2018020246 (print) | LCCN 2018029403 (ebook) | ISBN 9781351657587 (Adobe PDF) | ISBN 9781351657570 (ePub) | ISBN 9781351657563 (Mobipocket) | ISBN 9781138067240 (hardback) | ISBN 9781315158778 (ebook)
Subjects: LCSH: Water quality management--Technological innovations. | Water quality management--Social aspects.
Classification: LCC TD353 (ebook) | LCC TD353 .W346 2019 (print) | DDC 628.1--dc23
LC record available at https://lccn.loc.gov/2018020246

Visit the Taylor & Francis Web site at
http://www.taylorandfrancis.com

and the CRC Press Web site at
http://www.crcpress.com

MIX
Paper from responsible sources
FSC www.fsc.org FSC® C013056

Printed and bound in Great Britain by
TJ International Ltd, Padstow, Cornwall

Dedication

To our families and children.

Contents

SECTION III Wastewater Treatment: Membrane and Polymer Based Process

SECTION IV Wastewater Treatment: Oxidation and Electrochemical Process

SECTION V Wastewater Treatment: Adsorption Process

SECTION VIII Water Management

SECTION IX Water-Energy Nexus

Preface

Exponential growth in population and improved standards of living demand an increasing amount of freshwater and are putting serious strain on the quantity of naturally available freshwater around the world and this goes well with the Ancient Mariners' rime: "Water, water everywhere/Nor any drop to drink". Therefore, how to manage this resource carefully is not only a localised problem but globalised, too. The management of water includes: (a) cost-effective and sustainable production of freshwater from saline water by desalination; (b) wastewater (water being polluted due to industrial use) treatment and re-use; (c) efficient and cost-effective water network for distribution; and (d) effective use of water in agriculture and industries.

Science and engineering play a vital and increasing role in meeting the current and future needs of both society and the planet Earth: from water supply to waste management, by developing new technology, know-how, and practical solutions. This includes many technologies from membrane meparation using reverse osmosis (RO) to the use of nano-particles for adsorption of impurities from wastewater to the use of thermal methods for desalination. Increasing the efficiency of water use in industry, as well as in agriculture and domestic use, are also important challenges. They are all integrated into an efficient system of water production, usage and recycling.

With the above in mind, Professor Mujtaba and Professor Majozi organised the trilateral (UK–Egypt–South Africa) workshop in South Africa in September, 2016, which was funded by the British Council under the Newton Fund. The workshop focused on 'Water Management' and included both the social aspects and the technical aspects of water. This workshop explored the water-energy nexus, which is evolving as the main challenge in resources management. To this end, the workshop did not only focus on water management, but also considered the inextricable link between water and energy. Increases in environmental degradation and social pressures in recent years have necessitated the development of manufacturing processes that are conservative with respect to both these resources, while maintaining financial viability. The workshop had contributions from chemical engineers, civil engineers, mechanical engineers, biochemical engineers, water resource engineers, microbiologists, forensic scientists, social scientists and industrial chemists, clearly demonstrating that 'water is everybody's business'. Ten participants from the UK, 10 participants from Egypt, and 10 participants from South Africa (academics and industrialists) presented their stimulating and state-of-the-art research and knowledge transfer ideas for water over a period of 4 days, and the workshop was well attended by over 60 participants.

The developments in energy efficient water production, management, wastewater treatment, and social and political aspects related to water management and re-use of treated water were widely discussed in this workshop. The proposed book will give the opportunity to bring forward those social and technical discussions for wider public benefit around the globe.

The book has a total of 30 contributions (most from the workshop mentioned earlier and a few solicited) and is divided into 9 main sections:

- Section I: Social Perspective
- Section II: Freshwater by Desalination
- Section III: Wastewater Treatment: Membrane and Polymer Based Processes
- Section IV: Wastewater Treatment: Oxidation and Electrochemical Processes
- Section V: Wastewater Treatment: Adsorption Processes
- Section VI: Wastewater Treatment: Biological Processes
- Section VII: Water Networks
- Section VIII: Water Management
- Section IX: Water-Energy Nexus

Section I includes three contributions on social perspective covering security, sectarian conflict, diplomacy, economic growth, social well-being, socio-political and cultural complexities linked to water and water management.

Section II includes five contributions on making freshwater and irrigation water by desalination covering technologies such as hybrid forward and reverse osmosis, multistage flash, multi-effect evaporator and microbial cells.

Section III discusses membrane and polymer-based processes for wastewater treatment. This section includes three contributions covering model-based evaluation of pore-blocking behaviours of low pressure membranes, Sodalite- and Chitosan-based composite membrane materials for metal removal, modelling and optimisation of RO process for the removal of phenolic compounds from wastewater.

Section IV highlights oxidation and electrochemical processes for wastewater treatment. The four contributions in this section cover industrial three phase oxidation reactor, electrolytic method, ozone-based method and photocatalytic oxidation method for the treatment of wastewater.

Section V includes three contributions highlighting adsorption process for wastewater treatment. This section covers bio-sorption of methylene blue dye, laser-induced breakdown spectroscopy (LIBS) technique for evaluation of water quality, low cost adsorbent for nitrogen removal from wastewater.

Section VI includes three contributions on the use of biological processes for wastewater treatment and discusses application of natural zeolite in anaerobic digestion system, activated sludge process, and anaerobic degradation process. This section also adds potential for Hythane (hydrogen [H_2] and methane [CH_4]) production from petrochemical wastewater using anaerobic digestion process.

Section VII discusses water networks for water management and includes three contributions. Regeneration-recycling of industrial wastewater to minimise usage of freshwater, total site water integration and water re-use opportunities in dairy industry via process integration are highlighted.

Section VIII highlights issues with water management and includes three contributions discussing wastewater management modelling for coral reefs protection, multidisciplinary approach for integrated water resources management and water efficiency lapses and sustainable solutions.

Finally, **Section IX** includes two contributions on water-energy nexus highlighting optimisation of water and energy in integrated water and membrane networks and interaction of energy consumption, energy quality and freshwater production in multi-effect evaporative desalination process.

MATLAB® is a registered trademark of The MathWorks, Inc. For product information, please contact:

The MathWorks, Inc.
3 Apple Hill Drive
Natick, MA 01760-2098 USA
Tel: 508 647 7000
Fax: 508-647-7001
E-mail: info@mathworks.com
Web: www.mathworks.com

Editors

Iqbal M. Mujtaba, PhD, is a professor of computational process engineering in the School of Engineering at the University of Bradford. He earned his BSc Eng and MSc Eng degrees in Chemical Engineering from Bangladesh University of Engineering & Technology (BUET) in 1983 and 1984 respectively and earned his PhD from Imperial College London in 1989. He is a Fellow of the IChemE, a Chartered Chemical Engineer, and the current Chair of the IChemE's Computer Aided Process Engineering Subject Group. He was the Chair of the European Committee for Computers in Chemical Engineering Education from 2010–2013. Professor Mujtaba leads research into dynamic modelling, simulation, optimisation and control of batch and continuous chemical processes with specific interests in distillation, industrial reactors, refinery processes, desalination and crude oil hydrotreating focusing on energy and water. He has managed several research collaborations and consultancy projects with industry and academic institutions in the UK, Italy, Hungary, Malaysia, Thailand and Saudi Arabia. He has published more than 300 technical papers and has delivered more than 60 invited lectures/seminars/short courses around the world. He has supervised 29 PhD students to completion and is currently supervising 10 PhD students. He is the author of Batch Distillation: Design & Operation (textbook) published by the Imperial College Press, London, 2004 which is based on his 18 years of research in Batch Distillation. Professor Mujtaba also edited books titled *Application of Neural Network and Other Learning Technologies in Process Engineering*, Imperial College Press, London, 2001 and *Composite Materials Technology: Neural Network Applications*, CRC Press, USA, 2009.

Thokozani Majozi, PhD, is a full professor in the School of Chemical and Metallurgical Engineering at Wits University where he also holds the National Research Foundation (NRF) Chair in Sustainable Process Engineering. Prior to joining Wits, he spent almost 10 years at the University of Pretoria, initially as an associate professor and later as a full professor of chemical engineering. He was also an associate professor in computer science at the University of Pannonia in Hungary from 2005 to 2009. Majozi completed his PhD in Process Integration at the University of Manchester Institute of Science and Technology in the United Kingdom. He is a member of Academy of Sciences of South Africa and a Fellow for the Academy of Engineering of SA. He has served in various senior positions, including Vice-President of the Engineering Council of South Africa (2009–2012), Director of Pelchem (2007–2010) and Director of Necsa (2010–2013). He is currently the Chairperson of the Board at Council for Scientific and Industrial Research (CSIR). He has received numerous awards for his research including the Burianec Memorial Award (Italy), S2A3 British Association Medal (Silver) and the South African Institution of Chemical Engineers Bill Neal-May Gold Medal. He is also twice a recipient of the NSTF Award and twice the recipient of the NRF President's Award. Majozi is author and co-author of more than 150 scientific publications, including two books on Batch Chemical Process Integration published by Springer in January 2010 and CRC Press/Taylor & Francis in 2015. Majozi is a B1 NRF rated researcher.

Mutiu Kolade Amosa, PhD, is currently a research fellow in sustainable process engineering in the School of Chemical and Metallurgical Engineering at the University of the Witwatersrand in Johannesburg, South Africa. He is a trained chemical engineer with bachelor's and master's degrees in chemical engineering awarded by Ladoke Akintola University of Technology and Ahmadu Bello University, respectively. Dr. Amosa earned his PhD in environmental process engineering from the International Islamic University Malaysia and won the best doctoral award. He also won the most highly cited researcher award at the University of the Witwatersrand in 2017. Dr. Amosa is a permanent staff member of the Department of Petroleum Resources (DPR)—the Oil and Gas Regulatory Agency in Nigeria, where he works as a senior chemical engineer at the agency's headquarters in Lagos, Nigeria. He is also an international research expert and member of the Environmental Engineering and Management Research Group, Faculty of Environment and Labour Safety, at Ton Duc Thang University, Ho Chi Minh City, Vietnam.

Previously, Dr. Amosa served as a research fellow between 2011 and 2012 under the Petroleum Technology Development Fund (PTDF) endowment of the Ahmadu Bello University, Nigeria, where he worked on drilling fluid technologies and development of zeolites and molecular sieves for petroleum refining applications. He also served as research/teaching assistant between 2012 and 2015 under the Bioenvironmental Engineering Research Centre (BERC) at International Islamic University Malaysia, where he worked on several research projects related to environmental process engineering. He currently serves as an editor for Cogent Engineering Journal (Taylor & Francis Group) and he is an award-winning reviewer (awarded by Publons in 2017) for many reputable journals and conferences.

Dr. Amosa is a corporate member of the American Institute of Chemical Engineers (AIChE); Nigerian Society of Chemical Engineers (NSChE); Association of Environmental Engineering and Science Professors (AEESP); International Water Association (IWA); the Society of Petroleum Engineers (SPE); amongst others. Widely published and cited, he focuses his research on process design, modelling and optimization; environmental process engineering; sustainable process systems engineering; less common separation technologies and their wide applications; and development of micro- and nano-porous materials.

Contributors

Sulyman Age Abdulkareem
Department of Chemical Engineering
University of Ilorin
Ilorin, Nigeria

Adewale George Adeniyi
Department of Chemical Engineering
University of Ilorin
Ilorin, Nigeria

Fatai Alade Aderibigbe
Department of Chemical Engineering
University of Ilorin
Ilorin, Nigeria

Adnan Alhathal Alanezi
Department of Chemical Engineering Technology
College of Technological Studies
Kuwait City, Kuwait

Radhi Alazmi
Department of Chemical Engineering Technology
College of Technological Studies
Kuwait City, Kuwait

Ma'an Fahmi Alkhatib
Bioenvironmental Engineering Research Centre (BERC)
Department of Biotechnology Engineering
Kulliyyah of Engineering
International Islamic University Malaysia
Kuala Lumpur, Malaysia

Salih Alsadaie
University of Sirte
Sirte, Libya

Ali Altaee
School of Civil and Environmental Engineering
University of Technology Sydney
Sydney, Australia

Mutiu Kolade Amosa
NRF-DST Chair: Sustainable Process Engineering
School of Chemical and Metallurgical Engineering
University of the Witwatersrand
Johannesburg, South Africa

and

DPR Headquarters,
Department of Petroleum Resources
Lagos, Nigeria

and

Environmental Engineering and Management Research Group
Faculty of Environment and Labour Safety
Ton Duc Thang University
Ho Chi Minh City, Vietnam

Seth Apollo
Centre for Renewable Energy and Water
Vaal University of Technology
Vanderbijlpark, South Africa

Maryam Aryafar
Department of Chemical and Process Engineering
University of Surrey
Guildford, Surrey, United Kingdom

Asmaa Abdallah Awad
Chemical Engineering Department
Waterloo University
Waterloo, Ontario, Canada

I. Azreen
Universiti Malaysia Sabah
Kota Kinabalu, Malaysia

Santanu Bandyopadhyay
Department of Energy Science and Engineering
Indian Institute of Technology Bombay
Mumbai, India

Esther Buabeng-Baidoo
School of Chemical and Metallurgical Engineering
University of the Witswatersrand
Johannesburg, South Africa

Alasdair N. Campbell
Department of Chemical and Process Engineering
University of Surrey
Guildford, Surrey, United Kingdom

Franjo Cecelja
Department of Chemical and Process Engineering
University of Surrey
Guildford, Surrey, United Kingdom

Mahmoud Dahroug
Egyptian Water and Wastewater Regulatory Agency (EWRA)
Ministry of Housing, Utilities and Urban Development, Egypt

Michael O. Daramola
School of Chemical and Metallurgical Engineering
Faculty of Engineering and the Built Environment
University of the Witwatersrand
Johannesburg, South Africa

Ahmed Elreedy
Sanitary Engineering Department
Faculty of Engineering
Alexandria University
Alexandria, Egypt

Mohamed Elsamadony
Environmental Engineering Department
Egypt-Japan University of Science and Technology (E-JUST)
Alexandria, Egypt
Public Works Engineering Department
Faculty of Engineering
Tanta City, Egypt

Ahmad Fikri Ahmad Fadzil
Process Systems Engineering Centre (PROSPECT)
Research Institute of Sustainable Environment
and
Faculty of Chemical and Energy Engineering
Universiti Teknologi Malaysia
Johor Bahru, Malaysia

Giacomo Filippini
Chemical Engineering Department
School of Engineering
Faculty of Industrial Engineering
Politecnico di Milano
Milan, Italy

Mamdouh A. Gadalla
Department of Chemical Engineering
Port Said University
Port Fouad, Egypt

and

Department of Chemical Engineering
The British University in Egypt
El-Shorouk City, Egypt

Saba A. Gheni
Chemical Engineering Department
Faculty of Engineering
Tikrit University
Tikrit, Iraq

Adewale Giwa
Department of Chemical Engineering
Khalifa University of Science and Technology
Abu Dhabi, United Arab Emirates

Dan Green
Wessex Water Services Ltd
Claverton Down, Bath, United Kingdom

Yakun Guo
School of Engineering
Faculty of Engineering and Informatics
University of Bradford
Bradford, West Yorkshire, United Kingdom

Prashanth Reddy Hanmaiahgari
Department of Civil Engineering
Indian Institute of Technology
Kharagpur, India

Shadi W. Hasan
Department of Chemical Engineering
Khalifa University of Science and Technology
Abu Dhabi, United Arab Emirates

Khaled M. Hassan
Department of Chemical Engineering
The British University in Egypt
El-Shorouk City, Egypt

Alaa H. Hawari
Department of Civil and Architectural Engineering
Qatar University
Doha, Qatar

Tamer T. El-Idreesy
Department of Chemistry
Faculty of Science
Cairo University
Giza, Egypt

and

Department of Chemistry
School of Sciences and Engineering
The American University in Cairo
New Cairo, Egypt

Mohammed Saedi Jami
Bioenvironmental Engineering Research Centre (BERC)
Department of Biotechnology Engineering
Kulliyyah of Engineering
International Islamic University Malaysia
Kuala Lumpur, Malaysia

Aysar T. Jarullah
Chemical Engineering Department
Faculty of Engineering
Tikrit University
Tikrit, Iraq

Chakib Kara-Zaïtri
Chemical Engineering
Faculty of Engineering and Informatics
University of Bradford
Bradford, West Yorkshire, United Kingdom

Jiří Jaromír Klemeš
Sustainable Process Integration Laboratory – SPIL
NETME Centre, Faculty of Mechanical Engineering
Brno University of Technology
Brno, Czech Republic

Akash Kumar
Department of Chemical Engineering
Indian Institute of Technology Gandhinagar
Gandhinagar, Gujarat, India

Y. Lija
Universiti Malaysia Sabah
Kota Kinabalu, Malaysia

Nielsen Mafukidze
School of Chemical and Metallurgical Engineering
University of the Witswatersrand
Johannesburg, South Africa

Thokozani Majozi
National Research Foundation (NRF) Chair in Sustainable Process Engineering
School of Chemical and Metallurgical Engineering
University of the Witwatersrand
Johannesburg, South Africa

Zainuddin Abdul Manan
Process Systems Engineering Centre (PROSPECT)
Research Institute of Sustainable Environment
and
Faculty of Chemical and Energy Engineering
Universiti Teknologi Malaysia
Johor Bahru, Malaysia

Flavio Manenti
Chemical Engineering Department
School of Engineering
Faculty of Industrial Engineering
Politecnico di Milano
Milan, Italy

Mohammed J. Al-Marri
Gas Processing Center
College of Engineering
Qatar University
Doha, Qatar

Claudio Mascialino
Department of Chemical Engineering
University of Surrey
Guildford, Surrey, United Kingdom

Machodi Mathaba
School of Chemical and Metallurgical Engineering
Faculty of Engineering and the Built Environment
University of the Witwatersrand
Johannesburg, South Africa

and

Department of Chemical Technology
University of Johannesburg
Doornfontein, Johannesburg, South Africa

Marcelle McManus
Water Innovation Research Centre
Department of Mechanical Engineering
University of Bath
Claverton Down, Bath, United Kingdom

Achisa C. Mecha
Department of Chemical, Metallurgical and Materials Engineering
Tshwane University of Technology
Pretoria, South Africa

Awad E. Mohammed
Chemical Engineering Department
Faculty of Engineering
Tikrit University
Tikrit, Iraq

Maggy N. B. Momba
Department of Environmental, Water and Earth Sciences
Tshwane University of Technology
Pretoria, South Africa

Alireza Abbassi Monjezi
Department of Chemical and Process Engineering
University of Surrey
Guildford, Surrey, United Kingdom

Iqbal M. Mujtaba
Chemical Engineering
Faculty of Engineering and Informatics
University of Bradford
Bradford, West Yorkshire, United Kingdom

Ibrahim Hassan Mustafa
Biomedical Engineering Department
Helwan University
Cairo, Egypt

and

Chemical and Materials Engineering Department
King Abdulaziz University
Jeddah, Saudi Arabia

Muftah H. El-Naas
Gas Processing Center
College of Engineering
Qatar University
Doha, Qatar

Vincenzo Naddeo
Department of Civil Engineering
University of Salerno
Fisciano (SA) Italy

Linda B. Newnes
Water Innovation Research Centre
Department of Mechanical Engineering
University of Bath
Claverton Down, Bath, United Kingdom

Nicholas Nyamayedenga
Silchrome Plating LTD
Leeds, United Kingdom

Chrysoula Papacharalampou
Water Innovation Research Centre
Department of Mechanical Engineering
University of Bath
Claverton Down, Bath, United Kingdom

Mudhar A. Al-Obaidi
Chemical Engineering
Faculty of Engineering and Informatics
University of Bradford
Bradford, West Yorkshire, United Kingdom

and

Middle Technical University
Baghdad, Iraq

Aoyi Ochieng
Centre for Renewable Energy and Water
Vaal University of Technology
Vanderbijlpark, South Africa

Olawale R. Olaopa
Department of Political Science
Obafemi Awolowo University
Ile-Ife, Nigeria

Maurice S. Onyango
Department of Chemical, Metallurgical and Materials Engineering
Tshwane University of Technology
Pretoria, South Africa

Maruf Oladotun Orewole
National Centre for Technology Management
Federal Ministry of Science and Technology
Obafemi Awolowo University
Ile-Ife, Nigeria

Benton Otieno
Centre for Renewable Energy and Water
Vaal University of Technology
Vanderbijlpark, South Africa

Aghaegbuna O. U. Ozumba
School of Construction Economics and Management
Faculty of Engineering and Built Environment
University of the Witwatersrand
Johannesburg, South Africa

Jaan H. Pu
School of Engineering
Faculty of Engineering and Informatics
University of Bradford
Bradford, West Yorkshire, United Kingdom

Md. Arafatur Rahman
Faculty of Computer Systems & Software Engineering
Universiti Malaysia
Pahang, Malaysia,

L. N. S. Ricky
Universiti Malaysia Sabah
Kota Kinabalu, Malaysia

Afshin Shahi
Division of Peace Studies & Middle Eastern Politics
University of Bradford
Bradford, West Yorkshire, United Kingdom

Prashant Sharan
Buildings & Thermal Science Center
National Renewable Energy Laboratory
Golden, Colorado USA

Adel O. Sharif
Department of Chemical and Process Engineering
University of Surrey
Guildford, United Kingdom

and

Qatar Environment and Energy Research Institute (QEERI)
HBKU, Qatar Foundation
Doha, Qatar

Babji Srinivasan
Department of Chemical Engineering
Indian Institute of Technology Gandhinagar
Gandhinagar, Gujarat, India

Rajagopalan Srinivasan
Department of Chemical Engineering
Indian Institute of Technology
Madras, India

Nashwa Tarek El-Tahhan
Chemical Engineering Department
The Higher Technological Institute (HTI)
10th of Ramadan City, Egypt

Ahmed Tawfik
Environmental Engineering Department
Egypt-Japan University of Science and Technology (E-JUST)
Alexandria, Egypt

and

National Research Centre
Water Pollution Research Department
Giza, Egypt

Sarojini Tiwari
Department of Chemical Engineering
Indian Institute of Technology Gandhinagar
Gandhinagar, Gujarat, India

Hamad Al-Turaif
Chemical and Materials Engineering Department
King Abdulaziz University
Jeddah, Saudi Arabia

Maya Vachkova
Division of Peace Studies & Middle Eastern Politics
University of Bradford
Bradford, West Yorkshire, United Kingdom

Sharifah Rafidah Wan Alwi
Process Systems Engineering Centre (PROSPECT)
Research Institute of Sustainable Environment
and
Faculty of Chemical and Energy Engineering
Universiti Teknologi Malaysia
Johor Bahru, Malaysia

Renju Zacharia
Gas Processing Center
College of Engineering
Qatar University
Doha, Qatar

A. Y. Zahrim
Chemical Engineering Department
Universiti Malaysia Sabah
Kota Kinabalu, Malaysia

Section I

Social Perspective

1 Water Security and the Rise of Sectarian Conflict in Yemen

Afshin Shahi and Maya Vachkova

CONTENTS

In recent years, various explanations have been provided for the rise of sectarianism in the Middle East. Iran–Saudi rivalry, uneven development, horizontal inequality, top-down ethno-sectarian discrimination, state propaganda and utilization of sectarian narratives for regime survival, US-led invasion of Iraq and the failure of Arab Uprising and the Syrian civil war have been highlighted as influential factors behind the new wave of sectarian violence in the region. Although all these factors are important in explaining the situation, we believe that often the environmental factors are overlooked in explaining the emergence of political violence in fragmented societies such as Yemen. There is enough evidence to suggest that factors such as drought and desertification are instrumental in setting the stage for social eruption, particularly in societies with fragmented sense of national consciousness.

The link between environmental problems and conflict is already well established and given the extreme water shortages in the region, the Middle East is particularly vulnerable to environmentally induced instabilities. The Middle East can be characterized as semi-arid or arid and hence, vulnerable to climate change [1]. Recurring droughts are by no means a novelty in the region. Climate change has exacerbated the effects of droughts. The 1998–2012 period was the driest one in the Levant for the past five centuries [2]. Moreover, severe droughts are expected in the Middle East within the next few decades [2]. With the current population growth and rates of water consumption, by 2050 the global water demand may reach 100% of the available supply [3]. A recent study reviewing water availability and climate change issued a prediction for the driest countries in the next few decades. According to the World Resource Institute, of the thirty-three countries that are expected to face extreme water stress by 2040, more than half are situated in the Middle East. Moreover, seven out of the ten most water-stressed countries are in the Middle East.

This chapter briefly examines how extreme water shortages have paved the way for the escalation of sectarian violence in Yemen. Although one cannot reduce the complicated conflict in Yemen only to one factor, water shortages have played a very important role in the instigation and the continuation of the civil war.

1.1 CASE VIGNETTE: YEMEN

Yemen is often described as a conflict-prone, underdeveloped country stretched across an expanse of scorched earth. It is a home to a wide diversity of tribes, all intertwined in cooperation and hostilities. The Shiite-Sunnite ratio in Yemen is close to equilibrium with around 40%–47% Shiites and 53%–60% Sunnites [4][1]. The Shiite sects in the country are mostly Zaydis and Twelver Shiites,

[1] *Atlapedia*, s.v. 'Yemen', accessed March 3, 2016.

while the Sunnite sects are Shafi, Maliki, Hanbali, and Hanafi, with the Shafi sect constituting the majority [5]. The northern parts of the country are dominated by the Zaydis, while the south and southeastern territories are the home of the powerful Shafi Sunnites. In recent years, however, there has been a wave of Sunnite and a corresponding wave of Shiite radicalisation. These sub-state identities began to gain more political importance after the unification of the North and the South in 1990, which spawned multitudinous sectarian political parties [6].

Yemen is one of the most water-stressed places in the world [7]. Droughts and scarce water resources have characterised the Yemeni geography for centuries [8]. The civilisations sustained by this climate and terrain have relied on ingenious solutions, such as terraced supply and demand farming. Traditional agricultural practices of those types allowed for sparing consumption of water resources. The agricultural revolution in the 1970s heralded a significant shift away from traditional farming practices [7]. After the seventies, the country became more engaged with the international market. Along with modern technology and investment, foreign water-demanding crops entered the newly liberalised Yemeni market. Inadequate planning and regulation lead to over-exploitation of land and water resources. Nonetheless, the rise of cash crops did enrich some parts of society.

The piping network in the country is not efficient and many households are not supplied by the state, but use drill wells instead. There have been some efforts towards regulating ground water use and distribution [9]. There is ample evidence that illegal drilling was common even after the introduction of a drilling permit system [10]. Ninety per cent of fresh water in Yemen is used for agriculture. Most of it is groundwater from springs and wells, which is extremely unsustainable [11]. The land degradation and ground water depletion, however, went overlooked and unaddressed. Decades of water mismanagement left arid Yemen extremely vulnerable to climate change [10].

Water scarcity fosters not only environmental, but also social vulnerability. Competition over this vital resource over the past couple of years has cost more lives than the recent civil war [12]. The weak governance institutions are often unable to resolve these conflicts and to provide basic water and sanitation to all areas of the country. In the absence of functional services, alternative forms of government infiltrate the chasm left by the abdication of the state. Unsurprisingly, the most water-stressed regions in Yemen host the strongholds of extremist organisations. It seems impoverished farmers are more likely to succumb to sectarian narratives and to enrol in ethno-sectarian militias. In the absence of effective state intervention, social groups violently confront each other to secure water and they use their tribal, sectarian and other ancient identities to frame their grievances. In the climate of struggle over survival, the binaries of "us" versus "them" became stronger than before.

Yemen is infamous for the weakness of its institutions. State legitimacy is often contested by tribal law and armed militant groups. Governance in Yemen has continuously relied on a balance between official and shadow institutions [13]. The conflict in Yemen escalated in response to anti-government protests. Sectarian extremist groups made their way into the havoc of protest crackdowns. Finally, at the time of writing, a heavily armed coalition in support of the government, championed by Saudi Arabia, is bombarding key infrastructure. Since the beginning of the conflict, 6400 people have lost their lives [14]. The fighting and the foreign bombing campaign have severed basic infrastructure and as of January 2016, 80% of the population was food-aid dependent [15]. As of 2015, 2.3 million were internally displaced and at least 121,000 were reported to have left the country [16]. Around 14.1 million people need support to meet basic healthcare needs and about two million are currently acutely malnourished, including 1.3 million children—320,000 of whom are enduring severe acute malnutrition. As of October 2015, health services reported 32,307 casualties (including 5604 deaths), an average of 153 injuries or deaths per day [17]. While the conflict turns into a sectarian bloodbath, with all sides drawing upon sacral narratives of ancient strife, the country is facing an environmental catastrophe, and more specifically a severe depletion of ground water resources [18].

The vigour of the so-called "Arab Spring" did not bypass Yemen. In 2011, a group of women, children and men gathered in "Change Square" in a peaceful demonstration against the authoritarian 33-year rule of President Ali Abdullah Saleh. Protesters hoped to transform the poorest Arab nation into a modern democracy. The civil society dominated the protests for a few weeks, before the leader of the Muslim Brotherhood affiliated Yemeni organization al-Islah hijacked the protest with the assistance of General Ali Mohsen al-Amar, commander of an elite militia with strong ties to armed Sunnite extremists. President Saleh's violent response to the demonstrations caused a high death toll within the span of a few months: hundreds of protesters died, while thousands were injured [19]. After external mediation provided by the Gulf Co-operation Council, Saleh handed power over to his deputy, Abdrabbuh Mansour Hadi. The deal was signed in Riyadh and stipulated that Hadi was to govern provisionally until the next presidential elections, in exchange for immunity for Saleh [20]. The deal was dismissed by protesters in the capital, who rejected the idea of granting the ousted president political absolution. Five years later, the country had descended into chaos and sectarian violence.

The main fight was waged between president Hadi and the Houthis, an armed militia group founded in the 1990s by Hussein al-Houthi. The Houthis draw recruits from the Northern Zaydi population who practice a form of Shiite Islam. In 2004, they instigated a rebellion against President Saleh. This internal conflict lingered, with fluctuations in intensity, until the 2011 revolution rekindled it [21]. The Houthis, who come from Northern Yemen, managed to push the presidential forces out of the capital Sana'a in February 2015. Alongside other grievances, the Houthis have continuously contested the unfair distribution of water between North and South [22]. Aden, the temporary capital in the loyal South, quickly became the next Houthi target [23]. Some of the presidential security forces defected in order to support the Houthis. The well-connected ex-president, Ali Abdullah Saleh, was out of power but far from powerless. While initially supporting the new presidential regime, he switched alliances to the Houthi rebels in 2014, conveniently parading his renewed Zaydi identity [24]. Despite coming from a Zaydi tribe, some thirty years ago the president rose to power with the support of powerful Sunnite tribes [13]. Locally, Hadi received support from the mostly Sunnite south, the al-Islah and internationally from Saudi Arabia and the Gulf regimes. Both President Hadi and the Houthis were opposed by al-Qaeda in the Arabian Peninsula (AQAP), which had strong positions in the Sunnite south and southeast. The conflict became further entangled due to the emergence a Yemeni Islamic State offshoot that aimed to contest AQAP's influence and became the new Sunnite jihadi presence in Yemen.

The aggressive foreign intervention supports the claim that the Yemeni conflict is now a proxy war with a sectarian character. The proxy conflict is waged between the two oil and religious colossi in the Middle East: Saudi Arabia and Iran. In March 2015, Saudi Arabia declared war on Yemen, calling on the support of a broad military coalition. In partnership with Jordan, Morocco, Bahrain, Kuwait, Qatar, Egypt, Sudan, the United Arab Emirates (UAE), the United States, the European Union, and Pakistan, Saudi Arabia began bombing Yemeni territories. The aim of that campaign is the eradication of the Houthi threat [25]. Although the conflict has been portrayed as a sectarian conflict with regional dimensions, its causes are mainly rooted in the drastic ecological problems that were devastating the national economy. The primary trigger for this conflict is water security.

In the pre-war period, the Yemeni government regularly struggled to tackle annual budget shortfalls [26]. The country has suffered groundwater depletion that inspired many grim forecasts—for instance that the groundwater in the capital Sana'a would be completely exhausted by 2025 [27] or even by 2015 [28]. Despite the inaccuracy of the latter, it is undeniable that water in Yemen is scarce. The issue of water scarcity is multi-faceted and may be traced back to a number of causes. Hydro-management was never a public priority. After the 1970s, Yemen experienced an agricultural revolution, spearheaded by deep tube wells—an innovation that helped expand land cultivation. In addition, farmers switched from traditional crops to water-intensive cash crops [8]. A weak, poorly managed system was used to deliver water to the commercial farms. The piping infrastructure is poor, leaks often, and is thus wasteful; in addition, illegal wells are drained to pump groundwater [10].

Farmers use up 90% of the ground water resources to irrigate crops and 37% of it flows into the very water-demanding qat crops [8]. Qat is a plant with stimulating properties whose soft upper leaves are chewed by some 72% of Yemeni men and 35% of women [13]. Lastly, climate change plays a crucial role in water depletion. Due to recurrent droughts, aquifers struggle to recharge [13]. Insufficient effort has been expended on conservation. The government seems to have made a tentative commitment to water conservation in 2010; however, the tumultuous start of 2011 cancelled all ecological plans, if there truly had been any [29].

Yemen is among the leaders in population growth. Simultaneously, it is the most drought-affected country in the Middle East [10]. Yemen is the most populous country on the Arab peninsula and its population is predicted to double by 2033 [30]. Two-thirds of Yemenis are under the age of 24 and each Yemeni woman bears an average of six children [31]. Rapid population growth has intensified pressure on natural resources—especially water—as well as on public services. In 2011, water consumption from the Sana'a Basin exceeded the rate of annual recharge [32]. Moreover, in 2011 the Yemeni economy plummeted, weakened by fuel shortages, power outages and extreme water scarcity. During the month of Ramadan, water prices escalated by 200% and fuel prices by 900% [4]. To poor families, this turn of events meant they could no longer afford bare necessities. Juxtaposed with the bleak prospects of starvation, militia recruitment became ever so palatable.

Environmental scarcity exacerbates social tensions and allows the penetration of sectarian narratives, which reinforce the binaries of "us" versus "them." In the south, tribal formations have reportedly already been sheltering extremist figures and symbolically approving their zeal for social change through jihad [13]. In the north, both the perceived relative deprivation of Zaydis and the restrictions on worship and religious expression in general have already fomented dissent against the government.

It is undeniable that water shortages played a significant role in the militant rise of the Houthi movement, which have contributed to the sectarian tensions in the country. Indeed, the Houthi homeland, the Sa'dah Plain, was already suffering ground water depletion in the nineties [33]. Experts conclude that overexploitation of groundwater resources is the main problem in the Sa'dah Plain, a semi-arid highland basin of Yemen. Groundwater-irrigated agriculture is the predominant livelihood in the area: hence, water depletion imperils food security and threatens the socioeconomic balance. In the past three decades, qat became a staple crop for the northern Zaydi highlands and, while it did bring economic prosperity, qat farming worsened groundwater depletion [30]. Due to poor institutional arrangements, there has been no adequate government intervention and the water crisis of the nineties has remained unaddressed. What is more, the Houthis have continuously reported unfair water distribution and have gained support from the North based on this grievance [22]. Indeed, this pressing ecological problem has intensified divisive identity politics, solidified the politics of othering and hardened the socio-religious boundaries of desperate collectivities who have to compete tirelessly for vital resources in their struggle for survival.

While the northern highlands are a Shiite/Zaydi Houthi domain, the southern highlands and lowlands are under the control of AQAP. Not surprisingly, water scarcity plays a key role in AQAP's legitimacy—the group governs water distribution and resolves disputes over water, which are extremely common in the country. Reportedly, around 4000 Yemenis die every year in small-scale water disputes [12]. The central role of water in AQAP's strategy is evident in a document discovered by the Associated Press in 2013 that states, "by taking care of their daily needs like water . . . [w]e will have a great effect on people, and will make them sympathise with us and feel that their fate is tied to ours" [12]. In partnership with tribes, AQAP also provides drinking water, electricity and a form of protection for the regions under its control, thus utilising water as a political tool [34]. Reportedly, many Yemenis have supported various extremist organisations because they provide better governance, relative to the central government, such as education and water provision [35]. This is the case for the Houthis, as much as for AQAP [36]. The Saudi-led forces also utilise water as a weapon by strategically destroying hydro infrastructure [37]. The strategic blockade of humanitarian aid on behalf of both extremists and the Saudi-led interventionist forces further exacerbated the water plight of Yemenis caught in the middle [38].

Although sub-state actors such as AQAP may have some success in hydro management in certain areas, their ideological mission only feeds into the further fragmentation of Yemeni national consciousness. AQAP's religio-political mandate, coupled with their short-term practical success, has reinforced the sectarian narratives and deepened the sense of distrust towards the state and other competing groups who also want a share of the remaining water recourses. Thus, water insecurity feeds identity politics and fosters conflict.

1.2 CONCLUSION

Yemen is one of the most water-stressed countries in the world. Droughts and scarce water resources have characterised the Yemeni climate for centuries. Pressing environmental problems coupled with water mismanagement have made Yemen one of the most unstable countries in the region. Although extreme water shortage has been part of life for centuries, the country embarked on an unsustainable agricultural revolution in the 1970s that undermined traditional farming practices suitable for that dry environment. In this light, water-demanding crops entered the newly liberalised Yemeni market and started to change the agricultural landscape and destroy underground water resources beyond recognition. Hence, gradually water became scarcer and securing it became even more costly. Even before the start of the civil war, water disputes were a permanent feature of social life. Struggle over securing water became responsible for thousands of micro conflicts across the country. As the state was unable to respond to these serious environmental problems, the right conditions were created for the emergence of sub-state actors. Some of these extremist groups such as Al Qaeda thrived in this environment because they could offer a more effective water distribution model. Whoever could control water, could control hearts and minds.

At least partly, the rise of the Houthi movement can be seen as a reaction to the extreme water shortage and water mismanagement. The Sa'dah Plain, the Houthi homeland has been suffering from drought for many years. In the past three decades, qat became a lucrative crop for the northern Zaydi highlands, which significantly aggravated groundwater depletion. This was coupled with what the Houthis regarded as "unfair" water distribution by the state. These pressing issues and grievances played a very important role in mobilising the Houthi farmers, which at least partly led to the destructive war. The rise of the northern Houthi farmers who happened to be Shia reinforced sectarian binaries of "us" versus "them." These sectarian binaries were strengthened by the prolongation of the war and the involvement of regional actors such as Iran and Saudi Arabia. The seemingly sectarian nature of their regional policies turned Yemen to another arena for rivalry and proxy confrontation.

Since the Saudi-led intervention, the environmental state of Yemen has deteriorated. Twenty million Yemenis do not have access to clean drinking water. The intervention of regional actors not only physically destroys the already weak water infrastructure, it cements the sectarian divisions within the country. Although the Saudi-led coalition may withdraw from the country in the near future, the water scarcity will continue to breed violence and conflict. Although Yemen requires a conclusive political solution to end the war, water security remains the most fundamental challenge facing the country. It is extremely hard to envisage a stable future for Yemen without a comprehensive solution for water scarcity.

REFERENCES

1. R. T. Watson, M. C. Zinoyawera, R. H. Moss, Eds., *IPCC Special Report for Policy Makers, The Regional Impacts of Climate Change: An Assessment of Vulnerability*, (Cambridge, UK: Cambridge University Press, 1997).
2. B. I. Cook, K. J. Anchukaitis, R. Touchan, D. M. Meko, E. R. Cook, Spatiotemporal drought variability in the Mediterranean over the last 900 years, *Journal of Geophysical Research* 122(5), 2060–2074 (2016).

3. M. T. Klare, The new geography of conflict, *Foreign Affairs* 80, 49 (2001).
4. S. W. Day, *Regionalism and Rebellion in Yemen: A Troubled National Union*, (Cambridge, UK: Cambridge University Press, 2012).
5. S. S. Canton, ed., *Middle East in Focus: Yemen*, (Oxford, UK: ABC Clio, 2013).
6. F. al-Muslimi, How Sunni-Shia sectarianism is poisoning Yemen, Carnegie Endowment for International Peace (December 2015) http://carnegieendowment.org/syriaincrisis/?fa=62375 (accessed February 1, 2016).
7. C. Ward, Yemen's water crisis, The British-Yemeni society (2001) http://www.al-bab.com/bys/articles/ward01.htm (accessed February 13, 2016).
8. G. Lichtenthaeler, Water conflict and cooperation in Yemen, *Middle East Report* (*Running Dry*) 254, 40 (2010).
9. Law No. 33 of 2002 (Yemen), available at http://www1.umn.edu/humanrts/arabic/Yemeni_Laws/Yemeni_Laws72.pdf (in Arabic)
10. N. Glass, The water crisis in Yemen: Causes, consequences and solutions. *Global Majority E-Journal* 1(1), 17–30 (2010). https://www.american.edu/cas/economics/ejournal/upload/global_majority_e_journal_1-1_glass.pdf (accessed February 2, 2016).
11. A. Almas, M. Scholz, Agriculture and water resources crisis in Yemen: Need for sustainable agriculture. *Journal of Sustainable Agriculture* 28(3), 55–75 (2006).
12. J. Fergusson, Yemen is tearing itself over water, *Newsweek*, (January 20, 2015) http://europe.newsweek.com/al-qaida-plans-its-next-move-yemen-300782?rm=eu
13. V. Clark, *Yemen: Dancing on the Heads of Snakes*, (New Haven, CT: Yale University Press, 2010).
14. Oxfam, Oxfam chief calls Yemen 'Syria without the cameras', *Press Release*, July 1, 2016. http://www.oxfam.org.uk/media-centre/press-releases/2016/07/oxfam-chief-calls-yemen-syria-without-the-cameras (accessed May 1, 2017).
15. BBC, Six thousand die in Yemen's forgotten war, news release January 7, 2016. http://www.bbc.co.uk/news/world-35258965?SThisFB
16. Associated Press, Despite war, Yemen still draws opportunity-seeking migrants, news release October 27, 2015. http://www.atlanticbb.net/news/read/category/Europe%20News/article/the_associated_press-despite_war_yemen_still_draws_opportunityseeking_m-ap
17. UNOCHA, *Yemen Humanitarian Pool Fund Interim Report* (2015) http://reliefweb.int/sites/reliefweb.int/files/resources/yemen_humanitarian_pooled_fund_-_interim_report_2015.pdf
18. C. Ward, *The Water Crisis in Yemen: Managing Extreme Water Scarcity in the Middle East*, (London, UK: I.B. Tauris, 2014).
19. BBC News, Arab uprising: Country by country – Yemen, news release December 16, 2011. http://www.bbc.co.uk/news/world-12482293
20. BBC News, Yemeni president Saleh Signs deal on ceding power, news release November 23, 2011. http://www.bbc.co.uk/news/world-middle-east-15858911
21. B. Salmoni, Yemen's forever war: The Houthi rebellion, The Washington Institute Policy Analysis (2010). http://www.washingtoninstitute.org/policy-analysis/view/yemens-forever-war-the-houthi-rebellion
22. A. Martino, Water scarcity is helping radicalize the middle east militant and extremist groups exploit the region's escalating water shortage to gain power, loyalty, and legitimacy. VICE April 25, 2015. https://www.vice.com/en_us/article/exq45z/is-water-scarcity-radicalizing-the-middle-east-235
23. F. al-Jalal, Aden: A temporary capital for Yemen, *The New Arab*, news release February 23, 2015. http://www.alaraby.co.uk/english/politics/2015/2/23/aden-a-temporary-capital-for-yemen
24. F. Gardner, Torn in two: Yemen divided, *BBC*, news release December 24, 2015. http://www.bbc.co.uk/news/world-middle-east-35160532
25. C. Shakdam, Yemen at war: The new Shia-Sunni frontline that never was, *Foreign Policy Journal* (2015). http://www.foreignpolicyjournal.com/2015/04/10/yemen-at-war-the-new-shia-sunni-frontline-that-never-was/
26. Central Intelligence Agency, *The World Factbook*, s.v. 'Yemen' (2016).
27. O. Naje, Yemen's capital will run out of water by 2025, Science and Development Network (October 2010). http://www.scidev.net/en/news/yemen-s-capital-will-run-out-of-water-by-2025-.html (accessed March 8, 2016).
28. C. Giesecke, Yemen's water crisis: Review of background and potential solutions, *USAID Knowledge Services Center (KSC) Series*, (June 2012). http://www.yemenwater.org/wp-content/uploads/2014/10/pnadm060.pdf

29. World Bank, Sana'a basin water management project, 2003. http://www.worldbank.org/projects/P064981/sanaa-basin-water-management-project?lang=en (accessed April 7, 2016).
30. UN Population Fund, About Yemen: Situation analysis http://yemen.unfpa.org/en/about-yemen (accessed March 26, 2016).
31. Central Investigation Agency, *The World Factbook*, s.v. 'Yemen'. (accessed April 3, 2016).
32. A. Hefez, How Yemen chewed itself dry, *Foreign Affairs* (Special edition: The Best of 2013). http://www.washingtoninstitute.org/policy-analysis/view/how-yemen-chewed-itself-dry
33. R. al-Sakkaf, Y. Zhou, M. J. Hall, A strategy for controlling groundwater depletion, *International Journal of Water Resources Development* 15(3), 349–365 (1999).
34. S. Al Batati, Yemen: The truth behind Al-Qaeda's takeover of Mukalla, *Aljazeera*, news release September 16, 2015. http://www.aljazeera.com/news/2015/09/yemen-truth-al-qaeda-takeover-mukalla-150914101527567.html (accessed February 1, 2016).
35. N. al-Dawsari, Tribes and AQAP in South Yemen. *Atlantic Council*, 2014. http://www.atlanticcouncil.org/blogs/menasource/tribes-and-aqap-in-south-yemen
36. D. O'Driscoll, Violent extremism and terrorism in Yemen. K4D (2017).
37. M. Benjamin, America will regret helping Saudi Arabia bomb Yemen. *The Guardian*, June 19, 2017. https://www.theguardian.com/commentisfree/2017/jun/19/america-helping-saudi-arabia-bomb-yemen-consequences
38. REUTERS, Saudi coalition, Houthi rebels restricting Yemen aid access: U.N. February 16, 2016. https://www.reuters.com/article/us-yemen-war-saudi-un/saudi-coalition-houthi-rebels-restricting-yemen-aid-access-u-n-idUSKCN0VP2Q6

2 Water Diplomacy

Solving the Equations of Conflict, Economic Growth, Social Well-Being and Ecosystem Demand

Maruf Oladotun Orewole

CONTENTS

2.1 INTRODUCTION

Water is an important source of life and livelihoods. The sixth goal of the Sustainable Development Goals (SDGs) is to "ensure availability and sustainable management of water and sanitation for all." Efficient water management is strategic to the success of the remaining SDGs. About 60% of fresh water available globally comes from river basins that traverse national boundaries and are sources of various degrees of conflict due to competing claims over utilization for economic growth, social being of citizens and ecosystem services. Climate change, population growth and urbanization as well as wasteful use of water in some countries are negatively impacting the availability and access of this critical resource to millions of people spread across the world. By the year 2025, nearly 2 billion people will live in conditions of absolute water scarcity, and two thirds of the world's population will be in areas of water stress, which is already leading to situations of unbalanced distribution and tensions among users (UN Water, 2014).

In order to curtail conflict escalation over water, it can no longer be business as usual. It requires effective management strategy that will tackle issues such as water wastage in current systems, which has been estimated to be up to 30%; unsustainable use, institutional dysfunction, unethical practices, pollution, ecosystem degradation, sanitation and human health crisis, water-related disasters, unsustainable development and conflicts over transboundary water basins. Water diplomacy is a veritable tool in achieving transboundary water cooperation, peaceful coexistence among riparian nations, political stability and economic growth. Water diplomacy is therefore discussed as a solution to the complex equations of conflict, economic growth, social wellbeing and ecosystem demand.

2.2 WATER: ITS INDISPENSABILITY AND IMPLICATIONS OF ITS LIMITED SUPPLY

The important role of water in shaping human settlement distribution and socio-economic development cannot be overemphasized. Water is an intrinsic factor in every culture and society. It occupies a significant portion of human body systems as well as other living things thereby making survival and continuity critically dependent on it. Functionality of human society is based on the water availability and before settlements are made, access to abundant supply of freshwater for drinking and sanitation and agriculture are usually considered a priority.

Water is also a critical input in various industries including transport and tourism, and its role in sustaining the earth's ecosystem is simply irreplaceable. According to the United Nations World Water Development Report (WWAP, 2015), water resources stand at the core of sustainable development because of the range of services they provide. They underpin poverty reduction, economic growth and environmental sustainability.

This earth is abundantly rich in water resources, which occupy two-thirds of the entire planet, but access to useful quality is very low. However, the greater percentage of this water is locked in the glaciers, salty oceans and deep aquifers, while most of the freshwater available also suffers pollution. Out of the global water, only about 2.5% is fresh water out of which 0.5% is accessible, while the remaining 97.5% of the whole belongs to the saltwater (Shiklomanov, 1998). This situation makes access to clean water and sanitation costlier than when it used to be managed and protected by communities as a commons.

Shiva (2002) observed that when water was a commons before its shift to economic goods and subsequent privatization momentum, it was well managed and accessible to all. Consideration of water as an economic good has its merits and demerits. The famous proclamation of water as *an economic good* at the Dublin Conference (ICWE, 1992) was a compromise between the economists who see water like other private goods that should be subjected to allocation through competitive market pricing and others who see water as a basic human right that should not be subjected to competitive market pricing and allocation. In the present situation of drought in many poor countries of the world and growing inaccessibility to water, considering water as *an economic good* may result in efficient management, but it will make access to this critical resource elusive to millions of already impoverished people across the globe.

Depending on the situation of different countries, water policies should be formulated in a way to foster accessibility and sustainable use while considering the objectives or properties that make water both a private and public good (Perry et al., 1997). A sharp increase in population growth, climate change, urbanization and land use change, hydraulic infrastructure and pollution all result into changes in the fluxes, pathways and storage of water. The overall effect of hydrologic changes is too little or too much water across the globe.

Global water demand is on the rise in response to unprecedented population growth, urbanization, food and energy security policies, and macro-economic processes such as trade globalization and changing consumption patterns (WWAP, 2015). This demand could be classified as irrigation demand and non-irrigation demand, the latter further disaggregated into domestic, industrial and livestock water demand (Rosegrant et al., 2002). The truth is that meeting this demand especially with the impact of climate change is becoming increasingly difficult and the consequences are dire.

Cape Town, a coastal city in South Africa, has been facing a critical water crisis since 2017 due to the drought experienced recently and it is likely to become the third city to run dry globally. Apart from famine, diseases, impedance to economic growth and death, another serious consequence of water shortage is conflict based on demand and allocation that could be internal within a nation or transboundary among nations. Science alone cannot solve complex water conflicts that involve policy issues and transcend political, social, jurisdictional, physical, ecological and biogeochemical boundaries (Choudhry and Islam, 2015).

2.3 WATER CONFLICTS IN THE FACE OF INCREASING STRESS

About 23 years ago Ismail Serageldin, the Vice-President of World Bank, made his popular statement during an interview with *Newsweek* that, "Many of the wars in this century were about oil, but those of the next century will be over water." Although large-scale wars over water are not imminent in the next 10 years (NIC, 2012), water scarcity has the potential to cause regional tension and conflict. It could also engender border disputes, local tribal and ethnic warfare, as well as being the focus of terrorism. All these mostly exist because of the underlying competition for economic development (Gleick and Heberger, 2012), but could also be caused by compromises to water quality, quantity and timing by riparian states in shared waters. Table 2.1 illustrates examples of selected water conflicts and their causes.

This prediction of water conflict is not far from the truth, as many conflicts are already happening between nations over water. However, conflicts over water predate when Serageldin made his statement in 1995. The history of water conflicts is well documented (Toset et al., 2000; Wolf, 2007). Numerous studies have found that water, food and energy challenges are the primary contributors to international and domestic conflict (Brock, 2011; Gleick et al., 2014).

India had witnessed internal conflict over water stress between the states of Tamil Nadu and Karnataka over water from the Cauvery River in 1974 when Karnataka attempted to discontinue adequate water supply to the downstream state of Tamil Nadu, asserting that the 1924 agreement entailed a discontinuation of the water supply to Tamil Nadu after 50 years. The transboundary conflict between India and Pakistan as reported by Vaid and Maini (2012) and Qureshi (2017), is the result of disagreement over water allocation due to construction of the Kishangangan dam on the Jhelum river, and the Ratle and Baglihar dams on the Chenab River by India government in contravention of the Indus Waters Treaty (IWT). This conflict is further escalated by the Mumbai attack of November 26, 2008. The sharing of water from the Ganges River water for agriculture and drinking purposes is the main cause of conflict between India and Bangladesh. However, Hossain (1981) also noted the problem over the maritime belt and border security just as studies have shown that water conflicts are most commonly intertwined with other conflicts.

The conflict between Ethiopia and Egypt over the right to exploit the Nile, which is the world's longest river, running some 6853 km (4250 miles) through north and northeast Africa, is yet to be resolved (Abebe, 2014). The Nile's water volume has been declining, while demands for access to the Nile's water resources from upper riparian states (Egypt and Sudan downstream; South Sudan, Eritrea, Rwanda, Burundi and the Democratic Republic of Congo midstream; and Tanzania, Uganda, Kenya, and Ethiopia upstream) have been increasing. While Egypt insists on a guarantee of its "historic rights" to two-thirds of the river's flow, Ethiopia demands an "equitable" distribution of water among all of the riparian countries.

In 2011 and 2012 alone, violence over water was reported in every major developing region of the world, especially the Middle East, Africa and Asia, with additional important examples in Latin America (Gleick and Heberger, 2012). A critical analysis of this conflict trend presupposes that increased water stress is likely to escalate further future conflicts among nations. While conflicts within a national boundary may be easier to address, transboundary conflicts require more commitment and diplomacy. It is equally important to understand the contemporary stressors responsible for conflict over water.

TABLE 2.1
Selected Examples of Conflicts Over Water

Location of Conflict	Main Issue	Observation
Cauvery River	Quantity	The dispute on India's Cauvery River sprung from the allocation of water between the downstream state of Tamil Nadu, which had been using the river's water for irrigation, and upstream Karnataka, which wanted to increase irrigated agriculture. The parties did not accept a tribunal's adjudication of the water dispute, leading to violence and death along the river.
Okavango River	Quantity	In the Okavango River basin, Botswana's claims for water to sustain the delta and its lucrative ecotourism industry contribute to a dispute with upstream Namibia, which wants to pipe water passing through the Caprivi Strip to supply its capital city with drinking water.
Mekong River Basin	Quantity	Following construction of Thailand's Pak Mun Dam, more than 25,000 people were affected by drastic reductions in upstream fisheries and other livelihood problems. Affected communities have struggled for reparations since the dam was completed in 1994.
Incomati River	Quantity and Quality	Dams in the South African part of the Incomati River basin reduced freshwater flows and increased salt levels in Mozambique's Incomati estuary. This altered the estuary's ecosystem and led to the disappearance of salt-intolerant plants and animals that are important for people's livelihoods.
Rhine River	Quality	Rotterdam's harbor had to be dredged frequently to remove contaminated sludge deposited by the Rhine River. The cost was enormous and consequently led to controversy over compensation and responsibility among Rhine users. While in this case negotiations led to a peaceful solution, in areas that lack the Rhine's dispute resolution framework, siltation problems could lead to upstream-downstream disputes, such as those in Central America's Lempa River basin.
Syr Darya	Timing	Relations between Kazakhstan, Kyrgyzstan, and Uzbekistan—all riparian of the Syr Darya, a major tributary of the disappearing Aral Sea—exemplify the problems caused by water flow timing. Under the Soviet Union's central management, spring and summer irrigation in downstream Uzbekistan and Kazakhstan balanced upstream Kyrgyzstan's use of hydropower to generate heat in the winter. But the parties are barely adhering to recent agreements that exchange upstream flows of alternate heating sources (natural gas, coal, and fuel oil) for downstream irrigation, sporadically breaching the agreements.

Source: Wolf, A.T., *Annu. Rev. Environ. Resour.*, 32, 1–29, 2007.

2.3.1 Global Water Stressors

Water stressors are the activities or phenomena that negatively impact the quality, quantity and availability of water. The impacts of climate change are being predicted to affect water supply and use by several studies. About 1 billion people have been projected to experience water shortages by 2025 due to these impacts (Alcamo et al., 1997; Arnell, 1999; Vorosmarty, 2000). Apart from climate change resulting from high fossil energy consumption, other prominent stressors of global water include population, land use, urbanization and economic growth.

However, Rogers et al. (2004) posited that water crisis would not only be due to physical water scarcity but also mismanagement of water resources, or in other words, to poor governance of water resources. In order to manage our waters sustainably so that they will be able to provide utilities, especially in the transboundary basins, there is need for effective water governance, which could be achieved through water diplomacy.

2.4 WATER DIPLOMACY AND ITS RELEVANCE

Diplomacy has been a very important historical tool for maintaining good relations among people. Nicolson (1950) in his famous book, defines diplomacy as *the art of conducting dialogue between and among states.* A diplomat performs functions such as representation, negotiation, information, diplomatic protection, international cooperation and consular services (Iucu, 2010). However, diplomacy is not simply limited to between states, but wherever people live in different groups. Its functions, among many, include preventing conflicts or full-time war and bringing about peace between nations. Without diplomacy, the world will be an extremely difficult place to live.

Water diplomacy uses diplomatic techniques of negotiation, mediation and intercultural communication to promote sustainable development of water resources and transform the potential risks of competing demands and even conflict over water into forms of cooperation that extend beyond water and economics (Hefny, 2011).

On February 11, 2011, the United Nations General Assembly declared 2013 as the United Nations International Year of Water Cooperation. Water, as an important resource, often causes conflicts among groups of people or nations due to its indispensable nature and spatial distribution. The United Nations recognized this nature of water and therefore called for a concerted effort to address it. The Year of Water Cooperation was declared to highlight the history of successful water cooperation initiatives, as well as to identify burning issues on water education, water diplomacy, transboundary water management, financing cooperation, national/international legal frameworks and the linkages with the Millennium Development Goals (MDGs). In order to consolidate on the gains of the MDGs, the newly adopted SDGs also emphasize achieving greater access to clean water and sanitation, for which water diplomacy will be instrumental.

Adopting water diplomacy to prevent and resolve conflicts around the distribution, allocation and use of water in order to sustain peaceful coexistence is equally important. Therefore, water diplomacy is "*niche diplomacy*," with a focus on fruitful cooperation on water and its associated resources. It revolves around dialogue, cooperation, negotiation and reconciling conflicting interests among riparian states, and sometimes involves the institutional capacity and power politics of states (Hefny, 2011).

As societies grow, water management problems also grow from simple to complicated and then to complex. Solving these water problems requires a system of management that addresses the complexity of meeting competing water needs in a multisectoral manner and the claims to water by multiple stakeholders—including many nations and the needs of the ecosystems of such nations (Islam and Repella, 2015).

Water diplomacy is a framework that enables countries to negotiate agreements on water management for peaceful coexistence and sustainable development. Water diplomacy is a call to action beyond theoretical concepts, as it encourages the people whose lives are at stake to become involved in decision-making processes and negotiations and to understand the impact that choices made in the present can have on their lives and the lives of the coming generations. The relevance of water diplomacy puts sound scientific skills and the competencies of the diplomatic body to the benefit of the challenges posed by the decrease in per capita freshwater as a means to prevent or deter conflict and promote cooperation (UNITAR, 2013).

2.4.1 Framework of Water Diplomacy

Beyond the definition of water diplomacy provided in the literature, there is an agreement concerning its three features (Hefny, 2001) necessary to achieve the objectives of cooperation among nations and the resolution of conflicts. These include:

1. *The need to integrate multiple perspectives.* Because of the multifaceted nature of water demands and disputes, water diplomacy requires input from hydrologists, engineers, politicians, economists, sociologists and all stakeholders. These stakeholders include local communities, local governments, technical groups, non-governmental organizations, as well as representatives industries and water users.
2. *The importance of negotiation, mediation and intercultural communication.* No meaningful solution can be provided to water problems if these key tools are not properly utilized. Water diplomacy seeks negotiated solutions informed by adequate knowledge to resolve problems of water allocation and quality and competing water needs while taking the sensibilities of people and their interests into consideration.
3. *Support for the wider diplomatic process.* Bilateral or multilateral diplomatic engagement on transboundary water or basins can help redefine regional and international foreign policies among riparian states. It offers an opportunity to develop mutual relationships and international partnerships. There is also the possibility of scientific cooperation, knowledge transfer, expanded trade cooperation and capacity building.

2.4.2 Factors Promoting the Effectiveness of Water Diplomacy

Adoption of diplomacy may achieve little or no success in resolving conflicts between nations if cooperation is not entrenched between the stakeholders. It is therefore important to consider and address the factors that influence cooperation among groups in conflict for water diplomacy to be effective in solving transboundary water problems. Six factors identified by Huntjens and De Man (2017) are: the ability to build trust among competing stakeholders, the ability to organize multi-sector and multi-level interactions, the ability to manage a growing multi-actor policy environment, the ability to deal with uncertainties, sustainable financing and a sustainable legacy.

2.5 WATER DIPLOMACY TOOLS

Since the basic objective of water diplomacy is to achieve a sustainable water management framework that ensures access to equitable shares of water for human and ecosystem functioning, various tools are available to water diplomats to achieve it.

2.5.1 Negotiation

In order to achieve cooperation among nations on shared water resources, negotiation is very important to address the concerns of the parties involved. At the negotiation stage, employing convincing scientific knowledge along with knowledge about the socio-cultural, religious and political situations of the parties involved is paramount.

2.5.2 Cooperation

Another important tool of water diplomacy is cooperation. It is very effective in de-escalating conflicts between riparian countries (Genderen and Rood, 2011) and also promotes integrated water resource management principles (GWP, 2000). Prominent models of transboundary water cooperation include volumetric allocation, where a shared percentage of volume is agreed upon, and benefit sharing (Phillips et al., 2006). Such benefit sharing may be in the form of revenue sharing, hydropower trade, preferential electricity rates, compensation for costs or side payments (Daoudy, 2010).

2.5.3 Conventions, Treaties and Agreements

Global conventions, regional treaties and bilateral or multilateral agreements constitute legal tools in diplomacy and water governance. Iza (2004) compiled and analyzed both watercourse and non-watercourse international agreements that have relevance in protecting the earth's ecosystem and peaceful cooperation on transboundary water management.

The United Nations Convention on the Law of the Non-Navigational Water Uses of International Watercourses adopted by United Nations General Assembly on May 21, 1997 is an example of an international agreement. This convention re-emphasized the right of a watercourse state to utilize the watercourse within its territory in an equitable and reasonable manner, as well as the right to use and develop an international watercourse with the view of achieving optimal and sustainable utilization while taking into account the interests of other watercourse states and the protection of the watercourse.

The protocol on share watercourse systems in the Southern African Development Community (SADC) agreed to by all the regional actors and signed on August 28, 1995, is an example of regional as well as multilateral agreement with a focus on cooperation on shared water resources. The protocol made provision for maintaining a balance between resource development and environmental conservation. It also mandates members to guard against pollution or introduction of alien species that may alter the ecosystem and have deleterious effects in the shared watercourse system.

The agreement between the Federal Republic of Nigeria and the Republic of Niger concerning the equitable sharing in the development, conservation and use of their common water resources signed on July 18, 1990, in Maiduguri, Nigeria, is an example of a bilateral agreement. The primary focus of the agreement is on water quantity based on the equitable development, conservation and use of the shared water resources.

2.5.4 Scientific and Technical Knowledge

Another important tool in water diplomacy is scientific and technical knowledge. Scientific knowledge that is robust and convincing enough to create value for negotiating parties, such that trust is built to arrive at beneficial decisions, is important in water diplomacy. However instead of aiding negotiations, the mere use of scientific information to justify arbitrary (or political) decisions can be counterproductive (Susskind and Islam, 2012). Therefore, bringing scientific knowledge on board in all water negotiations will provide stakeholders with reliable information that can be used to formulate innovative ideas that can enhance cooperation.

2.6 CHALLENGES OF WATER DIPLOMACY

When diplomacy fails in resolving conflicts over competing interests about shared water, even after the deployment of all water diplomacy tools and considering all factors that engender cooperation, the objective of peaceful coexistence is jeopardized and sometimes leads to escalation of tension and arm conflict. The major challenge of water diplomacy is the inability to reach an agreement due to uncertainty in three domains (Susskind and Islam, 2012).

1. *Uncertainty of Information* exists when parties are unable to be confident about the likelihood of the occurrence of hydrologic events that can put their nations into an advantageous or disadvantageous position. For example, in an arid basin, it is quite difficult to predict

next year's rainfall by relying on historical records. This type of uncertainty might make it difficult for parties to shift ground and give away what is seen as theirs.

2. *Uncertainty of Action* exists when parties cannot predict a cause–effect relationship. Formulating a policy or adopting a program may or may not produce the desired results in the long run. This uncertainty also limits the adoption of a stand that is perceived as possibly limiting access to water.
3. *Uncertainty of Perception* exists when people do not see the reality. Rather, they see what they have in their minds, especially when issues have to do with either ideological or political matters. This type of uncertainty engenders mistrust and discourages consideration of the negotiating partner's concerns.

Since it is quite difficult to completely eliminate uncertainty during negotiation, the way out of the delicate situation is to involve experts among the representatives of the parties who will jointly undertake scientific analyses and come out with evidence-based information to bring about mutual trust that could lead to cooperation and the final agreement.

2.7 WATER DIPLOMACY AS SOLUTION TO THE CHALLENGES

As discussed previously, the competition for scarce water for economic growth, social well-being and ecosystem demand is a global challenge that often results in conflicts, especially in transboundary basins. This element of conflict further complicates the existing variables in a complex equation. The scale, urgency and complexity of the equation of water challenges require an inclusive, comprehensive and international approach combining diplomacy, innovation, partnerships and new funding mechanisms as its solutions (Huntjens and De Man, 2017).

Based on this approach, water diplomacy will be a formula to provide the needed solution to this complex problem. For example, diplomatic efforts can ensure implementation of the SDGs by all nations, as well as national evaluation based on the provided framework and indicators. Similarly, if water diplomacy is well engaged, it could turn conflict into cooperation, like former Secretary General of the United Nations, Kofi Annan, said (Wolf, 2007): "But the water problems of our world need not be only a cause of tension; they can also be a catalyst for cooperation . . . If we work together, a secure and sustainable water future can be ours."

Adopting multi-track water diplomacy, which considers multiple opportunities, will help build cooperation that will enable riparian states to undertake collaborative or joint investments in shared river basins, and address local or community-based conflicts and foster sustainable development. As mentioned earlier, the role of scientific and technical knowledge in reducing conflicts over access to water and its governance cannot be underestimated. Incorporating scientific synergy into water diplomacy could bring about technological innovations that can be deployed in the areas of water-efficient agriculture, wastewater minimization, pollution control, sustainable water desalination and water treatment and reuse techniques. All these will foster food and water security, good sanitation, healthy ecosystems, economic growth and sustainable development.

ACKNOWLEDGMENTS

I hereby acknowledge the support of the following institutions and people: The World Academy of Science (TWAS) and the Council of Scientific and Industrial Research (CSIR), Government of India, for granting me the CSIR-TWAS Postdoctoral Research Fellowship to pursue advanced research at the CSIR-National Institute of Science Technology and Development Studies (NISTADS), New Delhi, India; My supervisor, the Director of NISTADS, Prof. P. Goswami; and the Director General of National Centre for Technology Management (NACETEM), Prof. Okechukwu Ukwuoma for allowing me to take up the fellowship.

REFERENCES

Abebe, D. 2014. Egypt, Ethiopia, and the Nile: The economics of international water law. *Chicago Journal of International Law* 15(1). http://chicagounbound.uchicago.edu/cjil/vol15/iss1/4 (Accessed February 15, 2018).

Alcamo, J., Döll, P., Kaspar, F. and Siebert, S. 1997. *Global Change and Global Scenarios of Water Use and Availability: An Application of Water-GAP1.0*. Center for Environmental Systems Research, University of Kassel, Kassel, Germany.

Arnell, N. W. 1999. Climate change and global water resources. *Global Environmental Change* 9: 31–49.

Brock, H. 2011. *Competition Over Resources: Drivers of Insecurity and the Global South*. Oxford, UK: Oxford Research Group.

Choudhry, E. and Islam, S. 2015. Nature of transboundary water conflicts: Issues of complexity and the enabling conditions for negotiated cooperation. *Journal of Contemporary Water Research & Education* 155: 43–52.

Daoudy, M. 2010. Getting beyond the environment-conflict trap: Benefit sharing in international river basins. In *Transboundary Water Management: Principles and Practice*, Eds. Earle, A. Jägerskog, A. Öjendal, J., pp. 43–55. London, UK: Earthscan.

Genderen, R. V. and Rood, J. 2011. Water diplomacy: A niche for the Netherlands? https://www.clingendael.org/sites/default/files/pdfs/20111200_cling_report_waterdiplomacy_rgenderen_jrood.pdf (Accessed February 8, 2018).

Gleick, P. H., Ajami, N., Christian-Smith, J. et al. 2014. The World's water. *The Biennial Report on Freshwater Resources*, Vol. 8. Washington, DC: Island Press.

Gleick, P. H. and Heberger, M. 2012. Water and conflict: Events, trends, and analysis (2011–2012). In *The World's Water Volume 8: The Biennial Report on Freshwater Resources*, Ed. Gleick, P. H., pp. 159–171. Washington, DC: Island Press.

GWP. 2000. Integrated water resources management. Global Water Partnership (GWP) Technical Advisory Committee, Background Paper No.4. http://www.gwp.org/globalassets/global/toolbox/publications/background-papers/04-integrated-water-resources-management-2000-english.pdf (Accessed February 8, 2018).

Hefny, A. M. 2011. *Water Diplomacy: A Tool for Enhancing Water Peace and Sustainability in the Arab Region*. Technical Report, UNESCO.

Hossain, I. 1981. Bangladesh-India relations: Issues and problems. *Asian Survey* 21(11): 1115–1128.

Huntjens, P. and De Man, R. 2017. Water diplomacy: Making water cooperation work. The Hague Institute for Global Justice Policy Brief, pp. 5–10.

ICWE (International Conference on Water and the Environment). 1992. *The Dublin Statement and report of the conference*. Geneva, Switzerland: World Meteorological Organization.

Islam, S. and Repella, A. C. 2015. Water diplomacy: A negotiated approach to manage complex water problems. *Journal of Contemporary Water Research & Education* 155: 1–10.

Iucu, O. 2010. Diplomacy and diplomatic functions. *Manager* 11: 129–134.

Iza, Alejandro. 2004. *International Water Governance: Conservation of Freshwater Ecosystem*. IUCN International Agreements – Compilation and Analysis Vol. 1, pp. 1–26. Gland, Switzerland: IUCN.

NIC. 2012. Global trends 2030: Alternative worlds. National Intelligence Council https://cgsr.llnl.gov/content/assets/docs/Global_Trends_2030-NIC-US-Dec12.pdf (Accessed February 8, 2018).

Nicolson, H. 1950. *Diplomacy*. London, UK: Oxford University Press.

Perry, C. J., Rock, M. and Seckler. D. 1997. Water as an economic good: A solution, or a problem? Research Report 14. Colombo, Sri Lanka: International Irrigation Management Institute.

Phillips, D., Daoudy, M. Öjendal, J. Turton, A. and McCaffrey, S. 2006. *Transboundary Water Cooperation as a Tool for Conflict Prevention and for Broader Benefit-Sharing*, Stockholm, Sweden: Ministry for Foreign Affairs. http://www.protos.ngo/sites/default/files/library_assets/W_IWB_E16_Transboundary_Cooperation.pdf (Accessed February 8, 2018).

Qureshi, W. A. 2017. Water as a human right: A case study of the Pakistan-India water conflict. *Penn State Journal of Law & International Affairs* 5(2): 375–397.

Rogers, P. P., Ramón M. and Martínez-Cortina, L. 2004. *Water Crisis: Myth or Reality?* Leiden, the Netherlands: Taylor & Francis Group.

Rosegrant, M. W., Cai, X. and Cline, S. A. 2002. *World Water and Food to 2005: Dealing with Scarcity*. Washington, DC: IFPRI.

Shiklomanov, I. A. 1998. *World Water Resources: A New Appraisal and Assessment for the 21st Century*. Paris, France: UNESCO.

Shiva, V. 2002. *Water Wars: Privatization, Pollution, and Profit.* Cambridge, MA: South End Press.
Susskind, L. and Islam, S. 2012. Water diplomacy: Creating value and building trust in transboundary water negotiations. *Science & Diplomacy* 1(3): 1–7.
Toset, H. P. W., Gleditsch, N. P. and Hegre, H. 2000. Shared rivers and interstate conflict. *Political Geography* 19: 971–996.
UNITAR. 2013. Introduction to Water Diplomacy, Module 1: Water resources and their potential for conflict. UNESCO-IHE, University of East Anglia.
UN Water. 2014. Investing in water and sanitation: Increasing access, reducing inequalities. *UN-water Global Analysis and Assessment of Sanitation and Drinking-Water (GLAAS) 2014 – Report.* Geneva, Switzerland: WHO.
Vaid, M. and Maini, T. S. 2012. Indo-Pak water disputes: Time for fresh approaches. *South Asian Journal of Peacebuilding* 4(2): 1–14. http://wiscomp.org/peaceprints/4-2/4.2.5.pdf (Accessed February 15, 2018).
Vorosmarty, C. J. 2000. Global water resources: Vulnerability from climate change and population growth. *Science* 289: 284–288.
Wolf, A. T. 2007. Shared waters: Conflict and cooperation. *Annual Review of Environment and Resources* 32(3): 1–29.
WWAP (United Nations World Water Assessment Programme). 2015. *The United Nations World Water Development Report 2015: Water for a Sustainable World.* Paris, France: UNESCO.

3 Achieving Effective Water Management and Access within Africa's Socio-Political and Cultural Complexities

Issues and Policy Directions/Options

Olawale R. Olaopa

CONTENTS

3.1 INTRODUCTION

The issue of water supply systems and hygiene is becoming an increasingly popular topic within academia. Water to human existence is like blood in the circulatory system: it is a fundamental human requirement of life. Polluted, unhygienic water sources and indecorous or indiscriminate waste management have contributed to the spread of countless avoidable societal ailments (Godfrey, 2012). These facts, along with some global events, have proven wrong the belief and assumption that the entire quantity of water on earth has continued to remain unchanged. These events include the experienced global population explosion, the increase of irrigation-based agriculture and industrial development, which are mounting pressure on the amount and value of natural systems. Consequently, and as a result of the accompanying challenges, society has begun to realize that it can no longer follow a "use and discard" philosophy—either with water resources or any other natural resources (Cox, 1987). Hence, the apparent need for a consistent policy of management of water resources.

It is on this note that this chapter focuses on how effective water management systems can be realised within the milieu of African social, political and cultural complexities. Using as the basis or point of departure, "Culture Economy," a term coined by Mazzucato and Niemeijer (2002), to indicate the notion that a culture's practices, habits and lifestyles influenced and are also being influenced and determined by issues or events currently affecting them. The chapter will focus on the "culture economy" of Africa. In doing so Africa's cultural beliefs and relationship with water and how these impact water management and society water systems, as well as how, automatically, water systems impact and shift cultural practices are examined. The chapter concludes by providing a brief synthesis of the common cultural and religious uses of water in various cultural and religious belief systems in Africa with modern technologies and macro-level policies and agendas that could be used to foster, encourage and promote more sustainable approaches in future water-related policies, programs, projects and management.

3.2 WATER AND SOCIO-ECONOMIC DEVELOPMENT: ANY ROLE FOR AFRICA'S TRADITIONS, RELIGIONS, CULTURE AND VALUES

Socio-cultural, economic and political development processes cannot be easily separated from water resources due to the different varied interfaces between water and human activities. Water serves both as positive and negative inputs for many activities. It serves vital natural functions as a basic element of social and economic infrastructure, and as a natural facility contributing to mental and emotional welfare. Its negative roles are related to flooding and disease transmission. The significant magnitude and pervasive nature of these positive and negative attributes create a close relationship between water and human welfare (Cox, 1987; Godfrey, 2012). The history of man is replete with an unending sequence of water management activities aimed at improving the quality of life through augmentation of water's positive functions and reduction or mitigation of its negative functions.

The exact nature of the relationship between water and socio-cultural, economic and political development is hidden by several factors, however. A major factor, as argued by scholars including Cox (1987), Olokesusi (2006) and Mohsen (2013), is the difficulty of the development process itself. For instance, development is the outcome of several interrelated issues, and separation or seclusion of the effect of any sole variable may be challenging. Another difficulty in appreciating the general role of water is the mutually dependent nature of the numerous individual influences of water on human welfare, all of which often times become mainly substantial anytime water resource development activities are undertaken. While the valuable effects of different water project purposes can be aggregated, it may also include some level of engagement in which the accomplishment of a satisfactory level frequently requires trade-offs among the individual functions. Another basic aspect of the interactive relationship between water and socio-cultural, economic and political development is the possibility for development to impact water resources. Many socio-cultural, economic and political development activities involve modification of the quality and flow of water. As a result, the features of the water resource must be viewed as dynamic, along with the development process itself.

The above various arguments regarding the significance of water and its collaborative relationship in socio-cultural, economic and political development notwithstanding, there are a plethora of examples where water management activities, particularly irrigation, propelled and initiated by indigenous people's knowledge, technical skills and culture, played a central role in socio-economic development. A typical example can be found in the earliest known civilizations in the valleys of the Tigris, Euphrates, Nile, Indus and Yellow Rivers. In Mesopotamia, along the Tigris and Euphrates, irrigation was established at least as early as 4000 BCE. Although irrigated agriculture was the economic base of Mesopotamian civilization, navigation and flood control were also significant water management objectives (Fukuda, 1976, p. 21). Irrigation also played a major role in the development of Egyptian civilization along the Nile. The left bank of the Nile was under irrigation by 3400 BCE (Teclaff and Teclaff, 1973, p. 31), and this irrigation practice has been maintained until the 1980s with little modifications (Cox, 1987; Mazzucato and Niemeijer, 2002; Mohsen, 2013).

In the period between 400 and 200 BCE, floodwater diversion by means of dike construction has been described to have been a regular tool of war by the feudal states of the Yellow River Basin. This, to a certain extent, is an exceptional incorporation of water management into social processes (Greer, 1979, p. 25; Knutsson, 2014). In the western hemisphere, the Incas of Peru, which date back to 1000 BCE, provide an early example of irrigation civilization. Irrigation was practiced widely after 700 BCE, and watered agriculture was well developed at the time of the Spanish conquest in 1532 (Fukuda, 1976, pp. 29–30; Mazzucato and Niemeijer, 2002).

There is no doubt that water resource management, access and development are central to responsible socioeconomic growth and poverty reduction for enhanced sustainable development. However, sustainable development through the use of water for social and economic development is not only about guaranteeing people's access to water and sanitation. It also requires that they have a good quality of life where their cultures and values are recognised, respected and enhanced. Hence, cultural values and beliefs also have significant direct impacts on the governance institutions and stakeholders involved in water management. It is on this note that the debate around culture and development has been stimulated by a growing awareness that development programmes failed to consider the cultural environment and cultural factors influencing their sustainability. An added dimension to this is the issue of cultural rights and the recognition that people's cultural identity, beliefs and values can be a powerful ally as well as a barrier to development.

The role played by water is very significant in many religions and beliefs in Africa to the extent that societies and indigenous peoples have assigned religious and cultural beliefs and values to water for generations. This is reflected in the Indigenous Peoples Kyoto Water Declaration (2003), which states:

> We, the Indigenous Peoples from all parts of the world assembled here, reaffirm our relationship to Mother Earth and responsibility to future generations to raise our voices in solidarity to speak for the protection of water. We were placed in a sacred manner on this earth, each in our own sacred and traditional lands and territories to care for all of creation and to care for water. We recognise honor and respect water as sacred and sustains all life. Our traditional knowledge, laws and ways of life teach us to be responsible in caring for this sacred gift that connects all life. Our relationship with our lands, territories and water is the fundamental physical cultural and spiritual basis for our existence. This relationship to our Mother Earth requires us to conserve our freshwaters and oceans for the survival of present and future generations. We assert our role as caretakers with rights and responsibilities to defend and ensure the protection, availability and purity of water. We stand united to follow and implement our knowledge and traditional laws and exercise our right of self-determination to preserve water, and to preserve life.

Traditional management practices often reflect socially determined physical and spiritual norms for water allocation and sustainable practices. However, while countries have promoted equitable, efficient and social use of water resources, there is still very limited understanding about the use of water for cultural and religious activities and the value(s) attached to these uses and the manner in which these beliefs affect management decisions. Understanding the complex totality of these societal values, attitudes, beliefs and practices and their accompanying effects on water management strategies can be powerful drivers for social or economic growth. It equally engenders a sense of cultural identity and self-confidence obviously required for the design and the implementation of sound policy for effective water management systems (Godfrey, 2012; Knutsson, 2014). Surprisingly, these are poorly understood by African policymakers. To bridge this information gap there is need to showcase success stories in the utilisation of African values, attitudes, beliefs and practices in water management. This is with a view to providing a brief synthesis of the common cultural and religious uses of water of various cultural and religious belief systems in Africa for enhanced sustainable approaches in future water related policies, programs and projects.

3.3 SELECTED SUCCESS STORIES IN THE UTILISATION OF TRADITIONS, RELIGIONS, CULTURE AND VALUES IN WATER MANAGEMENT AROUND THE GLOBE

It should be noted that there are multifarious relations and interfaces in the way religious and cultural societies recognise and attach worth to water as a natural resource in Africa. This has implications for water management and development. The presence of different traditional religions as well as others make the concept of religion and its meaning a herculean task. The reason is that the concept can be subjected to different interpretation by different people given the difference in religious beliefs and affiliations. This notwithstanding, Gardner (2002) gave a definition that could be seen to receive global consensus and acceptability when he posited that religion is that which "offers a means of experiencing a sustaining creative force, whether a creator deity, an awe-inspiring presence in nature, or simply the source of all life" (Gardner, 2002, p. 10). In Africa and elsewhere, many different religions with different mode of worshiping and societal relationship exist. Regardless of all these, all religions possess some common features (UNESCO, n.d., as adapted from Bell and Hall, 1991). Some religion/worship is distinct and detached from day-to-day (domestic) life and their experiences could be personified or exemplified by an institution like a church, temple and mosque. In this circumstance, this religious experience is most of the time profoundly entrenched in the way societies live their lives. A major example is the traditional African religions, where the life of a society is incorporated with its natural resources.

On the other hand, the term "culture" is simply used to refer to a society and its way of life (Bodley, 1994). Many of the definitions of culture denote specific values and beliefs, while some others refer to the everyday life and behavior of people that flows from these beliefs (UNESCO, n.d.). Culture is very critical in any society, particularly in Africa, as it plays different roles in its different societies. The importance of culture was brought into perspectives when Mazrui (1980) lists several of its functions to include the provision of a lens of perception and a way of looking at reality (Mazrui, 1980, p. 47). Culture is all-encompassing, as it can be used to refer to customs and traditions, ritual, religion, music, dance, language, food, games, clothes and objects (Zenani and Mistri, n.d.; UNESCO, 2002). The implication of this definition is that in any activity, either implemented by the state or privately, due cognisance must be taken of local traditional and cultural frameworks whilst also challenging the untrue beliefs or impressions created about these local and cultural beliefs and values. Thus, as awareness of this potential loss of crucial knowledge spreads, innovative mechanisms are being sought by sociologists, anthropologists and development planners, to integrate indigenous and traditional knowledge systems into development planning and western science (Orange-Sengu, n.d.). In addition, the fact that culture has the propensity to shape people's way of life globally, the capacity to foster and bring about attitudinal changes and sustainable development is very high if utilised wisely. African indigenous and traditional knowledge and practices—as exemplified by their traditional religion and cultural values—are relevant to the water sector in the following areas: location; collection and storage of water; water resource management and irrigation methods; conservation strategies, hunting, fishing and gathering; medicinal plants and medicinal practices; and agricultural practices including: crop domestication, breeding and management; swidden agriculture; agro-ecology; agro-forestry; crop rotation; and pest and soil management, among others (Orange-Sengu, n.d.; Olaopa et al., 2014, 2016).

Indigenous people based on their traditions, religions and cultural values have successfully engaged in indigenous water resource management and irrigation systems, although there are variations in methods. These methods include, for example: canal, pond and well digging; *qanat*; open-surface irrigation; spate; under-surface and covered tunnels; buried clay pots; pitchers; wheels; wooden pivot methods; construction of bund around fields; and cultivation of low-moisture adaptive crops among societies. African farmers' knowledge of plant–soil–moisture relationships and the adaptive capability of domesticated plants also play a significant role in the evolution of indigenous irrigation and water resource management methods (Shajaat Ali, 2006).

Indigenous people have also noticed a causal relationship between events and their consequences for effective management of their resources, including water. Olaopa et al. (2016), for instance, argued that indigenous people have been able to predict the extent of a dry season or otherwise through their knowledge for easy water management and related activities. The sign or symbol used to predict the likelihood of a protracted dry season is when vulture is found gestating or hatching at a period when rainfall is generally anticipated (Olaopa et al., 2016; Domfeh, 2007).

Regarding water harvesting with respect to the precipitation pattern in semi-arid and arid areas where most total annual rainfall is low and often received in one or a few high-intensity storms, indigenous people have, over the centuries, developed local methods of water harvesting for conserving rain water in order to ensure their survival (Mohsen, 2013). In dry–wet and semi-arid regions with erratic rainfall, rainwater-harvesting practices, as asserted by Ferrand and Cecunjanin (2014), include, among other techniques, the construction of ponds, allowing the runoff to percolate through sand reservoirs. This method is referred to as *qanat,* although it has different names in different countries, and it is a traditional water extracting and transporting technique commonly used in Morocco, Spain, Syria, Iran and Central and Eastern Asia (Beshah et al., 2016). For instance, *qanat* is known as *khettara* in Morocco, *qanat* or *kārīz (kāhrez)* in Central and Eastern Asia including China, and *galerias* in Spain (Mohsen, 2013; Canavas, 2014). Not minding how dangerous and ordinary this technology could look compared to modern technologies, the fact remains that they have proven to be effective and perfectly sustainable for many years. The reason is not far-fetched. They were designed based on long-gathered experience and in a way that was compatible with local environment and local social systems (Ferroukhi and Chokkakula, 1996, p. 1; Chuvieco, 2012). However, the existence of these sources of water supply is faced with insignificant improvement compared to the developed countries where sophisticated modern water supply systems are in place, thus making the traditional water systems moribund (Knutsson, 2014).

The empirical work of Beshah et al. (2016) on Indigenous Practices of Water Management for Sustainable Services Case of Borana and Konso, Ethiopia, buttressed the earlier claim. The study showed that both communities have their own water source types, depending on local hydrogeological conditions. Specifically, Borana is known for the so-called Ella (wells) and Konso for Harta (ponds), which have been managed for more than five centuries. As argued by Arsano (2007), Coppock (1994) and UNESCO (2010), both have well-structured traditional institutions that have enabled their water systems to be sustained for centuries. The new technologies introduced failed to operate for long when it did not consider the local environmental, technical and social aspects—including norms, cultures, religions and traditional administrations (B. Bulee, personal interview, April 10, 2014; K. Garra, personal interview, April 12, 2014 cited by Beshah et al., 2016). All these go to show the propensity and efficacy of indigenous people's knowledge, technical skills and culture to fast-track water resource management in Africa and beyond.

At this juncture, it should be made clear that African indigenous and traditional knowledge and practices are not only limited to location, collection and storage of water, water resource management and irrigation methods, and conservation strategies alone. African traditional religion and cultural values also include elements of water purification, protection and hygiene. In some parts of Africa, water meant for drinking is subjected to one particular form of treatment or the other; for example, after collection, surface water is made to stand for some time to settle or precipitate before it is poured or transferred into clay pots for storage (Olokesusi, 2006).

Other water hygiene methods used to protect people from waterborne and other diseases in most Africa's homes or families include prohibition of the use of an individual's cup/calabash to get water from the family water pot and the provision of a single container or calabash tied to the pot as in the case of Southern Nigeria. Individuals are expected to use the tied calabash to draw water into their own containers whenever they want to drink.

In order to control or moderate water temperature and ensure that the water remains cool even when the outside temperature rises, the Yoruba of southwest Nigeria usually put the clay pots inside the home, as argued by Olokesusi (2006). The pot is then put on the locally processed shea butter before water collected from rooftops, streams or rivers is poured into them (Olokesusi, 2006). In some cases, given the beliefs among the people of southwest Nigeria, *Adenopus breviflorus* or *tagiri* (in Yoruba) is usually placed beside the family clay pot in the dry season to prevent evil spirits and "germs" causing measles (Olokesusi, 2006). According to Olokesusi (2006), the Uhabiri Ossah clan of Igbo stock in the Umuahia area of southeastern Nigeria endeavours to improve the taste of water for human drinking by heating the clay water pot with the hot smoke of Uhokiriho seeds.

To prevent environmental health hazards posed by livestock to humans, Sandford (1983 cited in Olokesusi, 2006) argued that in those areas where traditional *hafirs* and cisterns were most highly developed, drinking channels and other arrangements for dispensing water to livestock were carefully planned and enforced.

Examples of communities with daunting water challenges abound in Africa. The specific reasons for these problems are attributed to loss of environmental and traditional knowledge about water safety (Mahlangu and Garutsa, 2014). Some of these communities and people facing challenges include the Khambashe rural district in Buffalo City municipality, Eastern Cape and Cape Town, South Africa; the Dogon of Mali; Ader Doutchi Maggia area of Niger Republic; the Mossi farmers of Burkina Faso; and the Matengo and Luguru people in the Mbinga district of Ruvuma Uluguru regions in Tanzania. Others include the Mapan and Chingwan, in the Wokkos district of Pankshin in Plateau state and Shuwa and Kanuri of Kano state Nigeria as well as the Mossi, Kassena, Bifirfor and Dagari of Burkina Faso, among others. Given the earlier experience in water management and the apparent similarities in African culture, traditions and beliefs, the various methods detailed earlier with reference to individual communities' environmental, ethical and cultural peculiarities can be adopted and adapted in communities facing water challenges. Specifically, each community can utilise the indigenous knowledge within its culture and tribes to properly manage its water resources.

3.4 THE SCOPE AND IMPLICATIONS OF WATER USES AND MANAGEMENT IN AFRICAN TRADITIONAL RELIGIONS AND CULTURE

In Africa, the role played by traditions, religions and cultural values in the lives of the rural communities cannot be downplayed. Specifically, water has been, and still is, paramount to African traditional religions and culture. As a result of this, Africans have continuously preserved and sustained their relationships and association with water. Given the significant importance and role of water as found in many religions and beliefs in Africa, Africans frequently put in place regulations and instructions concerning the use and management of water. However, these regulations vary with religions, their teachings, values, philosophies and ideologies. According to Zenani and Mistri (n.d.), water as a "source of life" represents birth or re-birth and purity. The implication of this is that water is greatly symbolic and holy in status. Thus, water is a crucial resource in any ceremony and religious rite, as well as in endeavours like agriculture.

African societies present varieties of purposes of using and managing water. This includes agriculture, domestic consumption, and cultural and religious purposes (Olokesusi, 2006; Zenani and Mistri, n.d.). In religion, how water has been viewed as part of the holy life process and not simply another product for consumption is presented. It is important to note that, in the opinion of the west, water is viewed as a serviceable and marketable economic good. However, in the context of Africa and other similar societies, there are cultural and spiritual conceptions of water in addition to its social and economic importance. Thus, any simplified way of divorcing domestic and agricultural use of water from sacred and cultural cultural activities remains a difficult task.

Generally, there are a variety of practices performed by cultural and religious societies that entail the use of water from various sources including rivers, streams, dams and springs. Regardless of religious affiliation, some of these uses and their implication for water resource management and policy are discussed as follows.

In Africa, the traditional, Pentecostal, Anglican and Methodist churches all perform initiation ceremonies known as baptism that involve the use of water. Baptism is a type of initiation representing deliverance or emancipation from evil domination or persecution caused by immorality that separates man from his creator or the almighty (God). It is also referred to as a sacramental procedure or ceremony of proclaiming one's faith in Christ and is an indication of introduction to the church or religious group. Water is used as a cleansing and healing agent as it is taken to represent the spirit of God, and their preference is the use of fresh water sources such as rivers, streams, lakes and dams. In some cases baptism pools are built in secluded areas, particularly where communities no longer have access to natural water sources. The pool is nourished with fresh water and regarded as a sacred water feature in the African tradition. Baptism may be observed differently in different Christian denominations, although the practice signifies the same belief and remains consistent. Here, water used is mostly used and sourced from natural sources while very little or no water is extracted.

Related to the earlier is the use of water for initiation ceremonies. Here water is used in African culture to initiate young males and females into adulthood. These practices are more pronounced in rural areas. However, increasing rural–urban development has assisted in the spreading of the practice to the urban areas as well. It involves a situation where females and males in the society go to the river to bathe and cleanse their body for certain hours of a chosen day (Mutshimbwe Water Users Association, 2005). This cleansing ritual represents maturity and in the process, the person being initiated is introduced to river spirits and continuously bathes in the stream of the river (Zenani and Mistri, n.d.). Only water from natural sources—including rivers, streams, and lakes—is used on-site and very little water is extracted, while areas used by initiates for bathing should be isolated.

The use of water is important and vital in the performance of rites and rituals by most religions in Africa. Some religions in Africa, such as "Osun" and "Oya" worshipers in Nigeria and Hinduism in South Africa have tried to bring rain through the performance of certain water-related rituals. In this circumstance, good relations are maintained with ancestors by appeasing, pacifying and/or protecting the water spirits by making sacrifices to them whenever rain is needed either for agricultural purposes, to support their livelihood or for the purification of utensils to be used for rituals (Zenani and Mistri, n.d.).

Since water used for rites and rituals falls into two categories—rainmaking and cleansing—the sources and extent of use are different. For rainmaking, water is sourced and collected from both natural rivers and/or treated waterfalls believed to possess strong water spirits. Such water extracted may be used on-site or a small amount may be collected and taken away. The quantity of water used may not be easily determined. In the case of water for spiritual cleansing, only treated water is sourced and extracted while less than 5 L of water is used to clean utensils used for rituals (Zenani and Mistri, n.d.).

Apart from the traditional religions' use of water for rites and rituals, the Islamic faithful also use water for these purposes. The Muslims use water for ablution either for cleansing and purification. For instance, they must wash before standing to pray to the almighty Allah. To facilitate this, mosques are built with courtyard having a pool of clean water for this purpose. Ablutions are of two types, the *ghusl* (major ablution) and the *wudu* (minor ablution). The former is and must be done before praying and burying any deceased person, and must also be performed after sexual intercourse (*Janabat*), while the latter must always be performed before each five daily prayers. Pure water—though not necessarily from the river—is used to wash the face and rub the head, hands, elbows and feet up to ankles. Here, no preferred sites are required, and tap or treated water is adequate, while approximately 5 L of water per person is required for ablution (General Secretary of the Jamiet Ulama, 2005).

In addition, one of the traditional and cultural uses of water is to serve as the last resting place for the dead. An example of this is found in Lake Fundudzi in the Northern Province of South Africa, which serves as a sacred lake to the Vha-Venda people (Mutshimbwe Water Users Association, 2005).

One of the strongly held beliefs is that it is inhabited by the god of fertility in the form of a python, symbolic of the Vha-Venda ancestors and treated like a holy shrine (Zenani and Mistri, n.d.). After burying a deceased member of the tribe for a number of years, the bones are exhumed, cremated and thrown into the lake, which thus serves as the final resting place for the ancestors. This practice is also common among the Hindu faiths in South Africa as the conventional and preferred method of disposing of the deceased. Cemeteries and funeral grounds are usually located near these rivers and accessed by the wider community (Zenani and Mistri, n.d.).

Sites for holy places such as the Celestial Churches, Cherubim and Seraphim churches, Hindu temples and mosques are usually located on riverbanks, on the foot of a mountain or along the coastline. This is closely related to the use of water for medicinal purposes. Those who hold this view believe in the ideology of water animism, which posits that natural objects such as river water, lakes and springs are "living and possess souls" (Water page-water in animism cited in Zenani and Mistri, n.d.). African traditional healers collect water from rivers and streams to prepare healing medicine since they are believed to possess healing powers from the spirits residing in them. In this circumstance, less than 5 L is extracted of water from natural sources (Zenani and Mistri, n.d.). Other religious uses of water include livestock farming and food production.

From the earlier discussions, in quantifying the extent to which water is or can be used by traditional, religious and cultural communities in Africa, two conclusions can be drawn. One is that water can be used for spiritual and physical health, recreation in its broad sense, domestic uses, livestock and agriculture. Second, water is used for preservation purposes, including the protection of sacred spirits in the rivers, lakes, sea and natural springs; protection of family and village totems that dwell in these water sources; and protection of some of areas and landscapes that have historical values and importance to the endowed communities, including waterfalls, rocks hanging over rivers, stones protruding from the depths of the river or lake or the convergence of two or more rivers. The implication therefore is that most water policies do not include some of these in their definitions of water use and as such have no separate category for religious and cultural uses. This makes it difficult for both the policymakers and those responsible for the implementation of water-related policies and projects to identify the affected cultural and religious groups. This equally affects the determination of the quantity of water used, particularly for non-commercial purposes, as well as whether those categories of people using small quantities of water are expected to register or apply for licenses.

In situations where cultural and religious uses of water required very little water to be extracted, and water is generally used from the site accessed—as in the case of baptism, cleansing and rituals—and where activities of cultural and religious communities involve the use of water as in the case of the last resting place, this is capable of affecting water quality and compromising health standards. This therefore requires effective understanding and intervention from the policymakers.

The traditional and cultural use of water for livestock farming is likely to have a negative impact on water quality in that the defecation of cattle is likely to increase the nutrient content of fresh water and underground water stores, and could increase the growth of phytoplankton and algae blooms (Zenani and Mistri, n.d.).

It is equally important to state that the implications of traditional and cultural activities on the use of water and its management are not only limited to the quantity and quality of water available for use. Human traditional and cultural activities are capable of having adverse consequences for communities. Some of the consequences, as highlighted by Zenani and Mistri (n.d.), may include the destruction of river ecosystems, while tinkering with river streams and dams may cause flooding that may affect the social, economic and cultural activities of local residents. Most especially, flooding is capable of washing away their ancestors' graves, which in most cases are close to rivers and dams. This may cause water pollution and may constitute a health hazard. This notwithstanding, these traditional methods of water management and conservation are capable of promoting industrial development in that the water conserved through traditional technologies, for whatever purpose, can be used for irrigation to boost the mechanised farming system or for industrial use.

One way to easily achieve this is through effective education and awareness programmes that bring into the limelight the advantage of such action to the communities concerned. The recognition of society's cultural values for water use and related traditional technologies combined with modern technologies will facilitate this effort. This without prejudice to the fact that excessive mining of groundwater and aquifers for industrial and commercial purposes tend to affect rivers and dams that are also used for cultural and religious purposes.

3.5 ISSUES IN THE INTEGRATION OF TRADITIONS, RELIGIONS AND CULTURAL VALUES IN WATER USES AND MANAGEMENT AND CHALLENGES

The examples discussed earlier have clearly shown the age-long existence, use and practice of African indigenous knowledge systems. Colonialism, Western education and religion have caused the great neglect and stultification of its growth, thereby reducing its further development and Africans' capacity to resolve their societal challenges using locally developed technologies. In fact, this has been responsible for the paucity of pragmatic enquiry and records of this knowledge type in African countries, which has created knowledge gaps on a spatio-temporal scale (Pacey and Cullis, 1986, p. 127). This has resulted in the difficulty in tracing the developmental trend and improvement of any given technique or method over a long period of time in a methodical and organised way.

Consequently, an effective discussion of water use, management and policies must take cognisance of the intricate totality of the societal values, attitudes, beliefs and practices and their accompanying effects on water availability and allocation. Finding ways of connecting these has proven to be challenging, while at the same time remains critical to achieving effective water management and sustainability. To attain this requires the acknowledgement, the consideration and beliefs of cultural and religious societies about the sacredness of fresh water sources; recognising that people's relation to fresh water has resulted in increased identity for non-material culture through folklore, music, mythology, oral tradition and customs, and that these form the basis of social practices and the traditional forms of social institutions for managing water resources; and know that the cultural and religious values about water are of great importance to communities, serve as an expression of their identity (Oestigaard, 2009; Fisher and Rucki, 2017) and create a sense of belonging, loyalty and nationalism, the loss which may indicate isolation from the fresh water source and their ancestral origin.

Various stakeholders, relevant government ministries, departments and agencies have a leading role to play in assisting the religious and cultural communities to achieve their goals within the provisions of the law. In this regard, even though proactive efforts at promoting interests in indigenous knowledge systems was taken all over the world almost forty years ago (Warren and Cashman, 1988) and has led to the establishment of more than thirty indigenous knowledge resource centres around the world, including the Kenya Resource Centre for Indigenous Knowledge (KERIK) and the African Resource Centre for Indigenous Knowledge located in the Nigeria Institute for Social and Economic Research (NISER), Nigeria, the Centre for Indigenous Knowledge, North-West University, Mafikeng, South Africa, few success has been recorded. The challenges retarding their effective operation include shortage of funds on the part of governments in these and other African countries to address water and sanitation issues and research, as well as to fund existing institutions or establish suitable and appropriate institutions to manage the implemented systems. The reason for this was ascribed by Foster and Briceño-Garmendia (2010), cited in Hukka and Katko (2015), to the low and uncollectable tariff accrued and generated from such services, and the lack of external agents interested in assisting in water development compared to the number of service explorers and beneficiaries.

Another challenge to effective integration of African traditions and cultural values is the communities' total trust, adherence, patriotism and loyalty to these institutions and their resistance and lack or difficulty in trusting newly introduced systems.

Questions remain about how water policies will cater to development-integrated water management strategies that include cultural and religious systems in cases that lack information on how communities take care of this resource, as well as the impacts of contextually changing cultural and religious uses of water on the water sources among different communities and denominations in Africa. More specifically, how can a general framework that observes and guarantees the religious freedom of all peoples at the local level be designed? How will the rights of individuals belonging to a cultural, religious or linguistic community be preserved when other members of the same community enjoy their own culture, practice their own religion and use their own linguistics among other water-related rights without interference? How will a holistic approach to water use—environmental, cultural, spiritual, and aesthetic—be reflected in the water policy in a manner consistent with the other sections of the constitutional provisions? These remain very serious challenges that need critical consideration. Other multifaceted issues that the incorporation of indigenous and traditional knowledge raised include land tenure rights, genetic resource ownership, intellectual property rights and benefit sharing.

3.6 CONCLUSION

There is no doubt that the knowledge of African religious, traditional and cultural values is vital to the determination of the quantity, quality, access and use of water and water management. Focusing research on these issues would bring effective answers about the extent to which the local cultural values, religions and traditions contributed to the development of water management policy, as well as what the main factors were that contributed to the development of African societies in general, and how Africa will manage its declining water reserves, along with the relevance of the different belief systems and whether there is any need either to maintain these belief systems or to care for all African people and their respective societies in the same way in the process of managing water resources. These, and many other questions are among the many touchy and controversial problems arising around water management and sustainability. All these suggest the need for policymakers and stakeholders to be proactive in taking necessary action. This is especially so given the high rate at which the use of global water is increasing, the limited availability of water and the accompanying stress faced by a significant portion of the world population, given the United Nations' projection that the share of people who will be affected by water scarcity may double during the years to come.

3.7 RECOMMENDATIONS

On the earlier note several efforts have been taken, and mitigation measures have been suggested and put in place. This include: the need for and design or development of national water policies; the decentralisation of water management and the appropriate definition of its boundaries; the separation of the development and regulatory aspects of water supply and sanitation delivery functions; the entrenchment of financial discipline and accountability in many relevant government ministries, departments, agencies and utilities; sound planning and design of sustainable, economically efficient water projects; ensuring successful water management through proper understanding, application and maintenance of any introduced technologies; adequate consideration and establishment of appropriate fees and tariff structures, as well as other economic, environmental, financial and social issues; the elimination of adverse environmental effects on water resources through clear environmental policies and strategies that allow the best and efficient use of these water resources; the proper coordination of activities in the water subsectors to effectively address multisector, interrelated issues such as water-related health and environmental problems, and multi-purpose hydropower dam projects, among others (ADB and ADF, 2000).

In spite of the earlier suggestions and efforts, little success has been achieved with respect to water management, as there are still persistent challenges to effective water policy formulation and implementation. It is on this note that effort is required to direct the searchlight for a solution towards

African traditions, religions and cultural values hitherto neglected in the diagnosis of and prognosis for the problem. Specifically, the ineffectiveness of some proffered solutions and approaches to water management are traceable to their lack of meaningful inclusion of the existing approaches to water policy and planning processes (Oestigaard, 2009; Zenani and Mistri, n.d.). To be specific, traditional or customary access and rights to water is rarely recognised by policymakers and the state authorities that nowadays control indigenous areas and sources of water that are critical to the cultural, spiritual and physical wellbeing of the rural dwellers. There is also a lack of understanding or total neglect of the rationale for integrating Africa's indigenous knowledge systems in water use and management with Western science.

For effective water management, policy discussions and directions must therefore focus on issues of availability, efficiency, human equity and the needs of ecological systems, without mortgaging the well-being of future generations. For such policy to equitable, fair and just, it must be a bottom-up recognition of the inputs and contributions of all the principal stakeholders. This must involve proper consideration and systematic understanding of how water resources are sustained and used, and ethical appreciation of how different participants value water and understand the notion of equity. In doing this, special attention and credence must be given to the power of religion and culture (Oestigaard, 2009; Acquah, 2011; Zenani and Mistri, n.d.). This is so because, for at least 2–3 billion people living across the globe, religion shows how to deal with water and not scientific knowledge, and this influence of religion and tradition on water use and management is underestimated (Hans-Curt Flemming Biofilm Centre, 2008).

There is a need to sincerely acknowledge, respect and engage the social and cultural bases within which traditional knowledge is embedded with the aim of encouraging their role in public life and policy processes. Without doubt, religious and cultural organizations have the ability to make significant contributions to the effort to build a sustainable world. The recognition of the rights of indigenous people to own, access and realise benefits of their knowledge resources and systems, requires both: the full and equal participation of traditional knowledge holders during all stages of development plans, programs and policies; and that partnerships with traditional knowledge holders are only entered into with prior consent and that they are fully informed and understand the ramifications of partnerships. The promotion of training and capacity development programs to better equip scientists and traditional knowledge holders to conduct research on traditional knowledge (ICSU, 2002) will also motivate local residents and encourage people's acceptance of policies, reducing the chance of policy failure.

It is also important to ensure and promote environmental and sustainable governance institutions that guarantee effective integration of Africa's indigenous knowledge systems in water use and management with Western science. This can be resolved through the establishment of a mechanism for systematically acquiring data and knowledge of the water system at all institutional levels. This can be facilitated by adapting the Indian Digital Library of Traditional Knowledge and World Intellectual Property Organization guidelines on data acquisition, storage and codification (Adigun, 2014; Emeagwali and Sefa Dei, 2014). This promotes the availability of clean data and thereby encourages sound, evidence-based policy planning and design. It will also ensure sustainability and prevent water management policy decisions based on mere guesses.

REFERENCES

Acquah, F. (2011). *The Impact of African Traditional Religious Beliefs and Cultural Values on Christian Muslim Relations in Ghana from 1920 through the Present: A Case Study of Nkusukum-Ekumfi-Enyan area of the Central Region* (Unpublished Doctoral Thesis). PhD in Theology, University of Exeter.

Adigun, T. A. (2014). Application of competitive intelligence in indigenous knowledge: The role of the library professionals in Nigeria. *Journal of Research in Education and Society* 5(2).

African Development Bank (ADB) and African Development Fund (ADF). (2000). *Policy for Integrated Water Resources Management*. OCOD. African Development Bank/Africa Development Fund, Abidjan, Côte d'Ivoire.

Arsano, Y. (2007). *Ethiopia and the Nile: Dilemmas of National and Regional Hydropolitics* (Doctoral thesis). University of Zürich, Switzerland.
Behailu, B. M., P. E. Pietilä, and T. S. Katko. (2016). Indigenous practices of water management for sustainable services: Case of Borana and Konso, Ethiopia. *Sage Open* 6(4), 1–11.
Bodley, J. H. (1994). *Cultural Anthropology: Tribes, States and Global Systems*. Mountain View, CA: Mayfield Publishing.
Canavas, C. (2014). Public awareness and safeguarding traditional knowledge: Challenges and conflicts in preserving and representing kārīz/kǎnérjǐng in Xinjiang, PR China. *Water Science & Technology: Water Supply* 14, 758–765.
Chuvieco, E. (2012). Religious approaches to water management and environmental conservation. *Water Policy* 14, 9–20.
Cleaver, F. (2012). Development through bricolage: Rethinking institutions for natural resource management. London, UK: Routledge.
Coppock, D. L. (1994). *The Borana Plateau of Southern Ethiopia: Synthesis of Pastoral Research, Development, and Change, 1980–91* (Vol. 5). Addis Ababa, Ethiopia: International Livestock Research Institute (ILRI) (aka ILCA and ILRAD).
Cox, W. E. (1987). *The Role of Water in Socio-Economic Development*. Paris, France: Educational, Scientific and Cultural Organization (UNESCO).
Domfeh, K. A. (2007). Indigenous knowledge systems and the need for policy and Institutional Reforms. *Tribes and Tribals* 1, 41–52.
Emeagwali, G., and G. J. Sefa Dei (Eds.). (2014). *African Indigenous Knowledge and the Discipline*. Rotterdam, the Netherlands: Sense Publishers.
Ferrand, E. A., and F. Cecunjanin. (2014). Potential of rainwater harvesting in a thirsty world: A survey of ancient and traditional rainwater harvesting applications. *Geography Compass* 8, 395–413.
Ferroukhi, L., and S. Chokkakula. (1996). Indigenous knowledge of water management. In *Reaching the Unreached: Challenges for the 21st Century. Selected papers of the 22nd WEDC Conference*, Ed. Pickford, J. et al., New Delhi, India, Loughborough, UK: Intermediate Technology Publications in association with The Water, Engineering and Development Centre.
Fisher, J., and K. Rucki. (2017). Re-conceptualizing the science of sustainability: A dynamical systems approach to understanding the nexus of conflict, development and the environment. *Sustainable Development* 25, 267–275. doi:10.1002/sd.1656
Foster, V., and C. Briceño-Garmendia. (2010). *Africa's Infrastructure: A Time for Transformation*. Africa Development Forum. World Bank. https://openknowledge.worldbank.org/handle/10986/2692
Fukuda, H. (1976). *Irrigation in the World: Comparative Developments*. Tokyo, Japan: University of Tokyo Press.
Gardner, G. (2002). Invoking the Spirit: Religion and Spirituality in the Quest for a Sustainable World. World Watch Paper 164, December.
General Secretary of the Jamiet Ulama – KZN Council for Muslim Theologians (2005).
Godfrey, M. M. (2012). Water security and management in Burkina Faso: How socio-political and cultural complexities affect development, Independent Study Project (ISP) Collection. Paper 1350. http://digitalcollections.sit.edu/isp_collection/1350
Greer, C. (1979). *Water Management in the Yellow River Basin of China*. Austin, TX: University of Texas Press.
Hans-Curt Flemming Biofilm Centre. (2008). *Water: Religion, Mythology, Art and Beauty*. Lecture Delivered, July 8.
Hukka, J. J., and T. S. Katko. (2015). Appropriate pricing policy needed Worldwide for improving water services infrastructure. *Journal-American Water Works Association* 107, E37–E46.
ICSU. (2002). Science, traditional knowledge and sustainable development. ICSU Series on Science for Sustainable Development No. 4. International Council for Science, Paris, p. 19.
Indigenous People's Kyoto Water Declaration Presented at the Third World Water Forum in Kiyoto in March 2003. Appeared in 'Indigenous Water Initiative' 2009. https://www.activeremedy.org/wp-content/uploads/2014/10/indigenous_peoples_kyoto_water_declaration_2003.pdf.
Knutsson, G. (2014). The role of springs in the history of Sweden. *Vatten—Journal of Water Management and Research*, 70, 79–86.
Mahlangu, M., and T. C. Garutsa. (2014). Application of indigenous knowledge systems in water conservation and management: The case of Khambashe, Eastern Cape South Africa. *Academic Journal of Interdisciplinary Studies*, Vol. 3. Rome, Italy: MCSER Publishing.
Mazrui, A. A. (1980). *The African condition: A political diagnosis* [The Reith Lectures]. London: Heinemann Educational Books.

Mazzucato, V., and D. Niemeijer. (2002). The cultural economy of soil and water conservation: Burkina Faso. *Development & Change* 31(4), 831–855.

Mohsen, T.-J. T. H. (2013). Review of ancient wisdom of Qanat, and suggestions for future water management. *Environmental Engineering Research* 18, 57–63.

Mutshimbwe Water Users Association. (2005). Use of water for Initiation ceremonies in the South African culture. In *Vuyisile Zenani and Asha Mistri* (n.d.) A desktop study on the cultural and religious uses of water using regional case studies from South Africa. A Report. Department of Water Affairs and Forestry (DWAAF), South Africa.

Oestigaard, T. (2009). *Water, Culture and Identity: Comparing Past and Present Traditions in the Nile Basin Region*. Nile Basin Research Programme. BRIC Press.

Olaopa, O. R., O. Ayodele, Y. O. Akinwale, and I. O. Ogundari. (2014). Security and securitization in plateau state: The significance of indigenous knowledge and practices. *A Paper Accepted for Presentation at the International Conference organised by the Department of History and International Studies*, University of Jos, Nigeria, June 30 to July 2.

Olaopa, O. R., V. Ojakorotu, and O. Ayodele. 2016. Developing, utilising and promoting African indigenous knowledge practices in strategic sectors: Issues, challenges and policy options. *Presented at 6th International Conference on African Unity for Renaissance on The Knowledge, Spiritual and Struggle Heritage for Re-imagining Innovative Africa*. UNISA CAMPUS, Pretoria, South Africa at Saint George Hotel, May 22–25.

Olokesusi, F. (2006). Survey of indigenous water management and coping mechanisms in Africa: Implications for knowledge and technology policy. ATPS Special Paper Series No. 25. Kenya, African Technology Policy Studies Networks.

Orange-Sengu (n.d.). People and the river. River awareness kit. https://www.Indigenous Traditional Knowledge Orange-Senqu River Awareness%20Kit.html. Accessed in March 17, 2018.

Pacey, A., and A. Cullis. (1986). *Rainwater Harvesting*. London, UK: Intermediate Technology Publications.

Shajaat Alli, A. M. (2006). Indigenous knowledge of water resource management in dry lands: A conceptual framework. *International Symposium on Drylands Ecology and Human Security (ISDEHS)*, March 18, 2018.

Teclaff, L. A., and E. Teclaff. (1973). A history of water development and environmental quality. In *Environmental Quality and Water Development*, Eds. C. R. Goldman, J. McEvoy III, and P. J. Richerson. San Francisco, CA: W. H. Freeman and Company, pp. 26–77.

UNESCO. (2002). Universal declaration on cultural diversity: A vision, a conceptual platform, a pool of ideas for implementation, a new paradigm. Cultural Diversity Series No. 1. A document for the World Summit on Sustainable Development, Johannesburg, August 26–September 4.

UNESCO. (n.d.). UNEP International Environmental Education Program. Adapted from Bell R and Hall R (1991). Impacts: Contemporary issues and global problems.

United Nations Educational, Scientific and Cultural Organization. (2010). Cultural properties—Konso cultural landscape (Ethiopia) (34 COM). Retrieved from http://whc.unesco.org/archive/2010/whc10-34com-20e.pdf

Warren, D. M., and K. Cashman. (1988). Indigenous knowledge for sustainable agricultural and rural development. Gatekeepers Series No. SA10. IIED, London.

Zenani, V., and A. Mistri (n.d.). A desktop study on the cultural and religious uses of water using regional case studies from South Africa. A Report. South Africa: Department of Water Affairs and Forestry (DWAAF).

Section II

Freshwater by Desalination

4 The Capability of Forward Osmosis in Irrigation Water Supply

Alireza Abbassi Monjezi, Maryam Aryafar,
Alasdair N. Campbell, Franjo Cecelja, and Adel O. Sharif

CONTENTS

4.1 INTRODUCTION

The availability of clean water for agriculture has become a growing concern due to the increasing global water scarcity. The agriculture sector already accounts for around 70% of the total freshwater withdrawals in the world, whereas domestic and industrial water use constitute about 10% and 20% respectively (Alexandratos and Bruinsma, 2012). The estimated agriculture water demands for irrigation are projected to exceed the available freshwater resources by the year 2050, taking into account population growth and the effects of global warming (Alexandratos and Bruinsma, 2012). Moreover, water sources often suffer from quality deterioration due to flooding, reduced rainfall and the discharge of inadequately treated municipal and industrial effluents. Irrigation with salty water is limited due to a number of drawbacks, such as lower yields, limited choice of crops, damage to soil structure and salt accumulation in the root zone unless large volumes of water in excess of the plant requirements are provided (Sowers et al., 2011). Emerging methods to supply freshwater such as brackish or seawater desalination, wastewater reclamation and reuse are therefore key to sustainable solutions for the principal problems of water quality and quantity, energy consumption, water supply costs, environmental impact and the increasing demands for freshwater for irrigation.

As for water availability, 71% of the surface of the Earth is covered with water, but only 1% is suitable for human consumption; the remaining water is either saline or locked in ice. It is estimated that over 80% of water used worldwide is neither collected nor treated (Beltran and Oshima, 2006). Wastewater treatment and seawater desalination constitute potential sustainable sources of freshwater supply for domestic, industrial and agricultural use. It is generally accepted that it is less expensive to treat wastewater produced in agriculture than to desalinate seawater.

Along with the growing demand for freshwater, there is also an increasing quantity of wastewater that needs to be treated adequately to meet public health and environment discharge regulations. Desalinated water is of high quality and leads to less negative impact on soil. Irrigation with desalinated water could result in a 24% increase in crop yield and simultaneously lead to about a 45% reduction in the current freshwater volume used (Beltran and Oshima, 2006). However, desalination is an energy-intensive process and consequently an expensive option. On the other hand, wastewater reclamation and reuse is limited by the presence of heavy metals, pharmaceuticals and other substances that are difficult to remove by conventional wastewater treatment processes (Ghermandi and Messalem, 2009).

The desalination technologies currently in use are classified into two main categories: thermal and membrane based. The three thermal distillation methods most widely used for seawater desalination include multiple-effect distillation, multi-stage flash distillation and vapour compression distillation. Membrane-based technologies are commercialised in two important processes: reverse osmosis and electrodialysis. The current membrane and thermal desalination systems have numerous limitations, including high energy consumption and capital cost (especially for thermal methods), coupled with negative environmental impact due to the discharge of brine and chemicals. The cost of desalted water depends predominantly on the type of desalination process, the quality of the feed and product water, plant capacity and the available options for waste disposal. Thermal processes are suitable for desalinating high salinity waters. However, these methods are rather expensive because of the high operating and construction costs. In contrast, the reverse osmosis desalination process is less expensive because of the lower energy demands. In reverse osmosis, mechanical pressure is applied to the concentrated solution, the water molecules are forced through the semi-permeable membrane and the contaminants are left behind. The cost per cubic metre of water desalinated by various methods and plants demonstrated in practice are summarised in Table 4.1.

Table 4.1 shows that thermal desalination processes costs are generally higher than the costs of the reverse osmosis process. In addition, thermal desalination processes rely on fossil fuels as the main sources of energy, making the cost of water highly dependent on general energy prices in addition to the environmental impact of increased CO_2 emissions and higher quantities of concentrated brine discharged to the sea at high temperature. It is for these reasons that low-cost and renewable energy sources are more frequently considered for desalination processes. However, it has been

TABLE 4.1
Desalinated Water Cost of Various Desalination Plants

Desalination Process	Plant Location	Product Water Cost (US$/m³)	References
Multi-stage flash	Singapore	1.5	Ghermandi and Messalem (2009)
	Qatar	2.74	Saif (2012)
	Abu Dhabi	2.48	Saif (2012)
Multi-effect distillation	USA (California)	0.46	Ghermandi and Messalem (2009)
Thermal vapour compression multi-effect	Qatar	2.48	Darwish (2014)
Reverse osmosis	Ashkelon	0.53	Fritzmann et al. (2007)
	Qatar	1.1	Darwish (2014)
	Singapore	0.5	Fritzmann et al. (2007)

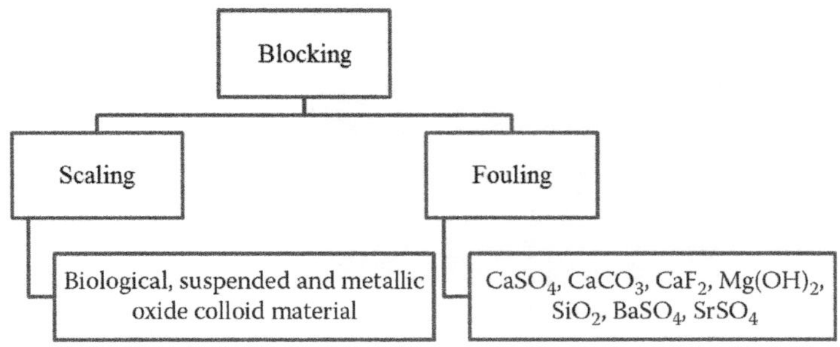

FIGURE 4.1 Limiting factors to membrane-based desalination by reverse osmosis. (Adapted from Fritzmann, C. et al., *Desalination*, 216, 1–76, 2007. With permission.)

demonstrated that the use of membrane-based desalination processes, such as reverse osmosis, is limited to lower salinity levels and result in a high level of scaling and fouling on the membrane. Figure 4.1 shows a number of salts and pollutants that are the main causes of blockage through scaling and fouling in reverse osmosis membranes in desalination (Fritzmann et al., 2007).

It has been established that an adequate pre-treatment process can reduce fouling and scaling on the membrane surface and hence extends the membrane lifetime and performance while reducing operating costs. Still, further development of innovative technologies to reduce the cost and environmental impact of treated water is needed to ensure a sustainable supply of water.

As an emerging technique, forward osmosis offers the possibility to provide a reliable and cost-effective method for producing freshwater through reducing the energy consumption and capital cost of desalination (Nicoll, 2013). The process is driven by the osmotic pressure difference between feed water (seawater, brackish water or wastewater) and the concentrated draw solution, which are separated by a semi-permeable membrane allowing water to permeate but retaining solutes. The diluted draw solution is then separated from the clean water by a thermal (Figure 4.2) or membrane (Figure 4.3) regeneration system and recycled to the forward osmosis stage.

The forward osmosis process is a comparatively new approach that offers the possibility for a substantial reduction in the energy consumption of desalination in comparison with the previously mentioned conventional processes. However, forward osmosis is still not widely industrialised because of (a) draw solution availability and regeneration cost and (b) membrane water permeability and rejection rate. It is assumed that the main energy-intensive stage in the forward

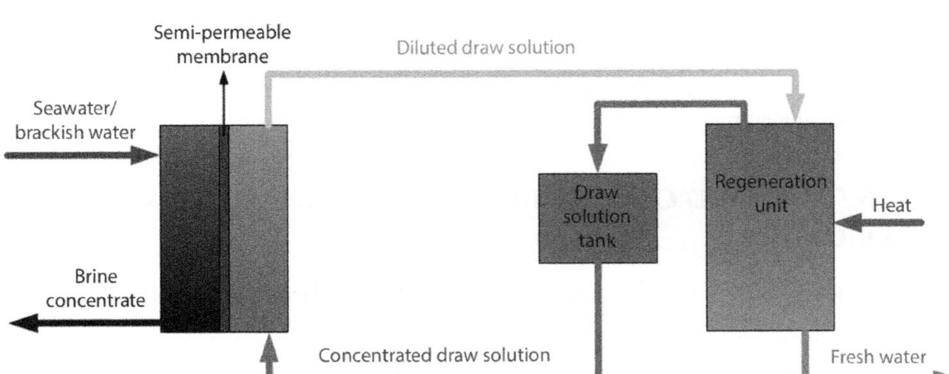

FIGURE 4.2 Forward osmosis process with a thermal regeneration system.

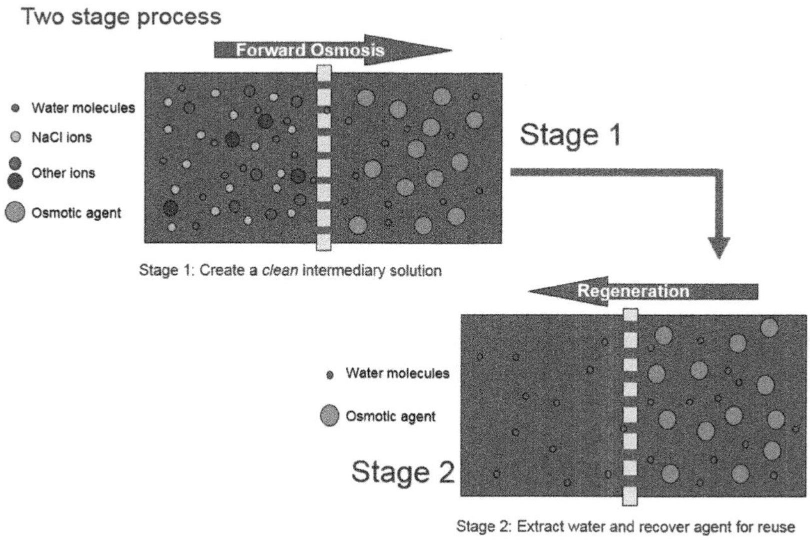

FIGURE 4.3 Forward osmosis process with a membrane regeneration system.

osmosis process is the separation of draw solute from pure water. Forward osmosis membrane fouling problems are less severe than those of reverse osmosis due to the lower operating pressure. Mi and Elimelech (2010) investigated the physical and chemical aspects of different organic foulants on forward osmosis membranes and found that suspended particles are not forced into the membrane pores, so any fouling at the forward osmosis step would be reversible. The recovery of permeate water flux after rinsing by water and without using chemicals was more than 98%. It is also expected that the effect of membrane compaction in the forward osmosis step would be minimal compared to the conventional reverse osmosis process (McCutcheon and Elimelech, 2006). In addition, the tendency for irreversible fouling during the forward osmosis process is much lower than in the reverse osmosis process. Thus, only periodic physical scouring and/or osmotic backwashing will be required to restore the desired flux levels in the forward osmosis process.

Desalination for irrigation is more energy-intensive than desalination for potable water because of stringent boron and chloride standards for agricultural water supply. Although boron is vital for plant growth, it reduces fruit yield at concentrations above 0.3–0.5 mg/L. Therefore, seawater desalination usually requires additional post-treatment steps such as ion exchange or multi-stage reverse osmosis units (Shaffer et al., 2012). Forward osmosis process could potentially lower the total energy consumption of desalination to produce irrigation water when used as an advanced pre-treatment or post-treatment to reverse osmosis plants. The recent investigations and commercial applications of the forward osmosis process in irrigation water supply and nutrient-rich water for direct fertigation from seawater, wastewater and brackish water are evaluated in the following section.

4.2 RECENT FORWARD OSMOSIS DESALINATION APPLICATIONS IN IRRIGATION

4.2.1 Forward Osmosis as a Pre-treatment for Reverse Osmosis Seawater Desalination

In these systems, forward osmosis is a pre-treatment stage whilst the reverse osmosis process acts as the freshwater extraction and draw solution regeneration stage. The forward osmosis–reverse osmosis process provides a double physical barrier for boron and hence meets the concentrations

TABLE 4.2
Forward Osmosis–Reverse Osmosis System in Al Khaluf, Oman

Technology	Unit	Seawater Reverse Osmosis	Forward Osmosis–Reverse Osmosis
Feed water recovery	%	25	35
Product water flow rate	m^3/d	71.4	100
Boron in clean water	ppm	NA	<0.5
Specific energy consumption	kWh/m^3	8.5	4.9

Source: Thompson, N.A. and Nicoll, P.G., *Forward Osmosis Desalination: A Commercial Reality*, IDAWC, Perth, Australia, 2011.

recommended by environmental guidelines. The feed of the reverse osmosis process is free of foulant materials that can improve salt rejection and eliminate the need for additional treatment (Thompson and Nicoll, 2011; Nicoll, 2013). Sharif (2007), Sharif and Al-Mayahi (2011), Al-Mayahi and Sharif (2007), and Al-Zuhairi et al. (2015) designed and tested a pilot unit of the combined forward osmosis–reverse osmosis desalination processes. The studies showed that significant savings of 30% can be achieved in both energy consumption and capital cost compared with the conventional reverse osmosis desalination process. Additional benefits of the forward osmosis–reverse osmosis system include minimum chemical treatment and reduced concentrated brine generation. The first industrial-scale installations of this method have been commissioned by Modern Water Plc. in Gibraltar and Oman beginning in 2008 (Thompson and Nicoll, 2011; Nicoll, 2013). Table 4.2 shows the operating data of the forward osmosis–reverse osmosis unit compared to the existing reverse osmosis desalination plant in Oman.

4.2.2 Integrated Forward Osmosis and Low-Pressure Reverse Osmosis Process for Water Reuse

Domestic treated wastewater is increasingly being considered as a water resource for the irrigation of crops as well as for landscaping, particularly in arid areas. The development of cost-effective wastewater treatment techniques could further stimulate the use of this potential resource. An innovative treatment technology, including desalination, to promote the use of treated wastewater could contribute towards addressing water scarcity. The forward osmosis process has the potential to be applied for harvesting clean water from wastewater sources when integrated with a low-pressure reverse osmosis system to produce irrigation water. Yangali-Quintanilla et al. (2011) coupled the forward osmosis process with a low-pressure reverse osmosis system for indirect desalination using a secondary wastewater effluent and Red Sea water as the feed and draw solutions respectively. Figure 4.4 shows the flow diagram of the integrated forward osmosis–low-pressure reverse osmosis system.

The results demonstrated that the integrated forward osmosis–low-pressure reverse osmosis technology is promising. The results are reported in Table 4.3.

The rate of fouling of the forward osmosis membranes was not a major issue during the experiments for 14 days. Furthermore, the initial water flux was recovered by 98.8% after cleaning the forward osmosis membranes with air scouring (Yangali-Quintanilla et al., 2011). The quality of water produced by the forward osmosis–low-pressure reverse osmosis process was satisfactory for water reuse and could be employed in irrigation.

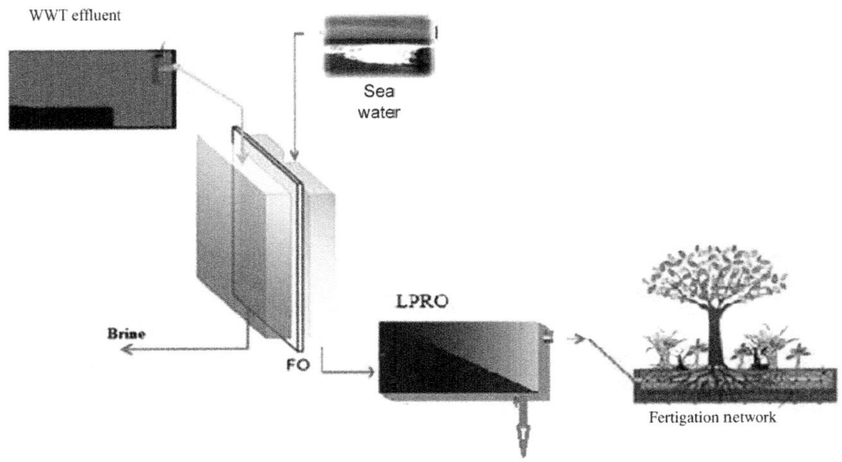

FIGURE 4.4 Integrated forward osmosis (FO) and low-pressure reverse osmosis (LPRO) system for irrigation.

TABLE 4.3
Forward Osmosis–Reverse Osmosis System Results

Technology	Unit	Seawater Reverse Osmosis	Forward Osmosis–Low-Pressure Reverse Osmosis
Product water quality	ppm	500	500
Boron in clean water	ppm	NA	<0.6
Specific energy consumption	kWh/m^3	2.5	1.5

Source: Yangali-Quintanilla, V. et al., *Desalination*, 280, 160–166, 2011.

4.2.3 Fertiliser Drawn Forward Osmosis Desalination

Dry climate, infertile soil, low water quality and scarce water resources contribute to low crop yields in arid and semi-arid regions. Fertiliser nutrients are widely used for agricultural developments; therefore fertilisers and water are crucial components of improving plant productivity. Forward osmosis desalination has the potential to provide nutrient-rich water for direct fertigation from any saline water source, because the diluted fertiliser draw solution can be used for crop irrigation after dilution with a freshwater source. Since the separation and recovery of the draw solution is eliminated, the specific energy consumption will be lower than the current desalination processes for potable water such as reverse osmosis (Hoover et al., 2011). Phuntsho et al. (2012) applied eleven soluble fertilisers as draw solutions to study the performance of fertiliser-drawn forward osmosis desalination under various operating conditions. They estimated that the energy required for direct fertigation using the forward osmosis process is less than 0.24 kWh/m^3. They also reported that most of the applied fertilisers achieved recovery rates of greater than 80% with seawater as the feed. This suggests that fertiliser-drawn forward osmosis desalination can be operated at very high feed recovery rates with a low energy consumption rate. Figure 4.5 shows the conceptual process diagram of fertiliser-drawn forward osmosis desalination using wastewater as the source of water for further dilution. Despite the forward osmosis process adding more value to irrigation water, the diluted fertiliser draw solution can be used directly for fertigation if the final fertiliser concentration meets the acceptable nutrient concentration requirements (Phuntsho et al., 2012).

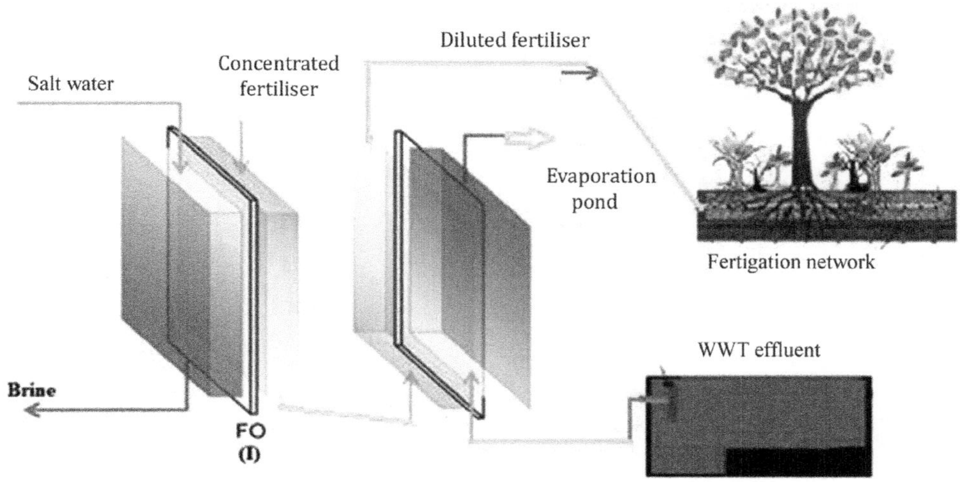

FIGURE 4.5 Forward osmosis (FO) desalination using a two-stage process with additional dilution water from a secondary wastewater effluent.

The two-stage forward osmosis process is used here as a multiple barrier for simultaneous seawater desalination (Phuntsho et al., 2012). The first forward osmosis stage desalinates the saline water using a concentrated fertiliser as the draw solution. Then the diluted fertiliser draw solution goes to the second forward osmosis stage to extract water from the wastewater source. The second forward osmosis stage achieves two objectives: treatment of wastewater effluent to the required irrigation standard and further dilution of the fertiliser draw solution so that it can be applied directly for fertigation.

An assessment of the existing seawater reverse osmosis and the aforementioned integrated forward osmosis–reverse osmosis, forward osmosis–low-pressure reverse osmosis and two-stage forward osmosis processes reveals that forward osmosis is a viable technology for irrigation water supply at low cost. Still, a trade-off between the value of savings in operation and maintenance costs as well as the extra capital cost due to the membrane area requirement in the regeneration system is a challenge for implementing the integrated forward osmosis processes.

4.3 MODIFIED FORWARD OSMOSIS PROCESS WITH THERMAL DEPRESSION REGENERATION

Typically, forward osmosis desalination is integrated with a regeneration process to extract fresh water from the diluted draw solution. The energy-intense stage in forward osmosis is the separation of the draw solution from freshwater.

The feasibility study of integrating the forward osmosis process with the thermal depression regeneration method for seawater desalination has shown the potential of reducing process-specific energy consumption and required membrane area. Draw solution selection can improve the performance of the forward osmosis process in producing cost-effective irrigation water. Here, the concept of employing liquefied gas compounds as the draw solution has been investigated among 137 gaseous compounds (Gas Encyclopaedia) by determining their solubility in water. The screening process identified liquefied gas dimethyl ether (DME) as a suitable draw solution for the forward osmosis desalination process. The DME-water solution is a polar and non-ideal with partial miscibility (AkzoNoble et al., 2010); it generates an osmotic pressure at maximum solubility around seven times greater than the osmotic pressure of seawater. In addition, there is a significant reduction in the solubility of liquefied DME in water when the external pressure on the DME-water

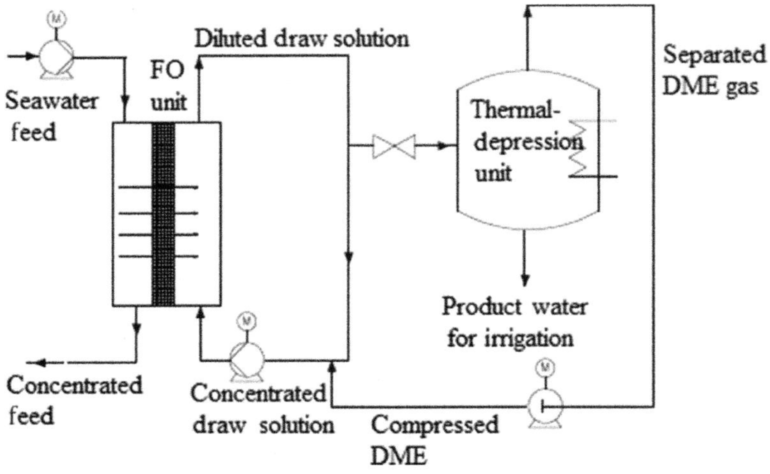

FIGURE 4.6 The forward osmosis (FO) process integrated with a thermal depression regeneration system for separating the draw solution from product water.

solution is reduced from 4 bar to atmospheric pressure. This suggests that the draw solute (DME) could be separated from the solution by thermal depression processes such as gas striping or atmospheric/vacuum flash methods. Figure 4.6 demonstrates the forward osmosis–thermal depression system for water desalination and irrigation water supply. To demonstrate the practicability of the forward osmosis–thermal depression system, the process was modelled and simulated for producing 1 m^3/h of potable water to estimate the required membrane area and specific energy consumption. The considered system recovery was 50% and the concentration of feed and draw solutions on the membrane wall were employed in the calculation of water flux using the modified external and internal concentration polarisation models (Tan and Ng, 2008; Li et al., 2011; Xiao et al., 2012; Tan and Ng, 2013).

Figure 4.6 presents the results of the forward osmosis system coupled with a single-stage flash column using 0.5 M NaCl to resemble seawater and 3 M DME-water as the draw solution in a closed loop process. Feed water is pumped into one side of the membrane and the pressurised DME-water draw solution is introduced to the other side of the membrane. Since the osmotic pressure of the concentrated DME-water draw solution is higher than the feed solution, water flows through the membrane by natural osmosis. After leaving the forward osmosis membrane, the diluted draw solution goes to the regeneration unit. The feed side is maintained at atmospheric or minimum hydrostatic pressure of less than 1 bar while the DME draw solution is operated under 4 bar hydraulic pressure. A fraction of the diluted DME draw solution, 1 m^3/h, is directed to the regeneration system and the remaining (2 m^3/h) is recycled into the forward osmosis system. The diluted DME-water at an elevated pressure is declined by passing through the regulator before entering the regeneration system. The concentrated DME draw solution is then separated from pure water by a gas depression flash tank. The solubility of the draw solute changes by altering the operating pressure; therefore the criteria of separating the draw solution from water in the flash tank is by depressurising the draw solution to atmospheric pressure. The upper stream in the separation process includes the concentrated DME draw solution whilst the bottom stream is pure water. The product water is suitable for drinking purposes or for irrigation water supply. The upper stream of the separating column comprising concentrated DME gas flows to the compressor to be liquefied at 4 bar and is then injected to the recycling diluted draw solution to provide a concentrated draw solution which returns to the forward osmosis membrane.

4.3.1 Model Parameters

The model assumes a 50% recovery rate based on the system design described in Figure 4.6. A tubular ceramic membrane with dimensions of 25 cm × 1 cm × 3 mm (L × D × t), coated in alumina with pore size 0.5 nm and supplied by Department of Biotechnology, Chemistry and Environmental Technology at Aalborg University (Farsi et al., 2014) was considered in the model. The membrane has a water permeability coefficient of 1.37×10^{-6} m/s bar which is determined by a reverse osmosis test at 25°C. Both feed and draw solution temperature and cross flow velocity were assumed to be 25°C and 22 cm/s respectively. The osmotic pressure of feed and draw solution were calculated using the OLI Stream Analyzer (2005) and Aspen HYSYS software. Table 4.4 summarises the process modelling conditions and Table 4.5 presents the modelling results of the

TABLE 4.4
The Integrated Forward Osmosis and Thermal Depression Process Modelling Conditions

Model Parameters	Value	Unit	Formula Calculation
Feed water (seawater) flow rate, Q_F	2	m³/h	
Feed water operating pressure, P	1	bar	
Draw solution (DME-water) flow rate, Q_D	2	m³/h	
Concentrated draw solution pressure, P	4	bar	
Feed water concentration, C_F	0.5	M	
Feed water osmotic pressure, π_{Fb}	27	bar	$\pi_{Fb} = 6.2971C^2 + 40.714C$ (Van't Hoff, 1887)
Draw solution concentration, C_D	3.26	M	
Draw solution osmotic pressure, Π_D	87	bar	$\Delta T_f = -i \cdot K_f \cdot C$ (Van't Hoff, 1887) $\pi = i \cdot C \cdot R_g T$ $\pi = -R_g T \ln(a_m)/V_m$
Forward osmosis process recovery, R	50	%	
Membrane structure parameters, thickness, t, porosity, ϵ, tortuosity, τ	3/40/1.75	mm	
Membrane water permeability coefficient, A	1.3E–6	m/s. bar	$J_w = A(\Delta - \Delta P)$ (Baker, 2004)
Membrane permeability coefficient for DME, B	1.4E–7	m/s. bar	$B = A(1-R)(\Delta\pi - \Delta P)/R$ (Baker, 2004)
Mass transfer coefficient solute resistance, K	8.3E+4	s/m	$K = \frac{t\tau}{D\varepsilon}$
Diffusion coefficient of DME in water, D	1.6E–5	cm²/s	$D = 7.4 \times 10^{-8} \frac{(xM)^{1/2} T}{\eta V^{0.6}}$ (Wilke and Chang, 1955)
Flash tank (DME separator) pressure, P	1	bar	
Flash tank operating temperature	50	°C	
Feed/draw solution operating temperature, T_F, T_{DS}	25	°C	
Relief valve differential pressure, ΔP	4	bar	
Compressor isentropic efficiency, η_{IS}	85	%	
Compressor pressure differential, ΔP	4	bar	
Ratio of specific heats of DME, $\gamma = Cp/Cv$	1.16		

Note: DME, dimethyl ether

TABLE 4.5
The Integrated Forward Osmosis and Thermal Depression Process Modelling Results

Model Parameters	Value	Unit	Formula Calculation
Clean water flow rate, Q_P	1	m³/h	$Q_P = Q_{\text{Din}} - Q_{\text{Dout}}$
DME concentration in clean water, C_D	0.48	M	Aspen HYSYS output
Chloride rejection in clean water, C_{Cl}	80	%	
Water flux, J_w	7.46	L/m²h	$\text{Sh}_{\text{ave}} = \frac{k_f d_h}{D_f} \frac{C_{dw}}{C_{db}} = \frac{k_D}{k_D - J_w}$ $K_m = \frac{t\tau}{D\varepsilon}$ (De and Bhattacharya, 1997; Lipnizki and Field, 2001; Gekas and Hallstrom, 1987) $J_w = \left(\frac{1}{k_f} + \frac{1}{k_m} + \frac{1}{k_D}\right)^{-1} \ln\left[\frac{A\pi_D + B - J_w \exp(J_w/k_D)}{A\pi_f + B}\right]$
Required membrane area, A	134	m²	$A = \frac{RQ_F}{J_w}$
Work required for compressor, W_S	2.65	kW	$W_S = \left(\frac{\gamma}{\gamma - 1}\right) \frac{P_{\text{in}} V_{\text{in}}}{\eta_{IS}} \left[1 - \left(\frac{P_{\text{out}}}{P_{\text{in}}}\right)^{\frac{\gamma-1}{\gamma}}\right]$ (Avlonitis et al., 2003)
Specific energy consumption, SEC	2.65	kWh/m³	$\text{SEC} = \frac{W_S}{Q_P}$

Note: DME, dimethyl ether.

integrated forward osmosis desalination process with thermal depression and a 50% overall system recovery.

The model was solved iteratively using the equations in Table 4.4. The viscosity, density, diffusion coefficient and osmotic pressure of the feed and draw solutions were calculated using the OLI Stream Analyzer and Aspen HYSYS software.

The results indicate that there is a direct relationship between the temperature of the diluted draw solution entering the flash tank and the quantity of recovered DME from the clean water produced in the regeneration process. The quality of the clean water in terms of DME concentration increases to reach less than 1 ppm either by raising the operating temperature or by using a second flash tank working under vacuum. On the other hand, the output temperature of the liquefied DME solution after compression was higher than the operating temperature in the forward osmosis process. Therefore, the hot liquefied DME solution can be used to heat up the diluted DME-water draw solution before going to the flash tank. This will increase the recovery rate of DME from the clean water. Fritzmann et al. (2007) reported the achievable energy consumption in seawater reverse osmosis desalination system applying a recovery system was as low as 2–4 kWh/m³ whereas energy consumption in the modified forward osmosis process is 2.65 kWh/m³ for seawater desalination (concentration of 35 g/L). This is considerably lower than current industrial desalination processes such as reverse osmosis. Moreover, the key potential advantages of the modified forward osmosis desalination process include high feed-water recovery, minimisation of brine discharge and separation of DME draw solution at atmospheric pressure. The estimated specific energy consumption of the process is considerably less than the current energy requirements for desalination, indicating the high potential of this method to produce freshwater for irrigation purposes.

4.4 FORWARD OSMOSIS PROCESS COUPLED TO SALINITY-GRADIENT SOLAR POND

Solar ponds are employed as sources of solar energy collection and storage, enabling the application of solar thermal energy in various processes. The theory behind the operation of solar ponds is very simple. A concentration gradient is artificially created where salts such as sodium chloride, magnesium chloride or sodium nitrate are dissolved in water with high concentrations of over 20% (often referred to as brine) at the bottom of the pond to almost entirely freshwater at the top. The concentration gradient created in the middle works as an insulator for the lower section of the pond (Velmurugana and Srithar, 2008).

Three zones exist in a salinity-gradient solar pond, namely the upper convective zone, the non-convective zone and the heat storage zone. The upper convective zone is a relatively thin layer containing freshwater. Water salinity increases in the non-convective zone, which is used as insulation for the heat storage zone. The solar radiation penetrates into the non-convective zone and the temperature increases with depth. The heat storage zone contains saturated brine in order to store solar thermal energy more efficiently. Hence, solar energy will predominantly be stored in heat storage zone making heat storage more convenient for extraction and use in various applications. A schematic view of a salinity-gradient solar pond including its three zones is depicted in Figure 4.7.

The early studies on the combination of solar ponds with desalination systems consisted of thermal desalination technologies, namely multi-effect, multi-stage processes. One of the most renowned studies in this context was conducted in El Paso (Engdahl, 1987; García-Rodríguez and Delgado-Torres, 2007) where a multi-effect, multi-stage desalination unit was operated on heat from a 3000 m^2 solar pond and the rejected brine from the desalination unit was recharged into the bottom of the pond in an effort to achieve zero discharge desalination (Ghermandi and Messalem, 2009).

Using heat from a solar pond, the performance of a flat-sheet air gap membrane distillation module was assessed by Walton et al. (2004) where the flow rate of the distillate reached a maximum of 6 L/m^2.hr operating with temperatures of 80°C on the hot side of the membrane, using feed water of similar concentration to seawater. This investigation mainly was mainly dedicated to the impact of the operational variables rather than the energy efficiency or cost.

Other plants using thermal energy from solar ponds to drive desalination technologies have been initiated in various countries (Delyannis, 1987; Safi, 1998; Szacsvay et al., 1999; Lu et al., 2000; Caruso et al., 2001). Considering the efficiency of the solar technology, solar fraction, thermal storage size and costs, García-Rodríguez (2007) claims that solar ponds offer the best solar technology to be coupled with desalination systems if the system capacity is not too small for a given annual demand of freshwater.

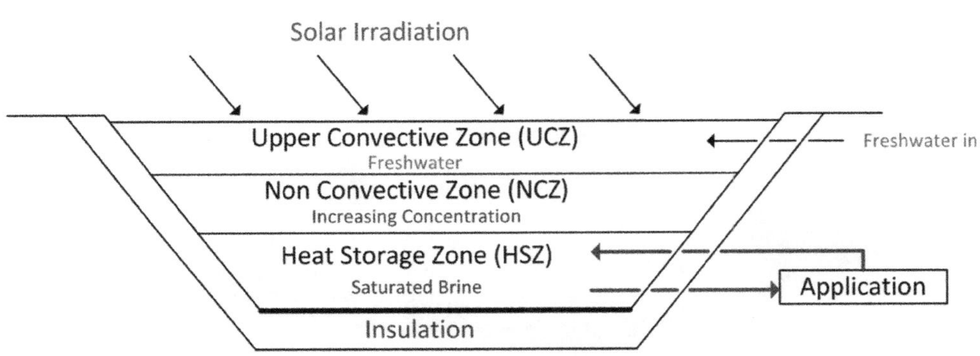

FIGURE 4.7 A typical salinity-gradient solar pond consisting of three zones and surrounded by insulation from the sides and the bottom. There is usually a black lining at the base of the pond to avoid reflection. (Adapted from Monjezi, A.A. and Campbell, A.N., *Appl. Therm. Eng.*, 120, 728–740, 2017. With permission.)

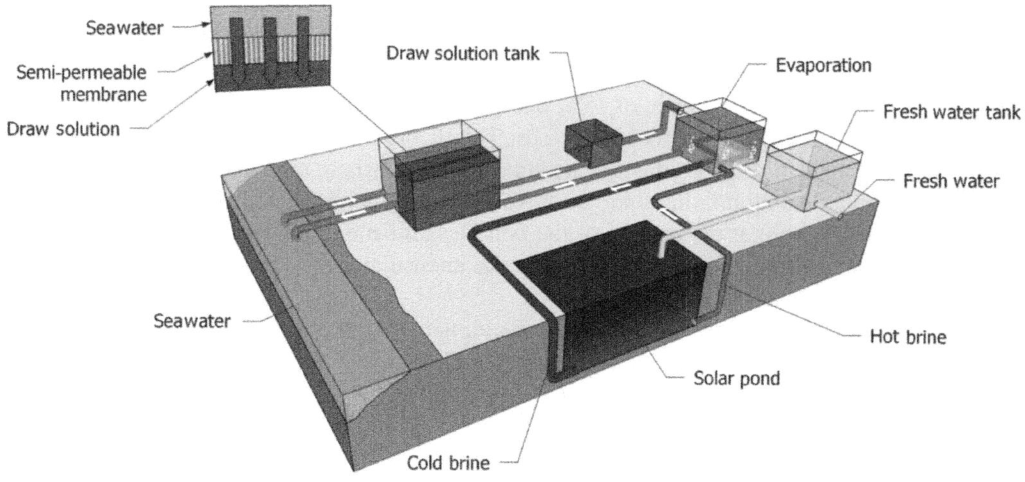

FIGURE 4.8 The proposed process for cost-effective seawater desalination coupling forward osmosis to a solar pond. (Adapted from Monjezi, A.A. et al., *Desalination*, 415, 104–114, 2017. With permission.)

This section presents a model for coupling forward osmosis desalination to solar ponds, where the regeneration of DME as the draw solution will be carried out using solar thermal energy from a simulated salinity-gradient solar pond. Thus, a sustainable solar desalination cycle is introduced and developed. This novel method, shown in Figure 4.8, can contribute towards supplying irrigation water particularly in arid and semi-arid regions of the world where there are high rates of solar irradiance.

A model has been previously developed and simulated for various places to predict temperature distributions and quantities of heat extracted from solar ponds (Monjezi and Campbell, 2016, 2017; Monjezi et al., 2017). A simulation was carried out for a solar pond of 10,000 m^2 in Chabahar (Iran), which benefits from high solar irradiance and access to an abundance of seawater from the Sea of Oman. The simulation results for the solar pond are shown in Figure 4.9.

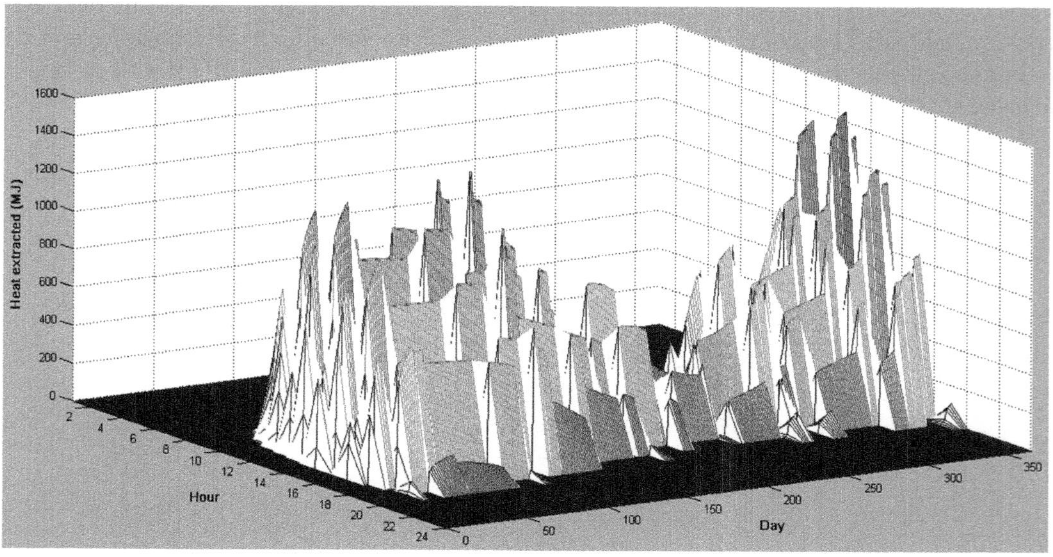

FIGURE 4.9 The quantities of heat extracted from the solar pond in Chabahar city during the first year of operation. (Adapted from Monjezi, A.A. et al., *Desalination*, 415, 104–114, 2017. With permission.)

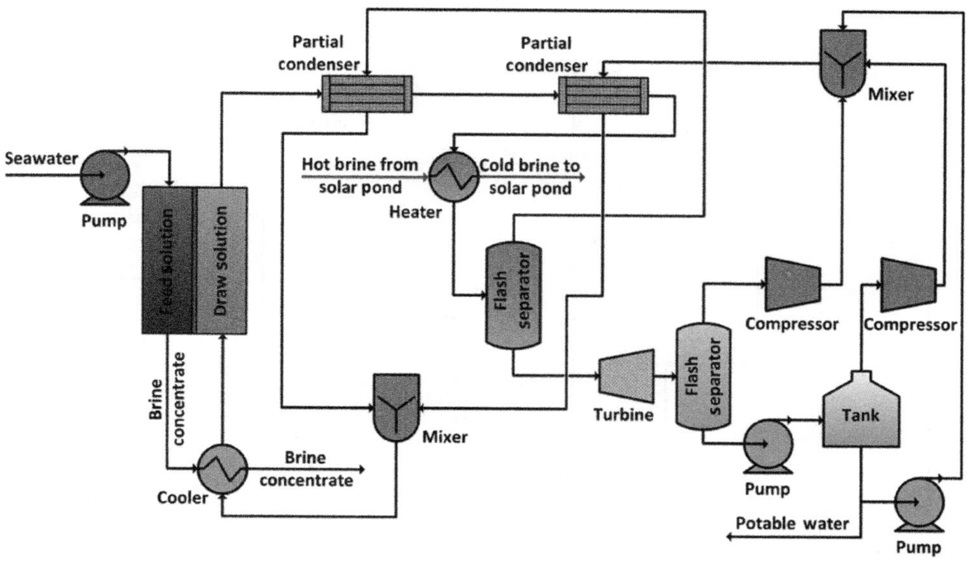

FIGURE 4.10 Process flow diagram of the proposed forward osmosis process using hot brine from the solar pond to heat the draw solution for thermal separation of dimethyl ether (DME) from water. (Adapted from Monjezi, A.A. et al., *Desalination*, 415, 104–114, 2017. With permission.)

Based on the previously mentioned forward osmosis system using DME as a draw solute, a process has been developed for the regeneration of the concentrated draw solution from the diluted draw solution using the hot brine available from the simulated solar pond. This process is illustrated in Figure 4.10.

This method aims to minimise the electricity consumption for desalination using forward osmosis. The employment of the two partial condensers ensures that the evaporated draw solution is condensed and returned to the mixer prior to entering the forward osmosis compartment. Additionally, it preheats the diluted draw solution, reducing the heat load of the heater which uses hot brine at 95°C supplied from the solar pond to heat the draw solution to 85°C. Cooled brine will then be transferred to the heat storage zone of the solar pond at 40°C. Moreover, a small portion of the freshwater produced will be supplied to the surface of the solar pond to maintain the upper convective zone. The solar pond model previously provided enables the calculation of the required supply of freshwater. This quantity is then subtracted from the freshwater produced using the forward osmosis process. The first flash separator is operated under the same pressure as the draw solution, and the top product containing mostly DME gas will not require pressurisation. However, the second flash separator operates at atmospheric pressure, but at 76°C after energy exchange in the turbine, in order to further recover the DME in the solution. The top product, also containing mainly DME gas, is then compressed to 4 bar to ensure solubility in water and provision of high osmotic pressure as a draw solution as specified in Table 4.4. The remaining DME in the bottom product is then transferred to an enclosed tank operated at atmospheric temperature and pressure. The findings of the study by Sato et al. (2014) showed that DME under atmospheric conditions would evaporate out of water within 24 h given its high volatility. Hence, a compressor connected to the tank is employed to raise the pressure of DME to the required level (4 bar).

Three stages of separation are carried out to fully recover the draw solution. The first stage uses heating, the second stage is merely a pressure reduction to atmospheric levels and the third is the reduction of temperature in a tank with a 24-h residence time allowing the remaining DME to evaporate. The only consumption of electricity in the process will then be in the compression of DME and pumping requirements. The inclusion of the turbine between the first and the second separators

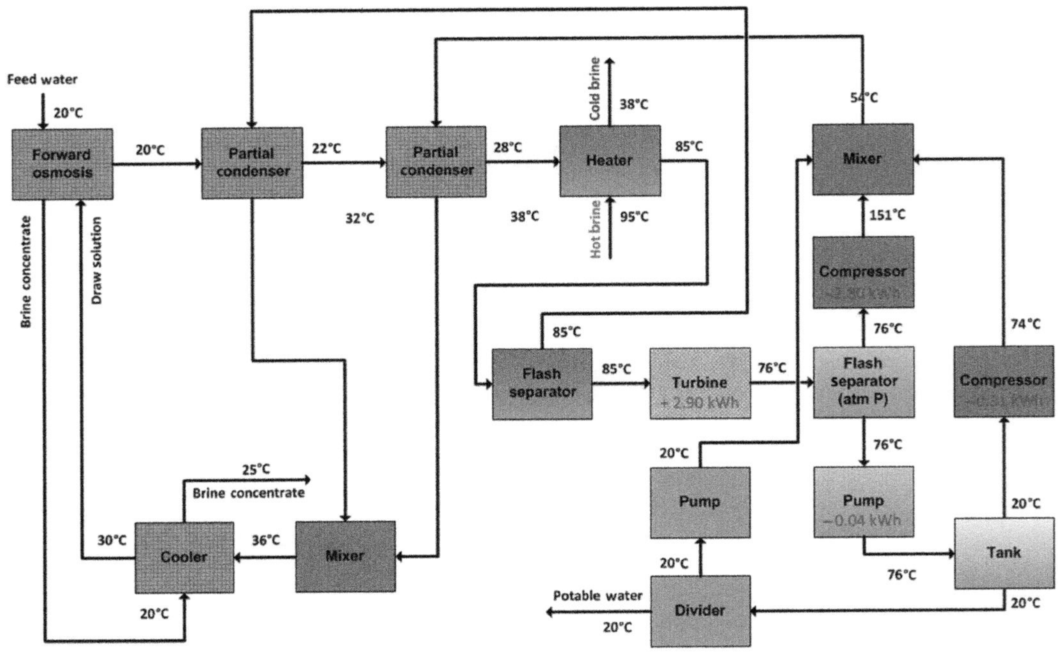

FIGURE 4.11 Block diagram of the proposed forward osmosis process using hot brine from a solar pond including an atmospheric tank indicating the temperatures of all streams and electricity consumption of pressure changing equipment.

enables a proportion of the energy consumed in compressing the draw solution to be reclaimed. Since the temperature of the product water can still be relatively high after 24 h, the concentrated brine is used to cool this stream. A proportion of the product water is used to make up the draw solution to the required concentration as shown in Table 4.4.

The proposed process was simulated on CHEMCAD 6.5.0 and the results are provided in Figure 4.11.

In addition to the electricity consumption rates indicated in Figure 4.12, it has been found that the electricity consumption of the pumping feed and draw solutions through the forward osmosis membrane process is 0.03 kWh per each cubic metre of desalinated water (Zhou et al., 2015).

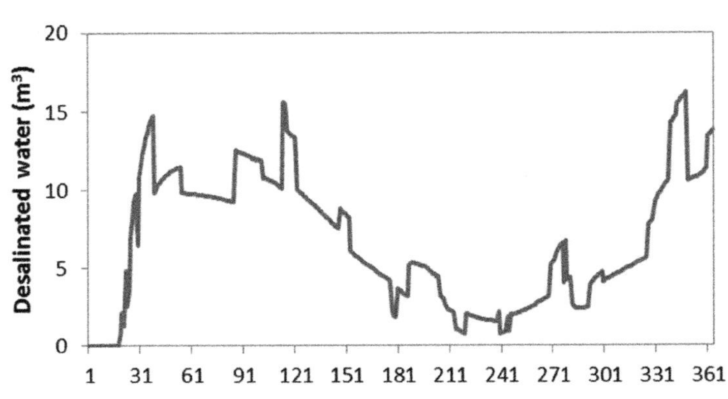

FIGURE 4.12 The quantity of desalinated water in the first year of operation, using the proposed process with dimethyl ether (DME) as draw solute, in Chabahar, Iran.

TABLE 4.6
The Electrical Duties of the Equipment Used in the Proposed Forward Osmosis Process Driven by a Salinity-Gradient Solar Pond

Equipment	DME Compressor-1	DME Compressor-2	Brine Pump	FO Pump	Turbine	Total
kWh/m³	−2.80	−0.31	−0.01	−0.03	+2.90	−0.25

Note: DME, dimethyl ether.

Although this value has been obtained for a different draw solution, it can be used as an estimation for the present work. Furthermore, the electricity consumption of the pumps in the system needs to be calculated. In order to pump the required quantity of brine (2.05 m^3) from the heat storage zone of the solar pond to the heater, a head difference of 1.1 m has to be overcome, assuming brine is extracted from the middle of the heat storage zone. The electrical energy consumption of this would be 0.01 kWh. The electrical duty of the equipment, obtained from the CHEMCAD simulations and pumping electricity consumption calculations are specified in Table 4.6.

It is shown that the total electricity consumption for each cubic metre of desalinated water using the proposed process is 0.25 kWh which is substantially lower than other desalination technologies (Ghalavand et al., 2014). Based on the aforementioned results on the performance of the solar pond modelled for Chabahar and the brine requirements for the heater in the forward osmosis regeneration process, the quantity of desalinated water can now be determined by dividing the quantity of thermal energy that can be extracted from the solar pond by the thermal energy requirements of the forward osmosis regeneration process. The values for the first year of the operation of the pond are shown in Figure 4.12.

The variations observed in Figure 4.12 correspond to those of Figure 4.9, which shows the quantities of heat that can be extracted from the solar pond. There is also a general agreement with the solar irradiation profile of Chabahar. It is also highlighted that there will be no water product for the first 19 days as the pond requires takes 19 days to reach the temperature of 95°C in the heat storage zone. Since the operation of the solar pond was set to start on the first day of May to exploit the highest solar radiation at the startup, it can be seen that the quantity of desalinated water increases sharply during the first 3 months. With the beginning of August, as the solar radiation reaching the pond starts to decrease, the thermal energy available to drive the desalination process is also reduced. The opposite trend is observed with at the start of April. The main drawback of this process is the inconsistent rates of freshwater production throughout the year.

Moreover, it is shown that a solar pond of a moderate size (10,000 m^2) can drive a forward osmosis plant to provide a total of approximately 2450 m^3 of freshwater in the first year of operation in the location considered in this study (Chabahar). This quantity will expectedly increase in the second year of operation since the initial heating-up period of the solar pond will not be required and therefore a higher quantity of heat can be harvested. The total quantity of desalinated water after 2 years of operation will be 5210 m^3 and the variations of the production rate are presented in Figure 4.13.

The availability of vast uninhabited coastal areas in many countries, particularly in the Middle East and North Africa region means that the proposed process can contribute towards addressing water scarcity in these countries. Further studies need to be carried out to evaluate the potential for electricity generation from solar ponds or solar photovoltaic panels in Chabahar to drive the compressors and pumps, which can complement the proposed method so that the forward osmosis desalination can be fully powered by solar energy and provide a fully off-grid method for freshwater production.

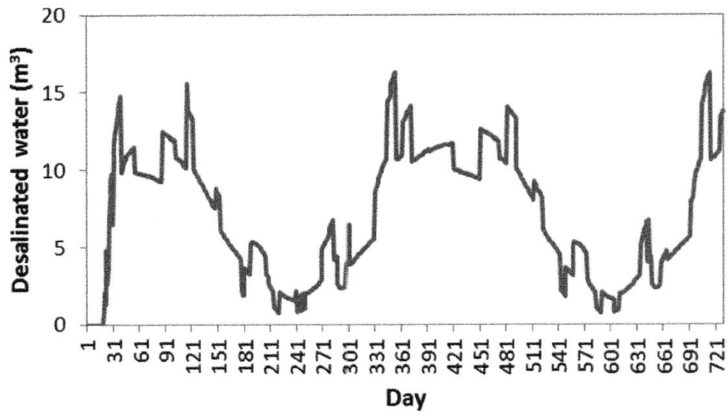

FIGURE 4.13 The quantity of desalinated water in the first two years of operation, using the proposed process with dimethyl ether (DME) as draw solute, in Chabahar, Iran.

4.5 ADVANTAGES OF THE INTEGRATED FORWARD OSMOSIS PROCESS WITH THERMAL REGENERATION

The proposed forward osmosis desalination process using DME as the draw solution with thermal regeneration offers a significant improvement in energy efficiency and cost over conventional desalination technologies. The values of specific energy consumption (kWh/m^3) available in the literature are listed in Table 4.7.

The proposed forward osmosis-thermal depression process with DME-water as the draw solution represents an effective approach to membrane-based seawater desalination, due to working at a pressure of less than 5 bar, which reduces the capital cost of the forward osmosis unit as there will be no requirement for high pressure pumps and pressure vessels. The solubility of DME in water decreases by lowering the operating pressure and increasing the temperature; therefore reverse-DME diffusion from the draw solution to the feed side can be purged and recycled back to the draw solution by decreasing the operating pressure of the feed solution. In addition, using a low-grade heat source, such as a solar pond, increases the cost-effectiveness of the system by reducing the electricity consumption of the process.

TABLE 4.7
Comparison of the Energy Requirement for Current Desalination Methods and the Proposed Forward Osmosis Desalination Processes

Technology	Specific Energy Consumption (KWh/m^3)	References
Multi-stage flash	5.66	Morin (1993)
Multi-effect distillation–Thermal vapour compression	4.05	Morin (1993)
Reverse osmosis	3.02	Avlonitis et al. (2003)
Forward osmosis–Reverse osmosis	2.93	Shaffer et al. (2012)
Forward osmosis–Thermal Depression	2.65	This work
Forward osmosis–Solar Pond	0.25	This work

4.6 CONCLUSION

The lack of freshwater resources, in conjunction with population growth, forms one of the main impediments to sustainable development, while agriculture is the main consumer of freshwater resources. Currently, the production of desalinated water is costly and its use has therefore been limited to the production of water for drinking purposes. Forward osmosis desalination promises to overcome most of the practical difficulties in the conventional reverse osmosis process such as scaling, chemical treatment, fouling and high power consumption. In addition, forward osmosis is capable of delivering higher throughput with lower environmental impact, including minimal chemical additives and less brine rejection. Hence, the forward osmosis process offers a more energy efficient, environmentally friendly and economically viable desalination method. This study presented the forward osmosis process with thermal depression regeneration, using DME as a draw solution, leading to a significantly lower energy consumption rate in comparison with conventional desalination processes. The process of modified integrated forward osmosis with thermal depression has the potential to achieve a sustainable and cost-effective desalination solution to address the increasing scarcity of water for irrigation. The specific energy consumption of the proposed process was estimated to be 2.7 KWh/m^3.

Furthermore, a process was introduced for the regeneration of DME as the draw solution in forward osmosis using thermal energy provided by a solar pond. The proposed process benefits from a very low electricity consumption rate of 0.25 KWh/m^3 but since there are varying rates of solar irradiation, the production of desalinated water throughout the year is inconsistent.

Further studies should be carried out to assess the performance of the proposed forward osmosis process on a pilot scale and verify the performance indicators such as energy consumption, water quality of the produced water as well as the associated safety and environmental concerns.

ACKNOWLEDGMENTS

The financial support of the Royal Society Brian Mercer Award for Innovation and Modern Water Plc. to undertake this work are gratefully acknowledged. The authors would like to sincerely thank Dr Ali Farsi from University of Aalborg, Denmark, for preparing a ceramic membrane for the project.

NOMENCLATURE

A	Pure water permeability coefficient (m/s bar)
A_m	Membrane area, m^2
B	Draw solute permeability coefficient (m/s)
C_f	Molar concentration of the feed solution (mol/L)
C_{fb}	Molar concentration at feed bulk (mol/L)
C_{fw}	Molar concentration of feed at membrane surface (mol/L)
C_D	Molar concentration of the draw solution (mol/L)
C_{Db}	Molar concentration of draw solute in the bulk solution (mol/L)
C_{Dw}	Molar concentration of draw solution at membrane surface (mol/L)
d_h	Hydraulic diameter (m)
D	Solute diffusion coefficient (m^2/s)
J_w	Water flux (m^3/m^2s)
k	Mass transfer coefficient (m/s)
Kf	Cryoscopy constant
k_f	Mass transfer coefficient in feed solution stream (m/s)
k_m	Mass transfer coefficient in support layer (m/s)
k_D	Mass transfer coefficient in draw solution stream (m/s)

k_c	Mean mass transfer coefficient (m/s)
K	Solute resistivity (s/m)
M	Molality of DS solution (mol/kg)
m	Mass flow rate (kg/s)
n	Number of dissolved species created by draw solute
P	Operating pressure (bar)
R	Recovery
R_g	Ideal gas constant (J/mol Kg)
Re	Reynolds number
S	Membrane structural parameter
Sc	Schmidt number
Sh	Sherwood number
T	Absolute temperature (°C)
ΔTf	Deviation in the freezing point
t	Thickness of membrane (mm)
V	Cross flow velocity (m^3/m^2s)
x	Mole fraction
ρ	Fluid density (kg/m^3)
δ	Boundary layer thickness (m)
τ	Membrane tortuosity
ε	Membrane porosity
π	Osmotic pressure (bar)
Δπ	Osmotic pressure difference across membrane (bar)
$\Delta\pi_{eff}$	Effective osmotic pressure difference across membrane (bar)
b	Bulk
d	Draw solution
f	Feed solution
m	Membrane
p	Permeate
s	Salt
w	Membrane wall

REFERENCES

AkzoNobel. Available at: http://www.akzonobel.com/ic/products/dimethyl_ether/product_specification/

Alexandratos, N. and J. Bruinsma, *World Agriculture Towards 2030/2050: The 2012 Revision*, ESA Working paper No. 12-03, FAO, Rome, Italy (2012).

Al-Mayahi, A. K. and A. O. Sharif, Osmotic Energy, 2007, European Patent No EP 1,660,772 B, 2010, Japan Patent No JP 4,546,473.

Al-Zuhairi, A., A. A. Merdaw, S. Al-Aibi, A. Hamdan, P. Nicoll, A. A. Monjezi, S. Al-Aswad, H. B. Mahood, M. Aryafar, and A. O. Sharif, Forward osmosis desalination from laboratory to market, *Water Science & Technology: Water Supply* 15(4), 834–844 (2015).

Avlonitis, S. A., K. Kouroumbas and N. Vlachakis, Energy consumption and membrane replacement cost for seawater RO desalination plants, *Desalination* 157, 151–158 (2003).

Baker, R. W. *Membrane Technology and Applications*, John Wiley & Sons, Chichester, UK (2004).

Beltran, J. M. and S. K. Oshima, *Water Desalination for Agricultural Applications*, Food and Agriculture Organization of the United Nations, Rome, Italy (2006).

Caruso, G., A. Naviglio, P. Principi, and E. Ruffmi, High-energy efficiency desalination project using a full titanium desalination unit and a solar pond as the heat supply, *Desalination* 136, 199–212 (2001).

Darwish, M. Qatar water problem and solar desalination, *Desalination and Water Treatment* 52, 1250–1262 (2014).

De, S. and P. K. Bhattacharya, Prediction of mass transfer coefficient with suction in the applications of reverse osmosis and ultra filtration, *Journal of Membrane Science* 128, 119–131 (1997).

Delyannis, E. E. Status of solar assisted desalination: A review, *Desalination* 67, 3–19 (1987).

Engdahl, D. Technical information record on the salt-gradient solar pond system at the Los Banos Demonstration Desalting Facility, State of California, Department of Water Resources (1987).

Farsi, A., V. Boffa, H. Qureshi, A. Nijmeijer, L. Winnubst, and M. L. Christensen, Modelling water flux and salt rejection of mesoporous γ-alumina and microporous organosilica membranes, *Journal of Membrane Science* 470, 307–315 (2014).

Fritzmann, C., J. Löwenberg, T. Wintgens, and T. Melin, State-of-the-art of reverse osmosis desalination, *Desalination* 216, 1–76 (2007).

García-Rodríguez, L. and A. M. Delgado-Torres, Solar-powered Rankine cycles for fresh water production, *Desalination* 212, 319–327 (2007).

García-Rodríguez, L., Assessment of most promising developments in solar desalination. In: Rizzuti L., Ettouney, H. M., and Cipollina, A. (eds) Solar Desalination for the 21st Century. NATO Security through Science Series. Springer, Dordrecht, 355–369 (2007).

Gas Encyclopaedia. Available at: http://encyclopedia.airliquide.com/Encyclopedia.asp?LanguageID=11&CountryID=19& Formula=C2H4F2&GasID=0&UNNumber=&btnFormula.x=8&btnFormula.y=11

Gekas, V. and B. Hallstrom, Mass transfer in the membrane concentration polarization layer under turbulent cross flow critical literature review and adaptation of existing Sherwood correlations to membrane operations, *Journal of Membrane Science* 80, 153–170 (1987).

Ghalavand, Y., M. S. Hatamipour, and A. Rahimi, A review on energy consumption of desalination processes, *Desalination and Water Treatment* 54, 1–16 (2014).

Ghermandi, A. and R. Messalem, Solar-driven desalination with reverse osmosis: The state of the art, *Desalination and Water Treatment* 7, 285–296 (2009).

Ghermandi, A. and R. Messalem, The advantages of NF desalination of brackish water for sustainable irrigation: The case of the Arava Valley in Israel, *Desalination and Water Treatment* 10, 101–107 (2009).

Hoover, L. A., W. A. Phillip, A. Tiraferri, N. Y. Yip, and M. Elimelech, Forward with osmosis: Emerging applications for greater sustainability, *Environmental Science and Technology* 45, 9824–9830 (2011).

Li, W., Y. Gao, and C. Y. Tan, Network modelling for studying the effect of support structure on internal concentration polarization during forward osmosis: Model development and theoretical analysis with FEM, *Journal of Membrane Science* 397, 307–321 (2011).

Lipnizki, F. and R. W. Field, Mass transfer performance for hollow Fibre modules with shell-side axial feed flow: Using an engineering approach to develop a framework, *Journal of Membrane Science* 193, 195–208 (2001).

Lu, H., J. C. Walton, and A. H. P. Swift, Zero discharge desalination, *The International Desalination and Water Reuse Quarterly* 103, 35–43 (2000).

McCutcheon, J. and M. Elimelech, Influence of concentrative and dilutive internal concentration polarization on flux behaviour in forward osmosis, *Journal of Membrane Science* 284, 237–247 (2006).

Mi, B. and M. Elimelech, Organic fouling of forward osmosis membranes: Fouling reversibility and cleaning without chemical reagents, *Journal of Membrane Science* 348, 337–345 (2010).

Monjezi, A. A. and A. N. Campbell, A comparative study of the performance of solar ponds under Middle Eastern and Mediterranean conditions with batch and continuous heat extraction methods, *Applied Thermal Engineering* 120, 728–740 (2017).

Monjezi, A. A. and A. N. Campbell, A comprehensive transient model for the prediction of the temperature distribution in a solar pond under Mediterranean conditions, *Solar Energy* 135, 297–307 (2016).

Monjezi, A. A., H. B. Mahood, and A. N. Campbell, Regeneration of dimethyl ether as a draw solute in forward osmosis by utilising thermal energy from a solar pond, *Desalination* 415, 104–114 (2017).

Morin, O. J. Design and operating comparison of MSF and MED systems, *Desalination* 93, Topsfield, USA: 69–109 (1993).

Nicoll, P. G. *Forward Osmosis: Ignore It At your Peril!* Vol. 22, International Desalination Association (2013).

OLI Stream Analyzer 2.0, OLI Systems Inc., Morris Plains, NJ (2005). Available at: http://www.olisystems.com/

Phuntsho, S., H. K. Shon, S. Hong, S. Lee, S. Vigneswaran, and J. Kandasamy, Fertiliser drawn forward osmosis desalination: The concept, performance and limitations for fertigation, *Reviews in Environmental Science and Bio/Technology* 11, 147–168 (2012).

Safi, M. J. Performance of a flash desalination unit intended to be coupled to a solar pond, *Renewable Energy* 14, 339–343 (1998).

Saif, O. The future outlook of desalination in the Gulf: Challenges and opportunities faced by Qatar and the UAE (2012). Available at: http://inweh.unu.edu/wp-content/uploads/2013/11/The-Future-Outlook-of-Desalination-in-the-Gulf.pdf

Sato, N., Y. Sato, and S. Yanase, Forward osmosis using dimethyl ether as a draw solute, *Desalination* 349, 102–105 (2014).

Shaffer, D. L., N. Y. Yip, J. Gilron, and M. Elimelech, Seawater desalination for agriculture by integrated forward and reverse osmosis: Improved product water quality for potentially less energy, *Journal of Membrane Science* 415, 1–8 (2012).

Sharif, A. O. and A. Al-Mayahi, Solvent removal method, US Patent 7,879,243 (2011).

Sharif, A. O. *Novel Manipulated Osmosis Water Purification and Power Generation Processes – A Pilot Plant Study*, The Royal Society, London, UK (2007).

Sowers, J., A. Vengosh, and E. Weinthal, Climate change, water resources, and the politics of adaptation in the Middle East and North Africa, *Climatic Change* 104, 599–627 (2011).

Szacsvay, T., P. Hofer-Noser, M. Posnansky, Technical and economic aspects of small-scale solar-pond powered seawater desalination systems, *Desalination* 122, 185–193 (1999).

Tallon, S. and K. Fenton, The solubility of water in mixtures of dimethyl ether and carbon dioxide, *Fluid Phase Equilibria* 298, 60–66 (2010).

Tan, C. H. and H. Y. Ng, Revised and internal concentration polarization models to improve flux prediction in Forward Osmosis process, *Desalination* 309, 125–140 (2013).

Tan, C. H. and H. Y. Ng, Modified models to predict flux behavior in forward osmosis in consideration of external and internal concentration polarizations, *Journal of Membrane Science* 324, 209–219 (2008).

Thompson, N. A. and P. G. Nicoll, *Forward Osmosis Desalination: A Commercial Reality*, IDAWC, Perth, Australia (2011).

van't Hoff, J. The function of osmotic pressure in the analogy between solutions and gasses, *Philosophical Magazine and Journal of Science* 26, 81–105 (1887).

Velmurugana, V. and K. Srithar, Prospects and scopes of solar pond: A detailed review, *Renewable and Sustainable Energy Reviews* 12, 2253–2263 (2008).

Walton, J., H. Lu, C. Turner, S. Solis, and H. Hein, Solar and waste heat desalination by membrane distillation, desalination and water purification research and development program, Report no. 81, U.S. Department of the Interior, Bureau of Reclamation, Denver, CO (2004).

Wilke, C. R. and P. Chang, Correlation of diffusion coefficients in dilute solutions, *AIChE Journal* 1, 264–270 (1955).

Xiao, D., W. Li, S. Chou, R. Wang, and C. Tan, A modelling investigation on optimizing the design of forwards osmosis hollow fibre modules, *Journal of Membrane Science* 392, 76–87 (2012).

Yangali-Quintanilla, V., Z. Li, R. Valladares, Q. Li, and G. Amy, Indirect desalination of red sea water with forward osmosis and low pressure reverse osmosis for water reuse, *Desalination* 280, 160–166 (2011).

Zhou, X., D. B. Gingerich, and M. S. Mauter, Water treatment capacity of forward osmosis systems utilizing power plant waste heat, *Industrial and Engineering Chemistry Research* 54, 6378–6389 (2015).

5 Recent Advancements in the Application of Microbial Desalination Cells for Water Desalination, Wastewater Treatment, and Energy Production

Adewale Giwa, Vincenzo Naddeo, and Shadi W. Hasan

CONTENTS

5.1 INTRODUCTION

The energy requirements of desalination are gaining importance because energy costs account for a significant proportion of the total cost of water production for most desalination processes. A self-sufficient process, in terms of energy requirements, would thus be an attractive option to achieve highly sustainable desalination systems. It is therefore essential to develop an approach for producing the energy required to operate the desalination system within the system itself. Bacterial cells can transfer electrons from organic molecules to electron acceptors during biodegradation. When these electrons are allowed to flow in an electric circuit and create a potential difference, electrical energy is generated. In addition, the positive and negative terminals created in this process could be used for the deionization of the saline feed through ion migration. Accordingly, the development of large-scale microbial desalination cells (MDCs) for desalination has been proposed. MDCs are devices derived from microbial fuel cells (MFCs) (Stoll et al., 2015), and are used for saline water desalination, wastewater treatment, and energy production (Cao et al., 2009; Kim and Logan, 2011; Youngge and Logan, 2013). MFCs are a bio-inspired electrochemical system that mimics the natural microbial interactions by driving an electric current through the system. They convert the chemical energy present in microbial cells to electrical energy through electron generation and transport. MFCs can be broadly classified into two categories: mediated and unmediated (Bhatnagar et al., 2011). Mediated MFCs employ a chemical substance as the mediator to carry electrons from bacteria to the anode (Bhatnagar et al., 2011), while in unmediated MFCs, electrons directly migrate from the cell to the anode via electro-active redox proteins, e.g., cytochromes, which are found on the outer membranes of the microbes (Kim et al., 2004; Pham et al., 2003). Electrochemically active microbes include *Shewanella putrefaciens* and *Aeromonas hydrophila,* and they can contribute to cheaper and cleaner electrical energy production (Tiquia-Arashiro, 2014). However, the majority of microbial cells are not electrochemically active, and the electron transfer from these cells to the anode can only be facilitated through chemical mediators, such as methyl viologen, thionine, methyl blue, and neutral red (Rahimnejad et al., 2012, 2011; Schröder, 2007).

MFCs consist of two chambers—the anode chamber (anaerobic) and the cathode chamber (aerobic)—as well as a semi-permeable membrane that separates the two chambers and regulates ion passage to prevent the potential slowdown in the redox processes, in addition to ensuring that charge balance is maintained across the system. The anode chamber contains organic matter that acts as an electron donor when utilized by the microbial biofilm and releases CO_2 during the process. After electrons are transferred to the anode via the electrochemically active microbes, they would flow from the anode to the cathode, where they would be exposed to air or aerated water, which act as electron acceptors. Electricity is produced through redox reactions, because a potential difference is created between the two electrodes (Moqsud et al., 2013).

MDCs combine the principles of MFC and electrodialysis. As shown in Figure 5.1, a conventional MDC has three chambers: anode, desalination, and cathode. In addition, two ionic exchange membranes (IEMs)—the anion exchange membrane (AEM) and cation exchange membrane (CEM)—separate the anode and cathode chambers, respectively, from the desalination chamber (Cao et al., 2009). In an operating MDC, the desalination chamber handles the saline water flow, while the anode chamber contains wastewater, and the cathode chamber contains either aerated water or another electron acceptor. The electricity generation process in MDCs is similar to that of the MFCs; however, in MDCs saline water desalination occurs when the IEMs get charged and attract oppositely charged ions (Logan and Rabaey, 2012). Therefore, MDCs can be used in

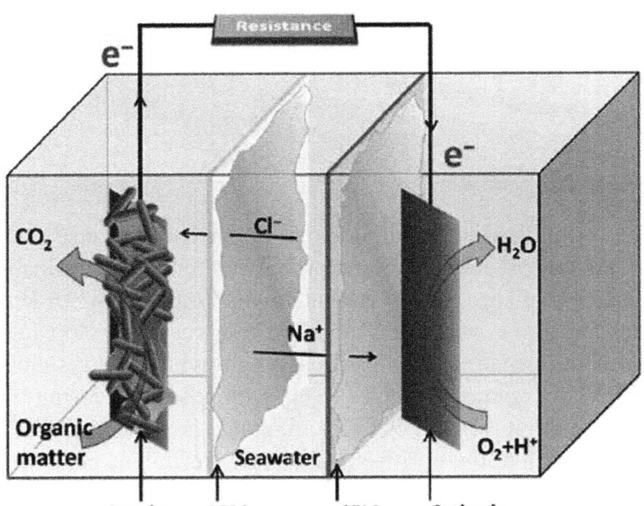

FIGURE 5.1 Schematic of the microbial desalination cell showing the anode chamber, desalination chamber, cathode chamber, and ion exchange membranes (AEM, anion exchange membrane; CEM, cation exchange membrane). (From Ping, Q. et al., *Desalination*, 325, 48–55, 2013.)

wastewater treatment and energy production, like the MFCs, in addition to water desalination (Chen et al., 2012b). The wastewater in the anode is the source of microorganisms that create a biofilm coat on the anode and utilize the organic matter. The electrons released by biodegradation of the organic matter in wastewater are carried by the microorganisms to the anode and then flow through an electric circuit to the cathode. The first MDC was introduced in 2009 (Cao et al., 2009), in which the cell held only 3 mL of saline water (NaCl) and removed 90% of the salts in a single desalination cycle.

Sustainable desalination via MDCs is discussed in this chapter with reference to recent developments and progress, viable opportunities, current challenges, and operating conditions. The major challenge associated with MDCs, which has been highlighted repeatedly in recent studies, is the material selection for the electrodes and the ion-selective membrane. This study focuses on this challenge while considering the recent improvements provided by scientists regarding the application of this technology, such as the optimization of process configuration and the hybridization of this technology with other desalination technologies.

5.2 RECENT PROGRESS IN MDC APPLICATIONS

5.2.1 Desalination

MDCs have previously been used for seawater desalination. Researchers were able to remove up to 99% of salts using upflow MDC (Jacobson et al., 2011a). The study was conducted over a period of 4 months, with a hydraulic retention time of 4 days and a feed solution with an initial concentration of total dissolved solids (TDS) of 30 g/L. The treated water was able to meet the standards set for drinking water in terms of TDS concentration. These results indicated that MDC can be used as an independent desalination unit or for pre-desalination treatment combined with other processes, such as reverse osmosis.

Another study on desalination used an air cathode MDC to reduce the salinity of a 20 g/L NaCl solution by 63% (Mehanna et al., 2010). However, the redox kinetics of air cathode MDCs are slow in ambient conditions and require expensive catalysts to accelerate the reactions, which in turn makes it economically unviable to upscale the system (Kokabian et al., 2013). On the other hand, Wen et al. (2012) used an aerobic biocathode to reduce the salinity of a 35 g/L NaCl solution by 92%, achieving a coulombic efficiency of 96.2% at a desalination rate of 2.83 mg/h. This approach was able to demonstrate the efficiency of another MDC configuration known as biocathode MDC (Wen et al., 2012). Moreover, a microbial capacitive desalination cell (MCDC) consisting of CEMs incorporated with an activated carbon cloth (ACC) and Ni/Cu mesh current collector has been used to desalinate a 10 g/L NaCl solution (Forrestal et al., 2012). The anode and cathode chambers were separated from the desalination chamber via ACC-CEMs. Results showed that 72.7 mg of TDS removal per gram of the activated carbon cloth was achieved due to the potential generated across the microbial anode and the air-cathode without using any external energy (Forrestal et al., 2012). The capacitive MDC proved to be 7–25 times more efficient than the capacitive deionization technology, and it prevented any significant pH changes from occurring in the system, because, in MDC, CEMs allow the transfer of protons from the suspension in the anode chamber (i.e., H^+) to oxygen gas in the cathode chamber to form water.

Zhang et al. (2012) used an osmotic microbial fuel cell, with wastewater in the anode chamber and salt water in the cathode chamber, coupled with MDC under high-power operation mode, and achieved 95.9% reduction in the conductivity and an energy production of 0.160 kWh/m^3. In most studies, electrodes were made of refined carbon. However in a recent study, granular activated carbon from biomass waste (coconut shell carbon) was used as the electrode material, and 84.2% removal of sodium and 58.25% removal of chlorine in one cycle was achieved while using an initial NaCl concentration of 20 g/L (Sophia and Bhalambaal, 2015).

Luo et al. (2017) investigated the treatment of RO brine using microbial electrolysis desalination and chemical production cell (MEDCC). After 18 h of operation, the maximum desalination rate of 86% ± 7% was achieved. No significant chemical oxygen demand (COD) removal in the RO concentrate was observed in the MEDCC, although the fulvic acid– and humic-like organic compounds could have potentially transferred from the desalination chamber into the acid-production chamber. The average total energy consumption was 6.51 ± 0.17–9.81 ± 0.23 kWh/m^3, through which 37%–61% was supplied by the bioenergy from substrate utilization.

MDC technology has also been shown to be applicable for the desalination of brackish water, such as groundwater, which is considered an important water resource. However, low salinity and conductivity of the feed solution is a problem as far as brackish desalination is concerned, as it causes an increase in the ohmic resistance. Researchers have therefore developed an ion-exchange resin MDC to overcome the problem of low conductivity and stabilize the ohmic resistance of the cell (Zhang et al., 2012). Morel et al. (2012) investigated this MDC type and reported a 1.5–8 times increase in the desalination rate of water with low salinity, specifically water with salt concentration <10 g/L, compared to conventional MDCs. In addition, the ohmic resistance range was also reduced to 3.0–4.7 Ω in ion exchange resins MDCs compared to 5.5–12.7 Ω in conventional MDCs (Morel et al., 2012). Ion exchange resins packed in the desalination chamber have also been employed by Zhang et al. (2012) to reduce the internal resistance associated with low salt concentration. By using gel-type ion exchange resins, the desalination profile followed a pseudo–first-order kinetics. As compared to the MDC without resin, the MDC with resin was able to increase the kinetic constant by 2.5 times (to 0.152 L/h) and 3.9 times (to 0.363 L/h) when the NaCl concentration in the desalination chamber was 0.7 and 0.1 mg/L, respectively (Zhang et al., 2012). Moreover, the infinite concentration of NaCl was reduced by 6.6 times (to 7.13 mg/L) and 2.6 times (to 10.6 mg/L), respectively. The resins were able to enhance ion transfer from the bulk solutions to the membranes. Further, to address the presence of nitrates in groundwater, Zhang et al. (2012) developed an MDC that was able to remove up to 90.5% of the nitrates from groundwater within 12 h (Zhang and Angelidaki, 2013).

These studies show the potential of using MDC either as a desalination unit or as a pretreatment unit for other processes, such as reverse osmosis, depending on the configuration used and the type of water (seawater or brackish water). It should be noted that although studies demonstrated high salt removal using MDC, they mainly focused on NaCl solution, while in reality seawater and brackish water contain different forms of ions and salts, which can affect MDC efficiency and the desalination rate when scaled up. In studies where artificial seawater and actual seawater were compared, the researchers observed a decrease of 27% in the MDC efficiency and of 47% in the current density, because of AEM fouling and CEM scaling as a result of microbe accumulation and magnesium and calcium deposition, respectively (Luo et al., 2012b).

5.2.2 Water Softening

Water hardness, caused by the presence of magnesium, calcium, and other divalent cations in water, is a major water quality problem. Therefore, it is important to remove water hardness before supplying water to consumers. Commonly, water softening is achieved by the addition of lime-soda or through ion exchange softeners. Lime-soda is added to hard water to precipitate calcium and magnesium ions as calcium carbonate and magnesium hydroxide, respectively. The drawbacks of this method are the huge amounts of lime, caustic soda, soda ash, and acids needed for the process, and the high sludge production that requires post-treatment (Brastad and He, 2013). There are other methods for water softening that are energy intensive and not sustainable, such as reverse osmosis, nanofiltration, carbon nanotubes, electrodialysis, and distillation (Kokabian et al., 2013).

Arugula et al. (2012) developed a device called enzymatic fuel cell to soften hard water using glucose oxidation by a dehydrogenase enzyme. It generated electricity that stimulated the softening process and achieved 46% removal in a 800 mg/L $CaCO_3$ solution (Arugula et al., 2012). This method could be improved and integrated into an MDC by genetically modifying a microorganism to

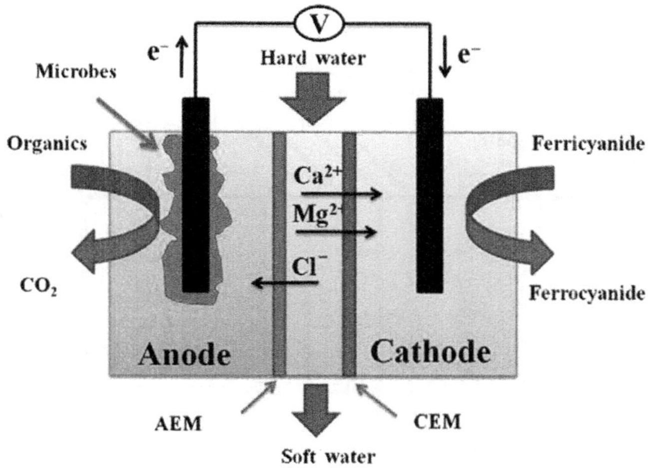

FIGURE 5.2 A microbial desalination cell used for water softening and removal of toxic heavy metals. (From Brastad, K.S. and He, Z., *Desalination*, 309, 32–37, 2013.)

overexpress the dehydrogenase enzyme, which might improve the performance by altering the enzyme kinetic rate to increase and accelerate the rate of glucose utilization. The disadvantage in this case is the high capital cost.

The bench-scale MDC shown in Figure 5.2 was used for water softening by Brastad and He (2013), and achieved 90% water softening using hard water samples collected from across the United States. In addition, 89% of the arsenic, 97% of the copper, 99% of the mercury, and 95% of the nickel were removed (Brastad and He, 2013). This proof of concept study shows the potential of using MDCs in water softening, because it is more sustainable than using excessive chemicals or energy-intensive methods to soften water. Nevertheless, MDCs for water softening must overcome some difficulties, such as the extended operation period, high capital cost, and membrane fouling and scaling.

5.2.3 Wastewater Treatment

Wastewater treatment could take place simultaneously with water desalination.

Compared to MFCs, MDCs are better for wastewater treatment since they offer a solution for the low conductivity of wastewater, which represents a difficulty for MFCs. The results of the study by Luo et al. (2012b) showed a twofold increase in wastewater conductivity, a fourfold increase in power generation, and a 52% reduction in COD per one batch cycle. The long-term performance of MDC in domestic wastewater treatment was monitored for 8 months in a study by Luo et al. (2012a), which reported successful organic removal regardless of the drop in the current density due to the biofouling of the AEM. This is a reasonable side effect as the AEM is facing the anodic chamber, which contains wastewater containing different kinds of microorganisms that causes a biofilm to develop on the AEM surface. This could be avoided if a specific membrane is developed to prevent the adherence of microorganisms on its surface, but not interfere with its ion exchange mechanism. It should also be mentioned that increasing the initial COD concentration decreases its removal efficiency by MDC (Zhang and He, 2015).

A recent study by Pradhan et al. (2015) investigated the removal of phenolic compounds from industrial wastewater simultaneously with saline water desalination. Figure 5.3 shows the stacked MDC used for phenol degradation, where the anodic chamber was inoculated with *Pseudomonas aeruginosa*, which degrades phenolic compounds, and other microbial consortia found in wastewater.

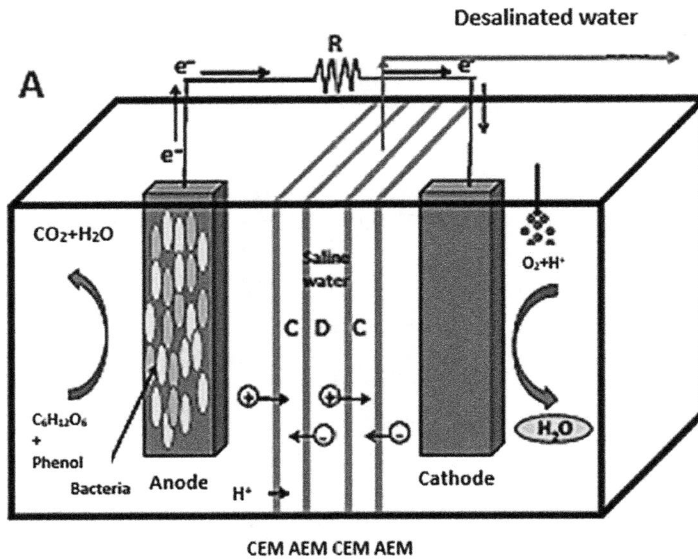

FIGURE 5.3 A stacked MDC (microbial desalination cell) for phenol degradation. (From Pradhan, H. et al., *Appl. Biochem. Biotechnol.*, 177, 1638–1653, 2015.)

A 90% degradation of phenolic compounds and 75% removal of TDS was reported. The study therefore recommended the use of MDCs in treating industrial wastewater because microorganisms have the capacity to utilize and degrade many pollutants under proper conditions.

Furthermore, Lu et al. (2016) introduced a single-stream MDC (SMDC) with four chambers for the treatment and desalination of domestic wastewater. A desalination rate of 12.2–31.5 mg/L.h, >90 and 75% removal of organics and ammonia were reported. A reduction in wastewater salinity from 1.45 to 0.75 mS/cm was also reported.

Therefore, the benefits of water treatment in MDCs are proven in cases of saline water desalination or produced water treatment (Stoll et al., 2015). Similar results are obtained when conducting water desalination and wastewater treatment simultaneously, and if wastewater contains environmental hazards, specific microorganisms can be used for their degradation.

5.2.4 Energy Production

The capability of MDCs to produce energy is one of their most important features. MDCs produce energy by converting chemical energy stored in organic matter in wastewater to electrical energy during anaerobic respiration (Cao et al., 2009). They can produce 4 times the power output of an MFC (Luo et al., 2012b). A study on a liter-scale cell revealed that an MFC connected to an RO plant produces enough electric power to drive the RO process, and it was also reported that the treatment of 1 m^3 of saline water produces 1.8 kWh energy (Jacobson et al., 2011b).

Since exoelectrogenic microorganisms are responsible for producing the current in MDCs, a pH imbalance in the anodic chamber can affect their performance. A maximum power density of 931 mW/m^2 was obtained when a phosphate buffer solution (PBS) was used to regulate the pH and anolyte/catholyte circulations (Qu et al., 2012). It should be mentioned that high COD concentrations do not result in high power densities (Kokabian and Gude, 2015). The energy produced by MDCs can be stored in capacitors or batteries, or used for many different purposes. The importance of energy production in MDCs is discussed later in this chapter, because the energy produced in MDCs is usually used to power downstream desalination plants such as RO.

5.3 CURRENT CHALLENGES, PROPOSED SOLUTIONS, AND OPPORTUNITIES FOR LARGE-SCALE MDC SYSTEMS

5.3.1 Optimization Through Different Configurations

Conducting desalination and wastewater treatment processes separately increases the capital cost (Yuan et al., 2015). MDCs offer a new low-cost and sustainable desalination system. Nevertheless, MDCs face some challenges, such as limited research on their scaling up, and the need for more field testing on their material performance in the long term and at large scale (Logan, 2010). So far, only bench-scale studies have been carried out on MDCs, and the largest total liquid capacity of MDCs to date was 105 L (Zhang and He, 2015); thus, it would be hard to assess their economic viability until they are applied at large scale with actual feed water (Subramani and Jacangelo, 2015). The number of electrodes used in MDCs is high, and it should be reduced in order to decrease the cost of scaling up (Logan, 2010). Currently, more cost-effective electrodes are under development with promising results (Sophia and Bhalambaal, 2015). For example, Liang et al. (2016) introduced a high-performance photo-microbial desalination cell (PMDC) based on photo-electrochemical interactions. Nanostructured α-Fe_2O_3 was used to modify the anode of the cell. Results reveal that the maximum current density of the PMDC during operation was 8.8 A/m^2 when an initial salt concentration of 20 g/L was used. Cyclic voltammetry and impedance spectroscopy analysis showed that the increase in current was controlled by the high electron transfer rate at electrode/biofilm interface.

The cost of implementing MDCs at large scale is still unpredictable, because MDCs are still under development, and the material used at bench scale experiments are too expensive for large-scale use (Qu et al., 2013). Therefore, the MDC technology will not be adopted at large scale until the cost of MDCs implementation is less than that of RO implementation. A recent study that investigated MDC life cycle and conducted cost analysis in the state of Ohio concluded that a conventional MDC is not profitable to be applied as a desalination device (Faze, 2015), because MDCs require high maintenance, cleaning, and membrane replacement. Thus, it is very important to conduct studies aimed at scaling up MDCs in order to understand how the cells would operate at a large scale, and to develop more cost-effective components and configurations.

Since MDC technology is still a fairly new approach to desalination, researchers are constantly developing new methods to modify and configure MDCs for better performance and to overcome the technical challenges in conventional MDCs. Currently, about twelve different configurations of MDCs have been developed (Saeed et al., 2015). Conventional MDC uses ferricyanide as a catholyte, and although it exhibits an excellent performance, ferricyanide is considered a toxic and expensive compound. To overcome this problem, an air cathode was designed, where oxygen is the terminal electron acceptor because of its high reduction potential (Kokabian et al., 2013). Oxygen is also more sustainable, because it is abundant in the atmosphere. A more favorable approach is the biocathode, which has proven to be more sustainable and self-generating than the other cathodes used in MDCs. A biocathode uses microorganisms found on the cathode to carry out the reduction reactions in the cathodic chamber.

In the stack structure MDC configuration, multiple IEMs are integrated in the MDC, creating what are called desalting/dilute cells and concentrate cells. The integration of multiple IEMs increases the charge transfer efficiency and the salt removal, because for every electron that passes through the circuit, an anion–cation pair is separated (Kim and Logan, 2011). In addition, the microbial community in the anode chamber recovers more energy, due to its stacked structure, as demonstrated by Shehab et al. (2013). The problem of increasing the number of IEMs is that the internal resistance of the cell increases, but the desalination rate also increases along with it, and the final result is that the cell would be more efficient and faster (Chen et al., 2012b).

The pH imbalance in the anode and cathode chambers is another drawback that affects the performance of microorganisms in the anode, which alters the desalination process and power generation (Cao et al., 2009; Kim and Logan, 2011). Therefore, a recirculation method of the anolyte and catholyte was introduced, and it was reported that a unit increase in the pH above neutrality

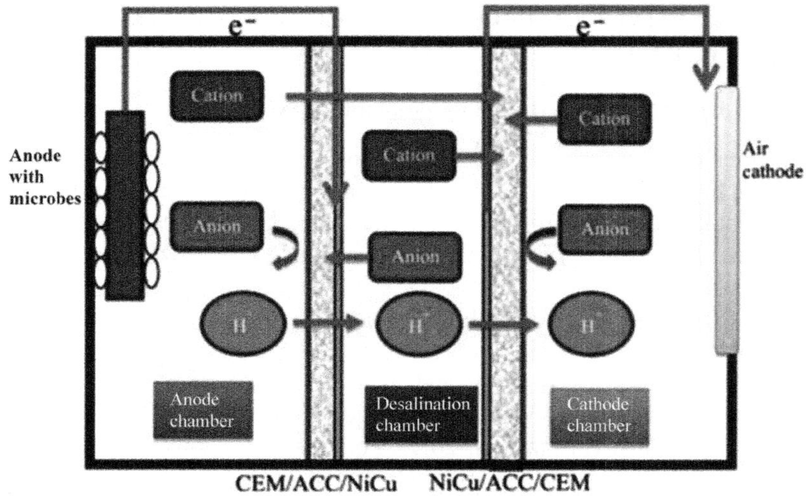

FIGURE 5.4 Capacitive MDC (microbial desalination cell) with cation exchange membrane (CEM)/activated carbon cloth (ACC)/NiCu assemblies that separate the chambers, allow free transport of H^+, and capture salts from saline water. (From Forrestal, C. et al., *Energy Environ. Sci.*, 5, 7161, 2012.)

leads to 59 mV reduction in the voltage (Kim and Logan, 2011). Further, a capacitive MDC was introduced to overcome the consequences of salt accumulation in the anode and cathode chambers. Forrestal et al. (2012) developed an MCDC based on the concept of capacitive deionization (Forrestal et al., 2012). The anode and cathode chambers were separated from the desalination chamber by 2 double-sided capacitors (Forrestal et al., 2012). The capacitors consist of CEMs integrated with ACC and Ni/Cu current collectors. When potential gradient is formed, water desalination takes place, because the ions in the salty water is adsorbed on the capacitors (Figure 5.4), and when the potential gradient is removed, the ions would be released back to water, which prevents salt from entering the anode and cathode chambers. This system also allows the free transport of H^+ protons across the chambers. Moreover, Meng et al. (2017) developed an MCDC with dewatered sludge as the anodic substrate to investigate salt migration issues and improve sludge recycling. Special designed CEM assemblies, layers of ACC, and nickel foam were used. Experimental results reported a power generation and a desalination rate of 2.06 W/m^3 and 15.5 mg/L.h, respectively.

An upflow microbial desalination cell with a tubular structure and two compartments separated by ion-exchange membranes has also been studied (Jacobson et al., 2011a). The inner compartment was the anode chamber filled with graphite granules and graphite rods that serve as current collectors. The anode chamber was surrounded by an AEM. The outer compartment was between this AEM and a CEM that was catalyzed at the outer surface by a platinum/carbon mixture. This compartment contained the salt solution, i.e., the desalination chamber. There was no cathode chamber in this MDC configuration. The cathode electrode was connected to an external electric field. This configuration was able to remove more than 99% of TDS from a saline solution having a concentration of 30 g/L. The unique structure of the upflow configuration is that mixing can be achieved without agitation because the contact between the biomass and substrate is ensured by the influent wastewater. This keeps the microorganisms in suspension and leads to a better oxidization of organic matter (He et al., 2006). The limitations of this configuration are the resistance that occurs within the anode and cathode, which can be eliminated with proper optimization of the cell (He et al., 2005), and the pH imbalance, which can be controlled by electrolyte recirculation (Zhang et al., 2010).

Osmosis MDC is proposed to replace AEM with a forward osmosis (FO) membrane (Kim and Logan, 2011; Zhang and He, 2013). In this type salty water is diluted instead of being desalinated, because FO membranes allow the migration of water molecules from wastewater in the anode chamber to the desalination chamber, which exposes the FO membranes to fouling (Kim and Logan, 2011).

The osmotic MDC is not efficient to replace desalination as a stand-alone unit, but it shows better performance when coupled with a microbial fuel cell (Zhang and He, 2012a). A bipolar membrane is another MDC configuration, in which the membrane is inserted close to the anode in the anode chamber, creating a fourth chamber (Kim and Logan, 2011). The desalination performance is affected by the membrane properties (Alvarez et al., 1997). The bipolar membrane splits water into hydrogen and hydroxyl, and then hydrogen moves into the fourth chamber where it reacts with chloride ions to form hydrochloric acid, while hydroxyl ions move to the anode chamber where it contributes to maintaining the pH (Kim and Logan, 2011). One disadvantage of this configuration is that it requires an external power source to facilitate the splitting of water by bipolar membranes (Ping and He, 2013).

Recent studies focused on integrating MDC systems with other methods to enhance wastewater treatment, desalination rates, and power generation. For instance, Li et al. (2017) investigated simultaneous removal of nitrogen and metals from municipal and industrial wastewaters, respectively, as well as the removal of salts from using an integrated microbial desalination cell-microbial electrolysis cell (MDC-MEC) system. Results showed 4.07 mg/L.h removal of nitrogen, 63.7% rate of desalination, and 293.7 mW/m^2 of power was generated. They also reported a positive energy balance of 0.0267 kWh/m^3, demonstrating a potential use of the MDC-MEC system. Also, Dong et al. (2017) investigated the performance of a combined system for simultaneous treatment of domestic wastewater and heavy metal wastewater, as well as high-salinity-water desalination. Four five-chamber MDCs with a desalination chamber and two concentrate chambers were hydraulically connected for desalination and alkalinity generation. Results reported 737.3 ± 201.1 mW/m^2, 53.6 ± 0.8 $kg/m^3.d$, and 1.84 ± 0.05 $kgCOD/m^3.d$ for average power density, salt removal, and COD removal rate, respectively. The use of electrodialysis revealed that 12.75 ± 1.26 V at switching time of 15 min, and $30.4\% \pm 2.6\%$ of maximum discharging voltage, and desalination efficiency could be achieved. Zuo et al. (2016) demonstrated the performance of a modularized filtration air cathode MDC (F-MDC) through which nitrogen-doped carbon nanotube membranes and Pt carbon cloths were used as filtration material and cathode. Their results reported 93.6% and 97.3% removal efficiency of salts and COD, and 68 ± 12 μS/cm and 0.41 NTU of final conductivity and turbidity, respectively.

A typical challenge associated with MDCs is the requirement of complete dissembling of the system when the variation of the liquids in the compartments is needed. Dissembling is also required for maintenance purposes. To overcome this challenge, a decoupled MDC configuration has been devised, such that the anode and cathode are placed in the salt solution, rather than inserting them in the anode and cathode chambers. The AEM is placed on the two sides of the anode. Likewise, the CEM is placed on the two sides of the cathode (Ping and He, 2014). This configuration highlights the advantage of the flexibility in controlling the volume and ratios as needed (Chen et al., 2012b). A separator coupled stacked MDC combines the stacked structure MDC with the recirculation MDC in addition to a glass fiber separator on the cathode on the side facing the desalination chamber (Chen et al., 2012b) to prevent biofouling and improve the coulombic efficiency. Ion exchange resin coupled MDC solves the problem of the increase in the ohmic resistance occurring when the salt concentration and conductivity decrease (Cao et al., 2009). In this configuration, the addition of ion exchange resins in the desalination chamber would increase the conductivity of the cell, which would decrease ohmic resistance and energy consumption (Morel et al., 2012). Table 5.1 summarizes all the mentioned configurations along with the most important features of the cells and their performance in terms of salt removal, energy production, and wastewater treatment.

TABLE 5.1
A summary of Different MDC Configurations in Terms of Features and Performance

		Performance			
	Key Features	**TDS Removed**	**Energy Produced**	**Wastewater Treated**	**Challenge**
Air cathode MDC	- Oxygen is the terminal acceptor - More sustainable	Conductivity: 43%–67% reduction (Mehanna et al., 2010)	PD[1]: 424–159 mW/m^2 CD[2]: 2.8–0.81 A/m^2 (Mehanna et al., 2010).	Acetate: 82.6%–77.3% reduction (Mehanna et al., 2010)	Slow redox kinetics
Biocathode MDC	- Sustainable and self-generating - Presence of microorganisms on the cathode	92% (Wen et al., 2012)	Electric potential: 609 mV (Wen et al., 2012) PD: 1.1 W/m^3 (Kokabian and Gude, 2015)	COD[3]: 82.17%–76.06% removal (Kokabian and Gude, 2015)	Further research needed
Stack Structure MDC	- More than one pair of IEMs - High efficiency of charge transfer	- 95.8% (Zuo et al., 2014) - 93%–100% (Kim and Logan, 2011)	Current: 3–5 mA CE: 86% PD: 800–1140 mW/m^2 (Kim and Logan, 2011)	N/A	The increase in IEMs increases the resistance overtime
Recirculation MDC	Recirculation of anolyte and catholyte between chambers	34%–37% (Shehab et al., 2013)	PD: 931–776 mW/m^2 (Qu et al., 2012)	N/A	Coulombic efficiency decreases
Capacitive MDC	Two sided capacitors on the anode and cathode	69.4% (Forrestal et al., 2012)	Electric potential: 0.25–0.28 V (Stoll et al., 2015)	DOC[5]: 27.5%–16.3% removed. (Zhang et al., 2010)	Not efficient compared to other configurations
Upflow MDC	- Tubular structure - Continuous flow	99% (Jacobson et al., 2011a)	Current: 62 mA PD: 30.8 W/m^3 (Jacobson et al., 2011a)	COD: 92% removal (Jacobson et al., 2011b)	Long retention time
Osmotic MDC	Use of FO membranes	Conductivity: 95.9% reduction (Zhang and He, 2013)	0.160 kWh/m^3 (Zhang and He, 2013)	COD: 82%–86% removal (Zhang and He, 2013)	FO membranes are highly susceptible to fouling
Bipolar membrane MDC	Use of bipolar membranes	86% (Chen et al., 2012a)	N/A	N/A	Lack of research
Decoupled MDC	Electrodes have AEM and CEM on both sides	0.143 g/d removal of salts (Ping and He, 2013)	CD: 187.3 A/m^3 PD: 82.6 W/m^3 (Ping and He, 2013)	N/A	Not efficient compared to other configurations

(Continued)

TABLE 5.1 (*Continued*)
A summary of Different MDC Configurations in Terms of Features and Performance

		Performance			
	Key Features	TDS Removed	Energy Produced	Wastewater Treated	Challenge
Separator coupled stacked circulation MDC	- Combines the stack type with the recirculation type - Cathode is covered with glass fiber to prevent biofouling	Desalination ratio: 81%–40% (Chen et al.,2012b)	N/A	COD: 64% removal (Chen et al., 2012b)	Not efficient compared to other configurations
Ion-exchange resin coupled MDC	- Use of ion exchange resins in desalination chamber - Overcomes the increase in resistance overtime - Excellent for low salinity water	TDS decreased from 700–7.13 mg/L (Zhang et al., 2012)	Current: 1.5–0.65 mA PD: 360–60 mW/m^2 (Zhang et al., 2012)	N/A	N/A
Microbial electrolysis desalination and chemical production cell (MEDCC)	–	Maximum desalination rate of 86% $\pm$ 7% (Luo et al., 2017)	N/A	Average total energy consumption was 6.51 $\pm$ 0.17–9.81 $\pm$ 0.23 kWh/m^3	No significant COD removal in the RO concentrate
Modularized filtration air cathode MDC (F-MDC)	- Nitrogen-doped carbon nanotube membranes and Pt carbon cloths were used as filtration material and cathode	93.6% removal of salt with 68 $\pm$ 12 μS/cm of dilute conductivity (Zuo et al., 2016)	N/A	97.3% of COD	Future research is needed
Single-stream MDC (SMDC)	- Four chambers were used for the treatment and desalination of domestic wastewater	12.2–31.5 mg/L.h desalination rate (Lu et al., 2016)	N/A	>90 and 75% removal of organics and ammonia	Future research is needed
Combined MDC system	- Four five-chamber MDCs with a desalination chamber and two concentrate chambers were hydraulically connected for desalination and alkalinity generation	53.6 $\pm$ 0.8 kg/m^3.d salt removal (Dong et al., 2017)	737.3 $\pm$ 201.1 mW/m^2	1.84 $\pm$ 0.05 $kgCOD/m^3$.d of COD removal	Future research is needed

(*Continued*)

TABLE 5.1 (*Continued*)
A summary of Different MDC Configurations in Terms of Features and Performance

	Key Features	Performance			Challenge
		TDS Removed	Energy Produced	Wastewater Treated	
Microbial capacitive desalination cell (MCDC)	- Dewatered sludge as anodic substrate	Desalinization rate was about 15.5 mg/L.h (Meng et al., 2017)	2.06 W/m^3	N/A	Conductivity of the anodic substrate affects salt accumulation
Microbial desalination cell-microbial electrolysis cell (MDC-MEC) system	–	63.7% rate of desalination (Li et al., 2017)	293.7 mW/m^2 with a positive energy balance of 0.0267 kWh/m^3	4.07 mg/L.h removal of nitrogen	Future research is needed
Photo-microbial desalination cell (PMDC)	- Nanostructured α-Fe_2O_3 modified anode	>96% salt removal (Liang et al., 2016)	8.8 A/m^2	N/A	Future research is needed

Note: PD: power density; CD: current density; COD: chemical oxygen demand; CE: current efficiency; DOC: dissolved organic carbon; MDC: microbial desalination cell; TDS: total dissolved solids.

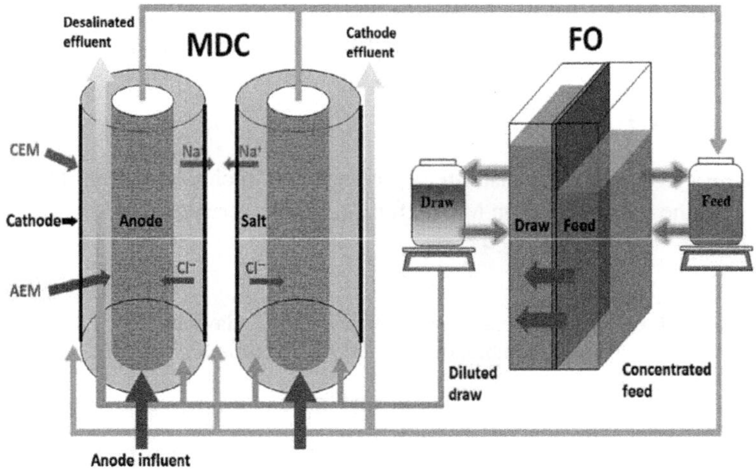

FIGURE 5.5 A schematic representation explaining the hybridization of forward osmosis (FO) with a microbial desalination cell. (From Yuan, H. et al., *Chem. Eng. J.*, 270, 437–443, 2015.)

5.3.2 Hybridization of MDCs with Other Desalination Techniques

Although a large scale MDC-RO hybrid system has not yet been realized, MDCs could be efficiently used as pretreatment units for seawater desalination, where they can lower the salt concentration, produce energy to power the actual desalination unit, and reduce energy costs (Mehanna et al., 2010). An upflow-MDC that was tested as a pretreatment for electrodialysis desalination (ED) has revealed that an upflow-MDC is capable of reducing ED energy consumption by 45% and desalination time by 25%, thereby supporting the great potential of MDCs as a pretreatment method for desalination plants (Zhang and He, 2012b). In another study, where an upflow MDC was used as a pretreatment for an RO plant, the MDC produced 1.8 kWh of energy to drive the RO plant, and therefore reduced the energy required for RO to 3.5 kWh/m^3, which lowered the energy cost by 22% (Jacobson et al., 2011b).

FO membranes have been integrated with MDCs by replacing the AEM with an FO membrane (Zhang and He, 2012a), but this combination underperformed because the replacement affected ion transfer. A recent study also focused on using FO membranes, not in MDCs, but rather as a supporting unit. FO membranes, placed between wastewater and the saline water feed, served to extract water from wastewater to dilute saline water simultaneously, thus recovering water and reducing the volume of wastewater. According to this study, the developed system could effectively treat wastewater that contains high levels of COD with no effect on desalination performance (Yuan et al., 2015), as shown in Figure 5.5. The diluted saline water can then be used in MDCs to reduce the inhibition of the microorganisms in the anode chamber by the Cl^- ions from the salts in the saline water (Zhang et al., 2012).

5.4 CONCLUSIONS

Since the introduction of MDCs in 2009, scientists have been developing and upgrading configurations to enhance MDC performance and overcome operational problems. As shown by recent research, MDC technology has a promising future. Several projects are currently underway in order to upscale the cells and to study their operational performance on a larger scale, which would provide data that is crucial for MDC research. In addition, promising results have been shown for using MDCs as a pretreatment unit for the RO process, because the cells can provide enough energy to drive the process. Moreover, many studies have shown the ability of the MDC to treat wastewater by significantly lowering the COD levels. Furthermore, microorganisms also have the capability to degrade hazardous phenolic compounds. Future studies should focus on the implementation and operation of large-scale MDCs that are economically viable, which will attract investment; this would accelerate the implementation of MDCs in the industry as an established desalination technology.

ACKNOWLEDGMENT

Authors are grateful to the Masdar Institute of Science and Technology, Abu Dhabi, United Arab Emirates, where all the publications on MDCs for this study were collected.

REFERENCES

Alvarez, F., R. Alvarez, J. Coca, J. Sandeaux, R. Sandeaux, and C. Gavach. 1997. Salicylic acid production by electrodialysis with bipolar membranes. *Journal of Membrane Science* 123(1): 61–69. doi:10.1016/S0376-7388(96)00197-4.

Arugula, M. A., K. S. Brastad, S. D. Minteer, and Z. He. 2012. Enzyme catalyzed electricity-driven water softening system. *Enzyme and Microbial Technology* 51(6–7): 396–401. doi:10.1016/j.enzmictec.2012.08.009.

Bhatnagar, D., S. Xu, C. Fischer, R. L. Arechederra, and S. D Minteer. 2011. Mitochondrial biofuel cells: Expanding fuel diversity to amino acids. *Physical Chemistry Chemical Physics: PCCP* 13(1): 86–92. doi:10.1039/c0cp01362e.

Brastad, K. S., and Z. He. 2013. Water softening using microbial desalination cell technology. *Desalination* 309: 32–37. doi:10.1016/j.desal.2012.09.015.

Cao, X., X. Huang, P. Liang, K. Xiao, Y. Zhou, X. Zhang, and B. E. Logan. 2009. A new method for water desalination using microbial desalination cells. *Environmental Science & Technology* 43(18): 7148–7152. doi:10.1021/es901950j.

Chen, S., G. Liu, R. Zhang, B. Qin, and Y. Luo. 2012a. Development of the microbial electrolysis desalination and chemical-production cell for desalination as well as acid and alkali productions. *Environmental Science and Technology* 46(4): 2467–2472. doi:10.1021/es203332g.

Chen, X., P. Liang, Z. Wei, X. Zhang, and X. Huang. 2012b. Sustainable water desalination and electricity generation in a separator coupled stacked microbial desalination cell with buffer free electrolyte circulation. *Bioresource Technology* 119: 88–93. doi:10.1016/j.biortech.2012.05.135.

Chen, X., X. Xia, P. Liang, X. Cao, H. Sun, and X. Huang. 2011. Stacked microbial desalination cells to enhance water desalination efficiency. *Environmental Science and Technology* 45(6): 2465–2470. doi:10.1021/es103406m.

Dong, Y., J. Liu, M. Sui, Y. Qu, J. J. Ambuchi, H. Wang, and Y. Feng. 2017. A combined microbial desalination cell and electrodialysis system for copper-containing wastewater treatment and high-salinity-water desalination. *Journal of Hazardous Materials* 321: 307–315. doi:10.1016/j.jhazmat.2016.08.034.

Faze, N. R. 2015. *Life Cycle and Economic Analysis Comparing Microbial Desalination Cell and Reverse Osmosis Technologies*. Columbus, OH: Ohio State University.

Forrestal, C., P. Xu, and Z. Ren. 2012. Sustainable desalination using a microbial capacitive desalination cell. *Energy & Environmental Science* 5(5): 7161. doi:10.1039/c2ee21121a.

He, Z., N. Wagner, S. D. Minteer, and L. T. Angenent. 2006. An upflow microbial fuel cell with an interior cathode: Assessment of the internal resistance by impedance spectroscopy. *Environmental Science and Technology* 40(17): 5212–5217. doi:10.1021/es060394f.

He, Z., S. D. Minteer, and L. T. Angenent. 2005. Electricity generation from artificial wastewater using an upflow microbial fuel cell. *Environmental Science and Technology* 39(14): 5262–5267. doi:10.1021/es0502876.

Jacobson, K. S., D. M. Drew, and Z. He. 2011a. Efficient salt removal in a continuously operated upflow microbial desalination cell with an air cathode. *Bioresource Technology* 102(1): 376–380. doi:10.1016/j.biortech.2010.06.030.

Jacobson, K. S., D. M. Drew, and Z. He. 2011b. Use of a liter-scale microbial desalination cell as a platform to study bioelectrochemical desalination with salt solution or artificial seawater. *Environmental Science and Technology* 45(10): 4652–4657. doi:10.1021/es200127p.

Kim, B. H., H. S. Park, H. J. Kim, G. T. Kim, I. S. Chang, J. Lee, and N. T. Phung. 2004. Enrichment of microbial community generating electricity using a fuel-cell-type electrochemical cell. *Applied Microbiology and Biotechnology* 63(6): 672–681. doi:10.1007/s00253-003-1412-6.

Kim, Y., and B. E. Logan. 2011. Series assembly of microbial desalination cells containing stacked electrodialysis cells for partial or complete seawater desalination. *Environmental Science and Technology* 45(13): 5840–5845. doi:10.1021/es200584q.

Kokabian, B., and V. G. Gude. 2015. Sustainable photosynthetic biocathode in microbial desalination cells. *Chemical Engineering Journal* 262: 958–965. doi:10.1016/j.cej.2014.10.048.

Kokabian, B., V. G. Gude, and V. Gadhamshetty. 2013. Beneficial bioelectrochemical systems for energy, water, and biomass production. *Journal of Microbial & Biochemical Technology*. doi:10.4172/1948-5948.S6-005.

Li Y., J. Styczynski, Y. Huang, Z. Xu, J. McCutcheon, and B. Li. 2017. Energy-positive wastewater treatment and desalination in an integrated microbial desalination cell (MDC)-microbial electrolysis cell (MEC). *Journal of Power Sources* 356: 529–538. doi:10.1016/j.jpowsour.2017.01.069.

Liang, Y., H. Feng, D. Shen, N. Li, Y. Long, Y. Zhou, Y. Gu, X. Ying, and Q. Dai. 2016. A high-performance photo-microbial desalination cell. *Electrochimica Acta* 202: 197–202. doi:10.1016/j.electacta.2016.03.177.

Logan, B. E. 2010. Scaling up microbial fuel cells and other bioelectrochemical systems. *Applied Microbiology and Biotechnology*. doi:10.1007/s00253-009-2378-9.

Logan, B. E., and K. Rabaey. 2012. Conversion of wastes into bioelectricity and chemicals by using microbial electrochemical technologies. *Science* 337(6095): 686–690. doi:10.1126/science.1217412.

Lu, Y., I. M. Abu-Reesh, and Z. He. 2016. Treatment and desalination of domestic wastewater for water reuse in a four-chamber microbial desalination cell. *Environmental Science and Pollution Research* 23(17): 17236–17245. doi:10.1007/s11356-016-6910-z.

Luo, H., H. Li, Y. Lu, G. Liu, and R. Zhang. 2017. Treatment of reverse osmosis concentrate using microbial electrolysis desalination and chemical production cell. *Desalination* 408: 52–59. doi:10.1016/j.desal.2017.01.003.

Luo, H., P. Xu, and Z. Ren. 2012a. Long-term performance and characterization of microbial desalination cells in treating domestic wastewater. *Bioresource Technology* 120: 187–193. doi:10.1016/j.biortech.2012.06.054.

Luo, H., P. Xu, T. M. Roane, P. E. Jenkins, and Z. Ren. 2012b. Microbial desalination cells for improved performance in wastewater treatment, electricity production, and desalination. *Bioresource Technology* 105: 60–66. doi:10.1016/j.biortech.2011.11.098.

Mehanna, M., T. Saito, J. Yan, M. Hickner, X. Cao, X. Huang, and B. E. Logan. 2010. Using microbial desalination cells to reduce water salinity prior to reverse osmosis. *Energy & Environmental Science* 3(8): 1114. doi:10.1039/c002307h.

Meng, F., Q. Zhao, X. Na, Z. Zheng, J. Jiang, L. Wei, and J. Zhang. 2017. Bioelectricity generation and dewatered sludge degradation in microbial capacitive desalination cell. *Environmental Science and Pollution Research* 24(6): 5159–5167. doi:10.1007/s11356-016-6853-4.

Moqsud, M. A., K. Omine, N. Yasufuku, M. Hyodo, and Y. Nakata. 2013. Microbial fuel cell (MFC) for bioelectricity generation from organic wastes. *Waste Management* 33(11): 2465–2469. doi:10.1016/j.wasman.2013.07.026.

Morel, A., K. Zuo, X. Xia, J. Wei, X. Luo, P. Liang, and X. Huang. 2012. Microbial desalination cells packed with ion-exchange resin to enhance water desalination rate. *Bioresource Technology* 118: 243–248. doi:10.1016/j.biortech.2012.04.093.

Pham, C. A., S. J. Jung, N. T. Phung, J. Lee, I. S. Chang, B. H. Kim, H. Yi, and J. Chun. 2003. A novel electrochemically active and Fe(III)-reducing bacterium phylogenetically related to aeromonas hydrophila, isolated from a microbial fuel cell. *FEMS Microbiology Letters* 223(1): 129–134. doi:10.1016/S0378-1097(03)00354-9.

Ping, Q., and Z. He. 2013. Improving the flexibility of microbial desalination cells through spatially decoupling anode and cathode. *Bioresource Technology* 144: 304–310. doi:10.1016/j.biortech.2013.06.117.

Ping, Q., and Z. He. 2014. Effects of inter-membrane distance and hydraulic retention time on the desalination performance of microbial desalination cells. *Desalination and Water Treatment* 52(7–9): 1324–1331. doi:10.1080/19443994.2013.789406.

Ping, Q., B. Cohen, C. Dosoretz, and Z. He. 2013. Long-term investigation of fouling of cation and anion exchange membranes in microbial desalination cells. *Desalination* 325(September): 48–55. doi:10.1016/j.desal.2013.06.025.

Pradhan, H., S. C. Jain, and M. M. Ghangrekar. 2015. Simultaneous removal of phenol and dissolved solids from wastewater using multichambered microbial desalination cell. *Applied Biochemistry and Biotechnology* 177(8): 1638–1653. doi:10.1007/s12010-015-1842-5.

Qu, Y., Y. Feng, J. Liu, W. He, X. Shi, Q. Yang, J. Lv, and B. E. Logan. 2013. Salt removal using multiple microbial desalination cells under continuous flow conditions. *Desalination* 317: 17–22. doi:10.1016/j.desal.2013.02.016.

Qu, Y., Y. Feng, X. Wang, J. Liu, J. Lv, W. He, and B. E. Logan. 2012. Simultaneous water desalination and electricity generation in a microbial desalination cell with electrolyte recirculation for pH control. *Bioresource Technology* 106: 89–94. doi:10.1016/j.biortech.2011.11.045.

Rahimnejad, M., G. D. Najafpour, A. A. Ghoreyshi, F. Talebnia, G. C. Premier, G. Bakeri, J. R. Kim, and S. E. Oh. 2012. Thionine increases electricity generation from microbial fuel cell using saccharomyces cerevisiae and exoelectrogenic mixed culture. *Journal of Microbiology* 50(4): 575–580. doi:10.1007/s12275-012-2135-0.

Rahimnejad, M., G. D. Najafpour, A. A. Ghoreyshi, M. Shakeri, and H. Zare. 2011. Methylene blue as electron promoters in microbial fuel cell. *International Journal of Hydrogen Energy* 36(20): 13335–13341. doi:10.1016/j.ijhydene.2011.07.059.

Saeed, H. M., G. A. Husseini, S. Yousef, J. Saif, S. Al-Asheh, A. A. Fara, S. Azzam, R. Khawaga, and A. Aidan. 2015. Microbial desalination cell technology: A review and a case study. *Desalination.* doi:10.1016/j.desal.2014.12.024.

Schröder, U. 2007. Anodic electron transfer mechanisms in microbial fuel cells and their energy efficiency. *Physical Chemistry Chemical Physics: PCCP* 9(21): 2619–2629. doi:10.1039/b703627m.

Shehab, N. A., B. E. Logan, G. L. Amy, and P. E. Saikaly. 2013. Microbial electrodeionization cell stack for sustainable desalination, wastewater treatment and energy recovery. *Proceedings of the Water Environment Federation* 2013(19): 222–227. doi:10.2175/193864713813667764.

Sophia, A. C., and V. M. Bhalambaal. 2015. Utilization of coconut shell carbon in the anode compartment of microbial desalination cell (MDC) for enhanced desalination and bio-electricity production. *Journal of Environmental Chemical Engineering* 3(4): 2768–2776. doi:10.1016/j.jece.2015.10.026.

Stoll, Z. A., C. Forrestal, Z. J. Ren, and P. Xu. 2015. Shale gas produced water treatment using innovative microbial capacitive desalination cell. *Journal of Hazardous Materials* 283: 847–855. doi:10.1016/j.jhazmat.2014.10.015.

Subramani, A., and J. G. Jacangelo. 2015. Emerging desalination technologies for water treatment: A critical review. *Water Research.* doi:10.1016/j.watres.2015.02.032.

Tiquia-Arashiro, S. M. 2014. Biotechnological applications of thermophilic carboxydotrophs. *Thermophilic Carboxydotrophs and Their Applications in Biotechnology*, 29–101. doi:10.1007/978-3-319-11873-4_4.

Wen, Q., H. Zhang, Z. Chen, Y. Li, J. Nan, and Y. Feng. 2012. Using bacterial catalyst in the cathode of microbial desalination cell to improve wastewater treatment and desalination. *Bioresource Technology* 125: 108–113. doi:10.1016/j.biortech.2012.08.140.

Youngge, K., and B. E. Logan. 2013. Microbial desalination cells for energy production and desalination. *Desalination* 308: 122–130. doi:10.1016/j.desal.2012.07.022.

Yuan, H., I. M. Abu-Reesh, and Z. He. 2015. Enhancing desalination and wastewater treatment by coupling microbial desalination cells with forward osmosis. *Chemical Engineering Journal* 270(June): 437–443. doi:10.1016/j.cej.2015.02.059.

Zhang, B., and Z. He. 2012a. Energy production, use and saving in a bioelectrochemical desalination system. *RSC Advances* 2(28): 10673. doi:10.1039/c2ra21779a.

Zhang, B., and Z. He. 2012b. Integrated salinity reduction and water recovery in an osmotic microbial desalination cell. *RSC Advances* 2(8): 3265. doi:10.1039/c2ra20193c.

Zhang, B., and Z. He. 2013. Improving water desalination by hydraulically coupling an osmotic microbial fuel cell with a microbial desalination cell. *Journal of Membrane Science* 441: 18–24. doi:10.1016/j.memsci.2013.04.005.

Zhang, F., and Z. He. 2015. Scaling up microbial desalination cell system with a post-aerobic process for simultaneous wastewater treatment and seawater desalination. *Desalination* 360: 28–34. doi:10.1016/j.desal.2015.01.009.

Zhang, F., K. S. Jacobson, P. Torres, and Z. He. 2010. Effects of anolyte recirculation rates and catholytes on electricity generation in a litre-scale upflow microbial fuel cell. *Energy & Environmental Science* 3(9): 1347. doi:10.1039/c001201g.

Zhang, F., M. Chen, Y. Zhang, and R. J. Zeng. 2012. Microbial desalination cells with ion exchange resin packed to enhance desalination at low salt concentration. *Journal of Membrane Science* 417–418: 28–33. doi:10.1016/j.memsci.2012.06.009.

Zhang, Y., and I. Angelidaki. 2013. A new method for in situ nitrate removal from groundwater using submerged microbial desalination-denitrification cell (SMDDC). *Water Research* 47(5): 1827–36. doi:10.1016/j.watres.2013.01.005.

Zuo, K., J. Cai, S. Liang, S. Wu, C. Zhang, P. Liang, and X. Huang. 2014. A ten liter stacked microbial desalination cell packed with mixed ion-exchange resins for secondary effluent desalination. *Environmental Science and Technology* 48(16): 9917–9924. doi:10.1021/es502075r.

Zuo, K., Z. Wang, X. Chen, X. Zhang, J. Zuo, P. Liang, and X. Huang. 2016. Self-driven desalination and advanced treatment of wastewater in a modularized filtration air cathode microbial desalination cell. *Environmental Science and Technology* 50(13): 7254–7262. doi:10.1021/acs.est.6b00520.

6 Effect of the Draw Solution on the Efficiency of Two-Stage FO-RO/BWRO for Seawater and Brackish Water Desalination

Ali Altaee, Adnan Alhathal Alanezi, Radhi Alazmi, Alaa H. Hawari, and Claudio Mascialino

CONTENTS

6.1 INTRODUCTION

The forward osmosis (FO) process has been suggested as an alternative to the conventional reverse osmosis (RO) process for seawater desalination [1–6]. Compared to RO, the FO process exhibits lower membrane fouling propensity and power consumption [5,7–10]. The driving force in the FO process is the osmotic pressure gradient between the draw and feed solutions. Fresh water transports from the feed to the draw solution side of the FO membrane. Water flux across the FO membrane ceases when the difference in the concentrations between the feed and the draw solutions is insignificant. The diluted draw solution requires further treatment for freshwater extraction and draw solution regeneration and reuse. Without draw solution recycling and reuse, the FO process is uneconomical [7–10]. Therefore, the draw solution should be carefully selected to maintain a low cost for the FO process. Ideally, the draw solution should be highly soluble in water, cheap, readily available, environmentally friendly, non-toxic and easily recycled [8–12]. Typical draw solutions used in the past included $MgSO_4$, NaCl, glucose, $MgCl_2$, magnetic particles and ammonium carbon dioxide [8,13].

Different techniques have been suggested for use to regenerate the draw solution. The techniques that have been used for the regeneration of the draw solution were RO and nanofiltration (NF), membrane distillation (MD) and thermal evaporation [9–12]. Elimelech et al. proposed a column distillation process to regenerate the draw solution [9–10]. Ammonium carbon dioxide was used as the draw solution because of its high osmotic pressure. In the MD regeneration system, fresh water is collected on the permeate side while concentrated brine is the regenerated draw solution. Using MD to regenerate the draw solution has a number of drawbacks, such as a lower recovery rate for the MD and membrane wetting [13,14]. Abdulsalam and Adel proposed using a two-stage FO-NF/RO hybrid system [15]. A multivalent ionic solution, such as $MgSO_4$ and $MgCl_2$, was proposed as a draw solution when the regeneration process was carried out by an NF membrane, while the RO membrane was suggested for use for the regeneration of a monovalent draw solution, such as NaCl or KCl. The NF/RO process is more energy efficient than the MD process and operates at a higher recovery rate than the MD process. Shung et al. used magnetic nanoparticles coated with hydrophilic polymers as the draw solution in the FO process. Although the magnetic nanoparticles exhibited a high osmotic pressure, regeneration was a problem due to the agglomeration of nanoparticles [16]. Hydrogel polymers were proposed as the draw solution in the FO process because of their high osmotic pressure. Water flux across the FO membrane increased when carbon nanoparticles were added, but excessive addition of carbon nanoparticles resulted in a reduction of the flux [17]. The NF process was used to regenerate the draw solution in the two-stage FO-NF desalination system [14,18]. $MgSO_4$ was used as the draw solution because of the high rejection rate of the NF membranes to divalent ions and the lower power consumption compared to the RO process. The FO-NF process was more efficient than a single RO process. However, the study was carried out for the desalination of brackish water, and no information was provided regarding the process efficiency for seawater desalination. Additionally, the power consumption in the FO-NF system was not calculated. It should be noted that the regeneration cost of the draw solution should be added to the total desalination cost.

In the current study, the FO-RO and FO-brackish water reverse osmosis (BWRO) processes were proposed for seawater and brackish water desalination, respectively. In the first stage, FO membranes were used for feed water pre-treatment, while in the second stage, RO or BWRO membranes were used to regenerate the draw solution and for fresh water extraction. Using RO and BWRO to regenerate the draw solution has a number of advantages, such as a high salt rejection rate, high recovery rate, and high membrane flux. Despite the advantages of the RO process, its major drawback is its high power consumption. The power consumption in the RO and BWRO processes is affected by a number of factors, such as the feed water salinity, feed temperature, membrane permeability, and feed pressure [8,11]. Typically, power consumption in the hyperfiltration process increases as the feed water salinity increases. In the current study, three seawater feed salinities and two brackish water feed salinities were evaluated for desalination by the FO-RO and FO-BWRO processes, respectively. In the FO-RO process, the total dissolved solids (TDS) of the seawater feeds were 32, 38, and 45 g/L, while the TDS of the feeds in the FO-BWRO process were 1.5 and 3 g/L. Three types of chemical compounds were investigated as the draw solution, $MgSO_4$, NaCl, and $MgCl_2$, because of their high osmotic pressure and applicability to RO membrane treatment. Pre-developed software models were used throughout this study to estimate the performance of the FO and RO membranes [11,12]. Reverse Osmosis System Analysis (ROSA6.1) was used to model the performance of the RO and NF membranes. Different membrane recovery rates were simulated in the current study. In practice, the recovery rate in the RO system did not exceed 50% for a seawater salinity of ~35,000 mg/L because of scaling problems. For brackish water desalination, the range of recovery rates varied between 50% and 75%. However, with FO pre-treatment, higher recovery rates can be achieved in the RO and BWRO membranes because of the high purity of the draw solution. Therefore, the recovery rate of the second stage FO-RO and FO-BWRO processes increased to over 50% and 75%, respectively.

6.2 METHODOLOGY

The water flux in the FO and RO membranes, J_w, is a function of membrane permeability and net driving pressure across the membrane. In the RO process, the driving force is the difference between the hydraulic and osmotic pressures of the feed solution, while in the FO process, it is the osmotic pressure gradient between the draw and feed solutions. It is assumed here that the FO process requires no or negligible hydraulic pressure for operation. In general, the water flux in the RO and FO processes can be estimated from the following equations [19]:

$$J_{w\text{-RO}} = A_{w\text{-RO}} \times (\Delta P - \Delta\pi) \tag{6.1}$$

$$J_{w\text{-FO}} = A_{w\text{-FO}} \times (\Delta\pi - \Delta P) \tag{6.2}$$

In Equations 6.1 and 6.2, $J_{w\text{-RO}}$ and $J_{w\text{-FO}}$ are the water fluxes of the RO and FO membranes, respectively (L/m^{2}h); $A_{w\text{-RO}}$ and $A_{w\text{-FO}}$ are the coefficients of the membrane permeability for the RO and FO membranes, respectively (L/m^{2}h.bar); ΔP is the feed pressure difference across the membrane (bar); and $\Delta\pi$ is the osmotic pressure difference across the membrane (bar). It was assumed that the FO membrane was operated in the PRO mode (active layer facing the draw solution) and a conservative A_w value of 0.79 L/m^{2}h.bar was used [9]. A Filmtec seawater membrane, SW30HRLE-400i, with an A_w of approximately 1.13 L/m^{2}h.bar was used for seawater desalination in the RO system. Previous studies showed that the experimental water flux in the FO process was lower than the theoretical water flux due to the concentration polarization (CP) phenomena [8,9]; CP is divided into dilutive CP on the draw solution side of the FO membrane and concentrative CP on the feed side of the FO membrane. Taking dilutive and concentrative CP, Equation 6.2 can be described as follows:

$$J_{w\text{-FO}} = A_{w\text{-FO}} \left(\frac{\pi_{\text{Db}} e^{\left(\frac{-J_{w\text{-FO}}}{k}\right)} - \pi_{\text{Fb}} e^{(J_{w\text{-FO}}K)}}{1 + \frac{B}{J_w}\left(e^{(J_{w\text{-FO}}K)} - e^{\left(\frac{-J_{w\text{-FO}}}{k}\right)} \right)} - \Delta P \right) \tag{6.3}$$

where π_{Db} and π_{Fb} are the osmotic pressures associated with the bulk draw and feed solution, respectively (in bar); B is the solute permeability coefficient (4×10^{-4} m/h); k is the bulk mass transfer coefficient (0.31 m/s); and K is the solute resistivity for diffusion within the porous support layer (1.4×10^5 s/m). K is the ratio of the salt diffusion coefficient (D) and membrane structure parameter (S): $K = S/D$. The lower the S, the less the CP effects the feed side. Equation 6.3 will be resolved by trial and error to obtain the value of $J_{w\text{-FO}}$ [20,21].

The salt flux, J_s, in the membrane filtration process was estimated from the following equation:

$$J_s = B(C_f - C_p) \tag{6.4}$$

where:

C_f is the concentration of the feed solution (mg/L)
B is the salt permeability coefficient (m/h)
C_p is the permeate concentration (mg/L)

The salt permeability coefficient, B, was estimated from the following equation:

$$B = \frac{(1 - Rj) \times J_w}{Rj} \tag{6.5}$$

Membrane rejection, Rj, is the ratio of the permeate concentration to the feed concentration: $Rj = 1 - (C_p/C_f)$. Typically, Rj increases with membrane selectivity. The concentration of the permeate in the FO, $C_{p\text{-FO}}$, was estimated from the following equation [22]:

$$C_{p\text{-FO}} = \frac{BC_{f\text{-in}}}{J_{w\text{-RO}} + B} \tag{6.6}$$

In Equation 6.6, $C_{f\text{-in}}$ is the concentration of the seawater feed to the FO (mg/L). In the case of the RO membrane, the permeate concentration, $C_{p\text{-RO}}$, was calculated from Equation 6.7 [22]:

$$C_{p\text{-RO}} = B \times C_{\text{fc}} \times CP \times Rj \times \frac{A_m}{Q_p} \tag{6.7}$$

where A_m is the membrane area (m^2). Altaee et al. explained the processes of RO and FO-RO for seawater desalination, as shown in Figure 6.1. The schematic diagram in Figure 6.1 was also applied for brackish water desalination, but instead of a RO membrane, a BWRO membrane was used in the second filtration stage. The BWRO process usually required less energy for saline water treatment because BWRO membranes have a higher water permeability coefficient than RO membranes [22]. In addition to BWRO membranes, an NF membrane was evaluated in this study for draw solution regeneration and fresh water extraction. NF membranes reject most divalent ions, such as magnesium and sulfate. In this study, an NF membrane was used to regenerate the $MgSO_4$ draw solution in the second filtration stage for brackish water. Filmtec RO and BWRO/NF membranes were used in the second filtration stage for seawater and brackish water desalination, respectively. The Filmtec RO membrane was type SW30HRLE 440i, while BW30-400 and NF90-400 membranes were used in the second FO-BWRO and FO-NF filtration processes, respectively.

Three inorganic metal salts were evaluated in this study as draw solutions, NaCl $MgSO_4$, and $MgCl_2$, because of their (i) wide availability, (ii) high osmotic pressure, (iii) high rejection by RO membranes, and (iv) high solubility in water. A number of seawater salinities were evaluated in the

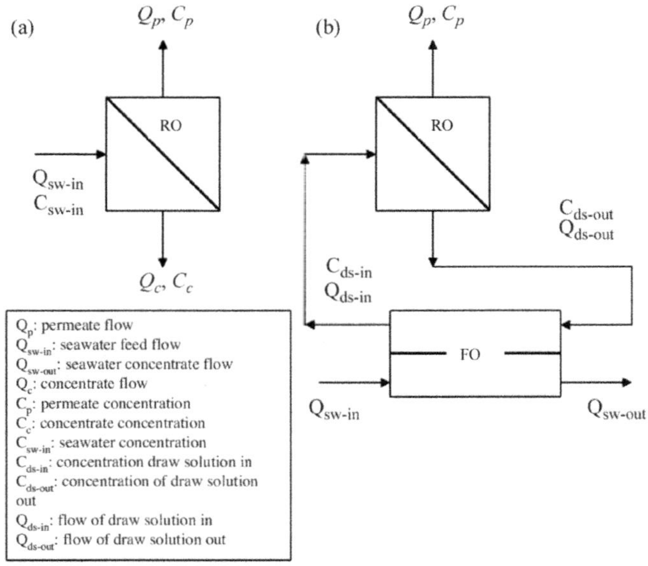

FIGURE 6.1 (a) RO and (b) FO-RO processes diagram. (From Altaee, A. et al., *Desalination*, 336, 50–57, 2014.)

TABLE 6.1
Seawater Composition

SW TDS	Ion Concentration mg/L							
(mg/L)	K	Na	Mg	Ca	HCO_3	Cl	SO_4	SiO_2
32,000	354	9,854	1,182	385	130	17,742	2,477	0.9
35,000	387	10,778	1,293	421	142	19,406	2,710	1.0
38,000	419	11,663	1,399	456	154	20,999	2,932	1.0
45,000	496	13,812	1,657	539	182	24,868	3,472	1.2

present study. Table 6.1 shows the salt concentrations and seawater compositions that were under investigation in this study [22]. It should be mentioned here that the feed and draw solution pressures for the FO process were assumed to be 1 bar and the pump efficiency, η, was 0.8. It should be mentioned that the Energy Recovery Device (ERD) was not considered in this study, which can be justified in small capacity desalination plants. For simplicity, it was assumed here that the composition of all of the brackish water feeds was only NaCl. The osmotic pressure of the feed and draw solutions was estimated using the van't Hoff equation as recommended by the FilmTec membrane company [22–24]:

$$\pi = 1.12\,(273 + T)\,\Sigma m_j \tag{6.8}$$

where:
m_j is the molal concentration of the jth ion species
T is the temperature (°C).

6.3 RESULTS AND DISCUSSION

6.3.1 Performance of the FO Process for Seawater Desalination

The water and salt fluxes in the FO membrane were evaluated for a number of seawater salinities. Figure 6.2 shows the water permeability in the FO membrane for seawater salinities of 32,000, 38,000, and 45,000 mg/L. The FO recovery rates varied from 16% to 60% at a 4% interval. The simulation results showed that the FO water flux decreased as the seawater salinity increased. This was due to the decrease in the net driving force across the FO membrane (Figure 6.2a–c). The water flux across the FO membrane was the highest when 1.2 M NaCl was used as the draw solution. However, the difference in the water flux between 1 M NaCl, 0.657 M $MgCl_2$, and 1.45 M $MgSO_4$ was insignificant due to the equal driving force generated by these draw solutions (Figure 6.2a–c). The osmotic pressures of 1 M NaCl, 0.657 M $MgCl_2$, and 1.45 M $MgSO_4$ were 47, 46, and 46.6 bar, respectively. The water flux increased significantly when 1.2 M $MgCl_2$ was used as the draw solution due to its higher osmotic pressure, which approached 57.5 bar. Using a high concentration draw solution would not only increase the water flux but also the salt flux from the feed to the draw solution. This phenomenon is shown in Figure 6.3a–c. The simulation results showed that the highest salt diffusion occurred when 1.2 M NaCl was used as the draw solution. This was due to the higher concentrative CP on the feed side of the membrane caused by the high solvent transport toward the membrane surface. The salt diffusion from the feed to the draw solution side of the FO membrane increased as the feed salinity increased (Figure 6.3). For example, the salt diffusion with a 1.2 M draw solution was 0.013 and 0.011 m/h at 32,000 and 45,000 mg/L of the feed salinities, respectively (Figure 6.3a and c). The higher salt diffusion at the higher feed salinity was due to the severe CP, which resulted in an increased solute accumulation in the support layer near the membrane surface.

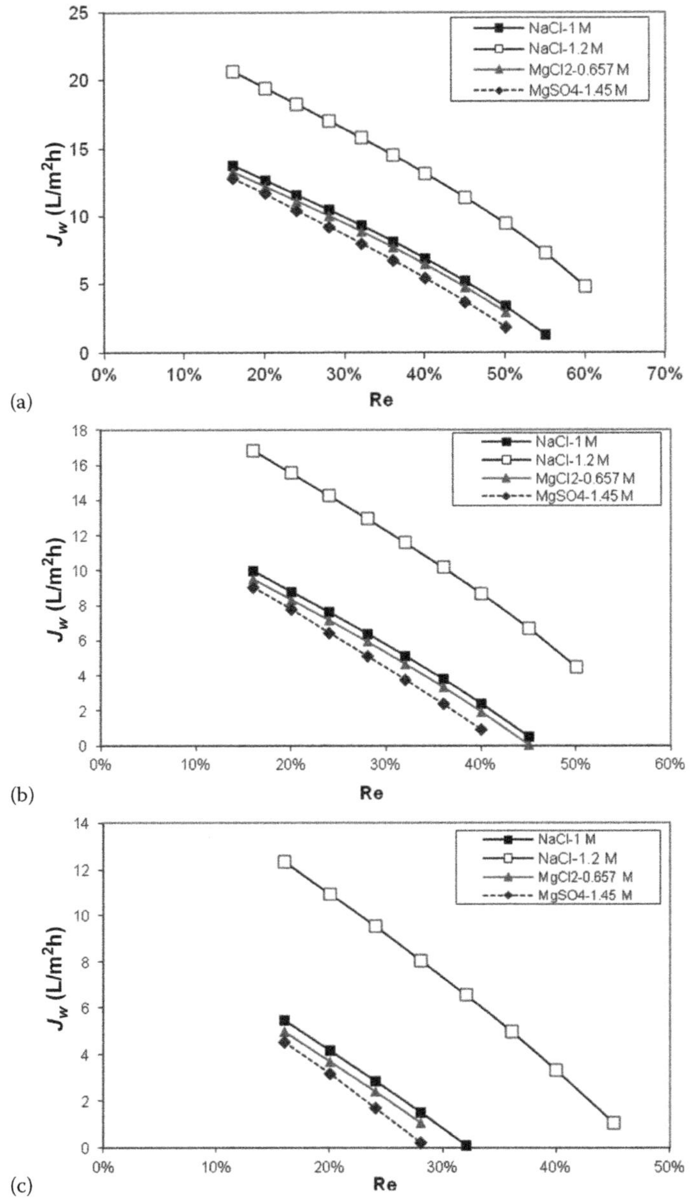

FIGURE 6.2 The water flux through FO at different membrane recovery rates (Re) with seawater salinities of (a) 32,000 mg/L, (b) 38,000 mg/L, and (c) 45000 mg/L.

The effect of the concentrative CP was obvious at all feed salinities and increased as the osmotic pressure of the draw solution increased. As shown in Figure 6.3a, J_s was 0.013 and 0.0087 m/h at 1.2 M and 1 M NaCl respectively. Interestingly, the differences in the salt diffusion between 1 M NaCl, 0.657 M $MgCl_2$, and 1.45 M $MgSO_4$ was insignificant because of the equal osmotic pressures that were generated by these draw solution concentrations (Figure 6.3a–c).

For seawater desalination, field studies carried out at pilot scale plants showed that FO fouling was negligible and reversible [25,26]. This will reduce the operating cost of the FO membrane system. However, it should be mentioned that the chemical and draw solution purging costs should be considered. These costs largely depend on the type of chemicals used in the FO process.

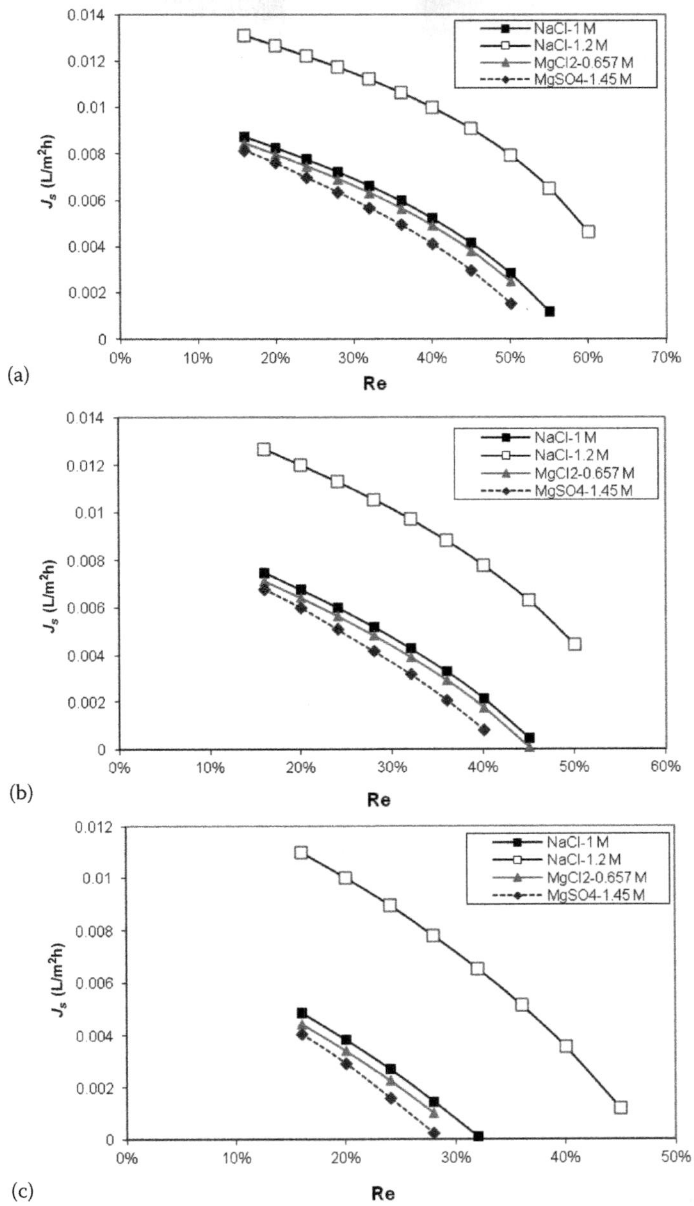

FIGURE 6.3 The solute flux across FO at different membrane recovery rates (Re) for seawater salinities of (a) 32,000 mg/L, (b) 38,000 mg/L, and (c) 45,000 mg/L.

6.3.2 Performance of the FO Process for Brackish Water Desalination

Three inorganic solutes were used as the draw solution in the FO membrane including NaCl, $MgCl_2$, and $MgSO_4$. Two brackish water salinities, 1.5 and 3 g/L, were evaluated as the feed solution for the FO membrane. The concentrations of the draw solutions were 0.33, 0.22, and 0.32 M for NaCl, $MgCl_2$, and $MgSO_4$, respectively, which gave equal osmotic pressures of approximately 15 bar. Typically, the recovery rate for brackish water in a conventional desalination plant was approximately 75%. A higher recovery rate may cause scale fouling in the BWRO membrane. However, in the FO process, a higher recovery rate can be reached because there was no hydraulic pressure involved, and in this study, FO recovery rates between 40% and 80% were evaluated. The results showed a gradual decrease in the membrane flux as the FO recovery rates increased from 40% to 80%. This observation applied to all of the osmotic agents and was mainly due to the decrease in the osmotic pressure across the membrane at higher FO recovery rates, as shown in Figure 6.4a and c. The results also showed that the water flux across the membrane was higher at lower feed concentrations or at a 1.5 g/L feed concentration. Obviously, this was due to the lower osmotic pressure driving force across the FO membrane at a higher feed concentration (3 g/L) (Figure 6.4a and c). It should be mentioned here that the water flux in the NaCl draw solution was slightly lower than that

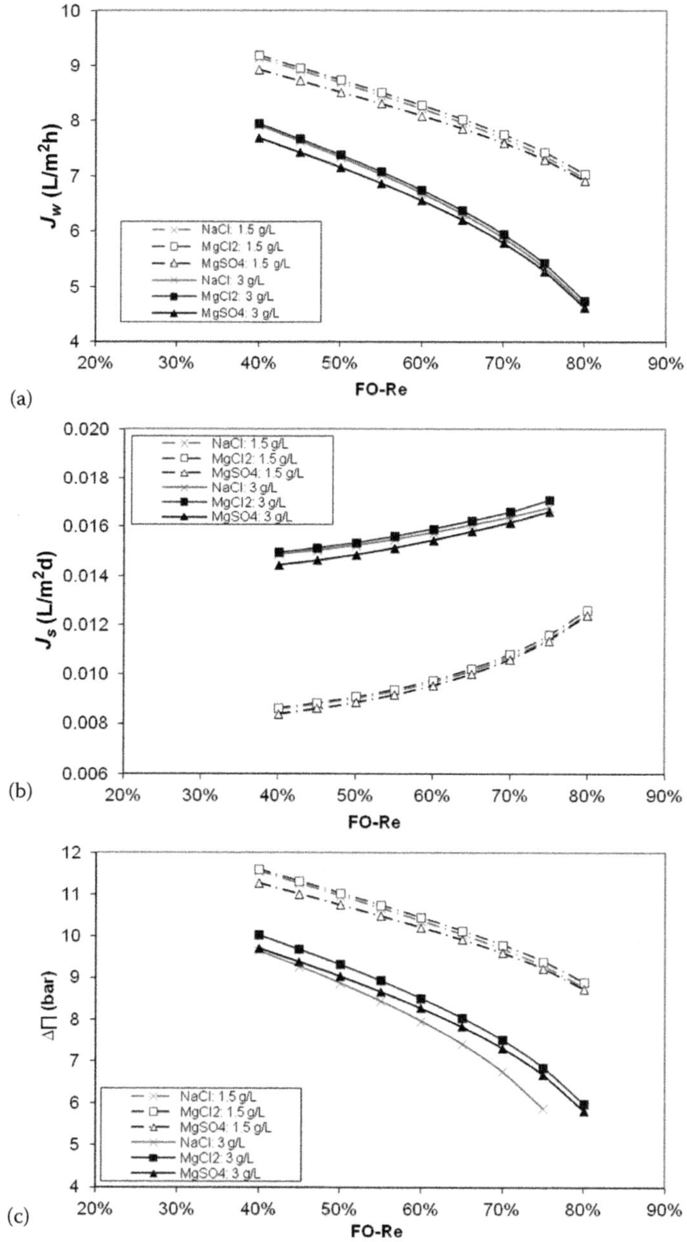

FIGURE 6.4 The FO performance at different recovery rates (FO-Re), and the impact on (a) the membrane flux, (b) solute flux, and (c) osmotic pressure difference across the membrane.

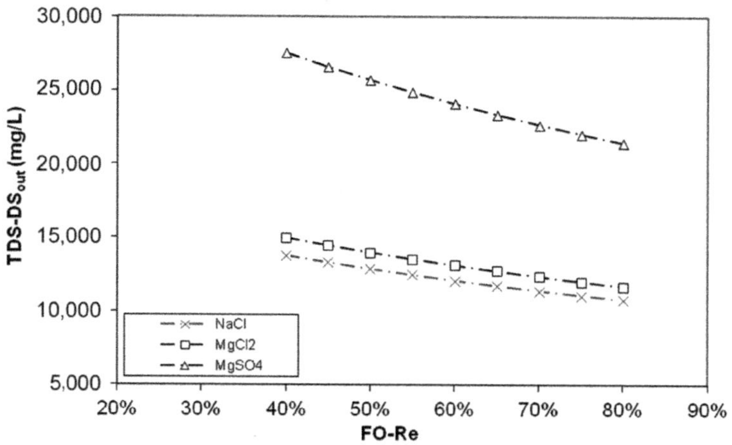

FIGURE 6.5 The concentration of the draw solution (TDS-DS_{out}) leaving the FO membrane at different FO recovery rates (FO-Re).

in the $MgCl_2$ and $MgSO_4$ draw solutions, which was mainly because of the lower membrane rejection rate for NaCl compared to $MgCl_2$ and $MgSO_4$. Finally, the solute flux from the feed to the draw solution side of the FO membrane is shown in Figure 6.4b. The results showed a higher solute flux at a 3 g/L feed TDS than at a 1.5 g/L feed TDS because of the higher concentration. The solute flux across the membrane also increased as the recovery rate of the FO membrane increased. Typically, the internal CP increased as the recovery rate increased, so a higher solute flux occurred.

In the FO-BWRO system, the feed to the second stage BWRO membrane is the diluted draw solution from the first stage FO process. The concentration of the BWRO feed solution affected the process performance in terms of energy consumption and permeate quality, which will be explained in the following section. Figure 6.5 shows the concentration of the draw solution out, $C_{ds\text{-}out}$, at different FO recovery rates. After leaving the FO membrane, the concentrations of the dilute draw solutions, $C_{ds\text{-}out}$, were in the following order: $MgSO_4$ > $MgCl_2$ > NaCl. Apparently, this was due to the higher initial concentration of $MgSO_4$ compared to the $MgCl_2$ and NaCl draw solutions. In general, the $MgSO_4$ solution exhibited a lower osmotic pressure than the $MgCl_2$ and NaCl solutions; therefore, a higher concentration of $MgSO_4$ was required to generate an osmotic pressure equal to that of the $MgCl_2$ and NaCl draw solutions.

6.3.3 Performance of the RO, BWRO, and NF Membranes

A Filmtec seawater RO membrane type SW30HRLE 400i was used for draw solution regeneration and fresh water extraction in seawater desalination. For brackish water desalination, a Filmtec BWRO membrane type BW30-400 was used for draw solution regeneration and fresh water extraction. In addition to the BWRO membrane type BW30-400, a Filmtec NF membrane type NF90-400 was used for $MgSO_4$ draw solution regeneration and fresh water extraction. Pretreatment of the seawater/brackish water was required to use the RO/BWRO membrane directly for desalination. However, in the FO-BWRO system, the feed water of the BWRO membrane did not require further treatment because the FO process works as a pretreatment for BWRO. Normally, conventional pretreatment of seawater yields a silt density index (SDI) of approximately 3, while in the case of FO pretreatment, the SDI can be as low as 1. As a matter of fact, this is one advantage of using FO membrane treatment to reduce the fouling propensity in the second stage RO membrane. It should be mentioned that the estimated energy consumption did not include the energy for FO pretreatment because it is insignificant [22]. The specific power consumption and permeate concentration of the desalinated seawater, TDS of 35 g/L, are shown in Figure 6.6a and b. The results show a decrease

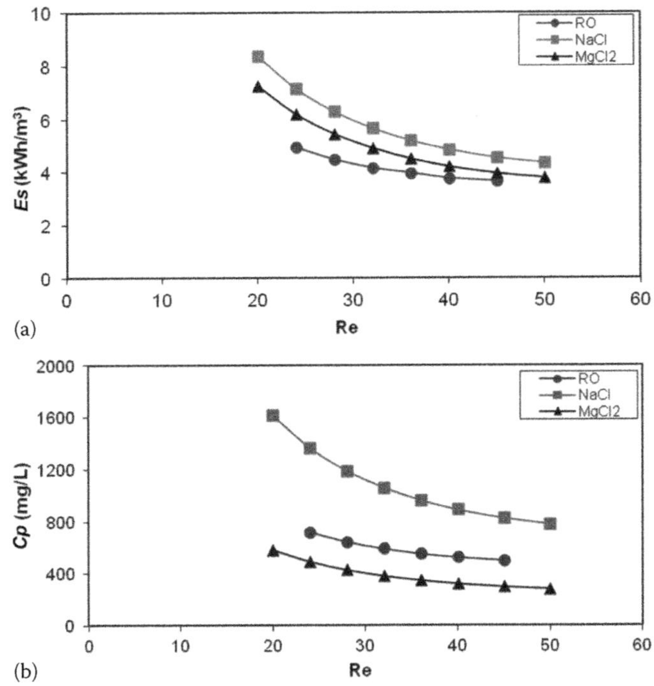

FIGURE 6.6 RO membrane performance for seawater desalination: (a) Specific power consumption in the RO and (b) Permeate TDS (feed TDS of 35 g/L).

in the specific power consumption at higher FO recovery rates (Figure 6.6a). The lowest power consumption was achieved in the conventional RO process without FO treatment followed by 0.657 M $MgCl_2$ and 1 M NaCl in the FO-RO system. Using the 1 M NaCl draw solution resulted in a slightly more concentrated feed solution for the RO membrane, which increased the power consumption of the RO process. However, power consumption decreased as the recovery rate of the RO system increased. Interestingly, the power consumption difference between the RO and the FO-RO system decreased as the recovery rate of the RO membrane increased, especially with the 0.657 M $MgCl_2$ draw solution. These results underline the importance of the draw solution in the FO process, which plays a significant role in improving process performance. Benefiting from the high purity draw solution, the FO-RO system achieved high recovery rates that could not be achieved in the conventional RO membrane system for seawater desalination. Most of the conventional RO desalination plants operate with recovery rates of less than 50%. However, a recovery rate over 50% can be achieved in the FO-RO system without causing major fouling problems in the RO system. For example, a 50% recovery rate was achieved in the 0.657 M $MgCl_2$ and 1 M NaCl FO-RO processes compared to 45% in the conventional RO process (Figure 6.6a). The other advantage of the FO-RO system was the high permeate water quality (Figure 6.6b). The lowest permeate concentration was exhibited by 0.657 M $MgCl_2$, followed by the conventional RO process and 1 M NaCl. This was due to the higher RO rejection rate for $MgCl_2$ than for NaCl, which constituted 80% of the seawater solution. Finally, the permeate concentration in the RO and FO-RO systems decreased as the recovery rate increased, which was due to the higher dilution of the permeate concentration at higher recovery rates.

In the case of brackish water, a Filmtec BW30-400 membrane was used for water desalination. Unlike seawater, brine disposal in brackish water treatment is problematic, especially in the case of inland desalination. Therefore, it is preferable to achieve high recovery rates in the brackish water desalination process. Depending on feed salinity and composition, a recovery rate of 75% or higher could be achieved in the brackish water desalination process. For a 3 g/L feed concentration,

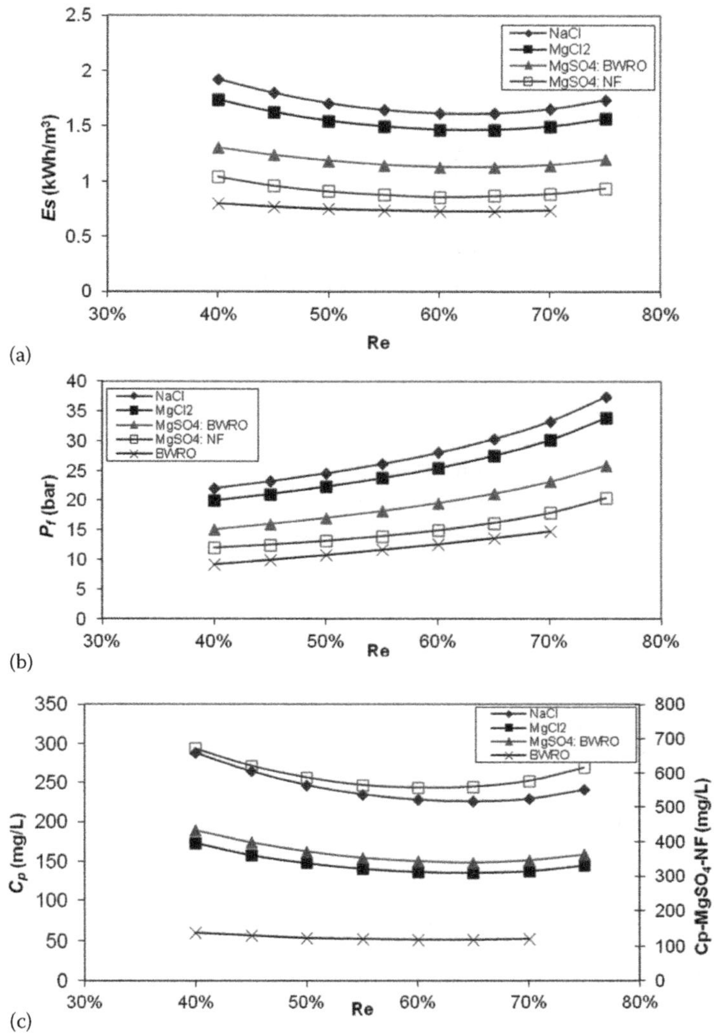

FIGURE 6.7 Performance of the RO and NF membranes for draw solution regeneration and fresh water extraction: (a) Feed pressure, (b) specific power consumption, and (c) permeate TDS.

the highest recovery rate that could be achieved by the BWRO membrane was 70%, while in the FO-BWRO system the recovery rate was approximately 75%. The higher recovery rate could also be achieved by the FO-BWRO system depending on the feed concentration. Figure 6.7a shows the power consumption in the BWRO system with different draw solutions. The results showed that the power consumption was lower in the conventional BWRO system than in the FO-BWRO system. In the case of the FO-BWRO system, the lowest power consumption was achieved by 0.32 M $MgSO_4$ followed by the 0.22 M $MgCl_2$ and 0.33 M NaCl draw solutions. However, the lowest achievable specific power consumption was determined using the NF Filmtec membrane NF90-400 to regenerate the $MgSO_4$ draw solution in the FO-BWRO system (Figure 6.7a). This was due to the high NF membrane permeability (9.9 L/m^{2}h.bar) compared to the BWRO membrane (3.3 L/m^{2}h.bar). The NF90-400 membrane rejection rate for $MgSO_4$ is >97%, which makes it suitable to regenerate the draw solution [14]. It should be mentioned here that the power consumption of the regeneration system increased slightly as the recovery rate increased (Figure 6.7a). For the BWRO and FO-BWRO systems, the lowest power consumption was achieved at a recovery rate of 65%, while in the FO-NF system, the recovery rate was 60%. The reason for this result is because as the feed pressure increased, the recovery rate of the BWRO/NF membrane system increased (Figure 6.7b). As the recovery rate increased for the brine concentration, its osmotic pressure also increased; therefore, a higher feed pressure was required to overcome the osmotic pressure of the feed solution. The lowest feed pressure was found in the conventional BWRO system because of the lower feed concentration (Figure 6.7b). In the case of the FO-BWRO system, the lowest feed pressure was found for the 0.32 M $MgSO_4$ draw solution, followed by the 0.22 M $MgCl_2$ and 0.33 M NaCl draw solutions. Nevertheless, more than a 31% energy reduction was achieved when the NF membrane was used to regenerate $MgSO_4$ in the FO-NF system (Figure 6.7b). Unfortunately, the major drawback to using the NF membrane for draw solution regeneration was the high permeate concentration due to the low solute rejection rate (~97% to $MgSO_4$). Figure 6.7c shows that at a 75% recovery rate, the permeate concentration of the FO-NF system was 615 mg/L compared to 159 mg/L for the FO-BWRO system (0.32 M $MgSO_4$ draw solution). An additional membrane treatment is often required for the FO-NF permeate if it is be used as drinking water. The permeate concentration from the conventional BWRO system, on the other hand, was only 53 mg/L at a 70% recovery rate because of the high rejection rate of the BWRO membranes. Although the FO-RO/BWRO system was less efficient in terms of power consumption and permeate quality compared to the conventional RO/BWRO system, it had a number of advantages, such as a higher recovery rate, lower RO/BWRO membrane fouling, and fewer antiscalants. Brine disposal in inland desalination is always a major problem that can be alleviated by increasing the recovery rate of the RO/BWRO process in the FO-RO/BWRO system. The high power consumption of the FO-RO/BWRO system can be resolved by using a combination of an engineered osmotic agent and a suitable regeneration membrane process such as the NF process, which can considerably reduce power consumption. However, using the NF process will adversely affect the permeate quality, but that is not a major problem if the end-use of the produced water is for domestic and human applications other than drinking water. It is also important to select a proper draw solution to reduce the desalination cost by improving the FO and RO filtration processes.

6.3.4 Cost of the FO-BWRO Membrane System

The cost and number of FO elements required for the FO-BWRO system are shown in Table 6.2. The values in Table 6.2 are based on the HTI, USA, FO membranes using 0.22 M MgC12 as the draw solution. Three feed concentrations were evaluated in this study: 1.5, 3, and 5 g/L. The cost

TABLE 6.2
Number and Cost of FO in the FO-BWRO System for Desalination with Different Seawater TDS Values

	1.5 g/L Feed			3 g/L Feed			5 g/L Feed		
Re	A (m^2)	No Elm.	Cost USD×10^6	A m^2	No Elm.	Cost USD×10^6	A m^2	No Elm.	Cost USD×10^6
40%	136,041	8,245	14.17	157,116	9,522	16.37	206,500	12,515	21.51
45%	139,408	8,449	14.52	162,853	9,870	16.97	218,845	13,263	22.80
50%	142,947	8,663	14.89	169,246	10,257	17.63	233,740	14,166	24.35
55%	146,735	8,893	15.29	176,565	10,701	18.39	252,480	15,302	26.30
60%	150,886	9,145	15.72	185,236	11,226	19.30	277,385	16,811	28.90
65%	155,581	9,429	16.21	195,979	11,878	20.42	313,084	18,975	32.62
70%	161,120	9,765	16.79	210,115	12,734	21.89	370,362	22,446	38.59
75%	168,059	10,185	17.51	230,388	13,963	24.00	481,602	29,188	50.17

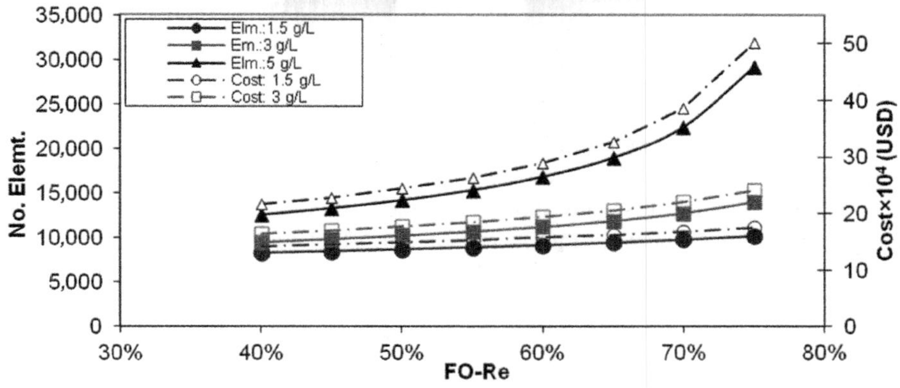

FIGURE 6.8 Number and cost of the FO elements at different FO recovery rates (FO-Re) and feed concentrations.

of the HTI FO membrane type 8040 FO-FS-P, with 16 m^2 of active membrane area, is approximately 1719 USD/element. These membranes are made of cellulose triacetate (CTA). The results in Table 6.2 and Figure 6.8 show that the number of FO membranes required for brackish water treatment increased as the feed salinity increased from 1.5 to 5 g/L. This was due to the decrease in the osmotic pressure difference and net driving force across the FO membrane at a higher feed salinity (Figure 6.4c). The cost of the FO membrane increased as the feed salinity increased, as shown in Figure 6.8. However, the difference in the number of FO elements and cost between 1.5 and 5 g/L of the feed concentration increased as the recovery rate of the FO membrane increased (Figure 6.8). This was mainly due to a reduction in the osmotic pressure driving force across the FO membrane (Figure 6.4c). Increasing the FO recovery rate would increase the CP at the feed side of the membrane, so the water flux decreases. For example, at a recovery rate of 40%, the cost of the FO membrane increased by 34% when the feed concentration increased from 1.5 to 5 g/L, while the cost increased by 65% at a recovery rate of 75%. However, it should be noted here that the expected FO membrane life is longer than the conventional RO membrane due to the lower fouling propensity and membrane compaction problems. Unfortunately, the cost of an FO membrane is currently rather high, but it is expected to drop in the near future, which will make the FO process more competitive for the conventional RO process.

6.4 CONCLUSION

The present study investigated the effect of the draw solution on the performance of the FO-RO and FO-BWRO systems for seawater and brackish water desalination, respectively. For seawater and brackish water desalination, the water flux decreased as the recovery rate of the FO process increased. This trend was also observed in the solute flux across the FO membrane for seawater desalination. However, for brackish water desalination, the solute flux increased as the FO recovery rates increased. The simulation results showed an increase in the RO power consumption for seawater desalination as the recovery rate increased. In general, seawater desalination power consumption was lower in the conventional RO process than in the FO-RO system. However, the difference in the power consumption between the RO and FO-RO systems decreased as the RO recovery rate increased. Interestingly, a higher recovery rate can be reached in RO with FO pretreatment than in the conventional RO process. In the case of brackish water desalination, the power consumption in the conventional BWRO process was lower than that in the FO-BWRO system, but the recovery rates of the conventional BWRO process were lower than the recovery rates that could be achieved by the FO-BWRO system. The FO-NF system was able to reduce the cost of brackish water desalination due to the higher NF permeability compared to the BWRO membrane but that depended on the cost of the permeate water TDS. The results from this study also showed that the energy requirements for FO desalination were affected by the type of draw solution; therefore, more research and experimental work needs to be done in that area. One advantage of the FO-BWRO process was that a high recovery rate could be achieved by the membrane treatment. In fact, the recovery rate is a key parameter in inland brackish water desalination due to increasing concerns about brine disposal. Finally, this study showed that the cost of the FO membrane increased as the feed salinity and/or the FO recovery rate increased. Nevertheless, the cost of the FO desalination process can be reduced if the cost of the FO membrane decreases in the near future.

REFERENCES

1. B. Peñate, L. García-Rodríguez, Current trends and future prospects in the design of seawater reverse osmosis desalination technology, *Desalination*, 284, 1–8 (2012).
2. N. Misdan, W. J. Lau, A. F. Ismail, Seawater Reverse Osmosis (SWRO) desalination by thin-film composite membrane—Current development, challenges and future prospects, *Desalination*, 287, 228–237 (2012).
3. A. A. Alanezi. A. O. Sharif, M. Sanduk, A. Khan, Potential of membrane distillation—A comprehensive review, *International Journal of Water*, 7(4), 317–346 (2013).
4. H. Abdallah, A. F. Moustafa, A. A. Alanezi, H. E. M. El-Sayed, Performance of a newly developed titanium oxide nanotubes/polyethersulfone blend membrane for water desalination using vacuum membrane distillation, *Desalination*, 346, 30–36 (2014).
5. A. AlTaee, A. O. Sharif, Alternative design to dual stage NF seawater desalination using high rejection brackish water membranes, *Desalination*, 273, 391–397 (2011).
6. M. M. Alhazmy, Multi stage flash desalination plant with brine–feed mixing and cooling, *Energy*, 36, 5225–5232 (2011).
7. V. Yangali-Quintanilla, Z. Li, R. Valladares, Q. Li, G. Amy, Indirect desalination of Red Sea water with forward osmosis and low pressure reverse osmosis for water reuse, *Desalination*, 280, 160–166 (2011).
8. R. L. McGinnis, M. Elimelech, Energy requirements of ammonia–carbon dioxide forward osmosis desalination, *Desalination*, 207, 370–382 (2007).
9. J. R. McCutcheon, R. L. McGinnis, M. Elimelech, A novel ammonia—Carbon dioxide forward (direct) osmosis desalination process, *Desalination*, 174, 1–11 (2005).
10. J. R. McCutcheon, M. Elimelech, Influence of concentrative and dilutive internal concentration polarization on flux behavior in forward osmosis, *Journal of Membrane Science*, 284, 237–247 (2006).
11. A. Altaee, A. Mabrouk, K. Bourouni, A novel Forward osmosis membrane pretreatment of seawater for thermal desalination processes, *Desalination*, 326, 19–29 (2013).
12. A. Altaee, G. Zargoza, A conceptual design of low fouling and high recovery FO-MSF desalination plant, *Desalination*, 343, 2–7 (2014).
13. A. Altaee, Forward osmosis: Potential use in desalination and water reuse, *Journal of membrane and Separation Technology*, 1, 79–93 (2012).
14. A. Al-Mayahi and A. Sharif, Solvent removal process, USPC Class: 210644.
15. T.-S. Chung, S. Zhang, K. Y. Wang, J. Su, M. M. Ling, Forward osmosis processes: Yesterday, today and tomorrow, *Desalination*, 287, 78–81 (2012).
16. D. Li, X. Zhang, G. P. Simon, H. Wang, Forward osmosis desalination using polymer hydrogels as a draw agent: Influence of draw agent, feed solution and membrane on process performance, *Water Research*, 47, 209–215 (2013).
17. S. Zhao, L. Zou, D. Mulcahy, Brackish water desalination by a hybrid forward osmosis–nanofiltration system using divalent draw solute, *Desalination*, 284, 175–181 (2012).
18. A. Altaee, Computational model for estimating reverse osmosis system design and performance: Part-one binary feed solution, *Desalination*, 291, 101–105 (2012).
19. Feed water type and analysis, Tech Manual Excerpt, http://www.dow.com/PublishedLiterature/dh_003b/0901b8038003b4a0.pdf?filepath=liquidseps/pdfs/noreg/609-02010.pdf&fromPage=GetDoc (accessed October 1 2013).
20. A. Altaee, P. Palenzuela, G. Zaragoza, A. A. Alanezi, Single and dual stage closed-loop pressure retarded osmosis for power generation: Feasibility and performance, *Applied Energy*, 191, 328–345 (2017).

21. A. Altaee, J. Zhou, A. A. Alanezi, G. Zaragoza, Pressure retarded osmosis process for power generation: Feasibility, energy balance and controlling parameters, *Applied Energy*, 206, 303–311 (2017).
22. A. Altaee, G. Zaragoza, H. Rost van Tonningen, Comparison between forward osmosis-reverse osmosis and reverse osmosis processes for seawater desalination, *Desalination*, 336, 50–57 (2014).
23. S. Goh, J. Zhang, Y. Liu, A. G. Fane, Fouling and wetting in membrane distillation (MD) and MD-bioreactor (MDBR) for wastewater reclamation, *Desalination*, 323, 39–47 (2013).
24. A. Altaee, Computational model for estimating reverse osmosis system design and performance: Part-one binary feed solution, *Desalination*, 291, 101–105 (2012).
25. N. A. Thompson, P. G. Nicoll, Forward osmosis desalination: A commercial reality, *Proceedings IDA World Congress*, Perth, Australia, September 2011.
26. P. G. Nicoll, N. A. Thompson, M. R. Bedford, Manipulated osmosis applied to evaporative cooling make-up water – Revolutionary technology, *Proceedings IDA World Congress*, Perth, Australia, September 4–9, 2011.

7 Freshwater Production Using Multiple-Effect Evaporator

Prashant Sharan and Santanu Bandyopadhyay

CONTENTS

ABBREVIATIONS

GCC Grand composite curve
MEE Multiple-effect evaporator
TVC Thermo-vapor compressor

7.1 INTRODUCTION

Of the total existing desalination plants in the world, 94% of them use a reverse osmosis (RO), multi-stage flash or multiple-effect evaporator system (Calle et al., 2015). RO is a membrane separation technology whereas multi-stage flash and multiple-effect evaporators are thermal desalination based systems. A thermal desalination system uses thermal energy for freshwater production. In a multiple-effect evaporator (MEE), evaporation takes place in stages. The vapor coming out from one effect acts as a heat source for the next effect, and this process continues until the last effect. After condensing, the vapor comes out as a distillate.

For lower salinity seawater, RO is the most cost-effective solution (El-Dessouky and Ettouney, 1999). With the increase in seawater salinity the energy cost of an RO system increases linearly, whereas for a thermal system there is hardly any variation in energy cost (Fiorenza et al., 2003). However, higher scale formation due to higher salinity may lead to higher pumping power and reduction in the heat transfer area (El-Dessouky and Ettouney, 2002). Therefore, appropriate cleaning of the tubes plays an important role in proper operation of an MEE system. Thermal desalination technologies are preferred for large-scale desalination (~500,000 m^3/d) (Karagiannis and Soldatos, 2008). In 1995, 69% of the desalination plants worldwide used multi-stage flash, whereas the MEE contribution was only 8% (Al-Shammiri and Safar, 1999). With the introduction of low temperature MEE, the system cost has been reduced and freshwater production has improved (Wang and Lior, 2011). The MEE generally requires low temperature heating (in range of 70°C–80°C), and are often integrated with power plants, a waste heat source, or a sensible heating source, etc. Such integration reduces the cost of thermal energy, which in turns reduces the cost of the fresh water produced in comparison to an RO plant. The benefits of an MEE system over multi-stage flash are lower energy consumption, lower maintenance costs, simpler geometry, and a higher heat transfer coefficient (Al-Mutaz and Wazeer, 2014). The benefits of MEE over an RO plant include robustness, less maintenance, lower electrical power requirements, and reduced operating cost, etc. However, the major disadvantage of MEE over RO is the high capital cost of the system.

7.2 FEED CONFIGURATION FOR MEE

The performance of MEE is strongly dependent on the feed configuration. Generally MEE is classified in four types based on feed flow: forward feed, backward feed, parallel feed, and parallel/cross feed. In a forward-feed MEE, all of the saline water enters the first effect of the MEE. External motive steam acts as a heat source for the first effect and the entering feed partially vaporizes. The brine produced in the first effect acts as a feed for the second effect, and the vapor produced in the first effect acts as a heat source for the second effect. This process continues until the last effect of the MEE. In a backward-feed configuration, the vapor and feed flow in opposite directions. The feed enters the MEE from the last effect, whereas brine is discharged from the first effect. In a backward-feed configuration, as the concentration of the feed increases, the temperature of the feed also increases, leading to a high scaling problem (Al-Mutaz and Wazeer, 2014). The third type of feed flow is a parallel flow. In this configuration, the feed enters every effect of the MEE with an almost equal flow rate, while the brine is discharged from each effect. Another variant of the parallel feed is the parallel/cross feed configuration. The brine in this case is discharged only from the last effect. A brief literature review of the various studies on MEE based on the parallel/cross feed configuration has been reported by Sharan and Bandyopadhyay (2016a). Figure 7.1 shows the schematic for a parallel/cross feed MEE. The parallel/cross feed configuration is commonly used for desalination on account of its high energy efficiency. This chapter deals with the study of parallel/cross feed flow MEE. A comparison table of the different feed configurations is shown in Table 7.1.

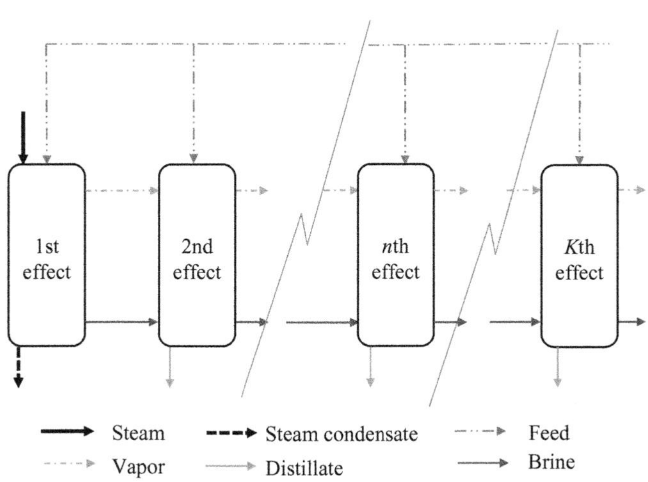

FIGURE 7.1 Schematic of a parallel/cross feed flow MEE.

TABLE 7.1
Comparison between Different Feed Flow for MEE

Parameters	Forward Feed	Backward Feed	Parallel Feed	Parallel/Cross Feed
Feed input	First	Last	Every	Every
Brine discharge	Last	First	Every	Last
Vapor and Brine flow direction	Co-current	Counter-current	–	Co-current
Energy efficiency	Moderate	Moderate	Least	Highest
Scaling	Least	Highest	Moderate	Moderate
Maintenance	Least	Highest	Moderate	Moderate
Area requirement	Moderate	Least	Highest	Moderate

7.3 SYSTEM DESCRIPTION AND MATHEMATICAL MODEL

A schematic for the *n*th effect evaporator is shown in Figure 7.2. The feed seawater, with mass flow rate F_n, temperature T_n, concentration X_f and enthalpy $h_{f,n}$ enters the *n*th effect. The brine with mass flow rate B_{n-1}, temperature T_{n-1}, concentration X_{n-1}, and enthalpy h_{n-1}, coming from *n*–1th effect, enters the *n*th effect. The vapor produced in (*n*–1)th effect acts as a heat source for the *n*th effect and loses its latent heat at saturation temperature $T_{\text{sat},n-1}$. Due to the addition of heat in the *n*th effect, vapor with mass flow rate V_n, temperature T_n and enthalpy H_n is produced.

Generally in the parallel/cross feed configuration, feed seawater enters each effect of the MEE with equal or almost equal flow rates. An equal feed flow rate might not necessarily give an energy-optimal solution. The first objective of the chapter is to calculate the optimal feed flow rate for each MEE effect to minimize the energy consumption of the desalination system. In order to further enhance the energy efficiency of MEE systems, they are often integrated with thermo-vapor compressors (TVCs). A TVC is basically a heat pump (explained in detail in Section 7.5). As a TVC is typically placed after the last effect of the MEE, the other objective of the chapter is to determine the optimal location of the TVC in an MEE system to further enhance the energy efficiency of the overall system.

A general mathematical model for the MEE system is explained in this section. The mass balance for the system is (Figure 7.2):

$$B_{n-1} + F_n = B_n + V_n \tag{7.1}$$

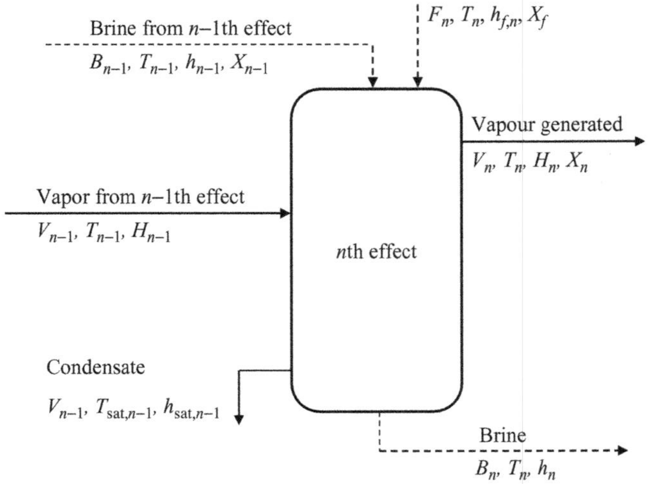

FIGURE 7.2 Schematic for the *n*th effect of the MEE.

Assuming the distillate is salt free, the salt load balance can be written as:

$$B_{n-1}X_{n-1} + F_nX_f = B_nX_n \tag{7.2}$$

Using Equation 7.2 the brine flow rate can be generalized as:

$$B_n = \left(\sum_{i=1}^{n} F_i\right)\frac{X_f}{X_n} \tag{7.3}$$

Inserting the brine flow rate in Equation 7.1 gives the vapor flow rate as:

$$V_n = \begin{cases} \left(1 - \dfrac{X_f}{X_1}\right)F_1 & n = 1 \\ \left(1 - \dfrac{X_f}{X_n}\right)F_n + \left(\displaystyle\sum_{i=1}^{n-1} F_i\right)\left(\dfrac{X_f}{X_{n-1}} - \dfrac{X_f}{X_n}\right) & n \geq 2 \end{cases} \tag{7.4}$$

The net fresh water produced by the system can be calculated by summing up the vapor produced from each effect. The energy balance for nth effect is:

$$V_{n-1}\lambda_{n-1} + B_{n-1}h_{n-1} + F_nh_{f,n} = V_nH_n + B_nh_n \qquad \forall\ n \geq 2 \tag{7.5}$$

Inserting the values of vapor and brine flow rate and rearranging the terms gives:

$$F_n(H_n - h_{f,n}) - F_n(H_n - h_n)\frac{X_f}{X_n} = F_{n-1}\lambda_{n-1}\left(1 - \frac{X_f}{X_{n-1}}\right) + \left(X_f\sum_{i=1}^{n-1} F_i\right)\left(\frac{h_{n-1}}{X_{n-1}} - \frac{h_n}{X_n}\right) + \left(\sum_{i=1}^{n-2} F_i\right)\left(\frac{X_f}{X_{n-2}} - \frac{X_f}{X_{n-1}}\right)\lambda_{n-1} - \left(\sum_{i=1}^{n-1} F_i\right)\left(\frac{X_f}{X_{n-1}} - \frac{X_f}{X_n}\right)H_n \quad n \geq 2 \tag{7.6}$$

For a detailed derivation of the earlier equation, please refer to Sharan and Bandyopadhyay (2016a). The assumptions made while deriving Equation 7.6 are as follows:

- Heat loss from the evaporator is neglected.
- Non-equilibrium allowance is neglected. Non-equilibrium allowance is the drop in pressure and temperature because of brine flashing.
- Distillate is salt free.
- Pressure drops in the evaporator, condenser, pipe, etc., are neglected.

The earlier set of equations can be solved using numerical methods to calculate the optimal feed flow rate for each MEE effect. In this chapter, the principles of process integration are applied to calculate optimal feed flow rate.

7.4 GRAND COMPOSITE CURVE GENERATION FOR MEE

Process integration came into existence in the late 1970s. Initially it was used to minimize the energy requirements for heat exchanger networks. Linnhoff and Flower (1978) developed the Problem Table Algorithm for targeting the external energy requirements of a heat exchanger network. Using the Problem Table Algorithm, a Grand Composite Curve (GCC) for the energy-integrated system can be generated. A GCC is a plot between the shifted temperature and feasible heat transfer. One of the key parameters for a GCC is the "pinch point." The point at which the GCC touches the *Y*-axis is called the pinch point. The pinch point is the bottleneck for the maximum possible energy integration. To minimize the steam requirement for an MEE, the principles of process integration have been used. Singh et al. (1997) calculated various energy saving opportunities using process integration for MEE. Khanam and Mohanty (2007) calculated the internal heat transfer possible to minimize the energy consumption of an MEE. Sharan and Bandyopadhyay (2016b) explained in detail the methodology for GCC generation of an MEE and suggested that, for minimum energy consumption of the MEE, all effects must be pinched. Figure 7.3 shows a typical GCC for a four-effect MEE. The net external energy requirement for the MEE is given by the hot utility requirement of the GCC, and the mass of steam required can be calculated accordingly.

Each MEE effect is represented in the form of a box. For a four-effect MEE, there are four boxes. The topmost box represents the first effect, whereas the bottom one is the last effect. Each box is represented in four thermal streams.

Energy required for evaporation: Each effect requires energy for evaporation. As the evaporation takes place at a constant temperature, it is represented by a horizontal line.
Vapor condensation: The vapor loses its latent heat and acts as a heat source. The condensation also takes place at a constant temperature, and is represented by a horizontal line.
Feed pre-heating: The feed entering in an effect is at a temperature less than the effect operating temperature. The feed is pre-heated up to its effect operating temperature.
Condensate cooling: The vapor after losing its latent heat comes out as condensate. This condensate is at higher operating temperature and has sufficient heat for energy integration. To further increase the energy efficiency of the system, the condensate heat can be reused.

The width of each box is given by summation of the minimum temperature driving force (ΔT_{min}) and boiling point rise (BPR). Because of the salt presence, seawater boils at a temperature higher than salt-free water, and this temperature difference is the BPR. The effect temperature (T_n) is the

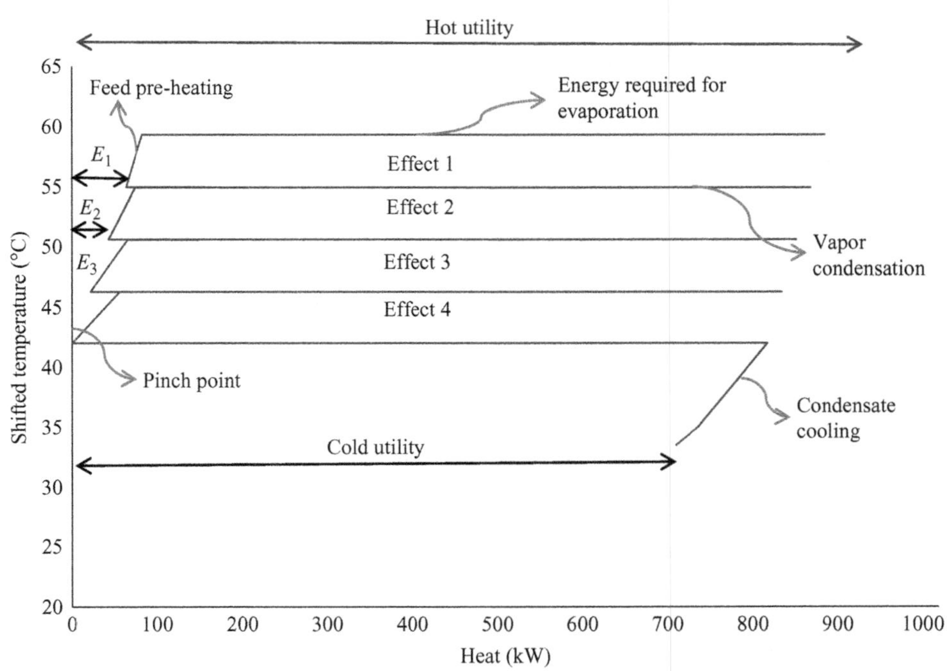

FIGURE 7.3 Typical Grand Composite Curve of a 4-effect MEE.

summation of the saturation temperature ($T_{sat,n}$) and the BPR (BPR_n). ΔT_{min} is the temperature difference between condensing vapor ($T_{sat,n-1}$) and the operating temperature of the next effect (T_n). The hot utility requirement signifies the total hot utility required by the system. Similarly the cold utility is the cooling required to condense the vapor of the last effect of the MEE in a condenser. From Figure 7.3 it can be observed that the fourth effect is pinched whereas the first, second, and third effect are shifted away from pinch by E_1, E_2, and E_3. For Figure 7.3, the feed flow rate for each effect is equal. For minimum energy consumption all MEE effects must be pinched (Sharan and Bandyopadhyay, 2016b). In order to pinch all the effects, the feed flow rate for each effect needs to be modified.

7.5 OPTIMAL FEED FLOW RATE CALCULATION

As discussed in the GCC generation section, if the MEE is modeled with equal feed flow rates, all MEE effects might not be pinched. In such a case the feed flow rate for each effect needs to be modified. Sharan and Bandyopadhyay (2016a) proposed a mathematical correlation for modifying the feed flow rate, which is given as:

$$dF_i \approx \frac{\dfrac{\left(K-(i-1)-(n-i)\right)\left(E_{n-1}-E_n\right)}{\lambda_i(1-X_f/X_i)} - \sum_{j=1}^{j=i-1} dF_j}{K-(i-1)} \qquad n \geq i+1 \tag{7.7}$$

where K is the total number of MEE effects, i is the effect for which the feed flow rate is calculated, n is the effect which is shifted away from pinch, λ is the latent heat of vaporization, and X is the brine concentration. The change in feed flow rate for Kth effect is given as:

$$dF_K = -\sum_{j=1}^{j=K-1} dF_j \tag{7.8}$$

Once the change of the feed flow rate is known, the optimal feed flow rate can be calculated as:

$$F_n^{new} = F_n + dF_n \tag{7.9}$$

The energy efficiency of MEE is often defined in terms of gain output ratio (GOR). GOR is defined as the ratio of the mass of fresh water produced to the mass of the steam supplied. The higher the GOR of the system, the more energy efficient is the system.

7.5.1 Methodology for Designing of MEE

For optimal energy integration of an MEE, the following methodology may be adopted:

1. Define the input parameters (number of effects K, incoming feed temperature T_f, last effect operating temperature T_K, minimum temperature driving force for each effect ΔT_{min}, feed concentration X_f, and maximum brine concentration allowed).
2. Calculate the effect operating temperature using ΔT_{min} and BPR.
3. Generate the GCC for the MEE, using an equal feed flow rate.
4. If each effect is pinched, the system is optimally designed. Otherwise, the feed flow rate for each is calculated using Equation 7.9.
5. Repeat the procedure from step 3 by replacing the equal flow rate with the new feed flow rate.

7.6 INTEGRATION OF MEE WITH TVC

The vapor produced in the last effect of MEE needs to be cooled in the condenser before discharge. This vapor can instead be recompressed in a TVC and reused. Reusing the vapor leads to simultaneous reductions in the hot and cold utility requirements, and the condenser duty is also reduced.

In a TVC, low-pressure vapor extracted from the MEE is compressed with a high-pressure motive steam to produce medium-pressure vapor. Figure 7.4 shows a typical TVC used in MEE. It consists of a nozzle section, mixing section, throat section and diffuser section. The high-pressure motive steam (V_0) enters the nozzle section. As the nozzle converges, it loses pressure and gains kinetic energy. The steam attains a speed greater than the speed of sound (Al-Juwayhel et al., 1997). As a result a partial vacuum is created and the entrained vapor ($V_{e,n}$) gets sucked into the TVC. The entrained vapor and motive steam are mixed in the mixing section. In the throat section the speed and pressure stabilizes, after which the mixture enters the diffuser. The pressure of the mixed vapor gradually rises in the diffuser section because of a decrease in momentum. The vapor comes out as a medium-pressure vapor and acts as a heat source for the MEE. TVC is very common in the desalination industry to increase the GOR of the system.

Hamed and Ahmed (1994) showed that by integrating a TVC with the MEE system, the GOR for the MEE could be increased up to 70%. Similarly, Darwish and El-Dessouky (1996) showed that by integrating a 4-effect MEE with TVC, the GOR obtained is equivalent to an 11-effect standalone MEE or a 24-stage flash desalination system. Generally, the TVC is placed after the last effect of the MEE. Kouhikamali et al. (2011) increased the GOR from 9 to 9.3 by moving the suction position of the entrained vapor from the last to the middle effect. Dahdah and Mitsos (2014) showed that by optimally locating the TVC suction position the GOR for the system increases and the area requirement decreases. Sharan and Bandyopadhyay (2016c) proposed a methodology for optimal location of the TVC suction position to maximize the GOR of the system. The methodology was capable to optimally locate the TVC suction position based on input parameters without requiring detailed simulation.

7.6.1 Optimal Location of the TVC

To maximize the GOR for an integrated MEE-TVC system, Sharan and Bandyopadhyay (2016c) proposed the following Theorem:

Theorem: For maximizing the GOR for an integrated MEE-TVC system with K number of effects, the optimal suction effect n for the TVC should be such that the following expression is maximized:

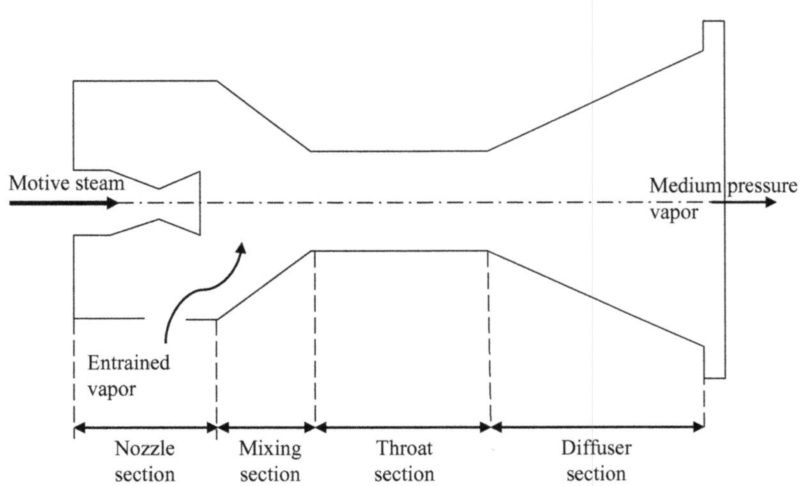

FIGURE 7.4 Schematic of TVC.

$$f(\text{TVC})_n = \frac{K(R_n+1)}{R_n} - \frac{(K-n)\lambda_n}{R_n\lambda_m} \tag{7.10}$$

where $f(\text{TVC})_n$ is the ratio signifying the GOR for the integrated system, λ_n is the latent heat of the low-pressure vapor (extracted vapor), λ_m is the latent heat of the mixed-pressure vapor and R_n is the mixing ratio. The mixing ratio is the ratio of the mass of steam supplied (V_0) to the mass of the extracted vapor ($V_{e,n}$). Many correlations exist in the literature for calculating the mixing ratio as a function of the motive-steam pressure, entrained-vapor pressure and mixed-vapor pressure. The TVC can be optimally located using the proposed Theorem.

7.7 ILLUSTRATIVE EXAMPLE

To demonstrate the methodology for the design of a parallel/cross feed MEE–based desalination system, an illustrative example is considered. The input parameters are listed in Table 7.2, and the correlation used is listed in Table 7.3. The number of MEE effects is varied from 4–12. To further enhance energy conservation, integration with a TVC is also considered.

For demonstration purposes, a four-effect MEE is considered. For ΔT_{min} equal to 3.8°C and last effect operating temperature (T_4) equal to 42°C, the other effect temperatures are calculated to be T_1 = 55.9°C, T_2 = 51.2°C, T_3 = 46.6°C and T_4 = 42°C. For an equal feed flow rate for each effect ($F_1 = F_2 = F_3 = F_4 = 1$ kg/s), the process stream data are generated and the GCC for the system is drawn. Figure 7.5 shows the GCC for a standalone MEE with equal feed flow rates. It can be observed that only the fourth effect is pinched. The total mass of the distillate produced is 1.33 kg/s whereas the mass of the motive steam required is 0.38 kg/s. Therefore, the GOR for the system is 3.53.

Using Equations 7.7 and 7.8, the change in the vapor flow rate for each effect is calculated to be dF and the values are 0.042, 0.014, −0.014, and −0.042 kg/s. The new feed flow rates F_1, F_2, F_3, and F_4 are updated as 1.042, 1.014, 0.986, and 0.958 kg/s. With the new feed flow rate the process stream data is generated and the methodology is repeated till the desired convergence ($dF_n < 10^{-5}$) is achieved. The final design parameters are listed in Table 7.4. The fresh water produced by the system is 1.33 kg/s, and the mass of steam required is 0.36 kg/s. The GOR for the system obtained is 3.67.

To further enhance the energy integration, the MEE can be integrated with a TVC. Using the Theorem for optimal location of the TVC, $f(\text{TVC}_1)$ = 7.9, $f(\text{TVC}_2)$ = 7.97, $f(\text{TVC}_3)$ = 7.7 and $f(\text{TVC}_4)$ = 7.64. For vapor extraction from the first or second effect, the compression ratio is less than 1.81. For stable operation of a TVC, El-Dessouky et al. (2000a) suggested that the compression ratio should be greater than 1.81, where the compression ratio is the ratio of the pressure of the medium-pressure vapor produced by the MEE to the pressure of the entrained (low-pressure) vapor. Therefore, using the proposed Theorem and the criterion of a compression ratio greater than 1.81,

TABLE 7.2
Input Parameters for Illustrative Example

Number of effects (K)	4–12
Feed temperature (T_f)	25°C
Feed concentration (X_f)	42,000 PPM
Maximum brine concentration (X_{max})	70,000 PPM
Top brine temperature	70°C
Last effect temperature (T_K)	42°C
Minimum temperature driving force (ΔT_{min})	1.6–3.8°C
Steam temperature for parallel feed (T_0)	75°C
Heat transfer coefficient for heat exchanger (U_{HE})	0.3 kW/m²K
Motive steam temperature for TVC ($T_{0,TVC}$)	250°C

TABLE 7.3
Correlations Used for Illustrative Example

Specific heat capacity of water (Cp_w) (El-Dessouky et al., 2000b)	$Cp_w = \left(A + BT + CT^2 + DT^3\right)\times 10^{-3}$ $A = 4206.8 - 6.6197Sa + 0.012288Sa^2$ $B = -1.2262 + 0.054178Sa - 0.00022719Sa^2$ $C = 0.012026 - 5.3566\times 10^{-4}Sa + 1.8906\times 10^{-6}Sa^2$ $D = 6.8777\times 10^{-7} + 1.517\times 10^{-6}Sa - 4.4268\times 10^{-9}Sa^2$ $Sa = 0.001X$
Boling point rise (BPR) (Al-Juwayhel et al., 1997)	$\text{BPR} = A.Sa^2 + B.Sa$ $A = (-4.584\times 10^{-4})T^2 + 0.2823T + 17.95$ $B = (1.536\times 10^{-4})T^2 + 0.05267T + 6.56T$
Maximum concentration (El-Dessouky et al., 2000b)	$X = 0.9(457628.5 - 11304.11T + 107.5781T^2 - 0.360747T^3)$
Evaporator heat transfer coefficient (El-Dessouky et al., 1998)	$U_{\text{evap}} = (1939.4 + 1.40562T - 0.0207525T^2 + 0.0023186T^3)10^{-3}$
Condenser heat transfer coefficient (U_{cond}) (Al-Mutaz and Wazeer, 2014)	$U_{\text{cond}} = (1719.4 + 32.063T - 0.015971T^2 + 0.00019918T^3)10^{-3}$
TVC mixing ratio (R) (Hassan and Darwish, 2014)	$\text{R} = -1.934 + 2.153\,\text{CR} + \frac{113.491}{\text{ER}} - 0.522\,\text{CR}^2 - \frac{14735.965}{\text{ER}^2} - 31.852\frac{\text{CR}}{\text{ER}} + 0.047\,\text{CR}^3 + \frac{900786.04}{\text{ER}^3} - 495.58\frac{\text{CR}}{\text{ER}^2} + 10.025\frac{\text{CR}^2}{\text{ER}} \quad (\forall\,\text{ER} \ge 100)$ $\text{R} = -3.208 + 3.933\,\text{CR} + \frac{27.236}{\text{ER}} - 1.192\,\text{CR}^2 - \frac{141.432}{\text{ER}^2} - 22.545\frac{\text{CR}}{\text{ER}} + 0.126\,\text{CR}^3 + \frac{348.507}{\text{ER}^3} + 41.796\frac{\text{CR}}{\text{ER}^2} + 6.44\frac{\text{CR}^2}{\text{ER}} \quad (\forall\, 10 \le \text{ER} \le 100)$ $\text{R} = -1.611 + 11.033\text{LN(CR)} + \frac{13.528}{\text{ER}} - 14.9334(\text{LN(CR)})^2 - \frac{34.44}{\text{ER}^2} - 48.477\frac{\text{LN(CR)}}{\text{ER}} + 6.462(\text{LN(CR)})^3 + \frac{29.97}{\text{ER}^3} + 70.811\frac{\text{LN(CR)}}{\text{ER}^2} + 46.959\frac{(\text{LN(CR)})^2}{\text{ER}} \quad (\forall 2 \le \text{ER} \le 10)$ where, $\text{CR} = \frac{P_m}{P_e}$ & $\text{ER} = \frac{P_0}{P_e}$

the third effect is the optimal location for the TVC. To produce 1.33 kg/s of fresh water, the mass of steam required is 0.186 kg/s. The GOR for such a system is 7.15. The final design parameters with the optimally located TVC are listed in Table 7.5.

For the illustrative example considered, the number of effects is varied from 4 to 12 and the GOR and specific area requirements are accordingly studied. A detailed sensitivity analysis of the various process parameters on the GOR and specific area requirements of the overall system has been reported in various research papers (Sharan and Bandyopadhyay, 2016a, 2016c).

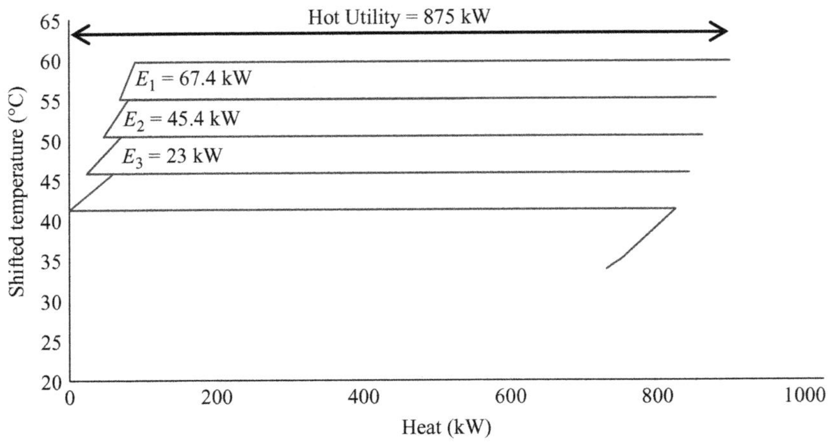

FIGURE 7.5 GCC for MEE with equal feed flow rate.

TABLE 7.4
Final Design Parameters with Optimal Feed Flow

Parameter	Value	Parameter	Value
F_1 (kg/s)	1.04	F_2 (kg/s)	1.01
F_3 (kg/s)	0.986	F_4 (kg/s)	0.96
B_1 (kg/s)	0.694	B_2 (kg/s)	1.37
B_3 (kg/s)	2.03	B_4 (kg/s)	2.67
Steam (kg/s)	0.36	Freshwater (kg/s)	1.33
GOR	3.67	A_s (m²/kg/s)	254.4

TABLE 7.5
Final Design Parameters for Integrated MEE-TVC System

Parameter	Value	Parameter	Value
F_1 (kg/s)	1.22	F_2 (kg/s)	1.18
F_3 (kg/s)	1.15	F_4 (kg/s)	0.445
$V_{e,3}$ (kg/s)	0.232	Optimal effect	3rd
B_1 (kg/s)	0.81	B_2 (kg/s)	1.6
B_3 (kg/s)	2.37	B_4 (kg/s)	2.67
Steam (kg/s)	0.186	Freshwater (kg/s)	1.33
GOR	7.15	A_s (m²/kg/s)	245

7.7.1 Gain Output Ratio (GOR)

Figure 7.6 shows the variation in the GOR with the number of MEE effects. For an equal feed flow rate the GOR for a four-effect MEE is 3.53, whereas with an optimal feed flow rate the GOR for the system increases to 3.67, an improvement of 4%. Similarly, for a 12-effect MEE the GOR for an equal feed flow rate is 9.13, whereas with an optimal flow it is 10.11.

To further increase the energy efficiency of the system, an MEE is integrated with a TVC. For an MEE with four- or five-effects, the optimal location for the TVC comes out to be the third effect,

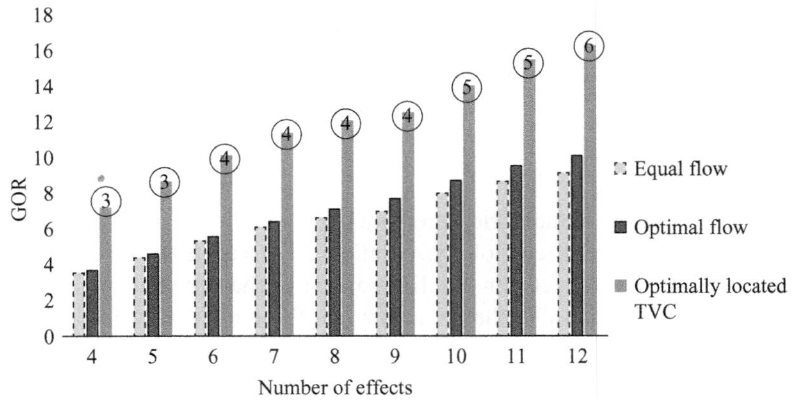

FIGURE 7.6 Variation in GOR for equal flow, optimal flow and TVC integration with number of effects.

(shown in Figure 7.6 with a blue histogram; an encircled number shows the optimal location of the TVC). As the number of effects is increased to nine, the optimal location for TVC shifts to the fourth effect. For an MEE with 10 or 11 effects, the fifth effect is the optimal TVC location and for a 12-effect MEE, the TVC should be located after the sixth effect. For a four-effect MEE, the GOR for the system improves from 3.67 to 7.15. For a 12-effect MEE, the GOR for system increases from 10.15 to 16.27: a 60.3% improvement in energy efficiency.

7.7.2 Specific Area (A_S)

The specific area is defined as the ratio of the summation of the evaporator area, condenser area and heat exchanger area to the total distillate produced. At an optimal feed flow rate, the feed flow rate for the effect operating at a higher temperature increases. An effect operating at a higher temperature has a higher heat transfer coefficient, leading to a reduction in the evaporator area. Due to the reduction in the GOR, the cold utility requirement decreases, leading to reduced condenser area. The heat exchanger area requirement remains almost constant. At the optimal feed flow rate the specific area requirement therefore decreases. For a 4-effect MEE, the area requirement decreases by 1.4%, whereas for a 12-effect MEE, the area requirement decreases by 2.2%, as shown in Figure 7.7.

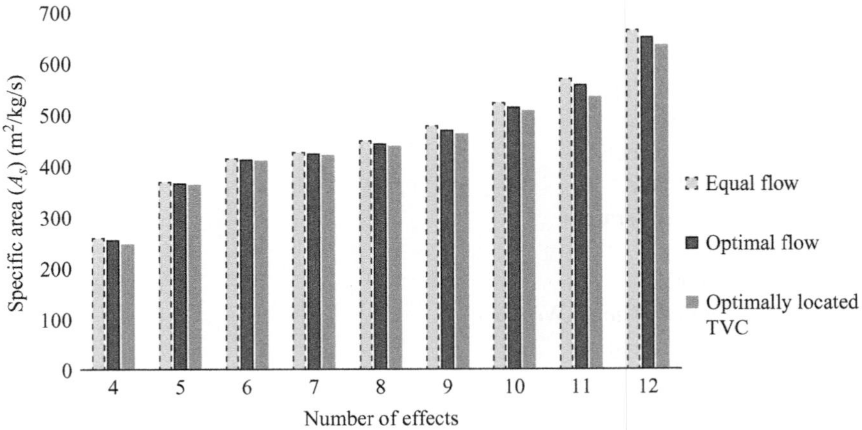

FIGURE 7.7 Variation in specific area for equal flow, optimal flow and TVC integration with number of effects.

Similarly for an MEE-TVC system, the feed flow rate for an effect operating at a higher temperature increases, and the cooling duty requirement decreases. The area requirement for a 4-effect MEE decreases by 3.8%, and for a 12-effect MEE it decreases by 2.62%.

7.8 CONCLUSION

With increasing water scarcity, industries are moving towards desalination to produce fresh drinking water. Due to improved technology, parallel/cross feed flow multiple effect evaporators (MEEs) are gaining importance for desalination. This chapter discusses the use of MEEs for fresh water production. Generally, the feed flow rate for each MEE effect in a parallel/cross feed is equal, which is not the most energy efficient solution. A methodology for calculating the optimal feed flow rate is discussed. At the optimal flow rate, the GOR for a 12-effect MEE increases by 10.7% with a simultaneous reduction in specific area by 2.2%. To further enhance energy recovery, an MEE is integrated with a TVC. The correlation for the calculation of the optimal location of the TVC is discussed. For a 12-effect MEE, the TVC should be located after the sixth effect. The GOR for the system improves by 60.3% and the area requirement for the system declines by 2.6%.

The methodology discussed in this chapter is based on the principles of process integration and mathematical optimization. The advantage of this methodology is that it gives fast convergence. Moreover, the correlation for the optimal location of the TVC is also proposed, and the advantage of the correlation is that there is no need to carry out out a detailed simulation to find the optimal location of the TVC.

NOMENCLATURE

A	area (m^2)
B	brine flow rate (kg/s)
BPR	Boiling point rise (°C)
CR	compression ratio
Cp	specific heat of water (kJ/kgK)
dF	small change in feed flow rate (kg/s)
E	distance by which effect is shifted away from pinch (kW)
ER	expansion ratio
f(TVC)	optimal ratio for TVC placement
F	feed flow rate (kg/s)
GOR	gain output ratio
H	vapor enthalpy (kJ/kg)
HU	hot utility (kW)
K	total number of effects
mCp	heat capacity (kW/K)
n	effect under consideration
R	mixing ratio
T	temperature (°C)
U	Heat transfer coefficient (W/m^2K)
V	Vapor flow rate (kg/s)
X	concentration (%)

Greek letter

Δ change
λ latent heat of vaporization (kJ/kg)

Subscript

e extracted vapor
f feed
m medium pressure vapor
n effect under study
max maximum
min minimum
0 motive steam

Superscript

new updated value

REFERENCES

Al-Juwayhel, F., El-Dessouky, H., Ettouney, H., 1997. Analysis of single-effect evaporator desalination systems combined with vapor compression heat pumps. *Desalination* 114, 253–275.

Al-Mutaz, I.S., Wazeer, I., 2014. Comparative performance evaluation of conventional multi-effect evaporation desalination processes. *Appl. Therm. Eng.* 73, 1194–1203.

Al-Shammiri, M., Safar, M., 1999. Multi-Effect disitillation plants: State of art. *Desalination* 126, 45–59.

Calle, A., Bonilla, J., Roca, L., Palenzuela, P., 2015. Dynamic modeling and simulation of a multi-effect distillation plant. *Desalination* 357, 65–76.

Dahdah, T.H., Mitsos, A., 2014. Structural optimization of seawater desalination: II novel MED-MSF-TVC configurations. *Desalination* 344, 219–227.

Darwish, M.A., El-Dessouky, H., 1996. The heat recovery thermal vapour-compression desalting system: A comparison with other thermal desalination processes. *Appl. Therm. Eng.* 16, 523–537.

El-Dessouky, H., Alatiqi, I., Bingulac, S., Ettouney, H., 1998. Steady-state analysis of the multiple effect evaporation desalination process. *Chem. Eng. Technol.* 21, 437–451.

El-Dessouky, H.T., Ettouney, H.M., 1999. Multiple-effect evaporation desalination systems: Thermal analysis. *Desalination* 125, 259–276.

El-Dessouky, H.T., Ettouney, H.M., 2002. *Fundamentals of Salt Water Desalination*. Amsterdam, the Netherlands: Elsevier BV.

El-Dessouky, H.T., Ettouney, H.M., Al-Juwayhel, F., 2000a. Multiple effect evaporation—Vapour compression desalination processes. *Chem. Eng. Res. Des.* 78, 662–676.

El-Dessouky, H.T., Ettouney, H.M., Mandani, F., 2000b. Performance of parallel feed multiple effect evaporation system for seawater desalination. *Appl. Therm. Eng.* 20, 1679–1706.

Fiorenza, G., Sharma, V.K., Braccio, G., 2003. Techno-economic evaluation of a solar powered water desalination plant. *Energy Convers. Manag.* 44, 2217–2240.

Hamed, O.A., Ahmed, K.A., 1994. An assessment of an ejectocompression desalination process. *Desalination* 96, 103–111.

Hassan, A.S., Darwish, M.A., 2014. Performance of thermal vapor compression. *Desalination* 335, 41–46.

Karagiannis, I.C., Soldatos, P.G., 2008. Water desalination cost literature: Review and assessment. *Desalination* 223, 448–456.

Khanam, S., Mohanty, B., 2007. A process integration based approach for the analysis of an evaporator system. *Chem. Eng. Technol.* 30, 1659–1665.

Kouhikamali, R., Sanaei, M., Mehdizadeh, M., 2011. Process investigation of different locations of thermo-compressor suction in MED-TVC plants. *Desalination* 280, 134–138.
Linnhoff, B.O.D., Flower, J.R., 1978. Synthesis of heat exchanger networks. *AIChE* 24, 633–642.
Sharan, P., Bandyopadhyay, S., 2016a. Energy optimization in parallel/cross feed multiple-effect evaporator based desalination system. *Energy* 111, 756–767. doi:10.1016/j.energy.2016.05.107.
Sharan, P., Bandyopadhyay, S., 2016b. Integration of multiple effect evaporators with background process and appropriate temperature selection. *Ind. Eng. Chem. Res.* 55, 1630–1641.
Sharan, P., Bandyopadhyay, S., 2016c. Integration of thermo-vapor compressor with multiple-effect evaporator. *Appl. Energy* 184, 560–573.
Singh, I., Riley, R., Seillier, D., 1997. Using pinch technology to optimise evapoorator and vapour configuration at Malelane mill, in: *Proceedings of the South African Sugar Technological Association*, Vol. 71, pp. 207–216.
Wang, Y., Lior, N., 2011. Thermoeconomic analysis of a low-temperature multi-effect thermal desalination system coupled with an absorption heat pump. *Energy* 36, 3878–3887.

8 Freshwater Production by the Multistage Flash (MSF) Desalination Process

Iqbal M. Mujtaba and Salih Alsadaie

CONTENTS

ABBREVIATIONS

ΔT	Temperature Difference
A	Surface Area
AEs	Algebraic Equation System
ACS	Advanced Control Strategy
BPE	Boiling Point Elevation
BR	Brine Heater
CAD	Computer Aided Design
CFD	Computational Fluid Dynamics
CMPC	Constrained Model Predictive Control
GCC	Gulf Cooperation Council
FLC	Fuzzy Logic Controller
FMRLC	Fuzzy Model Reference Learning Control
GMC	Generic Model Control
gPROMS	General Process Modeling System
HRJ	Heat Rejection Section
HRS	Heat Recovery Section

HS	Heuristic Selection
MED	Multiple-Effect Desalination
MIGD	Million Imperial Gallons per Day
MINLP	Mixed Integer Nonlinear Programming
MSF	Multistage Flash
MSF-BR	Multistage Flash Brine Recycle Desalination
MSF-OT	Multistage Flash Once Through Desalination
NCGs	Non-condensable Gases
NEA	Non-Equilibrium Allowance
NLMPC	Nonlinear Model Predicted Control
NN	Neural Network
ODEs	Ordinary Differential Equation System
OHTC	Overall Heat Transfer Coefficient
PID	Proportional Integral Derivative
PR	Performance ratio
RO	Reverse Osmosis
TBT	Top Brine Temperature
TDM	Tridiagonal Matrix
TE	Temperature Elevation
U	Heat Transfer Coefficient
Q	Heat flux
QDMC	Cascaded Quadratic Dynamic Matrix Control
WHO	World Health Organization

8.1 INTRODUCTION

Water is the most precious compound in the world and it is essential to humans and other life forms despite providing no calories or organic nutrients. Although access to safe drinking water has improved over the last few decades in almost every part of the world, almost 780 million people still lack access to safe drinking water and around 36% of the world's population (2.5 billion people) lack access to improved sanitation (Bennett, 2013). However, with the rapid increase in the world's population and improved standards of living, some observers have estimated that by 2030, the global needs for water would be 6900 billion m^3 compared to 4500 billion m^3 required in 2009 (Addams et al., 2009). A United Nations report (UN, 2015) estimated that the world population is expected to reach 9.7 billion by 2050, thus increasing the worldwide demand for fresh water and putting serious pressure on the quantity of naturally available freshwater. With most of the accessible water around us being saline (97% of the world's water) and 2.5% frozen (Fry and Martin, 2005), desalination technology has been recognized as one of the most sustainable sources for fresh water.

Desalination is a water treatment process that removes dissolved salts from saline water, thus producing fresh water from seawater or brackish water. According to World Health Organization (WHO) allowances, the permissible limit of salinity in drinking water is 500 ppm and up to 1000 ppm for special cases. Most of the water available on earth has a salinity up to 10,000 ppm and seawater normally has a salinity in the range of 35,000–45,000 ppm in the form of total dissolved salts (Tiwari et al., 2003).

Under these conditions, desalination of seawater and brackish water has become the only available solution to supply fresh water in regions where severe water shortages exist. Other alternative solutions, including transporting water from different zones, have proven to be more expensive, inadequate and less reliable (Al-bahou et al., 2007). More interestingly, 42 large cities out of 71 that do not have access to new fresh water resources are located along the coasts, and around 39% of the world's population lives at a distance of less than 100 km from the sea, making seawater desalination the only available option for some countries (Ghaffour et al., 2013). Also, being independent of climatic conditions and rainfall, desalination technology has become a more favorable option than other resources. In fact, adoption of this type of technique has resulted in an increase in the fresh water supply worldwide and a bridge for the gap between availability and demand for safe drinking water.

Desalination markets have grown significantly in the last few decades. Currently, there are more than 16,000 desalination plants in operation worldwide producing around 74.8 million m^3/ day. Between 40% and 50% of the world's desalinated water is produced in the Gulf countries (Bennett, 2013). Reverse osmosis (RO) and multistage flash (MSF) processes account for more than 86% of the total installed desalination capacity (IEA-ETSAP and IRENA, 2013). Due to the low cost of fossil fuels in the Gulf region and North African countries, MSF is the preferred choice, while in other parts of the world, where the fossil fuels cost is high, other desalination technologies such as RO are preferred.

Despite its higher cost compared to RO, the MSF desalting method is by far the most robust technology and does not require as intensive pretreatment as the RO process does (AlTaee and Sharif, 2011). However, the MSF process is sensitive to increases in energy prices and currently faces many challenges to reduce its costs and improve its market shares (profitability). Moreover, the high tendency of fouling and the release of the noncondensable gases (NCGs) are other serious problems that play an important role in increasing the cost of the MSF process.

In this chapter, a detailed description of the MSF process and a review of the studies on MSF modeling, including steady-state and dynamic analysis, are presented. The review provides an assessment and summary of the main features of these studies, including the objectives of each study. In addition, the effect of NCGs on MSF and the developed fouling models are also reviewed in this chapter. Moreover, a detailed review of the control strategies for MSF plants is also presented in this chapter.

8.2 PROCESS DESCRIPTION

The fundamental concept of thermal desalination relies on a phase change separation technique where saline water is heated to the boiling point to produce water vapor. The fresh water is then formed through condensation of the water vapor (UNEP, 2001). As mentioned before, the low cost of fossil fuels in the Gulf region and North African countries is the main reason for adopting thermal desalination technology in these countries. The thermal technology represents 70% of the total capacity in the Gulf Cooperation Council (GCC) countries, while RO represents only 30% (Sharif, 2016). In other countries, where the fossil fuels cost is high, other desalination technologies such as RO are preferred. Moreover, the opportunity to couple thermal desalination plants with power plants to produce water and electricity is another reason for the thermal process to hold a strong position in the water desalination market (Baig et al., 2011). The most well-known types of thermal desalination processes are multistage flash (MSF) and multiple effects (MED). Although MSF has been the most frequently used technique for large-scale commercial use until the late 1980s, many clients have been requesting the MED process during the past few years (De Gunzbourg and Larger, 1999).

Typically, the number of stages in an MSF plant can vary between 4 and about 40, and these stages normally operate at top brine temperatures (TBT) in the range of 100°C–110°C to produce 6–11 kg of distillate per kg of steam consumed (Mayere, 2011). MSF plants have been operating without problems for many years and have the highest capacity units (Darwish and Alsairafi, 2004). The success of the process has resulted in a dramatic increase in the unit production capacity from 6 MIGD during the 1980s and 1990s to 16.9 MIGD in 2004, when a MSF unit was built in the

United Arab Emirates (Al-bahou et al., 2007): this in turn has allowed a significant reduction in both the capital and operating costs (Borsani and Rebagliati, 2005).

Two types of MSF plants have been developed over the decades: the MSF once-through unit (MSF-OT) and the MSF brine recirculation unit (MSF-BR). In the MSF-OT, the MSF unit consists of two sections: a brine heater (BR) and heat recovery section (HRS). In the second type (MSF-BR), however, an extra section is added, called the heat rejection section (HRJ). The recovery and rejection sections are comprised of a series of stages where each stage has a flash chamber and condenser. The series of linked stages have a cold end and a hot end while the middle stages have intermediate temperatures.

8.2.1 Once-Through MSF Process (MSF-OT)

The MSF-OT process, which is illustrated in Figure 8.1, is the first version of the seawater desalination process, which is particularly known for its simplicity and the small number of components compared to the conventional MSF-BR. As shown, in the MSF-OT there is no specific heat rejection section (HRJ). The intake seawater is pumped into the condenser tubes of the last stage of the plant (the right side in Figure 8.1). As the seawater passes through the tubes from one stage to another, its temperature gradually increases by exchanging the heat from the flashing vapor in each stage. Leaving the tubes of the first stage at the highest possible temperature, the preheated seawater enters the brine heater for further increase to its temperature to the maximum allowable saturation temperature; however a care should be taken to avoid the tendency of scale formation in the brine heater tubes. The heat energy required to increase the brine temperature to the TBT is provided by surplus superheated steam that is supplied by a reboiler or electrical power plant. Thus, potable water and electricity are normally produced together in the same region. The saturated or supersaturated heating steam, with a temperature range of 97°C–117°C, flows on the outside surface of the brine heater tubes, while the brine stream flows on the inside of the tubes. As the steam releases its latent heat and condenses around the tubes, the brine stream gains that heat and its temperature is increased to reach the desired temperature (El-Dessouky et al., 1999).

At this point, the flashing brine enters the flash chamber of the first stage of the HRS through an orifice box, where the pressure inside the stage is reduced; this results in the water becoming superheated and flashed off to produce pure vapor. The vapor generated from the brine rises and passes through the demisters, where the entrained brine droplets are removed, and it condenses on the outside surface of the cooling tubes bundle that are located at the top of the stage. Since

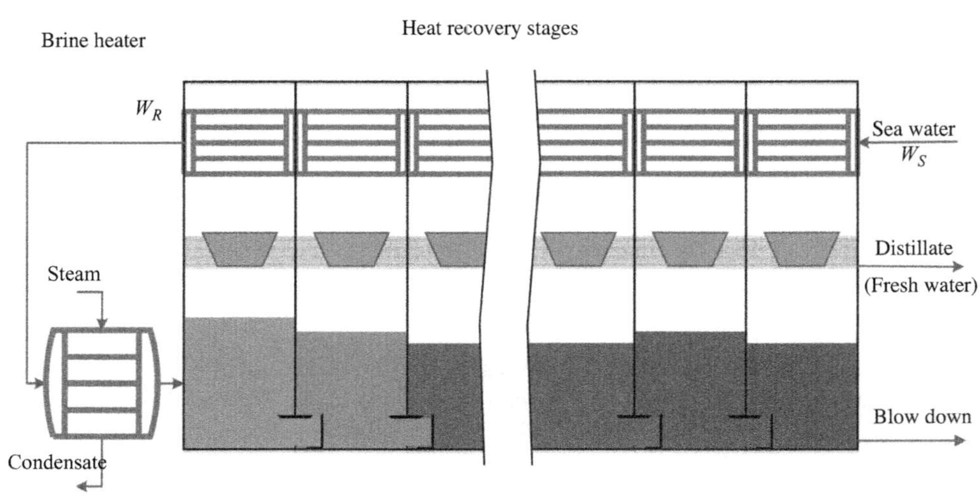

FIGURE 8.1 Schematic of once-through MSF desalination process (MSF-OT).

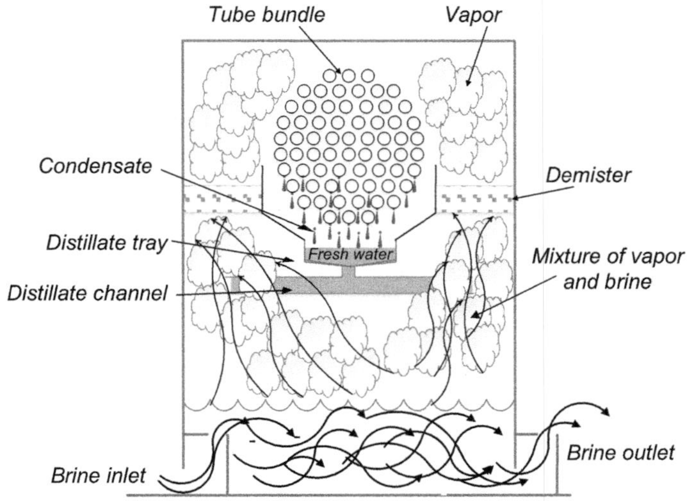

FIGURE 8.2 A single stage in the MSF desalination plant.

the cooling water entering the brine heater flows through the interior of the tube bundle, the vapor releases its latent heat and condenses whereas the cooling seawater gains the latent heat and it is preheated further. The heat exchange between the cooling seawater and the vapor increases the heat recovery as the cooling seawater temperature is increased incrementally to its maximum value, which minimizes the required energy for the brine heater. The condensed vapor is then collected in the distillate trays and pumped out as the final product. Figure 8.2 shows the internal cross-section of a stage. The large amount of latent heat required for the vaporization process leads to a small amount of brine being flashed off before the temperature of the brine drops below the boiling point (Gambier et al., 2002). As the brine would be still hot enough for possible boiling at reduced pressure, the brine flows through the orifice into the next stage with a lower pressure and another small fraction of the brine is flashed off to produce vapor. The flashing process is repeated as the flashing brine flows through a number of consecutive stages where pressure is decreased to allow the water to boil further at a lower pressure (Abdul-Wahab et al., 2012). The process is repeated until the last stage where the blowdown brine is discharged back to the sea. The distillate trays are connected to each other by a channel through which all the accumulated distillate flows. The distillate is finally collected in a distillate box at the last stage of the HRS and is then extracted by a distillate pump into product storage.

The main purpose of developing the MSF-OT system is to increase the performance ratio (kg of distillate per kg of steam) of a single stage. This can be done by dividing the flashing range over a larger number of stages. Increasing the number of stages leads to a reduction in the stage temperature drop and consequently decreases the temperature driving force between the flashing vapor and cooling water stream. This results in an increase in the total heat transfer area of the condensers (El-Dessouky and Ettouney, 2002). Though the once-through MSF process (MSF-OT) is simple and requires less capital investment compared to the MSF-BR, it consumes a large quantity of chemical additives due to the large amount of intake seawater (Helal and Odeh, 2004).

8.2.2 Brine Recirculation MSF Process (MSF-BR)

In the MSF-OT process, the entire seawater flow is heated to high temperature, and it has to be treated with anti-scale chemicals, which increases the operating costs. Moreover, the size of the stages must be designed for winter operation, leading to increased evaporator volume and thus

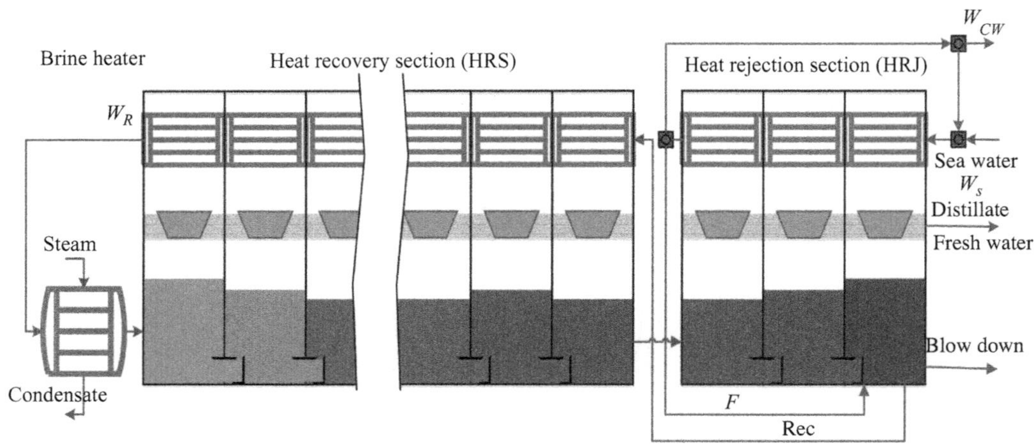

FIGURE 8.3 Schematic of the brine recirculation MSF desalination process (MSF-BR).

increased investment costs. These two points have led to the idea of separating the flashing stage into two parts (HRS and HRJ) and introducing the brine recycle process of MSF-BR. The MSF-BR process, also called the conventional MSF process, is illustrated in Figure 8.3. Normally this process involves recycling the brine from the reject section to the recovery section. The flashing stages are divided into a large number of heat recovery stages and a smaller number of heat rejection stages, commonly three. Although the HRJ and HRS are drawn separately, the two sections are integrated (El-Dessouky et al., 1995).

In the MSF-BR process, the intake seawater is fed into the condenser tubes of the HRJ. The outlet stream leaving the HRJ is then split into two parts; one stream enters the deaerator unit where the dissolved air is stripped out using steam. The deaerated seawater is then fed to the last stage of the MSF plant as make-up. The second stream is rejected back to the sea in the summer season while in the winter season this can be further divided into two parts: one is rejected back to the sea (thus rejecting part of the heat supplied) and the other part is mixed with the cold seawater to preheat it. It is worth noting that the temperature of the make-up stream should be the same as or very close to the temperature of the last stage of the HRJ to avoid thermal shocking that may lead to decomposition of the bicarbonate.

After the make-up enters the last stage of the HRJ, the recirculating brine is drawn from the last stage and then introduced to the last stage of the HRS where the recycled brine is gradually heated in the HRS as it flows through the condensing tubes from one stage to another, as described in the MSF-OT section. The remaining part of the concentrated brine is withdrawn from the brine pool and rejected to the sea as blow-down (El-Dessouky et al., 1995, 1999; Maniar and Deshpande, 1996; Al-Hengari et al., 2005; Al-shayji et al., 2005; Abdel-Jabbar et al., 2007; Bodalal et al., 2010; Abdul-Wahab et al., 2012).

The majority of the MSF plants are MSF-BR, which is superior to the MSF-OT design (El-Dessouky and Ettouney, 2002). The brine recirculation results in a decrease of the total flow rate of the intake seawater. As a result, the consumption rate of the chemical additive is decreased because only the make-up water is treated, rather than entire amount of cooling water. In fact, the MSF-OT requires about 70% more chemicals than needed for the traditional MSF-BR design, if it is assumed that the two plants are operated under the same conditions (Helal, 2004). Also, the recycle flow in the MSF-BR gives good control of the temperature of the feed seawater. The recycled brine also contains higher energy than the feed seawater. As a result, the thermal efficiency of the process is improved (El-Dessouky and Ettouney, 2002).

Although the MSF-OT is characterized by its simplicity over the conventional MSF-BR design, the latter has dominated the thermal desalination market (Helal and Odeh, 2004). The MSF-OT desalination plant is, however, more efficient for small plants and in areas where there is no large difference in the seawater temperature between summer and winter.

8.3 MATHEMATICAL MODELING

The high cost of thermal desalination processes such as MSF has led to more investigations and development by the computational community to reduce the cost by using mathematical models and process simulators. All desalination plants designers are interested in designing their units at minimum cost with high efficiency. For production of potable water by seawater desalination, the designers are challenged with the choice of the right configuration of the plant, as well as the other equipment needed to conduct the process at minimum cost. Thus, an interest in using computer technology to perform process design in a systematic way has been applied in the areas of simulation and optimal design.

Desalination modeling refers to formulating a set of mass and energy balance equations that mathematically describe the process units of an MSF plant. The main MSF units are the brine heater, flash stages, condensers, mixers and splitters. The model equations are supported by the physical and thermodynamic properties of brine, distillate and water vapor, as well as the heat transfer coefficients. When a process has to be simulated, the set of equations are formulated and then solved using an appropriate solution procedure. In addition, some independent process variables, mainly the conditions of the input streams, are given values to match the number of equation with the number of unknown variables. Due to the complexity of the process and the huge number of equations, the solution can be conducted with the aid of a computer (Husain et al., 1993). For design purposes, the model provides the engineers with the reasonable results for each task within a reasonable time and at a lower cost. The benefit of designing a piece of equipment on a computer is that it can be tested before it is purchased or constructed. It is much safer for the designers to make mistakes on the computer than on the plant. For operational purposes, computer models test the effects of different parameters, examine internal vapor and liquid loading, develop better insight into the working of the process and ultimately lead to optimal operation and control of the process.

Based on time-dependence, processes can be described by steady-state and dynamic models. In the steady-state model, the process is described by solving a set of algebraic equations. Steady-state models are useful at the design stage and for optimization purposes. Dynamic models, on the other hand, are time-dependent, and they are formulated and solved using differential equations with the support of algebraic equations. Dynamic models are useful for fault detection and implementation of control systems (Gambier and Badreddin, 2004). The difference between steady-state and dynamic models is the value of the accumulations. For example, the following is the general conservation equation for mass and energy balances.

$$\underset{\text{Input}}{\left\{\begin{array}{c}\text{Enters}\\ \text{through the}\\ \text{system}\\ \text{boundaries}\end{array}\right\}} - \underset{\text{Output}}{\left\{\begin{array}{c}\text{Leaves}\\ \text{through the}\\ \text{system}\\ \text{boundaries}\end{array}\right\}} + \underset{\text{Generation}}{\left\{\begin{array}{c}\text{Generated}\\ \text{by the}\\ \text{system}\end{array}\right\}} - \underset{\text{Consumption}}{\left\{\begin{array}{c}\text{Consumed}\\ \text{by the}\\ \text{system}\end{array}\right\}} = \underset{\text{Accumulation}}{\left\{\begin{array}{c}\text{Buildup}\\ \text{within the}\\ \text{system}\end{array}\right\}}$$

The earlier equation is applicable for a dynamic model; however, setting the value of accumulation to zero, the equation becomes valid for a steady-state model.

8.3.1 MSF Steady-State Model

Steady-state models are necessary at the design stage and are a useful tool to study the parameters of existing plants. Moreover, they are used to validate dynamic models and provide set points for control systems (Gambier and Badreddin, 2004). However, a steady-state model cannot be used for control purposes.

Several steady-state models have been developed in the last three decades. There are two types of steady-state models: simple and rigorous. Simple mathematical models of the MSF process are based on simplifying assumptions that are not sufficiently accurate because they generate a large discrepancy between the model's results and the actual data. However, they can be quite adequate to provide quick estimation for most of the process variables, whereas rigorous models provide more accurate and useful information for system design and simulation (Hamed et al., 2004).

From the literature, the following assumptions are made in the development of simple models for the MSF system:

- Steady-state operation.
- All the saline liquid droplets are removed by the demister and thus the distillate product is salt free.
- The sub-cooling of the condensate or superheating of the vapor is negligible.
- Constant heat transfer area in each section of the plant.
- Constant physical properties: this means assuming constant values (not as a function of temperature and salinity) for water density, heat capacity and the latent heat for evaporation.
- Constant overall heat transfer coefficient (OHTC): this assumption is based on the first assumption. As long as the physical properties are constant, then the OHTC is constant since the variation of the heat transfer coefficient depends on the physical properties.
- Constant thermodynamic losses: assuming the thermodynamics losses, such as boiling point elevation and non-equilibrium allowance, are constant makes the calculations of the flashed off temperature and condensed vapor easier.
- Constant latent heat of vaporization: this value is assumed constant and evaluated at the average temperature for the flashing brine.
- Negligible heat losses to the surroundings: this is a common assumption among all models since an adiabatic system is assumed. Moreover, calculations of heat losses are very difficult and require detailed information of the external heat transfer area of the system.
- Negligible heat of mixing: this is true since the released heat from mixing one cubic meter of water is much smaller than the heat required to vaporize the same amount of water.
- The heat and vapor losses are negligible due to the evacuated system: typically, a small fraction of the saturated water escapes with the NCGs through the venting system. Neglecting these values simplifies the calculations of the model.

A simple model was presented by Mandil and Ghafour (1970). They assumed constant physical properties, heat transfer coefficients and temperature drop in all stages. The results of the model provide a closed-form analytical solution that can be used to calculate the specific heat area and performance ratio. Coleman (1971) presented a stage-to-stage calculations model to formulate a method for cost optimization. He assumed a constant specific heat capacity for the water flowing through the condenser, linear boiling point temperature elevation (BPE) against salinity concentration and constant OHTC in the condenser tubes. The equations of the model were linearized and reformulated for ease of calculation. Moreover, he used a high temperature operation (steam temperature 300°F) and a seawater temperature of 70°F. Soliman (1981) relaxed the assumptions further by providing different values for different parameters in the different sections of the plant. The model assumed a linear temperature profile and a constant value for the heat transfer coefficient in each section. The latent heat of water vaporization is assumed constant and independent of temperature. The BPE losses are assumed to be constant for all sections; however, the non-equilibrium allowance (NEA) effect was neglected. The model was very fast in convergence and suitable for optimization.

Darwish (1991) developed a simple model of the MSF process in an attempt to arrive at a better quantitative evaluation of design and operating parameters on MSF process performance. The following assumptions were made to make the model simple: constant and average temperature drop of the flashing stream per stage, average latent heat of vapor and specific heat of distillate, constant and average values for the cooling water and brine streams. The fouling factor was included as different from section to section; however, its effect on the heat transfer coefficient was not mentioned. Another simple model was presented by El-Dessouky et al. (1998) to study the performance of the MSF process. The analysis is based on the performance characteristics for a number of simplified configurations starting from single stage flashing unit to a multistage flash unit with brine recirculation. The results showed that the MSF-BR is the most sufficient configuration, where the heat rejection section should not be less than three stages.

Despite the fact that the simple models are quite satisfactory in providing a quick estimate of the main process parameters, their main disadvantage is the deviation between the results of the model and the actual plant data due to the afore-mentioned assumptions. In addition, these models are unable to provide the whole picture for the entire process. Therefore, more detailed models are required, where the physical properties are calculated as a function of temperature and salinity, to provide more accurate and useful information for system design and simulation. Rigorous or detailed models include correlations for temperature losses, pressure drop, stage temperature profiles and heat transfer coefficients.

The earlier assumptions in the simple model are relaxed more in the development of detailed model of MSF process:

- Steady-state operation.
- The distillate product is salt free.
- The heat losses to the surroundings are negligible.
- As for simple model, the heat of solution or mixing is negligible.
- The sub-cooling of the condensate or superheating of the vapor is negligible.
- Constant heat transfer area in each section of the plant.
- All the saline liquid droplets are held in the demister.

Glueck and Bradshaw (1970) were among the earliest to present a model for MSF plants with a higher degree of rigor and few assumptions, taking into account the variation of the heat transfer coefficient. However, no results were presented and the correlations of the used properties were not included in their work. It is to be mentioned that this model was used for dynamic behavior as well. Beamer and Wilde (1971) developed a physical and economic model using stage-to-stage calculations to optimize an MSF plant. Their calculations started from the cold end to the hot end of the plant, with the brine heater as the last stage. The available data of the input stream variables and stage parameters on the cold side of a stage leads to the convergence of the solution procedure to obtain the output stream variables on the hot side of the stage. Barba et al. (1973) have developed simple and rigorous models for control and simulation of existing MSF plants. They carried out stage-to-stage calculations starting from the hot end of the plant using an online computer control system. The simple model solution was then used to provide the initial guesses for the rigorous model. The fouling factors were calculated as average values for each of the six stages in the recovery section. Rautenbach and Buchel (1979) developed a mathematical model with a modular structure to design and simulate different configurations of desalination plants, including the MSF process. The material properties within the model's equations were known functions of temperature and concentration. Their modular approach showed some robustness in handling the design and simulation of MSF plants. The solution of the model was done by stage-to-stage calculation starting from the hot end of the plant. However, to compute the output stream variables, the solution of the sequential modular approach required complete knowledge of the input streams and the equipment characteristics. Omar (1983) used his model, which was developed in 1981, to simulate and model an MSF using stage-to-stage calculations to solve the model equations. The model was computed into a Fortran IV with the aid of an IBM 370 computer machine. The program can be used to either design or accurately simulate plants in actual operation. Thermodynamic losses such as BPE and the NEA were included.

Helal et al. (1986) developed a rigorous method to solve a detailed steady-state model of an MSF plant. The method depends on decomposition of the large non-linear equations into smaller subsets equations for the ease of iterative sequential solution of these subsets. The new feature of the method was the formulation of energy balance and heat transfer equations, after linearizing them, into a tridiagonal matrix (TDM), which was then solved by the Thomas algorithm. The results showed that the method is stable and the convergence of the solution is fast over a wide range of initial conditions. The model included temperature losses across the demister and condenser tubes. Constant fouling and other thermodynamic properties are taken into account. Similarly to Helal et al. (1986), Al-Mutaz and Soliman (1989) developed a model to simulate an MSF plant using an orthogonal collection method to calculate the stream profiles across the stages. However, instead of solving material and energy balances for all stages, only a few stages were selected. The authors claim that the method is twice as fast as the TDM model and requires less convergence time. Steady-state models were developed by (Husain et al., 1993, 1994) using an advance computational platform (SPEEDUP) to solve the model equations. Another model was developed by Thomas et al. (1998). All of these models are described later in with the dynamic models. Their steady-state equations were obtained by setting the time derivative terms to zero.

Aly and Fathalah (1995) used the TDM technique to present a steady-state model to simulate an MSF system. However, more correlations describing the heat transfer process and physical properties were added. The model was used to study plant performance over extended ranges of TBT and cooling seawater temperatures. Results suggested that uprating is a promising technique to increase the production rate through elevated values of TBT, but fouling and scale formation were the main concern. El-Dessouky et al. (1995) described a detailed steady-state model to analyze the MSF process. Their model takes into account the effect of fouling factors and the presence of NCGs on the OHTC in the condensing tubes. However, the effect of NCGs on the overall heat transfer was neglected, assuming that all NCGs were extracted by the venting system. The model also considered the heat transfer losses from the stages and the brine heater to the surroundings and through rejection of the NCGs. In 1996, (El-Dessouky and Bingulac, 1996) presented an algorithm for solving a set of equations that describe the steady-state behavior of the MSF desalination process. The presented algorithm is a stage-to-stage calculation and starts from the hot end of the plant. The main advantages of the proposed algorithm were: (1) less sensitive to initial guess, (2) fewer iteration steps to obtain the required solution and (3) no derivative calculations were required. The developed algorithm was implemented using the computer aided design (CAD) interactive package L-A-S (Linear Algebra and Systems).

Rosso et al. (1997) presented a model similar to that previously developed by Helal et al. (1986). It is based on a detailed physicochemical representation of the process, including, in particular, the geometry of the stages, the mechanism of heat transfer and the role of fouling. The model was not only used for design purposes, but was also used to support the development of a dynamic model; it will be discussed further with the other dynamic models, where the time-dependent behavior of the plant can be reviewed. Helal et al. (2003) developed a mathematical model to study the feasibility of hybridization of RO and MSF to improve the performance of the MSF and reduce the cost of desalted water. The design equations representing the process models as well as the cost model were included in their work. The calculations were carried out using the SOLVER optimization tool in the Microsoft Excel® software to maximize plant capacity. The model was based on the one developed by Soliman (1981); several additions and modifications were, however, introduced to account for the different design variations. Aly and El-Figi (2003) presented a steady-state model to analyze both the MSF and MED processes. For the MSF process, the model accounts for the geometry of the stages, the heat transfer mechanism and the effect of fouling. The main goal of the study was to produce desalted water at a lower price by changing operating variables. The results obtained from the model were compared to actual data, with good agreement. Abdel-Jabbar et al. (2007) developed a mathematical model for the MSF process. This study was motivated by the need for an integrated model of the design for large-scale MSF units. Clean operation is assumed and consequently no fouling was taken into account. The model focused on evaluation of the weir loading, stage dimensions of the condenser tubes bundle, demister dimensions, as well as the flow rates and temperature profiles.

Tanvir and Mujtaba (2006a) presented steady-state model for MSF process using the gPROMS modeling tool. In their model, instead of using empirical correlations from literature, a neural network (NN)–based correlation developed earlier by Tanvir and Mujtaba (2006b) is used to determine the BPE. This correlation is embedded in the gPROMS-based process model. It was found that the NN-based correlations can predict the experimental BPE very closely. They obtained a good agreement between the results reported by Rosso et al. (1997) and those predicted by their model. In 2008, the same authors developed a model based on Helal et al. (1986) incorporating NN-based correlation to estimate physical properties. Using the MINLP technique within the gPROMS model builder, the number of flash stages as an integer variable and a few significant operating parameters, such as steam temperature, recycled brine flow and rejected seawater flow, are optimized while minimizing the total annual cost of the process. The results revealed the possibility of designing stand-alone flash stages that would offer flexible scheduling in terms of connecting new units and efficient maintenance of the units throughout the year. The NN technique had been used by Hawaidi and Mujtaba (2011) to develop a correlation that is used to calculate the dynamic fresh water demand–consumption profiles at different seasons and times of day.

Alasfour and Abdulrahim (2009) formulated and implemented a rigorous steady-state model using the process simulation software IPSEpro®. For accurate simulation, the flashing stage was decomposed into three main compartments: flashing pool, distillate tray and tube bundle. The thermo-physical correlations are considered as a function of temperature and salinity. Moreover, to account for the effect of small changes in temperature, the energy balance of the model was carried out using stream enthalpy instead of specific heat. The energy losses from the flashing stage and the brine heater were taken into consideration. The effect of NCGs on the OHTC was considered in their model. However, assuming a constant amount of NCGs in all stages resulted in some disagreement in the OHTC and heat flux between the actual results and the results of the simulation.

Hawaidi and Mujtaba (2011) developed a steady-state model for an MSF based on a combination of material and energy balances, along with the support of the correlations of physical properties. A linear dynamic fouling model was included to calculate the fouling factor at different operation times (e.g., seasonal operation). The gPROMS model building software was used for model development, simulation and optimization. The model was validated against the simulation results reported in the literature. The model was then used to study the impact of a changing brine heater fouling factor under the variation of seawater temperatures. In addition, for fixed plant parameters such as the freshwater production rate and steam and TBT, the impact of fouling on the performance ratio also studied. Said et al. (2010) described a steady-state model of the MSF process including correlations that consider the effect of the presence of NCGs and fouling factors on the OHTC. The simulation results showed a decrease in the OHTC as the concentration of NCGs increases. Also, compared to the results obtained by Rosso et al. (1997) and Tanvir and Mujtaba (2006a), which did not include NCGs, Said's results showed a 0.015 (wt%) increase in the steam flow rate due to the presence of NCGs.

Abdul-Wahab et al. (2012) also developed a mathematical model for an MSF plant. The model was solved using the TDM method suggested by Helal et al. (1986) and was based on the basic principles of physics and chemistry describing each stage in the desalination process. Most of the parameters that are known to affect the operation of the MSF plant were taken into account in building the model. No heat losses, fouling or NCGs were considered. The model was considered to be sufficient and accurate since its results matched well when compared with vendor simulation results and the actual operating plant data for the MSF desalination plant. A summary for most of the previous studies on steady-state modeling of the MSF process is shown in Table 8.1.

TABLE 8.1
Previous Work on Steady-State Models for MSF Desalination

Authors (Year)	Description of the Model	Objectives
Mandil and Ghafour (1970)	A simple model with constant thermo-physical properties.	Optimization of multistage flash desalination.
Gluek and Bradshaw (1970)	A more rigorous model with few assumptions. Includes heat transfer variation, fouling, BPE.	To provide an accurate representation of a typical MSF plant.
Coleman (1971)	Simple stage-to-stage calculation model, constant specific heat for cooling brine, linear TE correlation, no fouling/scaling.	Cost optimization.
Soliman (1981)	Simple model with linear temperature profiles, constant heat transfer coefficient, constant boiling point elevation (TE).	Examining the effect of cooling water temperature and flow rate.
Flower and Karanovic (1982)	A simple model with further relaxation assumptions. Vapor temperature for all stages is assumed.	Further relaxation of the assumptions.
Omer (1983)	Stage to stage simulation program. Thermodynamic losses and non-equilibrium were included.	Simulation and design of MSF.
Helal et al. (1986)	Rigorous stage-to-stage model. Physical properties are a function of temperature and nonlinear BPE correlation. Fouling and other properties were included.	Developing new techniques to solve the large system of nonlinear equations.
Al-Mutaz and Soliman (1989)	Steady-state model based on Helal et al. New method called orthogonal collect was used to calculate the stream profile.	Simulation of existing of MSF.
Darwish (1991)	Simple model. Constant average value for specific heat and latent heat. Fouling and NCGs were neglected.	Study the effect of some operating and design parameters.
Aly and Fathalah (1995)	Mathematical model using TDM technique. Additional correlations for heat transfer and physical properties.	Exploring MSF plant performance by extending TBT and cooling water temperature.
El-Dessouky et al. (1995)	Mathematical model including constant fouling factor, NCGs and heat losses to surroundings.	Analysis and predicting performance of existing MSF plants.
El-Dessouky and Bingulac (1996)	Model based on stage-to-stage approach. Less sensitive to initial guesses and few iteration steps.	Developing algorithm to solve huge number of equations.
Rosso et al. (1997)	Model similar to Helal et al. Including geometry of the stage, mechanism of heat transfer and role of fouling.	Simulation for design purpose and dynamic model development.
El-Dessouky et al. (1998)	Simple model with constant value for specific heat and OHTC. Fouling and NCGs not included.	Study the performance of different simplified MSF configurations.
Helal et al. (2003)	Mathematical model based on Soliman. Linear temperature profiles, constant heat transfer coefficient and latent heat.	Study the feasibility of hybridization of RO and MSF.
Tanvir and Mujtaba (2006a)	Mathematical model including NN-based correlation to determine the TE.	Developing new TE correlation instead of using from literature.
Abdel-Jabbar et al. (2007)	Model focused on evaluation of the weir loading stage dimensions of the condenser tube bundle and demister dimensions.	Modeling and simulation of the performance of large MSF units.
Tanvir and Mujtaba (2008)	Mathematical model including NN based correlation for physical properties.	Optimal performance at minimum cost.
Alasfour and Abdulrahim (2009)	Rigorous model includes energy losses, effect of NCGs and using stream enthalpy instead of specific heat.	Detailed simulation of MSF process.

(Continued)

TABLE 8.1 (*Continued*)
Previous Work on Steady-State Models for MSF Desalination

Authors (Year)	Description of the Model	Objectives
Hawaidi and Mujtaba (2010b)	Mathematical model including simple linear dynamic fouling factor profile.	Studying the changing brine heater–fouling factor with varying seawater temperature.
Said et al. (2010)	Steady-state model considered the effect of the NCGs and fouling on heat transfer coefficient.	Study the effect of NCGs on the MSF performance.
Abdul-Wahab et al. (2012)	Mathematical model based on Helal et al. heat losses, fouling and NCGs were not considered.	Study MSF performance under variation of some parameters.
Alsadaie and Mujtaba (2014)	Mathematical model considering the release of NCGs and their effect on the OHTC.	Study the optimum location of the venting points.

8.3.2 MSF Dynamic Model

Dynamic models are useful in troubleshooting, fault detection, reliability, start-up and shutdown conditions and to implement advanced control (Gambier and Badreddin, 2004).

Apart from the assumptions mentioned in detailed steady-state model, the following additional assumptions are stated in the development of dynamic model:

- The model is developed using lumped parameter analysis, the mass considered to be perfectly mixed and spatial variation were not explicitly considered.
- Neglecting the presence of non-condensable gases and blow-through phenomenon.
- For each stage the liquid and vapor are in equilibrium.
- Mass cooling brine in condenser tubes remains constant and there is no accumulation of salt in the condenser tubes.
- The non-equilibrium parameter depends on water temperature and salt content.

It is believed that Glueck and Bradshaw (1970) were the first to develop a dynamic model to provide an accurate representation of a typical MSF plant. In their model, the flash stage is divided into four compartments, with the streams and their capacities interacting materially and thermally among themselves. However, including a differential energy balance to the model combining the vapor space and distillate in the flash stage made the model over specified. According to Reddy et al. (1995a) and other authors, a second effort of transient modeling was made by Delene and Ball (1971). They designed a digital code to dynamically simulate a large MSF desalinating plant. They considered the MSF process as consisting of two well-mixed tanks to provide better representation of the holdup of cooling brine flowing inside the tube. Empirical correlations were used to calculate evaporation rates and interstage flow; however, the NCGs in the vapor were not considered. Reddy et al. (1995a) also mentioned that Ulrich in 1977 carried out a simulation of an MSF using Delene and Ball's model and found good agreement between the measured and simulated results; however, significant deviations in the cooling water rate were noted.

Yokoyama et al. (1977) described a dynamic and physical model to predict the start-up characteristics of an MSF plant. The time dependence characteristics of an MSF plant—such as flashing brine, coolant temperature, brine level (BL) and heating steam flow rate to the brine heater—were calculated by a HITAC 8450 Computer using the Runge–Kutta–Gill Method. Two cases were considered for analysis of BL behavior: flashing or no flashing in the chamber. They mentioned that the difference between the actual and numerical results was due to measurement error and the estimated value of the orifice coefficient. Furuki et al. (1985) developed a dynamic model similar to the previous models by (Glueck and Bradshaw, Delene and Ball, and Yokoyama et al.) but used different brine flow equations to develop an automatic control system for the MSF process. Hydraulic formulas were applied for the orifice flow equations. The model was used to study the start-up characteristics of the plant and to control the BLs to avoid blow-through or liquid pile up in the stages. The dynamic stability of the plant was also studied by a real-time dynamic simulator. Rimawi et al. (1989) solved a dynamic model for an MSF-OT plant. Nine stage variables were calculated by simultaneous solution of a set of energy and mass balance dynamic equations, as well as thermodynamic relations and flow rate correlations. For a duration of 15 seconds, they observed rapid and nonlinear variation in the heights of the brine and distillate pool.

Husain et al. (1993) developed a model with flashing and cooling brine dynamics. This model was improved later by the same authors (Husain et al., 1994) by considering distillate dynamics. The model was solved using two methods, one using the SPEEDUP package and the other using a specifically written program based on TDM formulation. For a period of 90 min, the steam flow rate to the brine heater was reduced by 26%, the simulation results obtained showed good agreement with the actual data for the same reduction of steam flow in a real plant. Moreover, their studies showed that the BLs in the flash stages are quickly affected by the steam flow rate and temperature. Reddy et al. (1995a) made more improvements to the model and included brine recycling, which gave more accuracy and faster convergence. Reddy et al. (1995b) have reported a holdup and interstage flow model for accurate estimation of liquid level upstream of the orifice.

A theoretical model simulating transient behavior was reported by Aly and Marwan (1995). The model was based on coupling the dynamic equations of mass and energy balances for the brine and product tray within the flash stages. The model was solved by combination of the Newton–Raphson and Runge–Kutta methods. A step increase in the feed water temperature was investigated, and the responses of the system variables in different stages were illustrated. Maniar and Deshpande (1996) carried out a dynamic model applying empirical correlations for the evaporation rates. The degrees of freedom based on a dynamic model were used to determine the number of controlled and manipulated variables. The SPEEDUP package was used to simulate the MSF process. A complete model for steady-state as well as for dynamic simulation was proposed by Thomas et al. (1998). The models, both steady state and dynamic, were based on the same set of equations and were of the same order. The flashing stage was divided into four compartments: flashing brine tray, product tray, vapor space and condenser tubes. The simulation code was written in C and implemented in a UNIX-based system. However, the absence of a controller in the simulation resulted in a discrepancy between the actual and predicted responses.

Falcetta and Sciubba (1999) described a dynamic simulation of an MSF plant using a modular simulator (CAMEL). Originally, the code was developed with the purpose of simulating thermal power plants only, but its structure was designed to simulate an MSF plant. The authors validated their model against experimental data and the results showed good agreement between the two. However, the lack of details for the mathematical model makes it difficult to critically assess the model. Mazzotti et al. (2000) developed a dynamic model of an MSF taking into account stage geometry and including variations of the physical properties as a function of temperature and concentration, as well as thermodynamic losses. The model equations are solved by LSODA routine. The results obtained by this model were not compared with previous results due to the lack of detailed information in the literature about the operating parameters adopted. However, some nonlinear dynamic features of the model make it useful in order to develop optimal control strategies. In 2001, Tarifa and Scenna presented a dynamic simulator for an MSF desalination plant. It takes into account the dynamics of heaters and stages, hydraulics, standard instrumentation and control systems. Using Delphi 5.0, a visual computer language, the simulator studied the effects of faults that may affect an MSF system, which might be caused by the failure of the pumps, heaters and controllers. Since the model was formulated using a large number of algebraic equation systems (AEs) combined with large ordinary differential equation systems (ODEs), Tarifa et al. (2004) developed a new method called heuristic selection (HS) to increase the solution conversion and the robustness of the model presented by Tarifa and Scenna (2001). For a given set of algebraic equations, the HS method uses a matrix structure to determine the calculation sequence for a set of decision variables and a set of subsystems. The purpose of developing such a method was to create a dynamic simulator for the MSF desalination process.

Shivayyanamath and Tewari (2003) developed a simulation program to predict the dynamic behavior of MSF plants. Using FORTRAN 95 and the Runge–Kutta technique, stage-to-stage calculations were carried out to simulate and model the start-up characteristics of an MSF plant. The thermodynamic properties of the seawater were considered as a function of temperature and salinity. Apart from neglecting the distillate holdups, the interstage brine flow rates are assumed to be constant. This reduced the model to simulation of the energy dynamics within the brine heater and flashing stages. Therefore, it was possible to determine the start-up time to reach steady-state conditions. Gambier et al. (2002) presented a hybrid dynamic model of the brine heater for an MSF plant with satisfactory simulation results. The nonlinear model was implemented in Matlab/Simulink, where algebraic equations were implemented as an S-function. The model could be used to study the simulated system behavior, supervisory control and fault handling. Due to the fact that most models are too large for control design as well as for fault detection, where models have to be computed in real time, Gambier and Badreddin (2004) presented a block-oriented library for Matlab/Simulink to simulate different plant configurations and test control algorithms. The main obstacle was the lack of real-time data.

A detailed dynamic model for the MSF process was presented by Sowgath (2007) to describe the physical behavior of the plant. The model as developed to solve a set of brine and distillate holdup equations, taking into account non-equilibrium effects and the pressure drop in the demister. Good results were obtained when the model was validated at steady state against literature results. David et al. (2007) used gPROMS to develop a detailed dynamic model of the MSF that considered "the blow-through" phenomenon inside brine orifices and the presence of NCGs, which have been neglected in previous studies. However, the model was applied for an experimental unit and was not validated against real plant operating data. Later, Al-Fulaij et al. (2010) extended the work of David et al. (2007) to simulate dynamic and steady-state performance of an MSF-OT using gPROMS. The model results showed good agreement against field data of industrial-scale MSF-OT units. Bodalal et al. (2010) presented a dynamic model to evaluate the performance of MSF plant using dynamic analysis. The model developed was based on coupling the dynamic equations of mass, energy and momentum. The model was solved using the fifth order Runge–Kutta method and was able to investigate the effects of some key parameters, such as sea water concentration and other thermal parameters that may affect the general performance of the MSF plant during transient as well as steady-state operation conditions.

Hawaidi and Mujtaba (2011) developed a dynamic model for a storage tank linked to the freshwater line of the MSF process, which helps to avoid dynamic changes in the operating conditions of the process. The model was solved using the gPROMS tool. For a given design configuration, the operation parameters were optimized at discrete time intervals (based on the storage tank level, which is monitored dynamically and maintained within a feasible level) while the total daily cost is minimized. Al-Fulaij et al. (2011) developed a mathematical model for an MSF-BR plant that includes demister losses, distillate flashing and NCGs. The model was coded and solved using gPROMS. Before using the model to predict the stability regimes of the plant and study plant behavior under operating conditions, it was validated for steady-state and dynamic operation. However, the model neglects heat losses to the surroundings, and the NCGs are also considered as a mass of vapor and their effect on the OHTC was not studied. Recently, Alsadaie and Mujtaba (2014) presented a dynamic model that included the release rate of the NCGs inside the flash chambers and their effect on the OHTC and the performance of the plant. Sowgath and Mujtaba (2015) developed an operational schedule for a particular spring day by carrying out a dynamic optimization formula using the MSF dynamic model. The steam temperature profile of the MSF process is optimized subject to variation in the intake seawater temperature for a fixed production rate of fresh water and maximum performance ratio. Most recent, Lappalainen et al. (2017) presented a new method for one-dimensional modeling and dynamic simulation of the MSF process. The proposed method combined the simultaneous mass, momentum and energy solution. The Rachford–Rice equation was used for flash calculation. A summary of most of the previous studies on the dynamics of the MSF process is shown in Table 8.2.

TABLE 8.2
Previous Work on Dynamic Models for MSF Since 1970

Authors (Year)	Description of the Model	Objectives
Gluck and Bradshaw (1970)	The flash stage was divided into four compartments, streams interacting materially and thermally among themselves.	To provide accurate representation of a typical MSF plant.
Delene and Ball (1971)	Designed a digital code to simulate MSF by considering the MSF process as consisting of two well-mixed tanks, holdup of cooling brine is included.	To provide better representation of the holdup of cooling brine flowing inside the tubes.
Yokoyama et al. (1977)	Model using Runge–Kutta–Gill Method to predict start-up characteristics such as flashing brine and coolant temperature.	To analyze the start-up dynamic characteristics of MSF.
Rimawi et al. (1989)	Transient model defined by nine variables for each stage. Line and Gears method was used to solve the equations.	To simulate transient behavior of MSF plant.
Hussain et al. (1994)	Dynamic model with flashing and cooling brine using SPEEDUP package and tridiagonal matrix method (TDM) formulation.	Evaluation of written program and simulation package for steady-state and dynamic simulations.
Reddy et al. (1995a)	Improvement of Hussain's model by including holdup for brine, distillate, vapor and NCGs.	To develop rigorous and fast model for plant simulation.
Aly and Marwan (1995)	Dynamic model solved by combination of Newton–Raphson and Runge–Kutta methods.	Development of computer simulation program for MSF.
Maniar and Deshpande (1996)	Using SPEEDUP package for dynamic model, empirical corrections for the evaporation rates, some controlled variables were investigated.	Investigate the MSF control system for improved operation.
Thomas et al. (1998)	Using simulation code written in C and implemented in a UNIX-system. The flashing stage was divided into four compartments.	Development of a detailed dynamic model for MSF plant.
Mazzoti et al. (2000)	Dynamic model including: stage geometry, variation of physical properties as a function of temperature and concentration.	Development of dynamic simulation model for the MSF process.
Shivayyanamath and Tewari (2003)	Using FORTRAN 95 and Runge–Kutta method, stage-to-stage calculations, variation of physical properties with temperature and salinity.	To analyze the start-up characteristics of MSF plants.
Gambier and Badreddin (2004)	Dynamic model for analysis and control design purpose. Using Matlab/Simulink.	Simulation and control design of MSF plant.
Sowgath (2007)	Model included a set of brine and distillate holdup equations, non-equilibrium effects and demister pressure drop.	Simulation and optimization of MSF plant.
Hawaidi and Mujtaba (2011)	Dynamic model for storage tank linked to fresh water line. gPROMS used to solve for temperature variation.	Flexible design to meet the demand/consumption of fresh water.
Al-Fulaij et al. (2011)	Dynamic model that included demister losses, distillate flashing and NCGs. The model was solved using gPROMS.	To analyze the dynamic response and test the operation stability.
Alsadaie and Mujtaba (2014)	Dynamic model including the effect of NCGs. gPROMS used to solve the equations of the model.	To analyze the effect of NCGs and design venting system.
Sowgath and Mujtaba (2015)	Dynamic model including operational schedule for specific day taking into account the variation in seawater temperature.	MSF optimization to develop an operational schedule.
Alsadaie and Mujtaba (2016)	A detailed dynamic model included GMC and hybrid Fuzzy-GMC and dynamic fouling.	For development of advanced control system.

8.4 FOULING AND SCALE FORMATION

As mentioned in the previous section, fouling on the surfaces of the tubes caused by formation of scale at high temperature is the main factor of concern in the thermal desalination industry. The mechanism in the MSF plants is a heat exchange process and plant performance is strongly affected by the condition of the heat transfer surfaces. Scale formation results from an accumulation of undesired micro-materials on the internal surface of the condensing tubes. Increasing the thickness of the fouling layer can decrease the heat flux through the tubes, due to low conductivity of the foulant, and thus leads to deterioration of plant performance (Al-Anezi and Hilal, 2007). In addition, increasing the thickness of the fouling layer leads to a significant reduction in the cross-section of the tubes and more energy is consequently required for the pumps systems to maintain a constant flow rate.

The most common materials that result in fouling are the precipitation of calcium carbonate ($CaCO_3$) and, at higher temperatures, magnesium hydroxide ($Mg(OH)_2$). Both are alkaline scales. Non-alkaline scales such as calcium sulfate ($CaSO_4$) are also common foulants in MSF plants. Normally, with the increase in the temperature of the seawater, HCO^-_3 ions decompose to produce CO_3 and, in the presence of Ca, the resulting carbonate reacts with calcium to cause precipitation of $CaCO_3$. OH ions can also be produced from the decomposition of HCO_3 and, in the presences of magnesium, hydroxide and magnesium react together causing precipitation magnesium hydroxide. El Din and Mohammed (1989) reported that $CaCO_3$ may start to precipitate at around 65°C, while $Mg(OH)_2$ precipitation starts around 75°C and increases as the temperature rises. Calcium sulfate, on the other hand, starts to precipitate at a higher temperature and most MSF plants operate below its precipitation limits (<120°C) (Hamed and Al-Otaibi, 2010). However, El-Dessouky and Khalifa (1985) checked the preheater tubes of a MSF-OT plant after 10 years of operation and found out that the most common scale was calcium sulfate. Wildebrand et al. (2007) reported some calcium sulfate growth at around 75°C with increase in the salinity. Unlike alkaline scales, which result from the decomposition of bicarbonates, $CaSO_4$ scale results from the reaction of components that already exist in seawater (Shams El Din et al., 2005).

Calcium carbonate can crystallize into three different forms: vaterite, calcite and aragonite, where the latter is more expected to form in high salinity water (Zhong and Mucci, 1989). Calcium sulfate also occurs in three different forms: anhydrite ($CaSO_4$), hemihydrate ($CaSO_4.1/2H_2O$) or dehydrate (gypsum: $CaSO_4.2H_2O$). Although anhydrite scale ($CaSO_4$) would be expected above 40°C due to its low solubility (Figure 8.4) (Al-Ahmad and Aleem, 1994), most of the calcium sulfate scale in thermal units is hemihydrate (Najibi et al., 1997; Al-Rawajfeh et al., 2014). However, Zhao and Chen (2013) reported that gypsum is more likely to form as scale on the surface at a temperature range of 40°C–98°C, while anhydrite and hemihydrate are precipitated above 98°C. Moreover, during their experimental study at around 60°C, Bansal et al. (2001) found out that only gypsum was formed as a deposit. In fact, there is a long history of controversy about the correct value of the saturation solubility of anhydrite. Marshall and Slusher (1968) experimentally determined the solubility product (*K*sp) of anhydrite to be between 100°C and 200°C and produced a correlation to calculate the solubility at lower temperatures. Though there are no solubility measurements for anhydrite at 100°C, the solubility line was obtained by extrapolating the solubility product from lower and higher temperatures. Thus, a solubility product above 100°C is more accurate than the same line at a lower temperature. Moreover, it is generally agreed that no formation of anhydrite occurs below 109°C (Freyer and Voigt, 2003). For gypsum, which also has a solubility concentration lower than hemihydrate (Figure 8.4), Partridge and White (1929) reported that gypsum is converted into hemihydrate in less than one day when in contact with water at 100°C. Freyer and Voigt (2003) also confirmed that if gypsum is heated above the transition temperature, it will be converted into hemihydrate by a dehydration solid-state reaction.

In MSF-BR, which has both recovery and rejection sections, the concentration of Ca^{2+}, Mg^{2+} and CO^{2-}_3 ions are higher in the recovery section than in the rejection section due to the mixing

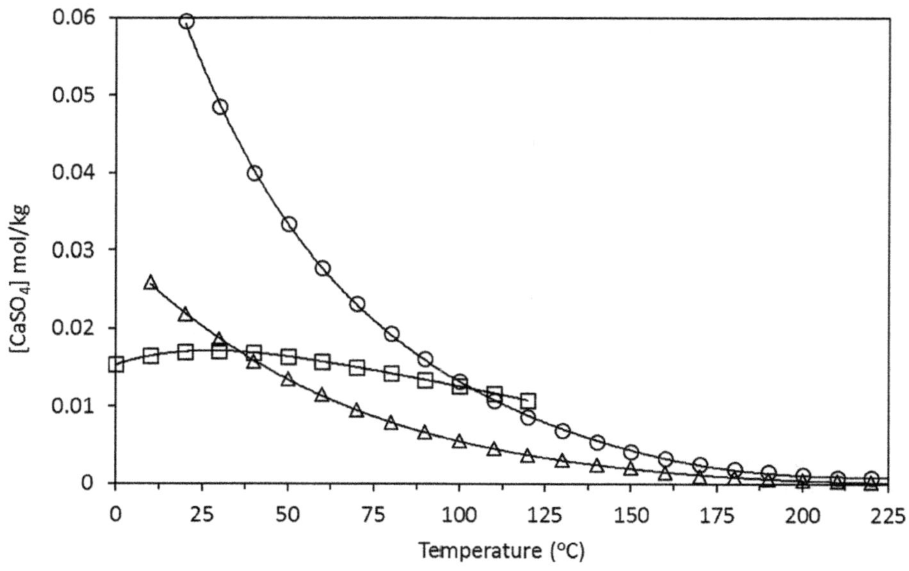

FIGURE 8.4 The solubility of $CaSO_4$ in its three different forms. (From Freyer, D. and Voigt, W., *Monatshefte für Chemie/Chemical Monthly*, 134, 693–719, 2003.)

of the recycle brine with water make-up, while HCO^-_3 is higher in the rejection section (Shams El Din and Mohammed, 1994). Shams El Din et al. (2005) conducted physicochemical analysis of MSF flash chambers that operate at TBT (112°C) and found that the first three stages were fouled completely by $Mg(OH)_2$, while stage four was mixed with $Mg(OH)_2$ and $CaSO_4$ and the scale in stage five was entirely $CaSO_4$. Again, $Mg(OH)_2$ appeared in the following stages (from stage six onward) up to stage nine with an increasing amount of $CaCO_3$. Besides the TBT, scale rate also depends on the concentration of bicarbonate and the partial pressure of CO_2 (Mubarak, 1998). Al-Sofi (1999) believed that scale formation would be expected at low temperatures without the need for CO_2 release.

The fouling process cannot be avoided under any circumstances, and scheduled cleaning is thus required. However, understanding and accurate simulation of the fouling process can help designers to reduce the often overestimated design fouling factor and thus reduce the cost of the extra surface area. Although several experimental studies have been carried out on the effect of scaling and corrosion, scale formation at the heat transfer surface is still not properly understood, and it is the weakest link in the design of heat transfer equipment. Moreover, the complexity and nonlinearity of the MSF process due to the continuous change in temperature and salinity makes the prediction of fouling behavior difficult.

One of the earliest attempts to model fouling was presented by Kern and Seaton (1959). In their model, both the rate of deposition and the removal rate were taken into account. While the rate of deposition was assumed to be constant, the removal rate was considered proportional to the accumulated mass. Although it was a simple model, it has led to the development of more accurate models. Taborek et al. (1972) extended Kern and Seaton's model to introduce the water characteristics into the deposition and removal rates. The rate of deposition was incorporated by the water quality factor and fouling probability, whereas the removal rate included the shear stress and scale strength factors. Although it seems to be a more pronounced model, the model was critiqued for its complexity and for having too many parameters. Hasson (1981) developed a fouling model that takes into account crystal nucleation and growth area, as well as water quality and sticking probability, to predict the crystallization rate of calcium sulfate. However, the model neglects the removal rate and has several unknown variables.

Though the removal rate mechanism is quite easy to model mathematically, the deposition mechanism is complex and difficult to describe based solely on particle attachment. The deposition of seawater involves both diffusion and reaction processes. Due to the concentration gradient between the bulk phase and the solid–liquid interface, all the ion species diffuse or transfer towards the hot surface of the tubes. The reaction process occurs at the surface, as most of the transported species accumulate around the tubes resulting in a crystal layer. To properly understand the fouling mechanism, the appropriate fouling model should consider the diffusion and reaction mechanisms of the fouling.

Hasson et al. (1968) developed a diffusion model to investigate the effect of velocity, temperature and water composition on the crystallization rate of $CaCO_3$. The results indicated that the effect of fluid velocity is more pronounced than the effect of temperature, and they concluded that the fouling control mechanism is the diffusion rate of the species toward the solid–liquid phase. Gazit and Hasson (1975) presented a basic kinetic fouling model to identify fouling control mechanisms by examining some parameters that may govern the deposition rate of $CaCO_3$. Their results indicated that the fouling rates can be treated in terms of kinetic parameters. Bohnet (1987) developed a physical model that considers the deposition and removal terms to predict the sedimentation and crystallization of $CaSO_4$ on hot surfaces. Due to the difficulty of identifying concentrations at the solid–liquid interface, the model combined diffusion and reaction mechanisms. For validation purpose, experimental procedures were carried out using aqueous $CaSO_4$ solutions and good agreement was obtained. Using four different heating modes in a double-pipe heat exchanger, Müller-Steinhagen and Branch (1988) investigated the rate of deposition of $CaCO_3$ using Hasson's ionic diffusion model. Scale and temperature profiles were obtained, and the results suggested that the rate of deposition increases as temperature and $CaCO_3$ concentration increase. Although the model neglects the removal rate, the results showed that the rate of deposition increases with increased flow velocity.

Mubarak (1998) mathematically described the decomposition of bicarbonate using a kinetic model. By applying MSF process conditions, the model predicted the decomposition of bicarbonate into hydroxide ions and carbon dioxide; the hydroxide ions then react with bicarbonate to produce carbonates, which in turn react with calcium to form $CaCO_3$. The results showed that, with continuous decomposition of bicarbonate, magnesium hydroxide is not formed and it is almost absent in the scale. Based on Krause's (1993) model, Brahim et al. (2003) simulated the dynamic growth rate of $CaSO_4$ using computational fluid dynamics (CFD) code FLUENT. The density of the fouling layer and the heat flux, along the heat transfer area, were considered as the distribution parameters. Mwaba et al. (2006) formulated a new semi-empirical correlation to predict the continuous development of the rate of $CaSO_4$ deposition. The surface reaction process was assumed as the controlled mechanism and the diffusion rate mechanism was thus neglected. Al-Rawajfeh (2008) correlated the rate of deposition of $CaCO_3$ to the release rate of CO_2 in a model that considers the diffusion and reaction mechanisms. The fouling resistance and thickness of the fouling layer were investigated at different TBT, pH and salinity.

The precipitation of $CaCO_3$ is not a one-step mechanism, but is in fact a very complex process involving several components moving from the bulk to the solid–liquid interface where they react and are produced simultaneously during the course of several reactions. A rigorous kinetic diffusion model was developed by Segev et al. (2012) to study the effect of velocity, pH and solution concentration on the deposition rate of $CaCO_3$. The model was a combination of diffusion mechanism, when all the species are transported towards the hot surface, and crystallization reactions of the species around the interface of the hot surface. Zhang et al. (2015) presented a generic CFD model to mathematically describe the transport, attachment and removal processes of the deposition of $CaSO_4$. In this model, the dynamic behavior of the fouling process was obtained by sequentially solving a set of steady-state processes with increases in the thickness of the fouling layer.

Notwithstanding the large number of developed models that predict the fouling process, most of them have been developed experimentally or simulated for several heat transfer fields, but have not been a part of the MSF process. Due to the complexity of the MSF process, with its harsh and changeable operating conditions related to temperature and brine concentration, such models may be considered meaningless and inadequate unless implemented under MSF operating conditions. Unfortunately, most of the MSF manufacturers or model developers represent fouling as a constant factor for safety considerations. However, there are a few publications that focus on modeling the fouling in the MSF process.

Cooper et al. (1983) criticized the commonly used linear relationship to predict fouling in MSF plants and reported that the fouling could reach a limit or steady-state value. Hence, Cooper et al. (1983) adopted the asymptotic model developed by Kern and Seaton to predict the fouling in an existing MSF. Cooper and his co-workers identified that the most changed parameter is the recycled brine flow rate in the brine heater and the first few stages in the heat recovery, which may be considered as a sign of fouling. Al-Ahmad and Aleem (1994) presented a brief review of the previous developed models and successfully applied the Kern and Seaton model to study fouling in an existing MSF plant. The authors compared the results obtained against actual data and reported that the assumptions made by Kern and Seaton are very simple and require more development. Hawaidi and Mujtaba (2010) developed a linear dynamic fouling model to predict fouling inside the brine heater. The model was then used to investigate the effect of fouling on plant performance with seasonal variation in seawater temperatures. Said et al. (2012) presented a regression dynamic fouling correlation as a function of temperature and implemented it into a steady-state MSF model to obtain a dynamic fouling factor. The obtained fouling factor is then used to investigate the fouling effect in the flash stages in the HRS.

However, similar to linear models, these regression models neglect several essential parameters that may have a strong effect on the fouling process; such models may thus not be adequate to predict fouling in MSF plants and might generate unacceptable results (Malayeri and Müller-Steinhagen, 2007). Al-Rawajfeh et al. (2014) extended Al-Rawajfeh's work (Al-Rawajfeh, 2008) to include the precipitation of $CaSO_4$ as well as $CaCO_3$ inside the MSF process. The model involves calculations of the carbonate species and sulfate in the intake seawater. As the temperature increases, the precipitation rate was calculated by calculating the reduction rate of carbonate and sulfate. However, the model neglected the removal rate and the effect of surface properties such as roughness.

Alsadaie and Mujtaba (2017b) developed very detailed fouling model to predict the precipitation of calcium carbonate and magnesium hydroxide inside the condensers tubes of the MSF process. The model considered the deposition and removal rates and took into account the effect of salinity, temperature and the flow rates of the intake seawater and the recycle brine stream. The model was incorporated into the MSF-BR process and, due to the recycle brine stream, the model considered the changes in the fouling rate to be caused by the dynamic variation of seawater salinity and temperature. Using their previous fouling model, Alsadaie and Mujtaba (2017a) studied the fouling phenomena inside the condenser tubes of the MSF-OT. The effects of several parameters (seawater temperature, flow velocity and TBT) were studied and it was found that the effect of TBT has a greater influence on the fouling rate than the effect of the seawater temperature. In a comparison of the two studies, it has been found that the fouling rate in the MSF-OT is less than the rate in the MSF-BR. This, as El-Dessouky and Khalifa (1985) pointed out, is due to the low concentration of the treated brine in the MSF-OT compared to the MSF-BR.

Complete prevention of scale formation is impossible. However, mitigation or control of fouling on heat transfer surfaces is possible and can be done chemically or mechanically. By chemical means, an acid such as H_2SO_4 can be added to cause reduction of the bicarbonate, or one of the commercial scale inhibiters derived from condensed polyphosphates, polyelectrolytes and organophosphonates can be used (Hamed and Al-Otaibi, 2010). However, improper control of the antiscalants dosing rate can lead to undesirable results. Salt can grow and build up around the antiscalant molecular chain resulting in the additive being less effective (Al-Sofi, 1999). Mechanical cleaning is another method of scale removal. There are two mechanical cleaning methods a used in desalination plants: offline cleaning (by brushes) while the plant is off and online cleaning (by balls) while the plant is operating (Al-Ahmad and Aleem, 1994). A summary for most of the previous studies on fouling models in the MSF process is shown in Table 8.3.

TABLE 8.3
Previous Work on Fouling Models in MSF

Authors (Year)	Description of the Model	Objectives
Cooper et al. (1983)	Similar to Kern and Seaton's model, included deposition and removal rates for steady-state operation.	To investigate the change in the operating condition on fouling behavior.
Al-Ahmed and Aleem (1994)	Applied Kern and Seaton's model in MSF plant and provide values for unknown parameters.	To test the capability of using Kern and Seaton model in MSF plant.
Mubarak (1998)	A kinetic fouling model to estimate the crystallization of $CaCO_3$.	To understand the complexity of $CaCO_3$ precipitation.
Al-Rawajfeh (2008)	Steady-state fouling model to estimate the deposition of $CaCO_3$ by calculating CO_2 release.	To simulate the crystallization of $CaCO_3$ in MSF plants.
Hamed and Al-Otaibi (2010)	A linear correlation to predict the dynamic fouling factor.	To investigate the effect of different types of antiscalant.
Hawaidi and Mujtaba (2010)	A linear dynamic fouling model for a brine heater with variation of seawater temperature.	Optimization of MSF under the effect of fouling.
Said et al. (2012)	Steady-state fouling model for MSF stages.	To study the effect of fouling on the performance ratio of the MSF.
Al-Rawajfeh et al. (2014)	Steady-state fouling model to estimate the deposition of $CaCO_3$ and $CaSO_4$.	To simulate and understand scale formation at high temperature.
Alsadaie and Mujtaba (2017b)	Dynamic fouling model considering deposit and removal rates to predict the crystallization of $CaCO_3$ and $Mg(OH)_2$.	To investigate the effect of temperature, velocity and salinity on the crystallization of $CaCO_3$ and $Mg(OH)_2$.

8.5 NON-CONDENSABLE GASES (NCGS)

As mentioned in the introduction, the tendency of scale formation and the release of NCGs inside the flash chambers play an important role in increasing the cost of the plants, and they are considered the main drawbacks of the thermal desalination process. The mechanism of the thermal desalination process is as any mechanism of heat exchange. This means that the internal and the external surfaces of the tubes are responsible for the heat transfer, and any accumulation of fouling or NCGs may reduce the effective surface area and consequently result in low performance. These gases have very low boiling points and thus, for any operating temperature in the MSF process, they are released at low temperatures and accumulate around the external side of the tubes, resulting in massive deterioration of MSF performance a (Al-Rawajfeh et al., 2003). The presence of NCGs can reduce the performance and decrease the efficiency of the whole desalination plant, and consequently increase the cost of most thermal desalination units. Trapping these gases (NCGs) inside the condensation zone can cause the following:

- The surface area taken up by NCGs will not be available for the steam to condense (low heat transfer area), and
- The NCGs will reduce the OHTC of the vapor inside the tube (low heat transfer coefficient).

According to the heat transfer equation $Q = U \times A \times \Delta T$, any decrease in the heat transfer area or OHTC will result in decrease in the amount of heat transferred between the steam and the cooling seawater. In order to get the same amount of transferred heat, the temperature difference has to be increased. Consequently a large amount of superheated steam is required, resulting in an increase in the cost and significant reduction of the performance ratio (PR) of the plant (El-Dessouky and Ettouney, 2002). Due to the law of conductivity of these gases, even a small amount of NCGs can result in a massive reduction of the OHTC and the overall performance of the desalination process (Al-Anezi and Hilal, 2007). According to Semiat and Galperin (2001), a half-percent of air in the vapor may cause a reduction in the heat transfer process by 50%. The NCGs inside the chambers of the MSF plants consist mainly of air (N_2 and O_2), argon and CO_2. The main causes of the presence of these gases are as follows:

- A small amount of all atmospheric gases is dissolved in the seawater feed.
- Due to the vacuum conditions inside the flash chambers, atmospheric air may leak into the stage through flanges, manholes and instrumentation nozzles.
- The high temperature of the brine causes a decomposition of the bicarbonates to produce carbon dioxide, which is transferred with the vapor towards the tube bundles.

When the hot brine leaves the brine heater and is fed into the flash chambers of MSF plants, the NCGs are released, due to their low boiling points and the decrease of CO_2 solubility at lower pressure, and carried with the steam into the vapor space of the stage (Glade et al., 2005). Steam releases its latent energy to the process and condenses on the surface area of the tubes. However, unlike steam, the NCGs are not able to condense and so accumulate around the tube bundle, effectively providing insulation material and thus a decrease in the efficiency of the heat transfer process. In the first few stages of the MSF, the brine temperature is very high and thus the solubility of the NCGs (mainly CO_2) decreases, resulting in a release of these gases (Al-Anezi et al., 2008). As a result, large amounts of most of these gases are transferred with the vapor in the first few stages of the MSF plant (Genthner et al., 1993). As the temperature drops through the stages, the increase in the salinity of the brine plays another factor in reducing the solubility of CO_2 (Al-Anezi et al., 2008). Under condensation conditions, the vapor is transported by diffusion toward the interface phase and condenses on the heat transfer surface of the tubes. The NCGs can also be transported with the vapor into the center of the tube bundle and, due to the condensation of the vapor, the concentration of the NCGs increases (Semiat and Galperin, 2001).

Carbon dioxide dissolves in the condensate and lowers its pH value causing, in the presence of O_2, serious corrosion problems in the vapor side of the flash zone, which leads to tube leakages and decreases the lifetime of the plant. Furthermore, accumulation of NCGs in an MSF plant can result in loss of the vacuum condition inside the stages, resulting in high BLs (Glade and Al-Rawajfeh, 2008). In addition, the release of CO_2 from the evaporating brine increases the pH and considerably influences the concentrations of $HCO^-{}_3$, $CO^{-2}{}_3$, H^+ and OH^- ions in the brine, thus playing an important role in alkaline scale formation (Glade and Ulrich, 2003; Al-Rawajfeh et al., 2005; Al-Anezi and Hilal, 2007).

Due to the low thermal conductivity of the NCGs, these gases create resistance and work as an insulator, thus affecting the efficiency of the heat transfer for the condensation process, increasing the energy consumption and consequently reducing the lifetime of desalination plants. Removing NCGs is thus essential to the efficient operation of all desalination plants. For NCGs to be removed from the feed seawater, the MSF-BR plants are usually equipped with a deaerator, where molecularly dissolved gases can be extracted almost completely from the make-up flow. However, CO_2 is hard to remove by simple deaeration, and an injection of a strong acid to the make-up stream and a decarbonator are therefore required. The additional acid results in a low pH value and speeds the conversion of bicarbonate and carbonate ions to CO_2, which is then expected to be released almost completely in the decarbonator (Al-Anezi and Hilal, 2007; Glade and Al-Rawajfeh, 2008).

For an MSF-OT, which is not equipped with a deaerator, the O_2, N_2 and Ar content entering the plant with the feed seawater can be removed almost completely in the first three stages. This can be done by an adequate venting system installed in the MSF plants. Also, in MSF-BR, the liberated

CO_2 and the remaining O_2, N_2 and Ar inside the flash chambers can be removed. Due to the high degassing rates, more than 95% of the air is expected to be released either in the deaerator or in the first stage. However, the situation with the CO_2 is quite different due to the high solubility of CO_2; the liberation of carbon dioxide is thus very slow compared to air. CO_2 is continuously liberated from the brine as the brine flows from one stage to another (Genthner et al., 1993). The main purpose of the venting system is to remove the NCGs and minimize their concentration around the tubes bundle. However, another use of the venting system in MSF plants is to reduce the pressure inside the flash chambers, particularly during start-up, and more steam is produced at low temperature.

In MSF plants, the extraction or venting points of the NCGs can be designed in series or in a parallel arrangement. In the series arrangement, the MSF stages are connected by venting points where the NCGs are transported directly from one stage to another. The whole amount of NCGs then accumulates in the last stage before they are extracted to the evacuated system. In the parallel arrangement, however, all the stages are connected directly to the evacuated system and the NCGs from each stage are transferred to the evacuated system directly (Darwish et al., 1995). Due to the drawbacks of using a single arrangement, most of MSF plants are usually designed to have a combination of both arrangements. As illustrated in Figure 8.5, the parallel arrangement is normally preferred in the first few stages where the concentration of NCGs is high. The remaining stages, however, are connected in a series because the concentration of NCGs is low.

Although the problem of NCGs in thermal desalination processes is partially solved using a venting system, there are a number of publications concerning the effect of these gases on the performance of the MSF plants. Seifert and Genthner (1991) and Genthner and Seifert (1991) developed a mathematical dynamic model taking into account the chemical and physical processes to predict the release rate of NCGs, mainly O_2, N_2 and CO_2, in the MSF chambers. The model was applied for the two types of MSF configuration (MSF-OT and MSF-BR) and the results showed that the concentration of NCGs in the MSF-OT is higher than in the MSF-BR due to the omission of the deaerator system in the MSF-OT. Genthner et al. (1993) provided a very detailed discussion about the behavior and effect of NCGs. The discussion included the influence of NCGs in the individual stages, the change of the heat transfer and saturation temperature profile and also the CO_2 release and distribution between the stages.

As mentioned in Section 8.3, several publications consider the modeling and simulation of the MSF process. However, the effect of NCGs was neglected in most of the MSF models. Only Reddy et al. (1995a) and Al-Fulaij et al. (2010) attempted to include mass balance equations in their models to predict the release rate of the NCGs. Recent studies by Alasfour and Abdulrahim (2009) and

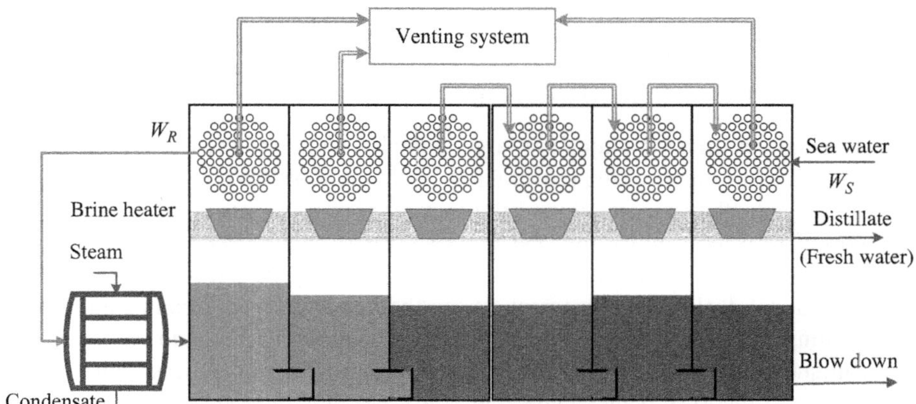

FIGURE 8.5 Typical venting arrangement in MSF plants.

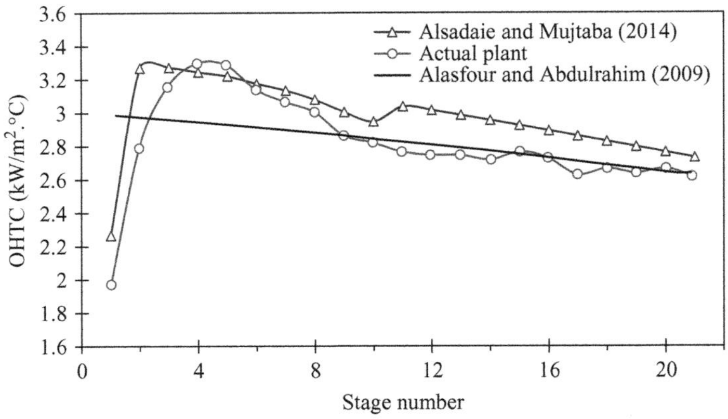

FIGURE 8.6 Overall heat transfer coefficient profile.

Said et al. (2010) included mass concentration of NCGs in their steady-state models to investigate the effect of NCGs on the OHTC. Alasfour and Abdulrahim (2009) assumed a constant value for the NCGs in all stages, whereas Said et al. (2010) attempted to test different values of NCGs, but for each test the concentrations were assumed to be constant for all stages. In industrial practice, the design and the arrangement of the extracted points of the evacuated system result in variation in the amount of NCGs between one stage and another. Most recently, Alsadaie and Mujtaba (2014) developed an MSF model to include the release of NCGs inside the chambers. The model used Henry's law in the mass equations to predict the release rate of NCGs and the varied NCG concentration was used to analyze the design of the venting system and the effect of NCGs on the OHTC. The equilibrium concentration of NCGs in the brine was calculated using Henry's law and the release rate was predicted based on the concentration driving force. The OHTC results obtained by Alsadaie and Mujtaba (2014) and Alasfour and Abdulrahim (2009) are plotted against actual data in Figure 8.6. As can be seen, the results obtained by Alsadaie and Mujtaba (2014) showed a close match to the actual data, as the OHTC varied as the concentration of the NCGs varied from one stage to another based on the location of the evacuated points. Thus, it is essential to model the release of NCGs as it provides a clear picture of the performance of the OHTC. Lappalainen et al. (2017) presented a dynamic model that includes the release rate of NCGs. The NCGs were modeled as nitrogen and oxygen (air), and the effect of temperature and water salinity was taken into account to predict the release rate of NCGs. Using Henry's law, the Henry's coefficient was calculated using the van't Hoff equation with the Setschenov salt effect. The effect of venting system failure on the stage pressure, BL and TBT was also studied.

8.6 ADVANCED CONTROL STRATEGIES

In general, thermal desalination plants are large, complex and consume energy and most importantly they are crucial to support life in several regions of the world. With the increase in the development of these plants, they operate at the limit to meet high standards of performance, including optimality, cost effectiveness, reliability, and above all, safety. Nowadays, enormous challenges are facing the MSF industry to reduce costs and increase profitability. As a result, MSF technology has experienced a gradual decline in investment during the last two decades compared to other techniques of desalting water. Additional enhancements are required to maintain the attractiveness of the MSF technique to investors. One interesting area of improvement is process control. Improving the control system of the MSF plant can decrease the consumption of energy and increase process performance.

In general, most of the chemical plants follow a nonlinear process; the nonlinearity of such plants varies according to the behavior of the physical and chemical properties of the materials and the parameters of the process. The MSF desalination process is a nonlinear process (Ismail, 1998; Ali et al., 1999; Lior et al., 2012). The nonlinearity of the process may cause variation in the TBT and the BLs in the flash chambers. Monitoring and controlling the TBT maintains the performance of the process at a desired value. In addition, a continuous monitoring of the BLs inside the stages is essential to maintain the required pressure and keep the continuity of the water evaporation process at a low temperature. Therefore, to increase the efficiency of MSF plants and maintain the operation at its optimum condition, an effective advance control system that handles nonlinear processes is required.

Due to its simplicity, the conventional proportional–integral–derivative (PID) controller is the preferred type of control used in MSF plants (Algobaisi et al., 1991). However, the PID controller was developed for a linear process and thus cannot reliably control highly complex systems containing nonlinear parameters. Another drawback of the PID controller system is the time consumed tuning the process of its parameters. Although much effort has been made to achieve an effective tuning process procedure, the procedure still relies heavily on the experience and the skills of the MSF plant operators. Moreover, due to the large difference in the temperature of seawater between summer and winter, most of MSF plants operate in two modes, summer and winter, and thus the fixed values of the PID parameters for one mode would not be optimal for the other mode (Al-Gobaisi et al., 1994). Presently, MSF plants are modern, very complex and operate at their limit, so more advanced control strategies (ACS) are required to handle such processes. The development of such ACS is possible nowadays due to the availability of powerful computing tools.

Several studies have been carried out in an attempt to implement ACS in the MSF desalination process. One of the earliest attempts to implement ACS in the MSF process was conducted by Maniar and Deshpande (1996) when they applied constrained model predictive control (CMPC) for the MSF process. Varieties of set points were obtained using CMPC and then implemented into existing PID controllers. Although excellent results were achieved, the linearity of the CMPC is still the main concern of this strategy. Later, Ali et al. (1999) applied a constrained nonlinear model predictive control (NLMPC) to control an MSF plant. Though the purpose of the study was to compare the performance of the control by using a reduced and full mathematical model, the results showed the ability of NLMPC to control MSF process with less computational time. The fuzzy logic controller (FLC) is another approach that has been implemented by several researchers to control the MSF process. The good ability of FLC to control very complex systems has maintained the use of this strategy at a full commercial scale in desalination plants (Alatiqi et al., 1999). Jamshidi et al. (1996) applied the FLC strategy to control the brine heater unit in an MSF plant. First a genetic algorithm (GA) approach was applied to the FLC of a brine heater unit and the NN technique was used to tune the fuzzy rule. The results showed a significant enhancement in controller response and tremendous reduction in oscillations and overshoot. The results also showed that the controller rules were properly tuned to perform towards the desired points. Ismail and AbuKhousa (1996) applied FLC to the MSF process to control one of the most important parameters in MSF, namely the TBT. A controller similar to PID (PID-like-FLC) was obtained by combination of a set of fuzzy rules. The proposed controller was found to give good results compared to the existing PID. Olafsson et al. (1999) presented the FLC to the brine heater process of an MSF desalination plant. Their proposed simple FLC was constricted to have two inputs and one output. Satisfactory results were obtained compared to the conventional PID controller.

One of the main drawbacks of FLC system is its inability to handle unpredicted high variations in the plants if the controller was developed for normal operations. This motivated Ismail (1998) to propose an adoptive fuzzy technique known as fuzzy model reference learning control

(FMRLC) to control the TBT. the proposed technique was compared to conventional PID as well as to non-adoptive fuzzy logic and the results showed significant improvements in overshoot and rise time.

Another ACS technique to handle very complex and nonlinear processes is the use of NNs. Ali et al. (2015) presented a very detailed review of the possible use of NNs as the base control technique in many engineering systems. The successful application of the NNs technique as a control strategy for a seawater-desalination solar-powered membrane distillation unit by Porrazzo et al. (2013) has motivated Tayyebi and Alishiri (2014) to adopt the same technique for the MSF desalination process. A neural network inverse model control strategy (NNINVMC) was proposed to control three parameters in MSF process: TBT, BL in the last stage and salinity. The proposed technique contained three layers that were trained with a descent gradient algorithm and identified from input and output data. The obtained results indicated that the proposed strategy is a robust tool for MSF plant control system. Alsadaie and Mujtaba (2016) applied the Generic Model Control (GMC) strategy to track the set points of the TBT and the BL in the last stage. The proposed strategy has been compared to the conventional PID controller and the obtained results showed the superiority of the GMC controller over the conventional PID controller. The optimistic results obtained using the GMC has encouraged (Alsadaie et al., 2016) to hybridize the GMC with the FLC to develop the GMC–Fuzzy controller. The proposed hybrid GMC–Fuzzy controller has been implemented to track the set points and handle the disturbances of the same loops used by Alsadaie and Mujtaba (2016). The developed control strategy has been compared to the pure GMC and the results revealed that the hybrid GMC–Fuzzy controller outperformed the pure GMC in terms of overshoot, rise time and steady-state settling time. A summary for most of the previous studies on control strategies implemented in the MSF process is presented in Table 8.4.

TABLE 8.4
Previous Work on Advanced Control Strategies Implemented in MSF

Authors (Year)	Description of the Model	Objectives
Maniar and Deshpande (1996)	Constrained Model Predictive Control (CMPC) was designed to control the TBT, BL and distillate flow rate.	Optimize plant performance and minimize energy consumption.
Jamshidi et al. (1996)	Genetic algorithm was applied to Fuzzy Logic Controller (FLC) to control the TBT.	To improve the convergence of the controller and reduce the oscillations.
Ali et al. (1999)	Nonlinear Model Predictive Control (NLMPC) was designed to control the TBT and distillate flow rate.	To validate the performance of MSF reduced model against large size model.
Ismail AbuKhousa (1996)	PID integrated with FLC to develop Hybrid PID-Like-FLC to control the TBT.	To improve the performance of the control system in MSF plants.
Ismail (1998)	Fuzzy Model Reference Learning Control (FMRLC) to regulate TBT	To improve the performance of the control system in MSF plants.
Olafsson et al. (1999)	FLC was developed to control the brine heater process.	To improve the operating performance of MSF plants.
Tayyebi and Alishiri (2014)	Designed NN inverse model to control the TBT, BL and salinity.	To improve the control system in MSF plants.
Alsadaie and Mujtaba (2016)	Generic Model control (GMC) designed to control the TBT and BL.	To reduce the operating cost by improving the control system in MSF plants.
Alsadaie et al. (2016)	Hybrid Fuzzy-GMC was developed to control TBT and BL.	To improve the pure GMC by introducing IF and THEN fuzzy rules.

8.7 CONCLUSIONS

Rapid increases in the world's population and the scarcity of natural water resources have created a major global challenge to overcome the water crisis. In response to this increase, many countries have focused on additional sources of water supply through desalination of sea water. Among different types of desalination techniques, the MSF desalting method is by far the most robust and reliable technology for the production of desalted water at large capacity. Hence, in this chapter, an attempt was made to cover all the aspects related to the MSF desalination process from its process description to mathematical models and problems facing this technology.

Although a lot has been done in mathematical modeling and simulation, more work is expected, especially in the area of fouling and new design configuration. Very detailed MSF mathematical models that consider the start-up and shutdown of the plant, fouling models and implementation of advance control can be further used to increase the efficiency of the MSF process and cut production costs.

Fouling and scale formation is a serious problem that requires further investigation. Although the crystallization of magnesium hydroxide is a well-known type of fouling in MSF plants, little attention has been paid to modeling its precipitation. Moreover, Shams El Din and Mohammed (1994) reported that part of the fouling in MSF is made of silica and clay and not just chemical species. Thus, a model that considers almost everything that may be expected to accumulate inside the MSF condensing tubes should be developed.

There are a number of studies in the literature on hybrid integration of a seawater RO unit with an MSF distiller (hybrid MSF/RO). Though, this can improve the performance of the MSF and reduce the cost of desalted water, a different configuration can result in a different cost reduction outcome. Hence, a superstructure study is required to optimize the best combination structure between the RO and MSF unit.

In addition, with the availability of powerful computer software, it is time to explore the feasibility of designing new MSF configurations through simulation studies. Several studies have focused on the development of new MSP designs for better performance and lower cost. For example, Mussati et al. (2003) presented different configurations of the MSF when the inlet and outlet of distillate, brine and the feed streams have been redesigned to minimize the total annual cost for a given water production. Al-Hamahmy et al. (2016) studied the effect of extracting part of the cooling brine from the water boxes and re-injected it directly to the flash chambers. The extracted part would not pass through the brine heater or high temperature stages and thus less surface area was required for the brine heater. Such new configurations may have a great impact on the total investment costs of MSF plants and more research in this area is urgently required.

Finally, most of the chemical processes have a specific simulation package that can be used easily for any given process. However, there is no a particular software that simulates the desalination process. In fact, most of the studies modeling and simulating the desalination process are conducted mathematically using software such as MATLAB, SPEEDUP and gPROMs. Although such programs have been proven to be quick in solution conversion and lower time consumption, building the model takes a long time because the user has to define the properties of each unit and stream manually. The process system engineering community is thus invited to put more interest in the water industry and develop software with a friendly user interface especially for desalination technology.

REFERENCES

Abdel-Jabbar, N. M., Qiblawey, H. M., Mjalli, F. S. and Ettouney, H. 2007. Simulation of large capacity MSF brine circulation plants. *Desalination*, 204, 501–514.

Abdul-Wahab, S. A., Reddy, K. V., Al-Weshahi, M. A., Al-Hatmi, S. and Tajeldin, Y. M. 2012. Development of a steady-state mathematical model for multistage flash (MSF) desalination plant. *International Journal of Energy Research*, 36, 710–723.

Addams, L., Boccaletti, G., Kerlin, M. and Stuchtey, M. 2009. *Charting Our Water Future: Economic Frameworks to Inform Decision-Making*. McKinsey & Company, New York.

Al-Ahmad, M. and Aleem, F. A. 1994. Scale formation and fouling problems and their predicted reflection on the performance of desalination plants in Saudi Arabia. *Desalination*, 96, 409–419.

Al-Anezi, K. and Hilal, N. 2007. Scale formation in desalination plants: Effect of carbon dioxide solubility. *Desalination*, 204, 385–402.

Al-Anezi, K., Johnson, D. J. and Hilal, N. 2008. An atomic force microscope study of calcium carbonate adhesion to desalination process equipment: Effect of anti-scale agent. *Desalination*, 220, 359–370.

Al-bahou, M., Al-Rakaf, Z., Zaki, H. and Ettouney, H. 2007. Desalination experience in Kuwait. *Desalination*, 204, 403–415.

Al-Fulaij, H., Cipollina, A., Bogle, D. and Ettouney, H. 2010. Once through multistage flash desalination: gPROMS dynamic and steady state modeling. *Desalination and Water Treatment*, 18, 46–60.

Al-Fulaij, H., Cipollina, A., Ettouney, H. and Bogle, D. 2011. Simulation of stability and dynamics of multistage flash desalination. *Desalination*, 281, 404–412.

Al-Gobaisi, D. M. K., Hassan, A., Rao, G. P., Sattar, A., Woldai, A. and Borsani, R. 1994. Towards improved automation for desalination processes, Part I: Advanced control. *Desalination*, 97, 469–506.

Al-Hamahmy, M., Fath, H. E. S. and Khanafer, K. 2016. Techno-economical simulation and study of a novel MSF desalination process. *Desalination*, 386, 1–12.

Al-Hengari, S., El-Bousiffi, M. and El-Mudir, W. 2005. Performance analysis of a MSF desalination unit. *Desalination*, 182, 73–85.

Al-Mutaz, I. S. and Soliman, M. A. 1989. Simulation of MSF desalination plants. *Desalination*, 74, 317–326.

Al-Rawajfeh, A. E. 2008. Simultaneous desorption–crystallization of CO_2–$CaCO_3$ in multi-stage flash (MSF) distillers. *Chemical Engineering and Processing: Process Intensification*, 47, 2262–2269.

Al-Rawajfeh, A. E., Glade, H. and Ulrich, J. 2003. CO, release in multiple-effect distillers controlled by mass transfer with chemical reaction. *Desalination*, 156, 109–123.

Al-Rawajfeh, A. E., Glade, H. and Ulrich, J. 2005. Scaling in multiple-effect distillers: The role of CO_2 release. *Desalination*, 182, 209–219.

Al-Rawajfeh, A. E., Ihm, S., Varshney, H. and Mabrouk, A. N. 2014. Scale formation model for high top brine temperature multi-stage flash (MSF) desalination plants. *Desalination*, 350, 53–60.

Al-shayji, K. A., Al-wadyei, S. and Elkamel, A. 2005. Modelling and optimization of a multistage flash desalination process. *Engineering Optimization*, 37, 591–607.

Al-Sofi, M. A.-K. 1999. Fouling phenomena in multi stage flash (MSF) distillers. *Desalination*, 126, 61–76.

Alasfour, F. N. and Abdulrahim, H. K. 2009. Rigorous steady state modeling of MSF-BR desalination system. *Desalination and Water Treatment*, 1, 259–276.

Alatiqi, I., Ettouney, H. and El-Dessouky, H. 1999. Process control in water desalination industry: An overview. *Desalination*, 126, 15–32.

Algobaisi, D. M. K., Barakzai, A. S. and Elnashar, A. M. 1991. An overview of modern control strategies for optimizing thermal desalination plants. *Desalination*, 84, 3–43.

Ali, E., Alhumaizi, K. and Ajbar, A. 1999. Model reduction and robust control of multi-stage flash (MSF) desalination plants. *Desalination*, 121, 65–85.

Ali, J. M., Hoang, N. H., Hussain, M. A. and Dochain, D. 2015. Review and classification of recent observers applied in chemical process systems. *Computers & Chemical Engineering*, 76, 27–41.

Alsadaie, S. M. and Mujtaba, I. M. 2017a. Crystallization of calcium carbonate and magnesium hydroxide in once-through multistage flash (MSF-OT) process desalination. In: Espuña, A., Graells, M. and Puigjaner, L. (Eds.) *Proceeding of the 27th European Symposium on Computer Aided Process Engineering–ESCAPE 27.* Barcelona, Spain.

Alsadaie, S. M. and Mujtaba, I. M. 2017b. Dynamic modelling of heat exchanger fouling in multistage flash (MSF) desalination. *Desalination*, 409, 47–65.

Alsadaie, S., Zanil, M. F. B., Hussain, A. and Mujtaba, I. M. 2016. Development of hybrid fuzzy-GMC control system for MSF desalination process. In: Kravanja, Z. and Bogataj, M. (Eds.) *Proceeding of the 26th European Symposium on Computer Aided Process Engineering–ESCAPE 26.* Portoroz, Slovenia.

Alsadaie, S. M. and Mujtaba, I. M. 2014. Modelling and simulation of MSF desalination plant: The effect of venting system design for non-condensable gases. *Chemical Engineering Transactions*, 39, 1615–1620.

Alsadaie, S. M. and Mujtaba, I. M. 2016. Generic model control (GMC) in multistage flash (MSF) desalination. *Journal of Process Control*, 44, 92–105.

AlTaee, A. and Sharif, A. O. 2011. Alternative design to dual stage NF seawater desalination using high rejection brackish water membranes. *Desalination*, 273, 391–397.

Aly, N. H. and El-Figi, A. K. 2003. Thermal performance of seawater desalination systems. *Desalination*, 158, 127–142.

Aly, N. H. and Marwan, M. A. 1995. Dynamic behavior of MSF desalination plants. *Desalination*, 101, 287–293.

Aly, S. E. and Fathalah, K. 1995. A mathematical model of a MSF system. In: *Proceedings of the IDA World Congress*, Abu Dhabi, United Arab Emirates, pp. 203–226.

Baig, H., Antar, M. A. and Zubair, S. M. 2011. Performance evaluation of a once-through multi-stage flash distillation system: Impact of brine heater fouling. *Energy Conversion and Management*, 52, 1414–1425.

Bansal, B., Müller-Steinhagen, H. and Chen, X. D. 2001. Comparison of crystallization fouling in plate and double-pipe heat exchangers. *Heat Transfer Engineering*, 22, 13–25.

Barba, D., Linzzo, G. and Tagliferri, G. 1973. Mathematical model for multiflash desalting plant control. In: *4th International Symposium on Fresh Water from the Sea*, Heidelberg, Germany, pp. 153–168.

Beamer, J. H. and Wilde, D. J. 1971. The simulation and optimization of a single effect multi-stage flash desalination plant. *Desalination*, 9, 259–275.

Bennett, A. 2013. 50th anniversary: Desalination: 50 years of progress. *Filtration+ Separation*, 50, 32–39.

Bodalal, A. S., Abdul_Mounem, S. A. and Salama, H. S. 2010. Dynamic modeling and simulation of MSF desalination plants. *JJMIE*, 4, 394–403.

Bohnet, M. 1987. Fouling of heat transfer surfaces. *Chemical Engineering & Technology*, 10, 113–125.

Borsani, R. and Rebagliati, S. 2005. Fundamentals and costing of MSF desalination plants and comparison with other technologies. *Desalination*, 182, 29–37.

Brahim, F., Augustin, W. and Bohnet, M. 2003. Numerical simulation of the fouling process. *International Journal of Thermal Sciences*, 42, 323–334.

Coleman, A. K. 1971. Optimization of a single effect, multi-stage flash distillation desalination system. *Desalination*, 9, 315–331.

Cooper, K. G., Hanlon, L. G., Smart, G. M. and Talbot, R. E. 1983. A model for the fouling of MSF plants based on data from operating units. *Desalination*, 47, 37–42.

Darwish, M. A. 1991. Thermal analysis of multi-stage flash desalting systems. *Desalination*, 85, 59–79.

Darwish, M. A. and Alsairafi, A. 2004. Technical comparison between TVC/MEB and MSF. *Desalination*, 170, 223–239.

Darwish, M. A., El-Refaee, M. M. and Abdel-Jawad, M. 1995. Developments in the multi-stage flash desalting system. *Desalination*, 100, 35–64.

David, I., Bogle, L., Cipollina, A. and Micale, G. 2007. Dynamic modeling tools for solar powered desalination process during transient operations. In: L. Rizzuti, H.M. Ettouney, and A. Cipollina (Eds.) *Solar Desalination for the 21st Century*. Springer, the Netherlands, pp. 43–67.

De Gunzbourg, J. and Larger, D. 1999. Cogeneration applied to very high efficiency thermal seawater desalination plants. *Desalination*, 125, 203–208.

Delene, J. G. and Ball, S. J. 1971. *Digital Computer Code for Simulating Large Multistage Flash Evaporator Desalting Plant Dynamics*. Oak Ridge, TN: Oak Ridge National Laboratory.

El-Dessouky, H., Alatiqi, I. and Ettouney, H. 1998. Process synthesis: The multi-stage flash desalination system. *Desalination*, 115, 155–179.

El-Dessouky, H. and Bingulac, S. 1996. Solving equations simulating the steady-state behavior of the multi-stage flash desalination process. *Desalination*, 107, 171–193.

El-Dessouky, H., Shaban, H. I. and Al-Ramadan, H. 1995. Steady-state analysis of multi-stage flash desalination process. *Desalination*, 103, 271–287.

El-Dessouky, H. T. and Ettouney, H. M. 2002. *Fundamentals of Salt Water Desalination*. Amsterdam, the Netherlands: Elsevier Science.

El-Dessouky, H. T., Ettouney, H. M. and Al-Roumi, Y. 1999. Multi-stage flash desalination: Present and future outlook. *Chemical Engineering Journal*, 73, 173–190.

El-Dessouky, H. T. and Khalifa, T. A. 1985. Scale formation and its effect on the performance of once through MSF plant. *Desalination*, 55, 199–217.

El Din, A. M. S. and Mohammed, R. A. 1989. The problem of alkaline scale formation from a study on Arabian Gulf water. *Desalination*, 71, 313–324.

IEA-ETSAP and IRENA. 2013. Water Desalination using renewable energy. Technology Policy Brief I12, January. www.etsap.org.

Falcetta, M. F. and Sciubba, E. 1999. Transient simulation of a real multi-stage flashing desalination process. *Desalination*, 122, 263–269.

Freyer, D. and Voigt, W. 2003. Crystallization and phase stability of $CaSO_4$ and $CaSO_4$—based salts. *Monatshefte für Chemie/Chemical Monthly*, 134, 693–719.

Fry, A. and Martin, R. 2005. Water facts and trends. *World Business Council for Sustainable Development*, 16.

Furuki, A., Hamanaka, K., Tatsumoto, M. and Inohara, S. 1985. Automatic control system of MSF process (ACSODE®). *Desalination*, 55, 77–89.
Gambier, A. and Badreddin, E. 2004. Dynamic modelling of MSF plants for automatic control and simulation purposes: A survey. *Desalination*, 166, 191–204.
Gambier, A., Fertig, M. and Badreddin, E. 2002. Hybrid modelling for supervisory control purposes for the brine heater of a multi stage flash desalination plant. In: *Proceedings of the American Control Conference*, Anchorage, AK, pp. 5060–5065.
Gazit, E. and Hasson, D. 1975. Scale deposition from an evaporating falling film. *Desalination*, 17, 339–351.
Genthner, K., Gregorzewski, A. and Seifert, A. 1993. The effects and limitations issued by non-condensible gases in sea water distillers. *Desalination*, 93, 207–234.
Genthner, K. and Seifert, A. 1991. A calculation method for condensers in multi-stage evaporators with non-condensable gases. *Desalination*, 81, 349–366.
Ghaffour, N., Missimer, T. M. and Amy, G. L. 2013. Technical review and evaluation of the economics of water desalination: Current and future challenges for better water supply sustainability. *Desalination*, 309, 197–207.
Glade, H. and Al-Rawajfeh, A. E. 2008. Modeling of CO_2 release and the carbonate system in multiple-effect distillers. *Desalination*, 222, 605–625.
Glade, H., Meyer, J.-H. and Will, S. 2005. The release of CO_2 in MSF and ME distillers and its use for the recarbonation of the distillate: A comparison. *Desalination*, 182, 99–110.
Glade, H. and Ulrich, J. 2003. Influence of solution composition on the formation of crystalline scales. *Chemical Engineering & Technology*, 26, 277–281.
Glueck, A. R. and Bradshaw, R. W. 1970. A mathematical model for a multistage flash distillation plant. In: *Proceedings of the 3rd International Symposium on Fresh Water from the Sea*, Dubrovnik, Croatia, pp. 95–108.
Hamed, O. A. and Al-Otaibi, H. A. 2010. Prospects of operation of MSF desalination plants at high TBT and low antiscalant dosing rate. *Desalination*, 256, 181–189.
Hamed, O. A., Al-Sofi, M. A. K., Mustafa, G. M., Imam, M., Ba-Mardouf, K. and Al-Washmi, H. 2004. Modeling and simulation of multistage flash distillation process. In: *Proceedings of the 4th SWCC Acquired Experience Conference*, Riyadh, Saudi Arabia.
Hasson, D. 1981. Precipitation fouling. In: E. F. C. Somerscale and Knudsen, J. G. (Eds.) *Fouling of Heat Transfer Equipment.* Washington, DC: Hemisphere Publishing Corporation.
Hasson, D., Avriel, M., Resnick, W., Rozenman, T. and Windreich, S. 1968. Mechanism of calcium carbonate scale deposition on heat-transfer surfaces. *Industrial & Engineering Chemistry Fundamentals*, 7, 59–65.
Hawaidi, E. A. M. and Mujtaba, I. M. 2010. Simulation and optimization of MSF desalination process for fixed freshwater demand: Impact of brine heater fouling. *Chemical Engineering Journal*, 165, 545–553.
Hawaidi, E. A. M. and Mujtaba, I. M. 2011. Meeting variable freshwater demand by flexible design and operation of the multistage flash desalination process. *Industrial & Engineering Chemistry Research*, 50, 10604–10614.
Helal, A. M. 2004. Once-through and brine recirculation MSF design—A comparative study. *Desalination*, 171, 33–60.
Helal, A. M., El-Nashar, A. M., Al-Katheeri, E. and Al-Malek, S. 2003. Optimal design of hybrid RO/MSF desalination plants Part I: Modeling and algorithms. *Desalination*, 154, 43–66.
Helal, A. M., Medani, M. S., Soliman, M. A. and Flower, J. R. 1986. A tridiagonal matrix model for multistage flash desalination plants. *Computers & Chemical Engineering*, 10, 327–342.
Helal, A. M. and Odeh, M. 2004. The once-through MSF design. Feasibility for future large capacity desalination plants. *Desalination*, 166, 25–39.
Husain, A., Hassan, A., Al-Gobaisi, D. M. K., Al-Radif, A., Woldai, A. and Sommariva, C. 1993. Modelling, simulation, optimization and control of multistage flashing (MSF) desalination plants Part I: Modelling and simulation. *Desalination*, 92, 21–41.
Husain, A., Woldai, A., Alradif, A., Kesou, A., Borsani, R., Sultan, H. and Deshpandey, P. B. 1994. Modeling and simulation of a multistage flash (MSF) desalination plant. *Desalination*, 97, 555–586.
Ismail, A. 1998. Fuzzy model reference learning control of multi-stage flash desalination plants. *Desalination*, 116, 157–164.
Ismail, A. and Abu-Khousa, E. 1996. Fuzzy TBT control of multi-stage flash desalination plants. In: *Proceedings of the 1996 IEEE International Conference on Control Applications*, September 15–18, New York, pp. 241–246.

Jamshidi, M., Akbarzadeh, M.-R. and Kumbla, K. 1996. Design and implementation of fuzzy controllers for complex systems–Case study: A water desalination plant. In: Soto, R., Sanchez, J. M., Campbell, M. and Francisco, J. (Eds.) *Proceedings of ISAI / IFIS 1996 Conference*. November, 12–15. Cantu, Cancun, Mexico, pp. 418–423.

Kern, D. Q. and Seaton, R. E. 1959. Surface fouling: How to calculate limits. *Chemical Engineering Progress*, 55, 71–73.

Krause, S. 1993. Fouling of heat transfer surfaces by crystallization and sedimentation. *International Chemical Engineering*, 33, 355–401.

Lappalainen, J., Korvola, T. and Alopaeus, V. 2017. Modelling and dynamic simulation of a large MSF plant using local phase equilibrium and simultaneous mass, momentum, and energy solver. *Computers & Chemical Engineering*, 97, 242–258.

Lior, N., El-Nashar, A. and Sommariva, C. 2012. Advanced instrumentation, measurement, control, and automation (IMCA) in multistage flash (MSF) and reverse-osmosis (RO) water desalination. *Advances in Water Desalination*, Ed. N. Lior, 453–658. Wiley, USA.

Malayeri, M. R. and Müller-Steinhagen, H. 2007. An overview of fouling mechanisms, prediction and mitigation strategies for thermal desalination plants. *Eleventh International Water Technology Conference, IWTC11 2007*, Sharm El-Sheikh, Egypt.

Mandil, M. A. and Ghafour, E. E. A. 1970. Optimization of multi-stage flash evaporation plants. *Chemical Engineering Science*, 25, 611–621.

Maniar, V. M. and Deshpande, P. B. 1996. Advanced controls for multi-stage flash (MSF) desalination plant optimization. *Journal of Process Control*, 6, 49–66.

Marshall, W. L. and Slusher, R. 1968. Aqueous systems at high temperature. Solubility to 200 degree of calcium sulfate and its hydrates in sea water and saline water concentrates, and temperature-concentration limits. *Journal of Chemical & Engineering Data*, 13, 83–93.

Mayere, A. 2011. Solar powered desalination. PhD thesis, University of Nottingham, Nottingham, UK.

Mazzotti, M., Rosso, M., Beltramini, A. and Morbidelli, M. 2000. Dynamic modeling of multistage flash desalination plants. *Desalination*, 127, 207–218.

Mubarak, A. 1998. A kinetic model for scale formation in MSF desalination plants. Effect of antiscalants. *Desalination*, 120, 33–39.

Müller-Steinhagen, H. M. and Branch, C. A. 1988. Influence of thermal boundary conditions on calcium carbonate fouling in double pipe heat exchangers: Einfluβ der thermischen Randbedingungen auf die Ablagerung von $CaCO_3$ in Doppelrohrwärmeübertragern. *Chemical Engineering and Processing: Process Intensification*, 24, 65–73.

Mussati, S. F., Aguirre, P. A. and Scenna, N. J. 2003. Novel configuration for a multistage flash-mixer desalination system. *Industrial & Engineering Chemistry Research*, 42, 4828–4839.

Mwaba, M. G., Golriz, M. R. and Gu, J. 2006. A semi-empirical correlation for crystallization fouling on heat exchange surfaces. *Applied Thermal Engineering*, 26, 440–447.

Najibi, S. H., Müller-Steinhagen, H. and Jamialahmadi, M. 1997. Calcium sulphate scale formation during subcooled flow boiling. *Chemical Engineering Science*, 52, 1265–1284.

Olafsson, T. F., Jamshidi, M., Titli, A. and Al-Gobaisi, D. 1999. Fuzzy control of brine heater process in water desalination plants. *Intelligent Automation & Soft Computing*, 5, 111–128.

Omar, A. M. 1983. Simulation of MSF desalination plants. *Desalination*, 45, 65–76.

Partridge, E. P. and White, A. H. 1929. The solubility of calcium sulfate from 0 to 200. *Journal of the American Chemical Society*, 51, 360–370.

Porrazzo, R., Cipollina, A., Galluzzo, M. and Micale, G. 2013. A neural network-based optimizing control system for a seawater-desalination solar-powered membrane distillation unit. *Computers & Chemical Engineering*, 54, 79–96.

Rautenbach, R. and Buchel, H. G. 1979. Modular program for design and simulation of desalination plants. *Desalination*, 31, 85.

Reddy, K. V., Husain, A., Woldai, A. and Al-Gopaisi, D. M. K. 1995a. Dynamic modelling of the MSF desalination process. In: *Proceedings of the IDA and WRPC World Congress on Desalination and Water Treatment*, Abu Dhabi, United Arab Emirates, pp. 227–242.

Reddy, K. V., Husain, A., Woldai, A., Nabi, S. M. and Kurdali, A. 1995b. Holdup and interstage orifice flow model for an MSF desalination plant. *Proceedings of the IDA World Congress on Desalination and Water Sciences*, Abu Dhabi, United Arab Emirates, pp. 179–197.

Rimawi, M. A., Ettouney, H. M. and Aly, G. S. 1989. Transient model of multistage flash desalination. *Desalination*, 74, 327–338.

Rosso, M., Beltramini, A., Mazzotti, M. and Morbidelli, M. 1997. Modeling multistage flash desalination plants. *Desalination*, 108, 365–374.

Said, S., Mujtaba, I. M. and Emtir, M. 2012. Effect of fouling factors on the optimisation of MSF desalination process for fixed water demand using gPROMS. In: *Proceeding of the 9th International Conference on Computational Management*. April 18–20, London, UK.

Said, S. A., Mujtaba, I. M. and Emtir, M. 2010. Modelling and simulation of the effect of non-condensable gases on heat transfer in the MSF desalination plants using gPROMS software. *Computer Aided Chemical Engineering*, 28, 25–30.

Segev, R., Hasson, D. and Semiat, R. 2012. Rigorous modeling of the kinetics of calcium carbonate deposit formation. *AIChE Journal*, 58, 1222–1229.

Seifert, A. and Genthner, K. 1991. A model for stagewise calculation of non-condensable gases in multi-stage evaporators. *Desalination*, 81, 333–347.

Semiat, R. and Galperin, Y. 2001. Effect of non-condensable gases on heat transfer in the tower MED seawater desalination plant. *Desalination*, 140, 27–46.

Shams El Din, A. M., El-Dahshan, M. E. and Mohammed, R. A. 2005. Scale formation in flash chambers of high-temperature MSF distillers. *Desalination*, 177, 241–258.

Shams El Din, A. M. and Mohammed, R. A. 1994. Brine and scale chemistry in MSF distillers. *Desalination*, 99, 73–111.

Sharif, A. O. 2016. Will reverse osmosis replace thermal desalination in GCC region. In: *Qatar Foundation Annual Research Conference Proceedings 2016: EEPP2725*, Doha, Qatar: HBKU Press.

Shivayyanamath, S. and Tewari, P. K. 2003. Simulation of start-up characteristics of multi-stage flash desalination plants. *Desalination*, 155, 277–286.

Soliman, M. A. 1981. A mathematical model for multistage flash desalination plants. *Journal of Engineering Sciences*, 7, 143–150.

Sowgath, M. T. 2007. *Neural Network based Hybrid Modelling and MINLP based Optimisation of MSF Desalination Process within gPROMS*. PhD Thesis, The University of Bradford.

Sowgath, M. T. and Mujtaba, I. M. 2015. Meeting the Fixed Water Demand of MSF Desalination using Scheduling in gPROMS. *Chemical Engineering Transactions*, 45, 6.

Taborek, J., Aoki, T., Ritter, R. B., Palen, J. W. and Knudsen, J. G. 1972. Predictive methods for fouling behavior. *Chemical Engineering Progress*, 68(7), 69–78.

Tanvir, M. S. and Mujtaba, I. M. 2006a. Modelling and simulation of MSF desalination process using gPROMS and neural network based physical property correlation. *Computer Aided Chemical Engineering*, 21, 315.

Tanvir, M. S. and Mujtaba, I. M. 2006b. Neural network based correlations for estimating temperature elevation for seawater in MSF desalination process. *Desalination*, 195, 251–272.

Tarifa, E. E., Humana, D., Franco, S. and Scenna, N. J. 2004. A new method to process algebraic equation systems used to model a MSF desalination plant. *Desalination*, 166, 113–121.

Tarifa, E. E. and Scenna, N. J. 2001. A dynamic simulator for MSF plants. *Desalination*, 138, 349–364.

Tayyebi, S. and Alishiri, M. 2014. The control of MSF desalination plants based on inverse model control by neural network. *Desalination*, 333, 92–100.

Thomas, P. J., Bhattacharyya, S., Patra, A. and Rao, G. P. 1998. Steady state and dynamic simulation of multi-stage flash desalination plants: A case study. *Computers & Chemical Engineering*, 22, 1515–1529.

Tiwari, G. N., Singh, H. N. and Tripathi, R. 2003. Present status of solar distillation. *Solar Energy*, 75, 367–373.

UN 2015. *World Population Prospects: The 2015 Revision, Key Findings and Advance Tables*. New York: United Nations.

UNEP. 2001. Desalination by distillation. Available: http://www.unep.or.jp/ietc/publications/techpublications/techpub-8c/distill.asp (Accessed February 1).

Wildebrand, C., Glade, H., Will, S., Essig, M., Rieger, J., Büchner, K.-H. and Brodt, G. 2007. Effects of process parameters and anti-scalants on scale formation in horizontal tube falling film evaporators. *Desalination*, 204, 448–463.

Yokoyama, K., Ikenaga, Y., Inoue, S. and Yamamoto, T. 1977. Analysis of start-up characteristics of commercial MSF plant. *Desalination*, 22, 395–401.

Zhang, F., Xiao, J. and Chen, X. D. 2015. Towards predictive modeling of crystallization fouling: A pseudo-dynamic approach. *Food and Bioproducts Processing*, 93, 188–196.

Zhao, X. and Chen, X. D. 2013. A critical review of basic crystallography to salt crystallization fouling in heat exchangers. *Heat Transfer Engineering*, 34, 719–732.

Zhong, S. and Mucci, A. 1989. Calcite and aragonite precipitation from seawater solutions of various salinities: Precipitation rates and overgrowth compositions. *Chemical Geology*, 78, 283–299.

Section III

Wastewater Treatment

Membrane and Polymer Based Process

9 Modeling of Pore-Blocking Behaviors of Low-Pressure Membranes during Constant-Pressure Filtration of an Agro-Industrial Wastewater

Mutiu Kolade Amosa, Mohammed Saedi Jami, Ma'an Fahmi Alkhatib, Thokozani Majozi, Adewale George Adeniyi, Fatai Alade Aderibigbe, and Sulyman Age Abdulkareem

CONTENTS

9.1 INTRODUCTION

Despite the widely documented excellent purification capacity of membranes, their main drawback—fouling—is still being extensively researched with a view to finding a sustainable solution. Fouling simply implies the process that results in the performance loss of a membrane due to the deposition of suspended or dissolved matters onto its external surface or the internal pore walls [1]. Fouling eventually leads to a reduction in the active area of the membrane and thereby results in a reduction in flux below the theoretical capacity of the membrane. Fouling or pore-blocking has been identified as the main reason limiting the adoption of membrane purification processes by many industries.

Consequently, an apt understanding of the pore-blocking mechanisms of membranes is imperative, as it is a pertinent factor dictating the overall performance of the filtration process. Pore blockage can occur in any of the two commonly known membrane operations: constant-pressure and constant-flux rate. In a constant-pressure operation, pore blockage usually leads to a sharp decline in permeate flux, while a severe pressure rise is usually encountered in a constant-flux rate operation. In principle, governing filtration models can facilitate the design of membrane processes more than any experiment or characterization can, yet data from experiments are usually required for validation purposes [2–5].

To properly control particulate fouling at the design stage, as well as appropriately monitor it during a plant operation, the methods utilized in evaluating the particulate content of feed-water in predicting membrane fouling are crucial. Soluble and colloidal materials are assumed to be responsible for membrane pore blockage, while suspended solids are mainly accountable for the cake layer resistance [4,6,7]. To accurately measure and predict particulate fouling, it is recommended that specific fouling mechanisms/indices be investigated with respect to specific membranes since the Modified Fouling Index (MFI), where a 0.45 μm membrane filter is used and usually represented as $MFI_{0.45}$, cannot represent all membrane types. This is due to the fact that some principal parameters such as retention of smaller particulates, the nature and concentration of solutes and solvents, proof of cake filtration, pore size distribution, surface morphology, module hydrodynamics and membrane type/material must be considered in such investigations [1,8].

Therefore, in any proposed membrane process with plans for sustainability, pore-blocking modeling is germane for the determination of some key factors necessary for the design of an efficient membrane system. These factors are: (1) the description of the extent of membrane fouling in terms of particle accumulation at the membrane surface or inside the membrane pores; (2) the prediction of the fouling potential of a specific feed with respect to a specific membrane; and (3) the identification of the most appropriate and sustainable cleaning method necessary for the membrane process.

In this study, a systemic investigation was carried out on high-strength agro-industrial wastewater to describe the successive steps involved in the flux decline of a membrane filtration process in terms of pore-blocking mechanism. Microfiltration (MF) and ultrafiltration (UF) membranes, which are the popular low-pressure membranes (LPMs), were utilized for the filtration of the high-strength wastewater. The wastewater is specifically a discharge of an end-of-pipe treatment process from the agro-industry palm oil milling process. An upstream adsorption process was applied to lower the feed strength and reduce its fouling effects on the membranes. The investigative experiments were conducted in a constant-pressure and cross-flow filtration mode through polyethersulfone (PES) MF (pore sizes: 0.1 and 0.2 μm) and UF (molecular weight cut-off: 1, 5 and 10 kDa) membranes at the transmembrane pressures of 40, 80 and 120 kPa. The examined results within the frame of the common blocking mechanisms revealed that the blocking index, η, decreased from 2 to 0 in all five membranes. The pore-blocking phenomenon was successively observed from the complete blocking mechanism (i.e., $\eta = 2$) down to the cake filtration mechanism (i.e., $\eta = 0$). Furthermore, there is an indication that the early blockage of the pores and formation of a cake resulted in a limiting cake height evident from the near-constant trend of the permeate flux. This means that cake filtration could be best used to explain the fouling mechanisms of the feed on the LPMs based on the coefficient of determination (R^2) values at all applied pressures. This further demonstrates that the fouling is primarily caused by the gradual reversible cake deposition, which could be easily removed by less onerous cleaning methods.

9.2 MECHANISM OF MEMBRANE FOULING OR PORE-BLOCKING

The fouling tendency, usually represented by the silt density index (SDI) and MFI, involves constant-pressure/constant-rate membrane filtration tests, so the indices are calculated from the experimentally determined relationship between filtration times and cumulative permeate volumes [8,9]. In both tests, feed-water is filtered through a 0.45 μm MF membrane in a dead-end flow at constant pressure. However, there are arguments that SDI cannot be used for mathematical modeling in predicting the flux decline due to particulate fouling, since it is not based on a distinction between the blocking and cake filtration mechanisms. On the contrary, MFI is based solely on the cake filtration mechanism and is dependent on particle size through the Kozeny–Carman equation for specific cake resistance [8,10,11].

To accurately measure and predict particulate fouling, it has been proposed that specific MFI be investigated with respect to specific membranes since $MFI_{0.45}$ cannot represent all membrane types. This is due to the fact that principal factors such as the retention of smaller particulates, proof of cake filtration, membrane pore size, surface morphology and material must also be considered in such investigations [8].

9.2.1 Blocking Filtration Models

Blocking phenomena studies, especially the complete and standard blocking laws [8], was pioneered in 1935 as reported by Hermans, Bredée [12] and Bowen et al. [13]. Nevertheless, the reformulation of the four blocking mechanisms in a common frame of power-law non-Newtonian fluids was revised by Hermia [14]. The four simplified fouling models have been used for evaluating the fouling of membranes by various types of water chemistries and matrices with multifarious contents [9]. Equation 9.1 is the general formula for all the blocking models [8,9,11] and Figure 9.1 depicts the mechanisms of each respective blocking model:

$$\frac{d^2t}{dV^2} = k_b \left(\frac{dt}{dV}\right)^{\eta} \tag{9.1}$$

where k_b and η are dimensionless numbers in relation to the particular pore-blocking phenomena. t and V are, respectively, time and total volume permeated through the membrane. Cake filtration, intermediate blocking, standard blocking and complete blocking filtration models will have respective characteristic η values of 0, 1, 1.5 and 2, as will be briefly reviewed in the subsequent sub-sections.

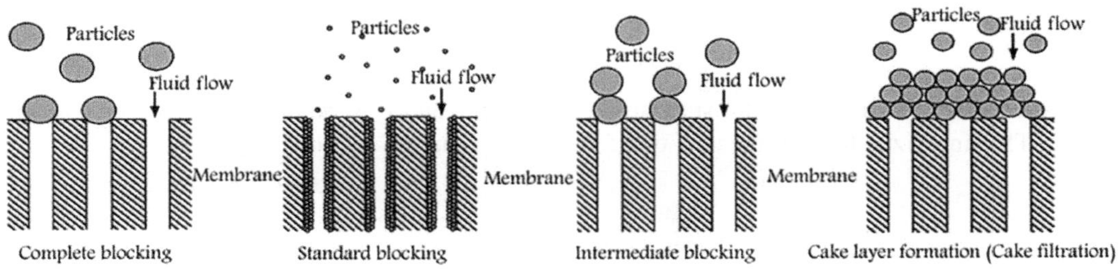

FIGURE 9.1 Pore-blocking filtration mechanisms.

9.2.1.1 Complete Pore-Blocking

The shutting of membrane pores by particles reaching the membrane indicates complete pore blocking. It is idealized that none of the particles are superimposed upon other particles during the filtration (devoid of any cake layer formation), or on top of the solid area of the membrane surface between the pores [9,13]. The blocked surface area is proportional to the filtrate volume in this type of pore blocking. This idealized scenario requires that all particles accumulate wholly on the membrane pores, in that all the participating particle sizes are larger than the membrane pores [10]. The mass transfer is affected by the reduction of the active membrane area, except with an enhanced feed velocity in which case the permeate flow rate might be increased by the increasing transmembrane driving force. Huang et al. [9] suggested that this is an unlikely scenario in membrane filtration, as model simulation reports and experimental evidence suggest that deposition of particles does not exclusively occur on the membrane pores, especially when LPMs are considered. The complete blocking phenomenon for filtration processes conducted at constant pressure is given by the characteristic equation in Equation 9.2 [13]:

$$\frac{d^2t}{dV^2} = k_{cb}\left(\frac{dt}{dV}\right)^2 \tag{9.2}$$

k_{cb} denotes the filtration constant for complete blocking and t is the cumulative filtration time. The linearized model follows the mathematical relation in Equation 9.3 [10,15]:

$$Q = Q_i - k_{cb}V \tag{9.3}$$

where:

Q is the volumetric permeate flow rate (L/h)
Q_i is the initial volumetric permeate flow rate (L/h)
V represents the permeate volume measured in liters.

9.2.1.2 Standard or Internal Pore-Blocking

This scenario involves the attachment of small particles to the internal walls of the pores, which eventually results in a narrowing of the membrane pores. The pore volume decreases as the permeate volume passing through the membrane pores increases, and the pore diameter is enormously larger than the particle diameter, which are thereby deposited on the pore walls and reduce the pore volume through the plugging of the pores. This results in an increasing membrane resistance due to the pore size reduction, with an effect on the mass transfer. Furthermore, this type of pore blocking is independent from feed velocity and no limiting might be observed [1]. Equation 9.4 is given as the characteristic equation that describes the standard blocking phenomenon [13]:

$$\frac{d^2t}{dV^2} = \frac{2k_{sb}}{\sqrt{J_v(0)}}\left(\frac{dt}{dV}\right)^{3/2} \tag{9.4}$$

k_{sb} represents the model constant for standard pore blocking and J_v is the measurement for volumetric flow. The linearized form of the equation follows as given in Equation 9.5 [10,15]:

$$\sqrt{Q} = \sqrt{Q_i} - \left(k_{sb}\frac{V}{2}\sqrt{Q_i}\right) \tag{9.5}$$

9.2.1.3 Intermediate or Particle Pore-Blocking

Here, it is assumed that each particle reaching the membrane surface may not only block pores in the way complete blocking does, but may also attach to other particles on the membrane surface. This postulates the occurrence of competition between the membrane pores and the existing particles on the membrane surface for the approaching particles, thus reducing the actual number of particles that can block the pores. This indicates that any particles reaching a pore might seal it over time and particles might bridge a pore and not block it completely. The effect on the mass transfer is also the reduction of active membrane area, which is similar to the complete pore blocking scenario but not so severe. Although the permeate volume is proportional to the blocked area, the blocking is less restrictive in that not every particle necessarily blocks the membrane pore. In this fouling type, membrane–contaminant attraction is favored, thereby leading to irreversible fouling as it is more resistant to the hydrodynamic forces applied during cleaning. Therefore, the characteristic equation is deemed as given in Equation 9.6 [13]:

$$\frac{d^2t}{dV^2} = \frac{k_{ib}}{J_v(0)}\left(\frac{dt}{dV}\right) \tag{9.6}$$

where k_{ib} represents the constant for intermediate pore-blocking law. The linearized version of the model is as follows in Equation 9.7 [10,15]:

$$\frac{1}{Q} = k_{ib}t + \frac{1}{Q_i} \tag{9.7}$$

9.2.1.4 Cake Layer Formation (Cake Filtration)

In this scenario, the formation of a cake on the membrane surface of particles that do not enter the pores is assumed [1]. The formation of a cake layer outside the external membrane surface is due to an increase in the hydraulic resistance encountered during membrane filtration. With cake filtration, it is assumed that the fouling type does not affect the membrane pore structure [9]. In terms of mass transfer effects, the overall resistance of the process becomes the resistance of the membrane plus the resistance of the cake formed. This is the universal type of membrane fouling in water treatment, as it does not depend upon the presence of favorable membrane–contaminant attractions that lead to reversible fouling, in contrast to intermediate pore blocking model [9]. The characteristic equation of the cake filtration process can be represented as shown in Equation 9.8 [13]:

$$\frac{d^2t}{dV^2} = \frac{k_{cf}}{2J_v^2(0)}\left(\frac{dt}{dV}\right)^0 \tag{9.8}$$

where k_{cf} represents the constant for cake filtration law; the simplified model follows the mathematical relation in Equation 9.9 [10,15]:

$$\frac{1}{Q} = \frac{1}{Q_i} + k_{cf}V \tag{9.9}$$

Since the wastewater employed for the pore-blocking study has a relatively uniform viscosity, a Newtonian behavior of the fluid was assumed. As such, filtration equations for constant-pressure filtration of Newtonian fluid-solid mixtures (as explained earlier) were fully considered.

9.3 MATERIALS AND METHODS

9.3.1 Unit and Membrane

A hybrid experimental set-up was employed in this study and its process schematic is shown in Figure 9.2 [7]. The hybrid system consists of a feedwater sedimentation tank, adsorption reactor, a micron filter, a peristaltic pump, a membrane module and a permeate tank. The geometrical configurations and rationale for each unit has been fully discussed elsewhere [5,7,16].

Five Sartocon® slice membrane cassettes made of PES material with different pore sizes and/or molecular weight cut-offs (MWCO) were investigated. Each of the membrane cassettes has an effective surface area of 0.1 m^2. The full characteristic details of the membranes are given in Table 9.1.

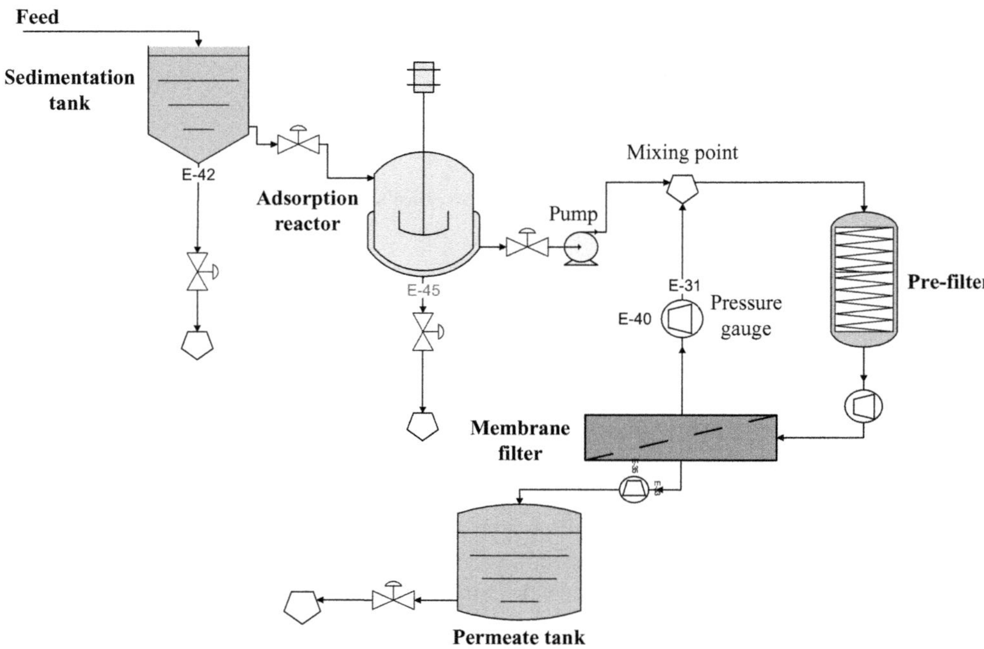

FIGURE 9.2 The bench-scale set-up for wastewater treatment. (Adapted from Amosa, M.K. et al., *Environ. Sci. Pollut. Res.*, 23, 22554–22567, 2016 [17].)

TABLE 9.1
Physical Characteristics of the LPMs

Characteristic Items	
Membrane material	Polyethersulfone (PES)
Membrane type	Cassette
Membrane brand name	Sartocon® slice cassettes
Effective filtration area (m^2)	0.1
MWCO of UF membranes (kDa)	1 (3051460901E-SG), 5 (3051462901E-SG), 10 (3051463901E-SG)
Pores of the MF membranes (μm)	0.1 (3051545801W-SG), 0.2 (3051860701W-SG)

TABLE 9.2
Feedwater Strength Before and After the Upstream Adsorption

Constituents	Raw Feed	Adsorbed Feed
Suspended solids (SS), mg/L	284	51
Silica, mg/L	58	20
Turbidity, NTU	840	53
Total dissolved solids (TDS), mg/L	970	820
Total alkalinity, mg/L $CaCO_3$	1860	880
Total hardness, mg/L $CaCO_3$	1680	440
Manganese as Mn, mg/L	2.14	0.1
Iron as Fe, mg/L	ND	ND
Chemical oxygen demand (COD), mg/L	1387	198
Sulfide as H_2S, mg/L	0.6	0.05
pH, units	8.56	9.2

ND: Not detected.

9.3.2 Feedwater Preparation

The raw feed quality was analyzed, and the concentrations of the contents were reduced with an upstream adsorption process before the filtration process. Powdered activated carbon (PAC) characterized by a high surface area of 886.2 m^2/g was employed in the adsorption process. A detailed characterization of the PAC and how the adsorption process proceeded have been reported elsewhere [18,19]. The strength of the feedwater, in terms of the concentrations of many of the pollutants, both before and after upstream adsorption, is given in Table 9.2 [5,7].

9.4 FILTRATION AND MEASUREMENTS

Prior to the feedwater filtration, a clean water filtration was conducted on each membrane with deionized water and the permeate volume (V) was recorded with filtration time (t) at the different transmembrane pressures (TMP) of 40, 80 and 120 kPa. Determination of the clean water flux rate allows the confirmation of the ability of the cleaning cycle to restore the filtration rate (flux) of the system. The flux-time data at each TMP were recorded and the initial flux capacity of each membrane was measured. These fluxes were applied as the initial permeate flux at $t = 0$ in the subsequent filtration of the feedwater, and were employed as the reference points for necessary model calculations. All the filtration experiments proceeded until there was no further change in flux (steady state) and this was achieved after 1 h filtration time. All fluxes in this study were calculated using Equation 9.10:

$$J = \frac{V}{A_m \times t} = \frac{Q_p}{A_m} \tag{9.10}$$

where:

J is the measured flux in $Lm^{-2}h^{-1}$
V is the volume of filtrate filtered in liters (L)
Q_p is the filtrate flow rate into the membrane in L/h
A_m is the active surface area of the membrane in m^2
t represents the filtration time in h.

The membrane filtration was conducted in cross-flow mode on all the LPMs at the applied constant pressures (TMPs) of 40, 80 and 120 kPa. The selected TMP range as applied in this study is well within the acceptable range for the two LPMs, together with the pore sizes, which were selected based on the reports from previous studies [8,20–27]. The permeate volume *V* at different filtration time *t* data were recorded at each TMP, fluxes were calculated, and the flux–time plots were employed in evaluating the fouling propensity of the feedwater [5,7]. Since the focus of the current study is to analyze the fouling or pore-blocking phenomena of the membranes with respect to the feedwater, the permeate quality from each membrane was not analyzed.

9.4.1 Analysis of Filtration Phenomena

The flux–time plots of the filtration experiments do not fully represent the fouling tendencies of the feed on the membranes. The fouling mechanism of the feed on the membranes were further evaluated using Hermia's revised blocking filtration laws [14] for complete blocking, standard blocking, intermediate blocking and cake filtration models at each TMP of 40, 80 and 120 kPa. The pore-blocking models given in Equations 9.1 through 9.9 were applied in modeling the description and quantification of blocking mechanisms controlling the membrane processes.

9.4.2 Membrane Cleaning

The membranes were cleaned after each filtration run to prevent a sediment layer caused by fouling from building up on the membranes, as well as ensuring a fair pore-blocking trend comparison among the membranes at each operating condition. A 1 N sodium hydroxide solution at a temperature of 50°C was utilized for the cleaning process, strictly adopting the manufacturer's cleaning guidelines [28]. As for the UF membranes, the cleaning solution was recirculated with the permeate valves open for 30 min with 200 and 0 kPa pressures at inlet and retentate points, respectively. The MF membranes were cleaned with the same cleaning solution as that of UF membranes, but with the permeate valves closed for 30 min. All the cleaning procedures were repeated twice for efficient cleaning. All the membranes were flushed with distilled water after the cleaning process. The clean water flux of the cleaned membranes was measured to be certain of proper recovery of the membranes from their fouled condition.

9.5 RESULTS AND DISCUSSION

9.5.1 Operation of the Hybrid System

The hybrid adsorption-membrane system was operated by first conducting the clean water flux (CWF) tests to ascertain the level of fouling in the LPMs when subjected to wastewater. The CWF revealed the flux capacities of all the LPMs at the tested TMPs. All the fluxes (J) are measured in $Lm^{-2}h^{-1}$. These were employed as the respective initial fluxes for similar operating conditions during the wastewater filtration.

Thereafter, the wastewater (10 L) was discharged into an adsorption reactor of 15 L volume capacity, followed by the membrane filtration of the adsorbed wastewater as described in our previous reports [5,7]. The fouling nature of the adsorbed wastewater was studied and modeled using Hermia's modified blocking mechanisms. Relatively constant fluxes were obtained at each applied pressure in relation to filtration time. This suggests that the distilled water employed in this study is devoid of particles that might have led to blockage of the membrane pores. Additionally, there was an observation that the permeate volume of water increased with the TMP. Relatively similar effects were observed from the CWF tests of all the LPMs.

Moreover, it was also observed that the permeate fluxes increased with increase in the membrane pore size. As for the MF membranes, 0.2 μm gave higher permeate fluxes (343, 445 and

532 $Lm^{-2}h^{-1}$) at the pressures of 40, 80 and 120 kPa when compared with the fluxes generated (230, 305, and 379 $Lm^{-2}h^{-1}$) by the 0.1 μm membrane. Similar trends were observed for the UF membranes as the membrane with highest MWCO (10 kDa) resulted in fluxes as high as 61 $Lm^{-2}h^{-1}$ at the highest TMP (120 kPa), while the lowest MWCO (1 kDa) generated fluxes as low as 5 $Lm^{-2}h^{-1}$ at the same pressure. The reason for this could be explained by the relationship between the permeate flux and the membrane resistance. As the pore of a membrane increases at a specified pressure, the resistance of the membrane decreases, and this leads to an increase in the permeate flux. This inverse proportionality between the resistance and flux is mathematically expressed in Equation 9.11:

$$J = \frac{\Delta p}{\mu_0 R} \tag{9.11}$$

where J, μ_0, and R respectively represent the flux or feed viscosity and the resistance of the clean membrane with respect to the CWF data [15,29].

9.5.2 Membrane Productivity and Performance

An investigation of the effect of TMP on the permeate flux and the fouling tendency of the feed was conducted on the LPMs. The plots of flux–time data for the MF membranes are depicted in Figures 9.3 and 9.4.

It can be observed from the two plots that higher TMPs generated higher fluxes for the whole filtration time of 60 min. It was also observed that the permeation flux (J) decreased sharply with time at the onset of the filtration, but the flux gradually attained stability and was nearly constant at longer filtration times, especially after 20 min of filtration. The flux declined from the initial flux by 50% for 0.2 μm MF and 40% for 0.1 μm MF membranes within the first 10 min of filtration. Similar trends were observed at all the TMPs, indicating that the feedwater still contained some constituents. These constituents may have accumulated on the membranes, resulting in fouling. Relatively similar flux–time fouling trends were observed in previous flux decline investigations [30,31].

The ability of the membranes to tend towards being stable at longer filtration times indicates that the membranes can still perform with continuous fluxing despite the fouling tendency of the feedwater. This is an indication that the fouling constituents of the feedwater may not completely foul

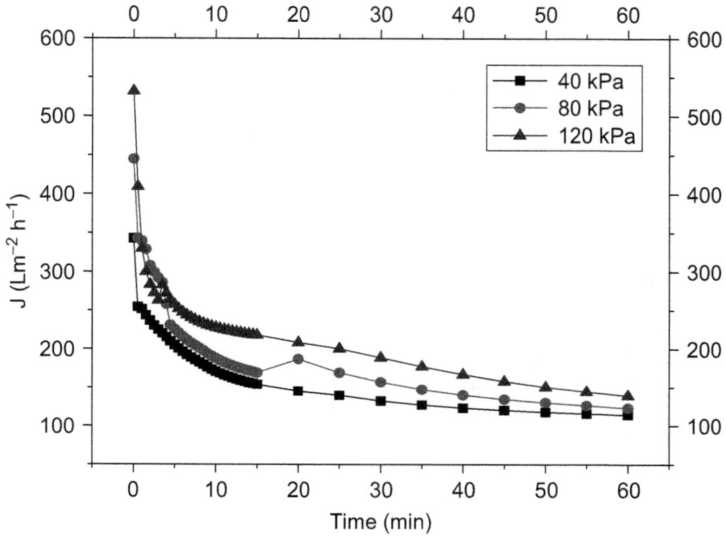

FIGURE 9.3 TMP influence on the flux of 0.2 μm MF membrane.

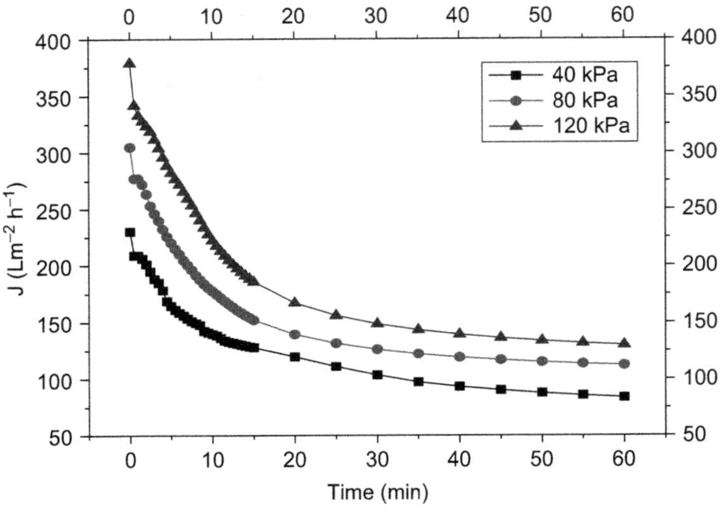

FIGURE 9.4 TMP influence on the flux of 0.1 μm MF membrane.

the pores of the MF membranes. The stability of the 0.2 μm MF membrane was more established than that of 0.1 μm MF membrane. This indicates that the 0.1 μm MF membrane experienced more accumulation of foulants with time compared to the 0.2 μm MF membrane, which may allow the passage of some of these foulants without much accumulation. These results are in agreement with previous studies [8,15].

Comparatively, the UF membranes have smaller pore sizes usually characterized by their MWCO, hence, lower fluxes are expected with feedwater of similar chemistry and applied pressure. The flux-time plots of the 10, 5 and 1 kDa UF membrane filtration data are presented in Figures 9.5 through 9.7, respectively.

Despite being subjected to feedwater of similar chemistry, the UF membranes were more affected by the foulants contained in the wastewater. This is evident from their very low fluxes in contrast to

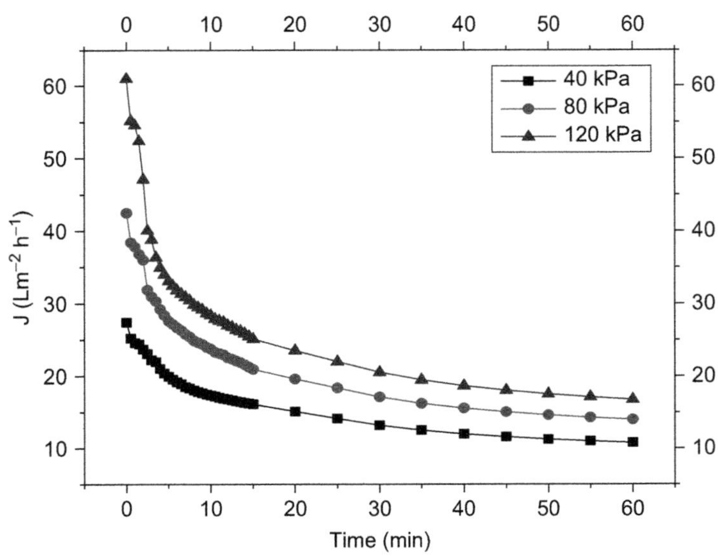

FIGURE 9.5 TMP influence on the flux of 10 kDa UF membrane.

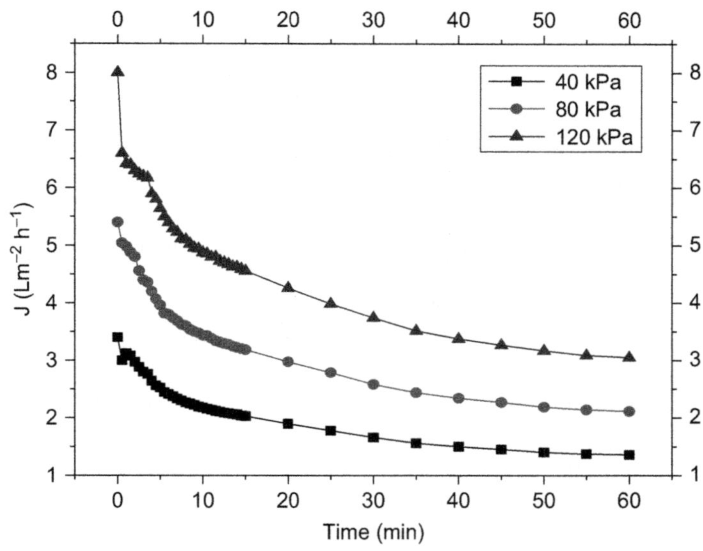

FIGURE 9.6 TMP influence on the flux of 5 kDa UF membrane.

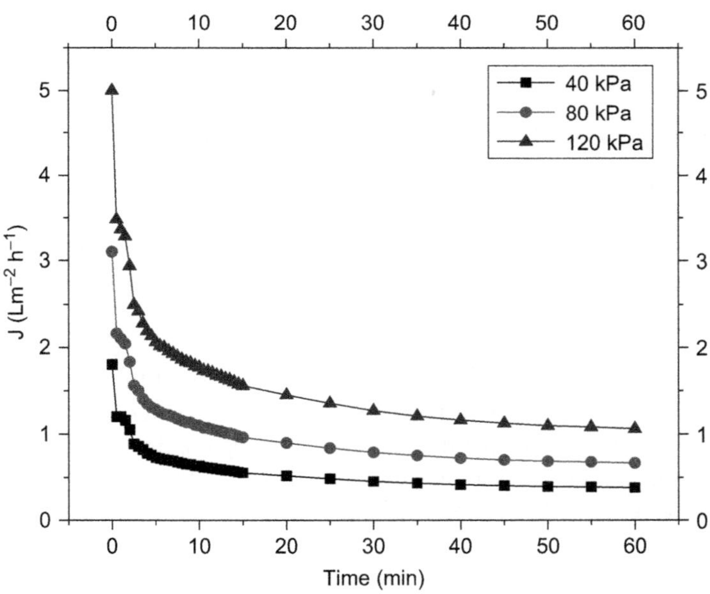

FIGURE 9.7 TMP influence on the flux of 1 kDa UF membrane.

the MF membranes. The reason for this may be attributed to the UF membranes' smaller pore sizes, which led to their higher resistances and retention capacities [21].

Furthermore, within the initial 5 min filtration time, the 1 kDa UF membrane exhibited a sharp flux decline of 59% (1.8–0.732 Lm^{-2}h^{-1}), 59% (3.1–1.284 Lm^{-2}h^{-1}) and 59% (5–2.064) at TMPs of 40, 80, and 120 kPa. However, the flux started tending towards sustainability after 10 min throughout the remaining filtration times. Similar observations were recorded in the cases of the 5 and 10 kDa membranes. The 5 kDa membrane witnessed a sharp flux decline during the first 5 min of filtration when it exhibited a 26% (3.4–2.52 Lm^{-2}h^{-1}), 27% (5.4–3.96 Lm^{-2}h^{-1}) and 30% (8–5.64 Lm^{-2}h^{-1}) reduction at TMPs of 40, 80 and 120 kPa. In the case of the 10 kDa UF membrane, the flux reductions at the initial 5 min of filtration were 27% (27.4–19.92 $Lm^{-2}h^{-1}$), 35% (42.5–27.6 $Lm^{-2}h^{-1}$) and 46% (61–33.12 $Lm^{-2}h^{-1}$) at the same TMPs as applied to the 1 and 5 kDa UF membranes.

It was observed that fluxes of the 1 and 5 kDa UF membranes were relatively more stable than the 10 kDa, as evident from their percent flux decline over time. The high initial flux indicates that the membrane performance was devoid of foulant deposition on the membrane surface [29,32–35]. Declining flux immediately after the initial flux stage suggests that foulants and particulate matter were deposited onto the membrane surface and this led to blockage of the membrane pores, thereby increasing the resistance of the flow of feedwater through the membrane [26].

Generally, it was observed that all the fluxes for the LPMs were pressure dependent because higher TMPs always brought about higher fluxes before realizing the stable or sustainable flux stages. The reason for this is that more colloids are brought onto the membrane surface in a specified filtration time, which leads to rapid growth of cake layers as has been observed in previous studies [30,31]. Though higher pressures did not always seem to result in an improved flux at steady states of filtration, a longer filtration time is a necessity in attaining stable and sustainable fluxes at high TMPs. In all the filtration experiments, it was observed that the feedwater indeed contained some particulates that were larger than the membrane pores as well as those that were smaller or had nearly the same size as the membrane pores. This is evident from the flux decline exhibited by all the membrane types due to the existence of particulates of a wide range of sizes and characteristics from the feedwater, which was complex. Lastly, since the flux–time plots are usually employed in predicting the flux condition at long filtration times, these results indicate that the permeate flux will approach a steady state at a prolonged filtration time, as suggested in earlier reports [3–5,7,17,36].

9.5.3 Analysis of the Pore-Blocking/Filtration Mechanisms

Declining flux with time is a phenomenon generally used in describing membrane fouling. The complete blocking, standard blocking, intermediate blocking and cake filtration mechanisms were used in studying the filtration mechanism for each of the membranes. The linear forms of the pore-blocking models were applied for the plots to depict the filtration mechanisms at each operating condition.

Previous investigators [8,36,37] asserted that it is phenomenal for the complete and standard blocking mechanisms to occur at the start of a filtration because of the accumulation of particles tending towards the complete sealing of both smaller and larger pores of the membrane, progressively. After this, it is expected that the intermediate blocking dominates the filtration mechanism for an extended period and serves as a transition phase between the previously occurring blocking mechanism (complete and standard) and cake layer stockpiling. This will continue until the establishment of a significant cake layer.

For each of the membranes utilized in this investigation, the filtration blocking models were fitted to the filtration experiments performed at the varying TMPs of 40, 80 and 120 kPa. Considering all the experiments, the best-fitting situations at these TMPs were selected based on the highest coefficients of determination (R^2) as the most stable for each membrane.

9.5.3.1 Filtration Mechanism for 0.1 μM MF Membrane

Table 9.3 shows the trend of the stability of 0.1 μm MF membrane at various TMPs as assessed by the coefficients of determination (R^2) of the pore-blocking models.

From the table, it is evident that cake filtration was the dominant mechanism with the highest R^2 values at all TMPs. The best stability was attained at the TMP of 40 kPa where the cake filtration model exhibited an R^2 value of 0.9835, which is the highest compared to all other models. The model plots representing the filtration mechanism at a TMP of 40 kPa are depicted in Figures 9.8 through 9.11.

TABLE 9.3
Fitting the Pore-Blocking Models to 0.1 μm MF Membrane

	R^2 of Pore-Blocking Models			
TMP (kPa)	Complete	Standard	Intermediate	Cake Filtration
40	0.8357	0.8864	0.9336	0.9835
80	0.7521	0.8022	0.8676	0.9194
120	0.8032	0.8466	0.8967	0.9388

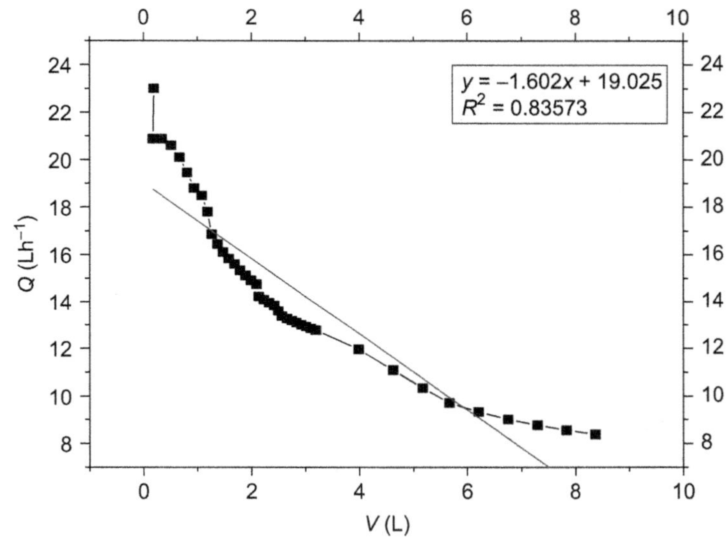

FIGURE 9.8 Plot of the complete blocking model at 40 kPa with 0.1 μm MF membrane.

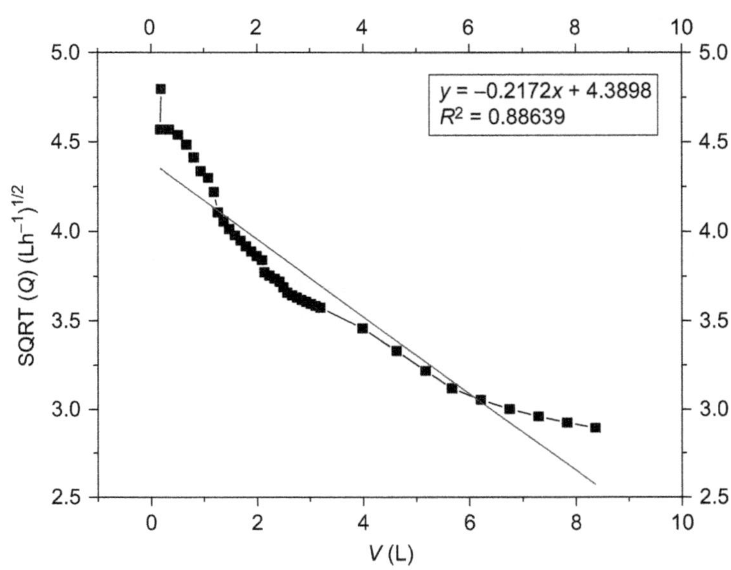

FIGURE 9.9 Plot of the standard blocking model at 40 kPa with 0.1 μm MF membrane.

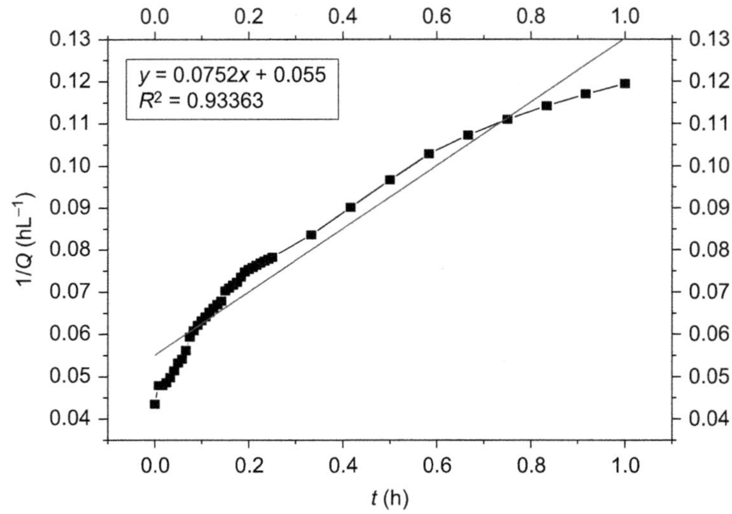

FIGURE 9.10 Plot of the intermediate blocking model at 40 kPa with 0.1 μm MF membrane.

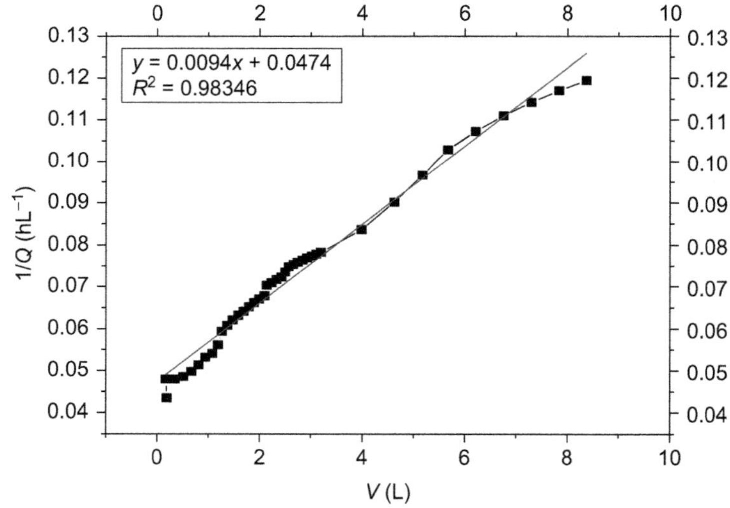

FIGURE 9.11 Plot of the cake filtration model at 40 kPa with 0.1 μm MF membrane.

It is evident from the plots that cake filtration dominated the filtration mechanisms with the highest coefficient of determination. It is thus intuitive that the intermediate blocking model should be the closest in stability to that of the cake filtration model from the principle, and this will serve as a confirmation for the establishment of cake layer formation in membrane filtration. The closest R^2 value after cake filtration appeared to be that of the intermediate blocking mechanism with a value of 0.9336. This indicates that intermediate blocking had nearly as strong an impact as cake filtration. This is because, based on principle, there exists a region where intermediate blocking plays a role in the transition phase from complete and standard blocking to the cake filtration mechanism. The intermediate blocking plot suggests that the transition phase is between 12 and 36 min of filtration time, which serves as the best straight line of the plot. The complete and standard blocking models are expected to be the dominant mechanism at the onset of the filtration experiment. It is evident from their plots that the two mechanisms dominated the onset of the filtration, as they both exhibited their somewhat best straight lines between 0.3–1.2 L, and 1.4–2 L of filtered volumes. This can also be observed from flux-time plot for 0.1 μm MF membrane in Figure 9.4. These results followed the cake filtration mechanism as reported in earlier filtration modeling investigations [8,11].

9.5.3.2 Filtration Mechanism for 0.2 μM MF Membrane

Table 9.4 shows the results of filtration or pore-blocking models as fitted to the filtration experiments performed using the 0.2 μm MF membrane.

It is once again evident from the table that cake filtration also dominated all other filtration mechanisms with the highest R^2 at all the applied TMPs. It was observed that filtration stability, according to the model values at 40 kPa, was relatively close to the one operated at 120 kPa, especially in the cases of cake filtration and intermediate blocking models. However, the best fitting condition (40 kPa) was selected based on the comparatively high R^2 values it possessed. This was done to maintain a balance between the fouling and flux of the membrane process. Additionally, higher TMPs require higher energy consumption.

From the model plots, it was clear that the mechanism followed similar trends as those of the 0.1 μm MF membrane. The complete and standard blocking models exhibited some dominance, as expected, at the beginning of the filtration. Figure 9.3 depicting the flux–time relationship of the 0.2 μm MF membrane filtration shows the sharp bent of the initial complete and standard blocking mechanisms in the membrane process. Furthermore, the model plots representing the complete and standard blocking models also showed that both models dominated at least the first filtration volumes of 0.2–3.5 L with a declining permeate flow-rate (Q), and their plots exhibited good straight lines around those filtration volumes (Figures 9.12 and 9.13). As for the intermediary stage between these two blocking models, the intermediate blocking plot showed that the mechanism occurred from 1 to 10 min, and then from 12 min and extending towards 35 min of filtration

TABLE 9.4
Fitting the Pore-Blocking Models to 0.2 μm MF Membrane

	R^2 of Pore-Blocking Models			
TMP (kPa)	**Complete**	**Standard**	**Intermediate**	**Cake Filtration**
40	0.6801	0.7537	0.8605	0.9167
80	0.5578	0.6346	0.7943	0.8388
120	0.561	0.6746	0.886	0.909

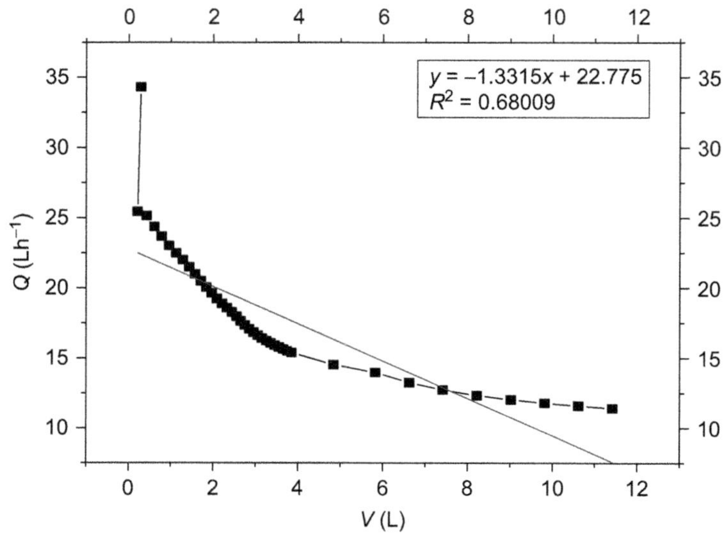

FIGURE 9.12 Plot of the complete blocking model at 40 kPa with 0.2 μm MF membrane.

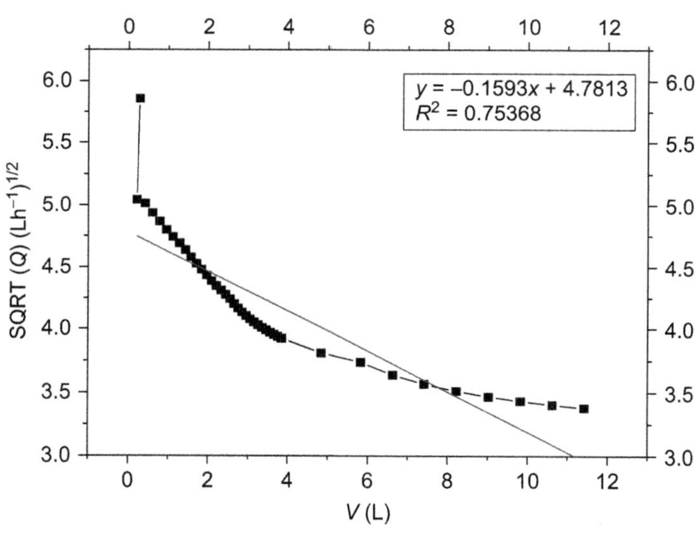

FIGURE 9.13 Plot of the standard blocking model at 40 kPa with 0.2 μm MF membrane.

time, with seemingly the best straight parts of the plot. This showed that the intermediate blocking mechanism existed before the dominant cake filtration mechanism took over completely and immediately after 3 L of filtered volume (Figures 9.14 and 9.15).

9.5.3.3 Filtration Mechanism for 1 kDa UF Membrane

Table 9.5 shows the results of the pore-blocking models as fitted to the filtration experiments performed using the 1 kDa UF membrane.

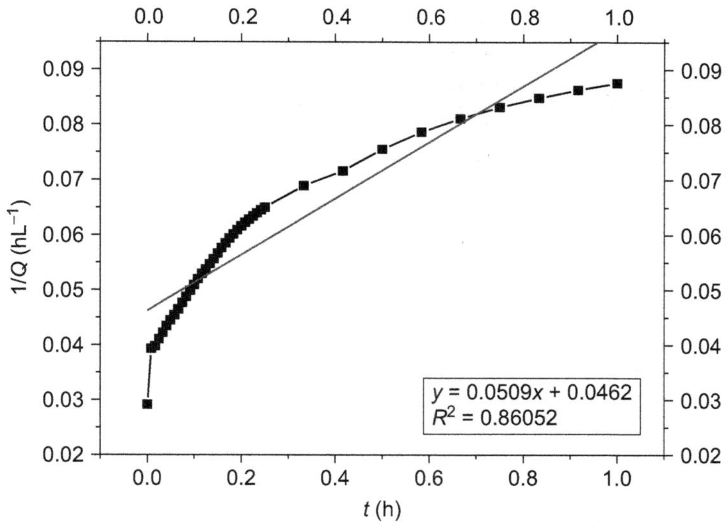

FIGURE 9.14 Plot of the intermediate blocking model at 40 kPa with 0.2 μm MF membrane.

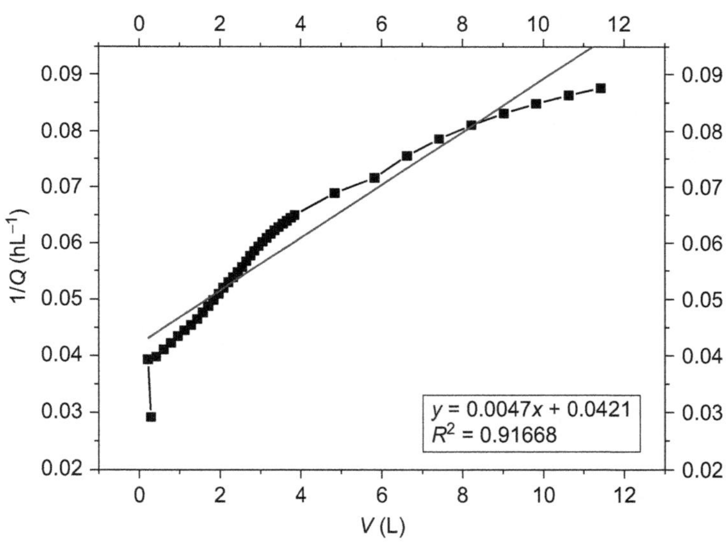

FIGURE 9.15 Plot of the cake filtration model at 40 kPa with 0.2 μm MF membrane.

TABLE 9.5
Fitting the Pore-Blocking Models to 1 kDa UF Membrane

	R^2 of Pore-Blocking Models			
TMP (kPa)	**Complete**	**Standard**	**Intermediate**	**Cake Filtration**
40	0.5384	0.6563	0.8567	0.9252
80	0.532	0.6462	0.8466	0.9155
120	0.5428	0.6591	0.8571	0.9264

Contrary to the conditions at which the MF membranes attained their stabilities, the 1 kDa UF membrane exhibited its best fitting at TMP of 120 kPa, although the values at the other TMPs of 40 and 80 kPa were relatively close to that of 120 kPa. Similar results have been reported in previous studies [4,8,38,39]. The model plots of the mechanisms at 120 kPa are depicted in Figures 9.16 through 9.19:

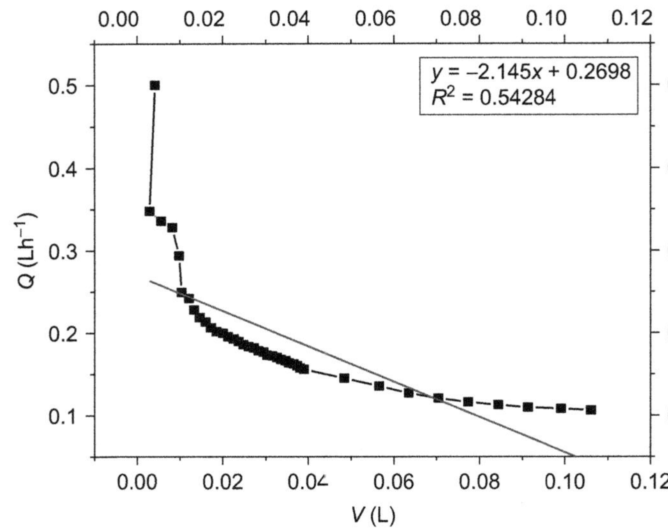

FIGURE 9.16 Plot of the complete blocking model at 120 kPa with 1 kDa UF membrane.

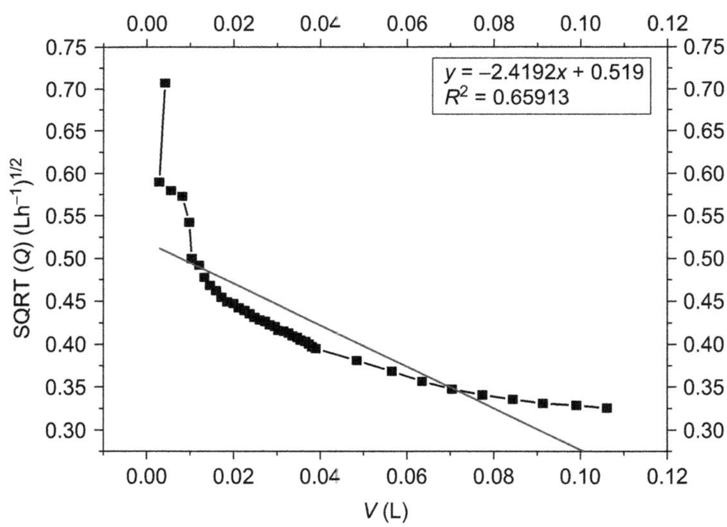

FIGURE 9.17 Plot of the standard blocking model at 120 kPa with 1 kDa UF membrane.

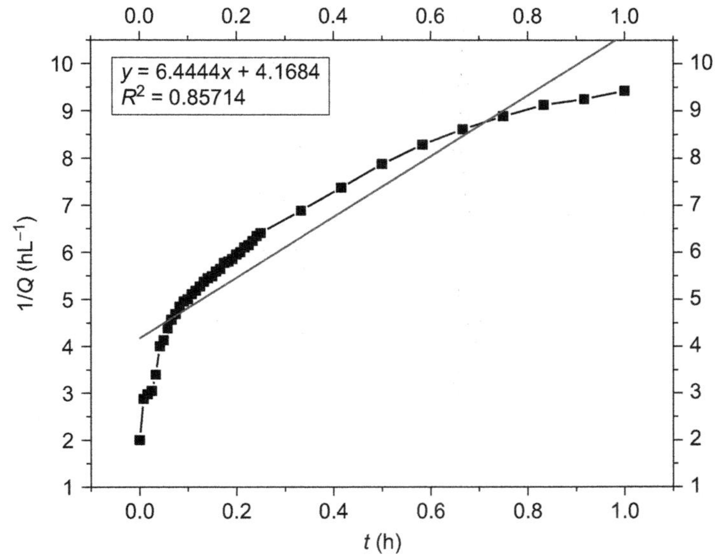

FIGURE 9.18 Plot of the intermediate blocking model at 120 kPa with 1 kDa UF membrane.

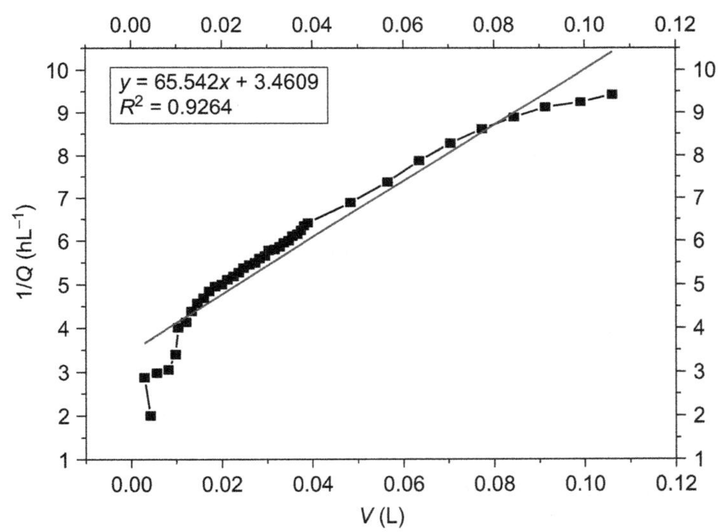

FIGURE 9.19 Plot of the cake filtration model at 120 kPa with 1 kDa UF membrane.

From the model plots, it also appeared the complete blocking model was more pronounced than the standard blocking model. The complete blocking mechanism was experienced at the filtered volumes of 0.0082–0.0104 L, then from 0.0212 to 0.039 L, whereas the standard blocking mechanism contributed to the fouling only after 0.016 L was filtered and up to a filtration volume of 0.0381 L, which roughly corresponds to the second phase at which the complete blockage occurred. This may be attributed to the fact that the feedwater contained more particulates larger than the membrane pores, thereby leading to more pronounced complete blocking. It also indicates that there is little accumulation of very small particles on the walls of the membrane pores, which might have produced the standard blocking mechanism [10,11,15]. The intermediate model plot suggests that the mechanism started at a filtration time of 8 min and extended until 30 min. The cake filtration mechanism started simultaneously with other mechanisms at 0.0172 L of filtered volume, as evident from the straightest part of its plot (Figure 9.19). In general, the cake filtration mechanism dominated the whole filtration process as suggested by the high value of its coefficient of determination.

9.5.3.4 Filtration Mechanism for 5 kDa UF Membrane

Table 9.6 shows the results of filtration models as fitted to the filtration experiments performed using the 5 kDa UF membrane.

Table 9.6 shows the coefficients of determination resulting from the filtration models as fitted to the experimental data for 5 kDa UF membrane filtration of wastewater. Just as with the 1 kDa UF membrane, it appeared that the 5 kDa UF membrane was also stable for the description of filtration mechanism, as evident from their comparatively close R^2 values. Though all the values are very close to one another at the operating TMPs, operation at a TMP of 120 kPa gave the best fitting model, and it also showed that cake filtration mainly dominated in the mechanism describing the filtration process with a relatively high R^2 value of 0.9849. Analyses of the filtration mechanisms based on the filtration obtained at a TMP of 120 kPa are presented in Figures 9.20 through 9.23.

As observed from the complete and standard blocking mechanisms, both mechanisms occurred simultaneously, as expected, with much better stability compared to the 1 kDa UF membrane. Complete and standard blocking started at first 0.0055 L of filtered volume and persisted until the volume reached 0.036 L, as observed from their linear plots. The intermediate blocking model, as observed from the plot, suggests that it occurred after the first 6.5 min and persisted until 15 min of filtration. The transition time for the intermediate blocking mechanism was observed to be longer compared to the 1 kDa UF membrane. This may be due to the principle [4,10,39,40] that the 5 kDa UF membrane has a larger pore than the 1 kDa and more filtration time and volume will be required before the particulate matter contained in the feedwater can accumulate. The cake filtration fully started after the filtration volume of 0.04 L, as observed from the possible straight plot that could be attained at that range.

TABLE 9.6
Fitting the Pore-Blocking Models to 5 kDa UF Membrane

	R^2 of Pore-Blocking Models			
TMP (kPa)	**Complete**	**Standard**	**Intermediate**	**Cake Filtration**
40	0.8532	0.8984	0.9401	0.9828
80	0.8503	0.8976	0.9407	0.9848
120	0.8582	0.905	0.9484	0.9849

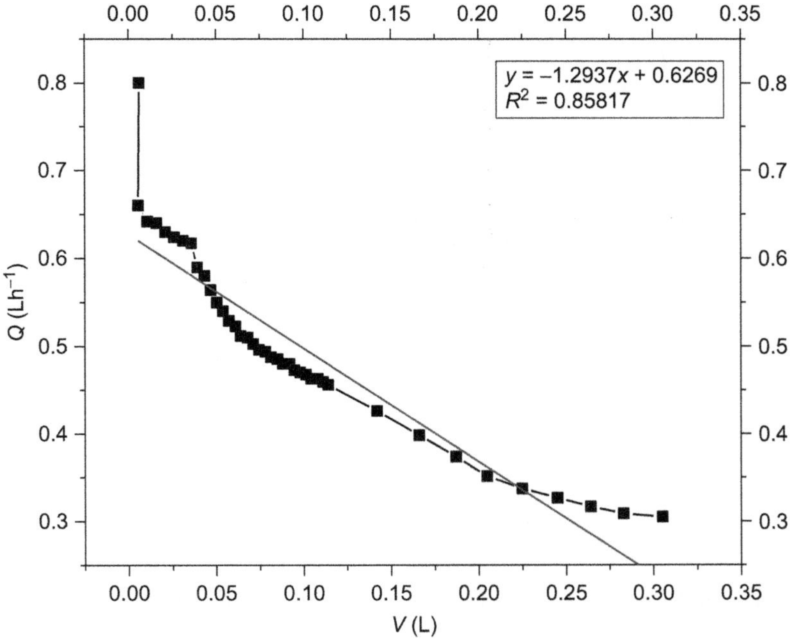

FIGURE 9.20 Plot of the complete blocking model at 120 kPa with 5 kDa UF membrane.

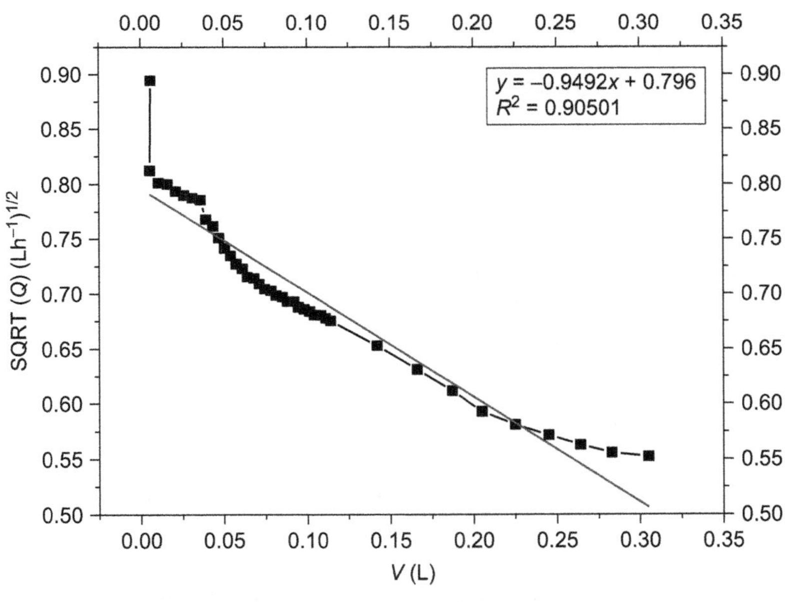

FIGURE 9.21 Plot of the standard blocking model at 120 kPa with 5 kDa UF membrane.

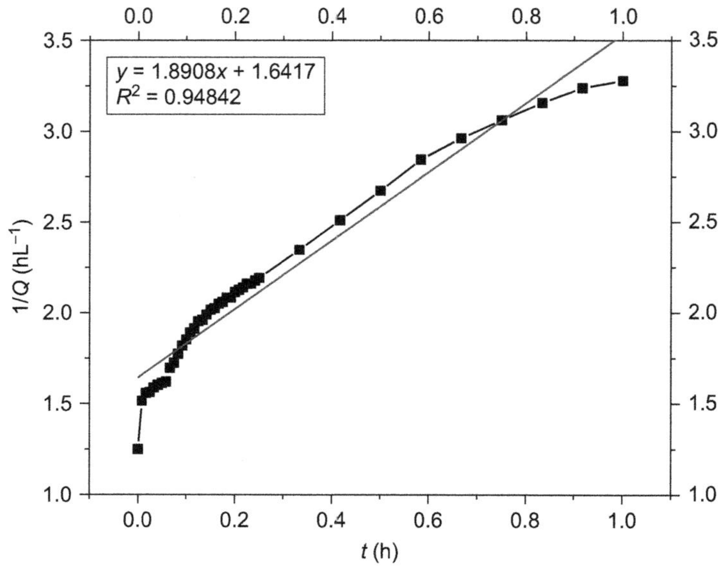

FIGURE 9.22 Plot of the intermediate blocking model at 120 kPa with 5 kDa UF membrane.

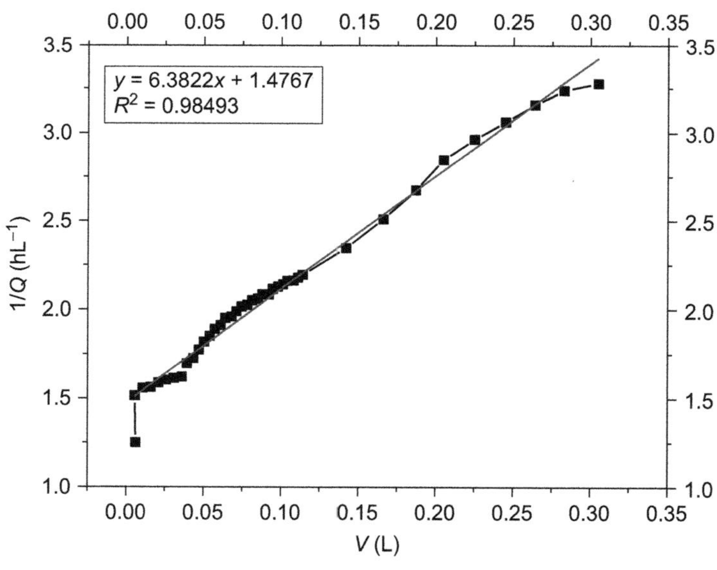

FIGURE 9.23 Plot of the cake filtration model at 120 kPa with 5 kDa UF membrane.

9.5.3.5 Filtration Mechanism for 10 kDa UF Membrane

Table 9.7 shows the results of the pore-blocking models as fitted to the filtration experiments performed using the 10 kDa UF membrane.

From Table 9.7, it was observed that the best stability was attained at a TMP of 40 kPa with a dominant cake filtration mechanism having the highest R^2 value of 0.9832. Just as experienced with all other membranes, cake filtration dominated at all the TMPs, and the closest in stability was the intermediate blocking mechanism. Figures 9.24 through 9.27 depict the filtration mechanism description based on the results obtained at the TMP of 40 kPa.

Observing the model plots, the usual complete and standard blocking models were the pioneering mechanisms in the 10 kDa UF membrane filtration, as they both simultaneously occurred between

TABLE 9.7
Fitting the Pore-Blocking Models to 10 kDa UF Membrane

	R^2 of Pore-Blocking Models			
TMP (kPa)	Complete	Standard	Intermediate	Cake Filtration
40	0.8465	0.8938	0.9383	0.9832
80	0.7858	0.8507	0.9212	0.9783
120	0.6497	0.74	0.8804	0.9483

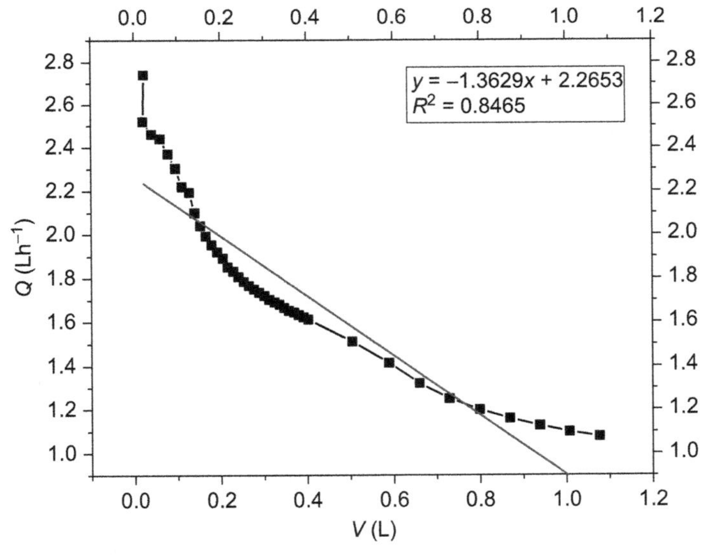

FIGURE 9.24 Plot of the complete blocking model at 40 kPa with 10 kDa UF membrane.

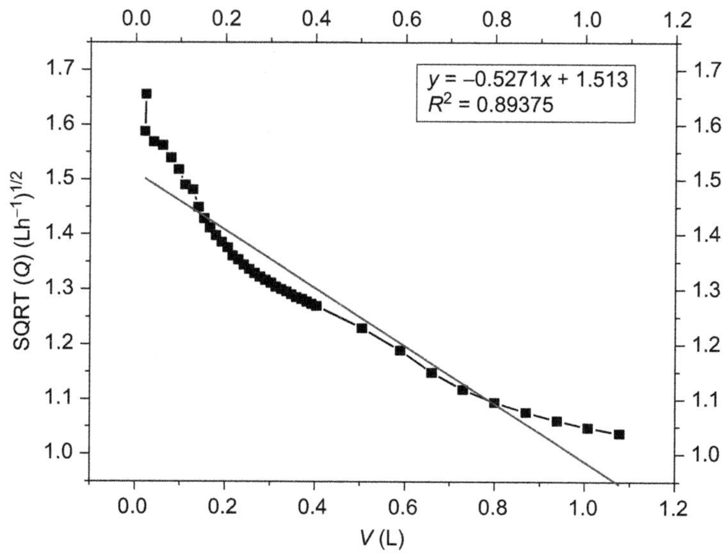

FIGURE 9.25 Plot of the standard blocking model at 40 kPa with 10 kDa UF membrane.

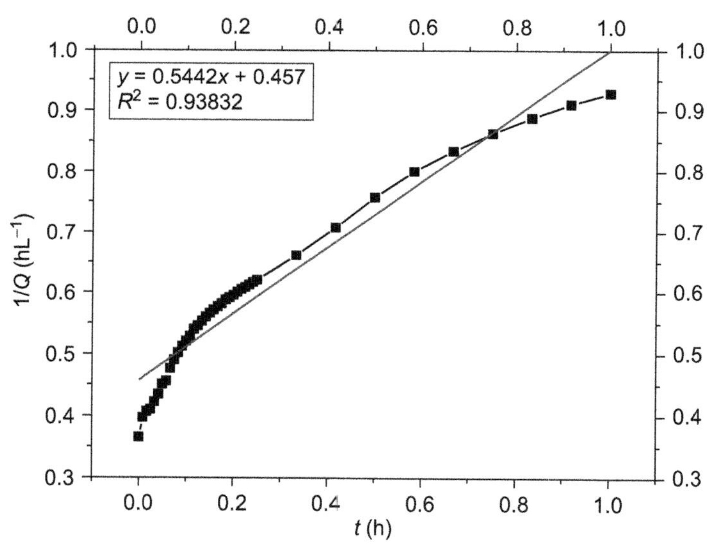

FIGURE 9.26 Plot of the intermediate blocking model at 40 kPa with 10 kDa UF membrane.

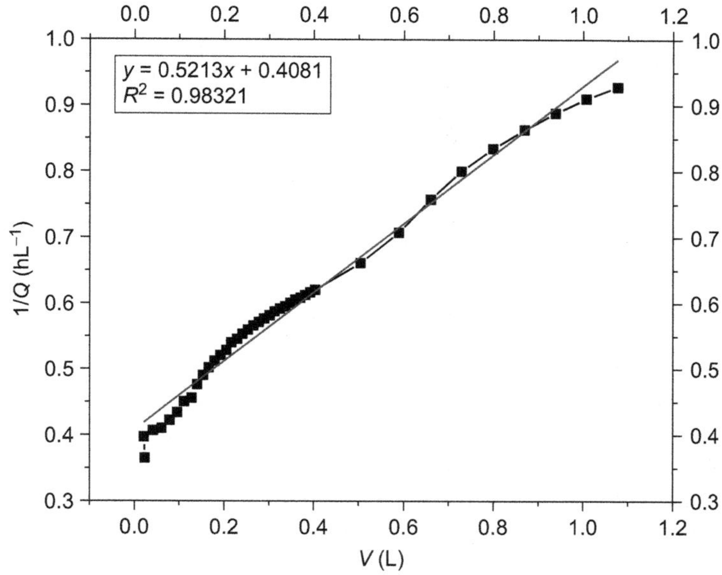

FIGURE 9.27 Plot of the cake filtration model at 40 kPa with 10 kDa UF membrane.

0.0055 and 0.036 L of filtered volume. The similarity of this phase to that of the 5 kDa UF membrane suggests that similar particulates are responsible for both blocking mechanisms. The intermediate blocking mechanism occurred around 6.5 min until 15 min of filtration, while the cake filtration fully occurred after the filtration of the first 0.192 L of permeate volume. All the phases occurred within those regions of straight plots from the models. Pore dimension similarity coupled with the distribution of pore size (such as heterogeneity) of the 5 and 10 kDa UF membranes may explain why some of the mechanisms occurred at the same phases as suggested by Boerlage et al. [8].

Generally, it was observed that with all the filtration experiments using LPMs, cake filtration dominated all other mechanisms, as it gave the best fitting (best linearity) when fitted with experimental data. The TMPs at which the cake filtration mechanism dominated for each of the LPMs are summarized in Table 9.8. In principle, all other mechanisms also occurred at a region in a model plot, but were not as stable for longer periods of time or for larger volumes as cake filtration.

TABLE 9.8
Summary of the Best-Fitting Mechanism

LPM	TMP (kPa)	R^2 for Cake Filtration Model
10 kDa UF	40	0.9832
5 kDa UF	120	0.9849
1 kDa UF	120	0.9264
0.2 µm MF	40	0.9167
0.1 µm MF	40	0.9835

Considering the behavior of all the LPMs as given in Table 9.8, the 5 kDa appeared to be the most stable for the filtration process at 120 kPa of TMP with the highest value of coefficient of determination. The closest values of R^2 were those of 10 kDa UF and 0.1 μm MF. With this, the three membranes could be proposed as promising reference membranes for use in the description of fouling mechanisms for wastewater filtration. However, the best conclusion could only be reached after these membranes are subjected to compressibility evaluation tests. The fitting of the cake filtration model as the most common model in describing the wastewater fouling mechanism on membranes has been reported in previous investigations [8,9,11].

9.5.4 Influence of the Upstream Adsorption on Fouling and Productivity

It was pertinent to analyze the extent of the feedwater fouling and productivity (flux) when the adsorption process is not integrated. This has led to the necessity of comparing the fluxes of the adsorbed feedwater with the raw feedwater based on the best model of cake filtration with an R^2 value of 0.9849, as shown in Table 9.8, with the operating conditions of 5 kDa of MWCO and 120 kPa. A comparison for the raw and adsorbed feed fluxes of the UF with the MWCO of 1 kDa was also evaluated, since it was operated at similar TMP (120 kPa). Plots of the two fluxes are presented in Figures 9.28 and 9.29.

From the plots, it was observed that membrane filtration without the adsorption process resulted in earlier and heavier fouling of both membranes. In the case of the adsorbed wastewater, the flux reduced from 8.0 to 5.64 $Lm^{-2}h^{-1}$, and from 5.0 to 2.13 $Lm^{-2}h^{-1}$ in the first 5 min of filtration for 5 and 1 kDa UF membranes, respectively. Whereas, the flux of the raw wastewater reduced drastically from 8.0 to 2.52 $Lm^{-2}h^{-1}$, and from 5.0 to 0.97 $Lm^{-2}h^{-1}$ at the same filtration time of 5 min for the 5 and 1 kDa UF membranes, respectively. This indicates that due to the heavy fouling of the raw wastewater, there was a flux reduction of 69% and 81% for 5 and 1 kDa UF membranes, respectively, while the adsorption process enhanced the adsorbed feedwater flux with only 30% and 57% flux reductions for 5 and 1 kDa UF membranes, respectively. The higher flux reduction in the 1 kDa UF membrane may be attributable to its smaller pore size, which might have led to the accumulation of

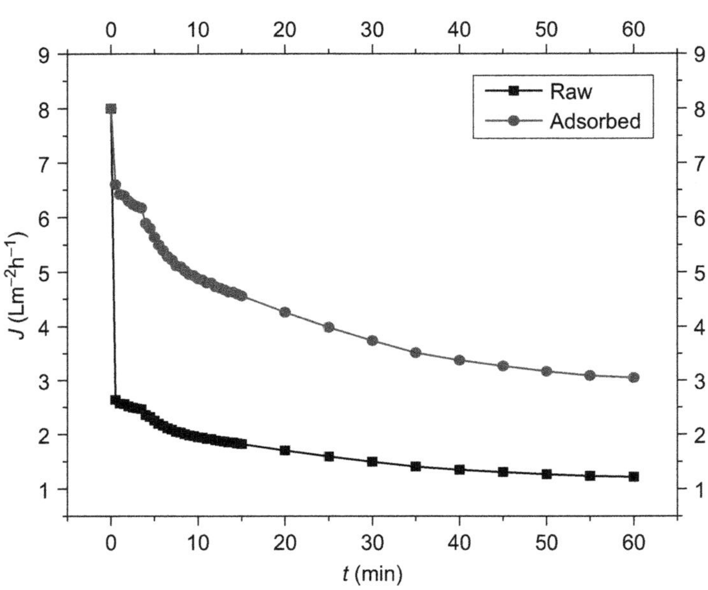

FIGURE 9.28 Comparison of the fluxes for the adsorbed and raw wastewater at 120 kPa with 5 kDa UF membrane.

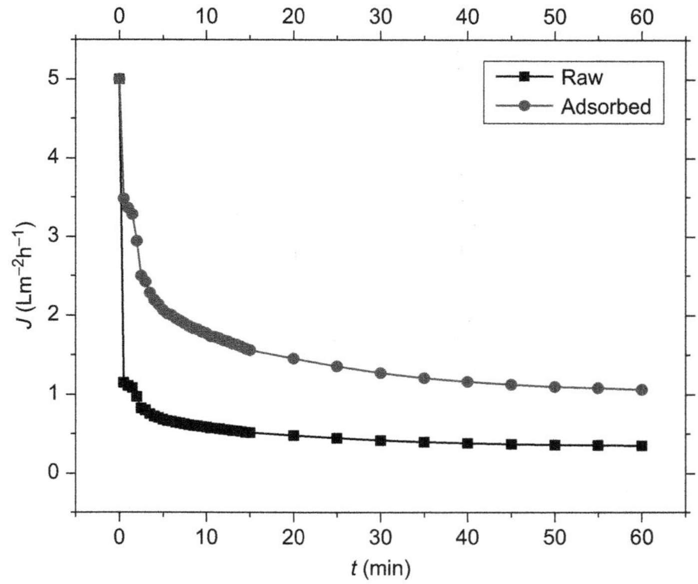

FIGURE 9.29 Comparison of the fluxes for the adsorbed and raw wastewater at 120 kPa with 1 kDa UF membrane.

bigger particulates on the membrane surface. These particulates block or limit the available pores responsible for the fluid flow, thus resulting in flux (productivity) reduction. The larger particulates resulting in pore blocking of the 1 kDa UF membrane might be able to pass through the larger pores of the 5 kDa, thereby resulting in its low reduction in flux as suggested in previous reports [3–5,7,41]. All other LPMs followed this trend regarding the flux enhancement by PAC addition. It is noteworthy to state that similar trends were also observed for all other membranes utilized in this study. Previous investigations showing similar trends have also been reported [21,42–50].

9.5.5 Effect of Chemical Cleaning on Membrane Filtration Flux

It was observed that chemical cleaning of the membranes yielded good results, as the CWF test after the first chemical cleaning led to at least a 75% restoration of the membrane flux. A reiterated chemical cleaning of the membrane proved sufficient, as it brought more than 96% restoration of CWF. This indicates that the fouling nature of the feedwater and blockage experienced by the membranes was neither severe nor permanent, as the chemical cleaning nearly brought the membrane back to its initial active strength (restoration rate close to 100%). This could be attributed to the upstream adsorption, which acted like a biofouling reducer prior to the downstream membrane process. The upstream step must have removed higher concentrations of some of the feedwater constituents that might be potentially responsible for biofouling or membrane pore blockage. Similar observations and conclusions were reported in previous reports [5,7,51,52]. The cake filtration phenomenon as characterized from all the tested membranes in this investigation also indicates that the residual foulants acted as secondary filters resulting from gradual particle bioaccumulation [4,7,41,53,54]. Reversible fouling must, therefore, have truly been the factor controlling the fouling of the membranes as suggested by the cake nature.

From the foregoing, this study has been able to evaluate the exact transport phenomena involved in the filtration of high-strength wastewater containing particulate solids and other contaminants of different concentrations using pore-blocking filtration laws. However, since the cake filtration mechanism is the best filtration law that could be applied in describing the effects of the particulate solids in the fouling of the membranes by the wastewater applied in this study, it is essential that the cake be further analyzed to completely grasp the fouling tendency of the wastewater with respect to the membranes employed. The build-up of solids on the membrane is called cake, and it increases the hydraulic resistance to flow, thereby limiting the flux in a constant-pressure filtration and increasing the pressure in a constant-rate filtration [27,55]. This could be well characterized by Darcy's law, which is the description of flow of liquid through a porous bed of solids allowing some fluids to flow through the apparently blocked pores [36], and is related to the driving pressure [56,57]. The cake analysis for the MF membranes employed in this study has been carried out in one of our previous studies with similar wastewater chemistry and operating conditions [41]. However, it is pertinent to carry out a similar analysis of the UF membranes to fully understand cake formation with respect to the wastewater employed in this study or any other wastewater with similar chemistry in terms of cake compressibility.

9.6 CONCLUSIONS

In this study, the pore blocking behaviors of some low-pressure membranes have been analyzed using the final discharge wastewater from the palm oil industry. The wastewater with an assumed Newtonian behavior has been evaluated to elucidate the fundamental mechanisms involved in membrane pore blocking with respect to different pore sizes of MF and UF membranes. It was observed that the blocking index η decreased from 2 to 0 successively as the blocking phenomenon started from complete blocking ($\eta = 2$) until a much more stable phase of the cake filtration phenomenon ($\eta = 0$) appeared. These consecutive steps of the fouling process were elucidated in terms of the successive or simultaneous presence of the four fouling stages of complete blocking, standard blocking, intermediate blocking and cake filtration phenomena. It was observed that the cake filtration phenomenon dominated all other fouling steps, but other blocking mechanisms (the complete, standard and intermediate blocking models) also participated slightly in pore blocking before the cake filtration eventually prevailed.

Since the cake deposition that resulted from particulate accretion on the membrane surface accounted for the major flux decline, this is an indication that the fouling did not really affect the pore structure of the membranes. Furthermore, with cake layer formation, it could be predicted that a steady-state filtration of the wastewater could be attained at longer filtration times without experiencing complete pore blocking since the phenomenon only acted as a secondary filter in the filtration process. Also, since fouling due to cake layer formation is reversible, this gives an idea of the cleaning methods necessary for the process at the design stage. The data presented here could be employed for upscaling designs in predicting the fouling behavior of similar effluents when subjected to the membrane filtration process.

REFERENCES

1. Cui, Z.F., Muralidhara, H.S. *Membrane Technology: A Practical Guide to Membrane Technology and Applications in Food and Bioprocessing*. Butterworth-Heinemann, Elsevier, Burlington, VT (2010).
2. Peinemann, K.-V., Nunes, S.P., Giorno, L. *Membranes for Food Applications, vol. 3. Membrane Technology*. WILEY-VCH Verlag GmbH & Co. KGaA, Weinheim, Germany (2010).
3. Amosa, M.K. Hybrid adsorption-membrane process for reclamation of bio-treated palm oil mill effluent for boiler-feed reuse. PhD Thesis, International Islamic University Malaysia, Kuala Lumpur (2015).
4. Amosa, M.K. Towards sustainable membrane filtration of palm oil mill effluent: Analysis of fouling phenomena from a hybrid PAC-UF process. *Applied Water Science* **7**(6), 3365–3375 (2017). doi:10.1007/s13201-016-0483-3.

5. Amosa, M.K., Jami, M.S., Alkhatib, M.F., Majozi, T. Studies on pore blocking mechanism and technical feasibility of a hybrid PAC-MF process for reclamation of irrigation water from biotreated POME. *Sep. Sci. Technol.* **51**(12), 2047–2061 (2016). doi:10.1080/01496395.2016.1192192.
6. Judd, S. *The MBR Book: Principles and Applications of Membrane Bioreactors for Water and Wastewater Treatment*, 2nd ed. Elsevier, Oxford (2011).
7. Amosa, M.K., Jami, M.S., AlKhatib, M.F.R. Cross-flow microfiltration of a high strength industrial wastewater: Modelling of membrane fouling mechanisms. *Qatar Foundation Annual Research Conference Proceedings* **2016**(1), EEPP1529 (2016). doi:10.5339/qfarc.2016.EEPP1529.
8. Boerlage, S.F., Kennedy, M.D., Dickson, M.R., El-Hodali, D.E., Schippers, J.C. *The modified fouling index using ultrafiltration membranes (MFI-UF): Characterisation, filtration mechanisms and proposed reference membrane. J. Membr. Sci.* **197**(1), 1–21 (2002).
9. Huang, H., Young, T.A., Jacangelo, J.G. Unified membrane fouling index for low pressure membrane filtration of natural waters: Principles and methodology. *Environ. Sci. Technol.* **42**(3), 714–720 (2008).
10. Iritani, E. A review on modeling of pore-blocking behaviors of membranes during pressurized membrane filtration. *Drying Technol.* **31**(2), 146–162 (2013).
11. Boerlage, S.F., Kennedy, M.D., Aniye, M.P., Abogrean, E., Tarawneh, Z.S., Schippers, J.C. The MFI-UF as a water quality test and monitor. *J. Membr. Sci.* **211**(2), 271–289 (2003).
12. Hermans, P.H., Bredée, H.L. Principles of the mathematical treatment of constant pressure filtration. *J. Soc. Chem. Ind.* **55**, 1–4 (1936).
13. Bowen, W., Calvo, J., Hernandez, A. Steps of membrane blocking in flux decline during protein microfiltration. *J. Membr. Sci.* **101**(1), 153–165 (1995).
14. Hermia, J. Constant pressure blocking filtration law: Application to power-law non-Newtonian fluid. *Trans. Inst. Chem. Eng.* **60**, 183–187 (1982).
15. Mohammadi, T., Kazemimoghadam, M., Saadabadi, M. Modeling of membrane fouling and flux decline in reverse osmosis during separation of oil in water emulsions. *Desalination* **157**(1), 369–375 (2003).
16. Jami, M.S., Amosa, M.K., Alkhatib, M.F.R., Jimat, D.N., Muyibi, S.A. Boiler-feed and process water reclamation from biotreated palm oil mill effluent (BPOME): A developmental review. *Chem. Biochem. Eng. Q.* **27**(4), 477–489 (2013).
17. Amosa, M.K., Jami, M.S., Alkhatib, M.F.R., Majozi, T. Technical feasibility study of a low-cost hybrid PAC-UF system for wastewater reclamation and reuse: A focus on feedwater production for low-pressure boilers. *Environ. Sci. Pollut. Res. Int.* **23**(22), 22554–22567 (2016). doi:10.1007/s11356-016-7390-x.
18. Amosa, M.K. Process optimization of Mn and H_2S removals from POME using an enhanced empty fruit bunch (EFB)-based adsorbent produced by pyrolysis. *Environ. Nanotechnol. Monit. Manage.* **4**, 93–105 (2015). doi:10.1016/j.enmm.2015.09.002.
19. Amosa, M.K., Jami, M.S., Alkhatib, M.F.R. Electrostatic biosorption of COD, Mn and H_2S on EFB-based activated carbon produced through steam pyrolysis: An analysis based on surface chemistry, equilibria and kinetics. *Waste Biomass Valor* **7**(1), 109–124 (2016). doi:10.1007/s12649-015-9435-7.
20. Baker, R., Fane, A., Fell, C., Yoo, B. Factors affecting flux in crossflow filtration. *Desalination* **53**(1), 81–93 (1985).
21. Lee, C.W., Bae, S.D., Han, S.W., Kang, L.S. Application of ultrafiltration hybrid membrane processes for reuse of secondary effluent. *Desalination* **202**(1), 239–246 (2007).
22. Li, Y., Wang, J., Zhang, W., Zhang, X., Chen, C. Effects of coagulation on submerged ultrafiltration membrane fouling caused by particles and natural organic matter (NOM). *Chin. Sci. Bull.* **56**(6), 584–590 (2011).
23. Agbekodo, K.M., Legube, B., Cote, P. Organics in NF permeate. *J. Am. Water Works Assn.* **88**(5), 67–74 (1996).
24. Fu, P., Ruiz, H., Thompson, K., Spangenberg, C. Selecting membranes for removing NOM and DBP precursors. *J. Am. Water Works Assn.* **86**(12), 55–72 (1994).
25. Jung, C.-W., Kang, L.-S. Application of combined coagulation-ultrafiltration membrane process for water treatment. *Korean J. Chem. Eng.* **20**(5), 855–861 (2003).
26. Springer, F., Laborie, S., Guigui, C. Removal of SiO_2 nanoparticles from industry wastewaters and subsurface waters by ultrafiltration: Investigation of process efficiency, deposit properties and fouling mechanism. *Sep. Purif. Technol.* **108**, 6–14 (2013).

27. Ohn, T., Jami, M., Iritani, E., Mukai, Y., Katagiri, N. Filtration behaviors in constant rate microfiltration with cyclic backwashing of coagulated sewage secondary effluent. *Sep. Sci. Technol.* **38**(4), 951–966 (2003).
28. Sartorius stedim biotech: User manual. In, vol. 6. pp. 1–56. Sartorius Stedim Biotech GmbH, Göttingen, Germany, (2012).
29. Foley, G. *Membrane Filtration: A Problem Solving Approach with MATLAB®*. Cambridge University Press, Cambridge, UK (2013).
30. Cai, M., Wang, S., Liang, H. Modeling and fouling mechanisms for ultrafiltration of Huanggi (Radix astragalus) extracts. *Food Sci. Biotechnol.* **22**(2), 407–412 (2013). doi:10.1007/s10068-013-0094-9.
31. Rashid, S.S., Alam, M.Z., Fazli, M.B.F.A. Separation of cellulase enzyme from fermentation broth of palm oil mill effluent by ultrafiltration process. *Int. J. Chem. Environ. Biol. Sci. (IJCEBS)* **1**(3), 501–506 (2013).
32. Crozes, G.F., Jacangelo, J.G., Anselme, C., Laine, J.M. Impact of ultrafiltration operating conditions on membrane irreversible fouling. *J. Membr. Sci.* **124**(1), 63–76 (1997).
33. Wong, J.M. Membranes for wastewater reclamation and reuse for petrochemical and petroleum refining industries. Proceedings of the Water Environment Federation **2011**(15), 1727–1738 (2011). WEF Technical Exhibition and Conference (WEFTEC) held on October 16–19. Los Angeles, California, USA.
34. Wu, T., Mohammad, A., Md Jahim, J., Anuar, N. Palm oil mill effluent (POME) treatment and bioresources recovery using ultrafiltration membrane: Effect of pressure on membrane fouling. *Biochem. Eng. J.* **35**(3), 309–317 (2007).
35. Suzuki, T., Watanabe, Y., Ozawa, G., Ikeda, S. Removal of soluble organics and manganese by a hybrid MF hollow fiber membrane system. *Desalination* **117**(1), 119–129 (1998).
36. Yuan, W., Kocic, A., Zydney, A.L. Analysis of humic acid fouling during microfiltration using a pore blockage–cake filtration model. *J. Membr. Sci.* **198**(1), 51–62 (2002).
37. Schippers, J.C., Verdouw, J. The modified fouling index, a method of determining the fouling characteristics of water. *Desalination* **32**(0), 137–148 (1980). doi:10.1016/S0011-9164(00)86014-2.
38. Grenier, A., Meireles, M., Aimar, P., Carvin, P. Analysing flux decline in dead-end filtration. *Chem. Eng. Res. Des.* **86**(11), 1281–1293 (2008). doi:10.1016/j.cherd.2008.06.005.
39. Said, M., Ahmad, A., Mohammad, A.W., Mohd Nor, M.T., Sheikh Abdullah, S.R. Blocking mechanism of PES membrane during ultrafiltration of POME. *J. Ind. Eng. Chem.* **21**, 182–188 (2015). doi:10.1016/j.jiec.2014.02.023.
40. Bacchin, P., Aimar, P., Field, R.W. Critical and sustainable fluxes: Theory, experiments and applications. *J. Membr. Sci.* **281**(1), 42–69 (2006).
41. Amosa, M.K., Jami, M.S., Alkhatib, M.F.R., Majozi, T., Abdulkareem, S.A. Cake compressibility analysis of BPOME from a hybrid adsorption-microfiltration process. *Water Environ. Res.* **89**(4), 292–300 (2017).
42. Ying, Z., Ping, G. Effect of powdered activated carbon dosage on retarding membrane fouling in MBR. *Sep. Purif. Technol.* **52**(1), 154–160 (2006).
43. Lee, J.-W., Choi, S.-P., Thiruvenkatachari, R., Shim, W.-G., Moon, H. Submerged microfiltration membrane coupled with alum coagulation/powdered activated carbon adsorption for complete decolorization of reactive dyes. *Water Res.* **40**(3), 435–444 (2006).
44. Thiruvenkatachari, R., Shim, W.G., Lee, J.W., Moon, H. Effect of powdered activated carbon type on the performance of an adsorption-microfiltration submerged hollow fiber membrane hybrid system. *Korean J. Chem. Eng.* **21**(5), 1044–1052 (2004).
45. Thiruvenkatachari, R., Shim, W.G., Lee, J.W., Aim, R.B., Moon, H. A novel method of powdered activated carbon (PAC) pre-coated microfiltration (MF) hollow fiber hybrid membrane for domestic wastewater treatment. *Colloids Surf. Physicochem. Eng. Aspects* **274**(1), 24–33 (2006).
46. Thiruvenkatachari, R., Shim, W.G., Lee, J.W., Moon, H. Powdered activated carbon coated hollow fiber membrane: Preliminary studies on its ability to limit membrane fouling and to remove organic materials. *Korean J. Chem. Eng.* **22**(2), 250–255 (2005).
47. Lesage, N., Sperandio, M., Cabassud, C. Study of a hybrid process: Adsorption on activated carbon/membrane bioreactor for the treatment of an industrial wastewater. *Chem. Eng. Proc* **47**(3), 303–307 (2008).

48. Satyawali, Y., Balakrishnan, M. Performance enhancement with powdered activated carbon (PAC) addition in a membrane bioreactor (MBR) treating distillery effluent. *J. Hazard. Mater.* **170**(1), 457–465 (2009).
49. Tomaszewska, M., Mozia, S. Removal of organic matter from water by PAC/UF system. *Water Res.* **36**(16), 4137–4143 (2002).
50. Löwenberg, J., Zenker, A., Baggenstos, M., Koch, G., Kazner, C., Wintgens, T. Comparison of two PAC/UF processes for the removal of micropollutants from wastewater treatment plant effluent: Process performance and removal efficiency. *Water Res.* **56**, 26–36 (2014).
51. Damayanti, A., Ujang, Z., Salim, M.R. The influence of PAC, zeolite, and *Moringa oleifera* as biofouling reducer (BFR) on hybrid membrane bioreactor of palm oil mill effluent (POME). *Bioresour. Technol.* **102**, 4341–4346 (2011).
52. Yuniarto, A., Ujang, Z., Zainon Noor, Z. Performance of bio-fouling reducers in aerobic submerged membrane bioreactor for palm oil mill effluent treatment. *Jurnal Teknologi* **2008**(49), 555–566 (2008).
53. Bugge, T.V., Jørgensen, M.K., Christensen, M.L., Keiding, K. Modeling cake buildup under TMP-step filtration in a membrane bioreactor: Cake compressibility is significant. *Water Res.* **46**(14), 4330–4338 (2012). doi:10.1016/j.watres.2012.06.015.
54. Kovalsky, P., Gedrat, M., Bushell, G., Waite, T.D. Compressible cake characterization from steady-state filtration analysis. *AIChE J.* **53**(6), 1483–1495 (2007). doi:10.1002/aic.11193.
55. Li, N.N., Fane, A.G., Winston Ho, W.S., Matsuura, T. *Advanced Membrane Technology and Applications*. John Wiley & Sons, Hoboken, NJ (2011).
56. Van den Berg, G.B., Smolders, C.A. Flux decline in ultrafiltration processes. *Desalination* **77**, 101–133 (1990).
57. Bessiere, Y., Abidine, N., Bacchin, P. Low fouling conditions in dead-end filtration: Evidence for a critical filtered volume and interpretation using critical osmotic pressure. *J. Membr. Sci.* **264**(1), 37–47 (2005).

10 Sodalite- and Chitosan-Based Composite Membrane Materials for Treatment of Metal-Containing Wastewater in Mining Operations

Machodi Mathaba and Michael O. Daramola

CONTENTS

10.1 INTRODUCTION

The major environmental challenge associated with mining industry is acid mine drainage (AMD), which forms when sulfide rocks are exposed to air and water for prolonged periods (Nordstrom et al., 2015). The formation of AMD is a natural process, but reaction caused by exposing sulfide-containing rocks to the environment through mining operations are often catalyzed by bacterial activity. The natural process of AMD formation takes close to 15 years in the absence of bacteria for ferric iron to produce acid, but the presence of bacteria shortens the reaction time down to 8 min (Metesh et al., 1998). Typical characteristics of AMD are very low pH (often from 3 to 2) and elevated concentration of metals and sulfates (Meschke et al., 2015). If left untreated, AMD has a significant negative environmental impact, including the mineralization of affected areas and acidification of receiving ground and surface waters. The solubility of transition metals is greater in low pH media, so AMD carries with it a high concentration of metals such as Al, As, and Mg and other transition metals such as Cu, Zn, Pb, Co, Mn, and Cd depending on the host rock (Hallberg, 2010). It uncontrollably enters the aquatic environment and poses a threat to humans, domesticated animals, and the ecosystem (Durand, 2012). According to Ford (2003), there are four commonly-accepted chemical reactions that represent the chemistry of pyrite weathering to form AMD (Equations 10.1 through 10.4); Equation 10.5 represents the oxidation of pyrite by unhydrolyzed ferric iron.

$$4\ FeS_2 + 15\ O_2 + 14\ H_2O \rightarrow 4\ Fe\,(OH)_3\downarrow + 8\ H_2SO_4 \tag{10.1}$$

$$2\ FeS_2 + 7\ O_2 + 2\ H_2O \rightarrow 2\ Fe^{+2} + 4\ SO_4^{-2} + 4\ H^+ \tag{10.2}$$

$$4\ Fe^{+2} + O_2 + 4\ H^+ \rightarrow 4\ Fe^{+3} + 2\ H_2O \tag{10.3}$$

$$4\ Fe^{+3} + 12\ H_2O \leftrightarrow 4\ Fe\,(OH)_3\downarrow + 12\ H^+ \tag{10.4}$$

$$14Fe^{+3} + FeS_2(s) + 8\ H_2O \rightarrow 2SO_4^{-2} + 15\ Fe^{+2} + 16\ H^+ \tag{10.5}$$

During AMD generation, pyrite is first oxidized under water and oxygen, with air being the predominant source of oxygen. The reaction generates two moles of acidity for each mole of pyrite oxidized; sulfur is oxidized to sulfate and ferrous ion is released. With more excess oxygen, the ferrous iron is further oxidized to ferric iron, which is soluble at very low pH (<3.5). In addition, ferric iron can not only undergo hydrolysis and precipitate as Fe $(OH)_3$, but it can become an oxidant itself and can further oxidize pyrite to generate more ferrous iron with high acidity. The abiotic oxidation rate of converting ferrous into ferric iron is slow in an acidic environment (pH > 3). However, acidophilic iron-oxidizing species such as *Acidothiobacillus ferroxidans* can generate energy by converting ferrous iron into ferric iron (Johnson, 1998). Drainage from coalmines is less acidic (in terms of proton acidity as opposed to mineral acidity) due to the moderately high carbonate content of the host rock, which provides buffering capacity. In contrast, drainage from metal mines and spoils are more acidic and contains a high concentration of metals (Hallberg, 2010).

The primary contributing factors perpetuating AMD formation include the pH, temperature and oxygen content in the gas phase. Other contributing features are oxygen concentration in the water phase, exposed surface area of metal sulfide, degree of saturation with water, chemical activity of Fe^{3+}, chemical activation energy required to initiate acid generation, and bacterial activity (Akcil and Koldas, 2006). Should the recipient water have insufficient neutralization capacity, a major concern associated with AMD is the proton acidity, which lowers the pH of the receiving water. An additional issue is the mineral acidity, which further lowers the pH of the water body far beyond the pH of the AMD source due to the oxidation and precipitation of metals contained in the AMD. Precipitation of these metals results in their accumulation on the surface of the receiving water's benthic zone, and living organisms capture these metals through bioaccumulation of free cations and plant or bacteria membranes, organically via the food chain by biomagnification, or directly through cell surfaces. AMD not only impacts surface or ground water, it is also responsible for soil quality degradation and dispersion of metals into the air.

While it is difficult to accurately assess the worldwide impact of AMD, it was estimated in 1989 that about 19,300 km of streams and rivers had been seriously damaged by mine effluents (Johnson and Hallberg, 2005). About 12,000 km of watercourses in the UK (Hallberg, 2010), USA, and Canada (DeNicola and Stapleton, 2002; Natarajan, 2008; Cole et al., 2011) had been negatively impacted by AMD. The Jaintia Hills district of Meghalaya in northeast India has traditionally depended on agriculture with paddy rice as the dominant product. However, the exploration of the 10^9 tons of coal reserves since the introduction of opencast mining in the 1970s has contaminated the soil with heavy metals and transformed the farmland into unproductive wasteland (Choudhury et al., 2017). The soil contamination has been attributed to overburden dumping from the mines and the associated AMD generated.

Gold mining in South Africa dates as far back as the eighteenth century, when gold was discovered in the Johannesburg Witwatersrand basin region, which is approximately 350 km long and 200 km wide (Naicker et al., 2003; Tutu et al., 2008). Mine tailings in the Witwatersrand precinct have been open to water and air, making the region vulnerable to oxidation of pyrite and other sulfide-containing rocks. Oxidation of the pyrite rocks causes acidification of rainwater penetrating though the dumps and entering the groundwater beneath (Naicker et al., 2003). The groundwater tables near Johannesburg have been reported to be heavily contaminated with AMD, and over time the underground mine voids have been filling up, until acidic water started decanting at 15–35 mL/day (Naicker et al., 2003; Ochieng et al., 2010) into the Tweelopies River, a tributary of the Bloubank, which in turn discharges into the Crocodile River, which supports close to 95,000 ha of agricultural irrigation (Lang, 2007). Due to population growth, industrialization, mechanization, and urbanization, the water supply continues to dwindle while demand rises (Ochieng et al., 2010). Over 70% of the water used in both rural and urban areas in South Africa is surface water drawn from rivers, streams, lakes, ponds, and springs (Department of Water Affairs and Forestry (DWAF), 2004) and mining wastewater currently contaminates these sources of water.

10.2 PREVENTION OF FORMATION OF AMD AND TREATMENT OF METAL-CONTAINING WASTEWATER

Since the formation of AMD requires sulfide-bearing rocks encountering water and oxygen, excluding either or both should prevent or at least minimize production of AMD. Interventions such as flooding the deep mines with water could help prevent AMD formation only if the present dissolved oxygen in the flooding water is consumed by the sulfide-bearing rocks and microorganisms present. Replenishment of dissolved oxygen through diffusion can be averted by sealing off the mine (Johnson and Hallberg, 2005). However, this is only effective when the location of the shafts and oxygen-containing water ingress is known; in most cases, it is difficult to identify these areas. Since iron and sulfate-oxidizing bacteria catalyze the AMD generation reaction, another intervention could be using anionic surfactant biocides such as sodium dodecyl sulfate (SDS) to kill this group of microorganisms. The challenge with this kind of intervention is that the effectiveness of biocide application is only short term and costly as it requires continuous application of the chemicals.

Because of probable economic and practical difficulties associated with preventing AMD formation at the source, an alternative option would be to collect the contaminated water, treat it and discharge it as clean water. Conventional treatment processes for AMD or water bodies affected with AMD have been divided into active and passive treatment. Active treatment involves neutralizing the acidity with alkaline substances such as lime and allowing formation of metal hydroxide precipitate, which could easily be removed by sedimentation (Elliott et al., 1998; Jong and Parry, 2003; Robinson-Lora and Brennan, 2009). Ion exchange technology, which explores the advantage of oppositely charged pollutants and employs solid resin to remove cation and anion from solutions, has been proposed as a treatment method as well. A high metal ion uptake capability of this resin makes ion exchange an attractive technology, but it is a desirable only for low metal ion concentrations; it becomes very expensive when dealing with high metal ion concentration AMD. Furthermore, the resins need to be regenerated when exhausted by chemicals, and this regeneration can pose a secondary pollution problem and elevates operational costs (Manahan, 2005). On the other hand, the passive treatment of AMD relies on biological, geochemical, and gravitation processes in a natural or constructed wetland ecosystem.

Conventional methods can only achieve partial treatment and have the disadvantages of producing sludge, requiring high energy consumption, and needing frequent maintenance (Ali et al., 2005). Growing global demand for clean water and increasing environmental concerns warrant the need to search for more sustainable and environmentally friendly technologies for metal ion removal from mining wastewater. Membrane technology has proven to be a promising alternative. Membrane technology, due to its easy operation, inexpensiveness, high separation efficiency, and low energy consumption, has emerged as a promising substitute to conventional methods for AMD treatment. A membrane is a thin layer of semi-permeable material that separates substances when a driving force is applied across it in a selective manner (Daramola et al., 2015). The mechanical strength, thermal stability, and chemical resistance of a membrane form part of the significant characteristics that define a good membrane and are highly dependent on the properties of material used in membrane construction (Masukume et al., 2014).

10.3 MEMBRANES FOR TREATMENT OF METAL-CONTAINING WASTEWATER IN MINING OPERATIONS

10.3.1 Organic and Inorganic Membranes

Water forms a major part of the required essentials needed during operation of a mine. There are, however, environmental concerns about the used water, which mostly is acidic and contains high concentrations of metals and sulfates. Mine wastewater must therefore be treated before being discharged into the environment or recycled back into the system (Sivakumar et al., 2013). Treatment of mine wastewater conforms to the sustainability concept in two ways: first is the removal of toxic metals and salts to reduce the water salinity for safe disposal, which will save money by avoiding penalties; second is the recovery of valuable metals from the wastewater, which could be sold to offset the operational costs of wastewater treatment processes (Jeppesen et al., 2009). Metals are contained in wastewater as ions and/or dissolved salts. Conventional treatment methods such as precipitation, coagulation and flocculation, ion exchange, flotation, and electrochemical processes have failed to effectively remove them from wastewater at reasonable costs and in an environmentally benign way (Fu and Wang, 2011). Membrane technology has proven its competitiveness towards these methods due to its high selectivity, low energy requirement, continuous operation, and minimal labor requirements (Chai et al., 1997; Kurniawan et al., 2006; Abhang et al., 2013). Most common pressure-driven membrane processes, which are distinguished by pore sizes, are microfiltration (MF), ultrafiltration (UF), nanofiltration (NF), and reverse osmosis (RO) (Mthethwa, 2015). Figure 10.1 provides an illustration of these membranes.

UF membranes are categorized by pore sizes of 5–20 nm and separate macro molecules with a molecular weight of 1,000–100,000 Da at a low transmembrane pressure. Since the size of the dissolved metal ions in their hydrated form is smaller than the pore size of UF membrane, they need to bind to large molecules of surfactants or form complexes with water-soluble polymers to increase their size before treatment (Landaburu-Aquirre et al., 2009). NF has pore sizes equivalent with a molecular weight of 200–1000 Da and can endure pressure between 10 and 34 bars, which is lower than RO. It is the intermediate membrane between UF and RO in terms of pore sizes and pressure requirement. Since all heavy metals are either bivalent or higher, NF is an effective process to remove them as it allows all monovalent salt such as NaCl to pass through,

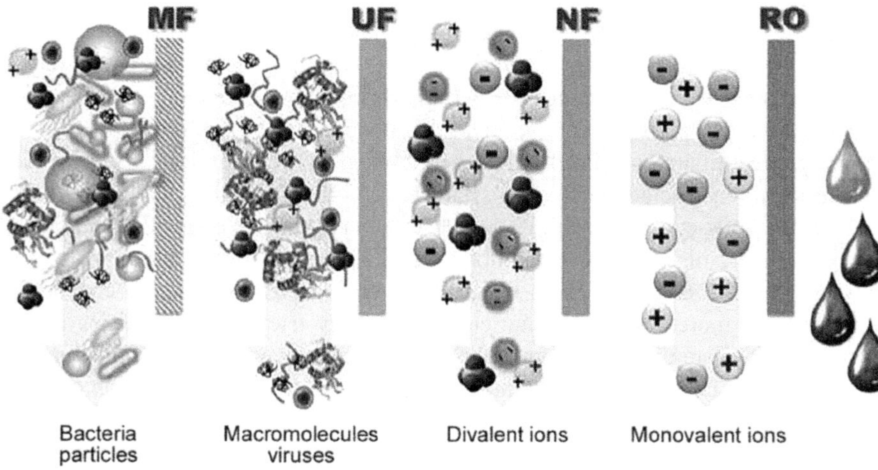

FIGURE 10.1 Pressure driven membranes. (From National Research Council, 2012, *Water Reuse: Potential for Expanding the Nation's Water Supply Through Reuse of Municipal Wastewater*, National Academies Press. With permission.)

but rejects all bivalent salts (Eriksson, 1988; Mthethwa, 2015). RO rejects all molecules including monovalent ions, but it requires a high operational pressure, which translates into high energy cost (Chan and Dudeney, 2008).

Membrane preparation employs various techniques depending on the material of construction and the final application. Normally, membrane materials are manufactured from synthetic polymers, although other forms such as ceramic (Sklari et al., 2015; Nędzarek et al., 2015) and metallic membranes have been used as membrane supports. The typical synthesis techniques used to prepare organic and inorganic membranes are sintering, stretching, and phase inversion (Drioli et al., 2006). The sintering technique normally produces symmetric ceramic or metallic membranes for UF and MF applications (Wu et al., 2013; Strathmann et al., 2010). The stretching technique normally produces polymeric membranes by stretching film or a hollow shaped homogeneous polymer with partial crystallinity perpendicularly to the axis of crystallite orientation (Bottino, 2009). Due to their intrinsic hydrophobicity, stretched membranes are used for vapor and gas separation rather than aqueous streams (Strathmann et al., 2010). The phase inversion method is most commonly used for the synthesis of polymeric membranes; it produces two distinct phases consisting of a polymer, solvent, and other additives by separating a homogeneous system.

The application of organic-based membranes for the treatment of AMD has been widely reported in literature. Georgiou et al. (2015) reported on As^{III} remediation uptake by hybrid polyethersulfone membrane decorated with zero valent iron (ZVI). Gherasim et al. (2013) investigated the removal of toxic Pb(II) ions from single salt and binary solutions using a polyamide thin-film composite NF membrane (AFC 80). Zhu et al. (2015) improved the performance of polyamide thin-film membrane using polyethersulfone-polyaniline membrane as supported and evaluated its separating performance using NaCl and $MgSO_4$. The preparation, characterization, and performance evaluation of thin film composite (TFC) NF membranes having porous polysulfone and polyvinyl alcohol as support base was reported (Gohil and Ray, 2009). A polyethersulfone (PES) membrane has been used widely for the removal of metal ions from solution due to its very attractive properties such as good thermal stability, chemical neutrality, high mechanical resistance, and wide range of pH tolerance (Wang et al., 2009; Shen et al., 2012). PES-based membranes always show asymmetric structural arrangements and are prepared via the phase inversion method (Zhao et al., 2013). The hydrophobic nature of PES membranes generated by the sulfonyl group leads to the attachment of particles in the membrane pores and surface and causes reduced permeability as a result of fouling (Arkhangelsky et al., 2007; Van der Bruggen, 2009; Khulbe et al., 2010). In 2015, Mthethwa investigated the treatment of AMD using a commercially purchased PES hollow fiber NF membrane. Even though the membrane had a larger surface area of 2 m^2, low flux was reported due to fouling and the hydrophobicity of the membrane. Different modifying agents such as surfactants (Al Malek et al., 2012), mineral fillers (Mousavi et al., 2012) or non-solvents (Vatanpour et al., 2012) could be used to increase the membrane's surface hydrophilicity (Rahimpour et al., 2010). Additionally, the PES polymer could be coated or grafted with other hydrophilic polymers to increase its hydrophilicity. PES membrane modification methods applied are plasma treatment, which adds different functional groups on the membrane surface, graft polymerization by chemically attaching hydrophilic monomers to the membrane surface, and introduction of hydrophilic components by physical preadsorption (Zhao et al., 2013).

Compared to organic membranes, inorganic membranes can withstand the effects of chemical corrosion, organic solvent attack, and high temperatures. Due to their thermally and chemically inert ceramic character, ceramic membranes can withstand a broad range in pH and temperature conditions without suffering irreversible structural changes that may affect their adsorption performance. Employment of inorganic membrane separation technology produces less sludge and does not require a high amount of energy. Advantages such as good thermal, mechanical, and chemical resistance, little pollution to the environment, and controllable micro-structure have attracted many researchers in the scientific community to the exploration of inorganic membranes for metal removal from wastewater (Chougui et al., 2014). Researchers have focused their attention on exploring easy and robust synthesis methods to fabricate inorganic membranes with low manufacturing costs and a long service life (Liu et al., 2004; Lu et al., 2007; Pendergast and Hoek, 2011). Ge et al. (2015) reported a novel free-standing and self-supporting geopolymer inorganic membrane for Ni^{2+} removal from mining wastewater using metakaolin and sodium silicate solutions. The overall cost for preparing this inorganic geopolymer membrane was lower than that of traditional commercial membranes. The morphology of the geopolymer (as obtained from scanning electron microscope (SEM) imaging) confirmed that the gels were converted into a three-dimensional network structure after dehydration, resulting in the membrane being able to capture, intercept, and fix heavy metal ions. Since the pH of the aqueous solution affects the surface charge of the membrane and Ni^{2+} precipitates at pH of 6.9, the initial pH of the feed solution during the study was <6 (Cui et al., 2008; Mihaly-Cozmuta et al., 2014). The results of the investigation by the authors confirmed that geopolymer-based membranes are not only of low cost, environmentally friendly, or convenient but are also a promising membrane material to remove Ni^{2+} and other heavy metal ions from mining wastewater.

In another report, the use of hydroxy sodalite membrane in pervaporative desalination of seawater to produce ultra-pure water was investigated and reported (Khajavi et al. 2010). In the investigation, a hydroxy sodalite membrane supported on ceramic material was synthesized and tested in pervaporative desalination of seawater at a pressure of 2.2 MPa and temperature of 303–473 K. Aqueous solutions of sodium chloride and sodium nitrate, which at times form part of the AMD constituents, were fed into the hydroxy sodalite membrane at a constant flow of 10 mL/min. The results of the investigation revealed that the hydroxy sodalite membrane effectively desalinated the sodium chloride and sodium nitrate aqueous solutions, indicating the potential of this membrane to treat AMD.

10.3.2 Mixed-Matrix Membranes

In terms of water selectivity, zeolite-based membranes display better performance when compared to polymeric membranes (Khajavi et al., 2010; Daramola et al., 2012), but the cost of the ceramic supports employed in fabricating zeolite-based membranes is very high, contributing to their huge costs overall. Although zeolite membranes are known to be thermally stable with good chemical resistance in comparison to polymer membranes, they are very fragile and brittle, which makes handling during module assembly difficult (Daramola et al., 2012). In addition, the technology for fabricating commercial zeolite membranes is still at the developmental stage, while the technology for commercial production of polymeric membranes is mature, with applications in a series of industrial processes (Morigami et al., 2001).

To overcome these shortcomings of the zeolite membranes, mixed-matrix membranes (MMMs) have been proposed. MMMs are composite membranes containing inorganic fillers (e.g., zeolites, carbon nanotubes, nanomaterials, etc.) within the matrix of polymer membranes, and MMMs aim to take great advantage of the processability, durability, permeability, and selectivity of polymers by offering the advantage of a unique surface chemistry and good mechanical properties (Goh et al., 2011; Daramola et al., 2015). The presence of crystals within the polymer chains/matrices improves the separation performance, mechanical strength, and thermal stability of polymeric membranes. The differences between polymer membranes and MMMs are shown in Table 10.1. The chemical structure of the inorganic fillers, the type of inorganic fillers, and the surface chemistry are, however, mitigating factors to obtaining high quality MMMs (Vu et al., 2003).

Various porous and non-porous inorganic fillers have been explored to improve the separation performance of polymeric membranes (Bastani et al., 2013). Carbon nanotubes, molecular sieves, zeolites, and metal organic frameworks are examples of porous inorganic fillers, while silica, titanium dioxide, and fullerene are potential non-porous fillers (Goh et al., 2011; Daramola et al., 2017). Introducing inorganic fillers to the polymeric matrix enhances the thermal and mechanical stability of MMMs and improves their performance compared to pure polymeric membranes (Aroon et al., 2010; Maphutha et al., 2013; Daramola et al., 2015, 2017).

TABLE 10.1
Differences between Pure Polymer Membranes and Mixed Matrix Membrane

Pure Polymer Membranes	Mixed Matrix Membrane
Poor mechanical properties	Good and desirable mechanical properties
Do not contain dispersed inorganic phase within the organic phase	Contain inorganic phase of unique structure dispersed within the organic phase
No inorganic phase within the organic phase to enhance the surface chemistry	Presence of inorganic phase enhances the surface chemistry
Chemical structure and type of organic phase dictate the membrane quality and separation performance	Chemical structure and type of inorganic phase and organic phase dictate the membrane quality and separation performance

For instance, a PES membrane was embedded with polyaniline-modified iron oxide nanoparticles to remove metal ions from solution (Daraei et al., 2012). The study suggests that the adsorptive performance of polymeric membranes can be improved by introducing nanoparticles during preparation as an additive (Ng et al., 2013). As observed by Daraei et al. (2012), iron oxide showed a greater affinity towards heavy metals compared to other metal oxides. The study also revealed improved adsorptive properties of Fe_3O_4 nanoparticles due to the presence of nitrogen atoms introduced by the N-H groups of the polyaniline. The hydrophobic nature of polyaniline, which caused a decrease in the water flux of the nanocomposite membrane, provoked Ghaemi et al. (2011) to investigate the use of hydrophilic materials to modify the iron oxide nanoparticles in order to improve the flux. A nano-enhanced PES MMM was fabricated by adding different types of iron oxide nanoparticles along with silica, metformin, and amine. The presence of modified Fe_3O_4 in the PES polymer resulted in an increase in the hydrophilicity and water flux of the membrane and consequently an increase in the membrane selectivity towards copper. The trend of research development and application of MMMs in mining wastewater treatment, as indicated by the progression in scientific publications, is depicted in Figure 10.2. According to Figure 10.2, there has been a dramatic increase in the number of scientific publication from 1974 to 2016, indicating active research efforts dedicated to the development and application of MMMs in wastewater treatment.

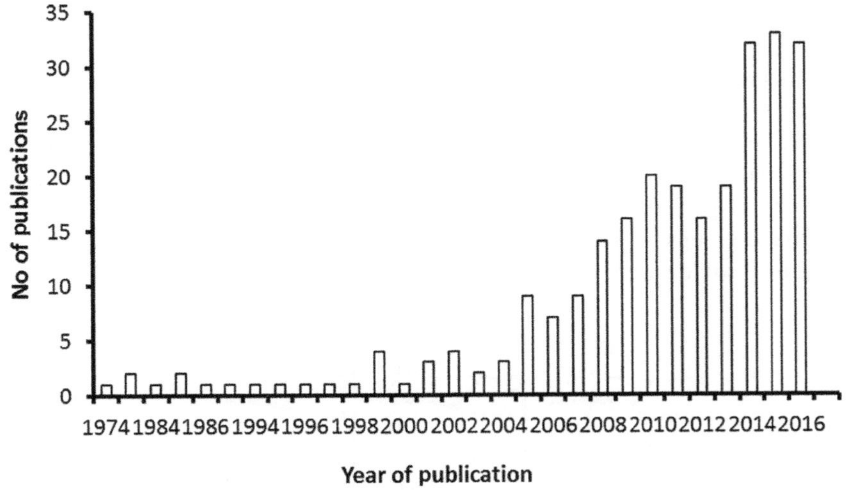

FIGURE 10.2 Scientific publications (books, book chapters, articles, conference proceedings) on mixed matrix membrane materials for wastewater treatment (Scopus 2017). (Obtained from www.scopus.com [accessed April 4, 2017].)

10.3.2.1 Chitosan-Based Composite Materials

The performance and success of membrane technology especially for wastewater treatment is highly dependent on the biodegradability and non-toxicity of the chosen materials, making natural polymeric materials such as chitosan a promising option (Wang and Wang, 2007).

According to Ma et al. (2015), chitosan is a naturally occurring biocompatible, biodegradable, bio-renewable, and non-toxic co-polymer composed of glucosamine and acetyl glucosamine. This polymer is derived from chitin, which is the second most abundant natural polysaccharide (Shukla et al., 2013), and is produced by many different living organisms such as insects, shellfish, crabs, fungi, arthropods, and crustaceans (Rinaudo, 2006). Chitin naturally occurs as three different structures namely α-, β-, and γ-structures. The γ-structure chitin is a combination of the α-structure and β-structure rather than a third allomorph. The α-chitin is the most abundant and it is found in the walls of yeast and fungal cells, shrimp cells, crab tendons and cells, as well as insect cuticle. The β-chitin is rare and is found in association with proteins in squid pens (Rinaudo, 2006). The structure of chitin, as shown in Figure 10.3, is compact and prevents it from being soluble in most solvents, making the transformation of chitin into chitosan very important (Rinaudo, 2006).

The use of chitosan as a material in wastewater treatment is economical because it is produced from a plentiful renewable resource (chitin). Other advantages of chitosan include its solubility in various media, ability to form films, and capability of binding with various metal ions (metal chelation), making it a suitable material for mining wastewater treatment (Shukla et al., 2013). Chitosan is produced from chitin under alkaline conditions or by enzymatic hydrolysis. Acetate moiety is removed from chitin under alkaline conditions through hydration to produce chitosan, a polymer that consists of amine and hydroxyl functional groups (Figure 10.3) (Shukla et al., 2013). Due to the presence of the aforementioned functional groups in chitosan, it can be applied in water and wastewater treatment. Due to its unique properties, application of chitosan-based materials has been showcased in industries such as wastewater treatment, food, cosmetics, biomedical, and agriculture. In the biomedical industry, chitosan has been found to be a good candidate for gene therapy, cell culture, and tissue engineering due to its excellent properties.

The presence of functionalities such as amine ($-NH_2$) and hydroxyl ($-OH$) in the chitosan molecules (Figure 10.4) provides the basis for the interaction with other materials. In recent times, the use of chitosan has received great attention in the development of novel functional composite materials produced either by chemical or physical modification. Physical modification is the easiest way, and is mostly done by physical mixing (Shukla et al., 2013).

Chitosan is widely known as a sorbent for the removal of heavy metals, transition metals, and dyes (Vieira and Beppu, 2006; Juang and Ju, 1997). Chitosan is the partially N-deacetylated derivative of chitin, and contains one primary amino and two free hydroxyl functional groups for each C3 and C6 building unit; chitosan can remove contaminants at near neutral solutions (Juang and Shao, 2002; Benavente et al., 2011). Chitin, which is a high molecular weight linear polymer of 2-acetamide-2-deoxy-glucopyranose units connected by 1,4-glycosidic bonds, is the second most abundant natural fiber after cellulose (Ngah and Isa, 1998). The high hydrophilicity of chitosan, generated by the large number of hydroxyl groups of glucose unit, the presence of a number of functional groups and their reactivity, and the flexible nature of the polymer chain make it an excellent biosorbent (Wan Ngaha

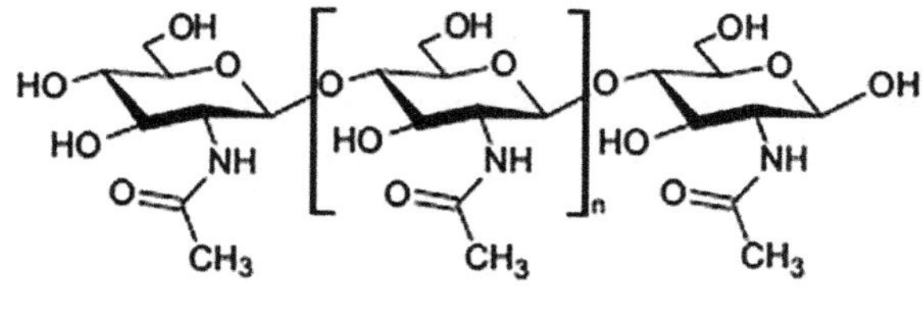

FIGURE 10.3 Structure of chitin. (From Shukla, S.K. et al., *Int. J. Biol. Macromol.*, 59, 46–58, 2013. With permission.)

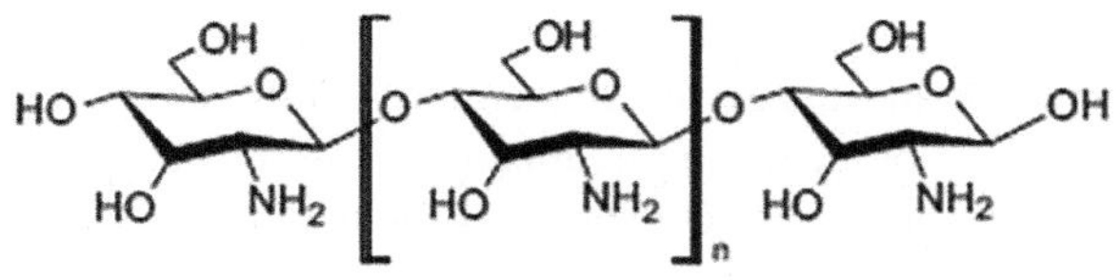

FIGURE 10.4 Structure of chitosan. (From Shukla, S.K. et al., *Int. J. Biol. Macromol.*, 59, 46–58, 2013. With permission.)

et al., 2011). The reactive amino groups bind to virtually all group III transition metals, but not to groups II and I. The protonation of amine groups of chitosan in acidic medium leads to adsorption of metal anions by ion exchange. Chitosan is characterized by its easy dissolution in many dilute mineral acids, so improving the chemical stability for better treatment of acidic solutions containing heavy metals is essential (Anirudhan and Rijith, 2012). There has been a growing trend in chemical modification of chitosan to improve its solubility and widen its application. Graft copolymerization is the most attractive modification method because it is a useful technique for modifying the chemical and physical properties of natural polymers (Jayakumar et al., 2005).

In the wastewater industry, the use of chitosan composite materials has been demonstrated. For example, Xie et al. (2013) studied chitosan-modified zeolite as an adsorbent for the removal of different wastewater pollutants. The authors synthesized the zeolite from coal fly ash and modified it with chitosan. The results obtained from the performance evaluation of the material in adsorptive treatment of wastewater indicate that the chitosan-modified zeolite enhanced the adsorption performance of zeolite material (Xie et al., 2013). In the same vein, Ngah et al. (2012) demonstrated the efficiency of a chitosan-zeolite composite as an adsorbent to remove copper from aqueous solution.

Chitosan-based membranes are preferred as membrane materials for the removal of metal ions from solution due to the ease of separation after process, high adsorption capacity, faster kinetics, better reusability, and biodegradability after use. Poor mechanical strength poses a serious concern preventing the wide application of chitosan-based membranes (Salehi et al., 2016). Figure 10.5 shows the progression of research and development in the fabrication and application of chitosan-based MMMs in mining wastewater treatment. There has been a dramatic increase in the number of scientific publications on chitosan-based MMMs from 1983 to 2016, indicating the active research efforts dedicated to the development and application of chitosan-based membranes for wastewater treatment.

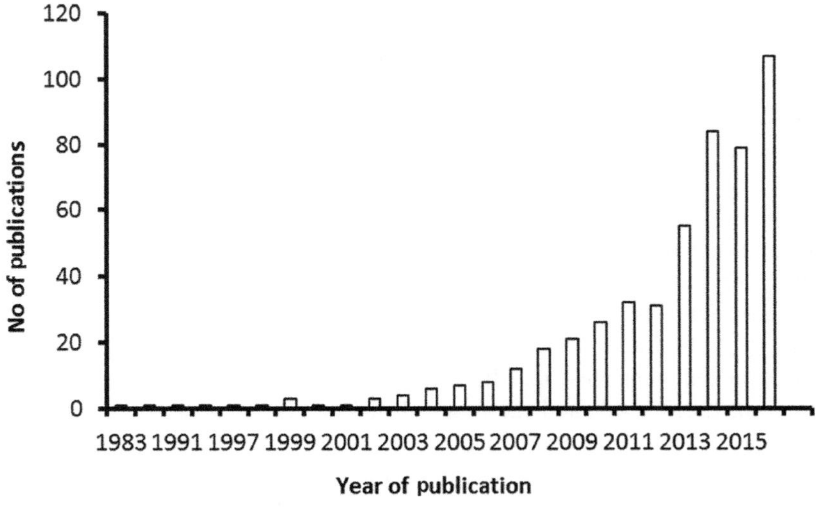

FIGURE 10.5 Scientific publications (books, book chapters, articles, conference proceedings) on chitosan based materials for wastewater treatment (Scopus 2017). (Obtained from www.scopus.com [accessed April 4, 2017].)

For example, Akbari et al. (2015) studied the influence of coating chitosan onto a polyamide/nanofiltration (PA/NF) membrane and evaluated the morphology, anti-fouling properties, and separation performance of the membrane during treatment of mining wastewater. Morphological characterization and the study of the surface chemistry of the membrane confirmed the occurrence of interfacial polymerization in the organic phase. In addition, the results show that amine monomers diffused through the polyamide (PA) layer already formed and reacted with acyl chloride in the organic side of the interface region (Kim et al., 2005; Akbari et al., 2015). The authors also observed that coating a PA/NF membrane with chitosan increased the thickness of the membrane and more PA amine monomers diffuse into the PA layer. This in turn led to fewer amine groups and more un-reacted acyl chloride groups on the organic phase side. When the membrane was exposed to water, the un-reacted acyl chloride groups were hydrolyzed into carboxylic acid. Although hydrolysis of acyl chloride is a relatively slower reaction compared to reaction of acyl chloride with chitosan (amine and hydroxyl groups), the occurrence makes modification of the PA surface layer possible by reacting the un-reacted acyl chloride with the amine and hydroxyl groups of chitosan.

In the same vein, adsorption of platinum and palladium from aqueous medium using ethylene-diamine-modified magnetic chitosan nanoparticles (EMCN) has been demonstrated (Zhou et al., 2010). The study revealed a series of mechanism involved in the modification step, including: (1) coordination of amino groups in a suspended manner or in combination with adjacent hydroxyl groups; (2) electrostatic attraction in acidic media; (3) ion exchange with protonated amine groups through proton or anion exchange, the counter ion being exchanged with the metal ion. According to the authors, the electrostatic attraction of anionic metal complexes was also induced by the protonation of the amine groups on the EMCN, which increased the available binding sites for the precious metal ion uptake. Furthermore, polyvinyl alcohol (PVA)/chitosan composite membrane was examined for the removal of Co^{2+} from radioactive wastewater (Zhu et al., 2014). In comparison with other adsorbents, the magnetic PVA-chitosan composite beads prepared by Zhu et al. displayed a maximum adsorption capacity of 14.39 mg/g (obtained from Langmuir model) (Zhu et al., 2014). In other studies, Wang et al. (2011) reported a maximum adsorption capacity of 8.85 mg/g for applying magnetic multi-walled carbon nanotube–iron oxide composites to the removal of cobalt from wastewater. In addition, Ma et al. (2015) found that the adsorption capacity of phosphate-modified montmorillonite (PMM) during the removal of cobalt from solution was 12.39 mg/g. These authors showed based on the results of Fourier Transform Infrared (FTIR) spectroscopy and SEM analyses that the functional groups, $-NH_2$ and $-OH$, participated in the adsorption process.

Uranium removal from wastewater by PES and polyamide NF membranes was studied and compared under various operating conditions (feed pH, uranium feed concentration, and transmembrane pressure) by Torkabad et al. (2017). It was found that the charge properties of both membranes and uranium constituents changed with pH. In addition, the membrane flux and selectivity toward uranium were better in the polyamide membrane than in the PES membrane. However, the authors concluded that both NF membranes could effectively be used for uranium removal from wastewater. In the same vein, Reiad et al. (2012) reported the performance of microporous chitosan/polyethylene glycol mixed matrix membrane during adsorptive removal of iron and manganese from wastewater. The polymeric blend of chitosan and polyethylene glycol (PEG) mixture was thoroughly stirred before adding 2% solution of glutaraldehyde, a cross-linking agent, that provided a link between the two polymers. According to the authors, the membranes displayed very good performance for the removal of iron and manganese ions from the wastewater sample tested and they are re-usable.

Gherasim et al. (2013) evaluated the removal of lead (II) using a polyamide NF membrane. A thin-film composite membrane consisting of aromatic polyamide skin-layer on a polysulfone substrate was used in the study. The authors attributed the 100% selectivity displayed by the membrane to the smaller average pore size of the membrane. In addition, the authors reported the isoelectric point (IEP) of the membrane to be at pH of 3.6. At a pH > IEP, the membrane was negatively charged and at pH < IEP the membrane was positively charged. Ammonium ($-NH_3^+$) and carboxyl ($-COOH$) groups were formed because of partial hydrolysis of polyamide, in which below the IEP the carboxyl groups are dissociated while the amine is protonated and the membrane becomes positively charged. In contrast, above the IEP, the carboxyl groups are dissociated and the membrane becomes negatively charged. It was observed that the membrane was positively charged in $Pb(NO_3)_2$ solutions with pH ≤ 6, indicating the ability of the membrane to remove lead from model solution at a near neutral pH.

Chethan and Vishalakshi (2013) achieved selective modification of chitosan through incorporation of ethylene-1,2-diamine molecule for the removal of divalent ions. The study employed chitosan derivative ethylene-1,2-diamine-6-deoxy-chitosan (CtsEn) and its pthaloylated precursor ethylene-1, 2-diamine-6-diamine-N-pthaloylchitosan (PtCtsEn) for the removal of divalent ions (copper, zinc, and lead) from a model solution used as the wastewater. The prepared chitosan derivatives were shown to have higher metal ion adsorption capacity compared to the parent chitosan due to the introduction of additional $-NH_2$ groups. Desorption studies conducted on the membrane also revealed possible successful regeneration of the adsorbents. The desorption tests carried out with an aqueous salt solution of pH 12 showed 92% of Cu and 97% of Zn desorbing from the loaded PtCtsEn, while 83% of Cu and 85% of Zn desorbed from the CtsEn samples.

A novel MMM containing chitosan beads for efficient removal of metal ions from a model aqueous solution used as wastewater was developed by Tetala and Stamatialis (2013). Ethylene Vinyl Alcohol (EVAL) was used as the polymeric material for the preparation of the MMM. A homogeneous precursor mixture containing the EVAL solution and chitosan beads was cast on a glass plate using a 400 μm casting knife and immersed immediately inside a coagulation bath containing water at 45°C to remove the solvent. SEM was used to check the morphology of the membrane. The images show that the membrane possessed highly interconnected pores that provided good access to the chitosan beads in the matrix. Performance evaluation of the membrane using a model solution as wastewater reveals high adsorption capacity and faster adsorption rate of the membrane compared to other chitosan membranes. The membrane displayed a maximum adsorption capacity of 225.7 mg/g adsorbed Cu^{2+} compared to other membranes that showed 5.9 mg/g Cu^{2+} (Ghaee et al., 2010), 10.79 mg/g Cu^{2+} (He et al., 2008), 200 mg/g Cu^{2+} (Steenkamp et al., 2002), and 35.3–48.2 mg/g Cu^{2+} (Liu and Bai, 2006). In another study, Habiba et al. (2017) fabricated a chitosan-PVA-zeolite membrane via the electrospinning technique and tested it for the removal of Cr^{6+}, Fe^{3+}, and Ni^{2+} adsorption from wastewater. Adsorption of Cr^{6+} on the material was well described by a pseudo-second-order kinetic model, while the adsorption of Fe^{3+} and Ni^{2+} satisfied the Lagergren first-order model. Comparison of the different types of chitosan-based membrane materials and their application in the treatment of wastewater as reported in literature is provided in Table 10.2.

10.3.2.2 Sodalite-Based MMMs

Hydroxy sodalite (HS) is a crystalline and hydrophilic zeolite type made of cubic array of β-cages as primary building blocks (Breck, 1974; Khajavi et al., 2007). HS has cage window dimension of ~0.265 nm, making it a good candidate for separating mixtures that contain small molecules such as water (~0.27 nm), helium (~0.26 nm), and ammonia (~0.25 nm) (Breck, 1974). Figure 10.6 depicts the structure of an HS crystal.

Previous studies on the synthesis of HS have demonstrated formation of pure HS crystals between 363 and 413 K (Khajavi et al., 2010). Consequently, HS membranes were prepared on asymmetric α-alumina-TiO_2 disks (average pore size distribution of the TiO_2 top layer 80 nm) at 413 K. The synthesized membranes were applied to a study series of industrial processes like the dewatering of alcohols, dehydrating of organic acids, and desalination of seawater to produce ultra-pure water (Khajavi et al., 2009, 2010, 2017). Furthermore, potential application of HS membranes was demonstrated in the esterification-coupling reaction and separation process for selective water removal from the product mixture (Khajavi et al., 2010). Also, potential application of the HS was showcased (via a mathematical modeling approach) in the Fischer–Tropsch (FT) process to remove water, the FT catalyst deactivation promoter, from the product mixture (Khajavi et al., 2010). In all the aforementioned applications of the synthesized

TABLE 10.2
Metal Ion Adsorption Performance of Chitosan Based Mixed Matrix Membranes During Wastewater Treatment

Membrane Material	Degree of Deacetylation	Preparation Technique	Metal Ions in the Feed	Operating Conditions	Rate Constant (L/g)	q_{max} (mg/g)	References
Chitosan/PVA/Zeolite	84%	Electrospinning	Cr^{4+}	Time (0–7 min)	69.68	8.84	
			Fe^{3+}	Temperature (N/A)	31.81	6.138	Habiba et al. (2017)
			Ni^{2+}	Concentration (1–60 mg/L)	855.26	1.76	
				pH (3,7, and 10)			
Chitosan/hydroxyapatite	92%	Electrospinning	Pb^{4+}	Time (0–30 min)	19.38	296.7	
			Co^{3+}	Temperature (298–318 K)	19.01	213.8	Aliabadi et al. (2014)
			Ni^{2}	Concentration (10–500 mg/L)	17.2	180.2	
Microporous chitosan	90%	Particulate Leaching	Cu^{2+}	Time 24 h	195.64	25.64	
			Ni^{2+}	Temperature 298 K	684.72	10.3	Ghaee et al. (2012)
				Concentration (50–800 mg/L)			
				pH 5			
Chitosan/PVA/MWCNT-NH_2	90%	Casting and evaporation	Cu^{3+}	Time 24 h	218.55	25.63	
				Temperature (293–313 K)			Salehi et al. (2012)
				Concentration (1–30 mg/L)			
				pH 5,5			
Cross linked chitosan	N/A	Crosslinking	Cr^{4+}	Time 24 h	0.62	1420	
				Temperature 298 K			Baroni et al. (2008)
				Concentration (250–1000 mg/L)			
				pH (2 and 6)			
Chitosan/Thiourea	90%	Water/Oil emulsion	Pt^{4+}	Time 4 h	389	129.9	
		Crosslinking	Pd^{2+}	Temperature (298–328 K)	978	112.4	Zhou et al. (2009)
				Concentration 30 mg/L			
				pH (1–9)			
Sulfated chitosan/PVA	90%	Thin film casting	Cu^{2+}	Time (30–180 min)	228.9	80.65	
			Ni^{2+}	Temperature (298–328 K)	709	35.46	Abu-Saied et al. (2017)
				Concentration (10–100 mg/L)			
				pH (2–6)			

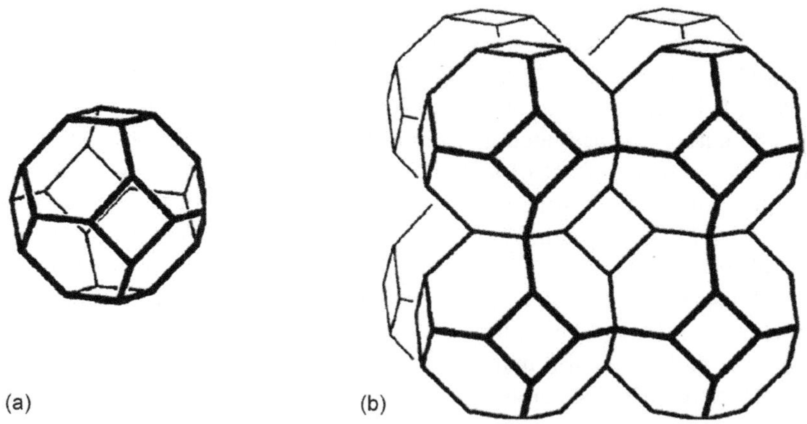

FIGURE 10.6 Crystal structure of hydroxy sodalite showing the primary building block (a) β-cage and (b) sodalite structure.

HS membranes, water flux (between 303 and473 K) and water-to-ethanol separation factor were 0.2–2.3 $kg{\cdot}m^{-2}{\cdot}h^{-1}$ and $>10^6$, respectively (Khajavi et al., 2009, 2010; Rohde et al., 2008; Khajavi 2010). Furthermore, other potential applications of HS membranes could be found in membrane reactors to remove water from product mixture during acetalization of ethanol with butyraldehyde to 1, 1 diethoxy butane (Agirre et al., 2012), during cyclization and methylation of γ-aminobutyric acid (GABA) to N-methylpyrrolidone (NMP) (Lammens et al., 2010) and as membranes for separating azeotropic mixtures in industrial processes (Khajavi et al., 2010). Apart from the use of HS as membrane, as a nanocrystalline material, HS crystals are explored in catalysis (Kimura et al., 2008; Makgaba and Daramola, 2015) and in pigment occlusion (Naskar et al., 2011a; Naskar et al., 2011b).

Application of HS-based materials (either as adsorbent or membranes) in wastewater treatment (even in mining wastewater treatment) is still limited. The progression of research and development in the development and application of sodalite-based MMMs in mining wastewater treatment, as indicated by the scientific publications, is depicted in Figure 10.7. According to Figure 10.7, there

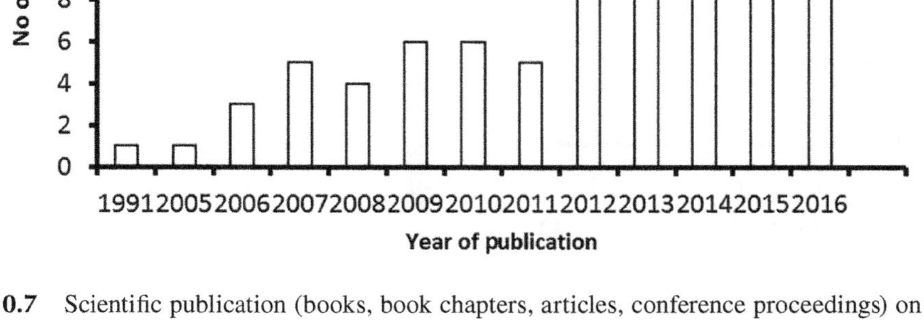

FIGURE 10.7 Scientific publication (books, book chapters, articles, conference proceedings) on HS based materials for wastewater treatment (Scopus 2017). (Obtained from www.scopus.com [accessed April 4, 2017].)

has been a dramatic increase in the number of scientific publications from 1991 to 2016, indicating the active research efforts dedicated to the development and application of HS-based MMMs.

As far as could be ascertained, the first open report on the synthesis and performance evaluation of HS-PES MMM for AMD treatment was from Daramola and his co-workers (Daramola et al., 2015). The authors investigated and reported on the synthesis of a HS-PES membrane for AMD treatment and the effect of sodalite loading in the membrane on the performance of the membranes for AMD treatment. The typical cubic shape of the HS crystals and their presence within the PES were confirmed from SEM images (Figure 10.8). In the study, the crystallinity and surface chemistry of the synthesized HS crystals were confirmed using X-ray diffraction and FTIR spectroscopy, respectively (Daramola et al., 2015). Maximum water flux through the membrane during the treatment of real mining wastewater (AMD) obtained from an abandoned mine in the area of Johannesburg in South Africa was about 2 $g \cdot cm^{-2} \cdot min^{-1}$ at HS loading of 10% (Figure 10.9). The authors attributed the low membrane flux to the thickness of the membrane, which was between 30 and 50 μ. The enhancement of the mechanical strength of the PES membrane with the incorporation of the HS crystals is depicted in Figure 10.9b. Figure 10.9a reveals that the mechanical strength of the HS-PES membrane increased with an increase in HS loading up to 15% loading. Furthermore, the membrane displayed a maximum membrane selectivity of about 60% to Pb^{2+} and a minimum selectivity of about 7% to Mn^{2+} in

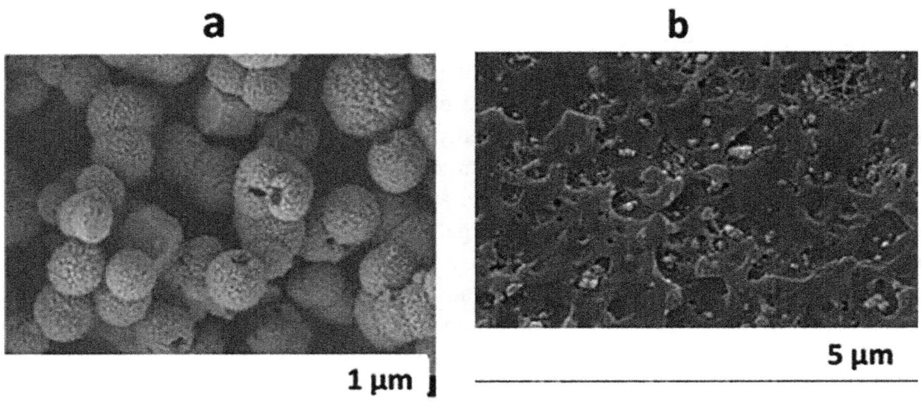

FIGURE 10.8 (a) SEM image of the HS crystal and (b) the HS-PES membranes with HS crystals dispersed within the polymer matrix (for 10% loading image). (Adapted from Daramola, M.O. et al., *J. South. Afr. Inst. Min. Metall.*, 115, 1221–1228, 2015.)

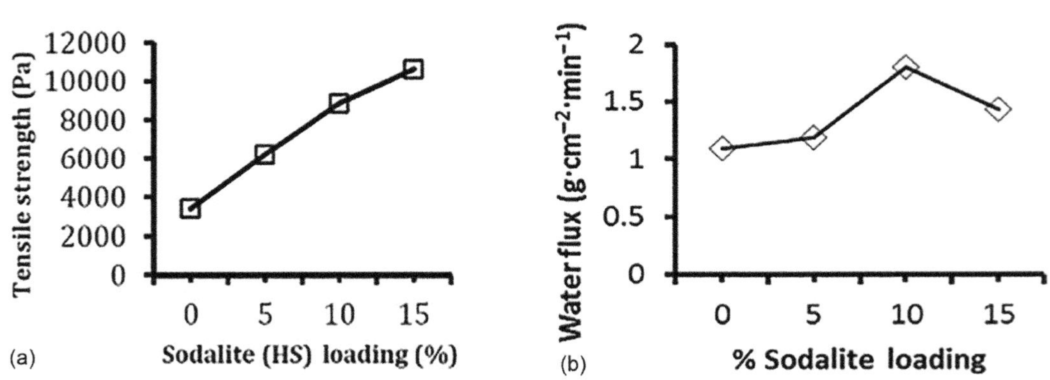

FIGURE 10.9 Tensile strength of the membrane (a) and membrane/water flux (b) as function of HS loading. (Adapted from Daramola, M.O. et al., *J. South. Afr. Inst. Min. Metall.*, 115, 1221–1228, 2015.)

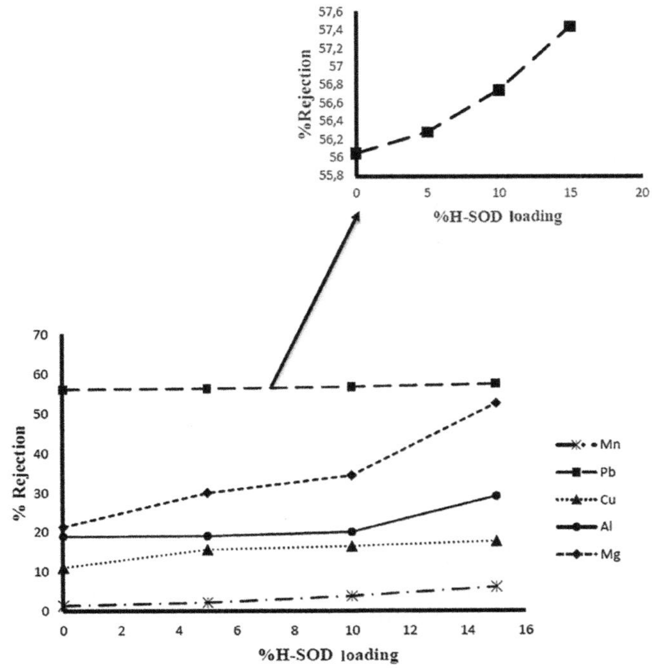

FIGURE 10.10 Performance of the membrane during AMD treatment as a function of HS loading. (Adapted from Daramola, M.O. et al., *J. South. Afr. Inst. Min. Metall.*, 115, 1221–1228, 2015.)

the AMD at 15% HS loading (Figure 10.10). The authors attributed this performance to the size of the metal ions present in the AMD. Since the average cage dimension of HS crystals is 0.26 nm, the separation performance of the membrane depends on size exclusion. Based on experimental results, the authors concluded that loading HS crystals within the PES polymer matrix enhanced the performance (selectivity and flux) of the pure polymeric membrane. It was also concluded, based on the characterization of the synthesized HS-PES MMM that the ratio of fractional free volume of the polymer to the amount of HS crystals within the PES polymer matrix decreased with increased HS loading.

10.4 CONCLUSIONS AND FUTURE OUTLOOK

Two commonly known principal difficulties associated with membrane operation are concentration polarization and fouling. Membrane fouling is a more serious concern than concentration polarization, but both need to be prevented during membrane system operation. During the membrane separation process, the natural consequence of semi-permeability and selectivity of a membrane result in the accumulation of rejected solutes or particles on the membrane surface. This process, which is reversible, is termed concentration polarization (Li et al., 2016). When the feed flow approaches the membrane surface, solvent molecules penetrate through the membrane, but the solutes are rejected and retained. These rejected molecules are relatively slow to diffuse back into the bulk solution and cause a concentration gradient just above the membrane surface. The concentration accumulation obstructs solvent flow through the membrane and creates osmotic backpressure that reduces the effective transmembrane pressure (TMP) of the system. Concentration polarization is an inevitable but reversible phenomenon that does not affect the fundamental properties of the membrane (Hughes et al., 2007).

Membrane fouling can be described as a phenomenon that takes place when suspended or dissolved matter in the feed solution migrate from the liquid phase to form deposits either on the membrane surface, at the pore openings, or within the porous structure of the membrane (Koros et al., 1996). It is possible to prevent membrane fouling before its occurrence through methods such as feed stream pre-treatment, chemical modification to improve the anti-fouling properties of the membrane, and optimization of operational conditions (Cheng et al., 2016). Commonly trouble-causing substances—termed foulants in membrane technology—are roughly divided into four categories: particulates, macromolecules, ions, and biological substances. Natural organic matter is the most problematic because it exists in natural water and can cause both reversible and irreversible fouling (Zularisam et al., 2007; Guo et al., 2009). Parameters affecting membrane fouling include the nature and concentration of solutes and solvents, membrane materials, surface characteristics, pore size, and the distribution and hydrodynamics of the membrane modules. In contrast to natural concentration polarization, which is reversible, fouling may cause irreversible loss of membrane permeability. The relative resistance to cleaning is a distinguishing factor between reversible and irreversible fouling. Reversible fouling can easily be removed with certain cleaning methods, while irreversible fouling remains even after cleaning. Fouling that cannot be cleaned by hydraulic means is called hydraulically irreversible fouling. Similarly, fouling left over after chemical cleaning is called chemically irreversible fouling (Kimura et al., 2004).

Another challenge in the application of MMMs for wastewater treatment, for instance the use of sodalite-based membranes for wastewater treatment, is the low membrane flux and membrane reproducibility (Daramola et al., 2012, 2015). Development of dependable and proven robust synthesis techniques for the fabrication of reproducible HS-PES membranes with high selectivity to metal ions such Mn^{2+} from the AMD is essential to the commercialization of HS-PES membranes for industrial application. In addition, developing ultra-thin membranes (of about 1 μm in effective membrane thickness) could enhance membrane flux. For industrial application, a tubular membrane configuration is preferred to a flat-sheet membrane, and the enhancement of the surface area–volume ratio of the tubular configuration could be instrumental to the efficient application of the membrane in industry. Instead of casting the membrane precursor solution into a flat-sheet HS-PES membrane, the precursor solution could be made into fibers using the electrospinning technique. Eletrospun fibers could be assembled into bundles, as illustrated in Figure 10.11, thereby enhancing the surface area–volume ratio of the membrane module. Lastly, parametric optimization studies and operational stability studies of the performance of the membrane using various real mining wastewater (AMD) might be necessary to benchmark the performance and integrity of the membrane.

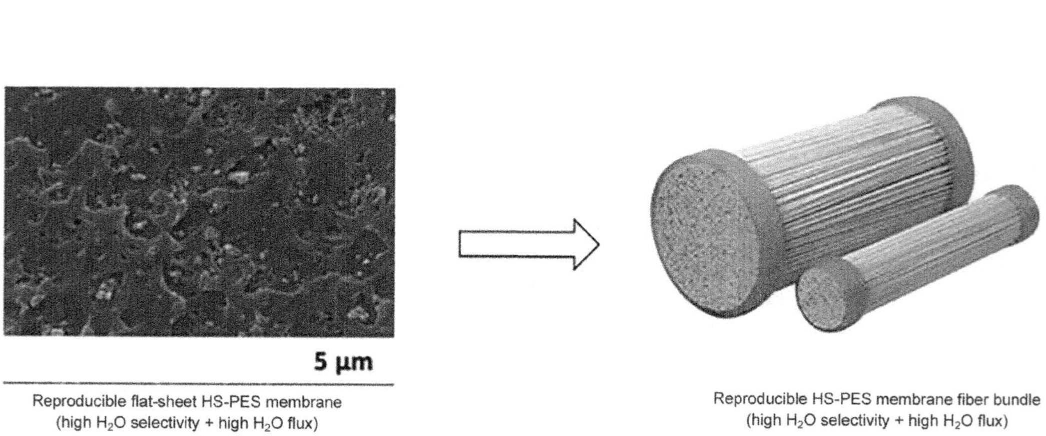

FIGURE 10.11 Proposed scale-up configuration from HS-PES membrane flat sheet to bundle of HS-PES fibers for industrial application.

REFERENCES

Abhanga, R., Wanib, K., Patilc, V., Pangarkara, B. and Parjanea, S., 2013. Nanofiltration for recovery of heavy metal ions from waste water—A review. *Liver, 1*, 0–60.

Abu-Saied, M.A., Wycisk, R., Abbassy, M.M., El-Naim, G.A., El-Demerdash, F., Youssef, M.E., Bassuony, H. and Pintauro, P.N., 2017. Sulfated chitosan/PVA absorbent membrane for removal of copper and nickel ions from aqueous solutions—Fabrication and sorption studies. *Carbohydrate Polymers, 165*, 149–158.

Agirre, I., Güemez, M.B., Motelica, A., van Veen, H.M., Vente, J.F. and Arias, P.L., 2012. The conceptual design of a continuous pervaporation membrane reactor for the production of 1, 1-diethoxy butane. *AIChE Journal, 58*(6), 1862–1868.

Akbari, A., Derikvandi, Z. and Rostami, S.M.M., 2015. Influence of chitosan coating on the separation performance, morphology and anti-fouling properties of the polyamide nanofiltration membranes. *Journal of Industrial and Engineering Chemistry, 28*, 268–276.

Akcil, A. and Koldas, S., 2006. Acid mine drainage (AMD): Causes, treatment and case studies. *Journal of Cleaner Production, 14*, 1139–1145.

Al Malek, S.A., Seman, M.A., Johnson, D. and Hilal, N., 2012. Formation and characterization of polyethersulfone membranes using different concentrations of polyvinylpyrrolidone. *Desalination, 288*, 31–39.

Ali, N., Mohammad, A.W., Jusoh, A., Hasan, M.R., Ghazali, N. and Kamaruzaman, K., 2005. Treatment of aquaculture wastewater using ultra low pressure asymmetric polyethersulfone (PES) membrane. *Desalination, 185*, 317–326.

Aliabadi, M., Irani, M., Ismaeili, J. and Najafzadeh, S., 2014. Design and evaluation of chitosan/hydroxyapatite composite nanofiber membrane for the removal of heavy metal ions from aqueous solution. *Journal of the Taiwan Institute of Chemical Engineers, 45*(2), 518–526.

Anirudhan, T.S. and Rijith, S., 2012. Synthesis and characterization of carboxyl terminated poly (methacrylic acid) grafted chitosan/bentonite composite and its application for the recovery of uranium (VI) from aqueous media. *Journal of Environmental Radioactivity, 106*, 8–19.

Arkhangelsky, E., Kuzmenko, D. and Gitis, V., 2007. Impact of chemical cleaning on properties and functioning of polyethersulfone membranes. *Journal of Membrane Science, 305*, 176–184.

Aroon, M.A., Ismail, A.F., Matsuura, T. and Montazer-Rahmati, M.M., 2010. Performance studies of mixed matrix membranes for gas separation: A review. *Separation and Purification Technology, 75*(3), 229–242.

Baroni, P., Vieira, R.S., Meneghetti, E., Da Silva, M.G.C. and Beppu, M.M., 2008. Evaluation of batch adsorption of chromium ions on natural and crosslinked chitosan membranes. *Journal of Hazardous Materials, 152*(3), 1155–1163.

Bastani, D., Esmaeili, N. and Asadollahi, M., 2013. Polymeric mixed matrix membranes containing zeolites as a filler for gas separation applications: A review. *Journal of Industrial and Engineering Chemistry, 19*(2), 375–393.

Benavente, M., Moreno, L. and Martinez, J., 2011. Sorption of heavy metals from gold mining wastewater using chitosan. *Journal of the Taiwan Institute of Chemical Engineers, 42*(6), 976–988.

Bottino, A., Capannelli, G., Comite, A., Ferrari, F., Firpo, R. and Venzano, S., 2009. Membrane technologies for water treatment and agroindustrial sectors. *Comptes Rendus Chimie, 12*(8), 882–888.

Breck, D.W., 1974. *Zeolite Molecular Sieves: Structure, Chemistry and Use*, New York: Wiley.

Chai, X., Chen, G., Po-Lock, Y. and Mi, Y., 1997. Pilot scale membrane separation of electroplating waste water by reverse osmosis. *Journal of Membrane Science, 123*(2), 235–242.

Chan, B.K.C. and Dudeney, A.W.L., 2008. Reverse osmosis removal of arsenic residues from bioleaching of refractory gold concentrates. *Minerals Engineering, 21*(4), 272–278.

Cheng, X., Liang, H., Ding, A., Qu, F., Shao, S., Liu, B., Wang, H., Wu, D. and Li, G., 2016. Effects of preozonation on the ultrafiltration of different natural organic matter (NOM) fractions: Membrane fouling mitigation, prediction and mechanism. *Journal of Membrane Science, 505*, 15–25.

Chethan, P.D. and Vishalakshi, B., 2013. Synthesis of ethylenediamine modified chitosan and evaluation for removal of divalent metal ions. *Carbohydrate Polymers, 97*(2), 530–536.

Choudhury, B.U., Malang, A., Webster, R., Mohapatra, K.P., Verma, B.C., Kumar, M., Das, A. and Hazarika, S., 2017. Acid drainage from coal mining: Effect on paddy soil and productivity of rice. *Science of the Total Environment, 583*, 344–351.

Chougui, A., Zaiter, K., Belouatek, A. and Asli, B., 2014. Heavy metals and color retention by a synthesized inorganic membrane. *Arabian Journal of Chemistry, 7*(5), 817–822.

Cole, M., Wrubel, J., Henegan, P., Janzen, C., Holt, J. and Tobin, T., 2011. Development of a small-scale bioreactor method to monitor the molecular diversity and environmental impacts of bacterial biofilm communities from an acid mine drainage impacted creek. *Journal of Microbiological Methods, 87*(1), 96–104.

Cui, X.M., Zheng, G.J., Han, Y.C., Su, F. and Zhou, J., 2008. A study on electrical conductivity of chemosynthetic Al_2O_3–$2SiO_2$ geoploymer materials. *Journal of Power Sources*, *184*(2), 652–656.

Daraei, P., Madaeni, S.S., Ghaemi, N., Salehi, E., Khadivi, M.A., Moradian, R. and Astinchap, B., 2012. Novel polyethersulfone nanocomposite membrane prepared by PANI/Fe_3O_4 nanoparticles with enhanced performance for Cu (II) removal from water. *Journal of Membrane Science*, *415*, 250–259.

Daramola, M.O., Aransiola, E.F. and Ojumu, V.T., 2012. Potential applications of zeolite membranes in reaction coupling separation processes, *Materials*, *5*, 2101–2136.

Daramola, M.O., Hlanyane, P., Sadare, O.O., Oluwasina, O.O. and Iyuke, S.E., 2017. Performance of carbon nanotube/polysulfone (CNT/Psf) composite membranes during oil-water mixture separation: Effect of CNT dispersion method, *Membranes*, *7*, 1–15.

Daramola, M.O., Silinda, B., Masondo, S. and Oluwasina, O.O., 2015. Polyethersulphone-sodalite (PES-SOD) mixed-matrix membranes: Prospects for acid mine drainage (AMD) treatment. *Journal of the Southern African Institute of Mining and Metallurgy*, *115*(12), 1221–1228.

DeNicola, D.M. and Stapleton, M.G., 2002. Impact of acid mine drainage on benthic communities in streams: The relative roles of substratum versus aqueous effects. *Environmental Pollution*, *119*(3), 303–315.

Department of Water Affairs and Forestry, 2004. *Olifants Water Management Area: Internal Strategic Perspective*. Report PWWA04/00/00/0340, Pretonia, South Africa.

Drioli, E., Curcio, E., Di Profio, G., Macedonio, F. and Criscuoli, A., 2006. Integrating membrane contactors technology and pressure-driven membrane operations for seawater desalination: Energy, exergy and costs analysis. *Chemical Engineering Research and Design*, *84*(3), 209–220.

Durand, J.F., 2012. The impact of gold mining on the Witwatersrand on the rivers and karst system of Gauteng and North-West province, South Africa. *Journal of African Earth Sciences*, *68*, 24–43.

Elliott, P., Ragusa, S. and Catcheside, D., 1998. Growth of sulfate-reducing bacteria under acidic conditions in an upflow anaerobic bioreactor as a treatment system for acid mine drainage. *Water Research*, *32*(12), 3724–3730.

Eriksson, P., 1988. Nanofiltration extends the range of membrane filtration. *Environmental Progress*, *7*(1), 58–62.

Ford, K., 2003. *Passive Treatment Systems for Acid Mine Drainage*. Technical Note 409. Bureau of Land Management, Denver, CO.

Fu, F. and Wang, Q., 2011. Removal of heavy metal ions from wastewaters: A review. *Journal of Environmental Management*, *92*(3), 407–418.

Ge, Y., Cui, X., Kong, Y., Li, Z., He, Y. and Zhou, Q., 2015. Porous geopolymeric spheres for removal of Cu (II) from aqueous solution: Synthesis and evaluation. *Journal of Hazardous Materials*, *283*, 244–251.

Georgiou, Y., Dimos, K., Beltsios, K., Karakassides, M.A. and Deligiannakis, Y., 2015. Hybrid polysulfone–Zero Valent Iron] membranes: Synthesis, characterization and application for as III remediation. *Chemical Engineering Journal*, *281*, 651–660.

Ghaee, A., Shariaty-Niassar, M., Barzin, J. and Matsuura, T., 2010. Effects of chitosan membrane morphology on copper ion adsorption. *Chemical Engineering Journal*, *165*(1), 46–55.

Ghaee, A., Shariaty-Niassar, M., Barzin, J. and Zarghan, A., 2012. Adsorption of copper and nickel ions on macroporous chitosan membrane: Equilibrium study. *Applied Surface Science*, *258*(19), 7732–7743.

Ghaemi, N., Madaeni, S.S., Alizadeh, A., Rajabi, H. and Daraei, P., 2011. Preparation, characterization and performance of polyethersulfone/organically modified montmorillonite nanocomposite membranes in removal of pesticides. *Journal of Membrane Science*, *382*(1), 135–147.

Gherasim, C.V., Cuhorka, J. and Mikulášek, P., 2013. Analysis of lead (II) retention from single salt and binary aqueous solutions by a polyamide nanofiltration membrane: Experimental results and modelling. *Journal of Membrane Science*, *436*, 132–144.

Goh, P.S., Ismail, A.F., Sanip, S.M., Ng, B.C. and Aziz, M., 2011. Recent advances of inorganic fillers in mixed matrix membrane for gas separation. *Separation and Purification Technology*, *81*(3), 243–264.

Gohil, J.M. and Ray, P., 2009. Polyvinyl alcohol as the barrier layer in thin film composite nanofiltration membranes: Preparation, characterization, and performance evaluation. *Journal of Colloid and Interface Science*, *338*(1), 121–127.

Guo, X., Li, Q., Hu, W., Gao, W. and Liu, D., 2009. Ultrafiltration of dissolved organic matter in surface water by a polyvinylchloride hollow fiber membrane. *Journal of Membrane Science*, *327*(1), 254–263.

Habiba, U., Afifi, A.M., Salleh, A. and Ang, B.C., 2017. Chitosan/(polyvinyl alcohol)/zeolite electrospun composite nanofibrous membrane for adsorption of Cr^{6+}, Fe^{3+} and Ni^{2+}. *Journal of Hazardous Materials*, *322*, 182–194.

Hallberg, K.B., 2010. New perspectives in acid mine drainage microbiology. *Hydrometallurgy*, *104*(3), 448–453.

He, Z.Y., Nie, H.L., Branford-White, C., Zhu, L.M., Zhou, Y.T. and Zheng, Y., 2008. Removal of Cu^{2+} from aqueous solution by adsorption onto a novel activated nylon-based membrane. *Bioresource Technology*, *99*(17), 7954–7958.

Hughes, T. and Taha, Z., 2007. Cui, Mass transfer: Membrane processes, in: S.S. Sablani (Ed.), *Handbook of Food and Bioprocess Modelling Techniques*, Boca Raton, FL: CRC Press, pp. 145–177.

Jayakumar, R., Prabaharan, M., Reis, R.L. and Mano, J., 2005. Graft copolymerized chitosan—Present status and applications. *Carbohydrate Polymers*, *62*(2), 142–158.

Jeppesen, T., Shu, L., Keir, G. and Jegatheesan, V., 2009. Metal recovery from reverse osmosis concentrate. *Journal of Cleaner Production*, *17*(7), 703–707.

Johnson, D.B. and Hallberg, K.B., 2005. Acid mine drainage remediation options: A review. *Science of the Total Environment*, *338*(1), 3–14.

Johnson, D.B., 1998. Biodiversity and ecology of acidophilic microorganisms. *FEMS Microbiology Ecology*, *27*(4), 307–317.

Jong, T. and Parry, D.L., 2003. Removal of sulfate and heavy metals by sulfate reducing bacteria in short-term bench scale upflow anaerobic packed bed reactor runs. *Water Research*, *37*(14), 3379–3389.

Juang, R.S. and Ju, C.Y., 1997. Equilibrium sorption of Copper (II) ethylenediaminetetraacetic acid chelates onto cross-linked, polyaminated chitosan beads. *Industrial & Engineering Chemistry Research*, *36*(12), 5403–5409.

Juang, R.S. and Shao, H.J., 2002. A simplified equilibrium model for sorption of heavy metal ions from aqueous solutions on chitosan. *Water Research*, *36*(12), 2999–3008.

Khajavi, S., Jansen, J.C. and Kapteijn, F., 2007. Preparation and performance of H-SOD membranes: A new synthesis procedure and absolute water separation. In: *From Zeolite to Porous MOF Materials*, R. Xu, Z. Gao, J. Chen, W. Yan (Eds.), The Netherlands: Elsevier, 1028–1035.

Khajavi, S., Jansen, J.C. and Kapteijn, F., 2009. Application of hydroxy sodalite films as novel water selective membranes. *Journal of Membrane Science*, *326*(1), 153–160.

Khajavi, S., Jansen, J.C. and Kapteijn, F., 2010. Application of a sodalite membrane reactor in esterification—Coupling reaction and separation. *Catalysis Today*, *156*(3), 132–139.

Khajavi, S., Jansen, J.C. and Kapteijn, F., 2010. Production of ultrapure water by desalination of seawater using a hydroxy sodalite membrane. *Journal of Membrane Science*, *356*(1), 52–57.

Khajavi, S., Sartipi, S., Gascon, J., Jansen, J.C. and Kapteijn, F., 2010. Thermostability of hydroxy sodalite in view of membrane applications. *Microporous and Mesoporous Materials*, *132*(3), 510–517.

Khulbe, K.C., Feng, C. and Matsuura, T., 2010. The art of surface modification of synthetic polymeric membranes. *Journal of Applied Polymer Science*, *115*(2), 855–895.

Kim, S.H., Kwak, S.Y. and Suzuki, T., 2005. Positron annihilation spectroscopic evidence to demonstrate the flux-enhancement mechanism in morphology-controlled thin-film-composite (TFC) membrane. *Environmental Science & Technology*, *39*(6), 1764–1770.

Kimura, K., Hane, Y., Watanabe, Y., Amy, G. and Ohkuma, N., 2004. Irreversible membrane fouling during ultrafiltration of surface water. *Water Research*, *38*(14), 3431–3441.

Kimura, R., Wakabayashi, J., Elangovan, S.P., Ogura, M. and Okubo, T., 2008. Nepheline from K_2CO_3/nanosized sodalite as a prospective candidate for diesel soot combustion. *Journal of the American Chemical Society*, *130*(39), 12844–12845.

Koros, W.J., Ma, Y.H. and Shimidzu, T., 1996. Terminology for membranes and membrane processes (IUPAC Recommendations 1996). *Pure and Applied Chemistry*, *68*(7), 1479–1489.

Kurniawan, T.A., Chan, G.Y., Lo, W.H. and Babel, S., 2006. Physico–chemical treatment techniques for wastewater laden with heavy metals. *Chemical Engineering Journal*, *118*(1), 83–98.

Ladhe, A.R., Frailie, P., Hua, D., Darsillo, M. and Bhattacharyya, D., 2009. Thiol-functionalized silica–mixed matrix membranes for silver capture from aqueous solutions: Experimental results and modeling. *Journal of Membrane Science*, *326*(2), 460–471.

Lammens, T.M., Franssen, M.C., Scott, E.L. and Sanders, J.P., 2010. Synthesis of biobased N-methylpyrrolidone by one-pot cyclization and methylation of γ-aminobutyric acid. *Green Chemistry*, *12*(8), 1430–1436.

Landaburu-Aguirre, J., García, V., Pongrácz, E. and Keiski, R.L., 2009. The removal of zinc from synthetic wastewaters by micellar-enhanced ultrafiltration: Statistical design of experiments. *Desalination*, *240*(1–3), 262–269.

Lang, S., 2007. Environment-South Africa, Radioactive Water, the Price of Gold. Available from: http://ipsnews.net/news.asp?idnews=40325 (Accessed on 24/02/2017).

Li, H., Xia, H. and Mei, Y., 2016. Modeling organic fouling of reverse osmosis membrane: From adsorption to fouling layer formation. *Desalination*, *386*, 25–31.

Liu, C. and Bai, R., 2006. Adsorptive removal of copper ions with highly porous chitosan/cellulose acetate blend hollow fiber membranes. *Journal of Membrane Science*, *284*(1), 313–322.

Liu, G., Liu, Y., Shi, H. and Qian, Y., 2004. Application of inorganic membranes in the alkali recovery process. *Desalination*, *169*(2), 193–205.

Lu, G.Q., Da Costa, J.D., Duke, M., Giessler, S., Socolow, R., Williams, R.H. and Kreutz, T., 2007. Inorganic membranes for hydrogen production and purification: A critical review and perspective. *Journal of Colloid and Interface Science*, *314*(2), 589–603.

Ma, J., Xin, C. and Tan, C., 2015. Preparation, physicochemical and pharmaceutical characterization of chitosan from Catharsius molossus residue. *International Journal of Biological Macromolecules*, *80*, 547–556.

Makgaba, C.P. and Daramola, M.O., 2015. Transesterification of waste cooking oil to biodiesel over calcined hydroxy sodalite (HS) catalyst: A preliminary investigation, In: *Proceedings of International Conference on Sustainable Energy and Environmental Engineering* (*SEEE2015*), October 25–26, Bangkok, Thailand, pp. 52–56.

Manahan, S.E., 2005. Heavy metal water pollutants, in: *Green Chemistry*. 2nd ed. Columbia (MO): Chem Char. Research, p.175.

Maphutha, S., Moothi, K., Meyyappan, M. and Iyuke, S.E., 2013. A carbon nanotube-infused polysulfone membrane with polyvinyl alcohol layer for treating oil-containing waste water. *Scientific Reports*, *3*, 1509.

Masukume, M., Onyango, M.S. and Maree, J.P., 2014. Sea shell derived adsorbent and its potential for treating acid mine drainage. *International Journal of Mineral Processing*, *133*, 52–59.

Meschke, K., Herdegen, V., Aubel, T., Janneck, E. and Repke, J.U., 2015. Treatment of opencast lignite mining induced acid mine drainage (AMD) using a rotating microfiltration system. *Journal of Environmental Chemical Engineering*, *3*(4), 2848–2856.

Metesh, J.J., Jarrell, T. and Oravetz, S., 1998. *Treating Acid Mine Drainage from Abandoned Mines in Remote Areas* (Vol. 9871). Missoula, MT: USDA Forest Service, Technology and Development Program.

Mihaly-Cozmuta, L., Mihaly-Cozmuta, A., Peter, A., Nicula, C., Tutu, H., Silipas, D. and Indrea, E., 2014. Adsorption of heavy metal cations by Na-clinoptilolite: Equilibrium and selectivity studies. *Journal of Environmental Management*, *137*, 69–80.

Morigami, Y., Kondo, M., Abe, J., Kita, H. and Okamoto, K., 2001. The first large-scale pervaporation plant using tubular-type module with zeolite NaA membrane. *Separation and Purification Technology*, *25*(1), 251–260.

Mousavi, S.M., Saljoughi, E., Ghasemipour, Z. and Hosseini, S.A., 2012. Preparation and characterization of modified polysulfone membranes with high hydrophilic property using variation in coagulation bath temperature and addition of surfactant. *Polymer Engineering & Science*, *52*(10), 2196–2205.

Mthethwa, V., 2015. *Investigation of Polyethersulfone (PES) Hollow Fiber Membrane for the Treatment of Acid Mine Drainage* (MSC Thesis). University of the Witwatersrand, South Africa.

Naicker, K., Cukrowska, E. and McCarthy, T. S., 2003. Acid mine drainage arising from gold mining activity in Johannesburg, South Africa and environs. *Environmental Pollution*, *122*(1), 29–40.

Naskar, M.K., Kundu, D. and Chatterjee, M., 2011a. Coral-like hydroxy sodalite particles from rice husk ash as silica source. *Materials Letters*, *65*(23), 3408–3410.

Naskar, M.K., Kundu, D. and Chatterjee, M., 2011b. Effect of process parameters on surfactant-based synthesis of hydroxy sodalite particles. *Materials Letters*, *65*(3), 436–438.

Natarajan, K.A., 2008. Microbial aspects of acid mine drainage and its bioremediation. *Transactions of Nonferrous Metals Society of China*, *18*(6), 1352–1360.

Nędzarek, A., Drost, A., Harasimiuk, F.B. and Tórz, A., 2015. The influence of pH and BSA on the retention of selected heavy metals in the nanofiltration process using ceramic membrane. *Desalination*, *369*, 62–67.

Ng, L.Y., Mohammad, A.W., Leo, C.P. and Hilal, N., 2013. Polymeric membranes incorporated with metal/metal oxide nanoparticles: A comprehensive review. *Desalination*, *308*, 15–33.

Ngah, W.S. and Isa, I.M., 1998. Comparison study of copper ion adsorption on chitosan, Dowex A-1, and Zerolit 225. *Journal of Applied Polymer Science*, *67*(6), 1067–1070.

Ngah, W.W., Teong, L.C., Toh, R.H. and Hanafiah, M.A.K.M., 2012. Utilization of chitosan–zeolite composite in the removal of Cu (II) from aqueous solution: Adsorption, desorption and fixed bed column studies. *Chemical Engineering Journal*, *209*, 46–53.

Nordstrom, D.K., Blowes, D.W. and Ptacek, C.J., 2015. Hydrogeochemistry and microbiology of mine drainage: An update. *Applied Geochemistry*, *57*, 3–16.

Ochieng, G., Seanego, E. and Nkwonta, O., 2010. Impacts of mining on water resources in South Africa: A review. *Scientific Research and Essays*, *5*(22), 3351–3357.

Pendergast, M.M. and Hoek, E.M., 2011. A review of water treatment membrane nanotechnologies. *Energy & Environmental Science*, *4*(6), 1946–1971.

Rahimpour, A., Madaeni, S.S. and Mansourpanah, Y., 2010. Fabrication of polyethersulfone (PES) membranes with nano-porous surface using potassium perchlorate ($KClO_4$) as an additive in the casting solution. *Desalination*, *258*(1), 79–86.

Reiad, N.A., Salam, O.E.A., Abadir, E.F. and Harraz, F.A., 2012. Adsorptive removal of iron and manganese ions from aqueous solutions with microporous chitosan/polyethylene glycol blend membrane. *Journal of Environmental Sciences*, *24*(8), 1425–1432.

Rinaudo, M., 2006. Chitin and chitosan: Properties and applications. *Progress in Polymer Science*, *31*(7), 603–632.

Robinson-Lora, M.A. and Brennan, R.A., 2009. Efficient metal removal and neutralization of acid mine drainage by crab-shell chitin under batch and continuous-flow conditions. *Bioresource Technology*, *100*(21), 5063–5071.

Rohde, M.P., Schaub, G., Khajavi, S., Jansen, J.C. and Kapteijn, F., 2008. Fischer–Tropsch synthesis with in situ H_2O removal–Directions of membrane development. *Microporous and Mesoporous Materials*, *115*(1), 123–136.

Salehi, E., Daraei, P. and Shamsabadi, A.A., 2016. A review on chitosan-based adsorptive membranes. *Carbohydrate Polymers*, *152*, 419–432.

Salehi, E., Madaeni, S.S., Rajabi, L., Vatanpour, V., Derakhshan, A.A., Zinadini, S., Ghorabi, S. and Monfared, H.A., 2012. Novel chitosan/poly (vinyl) alcohol thin adsorptive membranes modified with amino functionalized multi-walled carbon nanotubes for Cu (II) removal from water: Preparation, characterization, adsorption kinetics and thermodynamics. *Separation and Purification Technology*, *89*, 309–319.

Scopus, www.scopus.com, accessed 04-04-2017.

Shen, L., Bian, X., Lu, X., Shi, L., Liu, Z., Chen, L., Hou, Z. and Fan, K., 2012. Preparation and characterization of ZnO/polyethersulfone (PES) hybrid membranes. *Desalination*, *293*, 21–29.

Shukla, S.K., Mishra, A.K., Arotiba, O.A. and Mamba, B.B., 2013. Chitosan-based nanomaterials: A state-of-the-art review. *International Journal of Biological Macromolecules*, *59*, 46–58.

Sivakumar, M., Ramezanianpour, M. and O'Halloran, G., 2013. Mine water treatment using a vacuum membrane distillation system. *APCBEE Procedia*, *5*, 157–162.

Sklari, S.D., Plakas, K.V., Petsi, P.N., Zaspalis, V.T. and Karabelas, A.J., 2015. Toward the development of a novel electro-Fenton system for eliminating toxic organic substances from water. Part 2. Preparation, characterization, and evaluation of iron-impregnated carbon felts as cathodic electrodes. *Industrial & Engineering Chemistry Research*, *54*(7), 2059–2073.

Steenkamp, G.C., Keizer, K., Neomagus, H.W.J.P. and Krieg, H.M., 2002. Copper (II) removal from polluted water with alumina/chitosan composite membranes. *Journal of Membrane Science*, *197*(1), 147–156.

Strategic Perspective. Report P WMA 04/000/00/0340. Department of Water Affairs and Forestry, Pretoria, South Africa.

Strathmann, H., Giorno, L. and Drioli, E., 2010. Basic aspects in polymeric membrane preparation. *Comprehensive Membrane Science and Engineering*, *1*, 91–111.

Tetala, K.K. and Stamatialis, D.F., 2013. Mixed matrix membranes for efficient adsorption of copper ions from aqueous solutions. *Separation and Purification Technology*, *104*, 214–220.

Torkabad, M.G., Keshtkar, A.R. and Safdari, S.J., 2017. Comparison of polyethersulfone and polyamide nanofiltration membranes for uranium removal from aqueous solution. *Progress in Nuclear Energy*, *94*, 93–100.

Tutu, H., McCarthy, T.S. and Cukrowska, E., 2008. The chemical characteristics of acid mine drainage with particular reference to sources, distribution and remediation: The Witwatersrand Basin, South Africa as a case study. *Applied Geochemistry*, *23*(12), 3666–3684.

Van der Bruggen, B., 2009. Chemical modification of polyethersulfone nanofiltration membranes: A review. *Journal of Applied Polymer Science*, *114*(1), 630–642.

Vatanpour, V., Madaeni, S.S., Khataee, A.R., Salehi, E., Zinadini, S. and Monfared, H.A., 2012. TiO_2 embedded mixed matrix PES nanocomposite membranes: Influence of different sizes and types of nanoparticles on antifouling and performance. *Desalination*, *292*, 19–29.

Vieira, R.S. and Beppu, M.M., 2006. Interaction of natural and crosslinked chitosan membranes with Hg (II) ions. *Colloids and Surfaces A: Physicochemical and Engineering Aspects*, *279*(1), 196–207.

Vu, D.Q., Koros, W.J. and Miller, S.J., 2003. Mixed matrix membranes using carbon molecular sieves: II. Modeling permeation behavior. *Journal of Membrane Science*, *211*(2), 335–348.

Wan Ngaha, W.S., Teonga, L.C. and Hanafiah, M.A.K.M., 2011. Adsorption of dyes and heavy metal ions by chitosan composites: A review, *Carbohydrate Polymers*, 83, 1446–1456.

Wang, H., Yu, T., Zhao, C. and Du, Q., 2009. Improvement of hydrophilicity and blood compatibility on polyethersulfone membrane by adding polyvinylpyrrolidone. *Fibers and Polymers*, *10*(1), 1–5.

Wang, L. and Wang, A., 2007. Adsorption characteristics of Congo Red onto the chitosan/montmorillonite nanocomposite. *Journal of Hazardous Materials*, *147*(3), 979–985.

Wang, Q., Li, J., Chen, C., Ren, X., Hu, J. and Wang, X., 2011. Removal of cobalt from aqueous solution by magnetic multiwalled carbon nanotube/iron oxide composites. *Chemical Engineering Journal*, *174*(1), 126–133.

Wu, Z., Faiz, R., Li, T., Kingsbury, B.F. and Li, K., 2013. A controlled sintering process for more permeable ceramic hollow fibre membranes. *Journal of Membrane Science*, *446*, 286–293.

Xie, J., Li, C., Chi, L. and Wu, D., 2013. Chitosan modified zeolite as a versatile adsorbent for the removal of different pollutants from water. *Fuel*, *103*, 480–485.

Zhao, C., Xue, J., Ran, F. and Sun, S., 2013. Modification of polyethersulfone membranes—A review of methods. *Progress in Materials Science*, *58*(1), 76–150.

Zhou, L., Liu, J. and Liu, Z., 2009. Adsorption of platinum (IV) and palladium (II) from aqueous solution by thiourea-modified chitosan microspheres. *Journal of Hazardous Materials*, *172*(1), 439–446.

Zhou, L., Xu, J., Liang, X. and Liu, Z., 2010. Adsorption of platinum (IV) and palladium (II) from aqueous solution by magnetic cross-linking chitosan nanoparticles modified with ethylenediamine. *Journal of Hazardous Materials*, *182*(1), 518–524.

Zhu, S., Zhao, S., Wang, Z., Tian, X., Shi, M., Wang, J. and Wang, S., 2015. Improved performance of polyamide thin-film composite nanofiltration membrane by using polyetersulfone/polyaniline membrane as the substrate. *Journal of Membrane Science*, *493*, 263–274.

Zhu, Y., Hu, J. and Wang, J., 2014. Removal of Co^{2+} from radioactive wastewater by polyvinyl alcohol (PVA)/chitosan magnetic composite. *Progress in Nuclear Energy*, *71*, 172–178.

Zularisam, A.W., Ismail, A.F., Salim, M.R., Sakinah, M. and Ozaki, H., 2007. The effects of natural organic matter (NOM) fractions on fouling characteristics and flux recovery of ultrafiltration membranes. *Desalination*, *212*(1), 191–208.

11 The Removal of Phenol and Phenolic Compounds from Wastewater Using Reverse Osmosis

Process Modelling, Simulation, and Optimisation

Mudhar A. Al-Obaidi, Chakib Kara-Zaïtri, and Iqbal M. Mujtaba

CONTENTS

11.1 INTRODUCTION

Over the last few decades, there has been a sharp increase in the volume of industrial effluents and sewage, which are disposed haphazardly into rivers and oceans, and which continue to cause significant harm to the ecosystem. At the same time and with the continued population rise, there has also been a marked shortage of drinking water. Because of these competing aspects, there is an ever-increasing and urgent need to find better, faster, and cheaper methods for transforming wastewater into drinkable water, or at least cleaner water, that can be used in a plethora of applications.

Phenol and phenol compounds (aromatic compounds) can be found ubiquitously in several industrial effluents such as herbicides, pesticides, pharmaceuticals, disinfectants, and dyes [1]. Even low concentrations of such compounds in water have a highly toxic effect on public health, as recognised by a number of health agencies, including the United States Environmental Protection Agency (EPA). Indeed, the Agency of Toxic Substances and Disease Registry (ATSDR) limited the concentration of dimethylphenol at a maximum of 0.05 ppm in surface water [2]. It is not surprising therefore that several scientists have attempted to develop efficient methods for treating such industrial effluents before disposing of them into surface water. Although conventional methods—including microbial degradation, incineration, adsorption, irradiation, solvent extraction, catalyst wet air oxidation, and reverse osmosis—have been used with a degree of success to remove phenolic compounds from wastewater [3,4], there is still room for improvement in terms of finding a faster, better, and cheaper solution. Specifically, Mohammed et al. [4] addressed the main methods used to remove phenols from wastewater, which included UV/H_2O_2, catalytic wet air oxidation (CWAO), and reverse osmosis (RO). However, the use of UV requires high energy [5] and is therefore less economical. In contrast, the use of RO has proven to yield a cheaper solution for removing harmful compounds such as phenol from wastewater [6–8], but again, there appears to be room for improvement.

RO is a pressure-driven process, which can be specifically characterised by the idea of using a semi-permeable membrane (permeable to solvent, impermeable to solute) to separate two media of different solute concentration. Generally, in an osmotic process, the solvent spreads through the membrane due to its high osmotic pressure in order to achieve a balance in the chemical potential and equality of solute concentration on both sides of the membrane. The state of osmotic equilibrium can be reached by the gradual increase of concentrated side pressure, which finally inhibits solvent transfer. However, when applying a higher pressure than the osmotic pressure, the procedure is essentially reversed. This happens by forcing the solvent to flow from the concentrated side to the diluted side. The theory of solute and solvent flow through semi-permeable membranes can be schematically presented in three events as shown in Figure 11.1. This shows two solutions of different concentrations being separated by a semi-permeable membrane. Scenario A represents a natural phenomenon of water diffusion from the low concentration solution in to the high concentration solution, and this is referred to as osmosis. This process continues until osmotic equilibrium is achieved, which is characterised by the equivalent chemical potential of both solutions. The pressure difference between the two solutions is known as the osmotic pressure difference,

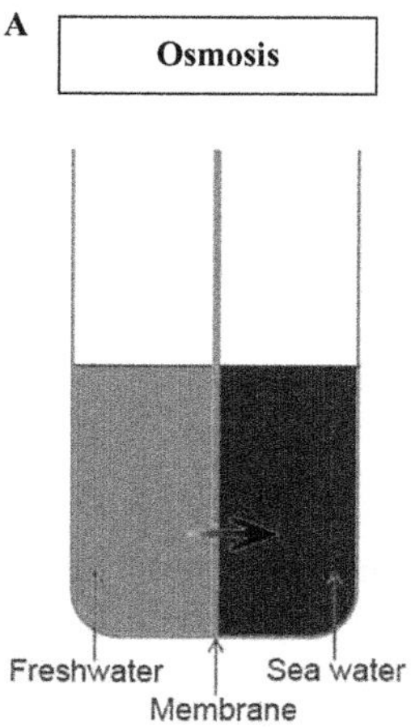

FIGURE 11.1 Scenarios A, B and C (A: Osmosis, B: Osmotic Equilibrium, and C: Reverse Osmosis). (Adapted from Al-Obaidi, M.A. et al., *Comput. Chem. Eng.*, 100, 48–79, 2017.)

which stops any further solvent flow from this point onward. This is referred to as the osmotic equilibrium state, i.e., Scenario B. Lastly, for the desalination of the concentrated salt water, the procedure can be reversed by applying pressure higher than the solution osmotic pressure, which forces water to pass over the membrane towards the low concentration solution side accompanied by salt rejection, i.e., Scenario C [9]. Usually, this process can happen under ambient temperature and without any phase change [10,11].

RO can therefore be readily considered as a prominent separation process in industrial applications due to its ability to separate impurities effectively and commensurate with environmental demands. In the past few decades, RO has been used for purifying sea and brackish water as well as for the treatment of effluents. It has also been extended to different types of industrial applications such as textile, paper, electrochemical, and biochemical industries [12–14]. Specifically, the use of the RO process offers a number of advantages, including minimum thermal damage, no chemical reactions, high packing density, low capital and operating costs, and low energy consumption. These are significant advantages in comparison with other thermal methods such as multistage flash desalination (MSF) and multi-effect distillation (MED) [15,16]. Moreover, RO processes usually yield lower pollutant concentration and therefore better water quality at an economical cost, which in turn provides significant opportunities for reclaiming water in a variety of different industrial applications [17]. This readily explains the large rise of RO applications in wastewater treatment. It is this rise that has motivated further research in developing an even cheaper solution—one that is based on an optimisation of associated mathematical models describing this complex process. For this to happen, a detailed and accurate process model (consisting of a set of ordinary algebraic and differential equations) is required to derive a reasonable prediction of membrane performance based on realistic operating. The fitness of such a model will be explored, together with its associated optimisation, to yield a validated cheaper solution.

Investigation and development of transport models for RO have attracted the attention of many scientists in recent years, who have researched a specific pattern of RO but assumed perfect operating conditions for the separation process. These models have also been used for the evaluation of the performance properties of the membrane with regard to its quality of separation [18]. Ideally, it would make sense to develop rigorous models requiring less experimental and pilot studies but also with a smaller number of parameters for characterising and evaluating membrane performance. These include the fluxes of solute and solvent and the efficiency of membrane rejection with regard to the operating conditions such as force per unit area and concentration driving forces [10].

The thorough literature review carried out by Al-Obaidi et al. [9] indicated that there are essentially two main models for estimating the mass transfer across the membrane. They are the irreversible thermodynamics model and the solution–diffusion model. The irreversible thermodynamics model depends on non-equilibrium thermodynamic equations, which consider the membrane as a black box where slow processes are occurring near the equilibrium. Unfortunately, there is insufficient information for describing the flow, transport mechanism, and structure of the membrane. Hence, the applicability of using the irreversible thermodynamics model for the accurate speculation of membrane separation is difficult. The starting fundamental formula for the irreversible thermodynamic model was established by Kedem and Katchalsky [19] and then by Spiegler and Kedem [20] for a dilute two-component non-electrolyte system of water and solute as linear equations and nonlinear equations, respectively, relating the fluxes of these components. One interesting aspect of this work is the idea that there should be a combination of three parameters, rather than two as assumed in the solution–diffusion model. For the first time, a third parameter consisting of the reflection coefficient will be added to properly model the broad criteria of the sensible interaction between the solute, solvent, and membrane, and to account for the phenomenological relations between them, thus providing an explanation for the underlying fluxes [11,21,22]. For a dilute single-solute system, the reflection coefficient is approximately equal to 1 for impermeable solute and less than 1 for permeable solute. The modification of this model to account for multi-component systems will also be explored. On the other hand, the solution–diffusion model can be considered one of the simplest non-porous or homogeneous models for transport mechanism criteria. This criterion is characterised by assuming that each solvent and solute are dissolved in the membrane separately on the high-pressure side. These are then diffused in individual fluxes through the membrane under the impact of pressure and concentration differences. The quality of permeate separation occurs due to the mobility of the dissolved solute and the rate of its diffusion [23]. Thus, the fluxes of solvent and solutes are effectively concerned with the values of solubility and diffusivity of solvent and each solute in the membrane. The solution–diffusion model assumes that the transfer of solute and solvent are largely dependent on the physical and chemical properties of both the solution and the membrane. Nevertheless, there are other types of mathematical models, referred to as pore models, which include the finely porous, preferential sorption-capillary flow, and surface force-pore flow models.

From the earlier, it can be stated that the solution–diffusion and the irreversible thermodynamic models are the most widely used to model the mechanism of transport phenomena in membrane technology. Murthy and Gupta [24] confirmed the accuracy of the Spiegler and Kedem model, while Mujtaba [25] confirmed the applicability of the solution–diffusion model to express the mechanism of water and solute transport in RO systems.

The RO membranes are placed in several types of modules of different geometries, such as spiral wound, hollow fiber, tubular, and plate and frame. Each type has several advantages and disadvantages.

Spiral-wound membrane modules, in comparison with other types of RO modules, are the most popular for use in desalination and industrial processes. This is because they are easy to operate and yield reasonable permeation rates with low fouling levels at low costs [26]. Table 11.1 shows the merits and limitations of each membrane module configuration. For example, the hollow fiber module has a high packing density and usually about 50–60% of the feed can be permeated as fresh water.

Spiral-wound modules contain an envelope consisting of a number of flat membranes glued and wrapped around a central tube. The brine water is pumped and forced to run along the membrane length where the fresh water is collected out in a tube at the permeate side. Figure 11.2 shows the spiral-wound module and the direction of flow inside the module [28]. The specific dimensions of the module are length (L) and width (W); x denotes the axial coordinate along the membrane length, and y denotes the tangential coordinate in the spiral direction starting from the sealed end of the leaf

TABLE 11.1
Comparison between the Reverse Osmosis Configurations

Element Type	Advantages	Limitations	Energy Consumption
Spiral wound	• Compact • Less expensive than others • Good packing density	• Plugging • Low fouling levels	Moderate
Hollow fiber	• Compact • Economical • High packing density	• Plugging	Low
Tubular	• Operating at high pressure • Easy to clean • Resistance to membrane fouling	• Plugging	High
Plate and frame	• Moderate packing density	• Plugging • Expensive • Difficult to clean	Low to moderate

Source: Baker, R.W., *Membrane Technology and Applications*, 2nd ed., Membrane Technology and Research, Newark, CA, 2004 [27].

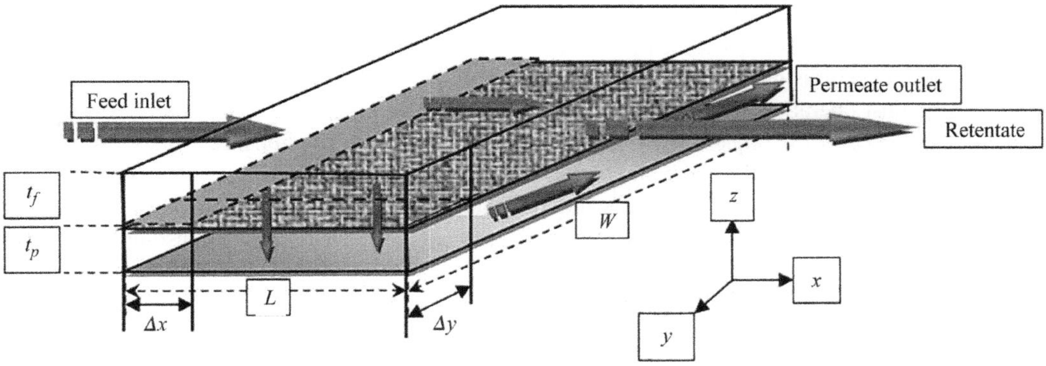

FIGURE 11.2 Schematic diagram of the spiral-wound module. (Adapted from Al-Obaidi, M.A. and Mujtaba I. M., *Comput. Chem. Eng.* 90, 278–299, 2016.)

to the end of membrane width. The feed and permeate spacer channels are t_f and t_p respectively. The effective membrane area can be calculated as $A_m = L \times W$. Specifically, the RO module (Figure 11.2) is composed of two sides: the feed side and the permeate side. The accumulated permeate water flows in the same direction as the feed and then flows in a spiral direction into a centrally perforated pipe, and the fresh water flows perpendicular from the feed side to the permeate side through the membrane region. To comply with the requirements of distributed modelling, the membrane area is split into 16 sub-sections of equal areas (Δx, Δy) using the method of discretisation used by gPROMS [29]. This in turn will be used to estimate the variation of the operating parameters along the two sides of the membrane. The area of each sub-section is given as, $\left(A_{sub-section} = \Delta x \; \Delta y\right)$, where $\Delta x = L/4$ and $\Delta y = W/4$.

The next section focuses on an extensive review of the current state of the art for the dimensional modelling, simulation, and optimisation of the RO process for the removal of phenol and phenolic compounds from wastewater. The main aim of this chapter is to enhance the existing RO process and improve its reliability in terms of achieving a better and cheaper solution for removing selected organic phenol and phenolic compounds from wastewater.

11.2 RO PROCESS MODELLING, SIMULATION, AND OPTIMISATION: THE STATE OF THE ART

This section discusses the latest published modelling, simulation, and optimisation methodologies used in the spiral-wound RO wastewater process.

11.2.1 Distributed Modelling of a Spiral-Wound RO Wastewater Process

In the past two decades, many models have been reported in the literature for spiral-wound RO processes for removing organic and non-organic compounds from aqueous solutions. These models explored the membrane performance for a spiral-wound module with different features and in different applications. Such models are based on a number of assumptions and have been generally validated with experimental data using sea and brackish water. More importantly, while the majority of these models took into consideration the average conditions of the membrane edges, they neglected the difference of operating parameters along the two sides of the membrane. In contrast, fewer models have been developed to explore the variation of operating parameters in one and two dimensions of the membrane. One of the main advantages of distributed modelling is the ability to consider the variance of all the operating parameters, the pattern of feed flow rate, solvent, and solute fluxes, as well as the mass transfer coefficient along the axial and tangential axes of the feed and permeate sides. This offers a more accurate prediction of the performance of the process in comparison to lumped modelling. A further advantage of using dynamic modelling is the ability to explore the behavior of the operating parameters against operational time. This helps to implement an efficient control method for enhancing the performance of the process. However, only a limited amount of published work on models of spiral-wound RO process used specifically for wastewater treatment is currently available in the public domain. Additionally, there are only a few validation studies of mathematical models using wastewater experimental data [18]. Finally, most of the suggested models have assumed constant pressure in the permeate side [30].

Analytical steady-state models were developed by Rautenbach and Dahm [31] for a spiral-wound module and worked out by Evangelista [26], for high-rejecting membranes and by Avlonitis et al. [32–34] and Boudinar et al. [35] for both Roga and Filmtec membrane types, respectively. These models considered the validity of the solution–diffusion model with fully axial flow of brine solution. These models assumed constant density and viscosity and ignored the concentration polarisation impact. Some of these models also neglect the pressure drop in the brine and permeate compartments. Based on the three-parameter model of Spiegler and Kedem, Senthilmurugan et al. [36] developed a model for turbulent flow by considering the pressure drop in both the channels as being proportional to an exponent of 1.2 to 2.7 in Darcy's law. Geraldes et al. [37] developed a one-dimensional steady-state model for spiral-wound RO membranes, but ignored both pressure drop in the permeate channel and diffusion flow in the feed side. Sagne et al. [38] developed a one-dimensional dynamic model to predict the performance of a spiral-wound RO process to remove dilute aqueous solution of five volatile organic compounds (acetic acid, butyric acid, 2-phenylethanol, 2,3-butanediol, and furfural) from brackish water that was used in fermentation industries. This model was based on the principles of the solution–diffusion methodology and ignored the impact of concentration polarisation. Oh et al. [39] settled a one-dimensional model based on the principles of the solution–diffusion model to predict the performance of a spiral-wound RO process. This model is characterised by assuming constant values for water flux and mass transfer coefficient. The solution–diffusion model is also used to develop a one-dimensional model by Kaghazchi et al. [40]. However, this model neglected the variation of the bulk flow rate and considered instead its average value at inlet and outlet edges.

Generally, all the earlier models are validated with sea and brackish water experimental data. Sundaramoorthy et al. [18,41] proposed a one-dimensional steady-state model by assuming the validity of the solution–diffusion model and constant values for both the permeate concentration and pressure along the permeate side. The model has been validated with the experimental data of chlorophenol and dimethylphenol solutes. Fujioka et al. [42] have developed a one-dimensional model based on the irreversible thermodynamic model and considered the variation of operating parameters while assuming zero permeate pressure. The model has been validated against experimental data of N-nitrosamine rejection. Most recently, Al-Obaidi and Mujtaba [28] and Al-Obaidi et al. [43–45] developed one- and two-dimensional steady-state and dynamic models of a spiral-wound RO process for the removal of phenol and phenolic compounds. Overall, most of the developed models are based on the principles of the solution–diffusion methodology and include the concentration polarisation mechanism. However, the concepts of the irreversible thermodynamics model have been used by Al-Obaidi et al. [45] to develop a one-dimensional model. The robustness of such models was appraised by a validation study against experimental data of phenol, chlorophenol, and dimethylphenol in the literature. Having said this, despite the investigation of other complex models, the models proposed by Al-Obaidi et al. are possibly the most advanced in that they relax the assumptions of constant physical properties and concentration of the fresh water on the permeate side, while using relatively simple concepts. The retentate concentration varies along the membrane length and width due to the impact of the plug and diffusion flow. Moreover, the process behavior along the operation time is analysed in the counter of the developed dynamic model of Al-Obaidi et al. Therefore, these models can be used to provide more accurate results for the design of the RO process.

The next section aims to present the two different one-dimensional distributed models of Al-Obaidi et al., which have been developed to simulate the rejection of phenol and phenolic compounds from an aqueous solution using a spiral-wound RO system. The models are themselves based on the principles of the solution–diffusion and irreversible thermodynamic models.

11.2.1.1 A Steady-State One-Dimensional Model of a Single Spiral-Wound RO System by Al-Obaidi et al.

Al-Obaidi et al. [44] developed a steady-state one-dimensional model, which includes the physical properties equations of a single spiral-wound RO system. The model consisted of a set of differential and algebraic equations (DAEs), which are solved using MATLAB, and where the filtration channel is divided into a number of segments of equal intervals (Δx). For a given inlet feed flow rate, pressure, solute concentration, and temperature, the proposed model can be used to predict the longitudinal variation of all parameters in the feed and permeate channels in the x-axis by using the estimated values of the membrane transport parameters.

The following assumptions were made to develop the proposed process model:

1. The solution–diffusion model is quantified to express the mass transport through the module.
2. The membrane characteristics and the channel geometries are assumed constant.
3. Validity of Darcy's law where the friction parameter is used to describe the pressure drop in the feed channel.
4. Constant atmospheric pressure (1 atm) at the permeate channel.
5. A constant solute concentration is assumed in the permeate channel, and the average value will be calculated from the inlet and outlet permeate solute concentrations.
6. The process is assumed to be isothermal.

11.2.1.2 Governing Equations

Table 11.2 shows the model equations used to estimate the performance of a single spiral-wound RO process to remove chlorophenol from aqueous solutions. Specifically, the model uses the experimental work of dilute chlorophenol aqueous solutions on a spiral-wound module from the literature. Therefore, the physical properties equations of the solution have been considered identical to the water equations proposed by Koroneos et al. [47].

11.2.1.3 Validation Analysis

11.2.1.3.1 Experimental Apparatus and Procedure

A pilot scale experiment has been designed consisting of a cross-flow RO filtration system of one commercial thin film composite RO membrane packed into a spiral-wound module of aqueous feed solutions of chlorophenol of specific concentration. The module was from the Ion Exchange Ltd. Company of India. This is the same as the one used by Sundaramoorthy et al. [41]. The characteristics of the spiral-wound module and the transport parameters of this model (A_w, B_s, and b) are given in Table 11.3. The feed was pumped at three different flow rates of 2.166E-4, 2.33E-4 and 2.583E-4 m^3/s. Also, for each feed flow rate, the solute concentrations vary from 0.778E-3 to 6.226E-3 $kmol/m^3$ with a set of pressures varying from 5.83 to 13.58 atm for each feed concentration.

TABLE 11.2
Equations Describing the Spiral-Wound RO Modelling

Model Equations	Specifications	Equation No.
$F_{b(x)} = \left\{ \begin{array}{l} F_{b(0)} - (W\theta \times \Delta P_{b(0)}) + \left(W\theta b \left(\frac{x^2}{2}\right) F_{b(0)} \right) + \\ \left(W\theta b \left(\frac{W\theta}{b}\right)^{0.5} \left(\frac{x^2}{2}\right) \left(\Delta P_{b(x)} - \Delta P_{b(0)}\right) \right) \end{array} \right\}$	Calculate feed flow rate at any point along the x-axis	11.1
$\theta = \frac{A_w B_s}{B_s + RT_b A_w C_{p(av)}}$	Parameter in Equation 11.1	11.2
$U_{b(x)} = \frac{F_{b(x)}}{t_f W}$	Calculate feed velocity at any point along the x-axis	11.3
$P_{b(x)} = \left\{ \begin{array}{l} P_{b(0)} - (b \times F_{b(0)}) + \left(bW\theta \left(\frac{x^2}{2}\right) (\Delta P_{b(x)}) \right) - \left[b^2 W\theta \left(\frac{x^3}{6}\right) F_{b(0)} \right] - \\ \left[b^2 W\theta \left(\frac{W\theta}{b}\right)^{0.5} \left(\frac{x^3}{6}\right) \left(\Delta P_{b(x)} - \Delta P_{b(0)}\right) \right] \end{array} \right\}$	Calculate retentate pressure at any point along the x-axis	11.4
$\Delta P_{b(x)} = \Delta P_{b(0)} - (b x F_{b(0)}) - \left[\left(\frac{W\theta}{b}\right)^{0.5} b \times \left(\Delta P_{b(x)} - \Delta P_{b(0)}\right) \right]$	Calculate pressure difference between the feed and permeate channels at any point along the x-axis	11.5
$\Delta P_{b(0)} = P_{b(0)} - P_p$	Calculates pressure difference between the feed and permeate channels at x = 0	11.6
$P_{loss} = P_{b(0)} - P_{b(L)}$	Calculate pressure loss along the membrane length	

(Continued)

TABLE 11.2 (*Continued*)
Equations Describing the Spiral-Wound RO Modelling

Model Equations	Specifications	Equation No.
$J_{w(x)} = \theta \left\{ \left[\Delta P_{b(0)} - \left(b \times F_{b(0)} \right) \right] - \left[\left(\frac{W\theta}{b} \right)^{0.5} bx \left(\Delta P_{b(x)} - \Delta P_{b(0)} \right) \right] \right\}$	Calculate water flux at any point along the x-axis	11.7
$J_{s(x)} = B_s \left(C_{w(x)} - C_{p(av)} \right)$	Calculate solute flux at any point along the x-axis	11.8
$\frac{\left(C_{w(x)} - C_{p(av)} \right)}{\left(C_{b(x)} - C_{p(av)} \right)} = \exp\left(\frac{J_{w(x)}}{k_{(x)}} \right)$	Calculate wall solute concentration at any point along the x-axis	11.9
$k_{(x)} de_b = 147.4 \ D_{b(x)} \ Re_{b(x)}^{0.13} \ Re_{p(x)}^{0.739} \ C_{m(x)}^{0.135}$	Calculate mass transfer coefficient at any point along the x-axis [41]	11.10
$C_{m(x)} = \frac{C_{b(x)}}{\rho_w}$	Calculate the dimensionless solute concentration at any point along the feed channel	
$Re_{b(x)} = \frac{\rho_{b(x)} de_b F_{b(x)}}{t_f W \mu_{b(x)}} \quad Re_{p(x)} = \frac{\rho_{p(x)} de_p J_{w(x)}}{\mu_{p(x)}}$	Calculate the Reynolds number at any point along the x-axis at the feed and permeate channels respectively	
$de_b = 2t_f \quad de_p = 2t_p$	Calculate the equivalent diameters of the feed and permeate channels	
$C_{b(x)} = \frac{F_{b(x-1)} \left(C_{b(x-1)} - C_{p(av)} \right)}{F_{b(x)}} + C_{p(av)}$	Calculate retentate concentration at any point along the x-axis	11.12
$C_{p(av)} = \frac{C_{p(0)} + C_{p(L)}}{2}$	Calculate average permeate solute concentration	11.13

(*Continued*)

TABLE 11.2 (*Continued*)
Equations Describing the Spiral-Wound RO Modelling

Model Equations	Specifications	Equation No.
$C_{p(0)} = \frac{B_s C_{b(0)} e^{\frac{J_{w(0)}}{k(0)}}}{J_{w(0)} + B_s e^{\frac{J_{w(0)}}{k(0)}}}$ and $C_{p(L)} = \frac{B_s C_{b(L)} e^{\frac{J_{w(L)}}{k(L)}}}{J_{w(L)} + B_s e^{\frac{J_{w(L)}}{k(L)}}}$	Calculate permeate solute concentrations at x = 0 and x = L [46]	11.14 and 11.15
$F_{p(x)} - F_{p(0)} + \left(W \times \theta \Delta P_{b(0)}\right) - \left[W\theta b\left(\frac{x^2}{2}\right)F_{b(0)}\right]$ $\left[W\theta b\left(\frac{x^2}{2}\right)\left(\frac{W\theta}{b}\right)^{0.5}\left(\Delta P_{b(x)} - \Delta P_{b(0)}\right)\right]$	Calculate permeated flow rate at any point along the x-axis	11.16
$Rec_{(\text{Total})} = \frac{F_{p(L)}}{F_{b(0)}} \times 100$	Calculate total water recovery	11.17
$Rej = \frac{C_{b(L)} - C_{p(av)}}{C_{b(L)}} \times 100$	Calculate solute rejection [48]	11.18

Source: Al-Obaidi, M.A. et al., *Chem. Eng. J.*, 316, 91–100, 2017.

TABLE 11.3
Membrane Characteristics and Geometry (Ion Exchange, India)

Property	Value
Membrane material	TFC polyamide
Module configuration	Spiral wound
Number of turns	30
Module length (L)	0.934 m
Module width (W)	8.4 m
Module diameter	3.25 inches
Feed spacer thickness (t_f)	0.8 mm
Permeate channel thickness (t_p)	0.5 mm
A_w*	9.5188E-7 $\left(\frac{\text{m}}{\text{atm s}}\right)$
B_s (chlorophenol)*	8.468E-8 $\left(\frac{\text{m}}{\text{s}}\right)$
b*	8529.45 $\left(\frac{\text{atm s}}{\text{m}^4}\right)$

Source: Al-Obaidi, M.A. et al., *Chem. Eng. J.*, 316, 91–100, 2017.
* Calculated by Sundaramoorthy et al [41].

11.2.1.3.2 Model Validation

Figures 11.3 and 11.4 depict a comparison between the theoretical values (y-axis) of the retentate chlorophenol concentration, average permeate concentration, chlorophenol rejection, retentate flow, outlet permeate flow rate, and retentate pressure and the experimental results (x-axis) and the model predictions (straight line) for three sets of inlet feed flow rate. Generally, the predicted values of the model are in good agreement with the experimental ones over the ranges of pressure and concentration. However, the assumption of constant values for the friction factor, water and solute permeability coefficients for all experiments had a negative impact on estimating the solute concentrations in both channels. This, in turn reduced the consistency of experimental and mathematical chlorophenol rejection results ($R^2 = 0.85$).

11.2.1.4 A Steady-State One-Dimensional Model Based on the Spiegler and Kedem Model by Al-Obaidi et al.

Following the principles of the irreversible thermodynamics theory, Al-Obaidi et al. [45] developed a new explicit one-dimensional steady-state model based on the Spiegler and Kedem model for a general spiral-wound RO system. This was resolved using the gPROMS software package (Process Systems Enterprise, PSE). The model can predict the variation of solute concentration, pressure, flow rate, mass transfer coefficient, solvent, and solute fluxes along the length of the feed channel. The gEST parameter estimation tool available in the gPROMS software has been used with experimental data to estimate the best values of the membrane and the friction parameters. Also, an equation for the mass transfer coefficient is investigated in line with experimental data to show the impact of solvent flux, flow rate, solute concentration, and both the solvent and solute properties. The actual data available in the literature about phenol removal from aqueous solutions was used to validate the model and show its robustness. The process model developed

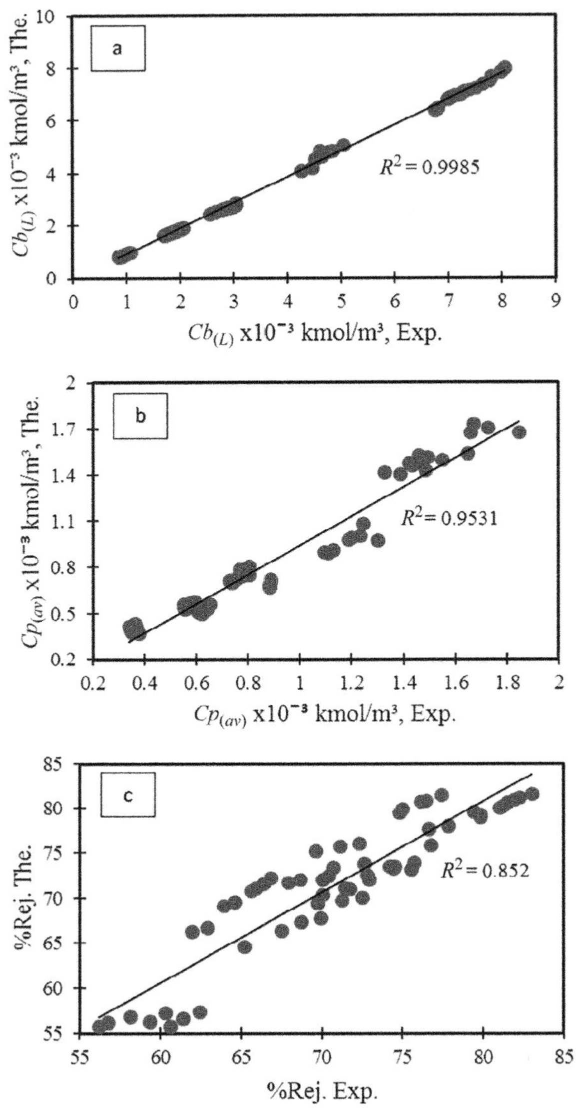

FIGURE 11.3 Experimental and model prediction of (a) retentate concentration, (b) average permeate concentration, (c) chlorophenol rejection (inlet conditions mentioned in Section 11.2.1.3.1). (Adapted from Al-Obaidi, M.A. et al., *Chem. Eng. J.*, 316, 91–100, 2017.)

was then used to study the variation of solute concentration, pressure, flow rate, solvent, and solute fluxes along the length of the feed channel.

The following assumptions were made to develop the proposed process model:

1. The flat membrane sheet has negligible channel curvature.
2. Validity of the Spiegler–Kedem model.
3. Validity of Darcy's law where the friction parameter is used to characterise the pressure drop.
4. A constant pressure of 1 atm is assumed at the permeate side.
5. A constant solute concentration is assumed in the permeate channel and the average value will be calculated from the inlet and outlet permeate solute concentrations.
6. The underlying process is assumed to be isothermal.

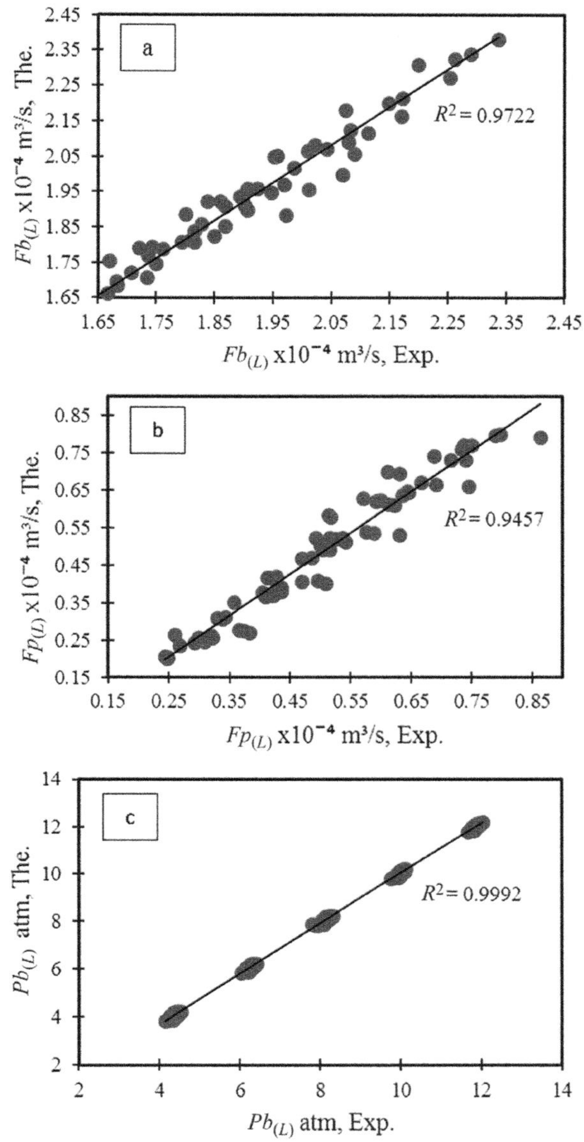

FIGURE 11.4 Experimental and model prediction of (a) retentate flow rate, (b) outlet permeate flow rate and (c) retentate pressure (inlet conditions mentioned in Section 11.2.1.3.1).

11.2.1.4.1 Process Modelling

Table 11.4 shows the baseline equations used in the model of Al-Obaidi et al. [45], which were used to estimate the performance of a single spiral-wound RO process for the removal of phenol from aqueous solutions. Specifically, the model used the experimental work of dilute phenol aqueous solutions from the literature. Therefore, the physical properties equations of the solution have been considered identical to the water equations proposed by Koroneos et al. [47].

11.2.1.4.2 Parameter Estimation

Before the proposed model can be applied to simulate the operation of a spiral-wound RO process, all the associated parameters must be assigned fixed values. The model has thus five parameters; namely L_p, ω, B_s, σ, and *b*.

TABLE 11.4
Equations Describing the Spiral-Wound RO Modelling

Model Equations	Equation No.
$F_{b(x)} = \dfrac{F_{b(L)}\left(e^{\sqrt{\frac{L_p}{Z}}x} - e^{-\sqrt{\frac{L_p}{Z}}x}\right) + F_{b(0)}\left(e^{\sqrt{\frac{L_p}{Z}}(L-x)} - e^{-\sqrt{\frac{L_p}{Z}}(L-x)}\right)}{\left(e^{\sqrt{\frac{L_p}{Z}}L} - e^{-\sqrt{\frac{L_p}{Z}}L}\right)}$	11.1
Where, $Z = \dfrac{1 + \dfrac{\sigma C_{p(av)} L_{p(T_b)}}{\omega_{(T_b)}} \quad \dfrac{C_{\tilde{s}(av)}(1-\sigma)L_{p(T_b)\sigma}}{\omega_{(T_b)}}}{W\,b}$	11.2
$\omega = \dfrac{B_s}{R(T_b + 273.15)}$	11.3
$C_{\tilde{s}(av)} = \dfrac{C_{\tilde{s}(0)} + C_{\tilde{s}(L)}}{2}$	11.4
Where, $C_{\tilde{s}(0)} = \dfrac{C_{s(0)} - C_{p(av)}}{\ln\left(\dfrac{C_{s(0)}}{C_{p(av)}}\right)}$ and $C_{\tilde{s}(L)} = \dfrac{C_{s(L)} - C_{p(av)}}{\ln\left(\dfrac{C_{s(L)}}{C_{p(av)}}\right)}$	11.5, 11.6
$F_{b(L)} = \dfrac{F_{b(0)}\left(e^{\sqrt{\frac{L_p}{Z}}L} + e^{-\sqrt{\frac{L_p}{Z}}L}\right)}{2} - \dfrac{\Delta P_{b(0)}\sqrt{\dfrac{L_p}{Z}}\left(e^{\sqrt{\frac{L_p}{Z}}L} + e^{-\sqrt{\frac{L_p}{Z}}L}\right)}{2b}$	11.7
$U_{b(x)} = \dfrac{F_{b(x)}}{t_f W}$	11.8
$P_{b(x)} = P_{b(0)} - \dfrac{b}{\sqrt{\dfrac{L_p}{Z}}\left(e^{\sqrt{\frac{L_p}{Z}}L} - e^{-\sqrt{\frac{L_p}{Z}}L}\right)}\left\{\left[F_{b(L)}\left[e^{\sqrt{\frac{L_p}{Z}}x} + e^{-\sqrt{\frac{L_p}{Z}}x} - 2\right] - F_{b(0)}\left[\left(e^{\sqrt{\frac{L_p}{Z}}(L-x)} + e^{-\sqrt{\frac{L_p}{Z}}(L-x)}\right) - \left(e^{\sqrt{\frac{L_p}{Z}}L} - e^{-\sqrt{\frac{L_p}{Z}}L}\right)\right]\right]\right\}$	11.9
$\Delta P_{b(x)} = \dfrac{\sqrt{\dfrac{L_p}{Z}}\,Zb\left\{\left[F_{b(0)}\left(e^{\sqrt{\frac{L_p}{Z}}(L-x)} + e^{-\sqrt{\frac{L_p}{Z}}(L-x)}\right)\right] - \left[F_{b(L)}\left(e^{\sqrt{\frac{L_p}{Z}}x} + e^{-\sqrt{\frac{L_p}{Z}}x}\right)\right]\right\}}{L_p\left(e^{\sqrt{\frac{L_p}{Z}}L} - e^{-\sqrt{\frac{L_p}{Z}}L}\right)}$	11.10
$J_{w(x)} = \dfrac{\sqrt{\dfrac{L_p}{Z}}}{W\left(e^{\sqrt{\frac{L_p}{Z}}L} - e^{-\sqrt{\frac{L_p}{Z}}L}\right)}\left\{\left[F_{b(0)}\left(e^{\sqrt{\frac{L_p}{Z}}(L-x)} + e^{-\sqrt{\frac{L_p}{Z}}(L-x)}\right)\right] - \left[F_{b(L)}\left(e^{\sqrt{\frac{L_p}{Z}}x} + e^{-\sqrt{\frac{L_p}{Z}}x}\right)\right]\right\}$	11.11

(Continued)

TABLE 11.4 (*Continued*)
Equations Describing the Spiral-Wound RO Modelling

Model Equations	Equation No.
$J_{s(x)} = J_{w(x)}(1-\sigma)C_{s(av)}^{\sim} + \omega_{(T_b)} RT_b (C_{w(x)} - C_{p(av)})$	11.12
$\frac{(C_{w(x)} - C_{p(av)})}{(C_{s(x)} - C_{p(av)})} = \exp\left(\frac{J_{w(x)}}{k_{(x)}}\right)$	11.13
$\frac{d\frac{(C_{s(x)}F_{b(x)})}{t_f W}}{dx} = -\frac{J_{w(x)}C_{p(av)}}{t_f} + \frac{J_{w(x)}C_{s(x)}}{t_f} + \frac{d}{dx}\left(D_{b(x)}\frac{dC_{s(x)}}{dx}\right)$	11.14
$C_{p(av)} = \frac{C_{p(0)} + C_{p(L)}}{2}$	11.15
$C_{p(0)} = \frac{B_{s(T_b)} C_{s(0)} e^{\frac{J_{w(0)}}{k_{(0)}}}}{J_{w(0)} + B_{s(T_b)} e^{\frac{J_{w(0)}}{k_{(0)}}}}$ and $C_{p(L)} = \frac{B_{s(T_b)} C_{s(L)} e^{\frac{J_{w(L)}}{k_{(L)}}}}{J_{w(L)} + B_{s(T_b)} e^{\frac{J_{w(L)}}{k_{(L)}}}}$	11.16, 11.17
$\frac{dF_{p(x)}}{dx} = WJ_{w(x)}$	11.18
$F_{p(\text{Total})} = F_{p(L)}$	11.19
$Rec_{(\text{Total})} = \frac{F_{p(\text{Total})}}{F_{b(0)}} \times 100$	11.20
$Rej = \frac{C_{b(L)} - C_{p(av)}}{C_{b(L)}} \times 100$	11.21

Source: Al-Obaidi, M.A. et al., *Desalin. Water Treat.*, 69, 93–101, 2017.

Generally, the parameters of the Spiegler–Kedem model can be predicted by fitting the experimental data with the predicted values for this model. Murthy and Gupta [49] used the non-linear parameter estimation method of the Box–Kanemasu to find the model parameters. Senthilmurugan et al. [36] achieved the same using the simplex search method.

Another way has been used to estimate the unknown parameters, which can be resolved within the gPROMS parameter estimation tool [29] for each set of experiments. Obviously, for any given set of values of the unknown parameters, the model equations can be solved to show the unit behavior at the experimental conditions, therefore yielding the objective function. The optimisation of these parameters is achieved by fitting the experimental data shown in Table 11.5 to the model predicted values. This was done by varying certain model parameters in order to maximise the probability that the model will closely predict the required values. The mathematical algorithm used to minimise the objective function is the sum of square errors (SSE) between the experimental retentate concentration, retentate flow rate, total permeated water, retentate pressure, and average solute rejection and the calculated values. This can be achieved by altering the model parameters from an initial guesstimate value to optimal values—usually referred to as the optimisation solver. gPROMS provides a mathematical solver tool called as MXLKHD, which is based on maximum likelihood optimisation. The optimisation problem is

TABLE 11.5
The Results of Parameter Estimation

$C_{s(0)} \times 10^3$	$P_{b(0)}$	T_b	$B_s \times 10^6$	$\omega \times 10^6$	$L_p \times 10^6$
2.125	4.93	32.5	1.7342	5.2388	1.4045
2.125	6.9	33.1	1.7342	5.2388	1.4045
2.125	8.9	33.0	1.7342	5.2388	1.4045
2.125	10.9	33.2	1.7342	5.2388	1.4045
2.125	14.8	34.0	1.7342	5.2388	1.4045
4.25	4.93	32.2	1.0707	0.67797	1.2483
4.25	6.9	32.8	1.0707	0.67797	1.2483
4.25	8.9	33.5	1.0707	0.67797	1.2483
4.25	10.9	33.9	1.0707	0.67797	1.2483
4.25	12.8	34.5	1.0707	0.67797	1.2483
4.25	14.8	34.5	1.0707	0.67797	1.2483
6.375	4.93	32.5	0.84163	1.5213	1.1314
6.375	6.9	33.0	0.84163	1.5213	1.1314
6.375	8.9	33.2	0.84163	1.5213	1.1314
6.375	10.9	33.5	0.84163	1.5213	1.1314
6.375	12.8	33.8	0.84163	1.5213	1.1314
6.375	14.8	34.0	0.84163	1.5213	1.1314
8.5	4.93	32.0	1.1476	1.8588	1.2090
8.5	6.9	32.5	1.1476	1.8588	1.2090
8.5	8.9	32.8	1.1476	1.8588	1.2090
8.5	10.9	33.0	1.1476	1.8588	1.2090
8.5	12.8	33.2	1.1476	1.8588	1.2090
8.5	14.8	33.5	1.1476	1.8588	1.2090
10.6	4.93	31.5	1.0972	0.58853	1.1184
10.6	6.9	32.2	1.0972	0.58853	1.1184
10.6	8.9	32.6	1.0972	0.58853	1.1184
10.6	10.9	32.8	1.0972	0.58853	1.1184
10.6	12.8	32.8	1.0972	0.58853	1.1184
10.6	14.8	33.0	1.0972	0.58853	1.1184
	b = 13,000 (atm s/m^4)			σ = 0.9075	

Source: Al-Obaidi, M.A. et al., *Desalin. Water Treat.*, 69, 93–101, 2017.

posed as a non-linear programming (NLP) problem and is solved using a successive quadratic programming (SQP) method within the gPROMS software [50]. The procedure for achieving this is explained as follows.

Given: Time invariant controls including, $C_{s(0)}$, $F_{b(0)}$, $P_{b(0)}$ and T_b
Measured variables data $\left(C_{s(L)}, C_{p(av)}, F_{b(L)}, F_{p(Total)}, P_{b(L)} \text{ and } Rej_{(av)}\right)$
The statistical variance models to be used for the measured variables.

The complete specification of a parameter estimation problem requires:

Obtaining: (L_p, ω, B_s, σ, and b).
Minimising: SSE.

For example, *SSE* for the retentate solute concentration is:

$$\text{SSE} = \sum_{i=1}^{N_{\text{Data}}} \left[C_{s,i}^{\text{Exp.}} - C_{s,i}^{\text{Cal.}} \right]^2 \quad (11.1)$$

With the earlier subject to process parameter constraints.

Results of the parameter estimation are given in Table 11.5 and show the variation of transport parameters with the inlet feed solute concentration.

It is worth noting the mass transfer coefficient is basically affected by the solvent flux, flow rate, solute concentration, and both the solvent and solute properties [41]. The mass transfer coefficient also varies along the x-axis dimension. So, the impact of all these factors can be correlated in Equation 11.2 as follows:

$$Sh_{(x)} = c_1 \left[Re_{p(x)}\, Re_{f(x)}\, C_{m(x)}\, Sc_{p(x)}\, Sc_{f(x)} \right]^{c2} \quad (11.2)$$

Where

$$Sh = \frac{2 t_f k_{(x)}}{D_{b(x)}} \quad (11.3)$$

$$Re_{p(x)} = \frac{2 t_p \rho_{p(x)} J_{w(x)}}{\mu_{p(x)}} \text{ and } Re_{f(x)} = \frac{2 \rho_{f(x)} F_{b(x)}}{W \mu_{f(x)}} \quad (11.4)$$

$$C_{m(x)} = \frac{C_{s(x)}}{\rho_m} \quad (11.5)$$

$$Sc_{p(x)} = \frac{\mu_{p(x)}}{\rho_{p(x)} D_{p(x)}} \text{ and } Sc_{f(x)} = \frac{\mu_{f(x)}}{\rho_{f(x)} D_{f(x)}} \quad (11.6)$$

Where $Sh_{(x)}, Re_{p(x)}, Re_{f(x)}, C_{m(x)}, Sc_{p(x)}, Sc_{f(x)}, \rho_{p(x)}, \rho_{f(x)}, \mu_{p(x)}, \mu_{f(x)}, D_{p(x)}$ and ρ_m are the Sherwood number, the permeate Reynolds number, the feed Reynolds number, dimensionless solute concentration, the permeate Schmidt number, the feed Schmidt number, the density of permeate, the density of feed, the viscosity of permeate, the viscosity of the feed, the solute diffusion coefficient of permeate at any point along the x-axis and the molal density of water respectively. In order to determine the values of $(c_1 \text{ and } c_2)$ mentioned in Equation 11.2, the mass transfer coefficients are calculated from the correlation reported by Wankat [51].

$$k_{(x)} = 1.177 \left(\frac{F_{b(x)} D_{b(x)}^{\;2}}{t_f^{\;2} W L} \right)^{0.333} \quad (11.7)$$

Then, the other parameters of Equation 11.2 in both ($x = 0$ and $x = L$) can be estimated from the experimental data of phenol, which were reported by Srinivasan et al. [48].

TABLE 11.6
Specification of Polyamide Membrane Module

Make	Membrane Material	Module Configuration	Number of Turns	Feed Spacer Thickness	Module Diameter	Module Width	Module Length
Ion Exchange, India	TFC Polyamide	Spiral-wound	13	0.85 (mm)	2.5 inches	1.6667 (m)	0.45 (m)

Source: Al-Obaidi, M.A. et al., *Desalin. Water Treat.*, 69, 93–101, 2017.

Finally, by plotting $Ln(Sh)$ vs. $Ln(Re_p\,Re_f\,C_m\,Sc_p\,Sc_f)$, the values of constants can be found. As a result, the last construction of mass transfer coefficient of phenol is:

$$k_{(x)} = \frac{6.5045\,D_{b(x)}^{0.9995}}{t_f}\left(\frac{F_{b(x)}\,t_p\,J_{w(x)}\,C_{s(x)}}{W\,\rho_m\,D_{p(x)}}\right)^{0.0005} \tag{11.8}$$

11.2.1.4.3 Experimental Procedure

The Perma-TFC polyamide RO membrane in a spiral-wound module (supplied by Permionics, Vododara, India) was used by Srinivasan et al. [48]. The characteristics of the spiral-wound module are given in Table 11.6. The experiments were carried out using binary mixtures of phenol compound in water at five different solute concentrations varying from 2.125×10^{-3} to 10.6×10^{-3} kmol/m^3. Also, the pressure was varied from 4.93 to 14.8 atm for each set of inlet feed solute concentrations under constant feed flow rate 3.333E-4 m^3/s. Lastly, the fluid temperature was kept between 31.5°C and 34.5°C.

11.2.1.4.4 Model Validation

Table 11.7 shows the experimental results of phenol removal and the model predictions for five groups of inlet feed solute concentration (each group holding six different inlet feed pressures). Tables 11.7 also depicts the percentage error between the experimental results and the model predictions for a number of model parameters, such as retentate pressure ($P_{b(L)}$), retentate flow rate ($F_{b(L)}$), average permeate solute concentration ($Cp_{(av)}$), the total volumetric permeated flow rate ($F_{p(\text{Total})}$), and the average solute rejection ($Rej_{(av)}$). The prediction of the values of the earlier parameters has been obtained by running the model for different inlet feed conditions. Figure 11.5 shows the comparison of experimental and theoretical results for average solute rejection and average permeate concentration. Generally, the predicted values of the theoretical model are in good agreement with the experimental ones over the ranges of pressure and concentration.

11.2.1.4.5 Steady-State Variation of Operating Parameters

In steady-state mode, Figure 11.6 shows the variation of feed pressure and concentration along the membrane length (x-axis). The feed pressure decreases due to pressure drop caused by friction. As a result, the pressure gradient is at its maximum point at the entrance of the membrane and at its minimum point at the end of the unit. The feed concentration progresses in the subsequent sub-sections of the feed channel since the solute is retained in the wall with the diffusion of water through the membrane.

It is worth noting that the feed flow rate decreases along the membrane channel as can be seen in Figure 11.7 and this can be attributed to the permeated water passing through the membrane, which reduces the velocity of the feed and increases the feed concentration along the membrane (Figure 11.6).

TABLE 11.7
Model Validation with Experimental Results for the Inlet Feed Flow Rate ($F_{b(0)}$ = 3.333 × 10^{-3} m^3/s)

No	$P_{b(0)}$ (atm)	T_b (°C)	$C_{s(0)} \times 10^3$ (kmol/m³)	$P_{b(L)}$ (atm)			$F_{p(Total)} \times 10^6$ (m³/s)			$Cp_{(av)} \times 10^3$ (kmol/m³)			$Rej_{(av)}$			$F_{b(L)} \times 10^4$ (m³/s)		
				Exp.	The.	%Error	Exp.	The.	%Error	Exp.	The.	%Error	Exp.	The.	%Error	Exp.	The.	%Error
1	4.93	32.5	2.125	2.99	2.99	0.01	3.53	3.02	14.4	0.831	0.806	3.02	0.6462	0.627	−3.14	3.30	3.30	−0.12
2	6.9	33.1	2.125	4.96	4.97	0.13	5.20	5.18	0.45	0.647	0.637	1.54	0.727	0.708	−2.66	3.28	3.28	−0.00
3	8.9	33	2.125	6.9	6.97	1.04	7.20	7.37	−2.30	0.580	0.574	0.96	0.7593	0.740	−2.62	3.26	3.26	0.03
4	10.9	33.2	2.125	8.9	8.98	0.87	9.60	9.55	0.48	0.524	0.551	−5.17	0.7861	0.753	−4.35	3.24	3.24	−0.06
5	14.8	34	2.125	12.9	12.88	0.15	12.2	12.2	0.00	0.349	0.368	−5.55	0.8745	0.838	−4.37	3.21	3.21	−0.07
6	4.93	32.2	4.25	2.99	2.99	0.00	3.30	2.68	18.7	1.24	1.24	−0.19	0.766	0.712	−7.61	3.30	3.31	−0.16
7	6.9	32.8	4.25	4.96	4.96	0.09	5.00	4.60	8.06	1.05	0.935	10.94	0.8132	0.785	−3.54	3.28	3.29	−0.12
8	8.9	33.5	4.25	6.9	6.97	1.00	7.00	6.54	6.54	0.80	0.81	−1.22	0.861	0.816	−5.46	3.26	3.27	−0.16
9	10.9	33.9	4.25	8.9	8.98	0.84	8.50	8.49	0.15	0.72	0.755	−4.84	0.8737	0.830	−5.21	3.25	3.25	−0.04
10	12.8	34.5	4.25	10.9	10.88	−0.17	10.25	10.3	−0.82	0.685	0.73	−6.62	0.8807	0.838	−5.14	3.23	3.23	−0.03
11	14.8	34.5	4.25	12.9	12.89	0.0	12.25	12.3	−0.23	0.718	0.734	−2.20	0.8766	0.839	−4.53	3.21	3.21	−0.07
12	4.93	32.5	6.375	2.99	2.99	0.00	3.20	2.43	20.4	1.40	1.65	−17.89	0.7983	0.744	−7.22	3.30	3.31	−0.21
13	6.9	33	6.375	4.96	4.96	0.00	4.33	4.17	3.70	1.24	1.21	2.47	0.8216	0.815	−0.86	3.29	3.29	−0.04
14	8.9	33.2	6.375	6.9	6.97	0.98	5.93	5.93	−0.06	1.176	1.03	12.15	0.8346	0.843	1.02	3.27	3.27	−0.02
15	10.9	33.5	6.375	8.9	8.97	0.81	7.00	7.70	−9.96	0.94	0.95	−1.05	0.868	0.857	−1.25	3.26	3.26	0.17
16	12.8	33.8	6.375	10.9	10.9	0.00	8.70	9.37	−7.73	0.87	0.913	−4.94	0.8798	0.864	−1.82	3.25	3.24	0.15
17	14.8	34	6.375	12.9	12.9	0.00	11.1	11.1	0.00	0.63	0.669	−6.12	0.9141	0.902	−1.39	3.22	3.22	−0.07
18	4.93	32	8.50	2.99	2.99	0.00	3.13	2.60	17.0	2.61	2.65	−1.60	0.7061	0.692	−2.00	3.30	3.31	−0.14
19	6.9	32.5	8.50	4.96	4.96	0.08	4.53	4.45	1.68	2.22	2.01	9.64	0.75	0.770	2.54	3.29	3.29	−0.02

(Continued)

TABLE 11.7 (*Continued*)
Model Validation with Experimental Results for the Inlet Feed Flow Rate ($F_{b(0)}$ = 3.333 × 10^{-3} m^3/s)

No	$P_{b(0)}$ (atm)	T_b (°C)	$C_{s(0)} \times 10^3$ (kmol/m³)	$P_{b(L)}$ (atm)			$F_{p(Total)} \times 10^6$ (m³/s)			$Cp_{(av)} \times 10^3$ (kmol/m³)			$Rej_{(av)}$			$F_{b(L)} \times 10^4$ (m³/s)		
				Exp.	The.	%Error	Exp.	The.	%Error	Exp.	The.	%Error	Exp.	The.	%Error	Exp.	The.	%Error
20	8.9	32.8	8.50	6.9	6.97	0.99	6.20	6.34	−2.22	1.93	1.75	9.58	0.7834	0.802	2.27	3.27	3.27	0.02
21	10.9	33	8.50	8.9	8.97	0.83	8.20	8.22	−0.26	1.60	1.63	−1.94	0.8261	0.816	−1.18	3.25	3.25	−0.03
22	12.8	33.2	8.50	10.9	10.9	0.00	9.30	10.0	−7.65	1.47	1.59	−8.16	0.8402	0.823	−2.11	3.24	3.23	0.16
23	14.8	33.5	8.50	12.9	12.9	0.00	11.5	11.9	−3.43	1.40	1.59	−13.31	0.8495	0.825	−2.96	3.22	3.22	0.04
24	4.93	31.5	10.6	2.99	2.99	0.00	2.66	2.40	9.94	3.09	3.35	−8.40	0.7112	0.688	−3.38	3.31	3.31	−0.06
25	6.9	32.2	10.6	4.96	4.96	0.00	4.13	4.11	0.47	2.52	2.51	0.39	0.7647	0.768	0.47	3.29	3.29	−0.00
26	8.9	32.6	10.6	6.9	6.97	0.97	5.86	5.85	0.14	2.02	2.16	−6.92	0.8164	0.803	−1.72	3.27	3.28	−0.02
27	10.9	32.8	10.6	8.9	8.97	0.81	7.50	7.59	−1.24	1.83	2.00	−9.16	0.8351	0.819	−1.95	3.26	3.26	−0.01
28	12.8	32.8	10.6	10.9	10.9	0.00	9.00	9.25	−2.73	1.69	1.94	−14.53	0.8462	0.826	−2.40	3.24	3.24	0.02
29	14.8	33	10.6	12.9	12.9	0.00	10.5	10.5	−0.44	1.40	1.50	−7.10	0.8739	0.867	−0.84	3.23	3.23	−0.06

Source: Al-Obaidi, M.A. et al., *Desalin. Water Treat.*, 69, 93–101, 2017.

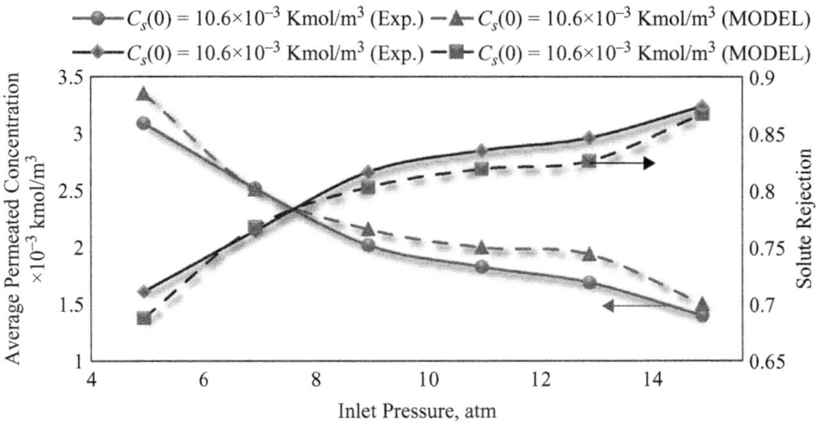

FIGURE 11.5 Comparison of theoretical and experimental results of solute rejection and average permeate concentration. (Adapted from Al-Obaidi, M.A. et al., *Desalin. Water Treat.*, 69, 93–101, 2017.)

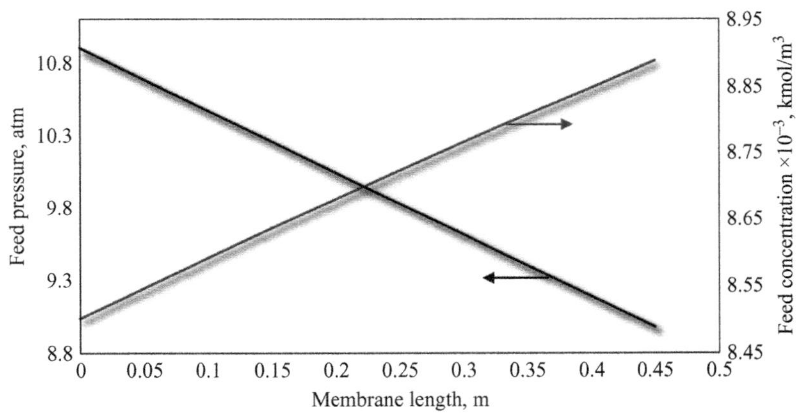

FIGURE 11.6 Feed pressure and concentration variation along the membrane length, inlet feed conditions ($3.333 \times 10^{-4}\,m^3/s$, $8.5 \times 10^{-3}\,kmol/m^3$, 10.9 atm and 32.2°C). (Adapted from Al-Obaidi, M.A. et al., *Desalin. Water Treat.*, 69, 93–101, 2017.)

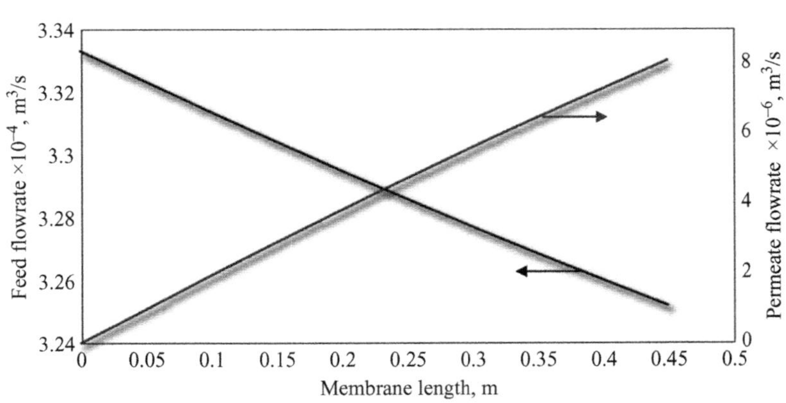

FIGURE 11.7 Feed and permeate flow rates variation along the membrane length, inlet feed conditions ($3.333 \times 10^{-4}\,m^3/s$, $8.5 \times 10^{-3}\,kmol/m^3$, 10.9 atm and 32.2°C). (Adapted from Al-Obaidi, M.A. et al., *Desalin. Water Treat.*, 69, 93–101, 2017.)

In addition, the water flux decreases along the membrane length as the pressure decreases due to friction, which decreases the net pressure driving force as shown in Figure 11.8. It is the usual expectation that the solute flux progresses in the subsequent sub-sections of the feed channel since the solute is retained in the membrane wall. Surprisingly, Figure 11.8 shows a steady decrease of solute flux along the membrane length. This can be attributed to the fact that the proposed solution is based on the principles of the thermodynamics model, which assumes that the solute flux is mainly driven by the water flux (Equation 11.12, Table 11.4). The assumption of constant values of the transport parameters along the membrane length also contribute to the main cause of solute flux reduction.

Figure 11.9 shows that the increasing operating pressure results when increasing solute rejection and total permeated flow rate are due to the increase in water flux for all the feed concentrations tested. Interestingly, for low inlet feed concentration conditions, the solute rejection increases due to the increase in the inlet feed concentration, and this may be attributed to the increase in the membrane solute isolation intensity. However, any further increase in inlet feed concentration can cause a reduction in solute rejection due to higher osmotic pressure, which causes a decrease in the driving force of water flux.

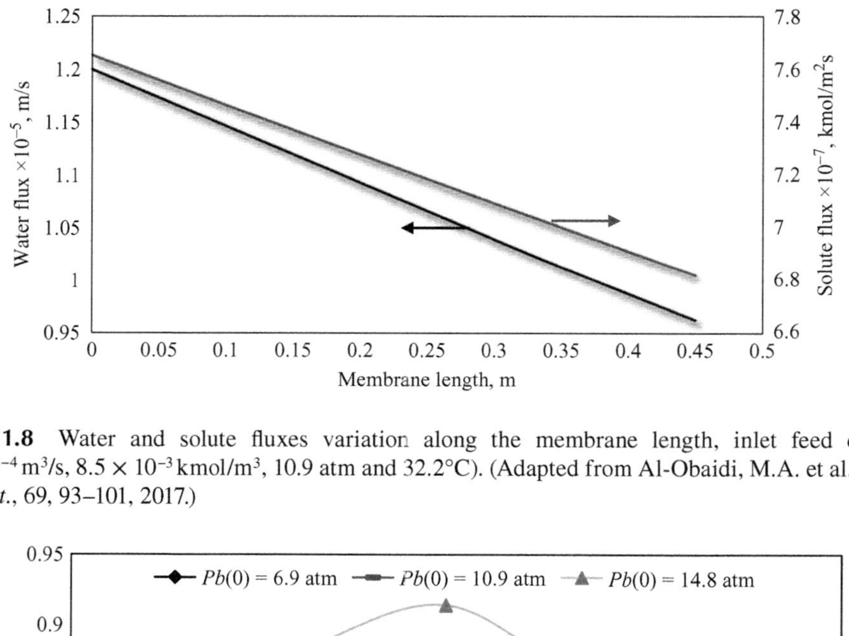

FIGURE 11.8 Water and solute fluxes variation along the membrane length, inlet feed conditions ($3.333 \times 10^{-4}\ m^3/s$, 8.5×10^{-3} kmol/m^3, 10.9 atm and 32.2°C). (Adapted from Al-Obaidi, M.A. et al., *Desalin. Water Treat.*, 69, 93–101, 2017.)

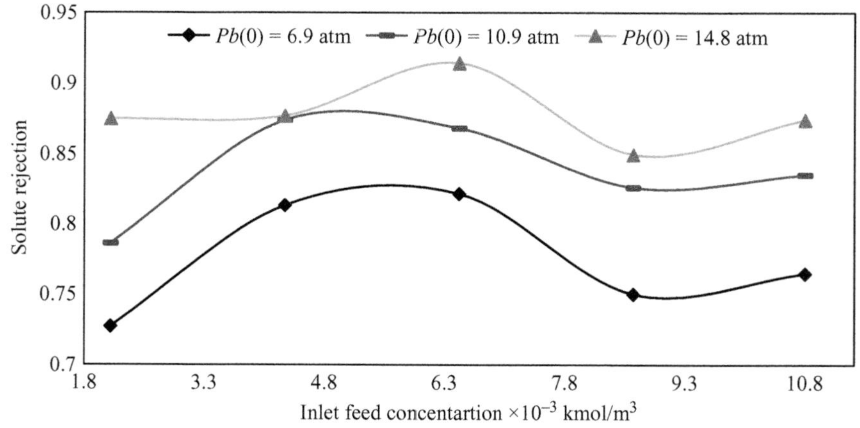

FIGURE 11.9 Solute rejection verses inlet feed concentration for three different inlet feed pressure, inlet feed conditions $3.333 \times 10^{-4}\ m^3/s$ and (32.2°C–34.5°C). (Adapted from Al-Obaidi, M.A. et al., *Desalin. Water Treat.*, 69, 93–101, 2017.)

11.2.2 Optimisation Approaches of RO Process Using GA (Genetic Algorithm)

In the past, the optimisation of seawater RO desalination systems has been carried out using different methods including global optimisation algorithm [52], SQP [53], mixed-integer non-linear programming (MINLP) [54], GA [55] and multi-objective optimisation and genetic algorithm (MOO+GA) [56].

In many optimisation applications, the use of GA has yielded better results in comparison to other conventional methods [57]. For instance, GAs have been used extensively in different areas of chemical engineering process design and operation, such as distillation systems [58], semi-batch reactors [59], multi-phase catalytic reactors (hydrogenation reaction system) [60], microchannel reactors (a novel technology for the synthesis of liquid hydrocarbons applications) [61], and steam reforming of hydrocarbons for the generation of hydrogen and synthesis gas [62]. Fang et al. [63] have combined an integrated neural network (NN) dynamic model and GA approach to optimise the performance of a full-scale municipal wastewater treatment plant with substantial influent fluctuations.

The use of GA to optimise the seawater RO desalination processes has already been implemented in a number of studies. Guria et al. [56] used the multi-objective optimisation and non-dominated sorting genetic algorithm (NSGA) technique for desalination of seawater using spiral-wound and tubular RO modules. The optimisation problem consisted of two or three objective functions for maximising the water flux in addition to minimising the permeate concentration and cost of filtration of an existing plant. Murthy and Vengal [55] used a single objective genetic algorithm technique (SGA) to optimise the rejection of NaCl in a laboratory scale RO desalination system of a disc-shaped flat cellulose acetate membrane. The experiments were carried out by varying the inlet feed flow rate and the overall water flux at a constant feed concentration. In this study, the mechanism of water and solute transport are measured using the Spiegler and Kedem model. Djebedjian et al. [64] implemented GA with a solution–diffusion model to optimise the performance of a real RO desalination plant and predicted the best operating pressure difference across the membrane, which enhances the water flux with low permeate concentration. Moreover, the modelling and prediction of the membrane-fouling rate in a microfiltration (MF) pilot-scale drinking water production system was achieved using the genetic programming of Lee et al. [65]. Park et al. [66] used GA to analyse the performance of a pilot-scale RO system. Bourouni et al. [67] used GA to optimise the configuration of a hybrid system of a small RO unit coupled with renewable energy source (photovoltaic and wind). Finally, Yuen et al. [68] used NSGA to maximise the removal of ethanol from beer and minimise the removal of the extract (taste chemicals) for hollow-fiber lab-scale beer dialysis module.

In contrast, and to the best of the authors' knowledge, the GA for RO-based wastewater treatment has not been widely used to find the optimal operation values achievable within the manufacturer's specifications. Okhovat and Mousavi [69] used GA to model the rejection of arsenic, chromium, and cadmium ions as a function of transmembrane pressure and initial concentration of pollutants in a nanofiltration (NF) pilot-scale system. Soleimani et al. [70] investigated the treatment of oily wastewaters with commercial polyacrylonitrile (PAN) ultrafiltration (UF) membranes by using artificial neural networks (ANNs) to predict the permeation flux and fouling resistance. GA was then used to optimise the operating conditions of transmembrane pressure, cross-flow velocity, feed temperature, and pH. The objective was to maximise the permeation flux while minimising the fouling behavior.

To the best of the authors' knowledge, there has not been any study that uses the GA optimisation technique in a distributed model for optimising the removal of organic compounds such as chlorophenol using a spiral-wound RO process. The following section presents the optimisation of the RO process for the removal of chlorophenol from wastewater based on the model developed by Al-Obaidi et al. [44].

11.2.2.1 Problem Description and Formulation

The objective function is to optimise chlorophenol rejection given the lower pressure drop constraint across the membrane with a lower than recommended value. The optimisation technique used here is based on a GA algorithm for the same pilot-scale RO wastewater system used by Sundaramoorthy et al. [41]. Experimental data show that a higher chlorophenol rejection of 83% is achieved at inlet concentration of 6.226E-3 kmol/m^3 and a pressure loss of 1.93 atm. This exceeds the maximum recommended manufacturer value for the module selected. The objective of this optimisation is to find the optimum solute rejection of chlorophenol for each inlet feed concentration (five cases) within the restricted operating conditions of inlet feed flow rate, pressure, and temperature. Also, the constraint of 1.38 atm as a maximum overall pressure loss across the membrane length (as declared by the Ion Exchange Ltd. Company, India) has been considered to represent the relative power consumption of each run. The optimisation iteration is carried out individually for five feed concentrations of chlorophenol used in the experiments, varying from 0.778E-3 to 6.226E-3 kmol/m^3. The model transport parameters (A_w, B_s, and b) have been considered constant along the optimisation procedure.

The optimisation problem is coded mathematically in two problems as follows:

Problem 11.1

Max $\quad Rej$

$F_{b(0)}, P_{b(0)}, T_b$

Subject to:

$$P_{\text{loss}} \leq P_{\text{loss}}{}^{d}$$

$$F_{b(0)}{}^{L} \leq F_{b(0)} \leq F_{b(0)}{}^{U}$$

$$P_{b(0)}{}^{L} \leq P_{b(0)} \leq P_{b(0)}{}^{U}$$

$$T_b{}^{L} \leq T_b \leq T_b{}^{U}$$

The model equation and the physical properties are given in Section 11.2.1.2.

The choice of the objective function is to achieve high chlorophenol rejection within the constraints of the decision variables. The limits of the decision variables of the inlet feed flow rate, pressure, and temperature and the recommended value of the pressure loss are given in Section 11.2.2.3.1.

In line with economic aspects of low energy consumption, Problem 11.2 is formulated as follows:

Problem 11.2

Max $\quad Rej$

$F_{b(0)}, P_{b(0)}, T_b$

Min $\quad P_{b(0)}$

$F_{b(0)}, P_{b(0)}, T_b$

Subject to:

$$P_{\text{loss}} \leq P_{\text{loss}}{}^{d}$$

$$F_{b(0)}{}^{L} \leq F_{b(0)} \leq F_{b(0)}{}^{U}$$

$$P_{b(0)}{}^{L} \leq P_{b(0)} \leq P_{b(0)}{}^{U}$$

$$T_b{}^{L} \leq T_b \leq T_b{}^{U}$$

The model equation and the physical properties are given in Section 11.2.1.2.

The choice of the first objective function is to secure optimal chlorophenol rejection, while the contribution of the second objective function is to maintain the process of filtration within an accepted consumption of energy (lower operating pressure).

There are two objectives in this problem, and the following penalty function is used to balance both objectives and transfer the problem into a single objective optimisation problem as follows:

$$L = W_1 \times Rej - W_2 \times P_{b(0)} \tag{11.9}$$

Weight factors W_1 and W_2 are used to balance the contributions of each objective. Then, the original problem will be altered to:

$$\max_{F_{b(0)},\, P_{b(0)},\, T_b} L$$

Subject to:

$$P_{\text{loss}} \leq P_{\text{loss}}{}^{d}$$

$$F_{b(0)}{}^{L} \leq F_{b(0)} \leq F_{b(0)}{}^{U}$$

$$P_{b(0)}{}^{L} \leq P_{b(0)} \leq P_{b(0)}{}^{U}$$

$$T_b{}^{L} \leq T_b \leq T_b{}^{U}$$

11.2.2.2 GA Modelling

GA was originally proposed by Holland [57] and is a stochastic and population-based optimisation technique constructed on the perceptions of natural evolution and the biological principles of natural selection. GAs have been successfully applied in various engineering optimisation problems [71,72].

The general GA procedure is shown in Figure 11.10. Initially, a population, which consists of a number of individuals, is randomly generated within the lower and upper limits of the decision variables.

There are two typical ways of representing individuals in a GA context. Traditionally, an individual is represented as a chromosome, i.e., as a bit string. This originated from the idea of stems derived from DNA work and fits well with integer decision variables. However, more and more researchers prefer a natural representation, i.e., one based on float numbers. In this section, an individual is therefore represented as a vector of real numbers of decision variables.

```
Begin
t=0;
initialize G(t);
Evaluate G(t);
while (not termination condition) do
Select G(t+1);
Crossover G(t+1);
Mutate G(t+1);
Evaluate G(t+1);
t=t+1;
End (while)
End
```

FIGURE 11.10 GA structure. (Adapted from Al-Obaidi, M.A. et al., *Chem. Eng. J.*, 316, 91–100, 2017.)

1. The concept of selection is used to select individuals from the current generations and use them to create a new generation based on their fitness. Usually, an individual with a higher fitness has a large probability of surviving in the next generation. In this study, a roulette-wheel method is used to select individuals from the current population. Suppose F_i is the fitness of individuals i, its probability of being selected can be calculated as:

$$P_i = \frac{F_i}{\sum_{j=1}^{N_p} F_j} \tag{11.10}$$

   Where N_p is the population size. An example of the roulette-wheel selection is shown in Figure 11.11. A proportion of a wheel is assigned to each individual based on their fitness or selected probability. A random number between 0 and 360 is generated to determine how many angles the wheel will rotate, and the individual to which the arrow points will be selected and copied to the next generation until N_p individuals have been selected. Figure 11.11 also shows that an individual with higher fitness enjoys a larger area and therefore has a higher probability of being selected.
2. The concept of crossover is used to select parents from the new formulated population to generate offspring with a probability of P_c. The crossover method depends on the representation of individuals. For example, if individuals are represented as chromosomes (i.e., a string of bites), the crossover method could be a one-point or two-point crossover. In this study, since individuals are represented as vectors of real numbers, the crossovers

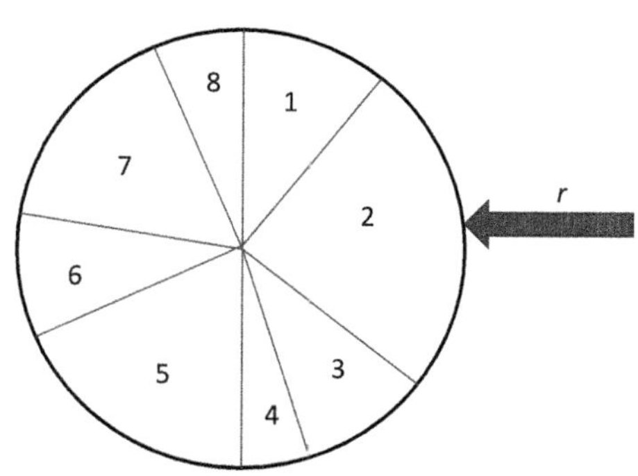

FIGURE 11.11 Roulette-wheel selection. (Adapted from Al-Obaidi, M.A. et al., *Chem. Eng. J.*, 316, 91–100, 2017.)

are performed as intermediate recombinations, so that an offspring O of randomly chosen parents S and T is expressed as follows:

$$x_0 = x_T + U\left(x_S - x_T\right) \tag{11.11}$$

Where U is a uniformly distributed random number over [0, 1].

3. The concept of mutation is used to allow a randomly chosen individual to move to a new position with a probability of P_m. The parent selection method is similar to the one of selecting parents for crossover. Suppose a parent has been selected, a random number j, between 1 and m (the dimension of decision variables) is used, and, the corresponding variable will be mutated by using the following uniform mutation will be applied:

$$X_j^{/} = x_j + R\left(x_j^u - x_j^l\right) \tag{11.12}$$

Where R is a uniformly distributed random number over [−1, 1] and x_j^u, x_j^l are the upper and lower limits, respectively, of variable j.

One of the most important characteristics of GAs is the generation of a number of different solutions for the specified problem at the end of each iteration, as opposed to a single solution. This is carried out without requiring a set of good initial guesses for the decision variables [56]. This approach therefore gives a wider choice for the optimised chlorophenol rejection for each input data of operation. GAs are considered to be global optimisation methods, while gradient-based methods can only find a local solution. This is why the GA technique has been applied to maximise the rejection in a wastewater treatment RO system.

11.2.2.3 GA Optimisation Numerical Results

This section describes the particular application of GA for optimising wastewater treatment using the developed RO system.

11.2.2.3.1 Inlet Feed Concentration

The optimisation study focuses on using GAs to locate the best operating parameters for the optimum rejection of chlorophenol using a single spiral-wound RO membrane element. The optimisation technique is implemented using the same experiments carried out by Sundaramoorthy et al. [41], this time using five cases of different inlet feed concentration of 0.778E-3, 1.556E-3, 2.335E-3, 3.891E-3 and 6.226E-3 kmol/m^3, and with the operating specification of maximum and minimum inlet feed flow rate, pressure, and temperature of 1E-4–1E-3 m^3/s, 4–24.77 atm and 15°C–40°C, respectively. Also, this optimisation will be based on the constraint of an allowable pressure drop (P_{loss}^d) across the membrane length of 1.38 atm. The specified bounds of the operating parameters are as recommended by the membrane manufacturer.

11.2.2.3.2 Parameter Settings

The RO model was coded in MATLAB and solved using the intrinsic modelling and optimisation tool.[1] The same can readily be implemented using MATLAB and Excel to build complex optimisation models. The GA parameters used to explore optimal solutions are given in Table 11.8.

11.2.2.3.3 Effects of GA Parameters

There are several GA parameters, which include weight factors, number of generations, and crossover and mutation probabilities. This section assesses the performance of the GA used to solve different cases of RO optimisation by analysing the effects of the parameters used.

---

[1] http://www.scholarpark.co.uk/mos (A modelling and optimisation platform developed by researchers).

TABLE 11.8
GA Parameters

Parameter	No.
Maximum generation, N_{gen}	500
Population size, N_{pop}	50
Mutation probability, P_m	0.1
Crossover probability, P_c	0.6

Source: Al-Obaidi, M.A. et al., *Chem. Eng. J.*, 316, 91–100, 2017.

The inlet feed concentration of 6.226E-3 kmol/m^3 in Problem 11.2 is chosen to analyse the GA performances and to identify the best parameter settings for solving the developed RO optimisation problems.

11.2.2.3.3.1 Effects of the Weight Factor There are two weight factors W_1 and W_2 as shown in Equation 11.9, where W_1 is always set to 1 to simplify the analysis, as the maximum value of rejection is 1 and the maximum value of P_{b0} is about 25. When, $W_2 = 1/25 = 0.04$, both objectives are of the same importance. The results of Table 11.9 show that a value of 0.04 for W_2 offers high chlorophenol rejection at an economical cost with reduced operating pressure and allowable temperature. Any further reduction in W_2 slightly increases the rejection parameter and requires higher operating pressure. Moreover, an increase of W_2 with a value larger than 0.04 has no significant impact, and the output results of optimum rejection and decision variables remain the same. This is because the operating pressure has reached its lower limit. The increase of W_2 cannot increase the contribution of the pressure to the objective function.

11.2.2.3.3.2 Effects of Generations Table 11.10 shows the influence of using different numbers of generations with a population size of 50 at the optimum weight W_2 of 0.04. As expected, the larger the generation, the more easily will the GA identify the solution. This is because there are more chances to explore the space and more fitness evaluations are therefore needed. It is observed that any generation between 80 and 200 can offer an optimum solution for the problem within the recommended decision variables.

This observation can be applied to the effect of population size. The larger the population size, the easier the solution, but more fitness evaluations are required. Generally, if a small population

TABLE 11.9
The Influence of the Weight Factor on GA Results

W_2	*Rej*	F_{b0}	P_{b0}	T_b
0.001	0.95	2.3698E-4	21.43	40
0.005	0.92	2.0418E-4	11.18	40
0.01	0.90	1.9235E-4	7.57	40
0.04	**0.90**	**1.9224E-4**	**7.53**	**40**
0.08	0.90	1.9224E-4	7.53	40
0.1	0.90	1.9224E-4	7.53	40
0.2	0.90	1.9224E-4	7.53	40

Source: Al-Obaidi, M.A. et al., *Chem. Eng. J.*, 316, 91–100, 2017.
Where $P_{loss} = 1.38$ (atm).

TABLE 11.10
The Influence of the Generation Number on GA Results

Generation	*Rej*	P_{loss}	F_{b0}	P_{b0}	T_b
10	0.87	0.82	1.2387E-4	7.53	38.1
20	0.89	0.70	1.0700E-4	7.53	40
50	0.90	1.36	1.8933E-4	7.53	40
80	**0.90**	**1.38**	**1.9224E-4**	**7.53**	**40**
100	0.90	1.38	1.9224E-4	7.53	40
200	0.90	1.38	1.9224E-4	7.53	40

Source: Al-Obaidi, M.A. et al., *Chem. Eng. J.*, 316, 91–100, 2017.

size is used, a larger generation is required; a large population size requires a small generation. In this context, the population size is set as 50 and the minimum generation as 80.

11.2.2.3.3.3 Effects of Crossover Probability and Mutation Probability Table 11.11 shows the influence of the crossover probability varying between 0.1 and 0.5 at a mutation probability (P_m) of 0.1 for W_2 of 0.04, 100 generations and a population size of 50 on the output results of GA. It is observed that an increase in the crossover probability has no influence. Furthermore, the impact of mutation probability variation is insignificant as can be shown in Table 11.12, where it is varied between 0.01 and 0.5 with constant values of crossover probability of 0.4. Nevertheless, it is reasonable to use P_m between 0.01 and 0.02 in order to achieve a lower pressure drop than the other tested values.

TABLE 11.11
The Influence of the Crossover Probability on GA Results

P_c	*Rej*	F_{b0}	P_{b0}	T_b
0.1	0.9	1.9224E-4	7.53	40.0
0.2	0.9	1.9224E-4	7.53	40.0
0.3	0.9	1.9224E-4	7.56	40.0
0.4	0.9	1.9224E-4	7.53	40.0
0.5	0.9	1.9224E-4	7.53	40.0

Source: Al-Obaidi, M.A. et al., *Chem. Eng. J.*, 316, 91–100, 2017.
Where P_{loss} = 1.38 (atm).

TABLE 11.12
The Influence of the Mutation Probability on GA Results

P_m	*Rej*	P_{loss}	F_{b0}	P_{b0}	T_b
0.01	0.90	1.37	1.9139E-4	7.53	40
0.02	**0.90**	**1.36**	**1.8990E-4**	**7.53**	**40**
0.05	0.90	1.38	1.9221E-4	7.56	40
0.4	0.90	1.38	1.9224E-4	7.53	40
0.5	0.90	1.38	1.9224E-4	7.53	40

Source: Al-Obaidi, M.A. et al., *Chem. Eng. J.*, 316, 91–100, 2017.

The earlier results of investigating the impact of GA computational parameters show the merits of obtaining a number of different results, which offer a number of solutions for a specific optimisation problem, especially for the case of altering the weight factor W_2.

11.2.2.3.4 Summaries of Effects of Parameters

It is therefore important to find the global solution of the underlying RO problem, with the population size set at 50 and the minimum generation at 80. Of course, the larger generation is ideal from a GA point of view, i.e., one that facilitates exploring global solutions. Since crossover and mutation probability have little influence on performance for this problem, any probability can, in effect, be used.

Weight factors have a big influence on the optimal result. In this case, W_1 set as 1 and W_2 set at 0.04 mean that both objectives are of the same importance and have therefore the same contribution to the system objective.

11.2.2.4 Discussion of Results

The optimisation problems of one- and two-objective functions were solved using the GA described in Section 11.2.2.3 in combination with the proposed model of Section 11.2.1.2. The optimisation results of Problem 11.1 for each inlet feed concentration and the optimised decision variables obtained are given in Table 11.13. The GA optimisation results show that a maximum chlorophenol rejection for all five cases can be achieved within the operating parameters of inlet feed flow rate, pressure, and temperature of 1.046–1E-4 m^3/s, 24.7717–16.09 atm and 15°C–40°C, respectively. It is readily observed that the removal efficiency of chlorophenol can be increased within the maximum allowable pressure. However, lower values of the operating feed flow rate are required to ensure lower pressure loss along the membrane length, which enhances the flux of water through the membrane and reduces permeate concentration. Also, the lower feed flow rate can secure the full rejection of chlorophenol from its aqueous solution by maintaining the solution for a longer resident time inside the unit. In contrast to Case 1, which has a higher concentration, the optimisation results of medium and low concentrations (Cases 2 to 5) are within a low feed temperature. One of the key outcomes of this is the fact that the optimiser can use such decision variables in cold areas too.

At medium and low feed concentration, it appears that there are competitive impacts of pressure, flow rate, and temperature, which determine the chlorophenol rejection. An increase in the temperature has two different impacts regarding the rejection parameter of RO wastewater treatment. The first one is to enhance water flux by decreasing the viscosity and density parameters and thermal expansion of the membrane [73], and at the same time increasing the organic solute diffusion and absorption through the membrane. This can readily plug the pores of the membrane [74] and deteriorate rejection. However, the combined impact of feed pressure and flow rate

TABLE 11.13
Optimal Values for Problem 11.1

Case	Feed Conc. $C_{b(0)} \times 10^3$, (kmol/m^3)	*Rej* (Max.) [41]	Experimental P_{loss}, [41]	GA P_{loss}	GA *Rej*	Decision Variables $F_{b(0)}$	$P_{b(0)}$	T_b
1	6.226	0.83	1.93	0.42	0.98	1.0464E-4	24.77	40
2	3.891	0.77	1.89	0.39	0.99	1E-4	21.64	15
3	2.335	0.75	1.84	0.39	0.99	1E-4	18.87	15
4	1.556	0.72	1.79	0.39	0.99	1E-4	17.48	15
5	0.778	0.66	1.74	0.39	0.99	1E-4	16.09	15

Source: Al-Obaidi, M.A. et al., *Chem. Eng. J.*, 316, 91–100, 2017.

has been used in this research to increase rejection at lower temperature. Also, for these cases, MATLAB has been used to test the rejection parameter at higher temperatures, and the results show a clear increase in solute flux and a decrease in chlorophenol rejection. From the results shown in Table 11.13, it appears there is a multimodal problem here, with the possibility of more than one solution. This is why for some cases, the temperature is near the upper limits and in other cases the proposed GA finds a solution where the temperature is near the lower limit.

For a better understanding of the temperature influence at high feed concentration, Case 1 is used to analyse the impact of varying the temperatures. Table 11.14 shows that the chlorophenol rejection has dropped slightly with the decrease of up to the temperature limit.

The optimisation results of Problem 11.2 for each inlet feed concentration and the optimised decision variables obtained are given in Table 11.15. The GA optimisation results show that a maximum chlorophenol rejection for all five cases can be achieved within the operating parameters of inlet feed flow rate, pressure, and temperature of 1.922–1.945E-4 m^3/s, 7.32–7.53 atm, and 40°C respectively. This is comparable to the optimisation results of Problem 11.1. Here, GA optimisation increases the chlorophenol rejection of five cases by 8.54%, 15.25%, 16.2%, 19.59%, 26.57% respectively, at the same time keeping an allowable pressure drop constraint. It also appears that considering the pressure drop as a constraint actually deviates the optimisation process to raise the rejection parameter by using a lower feed flow rate at a higher temperature. Table 11.15 shows that for each case, the temperature has achieved its maximum limits and the

TABLE 11.14
The Up Limit of Temperature on Results for Case 1 of Problem 11.1

Up Limit of Temperature	GA P_{loss}	GA Rej	Decision Variables $F_{b(0)}$	$P_{b(0)}$	T_b
40	0.42	0.98	1.0464E-4	24.77	40
39	0.43	0.97	1.0597E-4	24.77	39
38	0.45	0.96	1.0749E-4	24.77	38
37	0.46	0.94	1.0925E-4	24.77	37
36	0.41	0.95	1E-4	24.77	15
35	0.41	0.95	1E-4	24.77	15
34	0.41	0.95	1E-4	24.77	15

Source: Al-Obaidi, M.A. et al., *Chem. Eng. J.*, 316, 91–100, 2017.

TABLE 11.15
Optimal Values for Problem 11.2

Case	Feed Conc. $C_{b(0)} \times 10^3$, (kmol/m³)	Experimental P_{loss}, [41]	Rej (Max.) [41]	GA Rej	Decision Variables $F_{b(0)}$	$P_{b(0)}$	T_b
1	6.226	1.93	0.83	0.90	1.92243E-4	7.53	40
2	3.891	1.89	0.77	0.89	1.93143E-4	7.47	40
3	2.335	1.84	0.75	0.88	1.93936E-4	7.45	40
4	1.556	1.79	0.72	0.87	1.94295E-4	7.41	40
5	0.778	1.74	0.66	0.84	1.94572E-4	7.32	40

Source: Al-Obaidi, M.A. et al., *Chem. Eng. J.*, 316, 91–100, 2017.
Where GA P_{loss} = 1.38 (atm).

TABLE 11.16
The Up Limit of Temperature on Results for Case 1 of Problem 11.2

Up Limit of Temperature	GA *Rej*	Decision Variables		
		$P_{b(0)}$	T_b	$F_{b(0)}$
40	0.90	7.53	40	1.9224E-4
39	0.89	7.53	39	1.9199E-4
38	0.88	7.53	38	1.9172E-4
37	0.87	7.53	37	1.9142E-4
36	0.86	7.53	36	1.9110E-4
35	0.84	7.53	35	1.9076E-4
34	0.83	7.53	34	1.9039E-4

Source: Al-Obaidi, M.A. et al., *Chem. Eng. J.*, 316, 91–100, 2017.
Where GA P_{loss} = 1.38 (atm).

pressure has moved to the lower limits in order to achieve the two objective functions of the problem. This is a positive result when considering comparative results of raising the operating pressure.

The temperature influence is given in Table 11.16 for a high feed concentration (Case 1). Interestingly, Table 11.16 shows that the chlorophenol rejection drops with the decrease of the maximum limit of temperatures. This is due to choosing a minimum operating pressure as an objective function. This might cancel its impact and actually results in a reduction of the rejection parameter due to a decrease in temperature. This should be compared to using a higher operating pressure as in the case of Table 11.14, which slightly affects the rejection parameter.

11.3 CONCLUSIONS

It is clear from the research described in this chapter that RO remains the most promising and economically viable separation process for removing harmful contents from wastewater at levels commensurate with environmental legislation. The state of the art provided in this chapter indicates that significant progress has been made for removing phenol and phenolic compounds from wastewater, but there is still room for improvement in achieving a better and cheaper solution. The two- and one-dimensional distributed models presented by the authors in this chapter go a long way to realising this improvement. This is evidenced by the very small margin of error between the experimental results and the model predictions for a number of model parameters, including retentate pressure, retentate flow rate, average permeate solute concentration, the total volumetric permeated flow rate, and the average solute rejection. Finally, a further improvement has been achieved by the new GA-based optimisation methodology developed by the authors, which has yielded even greater chlorophenol rejection within the operating parameters of inlet feed flow rate, pressure, and temperature.

NOMENCLATURE

b	Feed channel friction parameter (atm s/m^4)
B_s	The solute permeability coefficients of the membrane, (the Solution–diffusion model) (m/s)
c_1, c_2	Constants in Equation 11.2
$C_{m(x)}$	Dimensionless solute concentration at any point along the membrane length
C_s, C_b	Retentate solute concentration in the feed channel (kmol/m^3)

$C_{\tilde{s}(av)}$	The mean solute concentration in the feed side (kmol/m³)
$C_{p(av)}$	Average permeate solute concentration in the permeate channel (kmol/m³)
$C_{w(x)}$	Solute concentration at the membrane wall at any point along the membrane length, (kmol/m³)
$D_{b(x)}$	Diffusivity coefficient of feed at any point along the membrane length (m²/s)
$D_{p(x)}$	Diffusivity coefficient of permeate at any point along the membrane length (m²/s)
d_{ep}	Equivalent diameter of feed channel (m)
d_{ef}	Equivalent diameter of permeate channel (m)
$F_{b(x)}$	Feed flow rate at any point along the membrane length (m³/s)
$F_{p(x)}$	Permeate flow rate at any point along the membrane length (m³/s)
$F_{p(\text{Total})}$	Total permeated flow rate of the permeate channel (m³/s)
$J_{s(x)}$	Solute molar flux through the membrane at any point along the membrane length, (kmol/m² s)
$J_{w(x)}$	Water flux at any point along the membrane length (m/s)
$k_{(x)}$	Mass transfer coefficient at any point along the membrane length (m/s)
L	Length of the membrane (m)
L_p	Solvent transport coefficient (m/atm s)
$P_{b(x)}$	Feed channel pressure at any point along the membrane length (atm)
P_p	Permeate channel pressure (atm)
P_{loss}	The pressure loss along the membrane length (atm)
R	Gas low constant $\left(R = 0.082 \dfrac{\text{atm m}^3}{\text{K kmol}}\right)$
$Re_{f(x)}$	The feed Reynolds number at any point along the membrane length (dimensionless)
$Rej_{(av)}$	Solute rejection coefficient (dimensionless)
$Re_{p(x)}$	The permeate Reynolds number at any point along the membrane length (dimensionless)
$Sc_{f(x)}$	The feed Schmidt number at any point along the membrane length (dimensionless)
$Sc_{p(x)}$	The permeate Schmidt number at any point along the membrane length (dimensionless)
$Sh_{(x)}$	Sherwood number at any point along the membrane length (dimensionless)
SSE	The sum of square errors
T_b	Feed temperature (°C)
t_f	Feed spacer thickness (m)
$U_{b(x)}$	The feed velocity at the feed channel (m/s)
W	Width of the membrane (m)
x	Any point along the membrane length
Z	Parameter defined in Equation 11.2 in Table 11.4

Subscript

$\rho_{f(x)}$	Feed density at each point along the membrane length (kg/m³)
ρ_m	The molal density of water (55.56 kmol/m³)
$\rho_{p(x)}$	Feed density at each point along the membrane length (kg/m³)
$\mu_{f(x)}$	Feed viscosity at each point along the membrane length (kg/m s)
$\mu_{p(x)}$	Feed viscosity at each point along the membrane length (kg/m s)
σ	The reflection coefficient (dimensionless)
ω	The solute permeability coefficients of the membrane (kmol/m² s atm)
Δx	Length of the sub-section (m)
$\Delta P_{b(x)}$	Trans-membrane pressure at each point along the membrane length (atm)
$\Delta\pi_{s(x)}$	The osmotic pressure difference at each point along the membrane length (atm)
θ	Parameter defined in Equation 11.1 in Table 11.2
ρ_w	Molal density of water (55.56 kmol/m³)

REFERENCES

1. Gami A. A., Shukor M. Y., Abdul Khalil K., Dahalan F. A., Khalid A. and Ahmad S. A. Phenol and its toxicity, *Journal of Environmental Microbiology and Toxicology* 2(1) (2014) 11–24.
2. Agency for toxic substances and disease registry (ATSDR), division of toxicology and human health sciences, 2015. Substance Priority List. https://www.atsdr.cdc.gov/spl/index.html.
3. Witek A., Koltuniewicz A., Kurczewski B., Radziejowska M., Hatalski M., Simultaneous removal of phenols and Cr^{3+} using micellar-enhanced ultrafiltration process, *Desalination* 191 (2006) 111–116.
4. Mohammed A. E., Jarullah A., Gheni S., Mujtaba I. M., Optimal design and operation of an industrial three phase reactor for the oxidation of phenol, *Computer and Chemical Engineering* 94 (2016) 257–271.
5. Fujioka T., Assessment and optimisation of N-nitrosamine rejection by reverse osmosis for planned potable water recycling applications, PhD thesis, University of Wollongong, Australia (2014).
6. Schutte C. F., The rejection of specific organic compounds by reverse osmosis membranes, *Desalination* 158 (2003) 285–294.
7. Bódalo-Santoyo A., Gómez-Carrasco J. L., Gómez-Gómez E., Máximo-Martin M. F., Hidalgo-Montesinos A. M., Spiral-wound membrane reverse osmosis and the treatment of industrial effluents, *Desalination* 160 (2004) 151–158.
8. Alzahrani S., Mohammad A. W., Hilal N., Abdullah P., Jaafar O., Comparative study of NF and RO membranes in the treatment of produced water—Part I: Assessing water quality, *Desalination* 315 (2013) 18–26.
9. Al-Obaidi M. A., Kara-Zaïtri C., Mujtaba I. M., Scope and limitations of the irreversible thermodynamics and the solution diffusion models for the separation of binary and multi-component systems in reverse osmosis process, *Computers and Chemical Engineering* 100 (2017) 48–79.
10. Jain S., Gupta S. K., Analysis of modified surface force pore flow model with concentration polarization and comparison with Spiegler–Kedem model in reverse osmosis systems, *Journal of Membrane Science* 232 (2004) 45–62.
11. Gauwbergen D van, Baeyens J., Modelling reverse osmosis by irreversible thermodynamics, *Separation and Purification Technology* 13 (1998) 117–128.
12. Bódalo-Santoyo A., Gomez-Carrasco J. L., Gómez-Gómez E., Máximo-Marttin F., Hidalgo-Montesinos A. M., Application of reverse osmosis to reduce pollutants present in industrial wastewater, *Desalination* 155 (2003) 101–108.
13. Nguyen V. T., Vigneswaran S., Ngo H. H., Shon H. K., Kandasamy J., Arsenic removal by a membrane hybrid filtration system, *Desalination* 236 (2009) 363–369.
14. Slater C. S., Ahlert R. C., Uchrin C. G., Applications of reverse osmosis to complex industrial wastewater treatment, *Desalination* 48 (1983) 171–187.
15. Fritzmann C., Löwenberg J., Wintgens T., Melin T., State-of-the-art of reverse osmosis desalination, *Desalination* 216 (2007) 1–76.
16. Moonkhum M., Lee Y. G., Lee Y. S., Kim J. H., Review of seawater natural organic matter fouling and reverse osmosis transport modelling for seawater reverse osmosis desalination, *Desalination and Water Treatment* 15 (2010) 92–107.
17. Blandin G., Verliefde A. R. D., Comas J., Rodriguez-Roda I., Le-Clech P., Efficiently combining water reuse and desalination through forward osmosis: Reverse osmosis (FO–RO) hybrids: A critical review, *Membranes: Open Access Separation Science and Technology Journal* 6(3) (2016) 1–24.
18. Sundaramoorthy S., Srinivasan G., Murthy D. V. R., An analytical model for spiral wound reverse osmosis membrane modules: Part I—Model development and parameter estimation, *Desalination* 280 (2011) 403–411.
19. Kedem O., Katchalsky A., Thermodynamic analysis of the permeability of biological membranes to non-electrolytes, *Biochimica et Biophysica Acta* 27 (1958) 229–246.
20. Spiegler K. S., Kedem O., Thermodynamics of hyperfiltration (reverse osmosis): Criteria for efficient membranes, *Desalination* 1 (4) (1966) 311–326.
21. Jonsson G., Overview of theories for water and solute transport in 9 UF/RO membranes, *Desalination* 35 (1980) 21–38.
22. Sapienza F. J., Gill W. N., Soltanieh M., Separation of ternary salt/acid aqueous solutions using hollow fiber reverse osmosis, *Journal of Membrane Science* 54 (1990) 175–189.
23. Wijmans J. G., Baker R. W., The solution–diffusion model: A review, *Journal of Membrane Science* 107 (1995) 1–21.
24. Murthy Z. V. P., Gupta S. K., Sodium cyanide separation and parameter estimation for reverse osmosis thin film composite polyamide membrane, *Journal of Membrane Science* 154 (1999) 89–103.

25. Mujtaba I. M., The Role of PSE community in meeting sustainable freshwater demand of tomorrow's world via desalination. In *Computer Aided Chemical Engineering*, I. A. Karimi and R. Srinivasan (Eds.), Vol 31, Amsterdam, the Netherlands: Elsevier (2012), pp. 91–98.
26. Evangelista F., An improved analytical method for the design of spiral-wound modules, *The Chemical Engineering Journal* 38 (1988) 33–40.
27. Baker R. W., *Membrane Technology and Applications*, 2nd ed., Newark, CA: Membrane Technology and Research (2004).
28. Al-Obaidi M. A., Mujtaba I. M., Steady state and dynamic modelling of spiral wound wastewater reverse osmosis process, *Computers and Chemical Engineering* 90 (2016) 278–299.
29. Process System Enterprise Ltd., *gPROMS Introductory User Guide*. London, UK: Process System Enterprise (2001).
30. Karabelas A. J., Koutsou C. P., Kostoglou M., The effect of spiral wound membrane element design characteristics on its performance in steady state desalination: A parametric study, *Desalination* 332 (2014) 76–90.
31. Rautenbach R., Dahm W., Design and optimization of spiral-wound and hollow fiber RO-modules, *Desalination* 65 (1987) 259–275.
32. Avlonitis S., Hanbury W. T., Boudinar M. B., Spiral wound modules performance. An analytical solution, part I, *Desalination* 81(1–3) (1991) 191–208.
33. Avlonitis S., Hanbury W. T., Boudinar M. B., Spiral wound modules performance an analytical solution: Part II, *Desalination* 89(3) (1993) 227–246.
34. Avlonitis S. A., Pappas M., Moutesidis K., A unified model for the detailed investigation of membrane modules and RO plants performance, *Desalination* 203 (2007) 218–228.
35. Boudinar M. B., Hanbury W. T., Avlonitis S., Numerical simulation and optimisation of spiral-wound modules, *Desalination* 86 (1992) 273–290.
36. Senthilmurugan S., Ahluwalia A., Gupta S. K., Modelling of a spiral-wound module and estimation of model parameters using numerical techniques, *Desalination,* 173 (3) (2005) 269–286.
37. Geraldes V., Escórcio Pereira N., Norberta de Pinho M., Simulation and optimization of medium-sized seawater reverse osmosis processes with spiral-wound modules, *Industrial and Engineering Chemistry Research* 44(6) (2005) 1897–1905.
38. Sagne C., Fargues C., Broyart B., Lameloise M. L., Decloux M., Modelling permeation of volatile organic molecules through reverse osmosis spiral-wound membranes, *Journal of Membrane Science* 330(1–2) (2009) 40–50.
39. Oh H., Hwang T., Lee S., A simplified model of RO systems for seawater desalination, *Desalination* 238 (2009) 128–139.
40. Kaghazchi T., Mehri M., Takht Ravanchi M., Kargari A., A mathematical modelling of two industrial seawater desalination plants in the Persian gulf region, *Desalination* 252 (2010) 135–142.
41. Sundaramoorthy S., Srinivasan G., Murthy D. V. R., An analytical model for spiral wound reverse osmosis membrane modules: Part II—Experimental validation, *Desalination* 277 (2011) 257–264.
42. Fujioka T., Khan S. J., Mcdonald J. A., Roux A., Poussade Y., Drewes J. E., Nghiem L. D., Modelling the rejection of N-nitrosamines by a spiral-wound reverse osmosis system: Mathematical model development and validation, *Journal of Membrane Science* 454 (2014) 212–219.
43. Al-Obaidi M. A., Kara-Zaïtri C., Mujtaba I. M., Wastewater treatment by spiral wound reverse osmosis: Development and validation of a two-dimensional process model, *Journal of Cleaner Production* 140 (2017) 1429–1443.
44. Al-Obaidi M. A., Li J.-P., Kara-Zaïtri C., Mujtaba I. M., Optimisation of reverse osmosis based wastewater treatment system for the removal of chlorophenol using genetic algorithms, *Chemical Engineering Journal* 316 (2017) 91–100.
45. Al-Obaidi M. A., Kara-Zaïtri C., Mujtaba I. M., Modelling of a spiral-wound reverse osmosis process and parameter estimation, *Desalination and Water Treatment* 69 (2017) 93–101.
46. Al-Obaidi M. A., Kara-Zaïtri C., Mujtaba I. M., Development of a mathematical model for apple juice compounds rejection in a spiral-wound reverse osmosis process, *Journal of Food Engineering* 192 (2017) 111–121.
47. Koroneos C., Dompros A., Roumbas G., Renewable energy driven desalination systems modelling, *Journal of Cleaner Production* 15 (2007) 449–464.
48. Srinivasan G., Sundaramoorthy S., Murthy D. V. R., Spiral wound reverse osmosis membranes for the recovery of phenol compounds-experimental and parameter estimation studies, *American Journal of Engineering and Applied Science* 3 (1) (2010) 31–36.

49. Murthy Z. V. P., Gupta S. K., Thin film composite polyamide membrane parameters estimation for phenol-water system by reverse osmosis, *Separation Science and Technology* 33 (16) (1998) 2541–2557.
50. Jarullah A. T., Mujtaba I. M., Wood A. S., Kinetic parameter estimation and simulation of trickle-bed reactor for hydrodesulfurization of crude oil, *Chemical Engineering Science* 66 (5) (2011) 859–871.
51. Wankat P. C., *Rate-Controlled Separations.* 1st ed., Dordrecht, the Netherlands: Springer (1990), pp. 873
52. Marcovecchio M. G., Aguirre P. A., Scenna N. J., Global optimal design of reverse osmosis networks for seawater desalination: Modelling and algorithm, *Desalination* 184 (2005) 259–271.
53. Villafafila A., Mujtaba I. M., Fresh water by reverse osmosis based desalination: Simulation and optimisation, *Desalination* 155 (2003) 1–13.
54. Lu Y-y., Hu Y-d., Xu D-m., Wu L-y., Optimum design of reverse osmosis seawater desalination system considering membrane cleaning and replacing, *Journal of Membrane Science* 282 (2006) 7–13.
55. Murthy Z. V. P., Vengal J. C., Optimization of a reverse osmosis system using genetic algorithm, *Separation Science and Technology* 41 (2006) 647–663.
56. Guria C., Bhattacharya P. K., Gupta S. K., Multi-objective optimization of reverse osmosis desalination units using different adaptations of the non-dominated sorting genetic algorithm (NSGA), *Computers and Chemical Engineering* 29 (2005) 1977–1995.
57. Holland J. H., *Adaptation in Natural and Artificial Systems.* Ann Arbor, MI: University of Michigan Press (1975).
58. Fraga E. S., Senos Matias T. R., Synthesis and optimization of a non-ideal distillation system using a parallel genetic algorithm, *Computers and Chemical Engineering* 20 (1996) S79–S84.
59. Gupta R., Gupta S. K., Multi-objective optimization of an industrial nylon-6 semi-batch reactor system using genetic algorithm, *Journal of Applied Polymer Science* 73 (1999) 729–739.
60. Victorino I. R. S., Maia J. P., Morais E. R., Wolf Maciel M. R., Filho R. M., Optimization for large scale process based on evolutionary algorithms, Genetic algorithms, *Chemical Engineering Journal* 132 (2007) 1–8.
61. Na J., Kshetrimayum K. S., Lee U., Han C., Multi-objective optimization of microchannel reactor for Fischer–Tropsch synthesis using computational fluid dynamics and genetic algorithm, *Chemical Engineering Journal* 313 (2017) 1521–1534.
62. Rajesh J. K., Gupta S. K., Rangaiah G. P., Ray A. K., Multi-objective optimization of steam reformer performance using genetic algorithm, *Industrial and Engineering Chemistry Research* 39 (2000) 706–717.
63. Fang F., Ni B.-J., Xie W.-M., Sheng G.-P., Liu S.-G., Tong Z.-H., Yu H.-Q., An integrated dynamic model for simulating a full-scale municipal wastewater treatment plant under fluctuating conditions, *Chemical Engineering Journal* 160 (2010) 522–529.
64. Djebedjian B., Gad H., Khaled I., Rayan M. A., Optimization of reverse osmosis desalination system using genetic algorithms technique, In *Twelfth International Water Technology Conference, IWTC12*, Alexandria, Egypt (2008).
65. Lee T.-M., Oh H., Choung Y.-K., Oh S., Jeon M., Kim J. H., Nam S. H., Lee S., Prediction of membrane fouling in the pilot-scale microfiltration system using genetic programming, *Desalination* 247 (2009) 285–294.
66. Park S.-M., Han J., Lee S., Sohn J., Kim Y.-M., Choi J.-S., Kim S., Analysis of reverse osmosis system performance using a genetic programming technique, *Desalination and Water Treatment* 43 (2012) 281–290.
67. Bourouni K., Ben-M'Barek T., Taee A. Al, Design and optimization of desalination reverse osmosis plants driven by renewable energies using genetic algorithms. *Renewable Energy* 36 (2011) 936–950.
68. Yuen C. C., Aatmeeyata, Gupta S. K., Ray A. K., Multi-objective optimization of membrane separation modules using genetic algorithm, *Journal of Membrane Science* 176 (2000) 177–196.
69. Okhovat A., Mousavi S. M., Modelling of arsenic, chromium and cadmium removal by nanofiltration process using genetic programming, *Applied Soft Computing* 12 (2012) 793–799.
70. Soleimani R., Shoushtari N. A., Mirza B., Salahi A., Experimental investigation, modelling and optimization of membrane separation using artificial neural network and multi-objective optimization using genetic algorithm, *Chemical Engineering Research and Design* 91 (2013) 883–903.
71. Leardi R., Genetic algorithms in chemometrics and chemistry: A review, *Journal of Chemometrics* 15 (2001) 559–569.

72. Kumar M., Husian M., Upreti N., Gupta D., Genetic algorithm: Review and application, *International Journal of Information Technology and Knowledge Management* 2 (2010) 451–454.
73. Thirugnanasambandham K., Sivakumar V., Loganathan K., Jayakumar R., Shine K., Pilot scale evaluation of feasibility of reuse of wine industry wastewater using reverse osmosis system: Modelling and optimization, *Desalination and Water Treatment* 57 (2016) 25358–25368.
74. Li Y., Wei J., Wang C., Wang W., Comparison of phenol removal in synthetic wastewater by NF or RO membranes, *Desalination and Water Treatment* 22 (2010) 211–219.

Section IV

Wastewater Treatment

Oxidation and Electrochemical Process

12 Industrial Three-Phase Oxidation Reactor for Wastewater Treatment

Aysar T. Jarullah, Saba A. Gheni, Awad E. Mohammed, and Iqbal M. Mujtaba

CONTENTS

12.1 INTRODUCTION AND WASTEWATER POLLUTION

At the beginning of the twentieth century, a few cities and industries began to recognize that the discharge of municipal and industrial sewage directly into natural watercourses caused health problems, and this led in the late 1930s to the development and construction of the first sewage treatment facilities. During the 1950s and 1960s, the U.S. government further encouraged the prevention of pollution by providing funds for the construction of municipal waste treatment plants, water pollution research and technical training and assistance in this field. After the establishment of the European Community in Western Europe, this organ actively influenced the environmental initiatives of its member states through the establishment of an environmental legislation frame (Safaa, 2009). In the 1980s, the greening all parts of industry and the sustainability of this greening were characterized. New procedures and enhanced strategies were created to treat sewage, dissect wastewater and assess the effect of pollution on the environment. Regardless of these endeavors, population increases and industrial and economic development brought about incremental increases in pollution (Spellman 2004). Satisfactory treatment, as well as protected transfer, of risky vaporous, fluid and solid wastes as therefore turned out to be significant issues for the foundation of future practical sustainable progress.

Such environmental concerns have recently been communicated by an ever-increasing amount of stringent legislation requiring lower toxicity and pollution, both natural and inorganic. One outcome is that, before being returned to the environment, industrial and municipal waste should typically undergo different physical, biological, chemical or combined treatment to reduce their potential danger. Industrial wastewater containing phenolic components is profoundly dangerous to aquatic life. Hence, phenolic wastewaters require uncommon treatment before sending off the effluents. In any case, the treatment cannot be done in traditional sewage plants in light of the fact that, even at low density, phenol has a bactericidal impact on smaller scale creatures. WAO (WAO) is one of the most widely used wet oxidation methods for the treatment of sewage and various moderately to highly concentrated modern effluents. The speculation and operation expenses of WAO plants are high as they use high temperatures ranging from 150–300°C and gaseous pressure up to 200 bar. In recent years, supercritical water oxidation (SCWO) has turned out to be an amazingly effective treatment for numerous types and concentration scopes of natural wastewater (Kritzer and Dinjus, 2001). The rising oxidation models performed well for drinking water filtration and low-level natural wastewater treatment at moderate temperatures (Neyens and Baeyens, 2003). For wastewater with low to medium natural charges, catalytic wet air oxidation (CWAO) yielded promising outcomes at the research-facility scale (Imamura, 1999; Pintar, 2008). These methods can be particularly alluring when they are combined with natural or physical–chemical end treatment. Consolidated treatment removes the requirement to finish toxin mineralization in CWO and is of increasing interest as a monetarily viable and environmentally encouraging system to treat difficult natural components, such as phenol, into CO_2 or innocuous intermediates, essentially unsaturated fats, which can later be dealt with organically. The procedure can also be done in moderate pressure and temperature conditions (Ohta et al., 1980; Tukac et al., 1995).

Water covers 70% of the earth's surface and undoubtedly is the most valuable resource on earth; 97% is in the form of seawater and only the remaining 3% is fresh water. Thus, a rapid 60% total increase in water demand for the coming 60 years being the higher projection resulting from agricultural activities. This increase in water demand is due to the fact that the world population is increasing at an alarming rate of 80 million people per year, requiring about 64 billion cubic meters of extra fresh water (Pardeep, 2010). The pattern of anticipated water demand for all human endeavors is shown in Figure 12.1 (Habtu et al., 2011).

Wastewater is composed of organic, inorganic compounds and dissolved gases. Among these groups, the aromatic compounds—especially phenol and its derivatives—are the most common

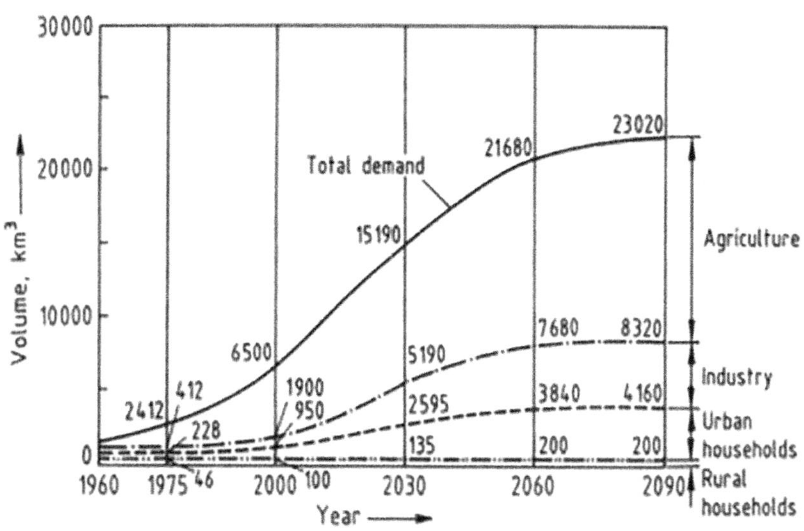

FIGURE 12.1 Projection of the increasing worldwide demand for water. (Adapted from Habtu, N.G. and Stuber, F., Catalytic wet air oxidation of phenol over active carbon in fixed bed reactor: Steady state and periodic operation, PhD. Universitat Rovirai Virgili, 2011.)

TABLE 12.1
Yearly Discharge in 2009 of Phenol and Substituted-Phenol in the USA

Compound	Phenol	Aniline	Hydroquinone	Chlorophenol	O-Cresol	Quinone
Emission (ton/yr)	2870	490	212.7	31	16.4	0.07

Source: Mohammed, A.E., *Optimal Design of Trickle Bed Reactor for Phenol Oxidation.* MSc thesis, Tikrit University, Tikrit, Iraq, 2016.

pollutants effluent in many industries (Singh et al., 2004). Large amounts of toxic compounds present in wastewater are discharged into the environment each year, as shown in Table 12.1.

Recently, the need for fresh, clean and uncontaminated water has become a problem of great importance, since many sources have been exhausted and others are likely to be contaminated because of rapid industrialization and the increasing world population, as well as the toxic waste being released into the environment and causing extensive environmental contamination such that many natural water reserves are damaged beyond repair (Habtu et al., 2011).

Water contamination arises from point or non-point sources. Point sources of water contamination occur when toxic or hazardous compounds are discharged directly into the water body at a single point of discharge. Most industrial effluents are considered point sources of water contamination. On the other hand, non-point sources arise across a wide, unspecified area, which makes water pollution difficult to control. Fertilizers and pesticides washed out from agricultural fields by rainfall are considered examples of non-point sources of water pollution. The quantity of wastewater produced varies in different communities and countries, depending on a number of factors such as water uses, climate, lifestyle and economics. It has also been noted that among the 900,000 tons of different dyes produced annually in the world, approximately (10%–15%) are lost in wastewater streams during manufacturing and processing operations, often into rivers and streams (Pardeep, 2010).

Wastewater pollutants can be generally grouped into eight classifications based on their nature and origin as shown in Table 12.2.

TABLE 12.2
Wastewater Pollutant Classification

Water Classification	Characterization
Oil products	Petroleum and chemicals utilized for fuel, oils, and so on
Pesticides and herbicides	Chemicals employed for killing undesirable creatures and plants
Heavy metals	Copper, lead, mercury, and selenium from industries
	Car fumes, mines, and even common soil
Dangerous wastes	Synthetic wastes: lethal (toxic), receptive (equipped for delivering unstable or dangerous gases), destructive (equipped for consuming steel), or ignitable (combustible)
Overabundance natural matter	Manures and different supplements that promote plant development
Residual	Solid molecule display in sufficiently extensive sums
Infectious organisms	Disease-creating living beings considered toxic when found in drinking water
Thermal contaminant	Indeed, even small temperature changes in a waterway can drive out fish and different species that were initially present, and support different species in their place

Source: Nathanson, J.A., *Basic Environmental Technology: Water Supply, Waste Disposal, Pollution Control*, John Wiley & Sons, New York, 1986.

Physically, wastewater is normally characterized by dim shading, smelly scent and total suspended solids (TSS) of around 0.1%, with 99.9% water content. Chemically, wastewater is made out of natural and inorganic components in addition to several dissolved gases. Natural materials may be made up of proteins, carbohydrates, fats and oils, surfactants, oils, pesticides, and phenols. Inorganic components may include substantial metals, nitrogen, phosphorus, sulfur, chlorides, alkalinity and others. Gases generally broken down in wastewater include H_2S, CH_4, alkali, O_2, CO_2 and N_2. The first three gases are produced from the deterioration of natural matter found in the wastewater. Wastewater may also naturally contain numerous pathogenic life forms that usually originate from human beings. (http://web.deu.edu.tr/atiksu/ana52/wtreat.html).

The U.S. Environmental Protection Agency (EPA) has recognized roughly 275 priority contaminants to be controlled by straight out release measures. Priority contaminants, both natural and inorganic, have been chosen on the basis of their carcinogenic nature, mutagenicity or intense poisonous quality (Metcalf and Eddy, 1991). Table 12.3 shows the positioning of some lethal natural components of effluents based on the EPA's priority contaminants ranking in the year 2001. The amount discharged every year into natural waterways in the USA in the year 2000 appears in Table 12.4.

TABLE 12.3
Extract from EPA Rundown of Priority Pollutants, 2001

Component	Risk Ranking
C_8H_{10}	56
C_6H_6O	162
C_7H_8O	169
$CH_3C_6H_4$	194
C_6H_5ClO	247
$C_6H_5NO_3$	256

Source: www.epa.gov/.

TABLE 12.4
Yearly Arrival of Lethal Phenol-Like Contaminants in the USA for the Year 2000

Component	Emission (ton/y)	Overall Ranking
C_6H_6O	22	35
$C_6H_4N_2O_5$	11	50
$C_6H_6O_2$	8.3	59
$C_6H_5NH_2$	5.8	70
$C_6H_6O_2$	1.9	95
$C_6H_4O_2$	0.64	115
C_6HCl_5O	0.55	120
C_6H_4ClOH	0.046	203
$C_6H_5O_3N$	0.026	213
$C_6H_5NO_3$	0.007	239

Source: www.epa.gov/tri.

TABLE 12.5
Utilize and Risk Ranking Details for Chemicals

Family	Structures	Poisonous Quality Ranking	Utilize
Phenols	C_6H_6O	3 Flawed carcinogen	Making disinfectants, chemicals, plastic, gums, elastic, refining oil, manure, coke, paints, removers, asbestos, fragrances, disinfectants, bactericides, fungicides
	$CH_3C_6H_4(OH)$	Toxic	Making disinfectant, fragrances, safeguarding operator or herbicides
Chlorinated hydrocarbons	C_6H_4ClOH	Flawed carcinogen, destructive	Making dyes, making other chemicals
Nitroaromatic hydrocarbons	$C_6H_5NO_3$	3	Making fungicides, pesticide, colors, and other chemicals
	$C_6H_5NO_2$	3 Likewise called oil of mirbane, poison, conceptive, impacts	Making shoe polish, dyes, explosives, floor and metal clean, different chemicals and paints
	$C_6H_5NH_2$	3 Suspected carcinogen, mutagen, allergen	Colors, shaded pencils, lithographic and other printing links, performs, pharmaceuticals, nylon, strands, saps, modern solvents, elastic handling
Sulfur compounds	$C_4H_8O_2S$	Harmful	Natural gas handling, making hardware and elastics

Source: Sax, N.I. and Lewis, R.J., *Dangerous Properties of Industrial Materials*. New York, Van Nostrand Reinhold, 1992; EPA, United State Environmental Protection Agency, 2015, www.epa.gov/tri.

Note: 3, serious poisonous quality, materials that can bring about harm of adequate seriousness to debilitate life, information for aggregate yearly discharge in the USA in 2001.

As can be noted from these tables, the important contamination families are aromatic compounds (C_6H_6O, $C_6H_4N_2O_5$) and substituted compounds (C_6H_4ClOH, $C_6H_5O_3N$, and, etc.). The significance of such contaminants is clear, as they include components previously reported as being highly poisonous and having a high yearly discharge.

Table 12.5 outlines the regular use and potential toxicity of these components. Generally, aromatic compounds are atoms dependent upon the benzene ring structure. They are essential segments in fuel and typically contain carcinogenic atoms like C_6H_6, C_6H_5-CH_3, C_8H_{10}, and C_8H_{10} (ortho-, meta-, and para-). Some of them have non-natural substituent such as nitrogen, sulfur, phosphorus, so they shape critical subgroups called substituted aromatic compounds (Bingham et al., 1979).

12.1.1 Phenol

The most widely recognized pollutants of wastewater are the phenols, which refer to the class of aromatic mixtures having a hydroxyl bunch and any extra natural substituent gathering on a six-carbon benzene ring. Phenols are dangerous and give water an undesirable smell even at low concentrations. The Ministry of Environment and Forests (MOEF), Government of India, and the EPA have placed phenol and phenolic components on the priority poisons list. Despite the fact that the enactment limitations have expanded, a large quantity of phenolic components are still released into the earth. For instance, in Europe 900,000 kg/yr of phenols are specifically or in a roundabout way released into natural bodies of water (Jordan et al., 2002).

Aqueous waste has a natural toxin stack in the scope of couple of hundred to a couple of thousand ppm. Phenol as natural toxin appears in the wastewater of different industries, such as refineries (6–500 mg/L), cooking operations (27–3900 mg/L), coal handling (9–6800 mg/L), fabrication of petrochemicals (28–1220 mg/L) and also in the pharmaceutical, lumber, paint and paper industries (0.1–1600 mg/L) (Busca et al., 2008).

The significant issues caused by phenol are listed as follows (Lin et al., 2006; Mohammed et al., 2016):

- Phenols are tenacious contaminants causing extraordinary harm to surroundings. They have been assigned as priority toxins.
- Phenols are extremely unsafe, destructive and dangerous, showing harmful consequences for aquatic life. Its impact relies upon the concentration of contaminants.
- Phenols may bring about destructive consequences for the focal sensory system and heart, bringing about dysrhythmia, seizures and coma.
- Phenol may also affect the kidneys. Extended exposure to the substance may effectively affect the liver and kidneys.
- Phenol and its gases are destructive to the eyes, skin and respiratory tract. Its destructive impact on the skin and mucous layers is a result of its protein-deteriorating impact.
- Prolonged or repeated skin contact with phenol may bring about dermatitis, or even second- or third-degree burns. Chemical burns from skin exposure can be flushed with polyethylene glycol, isopropyl liquor or maybe even bountiful measures of water.
- Systemic harm can happen notwithstanding in the neighborhood of burns because of the immediately retention of phenol in the skin.
- Phenols in waterways and seas diminish the light penetrating into the water, which has an impact on the flora and fauna found in these bodies of water.
- Phenol can give an obnoxious taste and scent to drinking water even at low concentrations. Phenol emits a sweet, bitter smell that is noticeable to most people at 40 ppb in air and at around 1–8 ppm in water.
- Phenol is also a reproductive poison, increasing the danger of premature birth and low birth weight, which demonstrates impeded advancement *in utero*.

The high quantity of contaminants releases to the ambient environment particularly for streams and seas causes tremendous issues; to remedy this situation, it is necessary to discover an appropriate strategy to treat these poisons—particularly phenols (Singh et al., 2004). In general, watery wastes have a natural toxin stack in the scope of 500–10,000 ppm and are excessively weakened, making it impossible to burn yet excessively poisonous. To detoxify these natural pollutants in groundwater or to separate the contaminants found in polluted water has become a noteworthy focus of interested for research and strategic discussions. The nearness of these pollutants in the water even at low quantities does not permit reuse of the water in the industrial processes (Mangrulkar et al., 2008). CWAO is the traditional procedure for removing these toxins from wastewater.

12.2 METHODS OF WASTEWATER TREATMENT

There is no doubt that water pollution, especially by a large number of different organic chemical species, is a continuing and even growing problem arising from human activities. No special arrangement appears to be feasible for treatment because of the heterogeneous structure of actual wastes and the differences of the characterizations of the contaminant chemical. Some waste treatment strategies only shift the harmful compound among phases; whereas this may serve to allow the waste to meet disposal guidelines, it does not actually change the science of the contaminant.

Different procedures utilize chemical reactions to change the waste into less dangerous by-products or innocuous final products, such as carbon dioxide and H_2O. Plainly the choice of the right procedure or mixture of treatments is a troublesome assignment that ought to be, for the most part, made based on the properties of the wastewater to be treated (e.g., concentration and grade of refractoriness of pollutants, flow rate) and the goal of the gushing (review of mineralization required). Notwithstanding, it should be said that exhaustive wastewater treatment advancement includes different viewpoints, mainly including logical science, energy, mass and heat exchange, reactor plans and process optimization, not all of which can be included in this review. Additional research is essential in these fields to develop the new treatment strategies to their maximum capacity. The various unit operations and procedures to expel wastewater pollutants are gathered together to give different levels of treatment. The primary treatment methods are listed in Table 12.6 divided into three categories: physical, biological and chemical methods.

Depending on the pollutants that need to be removed from wastewater, diverse treatment techniques are necessary, as summarized in Tables 12.6 (generally) and 12.7 (for lethal components). These tables additionally give a précis of the points of interest and downsides of common treatments (Guo and Al-Dahhan, 2003).

The utilization of traditional wastewater treatment forms, particularly on account of mild to higher natural loads, has turned out to be progressively tested with the identification of more natural and non-biodegradable pollutants. In other word, the development of wastewater treatment

TABLE 12.6
Advantages and Drawbacks of Physical, Biological and Chemical Treatment

Treatment	Physical, Physicochemical	Biological	Chemical
Kinds of contaminants	Ideally commercial wastewater, naturals, some in organics, substances	Commercial and local wastewater, little quantity naturals, some in organics	Ideally commercial wastewater, naturals in organics, substances
Strategies	Filtration, adsorption air floatation, extraction, flocculation, sedimentation	Anaerobic, aerobic, activated sludge	Warm oxidation, chemical oxidation, ion exchange, chemical precipitation
Advantages	Low capital expenses, moderately sheltered, simple to work	Troublesome upkeep, moderately protected, removal of dissolved, pollutants	High level of treatment, no auxiliary waste, removal of dissolved, pollutants
Disadvantages	Unpredictable outflows, high energy cost, troublesome upkeep	Unpredictable outflows, waste sludge disposal, susceptible to toxins	

TABLE 12.7
Operations Employed to Remove Poisonous Components

Operations	Removal Application
Activated carbon adsorption	Natural and industrial natural component involving insecticide, heavy metals
Activated residue powdered, activated carbon	Heavy metal, NH_3, chosen insusceptible, priority contaminants
Air stripping	Unpredictable natural mixtures and NH_3
Chemical coagulation, sedimentation and filtration	Heavy metals and PCBs
Chemical oxidative	NH_3, insusceptible, harmful halogenated aliphatic and fragrant components
Traditional biological treatment	Domestic wastewaters

TABLE 12.8
Techniques of Emerging Chemical Treatment of Wastewater

Wet air oxidation (WAO)	473–623 K 70.9–233 atm O_2 or air	Process of thermal oxidation
Catalytic wet air oxidation (CWAO)	<473 K <50.65 atm O_2 or air (and catalyst)	
Supercritical water oxidation (SCWO)	>648 K >223.87 atm O_2, air, or H_2O_2 (and catalyst)	
Wet peroxide oxidation (WPO)	>373 K >1 atm H_2O_2	Process of wet peroxide oxidation
Fenton (WPO)	298 K 1 atm $Fe^{+2} + H_2O_2$	
Advanced oxidation process (AOPs)	Radical - OH as intermediate (UV light, electrodes, O_3, or ultrasound pulses	AOPs
Collected treatment	UV + O_3 APOs + biological Adsorption via activated carbon + CWAO	Collected treatment

techniques are becoming increasingly widespread as they have demonstrated the capability of changing unsafe natural poisons into harmless components such as CO_2 and H_2O. A basic characterization of recent chemical treatment advances is given in Table 12.8.

12.2.1 Thermal Oxidation Process

There are three distinctive principle sorts of oxidation processes: CWAO, WAO and SCWO.

12.2.1.1 Wet Air Oxidation

WAO is a fluid-stage oxidation operation that happens when a disintegrated organic is blended with a vaporous source of O_2 at temperatures of 423–598 K and at pressures of 2–20 MPa (Copa and Gitchel, 1988).

The procedure is controlled by two stages; (1) exchange of O_2 to the fluid stage; and (2) reaction between disintegrated O_2 and natural. The review of oxidation is mainly based upon temperature, partial pressure, contact time and hardliner of the substrate (Mishra et al., 1995; Levec 1997; Kolaczkowski et al., 1999; Luck 1999, Mantzavinos et al., 1999; Lei and Wang, 2000). As detailed in the writing in Table 12.9, WAO has been studied for the treatment of numerous industrial and natural wastewaters.

WAO of natural toxins is the most part a free-radical chain response system (Li et al., 1991; Bachir et al., 2001; Robert et al., 2002).

12.2.1.2 Catalytic Wet Air Oxidation (CWAO)

12.2.1.2.1 Improvement and Industrial Application of CWAO

With the specific goal of decreasing the difficulties of the WAO processing conditions, CWAO was developed. The utilization of a catalyst is looked to upgrade general rate of reaction and eliminate intermediates unmanageable to non-reactant oxidation. Moderate states of temperature and pressure

TABLE 12.9
Wet Air Oxidation of Organic Contaminants and Commercial Effluents

Substrate	Reaction Conditions	Oxidant	Reactor Type	Removal (%)	References
C_6H_6O and substituted C_6H_6O	423–453 K 3–15.2 atm	O_2	Batch	>90% COD	Joglekar et al. (1991)
C_6H_6O and *m*-C_8H_{10}	473–548 K 69–138 bar (total)	Air	Batch	ns	Willms et al. (1987)
C_6H_6O o-C_6H_5ClO p-$C_6H_5NO_3$, etc.	423–593 K	Air	Batch	RE > 99%	Randall and Knopp (1980)
p-$C_6H_5NO_3$	423–453 K 3–9 bar (total)	O_2	Batch	ns	Gonzalez et al. (2002)

Note: ns, not specified; RE, removal efficiency; pressure given is the oxygen partial pressure, except where total pressure specified.

may bring down equipment and operation costs, however there are few contaminants obstinate enough to require CWAO, and quick catalyst deactivation is still a noteworthy issue to overcome (Hamoudi et al., 1999; Matatov and Sheintuch, 1998).

The industrial procedures of CWAO have been studied by Kolaczkowski et al. (1999). Some recent diverse and accessible innovations are those of the Osaka Gas operation (a blend of valuable and construct metals upon titanium or titanium-zirconium substrates, states of 523 K and 7 MPa, treats a coal gasifier effluents, wastewater from coke stoves, concentrated cyanide and sewage residue), the Kurita procedure (upheld platinum catalyst, for treatment of NH_3 at temperature over 373 K, utilizes nitrite as oxidant), and the Nippon Shokubai Kagaku operation (numerous heterogeneous catalysts in both pellet form and in a honeycomb frame, for such poisons as CH_3COOH, NH_3, C_6H_6O, CH_2O, at 493 K and 4 MPa) and with a homogeneous catalyst.

12.2.1.2.2 Operation Aspects of CWAO

The reaction mechanism of CWAO is thought to be similar to the mechanism of WAO, and the function of the catalyst is necessarily that of promoting the generation of free radicals. Hence, CWAO permits operation at low temperature and pressure (<5 MPa and <473 K) (Pintar and Levec 1992a; Fortuny et al., 1995; Duprez et al., 1996; Gallezot et al., 1999; Luck 1996; Lei et al., 1997; Fortuny et al., 1998, 1999).

Among the components previously reported, CWAO of phenol has been the subject of numerous studies (Pintar and Levec, 1994a; Duprez et al., 1996; Fortuny et al., 1998, 1999; Akyurtlu et al., 1998; Qin et al., 2001). Substituted phenols (Tukac and Hanika, 1995; Duprez et al., 1996; Pifer et al., 1999; Qin et al., 2001) and other components have drawn consideration, for instance, carboxylic acids (Imamura et al., 1986; Gomes et al., 2000; Lee and Kim, 2000), aniline (Oliviero et al., 2003), NH_3 (Huang et al., 2001), other nitrogenous components (Deiber et al., 1997) and commercial effluents (Donlagic and Levec, 1997; Harf et al., 1999; Pintar et al., 2001). Concerning the selection of reactors in research work, slurry reactors (SR) or fluidized bed reactors (FBR), operating in batch or continuous processes, are regularly utilized. Development of polymers has been considered for the CWAO of phenol, when utilizing batch SR (Pintar and Levec, 1992a, 1992b; Hamoudi et al., 1999; Stüber et al., 2001) and batch-recycle (Pintar and Levec, 1994a) processes. However introductory oxidation rates were substantially quicker and fewer intermediates were found in SR contrasted to the trickle-bed reactor (TBR) process (Stüber et al., 2001). In other words, the continuous reactors are perhaps preferable over batch and semi-batch reactors (Fortuny et al., 1995; Pintar and Levec, 1994b) for efficient treatment of phenol-contaminated wastewater. A synopsis of past work is shown in Table 12.10, which includes process conditions, oxidant, catalyst, removal and reactor type.

TABLE 12.10
Past Work of CWAO of Natural Contaminants and Commercial Effluents

Substrate	Reaction Conditions	Oxidant; Catalyst	Reactor	Removal (%)	References
C_6H_6O	423–483 K, 30 bar (total)	CuO, Zn, O_2, CO oxides	trickle bed reactor	>95% TOC	Pintar and Levec (1994a)
C_6H_6O C_6H_5ClO, $C_6H_5NO_3$	376 K, 6 atm (total)	Cu oxide, Zn, O_2; Co oxides	trickle bed reactor	ns	Pintar and Levec (1994b)
C_6H_6O, CH_3COOH	443–473 K, 20 atm O_2	Pt, O_2; Ru, Rh	Batch	<97% COD	Duprez et al. (1996)
C_6H_6O	308–338 K, 2–5 atm O_2	O_2; Pt-Ru	packed bed reactor	90% X	Atwater et al. (1997)
Substituted C_6H_6O	383–433 K, 20–50 atm (total)	O_2; Activate carbon	trickle bed reactor	45% X	Tukac and Hanika (1998)
C_6H_6O	363–423 K, 1–2 atm	O_2; Cu oxide+ Zn oxide, CuO+Al_2O_3	Batch	100% X	Akyurtlu et al. (1998)
C_6H_6O	393–433 K, 6–12 atm (O_2)	Air; AC; Cu/- γ-alumina	trickle bed reactor	<90% COD	Fortuny et al. (1998)
C_6H_6O	353–403 K, 2–25 atm O_2	O_2; MnO_2/CeO_2	Batch	98% TOC	Hamoudi et al. (1999)
$C_6H_5NO_3$	403 K, 7 atm O_2	O_2; Pd/C	Batch	>80% X	Pifer et al., (1999)
C_6H_6O, CH_3COOH $C_2H_2O_4$	393–433 K, 6–12 bar O_2	O_2; Cu/ γ-alumina	trickle bed reactor	<93% COD	Eftaxias et al. (2001)
P-C_6H_5ClO	453 K, 26 atm (total)	O_2; Pt, Pd, Ru	Slurry	<98% TOC	Qin et al. (2001)
C_6H_6O	393–433 K 6–12 atm (O_2)	O_2; Activate carbon, CuO_80_3/γ-Al_2O_3	trickle bed reactor	99%X 85% COD	Athanasios (2002)

(Continued)

TABLE 12.10 (*Continued*)
Past Work of CWAO of Natural Contaminants and Commercial Effluents

Substrate	Reaction Conditions	Oxidant; Catalyst	Reactor	Removal (%)	References
C_6H_6O	393–453 K, 0.1–8.1 bar (O_2)	O_2; Pt/G	Slurry CSTR	100% X	Masende et al. (2003)
C_6H_6O	393–433 K 2–9 atm (O_2)	O_2; Pt, Pd, Ru/CBC	trickle bed reactor	>90% X >80% COD	Janusz (2003)
C_6H_6O	298–823 K 10.13–345.5 atm (total)	O_2,H_2O_2; Activated carbon	trickle bed reactor	80% X >90%COD	Magdalena (2004)
C_6H_6O, o-C_7H_8O 314 K, m-C_9H_{10} 8.88 atm (O_2) o-C_6H_5ClO		O2; Activated carbon, Activated carbon +H_2O_2	trickle bed reactor	>99% X 85%COD	Stüber et al. (2005)
C_6H_6O, o-C_7H_8O C_6H_5ClO, p-$C_6H_5NO_3$	413 K, 12.83 atm (total)	O_2; Activated carbon	trickle bed reactor	30–55% X 15–50%TOC 12–45%COD	Suarez et al. (2005)
C_6H_6O, $C_6H_6O_2$, P-$C_6H_4O_2$	373–400 K, 7.89 atm O_2	O_2; Fe/Activated carbon	trickle bed reactor	>85% X >40%TOC	Quintanilla et al. (2006)
C_6H_6O	400 K, 7.88 atm (total)	O_2, HNO_3 Fe/ Activated carbon-N Fe/ Activated carbon-T	trickle bed reactor	95%X	Quintanilla et al. (2007)
C_6H_6O	413–433 K, 2 atm (O_2)	O_2; Activated carbon	trickle bed reactor	60% X	Aude et al. (2007)
C_6H_6O	<423 K <15.8 bar (total)	O_2,H_2O_2; Activated carbon	trickle bed reactor	100% X 95% COD	Alicia et al. (2007)

Notes: X, substrate conversion; TBR, trickle bed reactor; SDB, porous resin catalyst support; PBR, packed bed reactor; G, graphite; (total), total reaction pressure; AC, activated carbon; ns, not specified.

12.2.1.2.3 Type of Catalytic Wet Air Oxidation Catalysts

One parameter of CWAO is the development of the chosen catalyst. Noble metals are typically more efficient catalysts than metal oxides (Pirkanniemi and Sillanpaa, 2002). However, metal oxides are more appropriate for most applications, since they are more impervious to damage. Different noble metals (ruthenium, platinum, rhodium, iridium and palladium) and some metal oxides (copper, manganese, cobalt, chromium, vanadium, titanium, bismuth and zinc) have conventionally been utilized as heterogeneous catalysts in CWAO. A few reviews have ranked catalysts according to their reactivity; Imamura et al. (1986) ranked noble metal and metal oxide catalysts by the aggregate natural carbon transformation accomplished in 60 min, with the oxidation of polyethylene glycol (PEG) at 473 K and pH of 5.4. They found the accompanying request: ruthenium, platinum and rhodium > iridium > palladium, whereas the oxidation of C_6H_4ClOH catalyzed by noble metals bolstered on alumina (Al_2O_3) or titanium (TiO_2) at 423 K and under 0.3 MPa of O_2 pressure demonstrated the accompanying request of action: platinum >> palladium > ruthenium > rhodium (Okitsu et al., 1995). Such behavior concurs with Qin et al. (2001), who found the movement in all out natural carbon lessening of most basic respectable metals in CWO of chlorophenols is in taking after request: platinum > palladium > ruthenium, whether the support is activated carbon or Al_2O_3.

- *Supported metal oxides*: Metal oxides can be grouped by their physical and chemical characteristics; the steadiness of the metal oxide is the main characteristic. Metals with unsteady high oxidative case oxides (e.g., platinum, palladium, ruthenium and mercury) are unable to form stable oxides at ambient conditions (Besson and Gallezot, 2003).

  The most of the commonly utilized metal oxide catalysts (titanium, vanadium, manganese, zinc and aluminum) are more stable, while iron, cobalt and nickel are less stable (Pirkanniemi and Sillanpaa, 2002). Metal oxides are normally used in the form of fine molecules or powders, which can result in extreme concentrations in a particular area, while the dispersion of particles can make them less effective. A permeable support can be used to maintain the stability of the catalyst and also to generate a higher active phase. This also allows the contact time to be abbreviated, as the organic compounds to be oxidized can adsorb onto the support thus expanding the quantity of natural compounds at the surface. Normally, alumina or zeolites are used as a support, but the surface region of aluminum oxide is constrained and the pore size of zeolites is not reasonable for substantial natural mixes. The use of activated carbon (AC) as a substrate has been the subject of much enthusiasm, as it has substantial surface region and wide range of pore size (Hu et al. 1999).
- *CWAO base on AC catalyst/substrate*: activated carbon has been viewed favorably as a new catalyst because it shows great characteristics as an adsorbent for both organics and O_2 due to its porous frame and large surface area (Singh et al., 2004; Hu et al., 1999). Chemically, the stabilization of AC in exceptionally acidic and basic media is also of great significance. AC can also be easily prepared from renewable natural sources. In early studies AC appeared to act as a genuine catalyst for various reactions (Coughlin, 1969) and more recently has been considered as a substrate for other oxidation catalysts (Birbara and Genovese, 1995; Aksoylu et al., 2001). Nonetheless, if just CWAO procedures are viewed, AC has not been considered in many circumstances as a substrate for active metals (Duprez et al., 1996; Gallezot et al., 1999; Atwater et al., 1997; Hu et al., 1999; Alvarez et al., 2002a, 2002b) or reaction matter by catalyst (Fortuny et al., 1995, 1998, 1999; Qin et al., 2001; Nunoura et al., 2002; Gonzalez et al., 2002). AC is increasing in importance due to its great effectiveness as a result of its high adsorption limit and capacity to create oxygenated free radicals that promote the oxidation

response (Stüber et al., 2001). CWAO using AC can be an alternative and cost-effective method of wastewater treatment. AC can also catalyze polymerization reactions in the presence of oxygen via oxidative coupling. The phenol conversion is known to occur via complex pathways that could first lead to the formation of polymeric compounds that are adsorbed in part on the AC and finally limit the available AC surface (Stüber et al., 2001; Pintar and Levec, 1992); such polymers are especially hard to desorb from the AC (Cooney and Xi, 1994).

12.2.1.2.4 *Kinetic Mechanism of Catalytic Wet Air Oxygen*

As detailed by Sadana and Katzer (1974a, 1974b), during non-CWAO, the underlying rate of oxidation displays an acceptance period, which is trailed by a steady. The free radical initiation preceded by the proliferation step has also been reported by Akyurtlu et al. (1998). A stamped reliance of the acceptance time frame length on the underlying expansion of a free radical inhibitor focus basically demonstrates the inclusion of free radicals in the CWAO reaction. Another proof of free-radical investment is the dependence of rate on pH. Free radical reactions in fluid media are regularly pH-dependent and they commonly demonstrate a most extreme with pH (Sadana and Katzer, 1974a). It has also been noted that the acceptance period depends upon the temperature and conversely corresponds to the partial pressure of O_2 (Sadana and Katzer, 1974b).

12.2.1.2.5 *Reactivity of Organic Components*

Pintar and Levec (1994b) have noted during CWAO of some substituted C_6H_6O, that such compounds with electron-withdrawing substituents are more impervious to oxidation than C_6H_6O with electron-donating substituents. This can be clarified by the way that withdrawing electron thickness from the aromatic ring diminished its reactivity. The investigators indicated that the reactivity arrangement for oxidation with a catalyst containing CU, Zn and CO oxides has the following order: $C_6H_6O > C_6H_4ClOH > C_6H_5NO_3$. For the oxidative coupling reactions catalyzed by AC, the reactivity for substituted C6H6O introduced in the work of Cooney and Xi (1994) is quickened by: $CH_3OC_6H_4OH > C_7H_8O > C_6H_4ClOH > C_6H_6O > C_7H_6O_3$ acids $> C_7H_6O_2 > C_6H_5NO_3$. The request is indistinguishable with signifying whether the substituting bunch repulses or draws in electrons, hence making higher or bring down electron thickness on fragrant ring, separately. This request is in concurrence with the request reported by Joglekar et al. (1991) for non-CWAO. The noticed substituting bunch impact is shown in Figure 12.2.

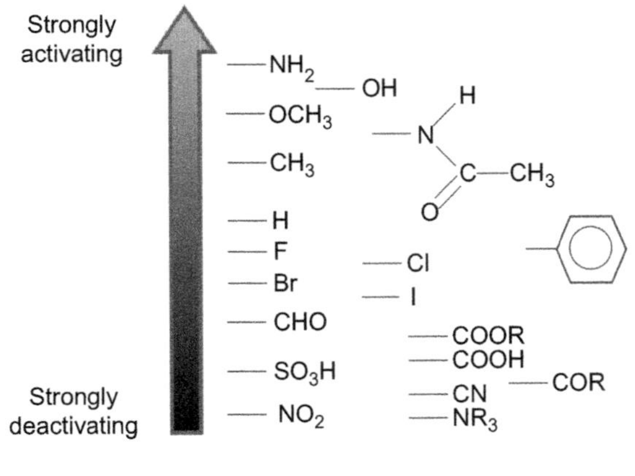

FIGURE 12.2 Substituent impact on the reactivity of aromatic rings. (From Joglekar, H.S. et al., *Water Res.*, 25, 135–145, 1991.)

12.2.1.2.6 Reaction Intermediates

As the component of catalytic wet air oxidation is accepted to be same without catalyst operation likewise comparable pathways and intermediate distribution ought to be noted. Alvarez et al. (2002a, 2002b) have discovered that CH_3COOH and $C_6H_4O_2$ to be the primary intermediates in catalytic wet air oxidation of C_6H_6O utilizing CUO bolstered over AC. The development of solid molecules that might be the organ cupric polymers was seen in the effluent coming by the reaction of R–COOH with CUO. They could block active sites and furthermore impact the adsorption of O_2 at the oxide sites.

Fortuny et al. (1999) oxidized C_6H_6O with an industrial Cu catalyst and they discovered that CH_2O_2, CH_3COOH and $C_2H_2O_4$ to be the fundamental intermediates in charge of right around 90% of all intermediates display in fluid phase at higher contact times. They are particularly undesired items as their poisonous quality is same or higher to that of phenol. The intermediates found are intelligent with the mechanism suggested by Devlin and Harris (1984). A comparable mid distribution has observed in the work of Duprez et al. (1996) over upheld Ruthenium, Platinum and Rhodium catalyst and Ohta et al. (1980) over upheld CuO. They have distinguished $C_6H_6O_2$, and $C_4H_4O_4$ that additionally oxidized to $C_3H_4O_2$ and CH_3COOH. They stated that CH_3COOH as just being an obstinate item. The reaction behavior of phenol catalytic wet air oxidation introduced by Devlin and Harris (1984) is displayed in Figure 12.3.

12.2.1.3 Supercritical Water Oxidation (SCWO)

Supercritical fluids are liquids conveyed to a temperature and weight higher than their basic temperature and pressure. Supercritical fluids have drawn consideration as a media for chemical reactions in light of the fact that their physiochemical characteristics alter drastically with small changes in temperature and pressure. The individual working conditions and references of these reviews are shown in Table 12.11.

12.2.1.4 Wet Peroxide Oxidation

Oxidative treatment with hydrogen peroxide (H_2O_2) has emerged as a practical option for wastewater treatment with natural load. Hydrogen peroxide is a strong oxidant and its implementation in the treatment of different inorganic and natural toxins is settled. Various utilizations of hydrogen peroxide are known in the expulsion of toxins from wastewater, such as $SO_2{-}_3$, ClO–, NO_2–, organic compounds and CL_2 (Neyens et al., 2003).

12.2.1.5 Fenton Promoted

Fenton detailed that ferrous particles emphatically promote the oxidation of $C_4H_4O_4$ by H_2O_2. It was found that the OH radical is a genuine oxidant in such frameworks. The Fenton catalyst (Fe2+/Fe3+ framework) causes the separation of H_2O_2 and the arrangement of exceedingly responsive OH radicals that assault and devastate the natural mixes. Fenton's reagent-based wastewater treatment procedures are known to be extremely viable in the removal of numerous dangerous natural toxins from H_2O. The component of assault of OH radicals on aromatic mixes is practically equivalent to electrophilic substitution, as also occurs in WAO.

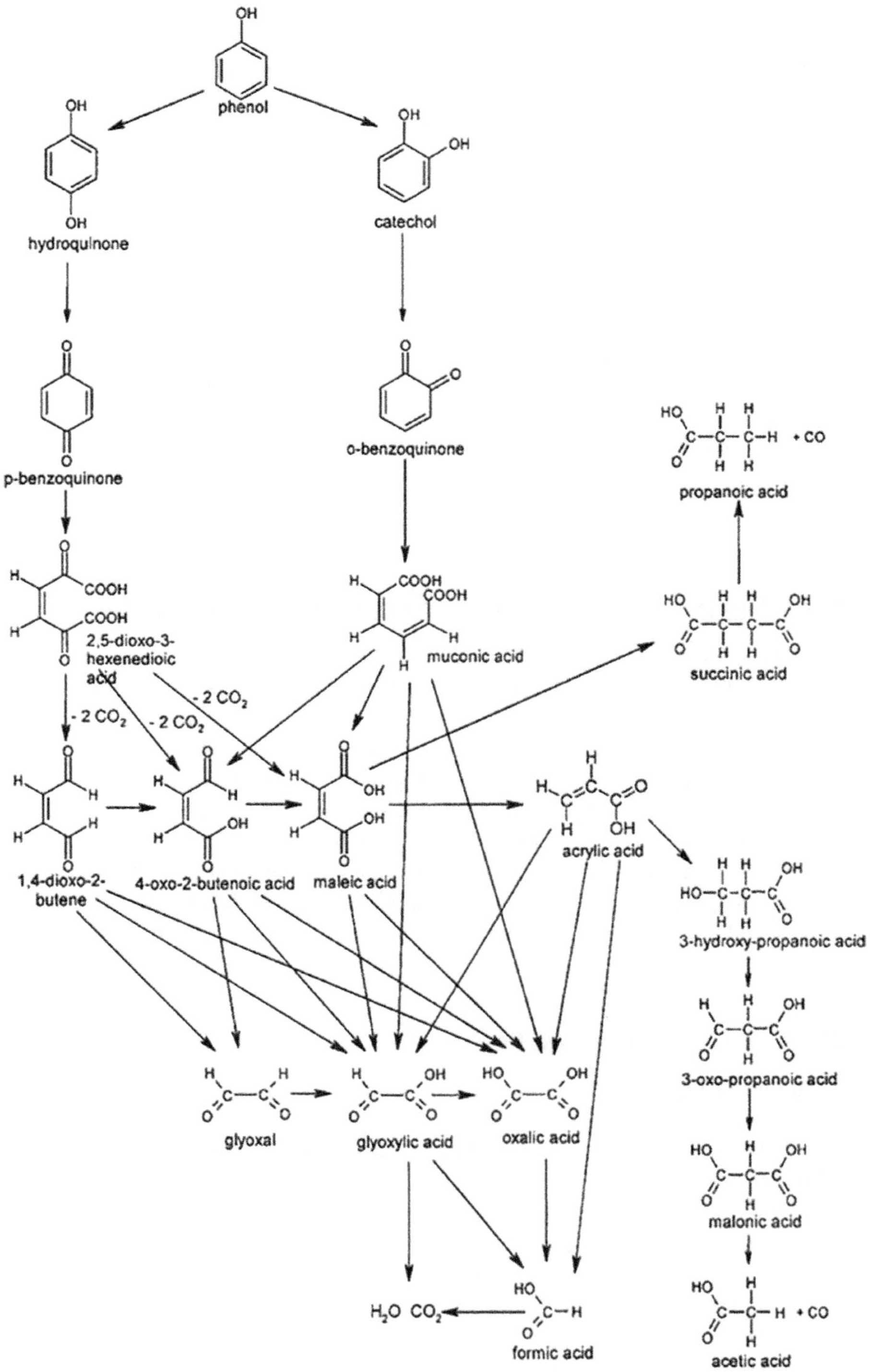

FIGURE 12.3 Reaction framework for phenol CWAO according to Devlin and Harris (1984).

TABLE 12.11
Supercritical Water Oxidation of Organic Contaminants and Commercial Effluents

Fluid	Ambient Water	Supercritical Water	Preheated Steam
Typical conditions			
Temperature (K)	298	723	723
Pressure (atm)	1	275.6	13.8
Properties and parameters			
Dielectric constant	78	1.8	1.0
Hydrocarbon solubility (mg/L)	Variable	∞	Variable
Oxygen solubility (mg/L)	8	∞	∞
Density, ρ (g/L)	998	128	4.19
Viscosity, μ (p)	8.90×10^{-3}	2.98×10^{-4}	2.65×10^{-7}
Particle Reynolds number	18.5	553	622
Effective diffusion coefficient, De (cm^2/s)	7.74×10^{-6}	7.67×10^{-4}	1.79×10^{-3}
Thiele modulus, φ	2.82	2.84×10^{-2}	1.22×10^{-2}

12.3 REACTORS IN CATALYTIC WET AIR OXIDATION TREATMENT

Three phase reactors (TPR) are required for responses composed of a strong stage (catalyst) with a relatively unpredictable gas phase (oxygen) and non-unstable fluid stage (contaminant in wastewater). TPRs utilized to reach these three phases have overwhelmingly been either fixed bed reactors (FBR) or slurry reactors. Two behaviors of liquid flow through a stationary bed of catalyst can be observed, simultaneously downwards (stream) and simultaneously upwards (bubble stream). The procedure plan and cost assessments for the catalytic oxidation in advanced wastewater treatment plants have been considered. Removal of natural toxins in the two-reactor frameworks has also been investigated for TBRs, notwithstanding the various mass exchange conditions. Diverse reactors have been utilized as a part of wastewater treatment and the determination of a legitimate reactor is vital and is considered to have a key impact upon the commercial advancement of wastewater treatment facilities.

12.3.1 Description of Three-Phase Reactors

Multiphase-catalytic packed-bed reactors (PBRs) are worked either with stream operation (consistent gas and disseminated fluid phase) when the mass exchange resistance is situated in the gas phase, or with bubble operation (nonstop fluid and dispersed gas phase) when the mass transfer is situated in fluid phase (Mederos et al., 2009).

As indicated by reactor configuration, gas and fluid move simultaneously downstream, or gas is encouraged to run upstream against the current. At present, the downstream model is the most well known reactor figuration employed in industrial applications. The fluid stage streams through the catalyst particles in the form of films, rivulets and beads as indicated schematically in Figure 12.4 (Mederos et al., 2009).

Multiphase PBRs can be characterized by the heading of the liquid stream into TBRs with co-current gas fluid downstream, while in TBRs with countercurrent gas–fluid stream and in packed-bubble reactors, the gas and fluid are reached in co-current up-stream. To appropriately select the catalyst and reactor and create the process plan, it is important to learn what the different sorts of reactors can and cannot do. When choosing an FBR, the most frequent potential inquiry is whether to use an upstream or downstream method of operation. The plan of multiphase reactors with catalysts is even more mind-boggling than that of homogeneous reactors

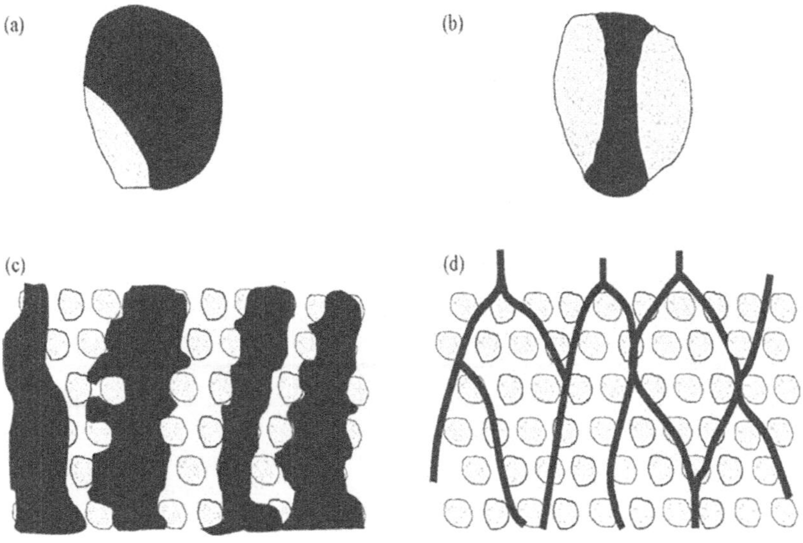

FIGURE 12.4 Schematic portrayal of fluid stream surfaces experienced during trickle-flow regime in TBRs: (a) molecule scale film stream, (b) molecule scale current flow, (c) bed-scale, and (d) bed-scale creek stream. (Adapted from Mederos, F.S. et al., *Appl. Cat. A: Gen.*, 355, 1–19, 2009.)

because of the concurrence of more than one stage present in the mass exchange resistance, as is shown in Figure 12.5.

TBRs are three-phase frameworks composed of a packed bed of catalyst with co-current downstream of gas and fluid. TBRs are broadly utilized for hydrotreating and hydrodesulfurization in refining operations and for hydrogenation, oxidation and hydrodenitrogenation applications in the chemical, biochemical and waste treatment industries (Augier et al., 2010). The TBR behavior is based on different parameters related to hydrodynamic administration, mainly, wetting efficiency, mass exchange among phases and radial and axial dispersion inside the bed. Upgrading and outlining the operation of TBRs is therefore not a simple task in the field of catalytic reactor engineering. Four implementations are possible in TBRs, contingent upon gas and fluid flow rates and liquid and packing characteristics (Mederos et al., 2009) as shown in Figure 12.6:

- Trickle Flow Regime: Low fluid and gas flow rates in this implementation, where the fluid streams over the packed catalyst and the gas stage fills the remaining voids.
- Pulsing Flow Regime: Higher fluid and gas loads, where the fluid occasionally hinders the channels between the packed catalyst particles, obstructing the structures.
- Spray Flow or (Mist Flow) Regime: High gas rates and low fluid flow rates, where the fluid stage is entrained in beads by the gas stage.
- Dispersed Bubble Flow Regime: Low gas stream rates and high fluid loads, where the gas streams descend as bubbles and the fluid becomes the dominant stage.

Moving from one zone into the next demonstrates that the transition between the trickle and pulse flow stream zones is sharp, whereas the transition between the spray stream and the scattered stream is more progressive. Understanding these transitions is imperative when planning TBRs. Many reviews have demonstrated that pressure drop, fluid blending, heat and mass exchange coefficients are influenced in every zone (Ruether et al., 1980; Rao et al., 1985). By and large, two most common methods for implementing the TBR process can be shown as follows.

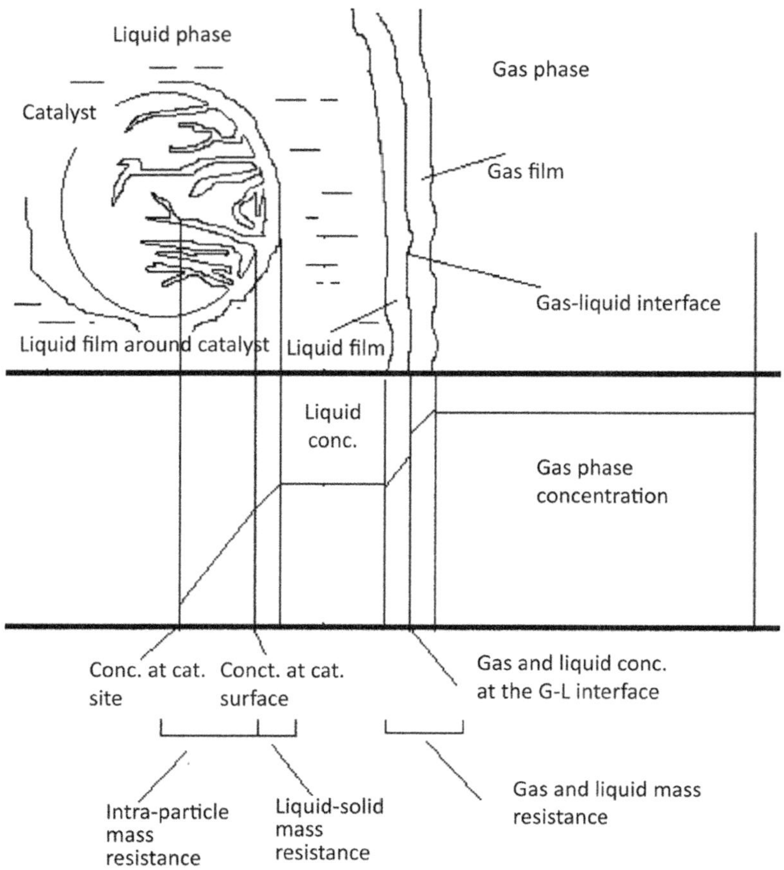

FIGURE 12.5 Gas-liquid-solid contact in three-phase reactors. (Adapted from Hedge, S.C., Simulation for polluting prevention: Gas-liquid reactors and sulfuric acid alkylation process, MSc Thesis, Lamar University, 1999.)

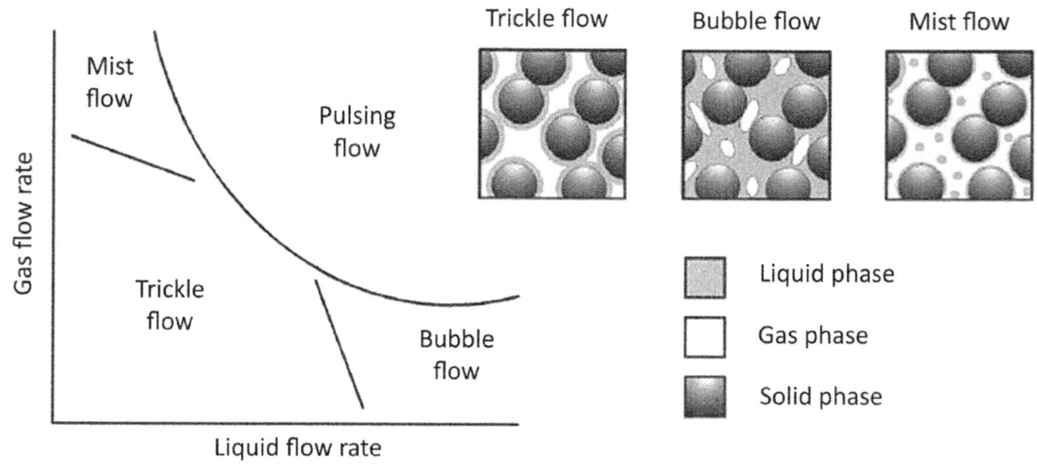

FIGURE 12.6 Various flow regimes in three phase packed beds as a function of flow rate of liquid and gas. (Adapted from Mederos, F.S. et al., *Appl. Cat. A: Gen.*, 355, 1–19, 2009.)

In the co-current gas-fluid downstream mode, gas is the consistent phase and the fluid holdup is lower. This process is the most frequently used as a part of the treatment process owing to serious impediments in the throughput rather than in countercurrent operation (Mederos et al., 2009). A down-stream reactor is favored for gas-constrained reactions (high fluid reactant flux to the catalyst molecule, low gas reactant flux to the molecule), particularly at partially wetted conditions, as it encourages the transport of the vaporous reactant to the catalyst (Dudukovic et al., 2002).

12.3.2 Reaction Pathway of Oxidation Mechanism and Kinetic

The mechanism for the oxidation of natural mixes is extremely complicated, even with an unadulterated compound, for example phenol, if the correct system or reaction pathway has not yet been built up. Much of the time, the oxidation experiences an extremely muddled pathway, which prompts the development of a wide range of intermediates, mostly lowering R-COOH (Zhang and Chuang, 1998). Regardless of whether the phenol was decimated under moderate working conditions, its mineralization was not completed as expected. The accompanying grouped schema suggested for phenol oxidation appears in Figure 12.7. Phenol oxidation takes place after a succession of parallel and sequential reactions (Devlin and Harris, 1984). The examination demonstrated that the primary fractional oxidation items must be light carboxylic acids, such as $C_2H_2O_4$ and CH_2O_2.

$C_3H_4O_4$, $C_4H_4O_4$ and its isomer were distinguished in the following sums. Quantifiable measures of $C_6H_6O_2$, $C_6H_6O_2$ and benzoquinones were achieved. The mediator prior to acceptance follows the ordinary patterns of a parallel sequential reaction pathway, which positively correlates with the interaction model shown (Devlin and Harris, 1984). This complicated reaction system is, for the most part, acknowledged to clarify the oxidation of phenol into CO_2. The exceptionally poor fitting acquired when checking kinetic systems, excluding an immediate oxidation venture from benzoquinones to $C_2H_2O_4$ and CH_2O_2, also supports this hypothesis. Direct oxidation of CH_3COOH into CO_2 and H_2O was not considered, as its reaction rate ought to be unimportant at temperatures below 293 K. The reaction pathway of the CWAO of phenol is exceptionally perplexing and these different pathways have been minimized to reduce the diverse characteristics. Although some of these schemes do not exist, many reaction systems are ambiguous since a few conceivable ways exist for the corruption of some of the partially oxidized molecules. For instance, benzoquinones can be debased through $C_6H_2Cl_2O_6$, i.e., without going through $C_4H_4O_4$. The former acid is very receptive and has not yet been identified as a mediator in phenol oxidation. Along these lines, benzoquinones could straightforwardly yield $C_2H_2O_4$ $C_2H_2O_2$, $C_2H_2O_3$. Thus, these can shape $C_2H_2O_4$, CH_2O_2, or could even be specifically changed over into CO_2. In other words, $C_4H_6O_5$ could be either decarboxylases to yield $C_3H_4O_2$ or debased by oxidation to $C_2H_2O_3$ and additionally $C_2H_2O_4$. $C_3H_4O_2$ was not distinguished despite the fact that the resulting oxidation items, $C_3H_4O_4$ and CH_3COOH, were without a doubt acquired. $C_2H_2O_4$ can be straightforwardly oxidized into CO_2, and H_2O however can likewise experience decarboxylation, yielding CH_2O_2. The last model accepts that phenol firstly experiences a –OH prompting $C_6H_6O_2$; a further oxidation yields the comparing $C_6H_4O_2$. At that point, $C_6H_4O_2$ is debased in two parallel pathways. In the primary way, $C_6H_4O_2$ is oxidized to $C_6H_6O_4$, which as such quickly disintegrates, stoichiometrically yielding $C_4H_4O_4$ and $C_2H_2O_4$. Such reactions have been noted even at room temperature (Zhang and Chuang, 1998). Figure 12.8 shows a simple pathway for the CWAO process.

FIGURE 12.7 Lumped scheme of natural components oxidation. (From Devlin, H.R. and Harris, I.J., *Ind. Eng. Chem. Fund.*, 23, 387–392, 1984.)

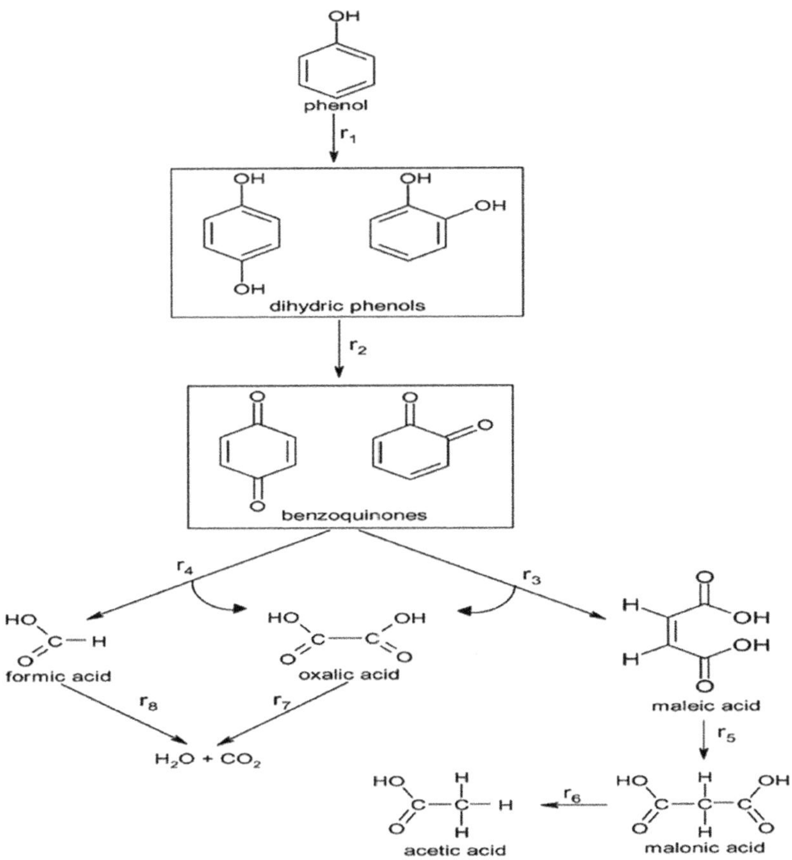

FIGURE 12.8 Simplified reaction pathway for CWAO of phenol. (From Zhang, Q. and Chuang, K.T., *Appl. Cat. B: Envir.*, 17, 321–332, 1998.)

More than 50 different reaction models have been designed and tested to select the best reaction. The following chemical reactions are specified for CWAO process reactions (Zhang and Chuang, 1998):

$$C_6H_4\left(OH\right)_2 + \frac{1}{2}O_2 \rightarrow C_6H_4O_2 + H_2O$$

$$C_6H_4O_2 + \frac{5}{2}O_2 + H_2O \rightarrow C_4H_4O_4 + C_2H_2O_4$$

$$C_6H_4O_2 + 5\,O_2 \rightarrow CH_2O_2 + C_2H_2O_4 + 3CO_2$$

$$C_4H_4O_4 + O_2 \rightarrow C_3H_4O_4 + CO_2$$

$$C_3H_4O_4 \rightarrow C_2H_4O_2 + CO_2$$

$$C_2H_4O_2 + \frac{1}{2}\,O_2 \rightarrow 2\,CO_2 + H_2O$$

$$CH_2O_2 + \frac{1}{2}\,O_2 \rightarrow CO_2 + H_2O$$

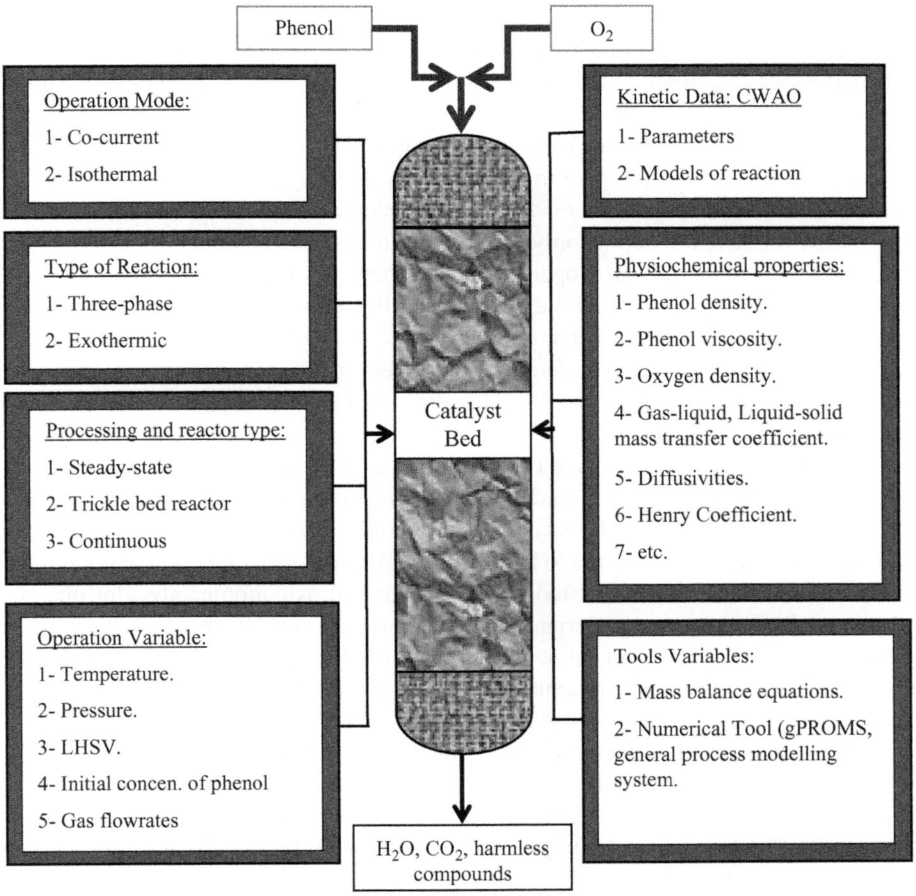

FIGURE 12.9 shows the required data and available tools for modeling and simulation of CWAO of phenol process. (Adapted from Mohammed, A.E. et al., *Comput. Chem. Eng.*, 94, 257–271, 2016.)

The mechanism of natural components is very complicated. Further steps have been reported in the literature related to the pathway mechanism from the first equation to the last equation, which include the kinetic expression as power law and the Langmuir–Hinshelwood kinetic model; these methods have been widely employed for the kinetic modeling of this process. The required data and available tools for the modeling and simulation of CWAO of phenol are shown in Figure 12.9.

12.3.3 Previous Works of Catalytic Wet Air Oxidation of Phenol

The kinetics of CWAO in commercial reactors for phenol oxidation has been considered by different authors, who have tentatively tried to develop the process. These reviews are summarized as follows:

Got and Smith (1975) investigated an experimental study for the WCO of CH_2O_2 and C_6H_6O by O_2 in a continuous TBR at temperature 485–813 K and pressure of 4 MPa. They suggested both axial dispersion and PF models to foresee the changing rate of oxidizing CH_2O_2 in the CWAO process. The disparities between the two models were discovered to be immaterial and the simulation results agreed with the experimental information. It has been confirmed in light of their results that there are four mass exchange resistances, which are, ranging from the most to least impact on the change rate: gas-to-fluid mass exchange, intra-molecule dissemination, fluid to-solid (molecule) mass exchange and axial dispersion.

Pintar and Levec (1994b) dealt with the CWAO of phenol in a TBR utilizing copper oxide, zinc and cobalt oxides as a heterogeneous catalyst and pure O_2 as oxidant with temperatures between 423–483 K and pressure around 3 MPa. This produced greater than 95% phenol expulsion of total organic carbon.

Pintar et al. (1997) concentrated CWAO of phenol in an isothermal TBR utilizing metal oxides as a heterogeneous catalyst at temperature 403–423 K and oxygen pressure of 0.7 MPa. Their results were close to that of a catalyst made out of upheld Cu, Zn, and Co oxides, which demonstrated that during the reaction course only small measures of fragrant and aliphatic hydrocarbons are gathered in the fluid stage. Such operations have been simulated utilizing one-dimensional axial dispersion and PF models, which has shown that the activity of the catalyst bed for phenol expulsion is affected by the mass-exchange rate of oxygen from the gas stage to the mass fluid stage and by resistance because of a surface response step. In the explored reactor framework, the response selectivity was equivalent to 90%. The phenol fixation took after the first-order model and 0.5 order for O_2.

Fortuny et al. (1998) examined CWAO of phenol in an FBR working in a trickle-flow regime utilizing AC and industrial CUO bolstered over γ-alumina as a catalyst. The oxidation was led in gentle states of gaseous tension (8–47 bar) and temperature of 393–413 K in the catalytic reactor. The phenol oxidation was demonstrated to happen through a first-order reaction based upon phenol. In this work, AC execution was contrasted with the industrial oxidation catalyst for phenol oxidation in 10-day trials. The catalyst experiences a fast deactivation by losing the active species as a result of its poor steadiness in the hot acidic watery medium. Hence, after a short acceptance period, the most noteworthy phenol transformation was of 78%, and after that it fell strongly to settle at an outstanding phenol change of 30%.

Zhang and Chuang (1998) concentrated the catalytic oxidation of industrial wastewater from paper and mash plants; dark alcohol (emanating from the pulping factory produced in the wake of cooking of wood or other crude materials called dark alcohol because its color) is oxidized in a SR utilizing Pt-Pd-Ce/alumina as a catalyst under various working conditions at temperatures of 433–463 K and pressure from 1.5–2.2 MPa. The response rate of oxidation is portrayed by an underlying quick response step taken after a moderate response step. The first-order reaction regarding the total organic carbon fixation in dark alcohol for both introductory and later response steps was noted, and the estimation error was around 5%.

Fortuny et al. (1999) investigated CWAO of phenol in fluid stage utilizing an FBR based on trickle flow regime using O_2 as an oxidant and accessible Cu upheld over γ-alumina as a heterogeneous catalyst, under temperature ranging from 393–433 K and oxygen partial pressure of 0.6–1.2 MPa. The phenol transformation obtained was 95% at 433K and 1.2 MPa of O_2. Phenol change and oxygen demand were influenced by temperature, whereas the oxygen partial pressure had only a minor impact. First-order kinetic of phenol focus and (0.5–/+0.1) regarding oxygen pressure were noted.

Christoskova and Stoyanova (2000) c examined the catalytic phenol oxidation (COO) found in wastewater utilizing Ni-oxide as a catalyst framework at low temperatures (288, 338, and 388 K) and pH 6–10 with differing quantities of catalyst (2–6 g/dm^3). The COO followed a pseudo-first-order kinetics with regard to phenol fixation. Since all tests were performed with a continuous flow of air through the liquid stage, it would be expected that the oxygen scope of the catalyst surface was consistent during the oxidation procedure. Phenol conversion around 100% can be accomplished at $6 < pH < 8$ and temperatures of 288–338 K.

Eftaxias et al. (2001) explored the CWAO of phenol using $CuO/\gamma\text{-}Al_2O_3$ as a catalyst in a non-stop TBR using air as oxidant at a temperature of 393–433 K and oxygen partial pressures of 0.6–1.2 MPa. They used the power law and Langmuir–Hinshelwood models to discover the kinetic factors. Non-linear multi-factor evaluations simulated the model precisely and correlate with all the test focus profiles with mean deviations beneath 8%.

Santos et al. (2005) attempted CWAO of phenol using heterogeneous industrial Cu as a catalyst at temperatures of 400–433 K and oxygen partial pressures (OPP) from 0.8 to 1.6 MPa. The suggested model was dependent on a heterogeneous free radical system, which considers the forager impact of the bicarbonate on the phenoxy radicals framed upon the catalyst surface by decreasing the active copper destinations. A high phenol change around 77% was achieved. Temperature had a slight positive impact on the kinetic equation for phenol oxidation, with no impact from OPP. These actualities can be clarified by the suggested free radical mechanism and the order for phenol fixation is found to be 1.5.

Aurora et al. (2006) have researched CWAO of phenol in a three-stage FBR with simultaneous up-flow at a consistent state utilizing industrial AC as catalyst at a temperature range of 400–433 K and OPP of 0.34–1.6 MPa. They successfully predicted the appearance and vanishing of phenol and the cyclic natural intermediates shaped during the course of phenol oxidation.

Ayude et al. (2007) have investigated phenol oxidation in a TBR, where the catalyst and oxidant utilized was AC and pure O_2, separately, under the accompanying conditions temperature (413–433 K) and OPP (0.2 and 0.9 MPa). A steady state resulted, which demonstrated that the execution of AC in CWAO emphatically relies upon the working conditions utilized, where directly expands the extent of the reaction with temperature and pressure, whereas, conversely, the extent of initial phenol fixation is acquired. At 433 K and 0.2 MPa of OPP, the conversion performed is higher than 90%, yet the AC consumed also improved, prompting a negative loss of the activity with working time.

Primo et al. (2007) considered tentatively the corruption of phenol in dirtied water by means of UV/H2O2. The test was conducted with an underlying centralization of phenol 1000 mg/L and a variable starting convergence of the oxidant in the range between 17,000 and 51,000 mg/L. A straightforward simulation model depending on the depiction of the kinetics of the fundamental operation factors for the photochemical oxidation handle, i.e., TOC, toxicity of the treated water and oxidant focus was suggested. The pseudo-first-arrange kinetic constants have improved for TOC expulsion, and zero-order kinetic expression was found to portray the debasement of H_2O_2 during the oxidation. The arrangement of standard differential correlations was explained utilizing gPROMS programming.

Mohammed and Abdullah (2008) concentrated the CWAO of phenol oxidized in an FBR based on a trickle flow regime utilizing a Cu-construct catalyst upheld in light of γ-alumina as a catalyst and air as oxidant. The basic power law as a model under moderate working conditions, temperature (393–433 K), OPP (0.9–1.2 MPa), LHSV (1–3 h^1) and stoichiometric abundance 80%, was employed. First-order concerning phenol and 0.5 kinetic order for O_2 were obtained.

Wu et al. (2009) dealt with the CWAO of phenol in a pilot plant TBR at steady state operation utilizing models to foresee the execution of TBR catalyzed by heterogeneous Cu catalyst at a temperature of 313–323 K and OPP of 1–3 MPa. It has been seen that 100% of fluid phenol can be corrupted when fresh catalyst is utilized, while 90% can be accomplished by means of the spent catalyst after 2 h of running.

Safaa (2009) considered the CWAO of phenol into a TBR utilizing Pt on alumina as a catalyst, pure oxygen as an oxidant at temperature (393–433 K), LHSV (1–3 h^{-1}), OPP (0.8–1.2 Mpa), initial phenol fixation (1–5 g/L), and gas flow rate (20%–100% S.E.). Increasing the temperature and diminishing LHSV increased the transformation to 97.40%.

Keav et al. (2010) attempted the CWAO of phenol in an FBR over Pt and Ru catalysts bolstered on cerium-based oxides at 433 K; fouling of the catalytic surface by a layer of adsorbed intermediates was considered a factor. They concentrated on the catalyst activity brought on by the development of a carbonaceous layer during the oxidation reaction.

Mohammed (2014) studied the oxidation of phenol in a TBR utilizing AC as a catalyst under various conditions (varying pH, gas flow rate, LHSV, T and OPP). The parameters of the model evaluated utilizing experimental information were achieved via a consistent TBR at various temperatures (393–433 K) and OPP (0.8–1.2 MPa). Basic power law and Langmuir–Hinshelwood expressions representing the adsorption impacts were checked to demonstrate the response organization, giving mean deviations under 10%. A non-straight multi-parameter estimation approach was utilized to assess the high number of the model parameters. A temperature of 433 K, 1.2 MPa and duration of around 1 h brought about phenol degradation around 100%. The model relations have been solved with a fourth-order Runge–Kutta technique to compute the hypothetical outlet phenol and middle concentrations of the goal work.

Most recently, Mohammed et al. (2016) have developed a full model to obtain the optimal design of a TBR to describe the performance CWAO of phenol found in wastewater with high efficiency, productivity and minimum cost under various operating conditions based on experiments (taken from literature). The mathematical model, simulation and optimization used phenol as the major pollutant, pure oxygen as an oxidant and Pt/γ-Al_2O_3 as a catalyst under various operating conditions (temperature (120, 140, and 160.), oxygen partial pressure (0.8, 1, and 1.2 MPa), liquid hourly space velocity (1, 2, and 3 h^{-1}), initial phenol concentration (1, 3, and 5 g/L) and gas flow rate (stoichiometric excess (20%, 40%, 80%, and 100%)). The best kinetic parameters of the relevant reaction were estimated based on the experimental results. Two approaches were used in estimating such parameters (linear and non-linear methods) and the sum of squared errors between the experimental and predicted concentration of phenol were minimized as an objective function. The results showed a good agreement between the experimental and predicted results with an average absolute error less than 5% among all the results. It is also observed, depending on the results, that the conversion increases by increasing the temperature (as shown in Figure 12.10), initial phenol concentration (as shown in Figure 12.11) and OPP (as shown in Figure 12.12) and decreasing the liquid hourly space velocity (as shown in Figure 12.13) as well as the gas flow rate (as shown in Figure 12.14).

The best kinetic parameters obtained from the small-scale TBR have been utilized to develop an industrial phenol oxidation reactor in order to predict the behavior of CWAO in industrial reactors. The optimal distribution of the catalyst bed has also been considered in scaling up. In this case, the optimal ratio of the reactor length to reactor diameter was calculated, taking into account the hydrodynamic factors (effect of radial, axial concentration and temperature

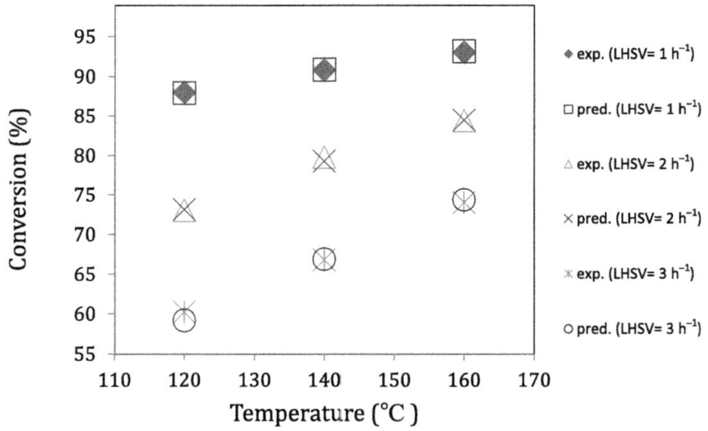

FIGURE 12.10 Effect of temperature on phenol conversion at different LHSV. Reaction conditions at initial phenol concentration of 5 g/L, OPP of 0.8 MPa and gas flow rate of 80%. (Adapted from Mohammed, A.E. et al., *Comput. Chem. Eng.*, 94, 257–271, 2016.)

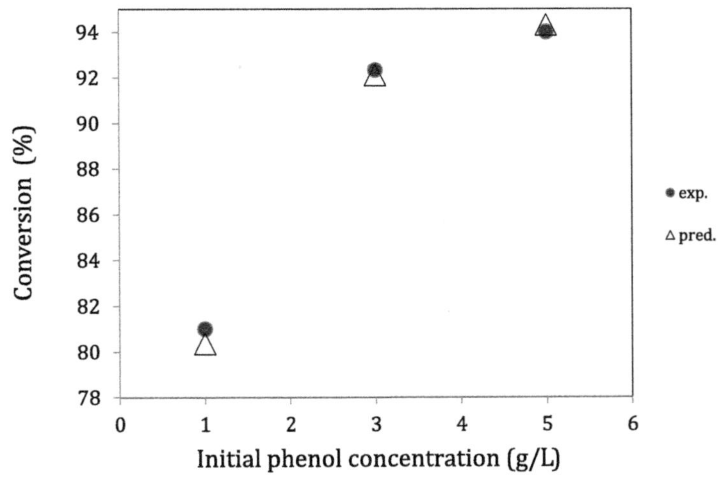

FIGURE 12.11 Effect of initial phenol concentration on phenol conversion. Reaction conditions (T = 160°C, LHSV = 1 hr^{-1} OPP = 1.2 MP and gas flow rate = 80%). (Adapted from Mohammed, A.E. et al., *Comput. Chem. Eng.*, 94, 257–271, 2016.)

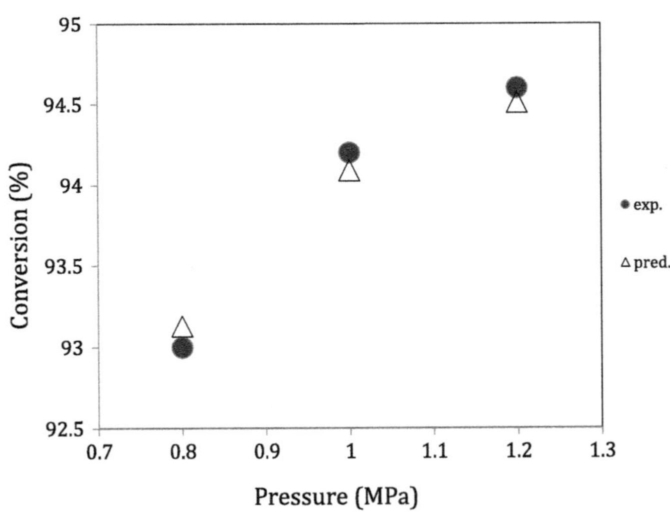

FIGURE 12.12 Effect of OPP on phenol conversion. Reaction conditions (T = 160°C, LHSV = 1 hr^{-1}, initial phenol concentration = 5 g/L and gas flow rate = 80%). (Adapted from Mohammed, A.E. et al., *Comput. Chem. Eng.*, 94, 257–271, 2016.)

distribution). The optimal results obtained based on the capital cost of the industrial TBR for CWAO is given Table 12.12.

Such parameters can effect the efficiency of the process, mainly the conversion and the productivity, so it is important to find the optimal values for the operating conditions used in the CWAO process; the optimal values for the operating conditions and the resulting cost parameters are shown in Tables 12.13 and 12.14, respectively.

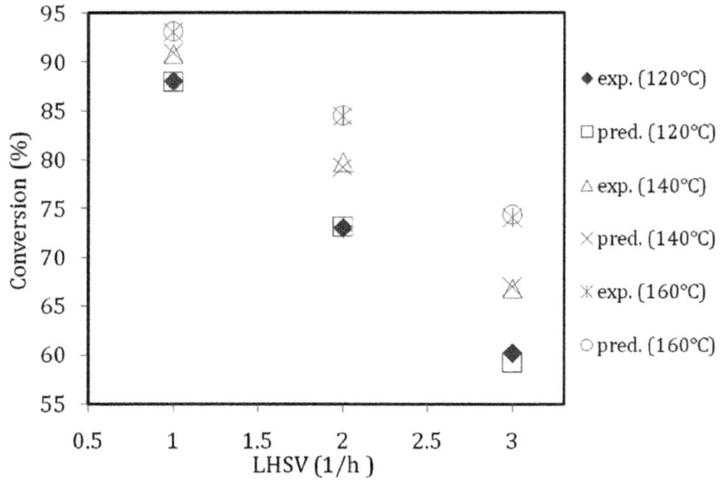

FIGURES 12.13 Effect of LHSV on phenol conversion at different temperatures. Reaction conditions (initial phenol concentration = 5 g/L, OPP = 0.8 MPa and gas flow rate = 80%). (Adapted from Mohammed, A.E. et al., *Comput. Chem. Eng.*, 94, 257–271, 2016.)

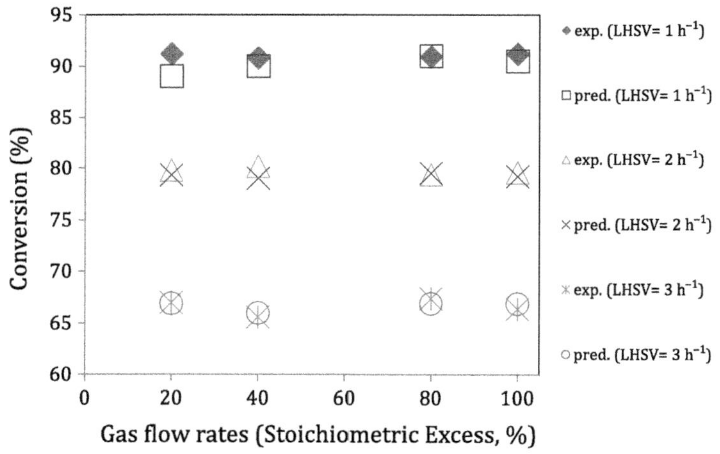

FIGURE 12.14 Effect of gas flow rates at different LHSV. Reaction conditions (T = 160°C., initial phenol concentration = 5 g/L and OPP = 5 MPa). (Adapted from Mohammed, A.E. et al., *Comput. Chem. Eng.*, 94, 257–271, 2016.)

TABLE 12.12
Optimal Results of Reactor Bed Length to Diameter with Addition 5% on Lr/Dr ratio

Decision Variable Type	Simulation Results
Lr/Dr	3.428
Lr (cm)	922.879
Dr (cm)	269.1983
Capital cost ($)	1801041.69

Source: Mohammed, A.E. et al., *Comput. Chem. Eng.*, 94, 257–271, 2016.

TABLE 12.13
Optimal Operating Conditions Obtained for Industrial Catalytic Wet Air Oxidation Process

Reaction Temperature	T_R	K	**472.87**
Partial pressure of oxygen	P	atm	5.92
Liquid hourly space velocity	LHSV	h^{-1}	0.50
Gas flow rate (stoichiometric excess)	*S.E.*	—	0.20
Initial phenol Concentration	$C_{ph,0}$	$mmol/cm^3$	1.0498E-2

Source: Mohammed, A.E. et al., *Comput. Chem. Eng.*, 94, 257–271, 2016.

TABLE 12.14
Optimal Values of the Cost Parameters Utilized by Mohammed et al. (2016)

Cost Parameter	Unit	Optimal Value
Oxygen cost	\$/yr	39,781.71
Phenol cost	\$/yr	347,559.32
Catalyst cost	\$/yr	485,251
Pumping cost	\$/yr	68,476
Compression cost	\$/yr	43,363
Energy cost	\$/yr	469,944.12
Conversion cost	\$/yr	155,158,210
Conversion	—	99.79
Cost function	\$/yr	190,507,980

The process of heat integration of CWAO in order to recover most of the external energy and reduce the environmental effects (in addition to maximizing the productivity at a minimum cost) has also been investigated here. The heat integration process has been undertaken based on experimental information from pilot-scale models, mathematical modeling and commercial processes. The final optimal configuration of the industrial TBR for CWAO with the optimal operating conditions, dimensions, oil feed and phenol conversion is shown in Figure 12.15.

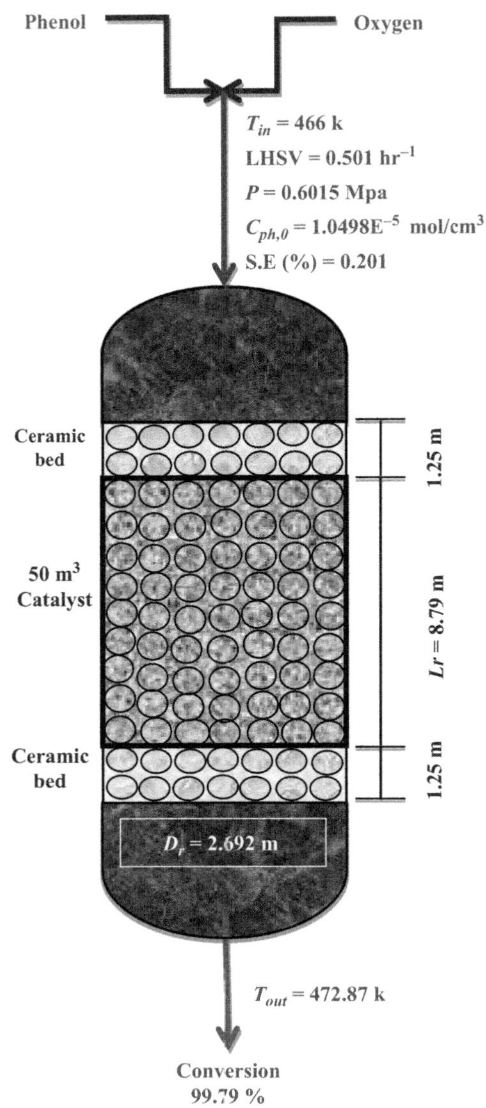

FIGURE 12.15 Industrial trickle bed reactor configuration for CWAO of phenol in wastewater with optimal conditions, operations and dimensions. (Adapted from Mohammed, A.E. et al., *Comput. Chem. Eng.*, 94, 257–271, 2016.)

REFERENCES

Aksoylu, A.E., Madalena, M., Freitas, A., Pereira, M.F.R., Figueiredo, J.L. 2001. The effects of different activated carbon supports and support modifications on the properties of Pt/AC catalysts. *Carbon*, 39, 175–185.

Akyurtlu, J.F., Akyurtlu, A., Kovenklioglu, S. 1998. Catalytic oxidation of phenol in aqueous solutions. *Catal. Today*, 40, 343–352.

Alicia, R., María, E., Julián, C., Josep, F., Frank, S., Christophe, B.A., Agustí, F., Azael, F. 2007. Biodegradability enhancement of phenolic compounds by hydrogen peroxide promoted catalytic wet air oxidation. *Catal. Today*, 124, 191–197.

Alvarez, P.M., McLurgh, D., Plucinski, P. 2002a. Copper oxide mounted on activated carbon as catalyst for wet air oxidation of aqueous phenol. 1. Kinetic and mechanistic approaches. *Ind. Eng. Chem. Res.*, 41, 2147–2152.

Alvarez, P.M., McLurgh, D., Plucinski, P. 2002b. Copper oxide mounted on activated carbon as catalyst for wet air oxidation of aqueous phenol. 2. Catalyst stability. *Ind. Eng. Chem. Res.*, 41, 2153–2158.

Athanasios, E. 2002. Catalytic wet air oxidation of phenol in a trickle bed reactor: Kinetics and reactor modeling. Ph.D thesis, Universitat Rovirai Virgili.

Atwater, J.E., Akse, J.R., McKinnis, J.A. Thompson, J.O. 1997. Low temperature aqueous phase catalytic oxidation of phenol. *Chemosphere*, 34, 203–212.

Augier, F., Koudil, A., Muszynski, L., Yanouri, Q. 2010. Numerical approach to predict wetting and catalyst efficiencies inside trickle bed reactor. *Chem. Eng. Sci.*, 65, 255–260.

Aurora, S., Pedro, Y., Sara, G., Gema, R., Felix, G.O. 2006. Reaction network and kinetic modeling of wet oxidation of phenol catalyzed by activated carbon. *Chem. Eng. Sci.*, 61, 2457–2467.

Ayude, A., Rodriguez, T., Font, J., Fortuny, A., Bengoa, C., Fabregat, A., Stüber. F. 2007. Effect of gas feed flow and gas composition modulation on activated carbon performance in phenol wet air oxidation. *Chem. Eng. Sci.*, 62, 7351–7358.

Bachir, S., Barbati, S., Ambrosio, M., Tordo, P. 2001. Kinetics and mechanism of wet-air oxidation of nuclear-fuel-chelating compounds. *Ind. Eng. Chem. Res.*, 40, 1798–1804.

Besson, M., Gallezot, P. 2003. Deactivation of metal catalysts in liquid phase organic reactions. *Catal. Today*, 81, 547–559.

Bingham, E., Trosset, R.P., Warshawsky, D. 1979. Carcinogenic potential of petroleum hydrocarbons: A critical review of the literature *J. Environ. Pathol. Toxicol.*, 3, 483–563.

Birbara, P.J., Genovese, J.E. 1995. Catalytic oxidation of aqueous organic contaminants. US Patent No. 5 362 405.

Busca, G., Berardinelli, S., Resini, C., Arrighi, L. 2008. Technologies for the removal of phenol from fluid streams: A short review of recent developments. *J. Hazard. Mater.* 160, 265–288.

Christoskova, Z., Stoyanova, M. 2000. Degradation of phenolic waste waters over Ni-oxide. *Wat. Res.*, 35, 2073–2077.

Cooney, D.O., Xi, Z. 1994. Activated carbon catalyzes reactions of phenolics during liquid-phase adsorption. *AIChE J.*, 40, 361–364.

Copa, W.M., Gitchel, W.B. 1988. *Wet Oxidation, Standard Handbook of Hazardous Waste Treatment and Disposal*, H.M. Freeman (Ed.). McGraw-Hill, New York.

Coughlin, R.W. 1969. Carbon as adsorbent and catalyst. *Ind. Eng. Chem. Prod. Res. Dev.*, 8, 12–18.

Deiber, G., Foussard, J.N., Debellefontaine, H. Removal of nitrogenous compounds by catalytic wet air oxidation. Kinetic study. 1997. *Environ. Pollut.*, 96, 311–331.

Devlin, H.R., Harris, I.J. 1984. Mechanism of the oxidation of aqueous phenol with dissolved oxygen. *Ind. Eng. Chem. Fund.*, 23, 387–392.

Donlagic, J., Levec., J. 1997. Oxidation of an azo dye in subcritical aqueous solutions. *Ind. Eng. Chem. Res.*, 36, 3480–3486.

Dudukovic, M.P., Larachi, F., Mills, P. 2002. Multiphase catalytic reactors: A perspective on current knowledge and future trends. *Cat. Revers. Sci. Eng.*, 44, 123–246.

Duprez, D., Delanoe, F., Barbier Jr. J., Isnard, P., Blanchard, G. 1996. Catalytic oxidation of organic compounds in aqueous media. *Catal. Today*, 29; 317–221

Eftaxias, A., Fonta, J., Fortuny, A., Giralt, J., Fabregat, A., Stüber, F. 2001. Kinetic modelling of catalytic wet air oxidation of phenol by simulated annealing. *Appl. Cat. Envir.*, 33, 175–190.

EPA, United State Environmental Protection Agency. 2015. (www.epa.gov/tri).

Fortuny, A., Bengoa, C., Font, J., Castells, A., Fabregat, A. 1999. Water pollution abatement by catalytic wet air oxidation in a trickle bed reactor. *Catal. Today*, 53, 107–114.

Fortuny, A., Ferrer, C., Bengoa, C., Font, J., Fabregat, A. 1995. Catalytic removal of phenol from aqueous phase using oxygen or air as oxidant. *Catal. Today*, 24, 79–83.

Fortuny, A., Font, J., Fabregat, A. 1998. Wet air oxidation of phenol using active carbon as catalyst. *Appl. Cat. Envir.*, 19, 165–173.

Gallezot, P., Lauarin, N., Isnard, P. 1999. Catalytic wet-air oxidation of carboxylic acids on carbon-supported platinum catalysts. *Appl. Catal. B*, 9, L11–L17.

Gomes, H.T., Figueiredo, J.L., Faria, J.L. 2000. Catalytic wet air oxidation of low molecular weight carboxylic acids using a carbon supported platinum catalyst. *Appl Catal B: Environ.*, 27, L217–L223.

Gonzalez, J.F., Encinar, J.M., Ramiro, A., Sabio, E. 2002. Regeneration by wet oxidation of an activated carbon saturated with p-nitrophenol. *Ind. Eng. Chem. Res.*, 41, 1344–1351.

Goto, S., Smith, J.M. 1975. Trickle-bed reactor performance: Part II reaction studies. *Appl. Chem. J.*, 21, 714–720.

Guo, J., Al-Dahhan, M. 2003. Catalytic wet oxidation of phenol by hydrogen peroxide over pillared clay catalyst. *Ind. Eng. Chem. Res.*, 42, 2450–2460.

Habtu, N.G., Stuber, F. 2011. Catalytic wet air oxidation of phenol over active carbon in fixed bed reactor: Steady state and periodic operation. PhD. Universitat Rovirai Virgili.

Hamoudi, S., Belkacemi, K., Larachi, F. 1999. Catalytic oxidation of aqueous phenolic solutions catalyst deactivation and kinetics. *Chem. Eng. Sci.*, 54, 3569.

Harf, J., Hug, A., Vogel, F., von Rohr, P.R. 1999. Scale-up of catalytic wet oxidation under moderate conditions. *Environ. Prog.*, 18, 14–20.

Hedge, S.C. 1999. Simulation for polluting prevention: Gas-liquid reactors and sulfuric acid alkylation process. MSc Thesis, Lamar University.

Hu, X., Lei, L., Chu., H.P., Yue, P.L. 1999. Copper/activated carbon as catalyst for organic wastewater treatment. *Carbon*, 37, 631–637.

Huang, T.-L., Macinnes, J.M., Cliffe, K.R. 2001. Nitrogen removal from wastewater by a catalytic oxidation method. *Water Res.*, 35, 2113–2120.

Imamura, S. 1999. Catalytic and non-catalytic wet oxidation. *Ind Eng Chem Res.*, 38, 1743–1753.

Imamura, S., Nakamura, M., Kawabata, N., Yoshida, J.-I. 1986. Wet oxidation of poly(ethylene glycol) catalyzed by manganese-cerium composite oxide. *Ind. Eng. Chem. Prod. Res. Dev.*, 25, 34–37.

Janusz, T. 2003. Noble metals supported on carbon black composites as catalysts for the wet-air oxidation of phenol. *Carbon*, 41, 1515–1523.

Joglekar, H.S., Samant, S.D., Joshi, J.B. 1991. Kinetics of wet air oxidation of phenol and substituted phenols. *Water Res.*, 25, 135–145.

Jordan, W., Van Barneveld, H., Gerlich, O., Kleine-Boymann, M., Ullrich, J. 2002. Phenol. In: *Ullmann's Encyclopedia of Industrial Chemistry*, 5th ed. VCH, Weinheim.

Keav, S., Martin, A., Barbier J., Duprez, D. 2010. Deactivation and reactivation of noble metal catalysts tested in the catalytic wet air oxidation of phenol. *Catal. Today,* 151, 143–147.

Kolaczkowski, S.T., Plucinski, P.F., Beltran, J., Rivas, F.J., McLurg, D.B. 1999. Wet air oxidation: A review of process technologies and aspects in reactor design. *Chem. Eng. J.*, 73, 143–160.

Kritzer, P., Dinjus, E. 2001. An assessment of supercritical water oxidation (SCWO): Existing problems, possible solutions and new reactor concepts. *Chem. Eng. J.*, 83, 207–214.

Lee, D.-K., Kim, D.-S. 2000. Catalytic wet air oxidation of carboxylic acids at atmospheric pressure. *Catal. Today*, 63, 249–255.

Lei, L., Hu, X., Chu, H.P., Chen, G., Yue, P.L. 1997. Catalytic wet air oxidation of dyeing and printing wastewater. *Water Sci. Technol.*, 35, 311–399.

Lei, L., Wang, D., Chin. J. 2000. Wet oxidation of PVA-containing desizing wastewater.*Chem. Eng.*, 8, 52–56.

Levec, J. 1997. Wet oxidation processes for treating industrial wastewaters. *Chem. Biochem. Eng. Q.*, 11, 47–58.

Li, L., Chen, P., Gloyna, E.F. 1991. Generalized kinetic model for wet oxidation of organic compounds. *AIChE J.*, 37, 1687.

Lin, T.M., Lee, S.S., Lai, C.S., Lin, S.D. 2006. Phenol burn. *Burns*, 32, 517–522.

Luck, F. 1996. A review of industrial catalytic wet air oxidation processes. *Catal. Today*, 27, 195–202.

Luck, F. 1999. Wet air oxidation: Past, present and future. *Catal. Today*, 53, 81–91.

Magdalena, A.P. 2004. Tailored chemical oxidation techniques for the abatement of bio-toxic organic wastewater pollutants: An experimental study. PhD diss. Universitat Rovirai Virgili, Tarragona, Spain.

Mangrulkar, P.A., Bansiwal, A.K., Rayalu, S.S. 2008. Adsorption of phenol and o-chlorophenol on surface altered fly ash based molecular sieves. *Chem. Eng. J.*, 138, 73–77.

Mantzavinos, D., Sahinzada, M., Livingston, A.G., Metcalfe, I.S., Hellgardt, K. 1999. Wastewater treatment: Wet air oxidation as a precursor to biological treatment. *Catal. Today*, 53, 93–106.

Masende, Z., Kuster, B., Ptasinski, K., Janssen, F., Katima, J. 2003. Platinum catalysed wet oxidation of phenol in a stirred slurry reactor: A practical operational window. *Appl. Cat. B: Envir.*, 41, 247–267.

Matatov-Meytal, Y.I., Sheintuch, M. 1998. Catalytic abatement of water pollutants. *Ind. Eng. Chem. Res.*, 37, 309–326.

Mederos, F.S., Ancheyta, J., Chen, J. 2009. Review on criteria to ensure ideal behaviors in trickle-bed reactors. *Appl. Cat. A: Gen.*, 355, 1–19.

Metcalf, L., Eddy, H.P. 1991. *Wastewater Engineering: Treatment, Disposal, and Reuse*, 3rd ed. McGraw-Hill, Boston, MA.

Mishra, V.S., Mahajani, V.V., Joshi, J.B. 1995. Wet air oxidation. *Ind. Eng. Chem. Res.*, 34, 2–48.

Mohammed, A.E. 2016. Optimal design of trickle bed reactor for phenol oxidation. MSc Thesis, Tikrit University.

Mohammed, A.E., Jarullah, A.T., Ghani, S.A., Mujtaba, I.M. 2016. Optimal design and operation of an industrial three phase reactor for phenol oxidation. *Comput. Chem. Eng.*, 94, 257–271.

Mohammed, W.T. 2014. Active carbon from date stones for phenol oxidation in trickle bed reactor, experimental and kinetic study. *J. Eng.*, 9, 170–173.

Mohammed, W.T., Abdullah, S.M. 2008. Kinetic study on catalytic wet air oxidation of phenol in a trickle bed reactor. *J. Eng.*, 9, 17–23.

Nathanson, J.A. 1986. *Basic Environmental Technology: Water Supply, Waste Disposal, Pollution Control.* John Wiley and Sons, New York.

Neyens, E., Baeyens, J. 2003. A review of classic Fenton's peroxidation as an advanced oxidation technique. *J. Hazard. Mater.*, 98, 33–50.

Nunoura, T., Lee, G.H., Matsumura, Y., Yamamoto, K. 2002. Modeling of supercritical water oxidation of phenol catalyzed by activated carbon. *Chem. Eng. Sci.*, 57, 3061–3071.

Ohta, H., Goto, S., Teshima, H. 1980. Liquid-phase oxidation of phenol in a rotating catalytic basket reactor. *Ind. Eng. Chem. Res.*, 19, 180–185.

Okitsu, K., Higasi, K., Nagata, Y., Dohmaru, T., Takenaka, N., Bandow, H., Maeda, Y. 1995. Decomposition of p-chlorophenol by wet oxidation in the presence of supported noble metal catalysts. *Nippon Kagati Kaishi*, 3, 208–214.

Oliviero, L., Barbier, J. Jr., Duprez, D. 2003. Wet air oxidation of nitrogen-containing organic compounds and ammonia in aqueous media. Review Article. *Appl. Catal. B*, 40, 163–184.

Pardeep, K., Gordon, H., Mehdi, N. 2010. Remediation of high phenol concentrations using chemical and biological technologies. Ph.D. Thesis. University of Saskatchewan.

Pinar, O. 2008. Catalytic wet air oxidation of mono azo dye orange II: Catalyst selection, reaction kinetics, and modeling. Ph.D. Thesis. University of Old Dominion.

Pifer, A., Hogan, T., Snedeker, B., Simpson, R., Lin, M., Shen, C., Sen, A. 1999. Broad spectrum catalytic system for the deep oxidation of toxic organics in aqueous medium using dioxygen as the oxidant. *J. Am. Chem. Soc.*, 121, 7485.

Pintar, A., Bercic, G., Levenic, J. 1997. Catalytic liquid-phase oxidation of aqueous phenol solutions in a trickle-bed reactor. *Chem. Eng. Sci.*, 36, 4143–4153.

Pintar, A., Besson, M., Gallezot, P. 2001. Catalytic wet air oxidation of Kraft bleach plant effluents in a trickle-bed reactor over a Ru/TiO_2 catalyst. *Appl. Catal. B*, 31, 275–290.

Pintar, A., Levec, J. 1992a. Catalytic liquid-phase oxidation of refractory organics in wastewater. *Chem. Eng. Sci.*, 47, 2395–2400.

Pintar, A., Levec, J. 1992b. Catalytic oxidation of organics in aqueous solutions: I. Kinetics of phenol oxidation. *J. Catal.*, 135, 345–357.

Pintar, A., Levec, J. 1994a. Catalytic liquid-phase oxidation of phenol aqueous solutions: A kinetic investigation. *Ind. Eng. Chem. Res.*, 33, 3070–3077.

Pintar, A., Levec, J. 1994b. Catalytic oxidation of aqueous p-chlorophenol and p-nitrophenol solutions. *Chem. Eng. Sci.*, 49, 4391–4407.

Pirkanniemi, K., Sillanpaa, M. 2002. Heterogeneous water phase catalysis as an environmental application: A review. *Chemosphere*, 48, 1047–1060.

Primo, O., Rivero, M.J., Ortiz, I., Irabien, A. 2007. Mathematical modelling of phenol photooxidation: Kinetics of the process toxicity. *Chem. Eng. J.*, 134, 23–28.

Qin, J., Zhang, Q., Chuang, K.T. 2001. Catalytic wet oxidation of p-chlorophenol over supported noble metal catalyst. *Appl. Catal.*, B, 29, 115–123.

Quintanilla, J.A., Casas, A.F., Mohedano, J.J., Rodríguez, J. 2006. Reaction pathway of the catalytic wet air oxidation of phenol with a Fe/activated carbon catalyst. *App. Cata. B: Env.*, 67, 206–216.

Quintanilla, J.A., Casas, A.F., Rodriguez, J. 2007. Catalytic wet air oxidation of phenol with modified activated carbons and Fe/activated carbon catalysts. *Appl. Cat. B: Envir.*, 76, 135–145.

Randall, T.L., Knopp, P.V. 1980. Detoxification of specific organic substances by wet oxidation. *J. Water Pollut. Contr. Fed.*, 52, 2117–2130.

Rao, V.G., Drinkenbur, A. 1985. A model for pressure drop in two-phase gas-liquid down flow through packed columns. *Appl. Chem. J.*, 31, 1010–1018.

Robert, R., Barbati, S., Ricq, N., Ambrosio, M. 2002. Intermediates in wet oxidation of cellulose: Identification of hydroxyl radical and characterization of hydrogen peroxide. *Water Res.*, 36, 4821–4829.

Ruether, J., Yana, C., Havduk, W. 1980. Particle mass transfer during cocurrent downward gas–liquid flow in packed beds. *Ind. Eng. Chem. Proc. Des. Dev.*, 19, 103–107.

Sadana, A., Katzer, J.R. 1974a. Involvement of free radicals in the aqueous-phase catalytic oxidation of phenol over copper oxide. *J. Catal.*, 35, 140–152.

Sadana, A., Katzer, J.R. 1974b. Catalytic oxidation of phenol in aqueous solutions over copper oxide. *Ind. Eng. Chem. Fundam.*, 13, 127–134.

Safaa, M.R. (2009). Catalytic wet air oxidation of phenolic compounds in wastewater in a trickle bed reactor at high pressure. MSc. Thesis. University of Tikrit.

Santos, A., Yustos, P., Quintanilla, A., Ruiz, G., Garcia-Ochoa, F. 2005. Study of the copper leaching in the wet oxidation of phenol with CuO-based catalysts: Causes and effects. *Appl. Cat. B: Envir.*, 61, 323–333.

Sax, N.I. and Lewis, R.J. 1992. *Dangerous Properties of Industrial materials*. New York, Van Nostrand Reinhold.

Singh, A., Pant, K.K., Nigam, K.D.P. 2004. Catalytic wet oxidation of phenol in a trickle bed reactor. *Chem. Eng. J.*, 103, 51–57.

Spellman F.R. 2004. *Handbook of Water and Wastewater Treatment Plant Operations*. Boca Raton, FL, Lowis Publisher, CRC Press.

Stüber, F., Font, J., Paradowska, A., Suarez, M.E., Bengoa, C., Fortuny, A., Fabregat, A. 2005. Chemical wet oxidation for the abatement of refractory non-biodegradable organic wastewater pollutants. *J. Ind. Chem. E.*, 83, 371–380.

Stüber, F., Polaert, I., Delmas, F., Font, J., Fortuny, A., Fabregat, A. 2001. Catalytic wet air oxidation of phenol using active carbon: Performance of discontinuous and continuous reactors. *J. Chem. Technol. Biotechnol.*, 76, 743–751.

Suarez-Ojedo, E., Stuber, M., Forntuy, A., Fabregat, A., Garrera, J., Font, J. 2005. Catalytic wet air oxidation of substituted phenols using activated carbon as catalyst. *App. Cata. B: Env.*, 58, 107–110.

Tukac, V., Hanika, J. 1995. Purification of phenolic waste waters by catalytic oxidation. *Collect. Czech.Chem. Commun.*, 60, 482–488.

Tukac, V., Hanika, J. 1998. Catalytic wet oxidation of substituted phenols in the trickle bed reactor. *J. Chem. Technol. Biotechnol.*, 71, 262–266.

Wadood, T.M., Sama, M.A. 2008. Kinetic study on catalytic wet air oxidation of phenol in a trickle bed reactor. *IJCPE*, 9, 17–23.

Willms, R.S., Balinsky, A.M., Reible, D.D., Wetzel, D.M., Harrison, D.P. 1987. Aqueous-phase oxidation: Rate enhancement studies. *Ind. Eng. Chem. Res.*, 26, 606–612.

Wu, Q., Xijun, H., Yue, J., Feng, X., Chen, H., Zhang, S. 2009. Modeling of a pilot-scale trickle bed reactor for the catalytic oxidation of phenol. *Sep. Purif. Technol.*, 67, 158–165.

Zhang, Q. and Chuang, K.T. 1998. Kinetics of wet oxidation of black liquor over a Pt-Pd-Ce/alumina catalyst. *Appl. Cat. B: Envir.*, 17, 321–332.

13 Electrolytic Treatment of Wastewater for Reuse Purposes

Case Study of the New Damietta Harbor Plants

Mahmoud Dahroug

CONTENTS

ABBREVIATIONS

HCWW	Egyptian Holding Company for Water and Wastewater
EO	Electrochemical oxidation
EOP	Electrochemical oxidation process
EFP	Electroflotation process
ECP	Electrocoagulation process
EC	Electrocoagulation
EF	Electroflotation
BOD	Biochemical oxygen demand
COD	Chemical oxygen demand
TSS	Total suspended solids
SPR	Special type of electrodes coated by $Ti/SnO_2/PdO_2/RuO_2$

13.1 INTRODUCTION

In Egypt, there is a great need for the development of cost-effective, context-appropriate and replicable small-scale sanitation systems for settlements not covered by present or future large-scale centralized schemes. By "small-scale" is meant "settlements or groups of settlements of up to 5,000 inhabitants." This need is reflected in some Egyptian projects, where solutions are currently needed for villages with a population up to 1,500 inhabitants. The development of a wide-scale replicable model is one of the ultimate goals of the Egyptian government. As such, the interface between small-scale systems and the utility (HCWW and Affiliates) and the integration of the model in its strategies is a key factor in Egyptian government's approach. In parallel, the Egyptian government intends to strengthen understanding of the particularities of sanitation in the Nile Delta villages and to develop a data baseline for planners and designers. In particular, tools and methodologies are being developed to quickly quantify and characterize the wastewater to be treated, on a site-specific basis. Those outcomes have the potential to benefit other large-scale sanitation projects and to support the National Rural Sanitation Strategy, currently under revision. One of these proposed technologies is treatment of wastewater by electrolysis.

Nowadays, various electrochemical treatments are methods used for removal of organic and inorganic impurities from fresh water, potable water and wastewater. The most usual methods are: electrocoagulation, electroflotation, electrochemical oxidation, electrochemical reduction and electrodeposition. Electrochemical technologies are a promising alternative for the treatment of wastewaters containing organic pollutants. The method consists of carrying out the oxidation reaction at the anode where pollutants are transferred into non-toxic substances by decomposition into simpler compounds or transfer into an oxidized form; it is used mostly for organic substances. Wastewater treatment by this procedure can be done through direct or indirect electrochemical oxidation (EO) [1], as shown in Figure 13.1.

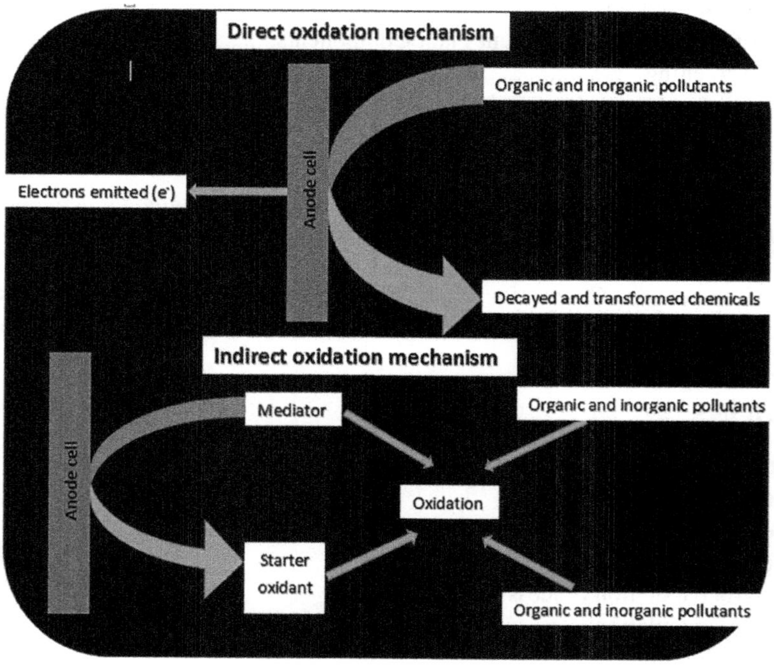

FIGURE 13.1 Pollutant removal pathway in the electrochemical oxidation process. (From Dimitrijević, S.B. and Vuković, M.D., Modern water treatment by electrochemical oxidation: A review. Mining and Metallurgy Institute Bor, Zeleni bulevar 35, 19210 Bor, Serbia. Innovation center of the Faculty of Technology and Metallurgy, Karnegijeva 4, 11000, 2013.)

EO is mainly used for degradation of aromatic compounds, pesticides, paints, industrial pollutants, pharmaceuticals waste and other organic substances. The advantage of this method is that the final products are mainly CO_2 and H_2O. This is the case for aliphatic organic matter, depending on stoichiometry. Even chlorine is not a problem, since it is converted in the form of the chloride ion. Therefore, the effluent of the wastewater treatment plant using this technology can be reused safely for many purposes [1].

13.2 THEORY OF EOP, EFP AND ECP

Toxic organic pollutants are usually destroyed by the generate oxidants such as Cl_2, ClO_2, O_3, OH, $O^{\cdot}$, ClOH and H_2O_2, that are produced from anodic oxidation during the electrolysis [2].

The absorbed hydroxyl radicals oxidize the organic matter:

$$H_2O + M \rightarrow M(^{\cdot}OH) + H^+ + e^- \tag{13.1}$$

$$R + M\,(^{\cdot}OH) \rightarrow M + RO + H^+ + e^- \tag{13.2}$$

Where RO refers to the organic matter that can be further oxidized by the hydroxyl radicals formed by electrolysis of water. M refers to the Anode metal [2].

An example of complete (electro) mineralization is the direct EO of the herbicide 2,4,5-Trichlorophenoxyacetic acid, also known as 2,4,5-T and Silvex (trade name) [3]. The whole process can be represented by the following reaction:

$$C_8H_5Cl_3O_3 + 13H_2O \rightarrow 8C_2O + 31H^+ + 3Cl^- + 28e^- \tag{13.3}$$

or by COD during electrochemical decomposition, the theoretical reaction of mineralization 2,4,5-T can be written as:

$$C_8H_5Cl_3O_3 + 7O_2 \rightarrow 8CO_2 + 3HCl + H_2O \tag{13.4}$$

The oxidation potential of the anode is directly dependent on the overpotential for oxygen separation and adsorption enthalpy of hydroxyl radicals on the anode surface.

It follows that for an anode material: the higher the overpotential for oxygen, the greater the oxidizing power of the anode material [4].

Electrocoagulation (EC) is a process that involves the generation of coagulants *in situ* from an electrode by the action of an electric current applied to these electrodes. This generation of ions is followed by electrophoretic concentration of particles around the anode. The ions are attracted by the colloidal particles, neutralizing their charge and allowing their coagulation. The hydrogen gas released from the cathode interacts with the particles causing flocculation and allowing the unwanted material to rise and be removed. The theory of EC has been discussed by several authors, and the phenomena involved can be summarized into three stages of operation:

1. Formation of a coagulating agent through the electrolytic oxidation of the sacrificial electrode, which neutralizes the surface charge, destabilizes the colloidal particles and breaks down emulsions (coagulation: EC step).
2. The particle agglutination promoted by the coagulating agent facilitates the formation and growth of flakes (flocculation: EF step).
3. Generation of micro-bubbles of oxygen (O_2) at the anode and hydrogen (H_2) at the cathode, which rise to the surface and are adsorbed when colliding with the flakes, carrying the particles and impurities in suspension to the top and thereby promoting the clarification of the effluent (flotation: electroflotation step). Figure 13.2 shows the mechanism of producing oxidants, floated and precipitates [5].

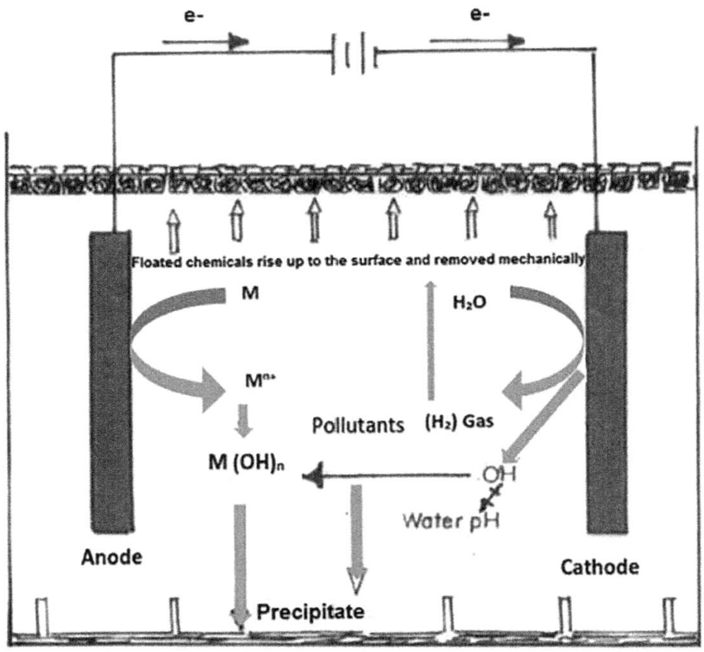

FIGURE 13.2 Mechanism of producing oxidants, floated and precipitates. (From Naumczyk, J., *Publ. Office Warsaw Univ., Techn., Res. Pap., Environment. Eng. (Warsaw)*, 37, 47–54, 2011.)

13.3 WHAT EC CAN REMOVE OR REDUCE

1. High BOD/COD, biomass and pesticides [7]
2. Bacteria, viruses [8], algae [9] and cysts
3. Fats, oils and grease (FOG)
4. Metals: Chrome 6, Nickel, Lead, Arsenic, Copper and Zinc
5. Nutrients and minerals: P, N, Ca and Mg
6. Suspended and colloidal solids
7. Hardness in drinking water
8. PCBs, textile dyes and silica

13.4 THE ADVANTAGES AND DISADVANTAGES OF TREATING WASTEWATER USING THE EC AND EF TECHNIQUES

The main advantages of these processes include environmental compatibility, versatility, energy efficiency, safety, selectivity, amenability to automation and cost effectiveness. However, the effectiveness of electrochemical approaches depends strongly on the electrode materials and cell parameters (mass transport, current density, water composition, etc.). The use of high performance anodic materials can also achieve high efficiency and lower the operating cost. Therefore, several research groups are studying the applicability of electrochemical technologies for treating real domestic and industrial effluents, with the aim of seeking a diversification of techniques and adapting the treatment to each situation as much as possible. In this context, this work presents the results concerning the application of electrochemical technologies to treat petrochemical effluents, emphasizing the new results obtained concerning the use of direct and indirect EO processes as an alternative to the pollution abatement of effluents generated by Brazilian petrochemical industries [10].

EC has a number of benefits [11]: compatibility, amenability to automation, cost effectiveness, energy efficiency, safety and versatility. Although EC received little scientific attention a decade ago, in the last couple of years this technology has been widely used for the treatment of dilute wastewaters contaminated by heavy metals [12], foodstuff [13,14], oil wastes [15,16], textile and dyes [17,18], fluorine [19], polymeric wastes [20], organic matter from landfill leachate [21], suspended particles [22,23], chemical and mechanical polishing wastes [24], aqueous suspension of ultrafine particles [25], nitrate [26], phenolic waste [27], arsenic [28] and refractory organic pollutants including lignin and EDTA [29]. EC is also applicable for drinking water treatment [30,31]. This process has the capability to overcome the disadvantages of other treatment techniques. Use of EC and EF techniques is advantageous because the plant is modular, the operation is continuous, it requires only electric power for operation, it does not require chemicals (except in a few cases in order to adjust pH), has a smaller footprint, does not create biological waste sludge, has an automated system, has no noise, has lower power consumption, the treatment is flexible and suitable for a wide range of contaminants at different levels, and the temperature and pressure required are low.

On the other hand, there are some drawbacks to using such techniques, including the probability of the formation of chlorinated organic compounds as a result of producing active chlorine ions; it is also necessary that the influent should have the conductivity properties to achieve electrolysis.

13.5 CASE STUDY: THE NORTHERN AND SOUTHERN WASTEWATER TREATMENT PLANTS LOCATED IN NEW DAMIETTA HARBOR

Egypt's Damietta port is located 8.5 km west of the Damietta branch of the Nile River westward of Ras El Bar; it is 70 km away from Port Said. The port installations extend on an area of 11.8 km^2; Figure 13.3 shows the location of the Northern and Southern treatment plants for wastewater and their discharge pathway. The port is bordered by an imaginary line connecting the eastern and western external breakwaters. In order to minimize the risk of pollutants on the environment, pilot wastewater treatment plants have been established in the northern and southern part (as shown in Figure 13.3) with a capacity of up to 500 m^3/day for each. The effluent of both treatment plants is discharged into the Mediterranean Sea. The wastewater in that area is characterized by a high content of domestic wastewater, oils and grease.

FIGURE 13.3 A map showing the location of the Northern and Southern Plants taken from Google Earth.

The Egyptian Ministry of Housing and Utilities and Urban development has been studying the applicability of installation of compact EF units in different areas in Egypt after the pilot demonstrated its success.

13.6 EF AND EC WASTEWATER TREATMENT PLANT COMPONENTS

13.6.1 Screening

Screening is the first unit operation used at wastewater treatment plants (WWTPs). Screening removes objects such as rags, paper, plastics and metals to prevent damage and clogging of downstream equipment, piping and appurtenances. Some modern wastewater treatment plants use both coarse screens and fine screens. The screen mesh of the coarse screen is 1 cm, and the mesh of fine screen is 0.5 cm.

13.6.2 Oil and Grease Removal

This unit works by aeration, the oils and grease float on the surface and are then removed by mechanical scum removal. This stage is efficient in reducing the concentration of oil and grease going to the electrolysis basins and consequently raising the electric conductivity of the treated wastewater media, which improves the electrolysis efficiency.

13.6.3 Electrochemical Basins

Electrolysis basins have two types of electrolysis basin according to purpose: an electro-oxidation basin and an EF basin, both of which are connected to a power source connected to sets of anodes and cathodes; this is illustrated in Figure 13.4. The anode and cathode are made of the same materials: they are made of titanium coated with precious metal oxides such as RuO_2, PdO_2, RuO_2 and SnO_2 to increase resistance to wastewater components and durability, in addition to decreasing the applied voltage to lower the cost of electricity. It is reported that a $Ti/SnO_2/PdO2/RuO_2$ (SPR) anode is the most effective [32].

These electrodes can, moreover, be reactivated by reversing the charge of electric current applied.

13.6.3.1 Electro-Oxidation Basin

In this basin, very active oxygen is formed, which oxidizes the dissolved matters so that many organic structures fall apart and precipitate; Figure 13.5 shows the flow direction of wastewater inside the electro-oxidation basin.

FIGURE 13.4 General view of electrolysis basin taken from the Northern treatment plant in Damietta harbor.

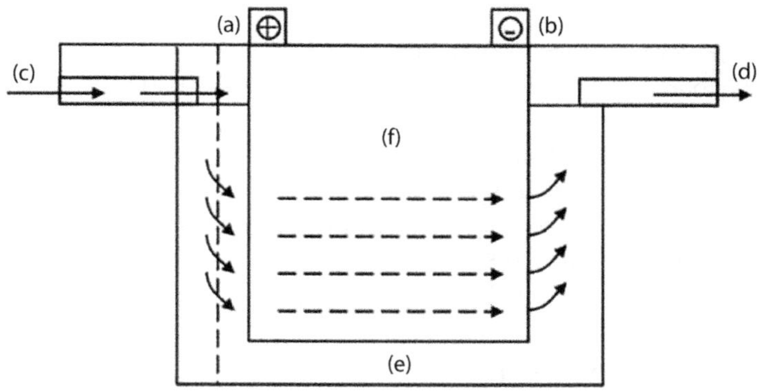

FIGURE 13.5 Electro-oxidation cell, showing the flow direction of wastewater inside the basin: (a) Anode, (b) Cathode, (c) untreated wastewater inlet, (d) treated wastewater outlet, (e) electrolysis chamber or basin and (f) Electroplate.

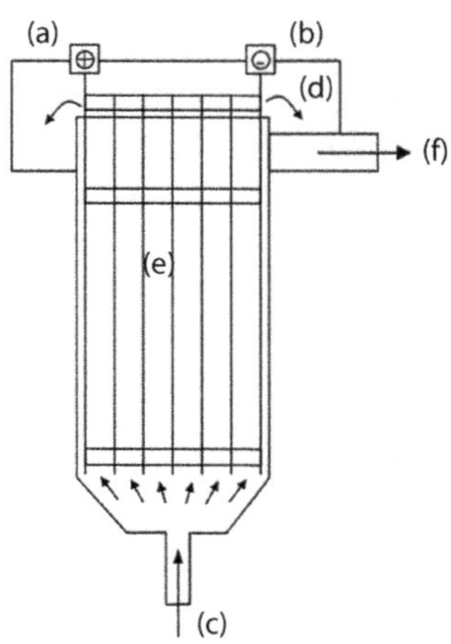

FIGURE 13.6 Electroflotation cell, showing the flow direction of wastewater inside the basin: (a) Anode, (b) Cathode, (c) untreated wastewater inlet, (d) scum removal chamber, (e) electrolytic cells and (f) treated wastewater outlet.

13.6.3.2 EF Basin

In this basin, hydrogen and oxygen gas bubbles are formed and suspended particles adsorbed and rise to the surface; Figure 13.6 shows the flow direction of wastewater inside the EF basin.

13.7 OPERATIONAL COST OF ONE M³ TREATED WASTEWATER

The operational cost relies on three main elements: chemicals, electricity and workers.

1. **The chemicals used**
   Chemicals are used to raise the pH to 9; the only chemical used throughout treatment is limewater, which costs around 0.03 L.E per $1m^3$.
2. **The electricity consumed per m³**
   Total energy consumed by cells = 5 K watt/h.
   Energy consumed by cells per $m^3 = 2.5 \text{ K watt/h}/10 \text{ m}^3\text{/h} = 2500 \text{ watt}/10 \text{ m}^3 = 250 \text{ W/m}^3$ 13.1
   Total energy consumed by pumps which operated 5 mins/hr. = 3 K watt/ hr.
   Energy consumed by pumps per $m^3 = 3000\text{watt/hr.}/22\text{hr.} \times 25\text{m}^3\text{/hr.} = 5.5 \text{ Watt/m}^3$ 13.2

→ Total consumption = 250 + 5.5 = 255.5 Watt/m^3. 13.3
Total cost = 255.5 watt × 46.5kwatt/hr./1000 = 11.88 L.E. 13.4
Therefore, the total cost of 500 m^{3}per day = 11.88 × 500 = 5940 L.E 13.5

3. **Workers**
   The required workers include one engineer, one technician and two unskilled workers. The total salary per one treated m^3 is almost 1.8 L.E.
   By adding the results of all the costs together, the total operation cost per one cubic meter treated is shown in equation 13.6.
   The total cost per m^3 = 0.03 + 11.88 + 1.8 = 13.71 L.E 13.6

13.8 SAMPLING AND ANALYSIS

Optimum operation conditions depend on many factors and could be adjusted throughout based on a set of samples taken from the influent during the operational period; by adjusting the pH, conductivity, flow rate, temperature and electric current applied, the voltammeter and retention time till optimum results can be reached. The optimum results are judged through the analysis of BOD_5, COD, TSS, N (total nitrogen), P (total phosphorous), oil and grease. The plant can be automated according to doses lifted from the plant. Figure 13.7 shows the three samples taken from influent (1), during electrolysis (2) and effluent (3) points; it is obvious from the figure the clarity of the effluent sample. Table 13.1 shows the results of effluent samples taken from the Northern and Southern plants. These results were obtained after several operational experiments that took place over the course of about 3 months.

The effluent of treated wastewater is discharged into the sea, so it has to comply with law number (4) from the 1994 Egyptian environmental protection laws, Annex (1). When comparing the results of parameters measured, the effluent of both wastewater treatment plants meets the limits specified in the law. It is obvious from Table 13.2 that all the results are below the maximum limits and, a comparison is shown in Table 13.3 with the results of the influent and effluent from a conventional wastewater treatment plant at Gabal Ataka in Suez, Egypt, based on a cluster sample taken on 26 November 2017.

FIGURE 13.7 Samples of influent (a), during electrolysis (b) and effluent (c).

TABLE 13.1
The Results of Effluent Samples Taken from the Northern and Southern Plants

WWPT		Northern Plant			Southern Plant		
Parameters	**Unit**	**Influent**	**Effluent**	**Removal %**	**Influent**	**Effluent**	**Removal %**
BOD_5	mg O_2/l	1500	14	99.1	24000	13	99.93
COD	mg O_2/l	4400	32	99.3	50000	38	99.92
TSS	mg /l	1800	16.2	99.1	21000	20	99.9
Total Nitrogen as N	mg /l	168	12	92.9	325	14	95.69
Total Phosphorous as P	mg /l	2.5	N.D.	100	6.5	0	100
Oil and Grease	mg /l	4600	5	99.9	12200	5	99.96

TABLE 13.2
Comparison of Results of Effluent Samples (Taken from the Northern and Southern Plants) to the Limits Stated in Law 4 for 1994

Parameters	Unit	Law 4 for 1994	Northern Plant Effluent	Southern Plant Effluent
BOD_5	mg O_2/L	60	14	13
COD	mg O_2/L	100	32	38
TSS	mg/L	60	16.2	20
Total Nitrogen as N	mg/L	Not applicable	12	14
Total Phosphorous as P	mg/L	5	N.D.	0
Oil and Grease	mg/L	15	5	5

TABLE 13.3
The Results of Influent and Effluent Sample of Suez Wastewater Treatment Plant (Gabal Ataka) Taken on November 26, 2017

Parameters	Unit	Influent	Effluent	Law 4 for 1994
BOD_5	mg O_2/L	128	38	60
COD	mg O_2/L	208	66	100
TSS	mg/L	198	108	60
Total Nitrogen as N	mg/L	56	28	Not applicable
Total Phosphorous as P	mg/L	2.4	0.53	5
Oil and Grease	mg/L	16.6	3.1	15

13.9 COMPARISON BETWEEN CONVENTIONAL AND ELECTROCHEMICAL TREATMENT OF WASTEWATER

Table 13.4 shows the advantages of electrochemical treatment technology over conventional treatment systems.

TABLE 13.4
Comparison between Conventional and Electrochemical Treatment of Wastewater

Comparison Item	Electrochemical Treatment Technology	Conventional Treatment Technology
Chemicals	Chemical-free in most cases	Feed chemicals lime, alum.
Operation	Can be operated as and when needed	Needs continuous operation
Biological treatment	Non biological process	Microbiological process
Disinfection efficiency	Gives pathogen-free output	Does not give disinfected water in most cases
Aeration	No need for aeration	Needs aeration
MLSS (Mixed Liquors Suspended Solids)	No need to maintain MLSS	Needs to maintain MLSS in aeration tanks
Reuse	Treated water can be recycled for many applications	Treated water cannot be used for most applications
Shock load	Shock load can be easily tackled	Shock load cannot be tackled.
Toxicity effect	Independent of toxicity	Toxicity destroys MLSS, restoration takes time.
Noise	Noiseless	Noisy because of aerators & its drives.
Sludge	Less sludge with better dewatering	Generates more sludge with lower dewatering
Plant size	Compact plant with small footprint	Large plant with larger footprint
Construction	Modular in construction	Fixed construction
Expandability	Easily expandable	Difficult to expand
Operation	Easy to operate	Difficult to operate
Moving parts	Less moving parts	More than 75% is moving parts
Maintenance	Easy to maintain the plant	Difficult to maintain the plant
Downtime	Low plant downtime	High plant downtime
Cost ratio	High performance to cost ratio	Low performance to cost ratio

REFERENCES

1. S.B. Dimitrijević, and M.D. Vuković, Modern water treatment by electrochemical oxidation: A review. Mining and Metallurgy Institute Bor, Zeleni bulevar 35, 19210 Bor, Serbia. Innovation center of the Faculty of Technology and Metallurgy, Karnegijeva 4, 11000, 2013.
2. B. Ramesh Babu, K.M. Seeni Meera, and V. Perumal, Removal of pesticides from wastewater by electrochemical methods: A comparative approach. *Sustainable Environment Research,* 21(3), 2011, 401–406.
3. B. Birame, E. Brillas, B. Marselli, P. Michaud, C. Comninellis, and D. Marième. Electrochemical decomtamination of waters by advanced oxidation processes (AOPS): Case of the mineralization of 2,4,5-T on BDD electrode bull. *Bull. Chem. Soc. Ethiop.,* 18(2), 2004, 205–214.
4. E. Brillas, I. Sirés, and M.A. Oturan, Electro-Fenton process and related electrochemical technologies based on Fentons reaction chemistry. *Chem. Rev.,* 109(12), 2009, 6570–6631.
5. A.A. Cerqueira, and M.R. da Costa Marques, Electrolytic treatment of wastewater in the oil industry, New technologies in the oil and gas industry, Dr. Jorge Salgado Gomes (Ed.), InTech, 2012. doi:10.5772/50712.
6. J. Naumczyk, Electro-oxidation of some impurities in its application to tannery wastewater, *Publ. Office Warsaw Univ., Techn., Res. Pap., Environment Eng. (Warsaw),* 37(3), 2011, 47–54.
7. L.S. Calvo et al., An electrocoagulation unit for the purification of soluble oil wastes of high COD. *Environmental Progress,* 22(1), 2003, 57–65.
8. B. Ramesh Babu et al., Removal of pesticides from wastewater by electrochemical methods: A comparative approach. *Sustainable Environment Research,* 21(3), 2011, 401–406.

9. John Wiley & Sons, Inc. Purification of wastewater by electrolysis. *Biotechnology and Bioengineering*, Vol. XXIII, 1981, 1881–1887.
10. K. Tumsri, and O. Chavalparit. Optimizing Electrocoagulation-electroflotation Process for Algae Removal. 2nd International Conference on Environmental Science and Technology IPCBEE Vol. 6 (2011), 452–456.
11. C.A. Martnez-Huitle, D.C. de Moura, and D.R. da Silva, Applicability of electrochemical oxidation process to the treatment of petrochemical effluents. *Chemical Engineering Reaction*, 41, 2014.
12. M.Y.A. Mollah, P. Morkovsky, J.A.G. Gomes, M. Kesmez, J. Parga, and D.L. Cocke, *J. Hazard. Mater.,* 114, 2004, 199.
13. M. Bier, in: M. Bier (Ed.), *Electrophoresis: Theory, Methods and Applications.* Academic Press, New York, 1959, p. 270.
14. N.F. Gray, *Water Technology: An Introduction for Scientists and Engineers.* Arnold, London, 1999.
15. E.C. Beck, A.P. Giannini, and E.R. Ramirez, *Food Technol.,* 22 (1974) 18/19.
16. J. Lawrence, and L. Knieper, *Ind. Wastewater,* 1–2, 2000, 20.
17. N. Biswas, and G. Lazarescu, *Int. J. Environ. Stud.*, 38, 1991, 65.
18. P. Stamberger, *J. Colloid Interface*, 1, 1946, 93.
19. Y. Xiong, P.J. Strunk, H. Xia, X. Zhu, and H.T. Karlsson, *Water Res.*, 35, 2001, 4226.
20. N. Mameri, A.R. Yeddou, H. Lounici, D. Belhocine, H. Grib, and B. Bariou, *Water Res.*, 32, 1998, 1604.
21. M. Panizza, C. Bocca, and G. Cerisola, *Water Res.*, 34, 2000, 2601.
22. T. Tsai, S.T. Lin, Y.C. Shue, and P.L. Su, *Water Res.*, 31, 1997, 3073.
23. J.C. Donnini, J. Kan, T.A. Hassan, and K.L. Kar, *Can. J. Chem. Eng.*, 72, 1994, 667.
24. J. Szynkarczuk, J. Kan, T.A. Hassan, and J.C. Donnini, *Clays Clay Miner.*, 42, 1994, 667.
25. M. Belongia, P.D. Harworth, J.C. Baygents, and S. Raghavan, *J. Electrochem. Soc.*, 146, 1999, 4124.
26. M.J. Matteson, R.L. Dobson, R.W. Glenn Jr., N.S. Kukunoor, H. Waits III, and E.R. Clayfield, *Colloid Surface, A*, 104, 1995, 101.
27. S. Koparal, and U.B. Ogutveren, *J. Hazard. Mater,* 89, 2002, 83.
28. W. Phutdhawong, S. Chowwanapoonpohn, and D. Buddhasukh, *Anal. Sci.*, 16, 2000, 1083.
29. N. Balasubramanian, and K. Madhavan, *Chem. Eng. Technol.*, 24, 2001, 855.
30. L.C. Chiang, J.E. Chang, and S.C. Tseng, *Water Sci. Technol.*, 36, 997, 123.
31. M.F. Pouet, and A. Grasmick, *Water Sci. Technol.*, 31, 1995, 275.
32. G. Chen, *Sep. Purif. Technol.*, 38, 2004, 11.

14 Inactivation of Waterborne Pathogens in Municipal Wastewater Using Ozone

Achisa C. Mecha, Maurice S. Onyango, Aoyi Ochieng, and Maggy N.B. Momba

CONTENTS

14.1 INTRODUCTION

14.1.1 HISTORY AND CHEMISTRY OF OZONE

Ozone was discovered by Carl Friedrich Schönbein as a gas generated during the electrolysis of water in 1840. However, its structure, O_3, as an allotrope of oxygen, was established much later by J. L. Soret in 1865. A detailed account of the early development and use of ozone is provided by Rubin (2001). Ozone has been used commercially since 1893 when the first full-scale drinking water treatment plant was implemented at Oudshoorn, Holland (Evans 1975). Ozone is a colorless gas at room temperature with a pungent smell; it is moderately soluble in water and is very reactive. The properties of ozone are shown in Table 14.1. It is usually generated on-site from dry air or pure oxygen through a high-voltage corona discharge process. Ozone is considered a health hazard at a concentration of 0.25 mg/L in air (by volume) with concentrations beyond 1 mg/L being extremely hazardous. Its corrosiveness means that only resistant materials like stainless steel and Teflon can be used to handle it (Scholz 2016). Upon dissolving in water, ozone undergoes complex self-decomposition and oxidation reactions because of its molecular electronic configuration, which makes it highly unstable (Gardoni et al. 2012, Zhou and Smith 2002).

TABLE 14.1
Physical Properties of Ozone

Property	Unit	Value
Molar mass	g mol^{-1}	48
Density at 1013 mbar, 0°C	kg m^{-3}	2.14
Solubility in water at 0°C	g L^{-1}	1.05
Redox potential	V	2.07
Boiling point	°C	−112
Melting point	°C	−193
Viscosity	Pa.s	0.0042
Odor threshold	ppm	0.02

Source: Sievers, M., 4.13: Advanced oxidation processes A2—Wilderer, Peter, in *Treatise on Water Science*, Elsevier, Oxford, UK, pp. 377–408, 2011; Gottschalk, C. et al., *Ozonation of Water and Wastewater—A Practical Guide to Understanding Ozone and its Applications,* 2nd ed., Wiley-VCH, Weinheim, Germany, 2010.

14.1.2 Applications of Ozone

Ozone has found numerous applications, including in the purification of drinking water, wastewater treatment, bleaching of pulp and kaolin, manufacture of semiconductors, air pollution control (odor abatement) and chemical synthesis (Sciences 1987). All of these applications are based on the reactivity and strong oxidizing properties of ozone. This high reactivity of ozone is derived from its inherent instability resulting from the molecule's eagerness to accept an electron, thus reducing ozone to O_2 and leading to the oxidization of the electron donor. Ozone has a comparatively high redox of 2.07 V (Table 14.2). Consequently, the use of ozone in wastewater treatment has seen wide application compared to all other oxidation processes. This is because of its effectiveness in: (1) control of waterborne pathogens; (2) oxidation of organic (phenols, detergents, pesticides, etc.) and inorganic (iron and manganese) compounds; (3) removal of taste and odor, algal toxins and color; and (4) particle (turbidity) reduction by oxidative flocculation (Gray 2014, Gottschalk et al. 2010). This chapter focuses on the application of ozone in wastewater treatment as a disinfectant.

TABLE 14.2
Standard Oxidation Potentials of Common Oxidants

Oxidant	Chemical Formula	Oxidation Potential (V)
Fluorine	F	3.03
Hydroxyl radical	•OH	2.80
Superoxide ion	$O_2^{\bullet -}$	2.42
Ozone	O_3	2.07
Hydrogen peroxide	H_2O_2	1.77
Chlorine dioxide	ClO_2	1.57
Hypochlorous acid	HOCl	1.49
Chlorine	Cl_2	1.36

Source: Pelaez, M. et al., *Appl. Catal. B: Environ.*, 125, 331–349, 2012.

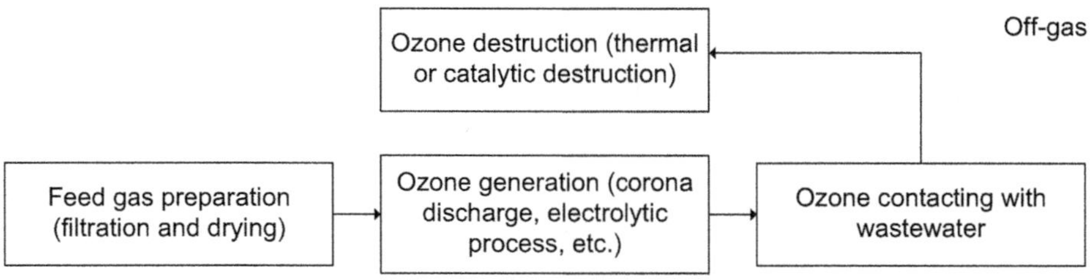

FIGURE 14.1 Schematic diagram of ozonation process.

14.2 THE PROCESS OF OZONATION

Ozone disinfection and the oxidation of organics is a four-step process (summarized in Figure 14.1):

1. *The preparation of air or the supply of oxygen*: Irrespective of the oxygen source, the gas needs to be cleaned and dried prior to use for ozone generation. The pre-treatment of the gas, especially when air is used, includes air compression, filtration and drying as well as regulation before entering the ozone generator.
2. *The generation of ozone using an ozone generator*: Ozone is formed when oxygen is broken down into radicals that react with oxygen to form ozone. This process requires high energy input and therefore high voltage. Plasma corona discharge, ultraviolet radiation and electrolytic processes are the common techniques for ozone generation. Among these, corona discharge is the most widely used process, especially for the large-scale production of ozone for water and wastewater treatment. This process involves the passage of oxygen or air between electrodes through which a high voltage potential is maintained. Ozone is therefore formed by the combination of ionized oxygen atoms and molecular oxygen. It is important that the process gas be sufficiently dried before being applied to the corona discharge (Langlais et al. 1991).
3. *Ozone transfer into the water to be treated*: Ozone is brought into contact with water using different methods, such as via bubble diffusers, injector dissolution or turbine mixers. Bubble diffuser systems are widely employed due to their relative simplicity and high effectiveness in terms of ozone transfer rates. Ozone transfer efficiencies between 70% and 95% have been reported (Blatchley et al. 2012, Mecha et al. 2016).
4. *The destruction of waste ozone in an ozone trap*: Since ozone is very toxic and because not all of the generated ozone is transferred to the water, residual ozone must be destroyed to prevent its accumulation at harmful levels in working environments. The destruction of off-gas is accomplished using heat and/or catalysts.

14.3 MECHANISM OF OZONE OXIDATION

Ozone and ·OH radicals are strong chemical oxidants, and both are involved in the destruction of microorganisms as well as the oxidation of a wide range of organic and inorganic compounds (Gray 2014). The degradation of pollutants in water using ozonation occurs through two mechanisms. The first is direct oxidation by molecular ozone, which occurs at low pH and involves highly selective reactions, such as electrophilic, nucleophilic or dipolar addition reactions with low reaction rates (Hoigne 1998). The other is an indirect mechanism, which occurs at basic pH through the ozone decomposition to produce ·OH radicals, which are non-selective and highly reactive. The formation

TABLE 14.3
Ozone Decomposition to Produce ·OH Radicals

Reaction	Rate Constant (k)	Equation Number
Initiation steps:		
$O_3 + OH^- \rightarrow O_2^{\bullet} + HO_2^{\bullet}$	$70\,M^{-1}s^{-1}$	(14.1)
$HO_2^{\bullet} \leftrightarrow O_2^{\bullet-} + H^+$	$1.6\times10^{-5}\,M^{-1}s^{-1}$	(14.2)
Propagation steps:		
$O_3 + O_2^{\bullet-} \rightarrow O_3^{\bullet-} + O_2$	$1.6\times10^{9}\,M^{-1}s^{-1}$	(14.3)
$O_3^{\bullet-} + H^+ \leftrightarrow HO_3^{\bullet}$	$5\times10^{10}\,M^{-1}s^{-1}$	(14.4)
$HO_3^{\bullet} \rightarrow OH^{\bullet} + O_2$	$1.4\times10^{5}\,M^{-1}s^{-1}$	(14.5)
Termination step:		
$OH^{\bullet} + HO_2^{\bullet} \rightarrow O_2 + H_2O$	$7.5\times10^{9}\,M^{-1}s^{-1}$	(14.6)

Source: Acero, J. L., and U. von Gunten., *Ozone Sci. Eng.*, 22, 305–328, 2000.

of ·OH radicals occurs in a number of reaction steps (Table 14.3). The first step of this process is the decomposition of ozone by hydroxide ions (Equation 14.1) to form superoxide radicals ($O_2^{\bullet-}$) and hydroperoxyl radicals ($HO_2^{\bullet}$). The formed hydroperoxyl radical is in an equilibrium state (Equation 14.2). The superoxide anion radical and ozone then react to form ozonide anion radical (Equation 14.3), which then immediately decomposes into oxygen and an ·OH radical (Equation 14.4) via hydrogen trioxide ($HO_3^{\bullet}$) (Gottschalk et al. 2010). The ·OH radical reacts with ozone to form oxygen and a hydroperoxyl radical (Equation 14.5) thus completing the chain reaction and starting it anew. Overall, three ozone molecules theoretically produce two ·OH radicals. Finally, the reaction is terminated (Equation 14.6).

14.4 FACTORS INFLUENCING OZONATION

The effectiveness of ozone treatment depends on factors such as ozone concentration, exposure (contact time), concentrations of pathogens and organic matter, as well as the ozone reactor. The ozone reactor should be designed such that: the ozone transfer efficiency is maximized, the system is free of leaks during operation and ozone resistant materials are employed for construction (Gonçalves and Gagnon 2011). Furthermore, mixing in ozone reactors depends on three variables: water flow, ozone gas flow and ozone contactor configuration. Consequently, high water and ozone flows, and an appropriate contactor configuration that minimizes flow short-circuiting and back mixing will result in better mixing (Schulz and Bellamy 2000). In addition, the following water quality parameters also affect the efficacy of ozone disinfection:

1. *Temperature:* Generally, as the temperature increases, the disinfecting power of ozone is enhanced. Whereas increased temperature reduces the solubility and stability of ozone, it does, however, increase the rates of diffusion of ozone through the cell wall of microorganisms and consequently the rates of reaction with the microorganism (Langlais et al. 1991).
2. *Solution pH*: Changes in water pH change the balance of available O_3 and OH radicals with an increase in pH from 6 to 9 resulting in a 40-fold decrease in ozone. Nevertheless, molecular ozone is principally responsible for the inactivation of microorganisms within most operational ozonation pH ranges (Elovitz and von Gunten 1999). In batch systems, increasing pH results in higher inactivation, but in systems with continuous ozone supply, there is limited effect of pH.

3. *Water matrix*: The presence of constituents other than the target microorganisms may increase ozone demand and reduce the disinfection capacity. Furthermore, turbidity indicates the presence of particulates that can shield microorganisms from the disinfectants. Particulate matter can also exert a disinfection demand, thus reducing the proportion of ozone available to inactivate the microbial target (Avery et al. 2013).

14.5 APPLICATION OF OZONE IN WASTEWATER TREATMENT

Municipal wastewater effluents contain a variety of pathogenic microorganisms that pose significant health risks to humans. Disinfection is therefore required for the inactivation of these pathogens to avert the spread of waterborne diseases to water users and the environment (Macauley et al. 2006). During the application of ozonation in wastewater treatment, direct oxidation by ozone and •OH radicals can take place depending on the nature of pollutants, solution pH and ozone dosage. However, the oxidation potential of •OH radicals is much higher and more efficient than that of ozone (Wu et al. 2008). Therefore, the mutual relevance of direct and indirect reactions depends on the wastewater matrix. If wastewater does not contain compounds that initiate the radical chain reaction (initiators such as OH^- ions), or if it contains many compounds that terminate the chain reaction very quickly (scavengers or inhibitors), then slower direct ozonation will be the prevalent mechanism.

By contrast, if the wastewater contains chain reaction promoters (such as superoxide radicals), the formation of •OH radicals will readily occur and a faster indirect reaction will be the prevalent mechanism (Gardoni et al. 2012). In terms of treatment of municipal wastewater, further benefits of ozonation include an increase in the dissolved oxygen levels, a decrease in the chemical oxygen demand (COD), turbidity and color (Singh et al. 2015). Ozone as a disinfectant is a strong germicide against bacteria, viruses and protozoa because of its high oxidizing capacity (Gray 2014, Silva et al. 2010). Its oxidizing power is superior to that of chlorine, chlorine dioxide and chloramines for the inactivation of different waterborne pathogens and therefore requires shorter contact times (typically four to five times less than chlorine) for antimicrobial action (Long et al. 1999, Zuma et al. 2009). Cysts of *Cryptosporidium parvum* and *Giardia lamblia* are known to be resistant to chlorine; however, they can be inactivated by ozone (Makky et al. 2011). Furthermore, ozone leaves no chemical residue and breaks down into molecular oxygen upon reaction or degradation (Priyanka et al. 2014).

There is considerable variability among studies on ozone inactivation of microorganisms, which makes a direct comparison of performance difficult. This is mainly due to the differences of experimental aspects such as: (1) the scale of the study (laboratory vs pilot vs full scale); (2) the water matrix studied (synthetic or real water sources); (3) the ozone supply (continuous vs discontinuous); and (4) the manner in which the values of ozone supply and utilization are reported (applied vs transferred and residual ozone doses). Nevertheless, the overall performance can be evaluated based on the log reduction of microorganisms as summarized in the case studies shown in Table 14.4.

14.6 MECHANISM OF OZONE DISINFECTION

Microbial inactivation by ozone is a complex process that predominantly involves the direct reaction with molecular ozone to inactivate microorganisms such as bacteria, viruses, fungi, yeast and protozoa. The process involves attack on the cell membrane and cell content constituents such as enzymes and nucleic acids, leading to the oxidation of sulfhydryl groups and amino acids of the cell constituents to short-chain peptides. Moreover, ozone also oxidizes polyunsaturated fatty acids to acid peroxides and also destroys viral ribonucleic acid (RNA) (Kim et al. 1999, Victorin 1992). Thus the microorganism is killed by the disruption of the cell envelope through oxidation of phospholipids and lipoproteins leading to leakage of cell contents (Pascual et al. 2007).

TABLE 14.4
Summary of Studies on the Inactivation of Microbiological Contaminants Using Ozonation

Contaminants	Water Matrix	Microbial Concentration	Treatment	Results	References
Total coliform	Anaerobic sanitary wastewater		5–10 mg/L for 5–15 min	2.00–4.06 log reduction	Silva et al. (2010)
Escherichia coli				2.41–4.65 log reduction	
Total coliform	Red-meat-processing wastewater	4600–7000 CFU/cm^3	23.09 mg/min/L for 8 min	99%	Wu and Doan (2005)
Eschericia coli		290–340 CFU/dm^3			
Bacillus Subtilis	Domestic well drinking water	4500 CFU/L	7–13 mg/min for 10 min	96%	Demir and Atguden (2016)
Escherichia coli		4000 CFU/L		98%	
Streptococcus faecalis		5000 CFU/L		98%	
Shigella sonnei	Lettuce wash water	100,000,000 CFU/mL	1.6–2.2 mg/L for 1 min	3.7–5.6 log reduction	Selma et al. (2007)
Total coliform	Secondary treated effluent	1,600,000 MPN/100 mL	30 mg/L	99.16	Verma et al. (2016)
Coliform bacteria	Municipal wastewater	860 MPN/mL	6.1 mg/L	>99%	Lee et al. (2008)
Total coliform	Secondary treated effluent	254,000 MPN/100 mL	10 mg/L for 5 min	>99%	Tripathi et al. (2011)
Fecal coliform		209,000 MPN/100 mL			
Total coliform	Municipal wastewater	28,000,000 MPN/100 mL	15.5 mg/L	99%	Zamudio-Pérez et al. (2014)
Fecal coliforms		8,480,000 MPN/100 mL		99%	
Helminth eggs		470 L^{-1}		99%	
Streptococcus faecalis	Synthetic wastewater	10,000–1,000,000 CFU/mL	0.3–2.5 mg/L for 10 min	90%	Lezcano et al. (2001)
Staphylococcus aureus					
Candida albicans					
Fecal coliforms, *Salmonella*, *Streptococcus faecalis* and *E. coli*	Anaerobically treated effluent		300 mg/L for 30 min	99%	Yasar et al. (2007)
Fecal coliforms	Municipal wastewater	52,000 MPN/100 mL	51.5 mg/L for 14 min	100%	Velásquez et al. (2008)
Salmonella typhi		68,000 MPN/100 mL			
Vibrio cholerae		3,500 MPN/100 mL			
Acanthamoeba sp.		23 MPN/100 mL			
Cryptosporidium parvum		1,250,000 oocysts		99%	Rennecker et al. (1999)
Giardia ssp	Municipal wastewater		11 mg/L for 15 min	75%	Passos et al. (2014)
			21 mg/L for 15 min	100%	
Bacillus subtilis	Swine wastewater	100,000,000 CFU/mL	100 mg/L	3.3–3.9 log reduction	Macauley et al. (2006)
Bacillus licheniformis					

Furthermore, ozone has good diffusion properties, so it can rapidly diffuse through the biological cell membranes, penetrate the microorganism and cause inactivation (Cullen et al. 2010, Makky et al. 2011). It has been reported that ozone at a low concentration of 0.1 mg/L, can inactivate bacterial cells, including their spores; ozone can inhibit cell growth in fungi and also in viruses by damaging the viral capsid thereby interrupting the reproductive cycle (Tiwari et al. 2017). On the other hand, the formation of hydroxyl radicals during ozonation at basic pH may enhance disinfection, since OH radicals have a greater oxidation potential (2.80 V) than ozone (2.07 V). Furthermore, OH radicals react unselectively with the cell membrane and other cell constituents, which may sometimes be resistant to ozone (Cho and Yoon 2006).

14.7 MODELING OZONE INACTIVATION KINETICS

The following models are commonly used to describe disinfection kinetics (Cho et al. 2003, Rennecker et al. 1999):

1. *Chick–Watson model*

$$In\frac{N}{N_o} = -kCT \tag{14.7}$$

where:
- C is the concentration of the disinfectant (mg/L)
- T is the exposure time
- k is the kinetic constant
- N and N_o are the final and initial microbial concentrations, respectively

The Chick–Watson model describes the inactivation of microorganisms based on first-order kinetics. It is based on the assumptions of: (1) the existence of a homogenous microbial population (all microbes are identical); and (2) "single-hit" inactivation, that is, one hit is sufficient for inactivation.

2. *Modified Chick–Watson model:*

$$In\frac{N}{N_o} = -\frac{kC_o^n}{nk'}\left[1-\exp\left(-nk't\right)\right] \tag{14.8}$$

where:
- C_o is the concentration of disinfectant in the water
- k' is the first-order ozone decay constant (sec^{-1})
- n is a coefficient of dilution, which is a function of disinfectant and pH of the medium (the value of n is usually close to unity)
- t is the exposure time.

3. *Modified Hom model*

$$In\frac{N}{N_o} = -\left(\frac{m}{nk'}\right)^m kC_o^n\left[1-\exp\left(\frac{-nk't}{m}\right)\right]^m \tag{14.9}$$

where m and n are model parameters for the modified Hom model. This model introduced an empirical coefficient (m) in the Modified Chick–Watson model. It reduces to the Modified Chick–Watson model for $m = 1$, produces shoulder and tail regions for: $m > 1$ and $m < 1$, respectively.

4. *Delayed Chick–Watson model*

$$In\frac{N}{N_o} = 0 \ if \ \bar{C}t \le \bar{C}T_{lag} = \frac{1}{k}In\frac{N}{N_o}$$

or

$$In\frac{N}{N_o} = -k\bar{C}t - k\bar{C}T_{lag} \ if \ \bar{C}t \ge \bar{C}T_{lag} = \frac{1}{k}In\frac{N}{N_o} \quad (14.10)$$

where:

N_o is the initial population (CFU/mL)
N is the remaining population at time t
C is the ozone concentration (mg/L)
$\bar{C} = \frac{\int_0^t Cdt}{t}$ is the time averaged ozone concentration (mg/L)
k is the inactivation rate constant with ozone (min^{-1}).

14.8 OVERALL QUALITY ASPECTS OF OZONE-TREATED WASTEWATER

The foregoing has suggested that ozonation has great potential in wastewater disinfection. However, it is essential to evaluate other aspects of wastewater treatment.

1. *Toxicity assessment*: Toxicity studies reported in the literature have generally shown that the toxicity of wastewater is reduced during ozonation for different wastewaters, such as: 2,4-dichlorophenol-containing wastewater (Van Aken et al. 2015); sewage effluent (Stalter et al. 2010); and water contaminated with phenolic compounds (Adams et al. 1997, Van Aken et al. 2015). All these studies imply that ozone degradation of wastewater pollutants results in treated water with reduced toxicity that is safer to discharge to the environment compared to the parent contaminants. However, despite this, there is need for a post-treatment process to remove the oxidation by-products before reusing the water (Stalter et al. 2010).
2. *Evaluation of organic matter content*: Ozonation also leads to the reduction of other global parameters such as COD, aromaticity (UV_{254}), biological oxygen demand (BOD) and the total organic carbon (TOC) as shown in Table 14.5. The TOC is the primary surrogate parameter for the measurement of the quantity of natural organic matter (NOM) in water supplies and UV absorbance at 254 nm (UV_{254}) is used to describe the type and character of NOM. A reduction in UV_{254} shows that ozone readily reacts with aromatics (Mecha et al. 2016, Penru et al. 2013).

TABLE 14.5
Effect of Ozonation on Global Wastewater Parameters

Water Type	Parameter	Removal Efficiency (%)	References
Secondary effluent	DOC	39	Crousier et al. (2016)
Secondary effluent	COD	57	Bataller et al. (2005)
	BOD	85	
Municipal wastewater	COD	75	Zamudio-Pérez et al. (2014)
	Turbidity	85	
Sea water	$SUVA_{254}$	30	Penru et al. (2013)
	TOC	10	
Secondary effluent	DOC	24.7	Mecha et al. (2016)

14.9 LIMITATIONS OF OZONE OXIDATION AND DISINFECTION

1. *Low mineralization of organic matter*: An important limitation of ozone is that ozonation of refractory organic compounds leads to their oxidation, but only minimal mineralization is achieved (Bashiri and Rafiee 2014). Low mineralization of contaminants results from the formation of ozone-recalcitrant intermediate compounds (Chin and Berube 2005). To overcome this limitation, ozonation is often combined with other processes such as heterogeneous catalysis and photocatalysis to increase the production of •OH radicals for mineralization processes (Agustina et al. 2005). Mecha and co-workers, for instance, recently showed that a combination of ozone and photocatalysis results in synergy, leading to faster and more efficient mineralization of organic compounds (Mecha et al. 2016).
2. *Microbial regrowth*: The low solubility of ozone in water greatly decreases its disinfection capacity. Furthermore, when used for disinfection purposes, regrowth of the microorganisms cannot be prevented after ozonation because of the difficulty in maintaining residual ozone. This may pose a health risk to water consumers, especially if ozonation alone is used for the treatment of wastewater intended for reuse or consumption purposes (Demir and Atguden 2016). A secondary disinfectant such as chlorine is therefore usually added if ozone is used as the primary disinfectant. In this case, the dosage of chlorine is reduced, which has the beneficial implications of decreasing the formation of halogenated disinfection by-products.
3. *High capital costs*: Comparing chlorination, chloramination, chlorine dioxide and ozonation based on capital, operation and maintenance costs, chlorination is the cheapest while chloramination is the most expensive. Ozone and chlorine dioxide processes are comparable in cost and fall between the other two disinfectants. Furthermore, ozone must be produced on-site since it is unstable and cannot be stored, and it requires more technical skills to use. In addition, ozone-resistant materials such as stainless steel, glass and Teflon should be used because ozone is very reactive and corrosive (Mecha et al. 2016).
4. *Production of disinfection by-products (DBPs)*: Chemical disinfectants such as chlorination, chloramination and chlorine dioxide produce DBPs of varying nature. Ozone also produces oxyhalides (bromate, chlorate, chlorite), aldehydes (formaldehyde and acetaldehydes) and carboxylic acids (butanoic acid, pentanoic acid and succinic acid) in treated water (Avery et al. 2013). Toxicological data on these DBPs shows that they are mutagenic and carcinogenic, although they are not legally regulated in most countries (Siddiqui et al. 1997). Furthermore, in the presence of bromide, bromate and bromine-substituted DBPs may be formed when ozone dosages above the instantaneous ozone demand are applied (Wert et al. 2007). This is the reason why a post-filtration system is required, such as the use of an activated carbon filter to remove DBPs. Siddiqui and co-workers showed that a combination of ozonation and biofiltration resulted in up to a 70%–80% reduction of the formation potentials of trihalomethanes, haloacetic acids and chloral hydrate (Siddiqui et al. 1997).

14.10 CONCLUSIONS AND FUTURE PERSPECTIVES

Ozonation is indeed a well-established process for the oxidation and inactivation of pollutants in wastewater. It is important to seek ways to address the limitations of this process. One of the strategies receiving wide attention is the coupling of ozone with other processes, such as ultrasound, UV, hydrogen peroxide (Priyanka et al. 2014) or even with advanced oxidation processes such as photocatalysis (Mecha et al. 2017) to improve the reaction kinetics and efficacy of ozone. The integration of ozone and other processes can significantly enhance ozonation performance resulting from synergistic effects, especially as a result of the production of hydroxyl radicals, which are powerful and indiscriminative oxidants. On the other hand, understanding the effect of water quality parameters on ozonation is fundamental to its effective use for the inactivation of microorganisms. Ozone can also be used as a primary disinfectant in combination with chlorine as a secondary disinfectant where detection of residual contamination is required. A further possibility is the coupling of biological and filtration processes with ozone treatment and disinfection for the removal of biodegradable matter.

REFERENCES

Acero, J. L., and U. von Gunten. 2000. Influence of carbonate on the ozone/hydrogen peroxide based advanced oxidation process for drinking water treatment. *Ozone Science & Engineering* 22 (3):305–328. doi:10.1080/01919510008547213.

Adams, Craig D., Randall A. Cozzens, and Byung J. Kim. 1997. Effects of ozonation on the biodegradability of substituted phenols. *Water Research* 31 (10):2655–2663. doi:10.1016/S0043-1354(97)00114-0.

Agustina, T. E., H. M. Ang, and V. K. Vareek. 2005. A review of synergistic effect of photocatalysis and ozonation on wastewater treatment. *Journal of Photochemistry and Photobiology C: Photochemistry Reviews* 6 (4):264–273. doi:10.1016/j.jphotochemrev.2005.12.003.

Avery, L., P. Jarvis, and J. Macadam. 2013. *Review of Literature to Determine the Uses for Ozone in the Treatment of Water and Wastewater.* Centre of Expertise for Waters, James Hutton Institute, Craigiebuckler, Aberdeen, AB15 8QH, Scotland, UK.

Bashiri, H., and M. Rafiee. 2014. Kinetic Monte Carlo simulation of 2,4,6-thrichloro phenol ozonation in the presence of ZnO nanocatalyst. *Journal of Saudi Chemical Society.* doi:10.1016/j.jscs.2014.11.001.

Bataller, M., E. Veliz, L. A. Fernandez, C. Hernandez, I. Fernandez, C. Alvarez, and E. Sanchez. 2005. Secondary effluent treatment with ozone. In *IOA 17th World Ozone Congress.* Strasbourg, France.

Blatchley, E. R., S. Weng, M. Z. Afifi, H.-H. Chiu, D. B. Reichlin, S. Jousset, and R. S. Erhardt. 2012. Ozone and UV254 radiation for municipal wastewater disinfection. *Water Environment Research* 84 (11):2017–2027.

Chin, A., and P. R. Berube. 2005. Removal of disinfection by-product precursors with ozone-UV advanced oxidation process. *Water Research* 39:2136–2144.

Cho, M., and J. Yoon. 2006. Enhanced bactericidal effect of O_3/H_2O_2 followed by Cl_2. *Ozone: Science & Engineering* 28:335–340. doi:10.1080/01919510600900316.

Cho, M., H. Chung, and J. Yoon. 2003. Disinfection of water containing natural organic matter by using ozone-initiated radical reactions. *Applied and Environmental Microbiology* 69 (4):2284–2291. doi:10.1128/AEM.69.4.2284-2291.2003.

Crousier, C., J.-S. Pic, J. Albet, S. Baig, and M. Roustan. 2016. Urban wastewater treatment by catalytic ozonation. *Ozone: Science & Engineering* 38 (1):3–13. doi:10.1080/01919512.2015.1113119.

Cullen, P. J., B. K. Valdramidis, B. K. Tiwari, S. Patil, P. Bourke, and C. P. O'Donnell. 2010. Ozone processing for food preservation: An overview on fruit juice treatments. *Ozone: Science & Technology* 32:166–179.

Demir, F., and A. Atguden. 2016. Experimental investigation on the microbial inactivation of domestic well drinking water using ozone under different treatment conditions. *Ozone: Science & Engineering* 38 (1):25–35. doi:10.1080/01919512.2015.1074534.

Elovitz, M. S., and U. von Gunten. 1999. Hydroxyl radical/ozone ratios during ozonation processes. I. The Rct concept. *Ozone: Science & Engineering* 21 (3):239–260. doi:10.1080/01919519908547239.

Evans, F. L. (Ed.) 1975. Ozone technology: Current status. *Ozone in Water and Wastewater Treatment.* Ann Arbor, MI: Ann Arbor Science Publishers.

Gardoni, D., A. Vailati, and R. Canziani. 2012. Decay of ozone in water: A review. *Ozone: Science & Engineering: The Journal of the International Ozone Association* 34 (4):233–242. doi:10.1080/01919512.2012.686354.

Gonçalves, A. A., and G. A. Gagnon. 2011. Ozone application in recirculating aquaculture system: An overview. *Ozone Science & Engineering* 33 (5):345–367. doi:10.1080/01919512.2011.604595.

Gottschalk, C., J. A. Libra, and A. Saupe. 2010. *Ozonation of Water and Wastewater—A Practical Guide to Understanding Ozone and its Applications.* 2nd ed. Weinheim, Germany: Wiley-VCH.

Gray, N. F. 2014. Ozone disinfection. In *Microbiology of Waterborne Diseases* (Second Ed.), Steven L. P., Marylynn V. Y., David W. W., Rachel M. C. and Nicholas F. G. (Eds.), pp. 599–615. London, UK: Academic Press.

Hoigne, J. 1998. Chemistry of aqueous ozone, and transformation of pollutants by ozonation and advanced oxidation processes. In *The Handbook of Environmental Chemistry Quality and Treatment of Drinking Water*, Jiri H. (Ed.), pp. 83–141. Berlin, Germany: Springer.

Kim, J. G., A. E. Yousef, and S. Dave. 1999. Application of ozone for enhancing the microbiological safety and quality of foods: A review. *Journal of Food Protection* 62 (9):1071–1087.

Langlais, B., D. A. Reckhow, and D. R. Brink, (Eds.) 1991. *Ozone in Water Treatment: Application and Engineering*. Chelsea, MI: Lewis Publishers.

Lee, B. H., W. C. Song, B. Manna, and J. K. Ha. 2008. Dissolved ozone flotation (DOF)—A promising technology in municipal wastewater treatment. *Desalination* 225 (1):260–273. doi:10.1016/j.desal.2007.07.011.

Lezcano, I., R. P. Rey, M. S. Gutiérrez, C. Baluja, and E. Sánchez. 2001. Ozone inactivation of microorganisms in water: Gram positive bacteria and yeast. *Ozone Science & Engineering* 23 (2):183–187. doi:10.1080/01919510108962001.

Long, B. W., R. A. Hulsey, and R. C. Hoehn. 1999. Complementary uses of chlorine dioxide and ozone for drinking water treatment. *Ozone: Science & Engineering: The Journal of the International Ozone* 21:465–476.

Macauley, J. J., Z. Qiang, C. D. Adams, R. Surampalli, and M. R. Mormile. 2006. Disinfection of swine wastewater using chlorine, ultraviolet light and ozone. *Water Research* 40 (10):2017–2026. doi:10.1016/j.watres.2006.03.021.

Makky, E. A., G.-S. Park, I.-W. Choi, S.-I. Cho, and H. Kim. 2011. Comparison of Fe(VI) (FeO_4^{2-}) and ozone in inactivating Bacillus subtilis spores. *Chemosphere* 83 (9):1228–1233. doi:10.1016/j.chemosphere.2011.03.030.

Mecha, A. C., M. S. Onyango, A. Ochieng, and M. N. B. Momba. 2016. Impact of ozonation in removing organic micro-pollutants in primary and secondary municipal wastewater: Effect of process parameters. *Water Science and Technology* 74 (3):756–765. doi:10.2166/wst.2016.276.

Mecha, A. C., M. S. Onyango, A. Ochieng, C. J. S. Fourie, and M. N. B. Momba. 2016. Synergistic effect of UV–vis and solar photocatalytic ozonation on the degradation of phenol in municipal wastewater: A comparative study. *Journal of Catalysis* 341:116–125. doi:10.1016/j.jcat.2016.06.015.

Mecha, A. C., M. S. Onyango, A. Ochieng, and M. N. B. Momba. 2017. Evaluation of synergy and bacterial regrowth in photocatalytic ozonation disinfection of municipal wastewater. *Science of the Total Environment* 601–602:626–635. doi:10.1016/j.scitotenv.2017.05.204.

Pascual, A., I. Llorca, and A. Canut. 2007. Use of ozone in food industries for reducing the environmental impact of cleaning and disinfection activities. *Trends in Food Science & Technology* 18 (Supplement 1):S29–S35. doi:10.1016/j.tifs.2006.10.006.

Passos, T. M., L. H. Moreira da Silva, L. M. Moreira, R. A. Zângaro, R. da Silva Santos, F. B. Fernandes, C. José de Lima, and A. B. Fernandes. 2014. Comparative analysis of ozone and ultrasound effect on the elimination of giardia spp. Cysts from wastewater. *Ozone Science & Engineering* 36 (2):138–143. doi:10.1080/01919512.2013.864227.

Pelaez, M., N. T. Nolan, S. C. Pillai, M. K. Seery, P. Falaras, A. G. Kontos, P. S. M. Dunlop et al. 2012. A review on the visible light active titanium dioxide photocatalysts for environmental applications. *Applied Catalysis B: Environmental* 125 (0):331–349. doi:10.1016/j.apcatb.2012.05.036.

Penru, Y., A. R. Guastalli, S. Esplugas, and S. Baig. 2013. Disinfection of seawater: Application of UV and ozone. *Ozone: Science & Engineering: The Journal of the International Ozone Association* 35 (1):63–70. doi: 10.1080/01919512.2012.722050.

Priyanka, B. S., N. K. Rastogi, and B. K. Tiwari. 2014. Chapter 19: Opportunities and challenges in the application of ozone in food processing In *Emerging Technologies for Food Processing*, Da-Wen S. (Eds.), pp. 335–358. San Diego, CA: Academic Press.

Rennecker, J. L., B. J. Mariñas, J. H. Owens, and E. W. Rice. 1999. Inactivation of Cryptosporidium parvum oocysts with ozone. *Water Research* 33 (11):2481–2488. doi:10.1016/S0043-1354(99)00116-5.

Rubin, M. B. 2001. The history of ozone. The Schönbein period 1839–1868. *Bulletin for the History of Chemistry* 26 (1):40–56.

Scholz, M. 2016. Chapter 19: Disinfection. In *Wetlands for Water Pollution Control* (Second Ed.), pp. 129–136. Elsevier, Amsterdam, the Netherlands.

Schulz, C. R., and W. D. Bellamy. 2000. The role of mixing in ozone dissolution systems. *Ozone Science & Engineering* 22 (4):329–350. doi:10.1080/01919510009408779.

Sciences, National Academy of. 1987. *Drinking Water and Health, Volume 7 Disinfectants and Disinfectant By-Products: Disinfectants and Disinfectant By-Products*. Vol. 7. Washington, DC: National Academy Press.

Selma, M. V., D. Beltrán, A. Allende, E. Chacón-Vera, and M. Isabel Gil. 2007. Elimination by ozone of Shigella sonnei in shredded lettuce and water. *Food Microbiology* 24 (5):492–499. doi:10.1016/j.fm.2006.09.005.

Siddiqui, M. S., G. L. Amy, and B. D. Murphy. 1997. Ozone enhanced removal of natural organic matter from drinking water sources. *Water Research* 31 (12):3098–3106. doi:10.1016/S0043-1354(97)00130-9.

Sievers, M. 2011. 4.13: Advanced oxidation processes A2—Wilderer, Peter. In *Treatise on Water Science*, pp. 377–408. Oxford, UK: Elsevier.

Silva, G. H. R., L. A. Daniel, H. Bruning, and W. H. Rulkens. 2010. Anaerobic effluent disinfection using ozone: Byproducts formation. *Bioresource Technology* 101 (18):6981–6986. doi:10.1016/j.biortech.2010.04.022.

Singh, S., R. Seth, S. Tabe, and P. Yang. 2015. Oxidation of emerging contaminants during pilot-scale ozonation of secondary treated municipal effluent. *Ozone: Science & Engineering* 37 (4):323–329. doi:10.1080/01919512.2014.998755.

Stalter, D., A. Magdeburg, M. Weil, T. Knacker, and J. Oehlmann. 2010. Toxication or detoxication? In vivo toxicity assessment of ozonation as advanced wastewater treatment with the rainbow trout. *Water Research* 44 (2):439–448. doi:10.1016/j.watres.2009.07.025.

Tiwari, S., A. Avinash, S. Katiyar, A. Aarthi Iyer, and S. Jain. 2017. Dental applications of ozone therapy: A review of literature. *The Saudi Journal for Dental Research* 8 (1–2):105–111. doi:10.1016/j.sjdr.2016.06.005.

Tripathi, S., D. M. Tripathi, and B. D. Tripathi. 2011. Removal of organic content and color from secondary treated wastewater in reference with toxic potential of ozone during ozonation. *Hydrology Current Research* 2 (1):1–6. doi:10.4172/2157-7587.1000111.

Van Aken, P., R. Van den Broeck, J. Degrève, and R. Dewil. 2015. The effect of ozonation on the toxicity and biodegradability of 2,4-dichlorophenol-containing wastewater. *Chemical Engineering Journal* 280:728–736. doi:10.1016/j.cej.2015.06.019.

Orta de Velásquez, M. T., M. N. Rojas-Valencia, and A. Ayala. 2008. Wastewater disinfection using ozone to remove free-living, highly pathogenic bacteria and amoebae. *Ozone: Science & Engineering* 30 (5):367–375. doi:10.1080/01919510802333738.

Verma, K., D. Gupta, and A. B. Gupta. 2016. Optimization of ozone disinfection and its effect on trihalomethanes. *Journal of Environmental Chemical Engineering* 4 (3):3021–3032. doi:10.1016/j.jece.2016.06.017.

Victorin, K. 1992. Review of the genotoxicity of ozone. *Mutation Research* 277:221–238.

Wert, E. C., F. L. Rosario-Ortiz, D. D. Drury, and S. A. Snyder. 2007. Formation of oxidation byproducts from ozonation of wastewater. *Water Research* 41 (7):1481–1490. doi:10.1016/j.watres.2007.01.020.

Wu, C.-H., C.-Y. Kuo, and C.-L. Chang. 2008. Homogeneous catalytic ozonation of C.I. Reactive Red 2 by metallic ions in a bubble column reactor. *Journal of Hazardous Materials* 154 (1–3):748–755. doi:10.1016/j.jhazmat.2007.10.087.

Wu, J., and H. Doan. 2005. Disinfection of recycled red-meat-processing wastewater by ozone. *Journal of Chemical Technology & Biotechnology* 80 (7):828–833. doi:10.1002/jctb.1324.

Yasar, A., N. Ahmad, H. Latif, and A. Amanat Ali Khan. 2007. Pathogen re-growth in UASB effluent disinfected by UV, O_3, H_2O_2, and advanced oxidation processes. *Ozone: Science & Engineering* 29 (6):485–492. doi:10.1080/01919510701617710.

Zamudio-Pérez, E., L. Gilberto Torres, and I. Chairez. 2014. Two-stage optimization of coliforms, helminth eggs, and organic matter removals from municipal wastewater by ozonation based on the response surface method. *Ozone Science & Engineering* 36 (6):570–581. doi:10.1080/01919512.2014.905194.

Zhou, H., and D. W. Smith. 2002. Advanced technologies in water and wastewater treatment. *Journal of Environmental Engineering & Science* 1:247–264. doi:10.1139/S02-020.

Zuma, F., J. Lin, and S. B. Jonnalagadda. 2009. Ozone initiated disinfection kinetics of Escherichia coli in water. *Journal of Environmental Science and Health. Part A: Toxic. Hazardous Substances and Environmental Engineering* 44 (1):48–56. doi:10.1080/10934520802515335.

15 Photocatalytic Oxidation of Non-Acid Oxygenated Hydrocarbons

Application in GTL Process Water Treatment

Renju Zacharia, Muftah H. El-Naas, and Mohammed J. Al-Marri

CONTENTS

15.1 INTRODUCTION

Gas-to-liquid (GTL) technology is a blanket term describing catalytic chemical processes that convert low-molecular weight hydrocarbons to long-chain hydrocarbon liquid fuels, which are then hydrocracked and fractionated to produce a wide variety of high-value products, such as motor oils, synthetic gasoline, naphtha, wax, motor lubrication base oil, kerosene, and jet fuel. The concept of the direct catalytic conversion of coal to liquid fuel was invented in Germany in 1913 as part of an effort to overcome the country's lack of oil reserves and to support industrial needs [1]. The Fischer–Tropsch synthesis (FTS), which supports most of today's commercial GTL facilities, was developed in 1925 [2]. This synthesis involves the reaction of a mixture of hydrogen and carbon monoxide (which is known as syngas) at an elevated temperature and pressure in the presence of a suitable transition metal catalyst. The drive for the production of synthetic liquid fuels from gaseous hydrocarbons peaked in Germany prior to World War II to meet the need for aviation fuel. The earliest commercial plant producing liquid hydrocarbons was commissioned in 1955 by Sasol in Sasolburg, South Africa. In a strict sense, the plant was not a GTL plant, as coal was used as the initial feedstock for production. In 2004, the plant began to use natural gas (NG) for its feedstock. The first commercial plant that used NG as the feedstock for FTS was commissioned in 1991 by PetroSA in the town of Mossel Bay, South Africa. Over the course of the following two decades, more complex and larger-scale GTL facilities have been established in Bintulu, Malaysia, by Royal Dutch Shell and in Ras Laffan Industrial City, Qatar, by QP/Sasol Partnership. Qatar's Pearl GTL plant, which was created in 2010 in a partnership between QP and Shell, is currently the world's largest GTL facility, with a full production capacity of 140,000 barrels per day (bpd) of petroleum liquids and 120,000 bpd of associated liquids. Large-scale GTL plants are capital intensive to build, and nearly twice as expensive as conventional oil refineries in energy-equivalent terms. The economic viability of large-scale GTL plants, which depends on a variety of factors, including the gas–oil price ratio, capital and operational costs, and the value and composition of the final products, is an important concern for final-investment decisions. Because of the high value of converted hydrocarbon liquids, GTL technology is highly regarded as a hedge against the volatility of the conventional NG market. Liquid fuels are easier to transport, as they required only standard bulk-liquid cargo vessels, not the cryogenic vessels used for liquified transport. Furthermore, the synthetic transportation fuels from GTL plants contain almost no sulfur or aromatic content, which makes them more environmentally friendly than conventional diesel and gasoline.

Over the last decade, several small- and medium-scale GTL players, such as Velocys, GasTechno, Greyrock Energy, Linc Energy CompactGTL, and Infra technology, have emerged [3]. They focus on developing modular GTLs, and they have capacities ranging from 100 to 15,000 bpd. These modular plants cost around $250 million to build, roughly 100 times lower than the cost of a large-scale facility. For geographical regions with stranded oil and gas reserves that are remote from the existing NG pipeline infrastructure, the utilization and monetization of gas are rather difficult. Modular GTL plants can easily be deployed in these reserves to improve the economics of the existing plants. Several mini-GTLs are in different phases of constructions in Austria, Australia, Brazil, the US, Kazakhstan, Russia, Algeria, China, Uzbekistan, and Turkmenistan. Timeline of various GTL facilities and the corresponding production capacities are shown in Figure 15.1.

Due to the very nature of the chemistry of the FTS, large quantities of water are produced during GTL processes. The process water from GTL plants contains dissolved hydrocarbons, owing to which it cannot be directly used within the plant and cannot be discharged to the environment. Non-acid oxygenated (NAO) hydrocarbons, such as aldehydes, ketones, alcohols, esters, and ethers are the principal contaminants in GTL process water. This chapter provides an overview of water generation in GTL process, current technologies of processes of water separation and treatment, and the application of TiO_2-mediated photocatalytic removal of NAO hydrocarbons. The interaction of TiO_2 with different families of NAOs is discussed here from a fundamental to unravel the mechanism of TiO_2-mediated reactions and photocatalysis.

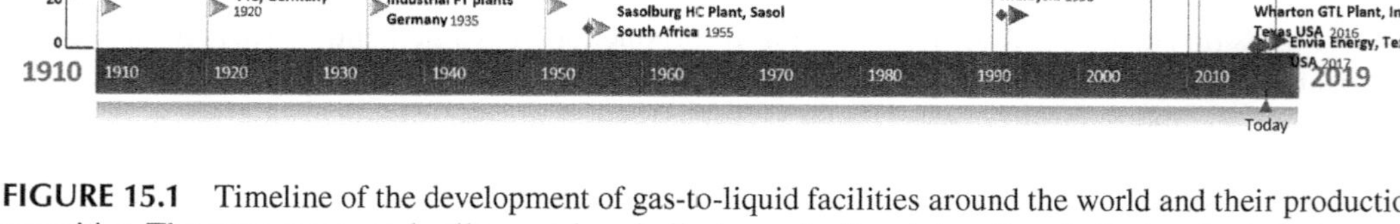

FIGURE 15.1 Timeline of the development of gas-to-liquid facilities around the world and their production capacities. The grey, green, and yellow markers indicate production status. Grey indicates plants that are not currently in production, green indicates plants are in production, and yellow and orange indicate plants in the construction or commissioning phase. Red indicates production capacity in 1000s of barrels per day. Data are collected from various web sources.

15.2 PRODUCTION OF WATER IN GTL PROCESSES

A simplified process diagram for liquid hydrocarbon production in GTL plants is given in Figure 15.2. The sweetened NG is first partially oxidized (methane reformation) to a mixture of CO and H_2. The H_2:CO ratio is adjusted to ensure the reaction is selective of the desired products. A ratio of 2:1 for H_2:CO yields alkenes and alcohols, while a ratio of 1.5:1 yields alkanes and C_1-C_2 alkanes are formed at ratios of 2.5–3:1 [4]. The mixture of gas reacts in an FT reactor at elevated temperature and pressure. The liquid hydrocarbons and water produced in the FT reactor are separated downstream of the reactor. Distillation, cracking, and fractionating columns are usually employed to separate and fine-tune the synthetic oils, fuels and lubricants. For most commercial GTL plants, FTS is employed in the synthesis of liquid hydrocarbons. One notable exception is the Shell's proprietary two-stage middle distillate synthesis, used in its Bintulu plant in Malaysia.

Examination of the fundamental reaction mechanisms of FTS has provided a good understanding of the quantity of water produced in GTL plants. One of the earliest mechanisms proposed involves the formation of chemisorbed enol species that reacts with adsorbed hydrogen, as shown in Figure 15.3 [5,6].

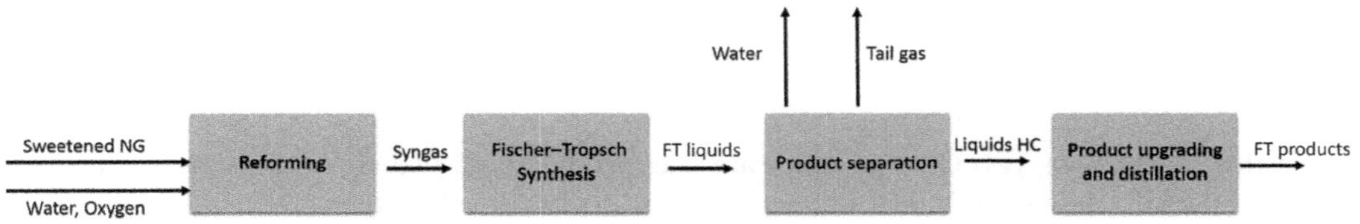

FIGURE 15.2 Generic block diagram of the gas-to-liquid process. Sweetened natural gas is initially reformed to produce syngas. The ratio of hydrogen to CO is adjusted to improve the selectivity and fed into the FT reactor. The liquid hydrocarbons and paraffin from the FT reactor are separated from the by-product. During downstream processing, the hydrocarbons are cracked and fractionated to yield FT products.

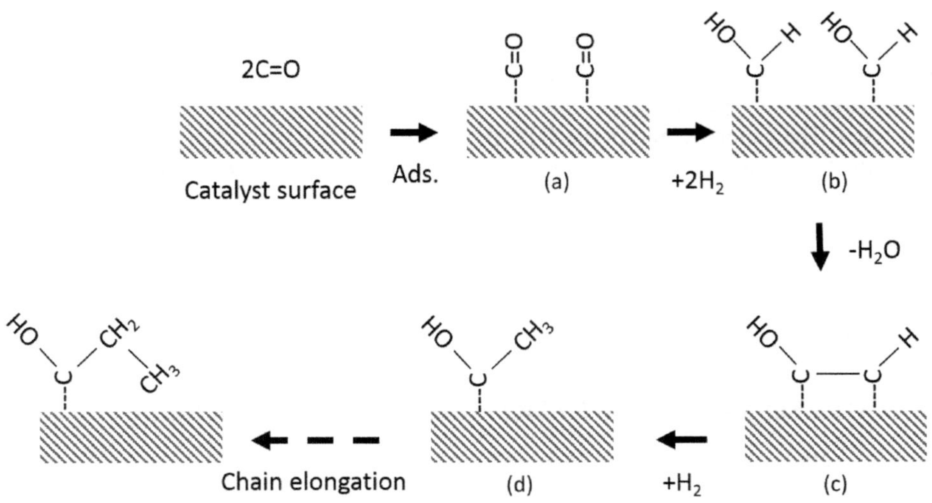

FIGURE 15.3 A proposed reaction mechanism of Fischer–Tropsch synthesis based on the Langmuir–Hinshelwood mechanism. (a) Adsorption of two CO molecules on vicinal catalytic active centers is followed by (b) hydrogenation of carbonyls, which (c) eliminates a water molecule to establish a C–C bond. (d) A second hydrogen molecule then is inserted and the reaction proceeds through steps with additional adsorbed CO to form long chain hydrocarbons. Through this mechanism, for each C–C bond formed, a molecule of water is eliminated. (From Davis, B.H., *Fuel Process. Technol.*, 71, 147, 2001.)

$$C{=}O + S \longleftrightarrow C{=}O\text{---}S$$

$$H_2 + S \longleftrightarrow 2H\text{---}S$$

$$C{=}O\text{---}S + H\text{---}S \longrightarrow HCO\text{---}S + S$$

$$HCO\text{---}S + H\text{---}S \longleftrightarrow C\text{---}S + H_2O\text{---}S$$

$$C\text{---}S + H\text{---}S \longleftrightarrow CH\text{--}S + S$$

$$CH\text{--}S + H\text{---}S \longleftrightarrow CH_2\text{--}S + S$$

SCHEME 15.1 FTS mechanism based on carbide (C–S) formation. Note that water is produced in the reaction in step 4. This mechanism is valid if the effect of adsorbed water on the FT kinetics is neglected. The water produced in the reaction is subsequently desorbed from the catalyst surface. (From Bhatelia, T. et al., *Chem. Eng. Trans.*, 25, 707, 2011.)

In this mechanism, two CO molecules are chemisorbed onto two adjacent catalytic sites, and they react with hydrogen molecules to form two enol structures, which then eliminate a molecule of water to form a C–C bond. Then, the new surface species successively eliminates a hydrogen molecule and reacts with another chemisorbed CO, resulting in chain elongation. This mechanism suggests that a molecule of water is produced for each C that is added to the chain. Another proposed mechanism for an FTS reaction is the carbide mechanism, given in Scheme 15.1, which suggests the elimination of two water molecules for each $-CH_2-$ added [7].

These mechanisms are consistent with the fact that nearly equal amounts of water and hydrocarbon products are produced during FTS. It is estimated that nearly 1.2 t of water is produced for 1 t of liquid hydrocarbon. Thus, water is considered a product of FTS and not just a by-product. For world-scale facilities such as the Pearl GTL plant in Qatar, which produces 260,000 barrels of synthetic hydrocarbons per day, the volume of GTL water produced amounts to hundreds of thousands of liters. The water created in the GTL process is largely acidic, due to the dissolved organic acids and alcohols in it and has a high chemical oxygen demand (COD), of up to 27 t/d [8]. On the other hand, GTL facilities, like many NG plants, require water for cooling and boiler feeds. Studies indicate that GTL and conventional NG plants consume among the most water in the petrochemical industry, with an average consumption of *ca.* 65 gal/MMBtu [9]. Water-treatment facilities are therefore an integral part of any GTL plant. The water-treatment plant of the Pearl GTL plant is designed to reuse 450,000 m^3 of spent process water every day.

15.3 WATER SEPARATION AND TREATMENT IN THE GTL PROCESS

As the FT reaction proceeds, liquid hydrocarbons and water slurry containing dissolved hydrocarbons and suspended catalyst particles are collected in the FT reactor. The effluent containing hydrocarbons and water is cooled to 137°C to condense as much water and liquid hydrocarbons as possible [10]. The stream is then passed through several three-phase separators to separate hydrocarbon vapor (tail gas), liquid (hydrocarbon condensate in the C_6–C_{20} range) and water. Some of the water that escapes with the tail gas is recovered through pressure reduction and cooling to 1.6°C in two successive three-phase separators [10]. The effluent water contains dissolved oxygenated hydrocarbons, including aliphatic alcohols, acids, aldehydes, and ketones. The aliphatic alcohols are typically removed through distillation or stripping columns [11] and are valorized using a saturator to be used as feedstock. The remaining water from the stripping/distillation column contains large quantities of carboxylic acids and other oxygenates. In Eni S.p.A.'s GTL water-treatment plant, acidic water is treated with an anaerobic digester and membrane reactors to yield reusable process water and sludge [12]. Pearl GTL's zero-liquid-discharge facility combines several technologies, including reverse osmosis and ultrafiltration to produce three streams of water: irrigation water, cooling water, and raw water [13]. The sludge from water–water treatment is evaporated and recrystallized. Sasol and GE introduced an anaerobic membrane bioreactor (AnMBR) technology at their Sasolburg GTL plant to convert the NAOs in the wastewater to methane [14]. While traditional aerobic treatment converts organics to CO_2, the AnMBR process converts them into methane-rich biogas, which can be used to generate power within the plant. In addition, AnMBR produces 80% less solid wastes than aerobic processes.

The use of a microbial membrane reactor for degrading mixtures of oxygenates is somewhat limited. While lower alcohols (C_1–C_5) are easily digested by anaerobic treatment, long-chain oxygenates with C_6–C_{10} carbons reduce microbial activity. This has been reported by Carlsen et al., who observed the adverse effects of long-chain aliphatic alcohols on the respiration and fermentation of baker's yeast [15]. Likewise, Dionisi et al. found that both acidogenic and methanogenic bacteria are inhibited by the presence of long-chain alcohols [16]. Moreover, GTL process water often contains leached out metal catalysts, which are toxic to microbes. As effluent water is acidic or alkaline and detrimental to microbial activity, it must be neutralized [17]. The application of activated carbons (AC) for the adsorptive removal of organic pollutants from process water has also been discussed. The volume of water produced in FTS limits the use of ACs. Furthermore, the recovery or disposal of the spent ACs that contain toxic chemicals is problematic and expensive.

The composition and the quantity of the dissolved hydrocarbons in GTL water depend on the type of catalysts used and the reaction temperatures. Organic analysis has shown that hydrocarbon contents vary between 1.17 wt.% for low-temperature cobalt-catalyzed FTS and 6.2 wt.% for high temperature iron-catalyzed FTS [18]. Figure 15.4 shows the fractions of different hydrocarbons present in effluent water from low-temperature and high-temperature FT reactions.

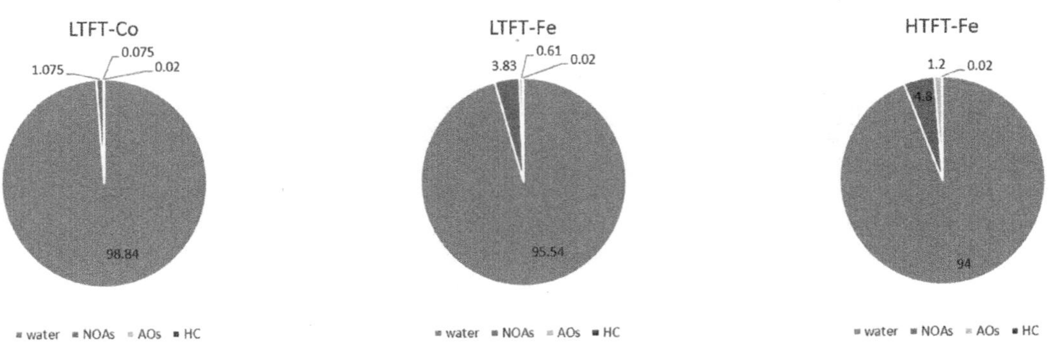

FIGURE 15.4 Comparison of the composition of hydrocarbons dissolved in water after low-temperature (Co- and Fe-catalyzed) and high-temperature (Fe-catalyzed) FT reactions. Fe-catalyzed high-temperature FT reaction produces as much as 4.8 mass% nonacid oxygenates and 1.2 mass% of acidic oxygenates. Low-temperature FT with Fe and Co catalysts produces less dissolved hydrocarbons. (Adapted from Dancuart, K.L.P.F. et al., Patent WO2003106351, 2003. With permission.)

15.4 NAOs

As mentioned previously, effluent water from the GTL stripper or distillation columns contains largely NAO hydrocarbons, which belong to C_1–C_7 hydrocarbons, which are polar and have high solubility in water. Analyses have indicated that the most common NAOs in this context are acetaldehyde, propionaldehyde, butyraldehyde, acetone, methyl propyl ketone, methanol, ethanol, propanol, butanol, pentanol, hexanol, and heptanol. Figure 15.5 shows the structures of the NAOs commonly found in GTL-process water. In Figure 15.6, the solubility of the NAOs is compared with that of a common non-polar solvent, n-hexane. With the exception of C_6+ alcohols, all NAOs are highly soluble in water. C_1–C_3 aldehydes and alcohols are completely miscible in the water and are not shown in Figure 15.6. Alcohols are the largest component of NOAs and are the ones that contribute the most to COD.

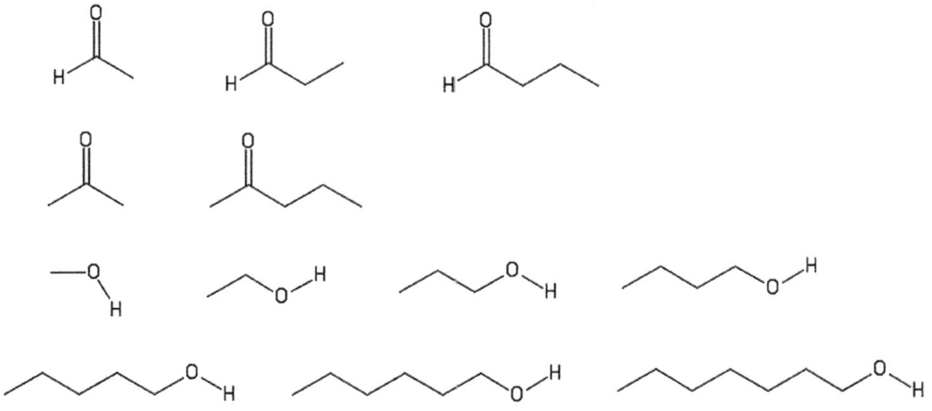

FIGURE 15.5 Common non-acid oxygenated hydrocarbons in GTL process water. In the first row (from left to right) there is acetaldehyde, propionaldehyde, and butyraldehyde. In the second row, acetone and methyl propyl ketone. In the third and fourth rows, methanol, ethanol, propanol, butanol, pentanol, hexanol, and heptanol are found.

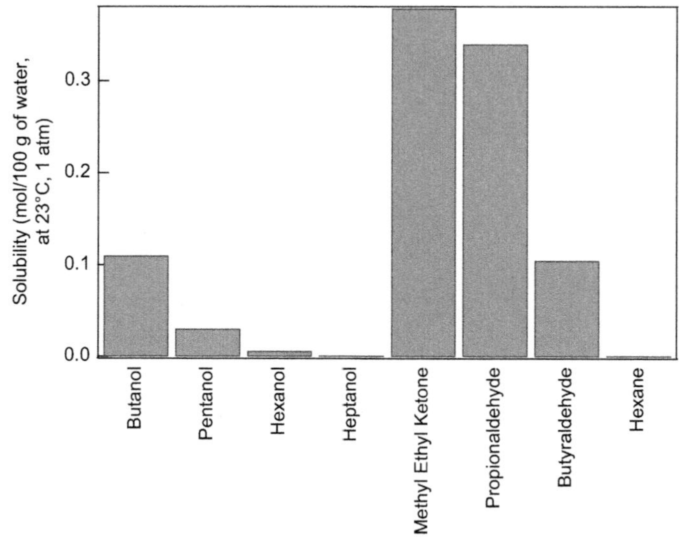

FIGURE 15.6 Comparison of room-temperature water solubility of non-acid oxygenated hydrocarbons in GTL process water with that of non-polar solvent n-hexane. Note that the NAOs acetone, methanol, ethanol, and propanol are completely miscible with water and are not plotted. Solubility values are collected from various online sources.

15.5 PHOTOCATALYTIC MINERALIZATION ON NAO HYDROCARBONS

Since the discovery of TiO_2-mediated photocatalytic splitting of water by Honda and Fujishima in 1972, significant strides have been made in utilizing TiO_2 and other semiconductor oxides and sulfides for the photocatalytic degradation of organic molecules [19]. Although many transition metal oxides and sulfides show electronic bandgaps in the UV–Visible region, which is necessary for their photoactivity, of these, TiO_2, ZnO, CdS, and ZnS are the most studied. Its high abundance, low cost, non-toxic nature, and high chemical stability make TiO_2 a particularly attractive photocatalyst for industrial applications. As TiO_2 becomes photoactive when irradiated with solar radiation, photocatalysis can be combined with solar technology to create photodegradation of organic pollutants. When a semiconductor oxide with a wide band gap is irradiated with photons having greater energy higher than the bandgap, electrons from the valence bands are excited and form an electron–hole pair. For TiO_2, the bandgap E_g between the valence and conduction bands is around 3.2 eV, as shown in the band structure (Figure 15.7, left panel [20]). The wavelength of the photons required for exciting the electrons over this bandgap is ~390 nm. Consistent with this excitation, the UV-Visible spectrum of TiO_2 exhibits a characteristic peak around 390 nm (Figure 15.7, right panel [21]).

A schematic illustration of the photooxidation mechanism using a wide bandgap semiconductor is shown in Figure 15.8. Upon irradiation with UV light, the excitation of the electron from valence band to conduction band creates an electron–hole pair (e^-h^+). The photogenerated hole (h^+) is a powerful oxidizing agent, which can oxidize organic pollutants to CO_2 and H_2O. In the presence of water, h^+ combines with water to form the free radical $OH^\cdot$. The excited electron (e^-) on the other hand, reduces the oxygen from the ambient atmosphere to form peroxide free radical (O_2^-) species. This reduced oxygen combines with water to produce the $OH^\cdot$ free radical. These radicals attack the organic pollutants and degrade them into CO_2 and water.

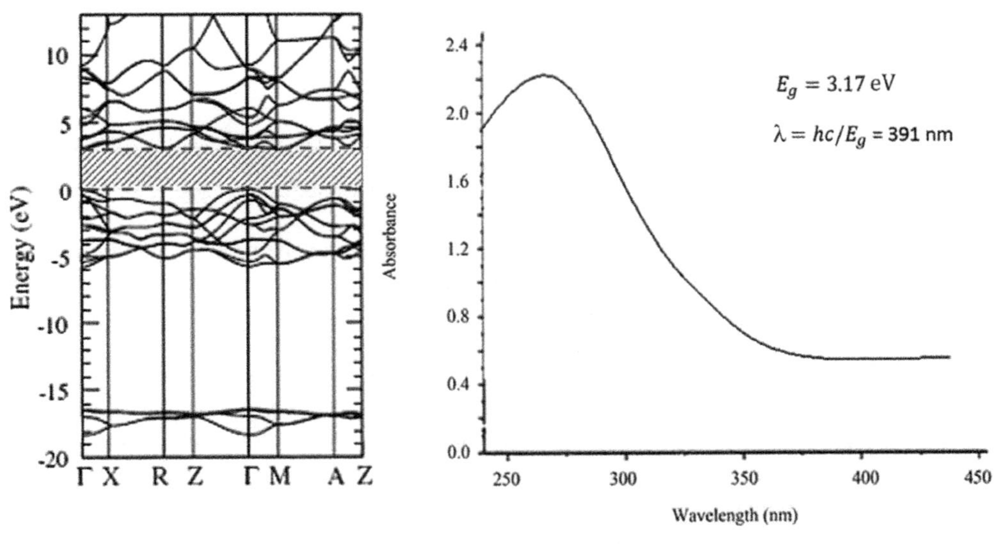

FIGURE 15.7 The calculated band structure of TiO_2. (Adapted from Sanjines, R. et al., *J. Appl. Phys.*, 75, 2945, 1994.) The bandgap, E_g is the difference between the highest level of valence band (Fermi level marked by broken red line in the left panel) and the lowest energy of the conduction band. The E_g from the band structure is 3.17 eV. The equivalent electron excitation wavelength can be computed from Planck's equation, $\lambda = hc/E_g$, where h is the Planck's constant and c is the speed of light. The equivalent wavelength is 391 nm, which falls in the UV region. On the right panel, a typical UV–Visible spectrum of TiO_2 clearly showing excitation around 391 nm is shown. (Adapted from Gouda, M. and Aljaafari, A.I., *Adv. Nanopart.*, 1, 29, 2012.)

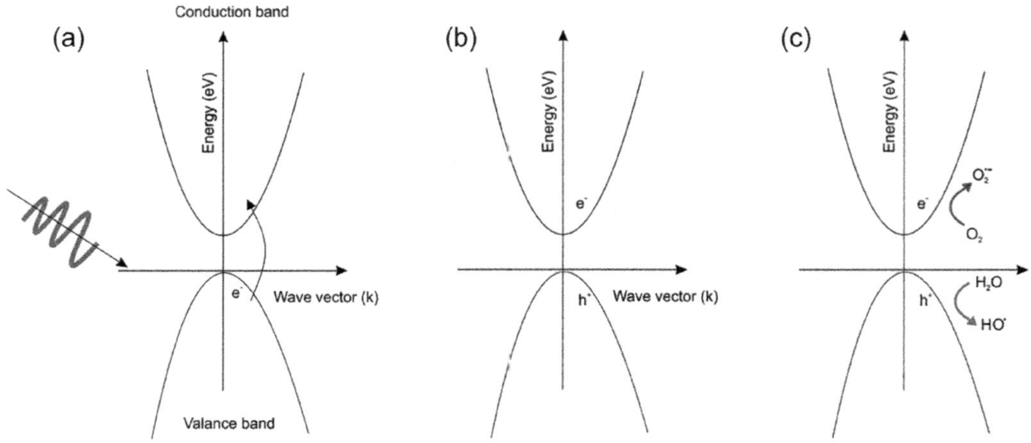

FIGURE 15.8 Mechanism of photocatalytic oxidation using a direct bandgap semiconductor. (a) The UV -visible photon excites an electron from the valence band to the conduction band, (b) creating a hole in the valence band. (c) The photogenerated hole combines with the water to form an OH free radical and the electron reduces ambient oxygen to form peroxide ion, which combines with water to form an OH free radical.

The kinetics of photocatalytic reactions between most NAOs considered here can be described with the classical Langmuir–Hinshelwood mechanism. In this mechanism, two molecules A and B adsorb onto vicinal catalytic sites. The adsorbed species AS and BS then combine, in a rate-determining step, to form products.

$$A + S \rightleftharpoons AS$$

$$B + S \rightleftharpoons BS$$

$$AS + BS \rightarrow \text{Products}$$

The adsorption of NAOs on TiO_2 is of fundamental importance for their photocatalytic oxidation. Even in the absence of UV photons, TiO_2 can catalyze a reaction between adsorbed NAO molecules. The titanium atoms on the surface of TiO_2 show two types of co-ordinations: fivefold and sixfold co-ordinations [22]. In Figure 15.9, the Ti^{n+} that is exposed to the ambient is fivefold coordinated and are marked Ti_{5c}. Constant-current topography reveals Ti_{5c} as bright lines in the figure (right panel, [23]). The sixfold coordinated Ti^{n+} appears below the surface layer oxygen and is marked Ti_{6c}. These surface atoms are Lewis acid sites. The surface species on TiO_2, including terminal Ti_{nc} Lewis acid sites, terminal Ti–O and bridging Ti–O–Ti, bind to NAO hydrocarbons and participate in thermal and photocatalytic oxidation reactions. To follow the intermediates and mechanisms of the photocatalytic oxidations, most studies employ spectroscopic and microscopic signatures for these species. In the FTIR spectra of TiO_2, the terminal Ti–O and bridging Ti–O–Ti are observed as two sharp bands below 1000 cm^{-1} (Figure 15.10, [24]).

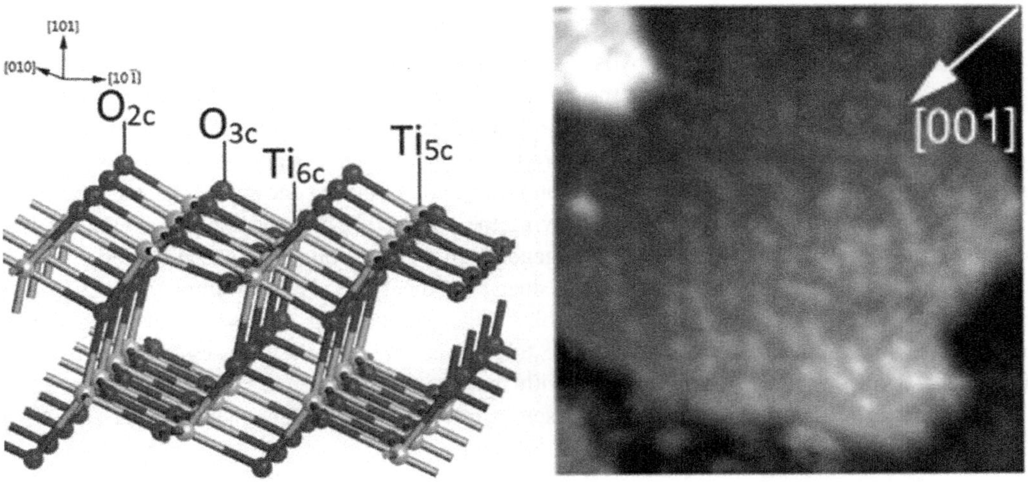

FIGURE 15.9 Illustration of Ti_{5c} and Ti_{6c} sites on the surface of TiO_2 [22]. These species act as Lewis acid centers on which the NAOs initially adsorb. On the right panel, a constant current topography image of TiO_2 is shown [23]. The bright parallel line indicates the rows of Ti_{5c}.

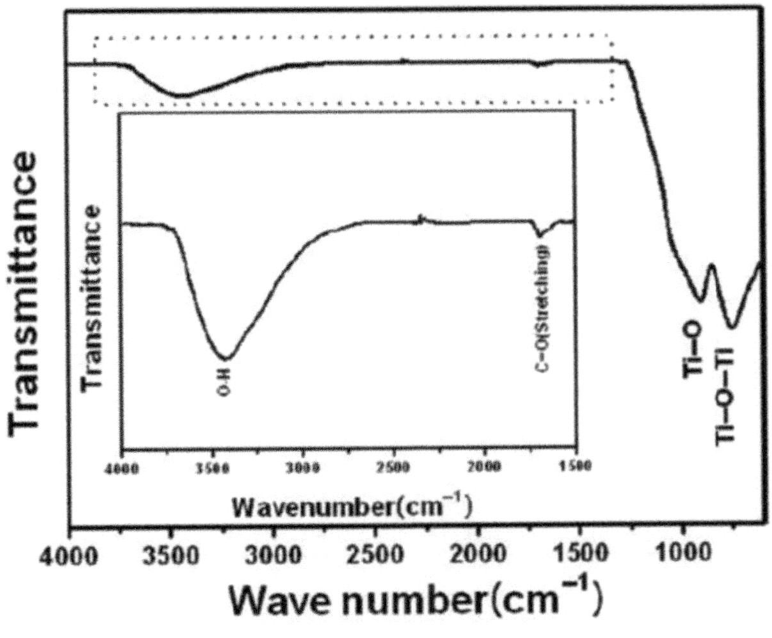

FIGURE 15.10 A typical FT-IR spectra of TiO_2 nanofilms showing two peaks below 1000 cm^{-1} that characterize the terminal Ti–O and bridging Ti–O–Ti. The broad band at 3600–3200 cm+ arises due to the linear OH group on the surface. (From Nam, S.H. et al., *Nanoscale Res. Lett.*, 7, 89, 2012.)

15.5.1 Aldehydes

The surface interaction of lower aldehydes (C_1–C_4) with TiO_2 and their photocatalytic decomposition has been studied in details. Even in the absence of UV irradiation, TiO_2 is known to catalyze the condensation of aldehydes to form aldols. Evidence suggests that the simplest aldehyde, acetaldehyde, interacts with TiO_2 via chemisorption, physisorption, and hydrogen bonding. In a low-pressure regime, acetaldehyde forms a chemisorbed monolayer with a binding energy of ~48.3 kJ mol^{-1} [25]. Singh et al. [26] proposed a mechanism involving the abstraction of an alpha hydrogen atom from an initially adsorbed acetaldehyde molecule, generating a carbanion (IR spectral evidence of aldol condensation: acetaldehyde adsorption on the surface of the TiO_2). This carbanion attacks a neighboring acetaldehyde to form 3-hydroxy butanol, which upon heating, removes a molecule of water to form the final 2-butenal product. Evidence for this is given by the progressive disappearance of the FT–IR band at 3718 and 3651 cm^{-1} with increasing acetaldehyde dosages (Figure 15.11). Concomitantly, the peak at 3350 cm^{-1} increases in intensity. The mechanism proposed by Singh et al. is shown in Figure 15.12.

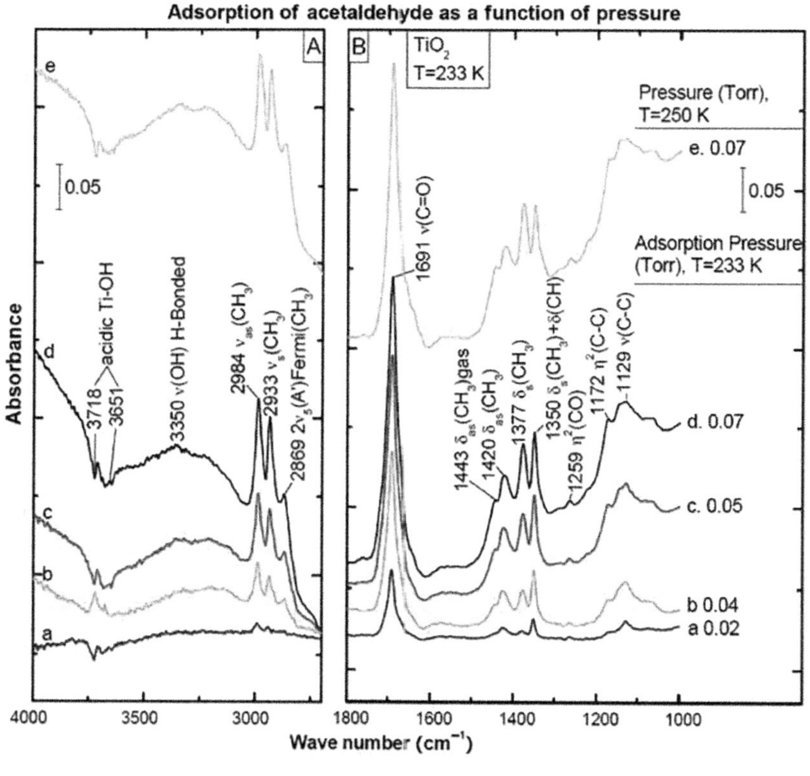

FIGURE 15.11 FT–IR spectra of TiO_2 on exposure of acetaldehyde [26]. With increasing partial pressure of acetaldehyde, the acidic peaks of Ti-OH at 3819 and 3651 cm^{-1} decreases in intensity indicating the binding of acetaldehyde on these site. (From Singh, M. et al., *J. Catal.*, 260, 371, 2008.)

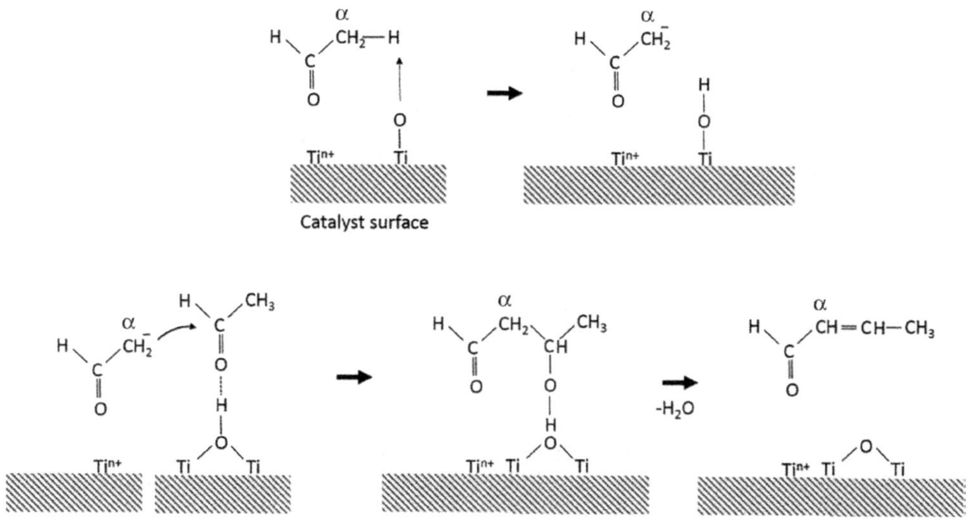

FIGURE 15.12 Scheme of the mechanism of aldol condensation of acetaldehyde on TiO_2. This mechanism is similar to the classical base-induced carbanion formation in the aldol condensation of gas-phase acetaldehyde. (From Singh, M. et al., *J. Catal.*, 260, 371, 2008.)

In the presence of light, acetaldehyde and propionaldehyde are completely mineralized to CO_2 [27]. Stengl et al. showed that acetaldehyde is an intermediate in the photocatalytic degradation of acetone [28]. Raillard et al. proposed a decomposition pathway for acetaldehyde proceeds through the formation of formaldehyde, formic acid, acetic acid, and finally CO_2, as shown in Figure 15.13 [29].

A comparison of the photooxidation of C_1–C_3 aldehydes showed that their activity increases from C_1 to C_3, which is attributed to the reactivity of the carbonyl functional group and steric factors (Figure 15.14, [30]).

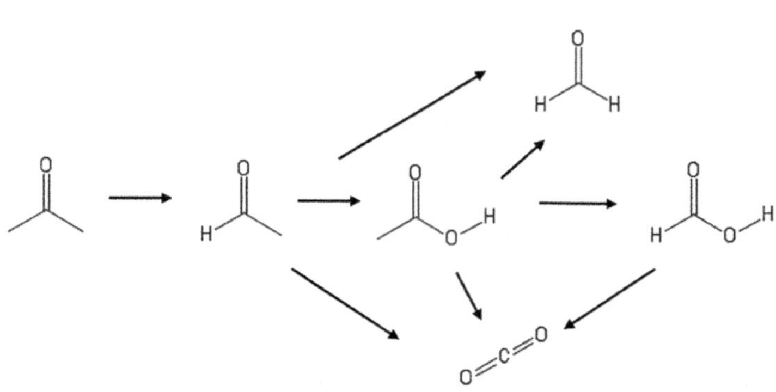

FIGURE 15.13 Intermediate molecules that are formed during the photocatalytic oxidation of acetone. (From Raillard, C. et al., *Water Sci. Tech.*, 53, 107, 2006.)

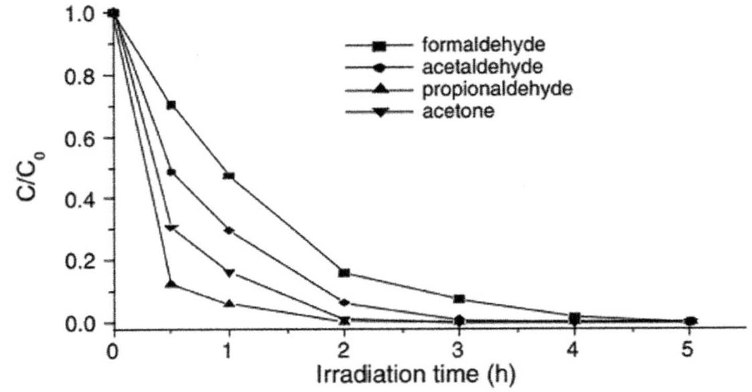

FIGURE 15.14 Comparison of the kinetics of photocatalytic degradation of mixed gaseous carbonyl compounds suggesting that decomposition is quickest for propionaldehyde. The observed kinetics is attributed to the enhanced reactivity of alpha-hydrogen atom of the ketone as compared to that of aldehydes. (From Zhang, M. et al., *Chemosphere*, 64, 423, 2006.)

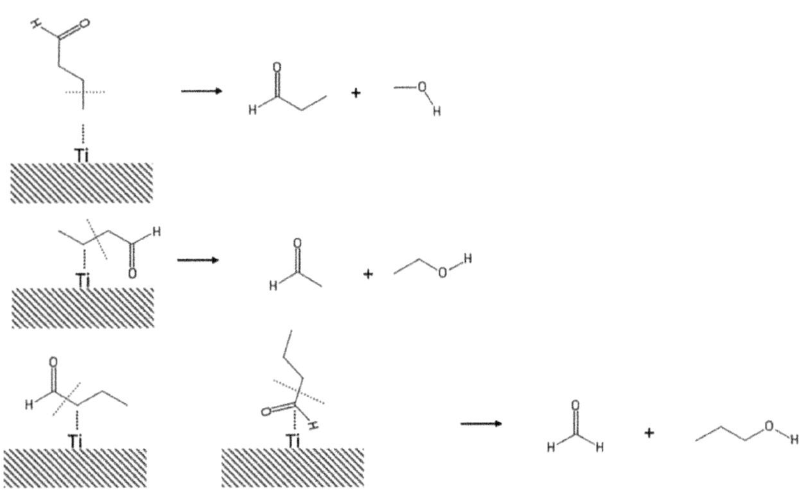

FIGURE 15.15 Different organic molecules are formed from photocatalytic oxidation of butyraldehyde on TiO_2 depending on the carbon atom that binds to the catalyst surface. (From Huang, C. et al., *Chem. Eng. Commun.*, 190, 373, 2003.)

The photodegradation of butyraldehyde, although CO_2 and water are the major products, provides others, including C_2–C_3 aldehydes and C_2–C_3, as well. The formation of by-products in this reaction can be explained through consideration of the adsorption of butyraldehyde onto TiO_2. Four carbon atoms of butyraldehyde attach to TiO_2 at carbons 1–4 (Figure 15.15). The dissociation of the C–C bonds followed by hydrolysis yields methanol, ethanol, propanol, formaldehyde acetaldehyde, and propionaldehyde [31].

15.5.2 Ketones

The biodegradation mechanism of ketones resembles that of acetone in many aspects, due to the similarity of the functional groups. For instance, as noted in the previous section, acetone photocatalytic oxidation proceeds through the formation of acetaldehyde. The geometry and adsorption sites of the acetone of TiO_2 have been studied using density-function theory [32]. $^1\eta$ configuration is the

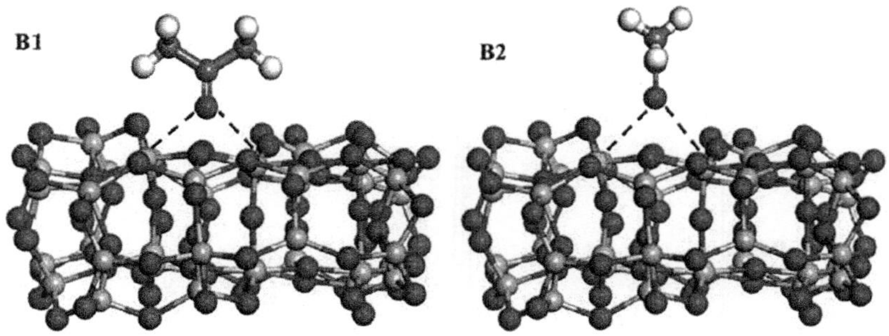

FIGURE 15.16 Adsorption sites and geometries of the adsorption of acetone on $(TiO_2)_{38}$ clusters. On a bridge position between two Ti atoms: (left) the acetone symmetry plane is parallel to the bridge oxygen rows; (right) the acetone symmetry plane is perpendicular to the bridge oxygen rows. (From Chen, Q. et al., *Vacuum*, 119, 123, 2015.)

most stable when an acetone molecule is relaxed in a $(TiO_2)_{38}$ cluster: in this case, the carbonyl group binds to the Lewis acid site of Ti^{n+}. Acetone adsorption prefers to take place on the top Ti atom, with the symmetry plane of acetone perpendicular to the bridge oxygen rows (Figure 15.16) [32].

The photocatalytic decomposition of acetone has been studied in greater detail, as acetone is widely considered a significant indoor pollutant. Loading dependent in situ FT–IR experiments performed by El-Maazawi et al. [33] showed that acetone adsorbs molecularly onto TiO_2. In the absence of light, at high coverage, IR bands are due to mesityl oxide, which is formed by the aldol condensation. The mechanism of the formation of mesityl oxide through aldol condensation proceeds through enolate formation. The reaction starts through the adsorption of acetone on the Lewis acid site, the exposed Ti^{n+} cation. In a typical enolate mechanism, as shown in Figure 15.17, the abstraction of alpha hydrogen is initiated by the surface Ti–O species. The resulting carbanion attacks the carbonyl group of the neighboring acetone molecule to form a hydroxyl ketone. Upon heating, the hydroxyl ketone eliminates a molecule of water to form mesityl oxide.

Relative to the reaction of acetaldehyde, reaction rates increase for acetone. At coverages lower than 50%, acetone undergoes photooxidation, leading to the formation of CO_2 and bidentate formate

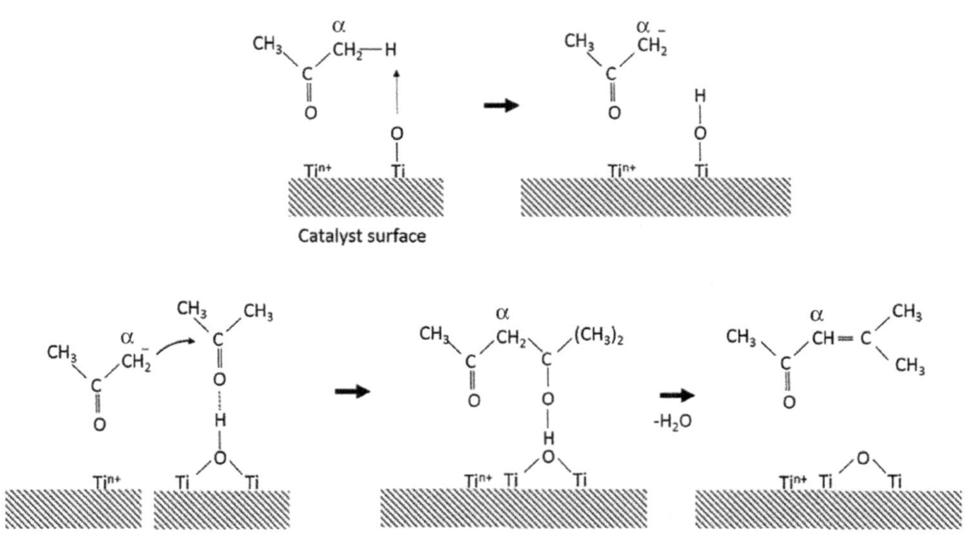

FIGURE 15.17 Formation mechanism of mesityl oxide on TiO_2 through the aldol condensation of acetone.

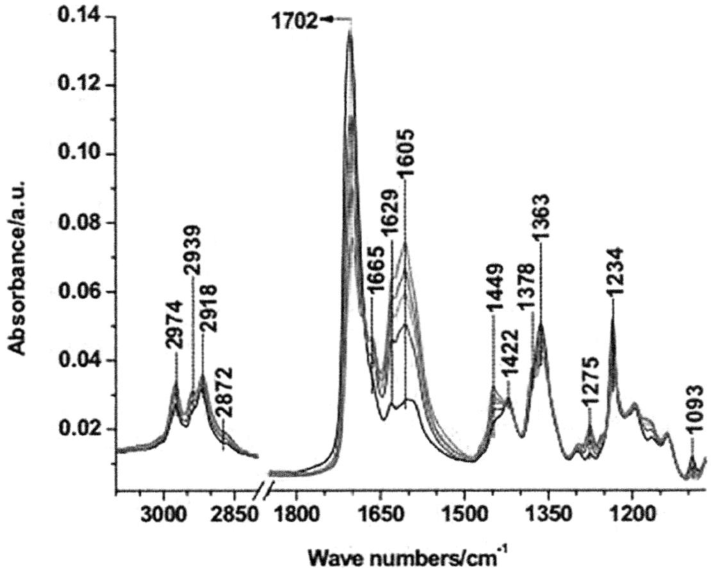

FIGURE 15.18 Series of IR spectra recorded during the adsorption of acetone on TiO_2 at 300 K. The band at 1605 cm^{-1}, corresponding to the mesityl oxide increases with the acetone dosing. (From Szanayi, J. and Kwak, J.H., *J. Mol. Catal. A Chem.*, 406, 213, 2015.)

species. At coverages above 50%, on the other hand, the acetone forms largely mesityl oxide, which is somewhat difficult to photooxidize, relative to the acetone; the photooxidation of mesityl oxide forms acetaldehyde [33]. The formation of mesityl oxide is confirmed by the increasing intensity of the C=C peak at 1605 cm^{-1}. When it is irradiated with UV light and in the presence of a large excess of oxygen, the peak at 1605 cm^{-1} begins to diminish and concomitantly, a new peak at 1730–1740 cm^{-1} begins to grow (Figure 15.18). This peak suggests the formation of acetaldehyde, while the acetone peak is less affected [34]. The formation of mesityl oxide is a thermal process and requires no UV radiation.

15.5.3 Alcohols

At room temperature and in the absence of UV irradiation, methanol, the simplest aliphatic alcohol, undergoes both dissociative and molecular adsorption onto TiO_2. This physisorption follows the Langmuir isotherm model, and the physisorbed molecules leave the surface without dissociation. The two distinct types of adsorbed forms found on the surface are confirmed by FT–IR experiments [35]. One form is a dissociated form (methoxy), and it has no signature in its OH bonds in the FTIR spectra (1337 cm^{-1}, 3400 cm^{-1}). The FT–IR spectrum of methanol on TiO_2 from Rossi and Busca [35] is presented in Figure 15.19. Methanol dissociation occurs primarily on bridge-bonded oxygen vacancies in TiO_2. First-principles calculations show that on a stoichiometric surface, the dissociative adsorption of CH_3OH can occur via both O–H and C–O ruptures and that this is favored over molecular adsorption.

It has been observed that methanol molecules are dominantly present on the top of the Ti_{5c} sites at low coverage and, as coverage increases, tend to present as chains along Ti_{5c} rows . Under irradiation, the methanol molecules adsorbed on Ti_{5c} dissociate to form formaldehyde. Studies of nanocrystalline TiO_2 has shown that in the absence of photo irradiation, methanol mostly adsorbs and desorbs as a molecule up to 450 K. Upon photoirradiation, a coupling reaction

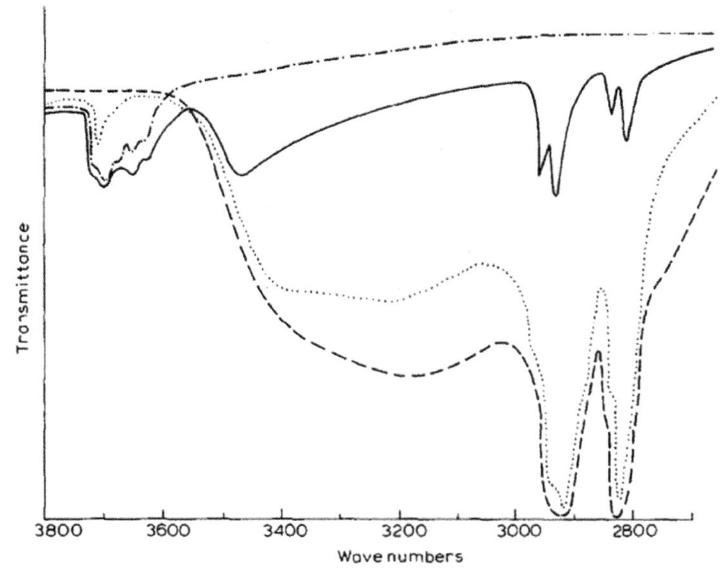

FIGURE 15.19 FT–IR spectra of methanol on TiO_2 showing two different types of forms of methanol. After the admission of methanol (undetectable equilibrium pressure and full and point lines), in equilibrium with 5 Torr methanol vapor (broken line). (From Rossi, P.F. and Busca, G., *Colloids Surf.*, 16, 95, 1985.)

occurs to form methyl formate, and small quantities of dimethyl ether [36]. Although it is feasible for the OH free radicals formed in the photoexcitation mechanism to oxidize methanol, the quantum efficiency of OH free radical formation is 10 times slower than that for CO_2 formation. This suggests that methanol is rather directly oxidized by photogenerated hole [36,37]. Site-specific photoreaction studies of EtOH species on TiO_2 in different oxidation states have shown that the photoreaction of EtOH on TiO_2 proceeds most efficiently when the reactants are adsorbed on regular surface Ti sites with O adatoms and form acetaldehyde [38]. However, strongly bound ethoxides, either adsorbed in bridged O vacancies or at defect sites associated with step edges, photoreact very slowly. Low-coordinated defect sites on the terraces and at the step edges, on the other hand, are least active and the presence of oxygen is required to induce photoreactivity of molecules at these sites.

15.6 CONCLUSIONS AND OUTLOOK

This chapter provided an overview of water production in the GTL process and the potential for photocatalytic water treatment. It focused on the removal of non-oxygenated hydrocarbons, because they are the most predominantly found organic contaminant in water produced by the GTL process. The current treatment of NAOs in commercial GTLs is conducted using mostly microbial methods. TiO_2-mediated photooxidative removal of NAOs is attractive; it allows coupling of solar technologies with water technologies of the GTL process. The clear conclusion drawn in many studies is that TiO_2 can oxidize almost all NAOs. However, because most were performed using pure NAOs and in the absence of any other contaminants or large excess of water, the extrapolation of their efficiency to predict the performance of water treatment in the real GTL process is rather difficult. For some NAOs, such as ketones, even the presence of water could adversely affect the reactivity of TiO_2. A potential way to overcome this is to tailor the hydrophobicity of the TiO_2 surface. Composites formation of TiO_2 with high surface area materials, such as graphene, will also be an effective method of overcoming the slow reactivity of bulk TiO_2.

REFERENCES

1. O. Glebova, Gas to liquids: Historical development and future perspectives, Oxford Institute for Energy Studies (2013).
2. A. N. Strages, *Energia*, **12** (2001), 1.
3. T. H. Fleisch, Associated gas monetization via miniGTL, Report III, World Bank-Global Gas Flaring Reduction Partnership (2015).
4. R. Rauch, H. Hofbauer, S. Sacareanu, A. Chiru, *Bulletin of Transylvania University of Brasov*, **5** (2010) , 33.
5. O. O. James, B. Chowdhury, M. A Mesubi, S. Maity, *RSC Advances*, **2** (2012), 7347.
6. B. H. Davis, *Fuel Processing Technology*, **71** (2001), 147.
7. T. Bhatelia, W. Ma, B. Davis, G. Jacobs, D. Burkur, *Chemical Engineering Transactions*, **25** (2011) 707.
8. Pearl GTL (Shell) Qatar, Case study, available at http://technomaps.veoliawatertechnologies.com/vwst-northamerica/ressources/documents/1/28711,Case-Study-Shell-Qatar_Corporate.pdf.
9. A. Siddiqi, L. D. Anadon, *Energy Policy*, **39** (2011), 4529.
10. A. D. Hix, M. Moor, R. Kendall, R. Svoboda, W. Maningas, Gas to Liquids (GTLs), University of Tennessee Honors Thesis Projects (2012).
11. A. Behroozsarand, A. Zamaniyan, *Journal of Cleaner Production*, **142** (2017), 2315.
12. R. Miglio, R. Zennaro, A. de Klerk, in *Greener Fischer-Tropsch Processes for Fuels and Feedstocks*, P. M. Maitlis and A. de Klerk (Eds.), Wiley VCH, Weinheim, Germany, 2013.
13. O. Onel, A. M. Niziolek, H. Butcher, B. A. Wilhite, C. A. Floudas, *Computers and Chemical Engineering*, **105** (2017), 276–296.
14. M. Rycroft, *Energize*, **55** (2013).
15. H. N. Carlsen, H. Degn, D. Lloyd, *Journal of General Microbiology*, **137** (1991), 2879.
16. D. Dionisi, M, Beccari, E. N. D'Addario, A. Donadio, M. Majone, R. Sbardellati, Anaerobic biotreatability of an industrial wastewater containing alcohols at different chain length, presented at *11th World Congress on Anaerobic Digestion*, Brisbane, Australia (2007).
17. M. Majone, F. Aulenta, D. Dionisi, E. N. D'Addario, R. Sbardellati, D. Bolzonella, M. Beccari, *Water Research*, **44** (2010), 2745.
18. K. L. P. F. Dancuart, P. G. H Du, T. F. J. Du, E. D. Koper, T. D. Philips, D. W. J. Van, Patent WO2003106351 (2003).
19. A. L. Linsebigler, G. Lu, J. T. Yates Jr., *Chemical Reviews*, **95** (1995), 735.
20. R. Sanjines, H. Tang, H. Berger, F. Gozzo, G. Magaritondo, F. Levi, *Journal of Applied Physics*, **75** (1994), 2945.
21. M. Gouda, A. I. Aljaafari, *Advances in Nanoparticles*, **1** (2012), 29.
22. J. G. Ma, C. R. Zhang, J. J. Gong, B. Yang, H. M. Zhang, W. Wang, Y. Z. Wu, Y. H. Chen, H. S. Chen, *Journal of Chemical Physics*, **141** (2014), 234705.
23. M. Yasuo, A. Sasahara, H. Onishi, *Journal of Physical Chemistry C*, **114** (2010), 14579.
24. S. H. Nam, S. J. Cho, J. H. Boo, *Nanoscale Research Letters*, **7** (2012), 89.
25. J. J. Plata, V. Collico, A. M. Marquez, J. F. Sanz, *Journal of Physical Chemistry C*, **115** (2011), 2819.
26. M. Singh, N. Zhou, D. K. Paul, K. J. Klabunde, *Journal of Catalysis*, **260** (2008), 371.
27. N. Takeda, M. Ohtani, T. Torimoto, S. Kuwabata, H. Yoneyama, *Journal of Physical Chemistry B*, **101** (1997), 2644.
28. V. Stengl, V. Houskova, S. Bakardjieva, N. Murafa, *New Journal of Chemistry*, **34** (2010), 1999–2005.
29. C. Raillard, V. Hequet, P. L. Cloirec, J. Legrand, *Water Science and Technology*, **53** (2006), 107.
30. M. Zhang, T. An, J. Fu, G. Sheng, X. Wang, X. Hu, X. Ding, *Chemosphere*, **64** (2006), 423.
31. C. Huang, D. H. Chen, K. Li, *Chemical Engineering Communications*, **190** (2003), 373.
32. Q. Chen, W. Zhu, X. Hou, K. Xu, *Vacuum*, **119** (2015), 123.
33. M. El-Maazawi, A. N. Finken, A. B. Nair, V. H. Grassian, *Journal of Catalysis*, **191** (2000), 138.
34. J. Szanayi, J. H. Kwak, *Journal of Molecular Catalysis A: Chemical*, **406** (2015), 213.
35. P. F. Rossi, G. Busca, *Colloids and Surfaces*, **16** (1985), 95.
36. D. A. Bennett, M. Cargnello, B. T. Diroll, C. B. Murray, J. M. Vohs, *Surface Science*, **654** (2016), 1.
37. J. Zhang, Y. Nosaka, *Applied Catalysis B: Environmental*, **166–167** (2015), 32.
38. J. Ø. Hansen, R. Bebensee, U. Martinez, S. Porsgaard, E. Lira, Y. Wei, L. Lammich et al., *Scientific Reports*, **6** (2016), 21990.

Section V

Wastewater Treatment

Adsorption Process

16 Biosorption of Methylene Blue Dye Using Anise Tea Residue*

Khaled M. Hassan, Mamdouh A. Gadalla, and Tamer T. El-Idreesy

CONTENTS

16.1 INTRODUCTION

Water is an essential material for human health and life. Unfortunately, polluted water and air are prevalent throughout the world (European Public Health Alliance, 2009). Polluted water can be very deceiving, as it can look muddy, smell bad, and have rubbish floating in it, which means that it is obviously unhealthy, but it could also look clean, although it is filled with harmful unseen chemicals. Human activity contributes to water pollution. Factories also dump chemicals and oils into waterways, and such chemicals are called runoffs (Ghouri, 2010). Nowadays, man-made dyes are utilized in a wide range of industries, such as the paper, pharmaceutical, cosmetics, paint, leather, food, and textile industries. Currently, more than 100,000 dyes are commercially available and greater than $8(10)^5$ tons of dye are produced annually to supply these industries (Pezoti et al., 2016). Azo dyes are the largest class of dyes, corresponding to more than 60% of manufacturing synthetic dyes; among the azo dyes, methylene blue (MB) is the most commonly utilized substance in paper coloring and in the textile industries that are the main users of fresh water and generators of wastewater.

* The work done in this chapter was part of Khaled M. Hassan's graduation project presented to The British University in Egypt in partial fulfillment of the requirement for the Bachelor Degree in Chemical Engineering.

Due to the presence of assorted toxic chemicals that have been extensively noticed at hazardous levels, drinking water has recently become a significant health concern to human beings in many parts of the world (Bhatnagar et al., 2010). Usually dye industries discharge their wastewater into natural water bodies. This discharged wastewater contains dyes that result in serious environmental problems, causing risk to ecosystems, living organisms and human health due to their toxic and carcinogenic properties. These dyes must therefore be treated properly before being dumped into the environment. This is not, unfortunately, what is currently happening (Pezoti et al., 2016).

There are countless technologies available that have a varying degree of success in overcoming water pollution. Unfortunately, there are many limitations to the majority of these technologies, such as elevated operational and maintenance prices, toxic sludge creation, and the complexity of the method encompassed in the treatment. Therefore, more alternatives have to be discovered. As a result, the scientific community commenced thinking of ecological and effectual methods, such as precipitation, chemical degradation, biodegradation, chemical coagulation, and adsorption. Adsorption is one of the most effective processes of wastewater treatment, because of its simplicity, convenience, ease of operation, and low cost (Pezoti et al., 2016).

Adsorption using solid materials is catching the world's attention, as its simple concept is the transfer of particular undesired substances from the solution onto the adsorbent surface. Many adsorbents can be used; some adsorbents can be expensive, such as activated carbon, and some can be cheap, including orange or lemon peels. Needless to say, extra attention has been given to the use of low-cost biosorbents. When agricultural wastes are being used for adsorbing pollutants and impurities from wastewater, it is called biosorption. For the past few decades, biosorption has emerged as a competitively priced and useful option for water and wastewater treatment. Biosorption uses naturally occurring and agriculture waste materials as biosorbents, as these are renewable, inexpensive, and plentifully available. Meanwhile, numerous studies have been published in a number of journals, expanding the biosorption field. The removal of assorted kinds of pollutants has been tested by different biosorbents (Pezoti et al., 2016). Biosorption is an emerging technique for water treatment utilizing abundantly available biomaterials.

A dye is a soluble, colored substance that has an affinity for a fiber or other substance, meaning that it will adhere to the surface. Plant, animal, and mineral substances can all be processed to make dyes. These dyes can absorb light of different wavelengths and the human eye detects this absorption, and responds to the colors (Alves, 2012). Dyes may be natural or synthetic based on their origin. Natural dyes are obtained from natural products, such as plants and insects and may therefore be more environmentally approachable and healthier for the consumer. The first man-made dye was discovered in the 1800s, and ever since, thousands of synthetic dyes have been prepared. Synthetic dyes are cheaper to produce, brighter, more colorfast, come in a wider variety of colors, and are easy to apply to fabric. They are utilized everywhere in everything from clothes to paper, from food to wood. All of these advantages cannot make one ignore the disadvantages of synthetic dyes, which are produced from chemicals that are exceedingly toxic, carcinogenic, or even explosive (Kolorjet Chemicals Pvt. Ltd., 2015).

On the other hand, industrial residual dyes from different origins (e.g., textile industries, paper and pulp industries, pharmaceutical industries) are believed to form a large collection of the organic pollutants running into the natural water resources or wastewater treatment systems. One of the biggest polluters in the world is the textile industry. The World Bank estimates that nearly 20% of world manufacturing water pollution comes from the treatment and dyeing of textiles. This includes 10,000 different textile dyes with an approximated annual creation of 7.105 metric tons are commercially obtainable worldwide; 30% of these dyes are utilized in excess of 1,000 tons each annum, and 90% of the textile products are utilized at the level of 100 tons each annum or less. The color and the breakdown products of many dyes are the reason behind the undesirability of sending dye-containing effluents into the water environment. These products are toxic, mutagenic, or carcinogenic to life forms generally because of carcinogens such as naphthalene and other aromatic compounds. If these products are not treated properly, these dyes can remain in the environment for a long time.

16.2 BIOSORPTION PROCESS

Biosorption is defined in numerous ways by different scientists and chemists. In this research chapter, the term will be defined as the ability of biological materials to remove dyes from wastewater through physical and chemical uptake pathways (Forest and Roux, 1992). Biosorption involves two processes:

- Solid phase (sorbent or biosorbent biological treatments)
- Liquid phase (consists of a solvent, usually water)

In the biosorption process, the dissolved (dye) particles are attracted and bounded to the adsorbent by various mechanisms. The process continues until equilibrium between the dye particles and the adsorbent is reached and saturation has occurred.

16.2.1 Biosorption Models

Adsorption isotherm gives the extent of an adsorbate adsorbed on the surface of an adsorbent at constant temperature, hence the name isotherm. It is the thermodynamic basis of the biosorption separation processes and determines the extent to which a material can be adsorbed onto a particular surface (Chowdury, 2012). According to many experimental researchers, it has been found that the Freundlich and Langmuir isotherms are fruently used (Wanyonyi et al., 2014). The Freundlich isotherm is applicable for non-ideal sorption on heterogeneous surfaces, as well as multi-layer sorption (Chowdury, 2012). The non-linear and linearized forms of the equation are as follows:

$$q_e = K_F C_e^{1/n} \tag{16.1}$$

$$\ln q_e = \ln K_F + \left(\frac{1}{n}\right) \ln C_e \tag{16.2}$$

Where, K_F (L/mg) represents the Freundlich constant that determines the adsorption capacity of the adsorbent. The term $1/n$ is an indication of the favorability of adsorption. However, the values of constant n, if $n > 1$, represent favorable adsorption conditions. Values of K_F and n are calculated from the intercept and the slope of the plot of ln q_e against ln C_e.

On the other hand, the non-linear and linearized forms of the Langmuir isotherm are as follows:

$$q_e = \frac{(Q_o b C_e)}{(1 + b C_e)} \tag{16.3}$$

$$\frac{1}{q_e} = \frac{1}{Q_o} + \left(\frac{1}{Q_o b}\right)\left(\frac{1}{C_e}\right) \tag{16.4}$$

where:

q_e (mg/g) is the amount of dye adsorbed at equilibrium
C_e (mg/L) is the equilibrium dye concentrations
Q_o is the maximum monolayer coverage capacity (mg/g)
b (L/mg) is the Langmuir isotherm constant related to the affinity of the binding sites

When $1/q_e$ is plotted against $1/C_e$, a straight line with slope $(1/bQ_o)$ and intercept $(1/Q_o)$ is obtained. A further analysis of this equation was made to obtain the dimensionless equilibrium parameter, R_L, also called the separation factor, given by:

$$R_L = \frac{1}{(1+bC_o)} \tag{16.5}$$

An indication of whether the shapes of the isotherm are either favorable or linear or unfavorable is by using the separation factor, R_L. The shapes are un-favorable when $R_L > 1$, linear when $R_L = 1$, favorable when $0 < R_L < 1$ or irreversible ($R_L = 0$).

16.2.2 Adsorption Kinetics

Adsorption kinetics depends on the adsorbate–adsorbent interaction and system condition. Kinetic analysis enumerates the solute uptake rate. Now a days, adsorption reaction models have been widely employed to describe the kinetic process of adsorption, using pseudo-first-order or pseudo-second-order, as follows (Ho, 2004):

1. Pseudo-first-order rate equation

$$\frac{dq_t}{dt} = K_1(q_e - q_t) \tag{16.6}$$

$$\mathrm{Log}(q_e - q_t) = \log q_e - \left(\frac{K_1}{2.303}\right)t \tag{16.7}$$

Where, q_e and q_t (mg/g) are the adsorption capacities at equilibrium and time t (min), respectively. K_1 (min^{-1}) is the pseudo-first-order rate constant for the kinetic model. Plotting a graph between log $(q_e - q_t)$ and time (t) gives a linear line, from which k_1 and q_e can be determined from the slope and intercept, respectively (Ho, 2004).

2. Pseudo-second-order rate equation

$$\frac{dq_e}{dt} = K_2(q_e - q_t)^2 \tag{16.8}$$

$$\frac{1}{q_t} = \left(\frac{1}{K_2 q_e^2}\right) + \frac{t}{q_e} \tag{16.9}$$

Where, K_2 is the pseudo-second-order rate constant (g/(mg·min)), q_e and q_t are the amounts of MB adsorbed (mg/g) at equilibrium and time t (min), respectively. The plot of t/q_e versus t gives a linear relationship, from which, q_e and K_2 can be determined from the slope and intercept, respectively.

16.3 MATERIALS AND METHODS

16.3.1 Methylene Blue as the Adsorbate

MB dye was first discovered in 1878. MB is a basic cationic dye, an aromatic chemical compound that is used widely around the world; it is scientifically used, for example, by biologists to help them see life under a microscope. MB dye ($C_{16}H_{18}ClN_3S.3H_2O$; 82% purity, $\lambda_{max} = 661$ nm) used in this study was purchased from NICE Chemicals and used without any further treatment or purification. A series of various concentrations of MB dye ($1(10)^{-5}$, $8(10)^{-6}$, $6(10)^{-6}$, $4(10)^{-6}$ M in distilled water) was prepared and their maximum absorption at 661 nm was recorded; then a calibration curve was plotted and used later for the estimation of the unknown dye concentration after the biosorption process with anise tea residue as adsorbent.

16.3.2 Anise Tea Residue as New Biosorbent

Firstly the anise tea was bought from a super market. The anise tea was then soaked in boiled water for 10 min in order to use it up, then filtered then dried. This process was repeated 3 times to ensure the preparation of the anise tea residue, which was then left to dry. The dry residue was extensively washed with tap water to remove any water-soluble impurities and other unwanted solids. Afterwards the anise tea residue was put on a flat plate and left for 48 h to dry completely. The residue was then ground to a fine powder using a pestle and mortar, as shown in Figure 16.1. The residue was then washed several times with distilled water before being used for adsorption purposes. When the residue was not being used, it was kept in a beaker closed by parafilm.

16.3.3 Batch Biosorption

Many variables were considered in this study, including the initial dye concentration, initial biosorbent dose, contact time, temperature, and pH. For each variable different known concentrations of the dye solutions were used to confirm the results obtained. Triplicates of each concentration were done and the average values were reported.

Batch biosorption experiments were performed in 250 mL Erlenmeyer flasks of 25 mL of known initial concentration of the dye ($1(10)^{-5}$ M). As a start 0.1 g of the anise tea residue was added to the solution. The flask was then agitated by an incubator shaker at a speed of 200 rpm for 15 min at 295 K (22°C). The samples were collected at regular time intervals and then residual dye concentration in the solution was analyzed by monitoring the change in absorbance values at maximum wavelength (λ_{max}) of 661 nm using spectrophotometer. All experiments were run in triplicate to enhance reproducibility. Table 16.1 shows a summary of the variables examined. Once the absorbance is measured, calculations can be done. These calculations are the percent removal of the dye and the amount of dye adsorbed at equilibrium onto the anise tea residue (amount of dye adsorbed per unit biosorbent) can be calculated using the following equations:

$$\text{Removal\%} = \left[\frac{(C_o - C_e)}{C_o}\right](100) \tag{16.10}$$

FIGURE 16.1 Anise tea residue being ground to fine materials by a pestle and mortar.

TABLE 16.1
Summary of the Variables Examined

Variables	Other Factors Kept Constant
Dose of biosorbent (0.1, 0.2, and 0.3 g)	Contact time 15 min, temperature 295 K, agitation 200 rpm, and pH 7.
Contact time (15, 30, and 45 min)	Biosorbent dose 0.2 g, temperature 295 K, agitation 200 rpm, and pH 7.
Temperature (303, 313, and 323 K)	Biosorbent dose 0.2 g, contact time 30 min, agitation 200 rpm, and pH 7.
Initial dye concentration	Biosorbent dose 0.2 g, contact time 30 min, agitation 200 rpm, temperature of 303 K, and pH 7.
pH (2–10)	Biosorbent dose 0.2 g, contact time 30 min, temperature of 303 K, and agitation 200 rpm.

$$q_t = \frac{[(C_o - C_t)V]}{M} \tag{16.11}$$

where:
C_o and C_t are the initial and final concentrations (mg/L)
V (L) is the volume of the solution
M (g) is the mass of the biosorbent

16.4 RESULTS AND DISCUSSION

16.4.1 Effect of Biosorbent Dose

One of the most important variables is the dose of the biosorbent used, since it determines the capacity of the biosorbent. The biosorbent/biosorbate equilibrium in the system can help predict the treatment cost of biosorbent per unit of dye solution, so this effect was investigated. The amount of anise tea residue (biosorbent) was varied from 0.1 to 0.4 g in 25 mL of dye solution, while the other variables such as the pH, contact time, initial dye concentration and temperature were kept constant and unchanged.

As the dose increases from 0.1 to 0.2 g, the percentage removal increased from 78% to 87%, which is probably due to an increase in the quantity of binding sites available for the biosorption process and also due to an increased surface area of the biosorbent. As the dose increases to 0.3 and 0.4 g, the percentage removal decreased. This could be due to the saturation of the dye binding sites due to aggregation, for example. From the following data, it is shown that the optimum biosorbent dose is 0.2 g. Table 16.2 shows the results obtained from this effect and the recorded concentration from the calibration curve. Table 16.3 shows the results obtained from the biosorbent dose effect. The graphical representation of the results is shown in Figure 16.2.

TABLE 16.2
Results Obtained from This Effect and the Concentration Calculated from the Calibration Curve

$C_1 = 1(10)^{-5}$ M			$C_2 = 8(10)^{-6}$ M		
Loading (g)	Absorbance	Concentration (M)	Loading (g)	Absorbance	Concentration (M)
0.1	0.180	$2.25(10)^{-6}$	**0.1**	0.155	$1.85(10)^{-6}$
0.2	0.105	$1.25(10)^{-6}$	**0.2**	0.085	$1.00(10)^{-6}$
0.3	0.161	$1.9(10)^{-6}$	**0.3**	0.122	$1.40(10)^{-6}$
0.4	0.172	$2.1(10)^{-6}$	**0.4**	0.141	$1.69(10)^{-6}$

TABLE 16.3
Results Recorded from Effect of Biosorbent Dose

$C_1 = 1(10)^{-5}$ M		$C_2 = 8(10)^{-6}$ M	
Loading at C_1	**% Removal**	**Loading at C_2**	**% Removal**
0.1	79	0.1	76.9
0.2	87.5	0.2	87
0.3	81	0.3	80
0.4	79	0.4	78.5

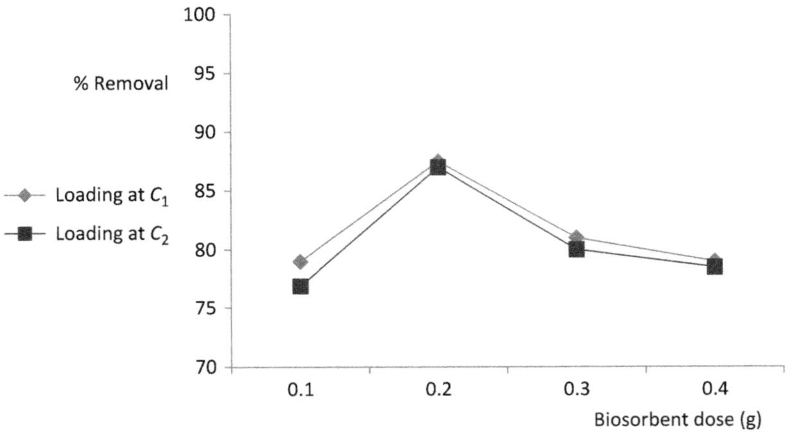

FIGURE 16.2 Effect of biosorbent dose (g) on biosorption of methylene blue by anise tea residue. Other variables: contact time = 15 min, pH = 7, temperature 295 K, and agitation speed = 200 rpm.

16.4.2 Effect of Contact Time

Once the biosorbent dose was examined, contact time was the second effect to examine. Contact time is important because it could lead to greater efficiency, if for example, 30 min would be enough to get the optimum results rather than 45 min. Longer contact time does not necessarily mean better the biosorption. The stock solutions, C_1 and C_2, were used in order to find out the optimum contact time. As the contact time is the variable to be computed, the temperature was kept constant at 295 K, agitation speed kept constant at 200 rpm, pH at 7 and the biosorbent dose was found to be optimal at 0.2 g, which was also kept constant. In order to get accurate results, triplicates were done and the average was calculated and recorded. In Figure 16.3 the results of this experiment are shown. It is obvious that the optimum contact time is 30 min (half an hour). The results are shown in Tables 16.4 and 16.5. Graphically, the results are plotted and presented in Figure 16.3. The percentage removal increased from 82% to 88% for concentration 1 ($1(10)^{-5}$ M) due to the large amount of surface area available for biosorption of the dye but it decreased after 30 min and when kept for a longer period of time. This could be because the vacant sites became difficult to occupy due to the repulsive forces between the MB particles on the solid and bulk phases when equilibrium was reached. In other words, saturation of binding sites occurred and slowed down the biosorption process.

16.4.3 Effect of Temperature

Temperature has a pronounced effect on the sorption process to remove the dye from solution or wastewater. Nevertheless, the effect of temperature was computed in this study in the range of

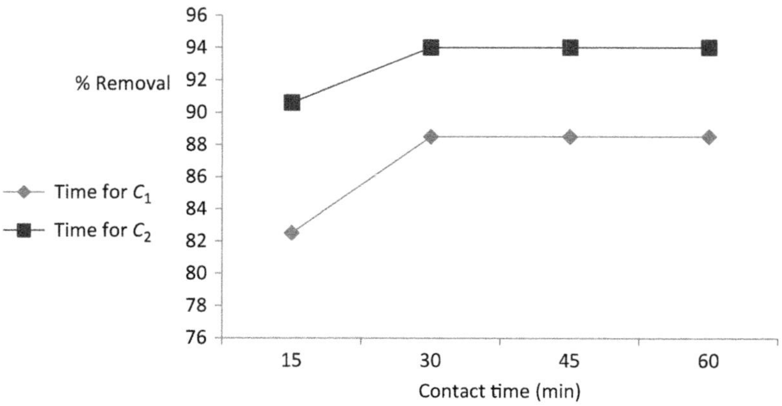

FIGURE 16.3 Effect of contact time on biosorption of methylene blue by anise tea residue. Other variables, biosorbent dose = 0.2 g, pH = 7, temperature 295 K and agitation speed = 200 rpm.

TABLE 16.4
Results Obtained from This Effect and the Concentration Calculated from the Calibration Curve

$C_1 = 1(10)^{-5}$ M			$C_2 = 8(10)^{-6}$ M		
Contact Time (min)	Absorbance	Concentration (M)	Contact Time (min)	Absorbance	Concentration (M)
15	0.145	$1.75(10)^{-6}$	15	0.075	$0.75(10)^{-6}$
30	0.099	$1.15(10)^{-6}$	30	0.050	$0.50(10)^{-6}$
45	0.010	$1.15(10)^{-6}$	45	0.050	$0.50(10)^{-6}$
60	0.011	$1.15(10)^{-6}$	60	0.0.51	$0.50(10)^{-6}$

TABLE 16.5
Results Recorded from Contact Time Effect

Time for C_1	% Removal	Time for C_2	% Removal
15	82.00	15	87.00
30	88.20	30	92.10
45	88.21	45	92.11
60	88.21	60	92.11

303–323 K, using an oven. Results are presented in Tables 16.6 through 16.9 and in Figure 16.4 shown. The figure shows that as the temperature increases the removal percentage decreases. This suggests that the MB sorption process is exothermic (releases heat) in nature. This correlation could be because the bonds between the active site of the biosorbent and the MB are being weakened by the increase of temperature. Not only that, but as the temperature increases, the solubility of MB also increases, which means that the solute–solvent interaction is stronger than that of the solute–biosorbent interaction. As a result, the solute remains in the solution and is not adsorbed by the anise tea residue (biosorbent).

From Figure 16.4, it is observed that the optimum operating temperature is 303 K (30°C), as the percentage removal is optimal, keeping in mind that the other variables are kept constant: agitation speed = 200 rpm; biosorbent dose = 0.2 g; contact time = 30 min; and pH = 7.

TABLE 16.6
Results Obtained from the Temperature Effect Study for C_1 and C_2

$C_1 = 1(10)^{-5}$ M			$C_2 = 8(10)^{-6}$ M		
Temperature (K)	Absorbance	Concentration (M)	Temperature (K)	Absorbance	Concentration (M)
303	0.160	$0.95(10)^{-6}$	303	0.052	$0.5(10)^{-6}$
313	0.159	$1.9(10)^{-6}$	313	0.087	$0.95(10)^{-6}$
323	0.201	$2.5(10)^{-6}$	323	0.122	$1.45(10)^{-6}$

TABLE 16.7
Results Obtained from the Temperature Effect Study for C_3

$C_3 = 6(10)^6$ M		
Temperature (K)	Absorbance	Concentration (M)
303	0.114	$1.25(10)^{-6}$
313	0.148	$1.8(10)^{-6}$
323	0.172	$2.1(10)^{-6}$

TABLE 16.8
Removal Percentages of the Dye from Solution at Different Temperatures

Temperature for C_1	% Removal	Temperature for C_2	% Removal	Temperature for C_3	% Removal
303	90.5	303	93.8	303	79
313	81	313	88	313	70
323	75	323	82	323	65

TABLE 16.9
Results Obtained from Studying the pH Effect

$C_1 = 1(10)^{-5}$ M			
pH	Absorbance	Concentration (M)	% Removal
2	0.540	$6.9(10)^{-6}$	31
3	0.370	$4.7(10)^{-6}$	53
4	0.255	$3.2(10)^{-6}$	68
5	0.130	$1.6(10)^{-6}$	84
6	0.100	$1.2(10)^{-6}$	86
7	0.850	$1.0(10)^{-6}$	90
8	0.850	$1.0(10)^{-6}$	90
9	0.850	$1.0(10)^{-6}$	90
10	0.850	$1.0(10)^{-6}$	90

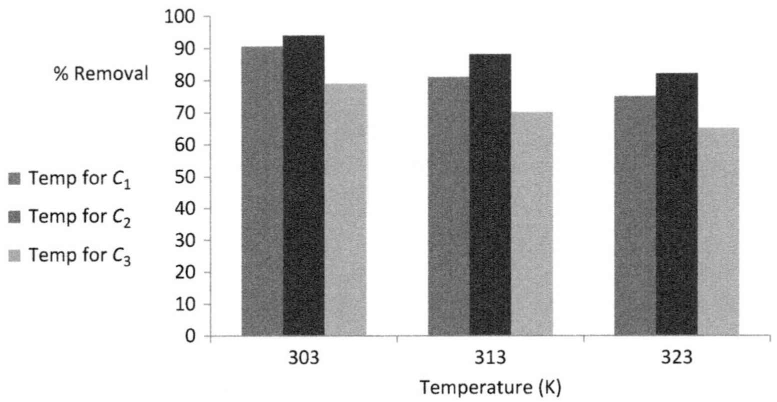

FIGURE 16.4 Effect of temperature on biosorption of methylene blue by anise tea residue. Other variables, biosorbent dose = 0.2 g, pH = 7, contact time = 30 min, and agitation speed = 200 rpm.

16.4.4 pH Effect

To look into the effect of pH on MB adsorption by anise tea residue, a series of batch processes were carried out over a range of pH 2–10. The pH value is one of the factors that affect the sorption behavior of the adsorbate, MB, onto the biosorbent surface, the anise tea residue, due to its impact on the surface binding site of the biosorbent and the dye solution. Results obtained from this experiment are recorded in Table 16.9. The results are presented graphically in Figure 16.5.

The percentage removal of the dye at equilibrium increases as the pH increases, considerably until pH 6. At pH 2 to pH 6, the percentage removal increases from 31% to 86%. However, the maximum removal (90%) occurred at pH 7. Hence, all further experiments were carried out at pH 7. This can be described by the change in the degree of ionization of the dye molecule and the surface charge of the biosorbent material. Any polar groups that become protonated (gain H^+ ion) at low pH will result in a net positive charge and deprotonation at high pH will result in a net negative charge. In other words, in acidic mediums, the negative charge from the biosorbent and the biosorbate will cause repulsion, hence the low biosorption uptake. However, at a higher pH, basic medium, the biosorbate and the biosorbent will attract each other causing a high sorption uptake.

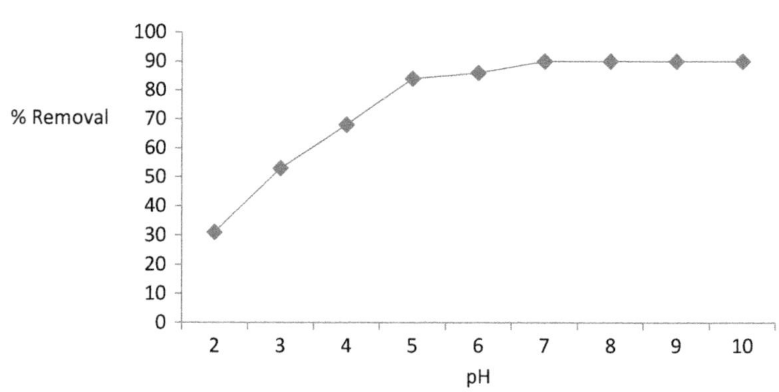

FIGURE 16.5 Effect of pH on sorption of methylene blue by anise tea residue. Other conditions, initial dye concentration = $1(10)^{-5}$ M, contact time = 30 min, biosorbent dose = 0.2 g and temperature = 303 K.

16.4.5 Initial Dye Concentration Effect

Initial dye concentration plays an important role when removing dye from the solution using a biosorbent, so tests were therefore conducted. Four different initial dye concentrations were prepared, $1(10)^{-5}$, $8(10)^{-6}$, $6(10)^{-6}$ and $4(10)^{-6}$ M. The results from this experiment are shown in Tables 16.10 and 16.11 and Figures 16.6 and 16.7.

The percentage removal of MB from the solution has increased from 87.5% to 93.8% when the initial dye concentration increased from $4(10)^{-6}$ to $8(10)^{-6}$ M. This may be attributed to an increasing concentration gradient providing an essential driving force to overcome all resistances of mass transfer of the dye molecules between the solid and aqueous phases, increasing the uptake of the dye molecules until saturation is achieved. However, when the initial concentration increased to $1(10)^{-5}$ M, a decrease in the percentage removal was observed, from 93.8% to 90.5%; this can be explained by the saturation of the adsorption sites at higher concentrations, whereas, at lower dye concentrations, the ratio of solute concentration to biosorbent sites is higher, causing an increase in color removal. The optimum initial dye concentration was recorded to be $8(10)^{-6}$ M.

TABLE 16.10
Results Recorded from the Experimental Work Done on the Initial Dye Concentration Effect and the Calculations Related to It

Initial Concentration (M)	Absorbance	Treated Concentration (M)	% Removal
$1(10)^{-5}$	0.160	$0.95(10)^{-6}$	90.5
$8(10)^{-6}$	0.052	$0.50(10)^{-6}$	93.8
$6(10)^{-6}$	0.061	$0.65(10)^{-6}$	89.2
$4(10)^{-6}$	0.049	$0.50(10)^{-6}$	87.5

TABLE 16.11
Converting Concentration from mol/L to mg/L

Concentration (mol/L) (M)	Concentration (mg/L)
$1(10)^{-5}$	3.74
$8(10)^{-6}$	2.99
$6(10)^{-6}$	2.24
$4(10)^{-6}$	1.50

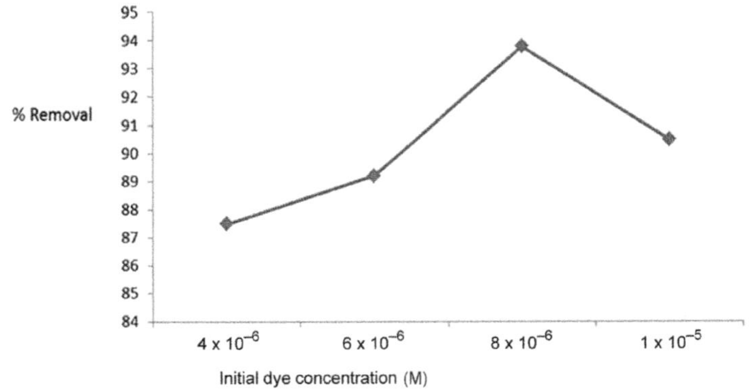

FIGURE 16.6 Effect of initial dye concentration on sorption of methylene blue by Anise tea residue. Other conditions, contact time = 30 min, pH = 7, biosorbent dose = 0.2 g and temperature = 303 K.

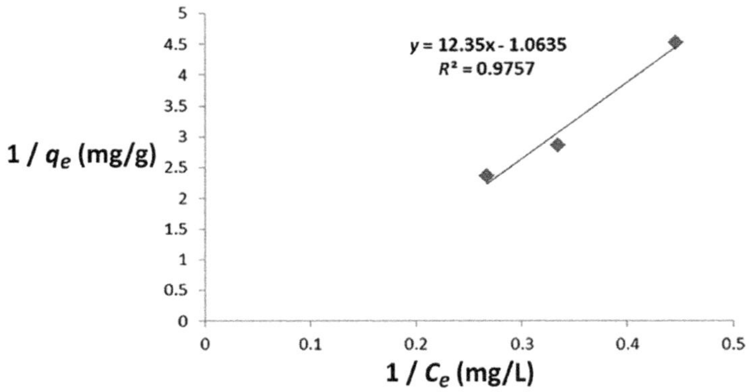

FIGURE 16.7 Langmuir adsorption isotherm for methylene blue dye on 0.2 g anise tea residue, 303 K temperature, contact time 30 min, and pH 7.

16.4.6 Adsorption Isotherms

In order to use the equations of adsorption isotherms, the concentration must be in mg/L. Throughout this research, the concentration was measured in mol/L, or in other words, molarity. Therefore, in order to convert the concentration from mol/L to mg/L, we used the following equation:

$$\text{Cocentration (mg/L)} = \text{concentration (M)} \times \text{Molecular Weight (374)} \times 1000 \quad (16.12)$$

The results obtained are shown in Table 16.11. The data to plot the curves for the Langmuir and Freundlich isotherms are shown in Table 16.12. The Langmuir and Freundlich model parameters determined using non-linear regression analysis are summarized in Table 16.13. To quantitatively compare the accuracy of the models, the correlation coefficients (R^2) were also calculated and are listed in Table 16.13. The graphical representations of the Langmuir and Freundlich isotherms are shown in Figures 16.7 and 16.8, respectively.

TABLE 16.12
Data to Plot the Curves for the Langmuir and Freundlich Isotherms

q_e	C_e	ln q_e	ln C_e	$1/q_e$	$1/C_e$
0.423	3.74	−0.86	1.319	2.36	0.267
0.35	2.99	−1.05	1.095	2.86	0.334
0.221	2.24	−1.51	0.806	4.52	0.446

TABLE 16.13
Langmuir and Freundlich Isotherm Constants for Adsorption of MB Dye Using Anise Tea Residue

Langmuir Adsorption Isotherm				Freundlich Adsorption Isotherm		
Q_o	B	R^2	R_L	K_F	n	R^2
0.940	0.086	0.9757	0.7567	0.923	0.780	0.9739

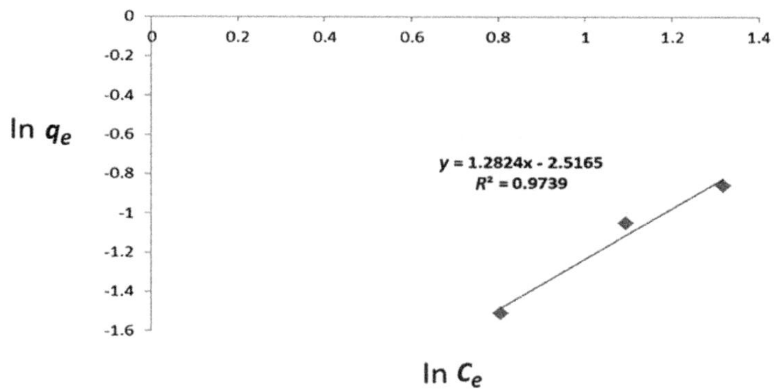

FIGURE 16.8 Freundlich adsorption isotherm for methylene blue dye on 0.2 g anise tea residue, 303 K temperature, contact time 30 min, and pH 7.

The high correlation coefficients ($R^2 = 0.9757$) indicate that the Langmuir isotherm model was the best fit applicable for describing the adsorption of MB onto the anise tea residue, as well as that R_L, separation factor, ranges between 0 and 1, which is consistent with the requirement for a favorable adsorption process. Moreover, $1/n$, the adsorption intensity for the Freundlich isotherm is less than one, and that does not favor this type of isotherm. This is because values of $n > 1$ represent favorable adsorption conditions. The suitability of the Langmuir isotherm model suggests monolayer coverage of dye molecules on the biosorbent surface (Chowdury, 2012).

16.4.7 Adsorption Kinetics

These calculations were carried out for concentration 1, which is $1(10)^{-5}$ M or 3.74 mg/L, in order to determine the kinetic model for this adsorption process. Other conditions such as 0.2 g of anise tea residue, temperature of 303 K, pH of 7, and agitation speed of 200 rpm were studied. For the pseudo-first-order model, Figure 16.9 and Table 16.14 show the results obtained from this investigation. On the other hand, Figure 16.10 and Table 16.15 show the results obtained from this investigation for the second-order model. The rate constants, predicted equilibrium uptakes, and corresponding correlation coefficients for the concentration (C_1 = 3.74 mg/L) tested have been calculated and summarized in Table 16.16. It can be seen that the correlation coefficients for the pseudo-second-order kinetic model were higher ($R^2 = 0.9997$) than that of the pseudo-first-order kinetic model ($R^2 = 0.935$). Also the theoretical (calculated), q_e

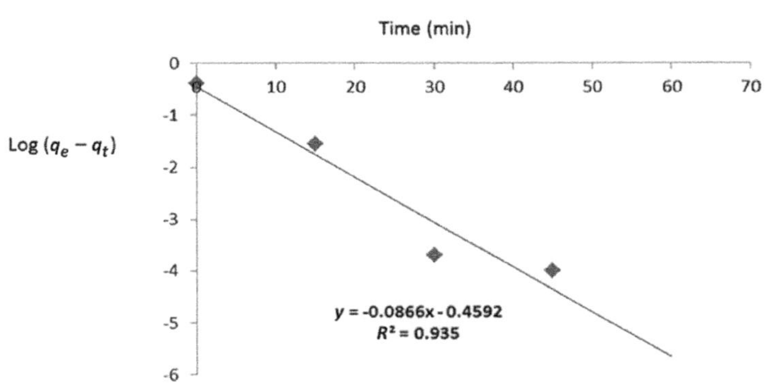

FIGURE 16.9 Pseudo-first-order kinetics plot for the adsorption of methylene blue by anise tea residue.

TABLE 16.14
Results and Calculations for First Order Model

Time (min)	q_e (mg/g)	q_t (mg/g)	Log ($q_e - q_t$)
0	0.414	0	−0.383
15	0.414	0.3860	−1.553
30	0.414	0.4138	−3.699
45	0.414	0.4139	−4.000
60	0.414	0.4140	—

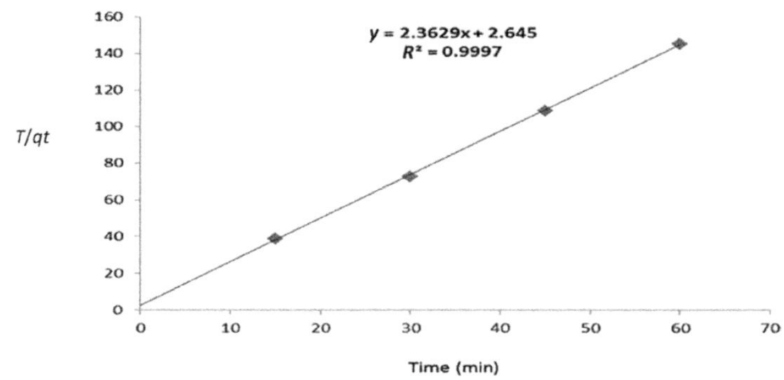

FIGURE 16.10 Pseudo-second-order kinetics plot for the adsorption of methylene blue by anise tea residue.

TABLE 16.15
Results and Calculations for Second Order Model

Time (T) (min)	$T^{1/2}$ (min)	q_t (mg/g)	T/q_t (min·g/mg)	T/q_e (min·g/mg)
0	0	0	—	—
15	3.873	0.3860	38.86	36.23
30	5.477	0.4138	72.50	72.46
45	6.708	0.4139	108.72	108.70
60	7.746	0.4140	144.93	144.93

TABLE 16.16
Pseudo First and Second Order Models Parameters for the Adsorption of MB by Anise Tea Residue

	Pseudo-First-Order Kinetic Parameters				Pseudo-Second-Order Kinetic Parameters			
Concentration (mg/L)	q_e Calculated (mg/g)	q_e Experimental (mg/g)	K_1 (min^{-1})	R^2	q_e Calculated (mg/g)	q_e Experimental (mg/g)	K_2	R^2
3.74	0.632	0.414	0.199	0.935	0.423	0.414	2.111	0.9997

calculated, value was closer to the experimental, q_e experimental, values at the studied initial MB concentration, indicating that the adsorption of MB by anise tea residue perfectly follows the pseudo-second-order kinetic model. These results suggest that the pseudo-second-order adsorption mechanism was predominant and that the overall rate of the MB sorption process appeared to be controlled by chemical process involving valence forces though the sharing of electrons between the MB dye (adsorbate) and the anise tea residue (adsorbent). Similar kinetics was also observed in the adsorption of Congo red dye on *Eichhornia crassipes* (roots) (Wanyonyi et al., 2014).

16.5 CONCLUSION

This study demonstrated that anise tea residue is indeed an efficient and viable biosorbent for the removal of MB dye from polluted wastewater. The efficiency of anise tea residue for removing MB dye from aqueous solutions was computed. Batch experiments indicated that biosorption significantly depends on adsorbent dosage, contact time, temperature, initial dye concentration, and pH. The percent removal was found to be optimum at 0.2 g of anise tea residue when added to $8(10)^{-6}$ M of initial dye solution for 30 min at a temperature of 303 K and a pH of 7.

The results indicated that the Langmuir isotherm model fitted the removal of the dye from aqueous solution better with $R^2 = 0.9757$, indicating monolayer biosorption on a homogeneous surface. Also the separation factor, R_L, proved that the biosorption process of MB by anise tea residue supported the Langmuir model rather than the Freundlich model. Kinetic studies confirmed that the rate of sorption was following the pseudo-second-order model with a high correlation coefficient, R^2, of 0.9997. Consequently, after calculating the theoretical q_e value from pseudo-second-order model, the results were almost the same as the experimental q_e value, which proves that the process is of second order. The study reveals that anise tea residue can be used as a widely available and economical alternative material for MB dye removal from wastewater, thus treating this polluted wastewater without any laborious pre-treatment.

REFERENCES

Alves, M. and Pereira, L. (2012). Dyes–Environment impact and remediation. In Malik, A. and Grohmann, E. (Eds.), *Environmental Protection Strategies for Sustainable Development*, pp. 111–162. New York: Springer.

Bhatnagar, A. et al. (2010). Coconut-based biosorbents for water treatment—A review of the recent literature. *Advances in Colloid and Interface Science* 160:1–15.

Chowdury, S. (2012). Biosorption of methylene blue from aqueous solutions by a waste biomaterial: Hen feathers. *Applied Water Science* 2:209–219.

European Public Health Alliance (2009). Air, water pollution and health effects. Retrieved from: http://www.epha.org/r/54.

Forest, E. and Roux, J.C. (1992). Heavy metal biosorption by fungal mycelial by products: Mechanisms and influence of pH. *Applied Microbiology and Biotechnology* 37:399–403.

Ghouri, D. M. (2010). Environmental pollution: Its effects on life and its remedies. *Journals of Arts, Science and Commerce* 2(2): 276–285.

Ho, Y.-S. (2004). Citation review of Lagergren kinetic rate equation. *Scientometrics* 59:171–177.

Kolorjet Chemicals Pvt. Ltd. (2015). *Synthetic Dyes*. Retrieved from Kolorjet Chemicals: http://www.dyes-pigments.com/synthetic-dyes.html.

Pezoti, O. et al. (2016). Percolation as new method of preparation of modified biosorbents for pollutants removal. *Chemical Engineering Journal* 283:1305–1314.

Wanyonyi, W. C. et al. (2014). Adsorption of Congo Red from aqueous solutions using roots of Eichhornia crassipes: Kinetic and equilibrium studies. *Energy Procedia* 50:862–869.

17 Laser-Induced Breakdown Spectroscopy (LIBS) as an Evaluation Tool for Improving Industrial Wastewater Quality by Different Adsorption Methods

Nashwa Tarek El-Tahhan

CONTENTS

17.1 INTRODUCTION

Industrial wastewater is one of the most important pollutant sources for environmental water pollution. During the last century, a large amount of industrial wastewater was drained into lakes, rivers and coastal areas. There are many types of industrial wastewater based on different contaminants and industries; each sector produces its own combination of pollutants. This resulted in serious pollution problems and caused negative effects on human life and the ecosystem [1].

17.1.1 Heavy Metal Ions

Heavy metals are elements having atomic weights between 63.5 and 200.6, and a specific gravity greater than 5.0. Heavy metals in industrial wastewater—including lead, chromium, mercury, uranium, selenium, zinc, arsenic, cadmium, silver, gold and nickel—are very dangerous to human health and the environment. Some industrial effluents contain high levels of several heavy metal ions. Unlike organic pollutants, the majority of which are liable to biological degradation, heavy metals will not degrade into harmless end products, so it is needful to remove them before drainage into the environment.

17.1.1.1 Chromium Toxicity

Chromium has been found to be one of the most toxic metals present in water obtained from industrial effluents. Chromium, a highly reactive metal, is widely used in electroplating, metal finishing and in the textile industries. Tannery wastewater generally contains a high concentration of solids, organic matters, chromium, sulfides and sulfate. The contamination levels of chromium in tannery wastewater are observed to be many times higher than the values permissible by the Egyptian Environmental Quality standards. Chromium has many troubling effects on human health, including: (a) breathing high levels can cause irritation to the lining of the nose, resulting in running nose,

nose ulcers and breathing problems such as asthma, cough, shortness of breath, or wheezing; (b) skin contact can cause skin ulcers and allergic reactions including severe redness and swelling of the skin; and (c) long-term exposure can cause damage to liver, kidney, circulatory and nervous tissues, as well as skin irritation [2].

It is therefore necessary to discover a non-toxic adsorbent from a natural source to combat the menace of chromium water pollution.

17.1.2 Adsorbents

17.1.2.1 Rice Husk

Rice husks are a non-toxic and environmentally friendly agricultural waste material generated in rice-producing countries, including Egypt. The annual world rice production is approximately 500 million metric tons, of which 10%–20% consists of rice husks. At present, researchers have focused on the use of modified or unmodified rice husk as an adsorbent for the removal of pollutants. Rice husk is considered a very attractive and alternative adsorbent due to its high adsorption capacity and rapid uptake. Rice husk for an adsorption experiment was sifted to less than 2 mm, which was followed by washing the rice husk several times in tap water, followed by washing with deionized water. After washing, it was soaked in NaOH with concentration of 2N for 45 min, the treated rice husk (TRH) was sun dried at temperature 30°C ± 2°C for about 30 h.

17.1.2.2 Activated Charcoal

Adsorption using activated carbon is an effective method for the treatment of wastewater rich in heavy metal ions. Cost is, however, an important parameter when comparing sorbent materials.

Commercial activated carbon is very expensive, because of high material costs; there is a need to explore less expensive cleanup methodologies for the eradication of metal ions from aqueous systems. In this study a ready-for-use activated charcoal (AC) "AquaSorb 1000 (Granular coal-based activated carbon)" was purchased. AC was used in some of the fixed-bed experiments, for comparison with the effectiveness of TRH.

17.1.3 Detection Method

Most of the detection techniques of heavy metal ions in industrial effluent are carried out by conventional analysis techniques like atomic absorption spectroscopy (AAS) and UV-Visible spectrophotometry. Both analytical techniques require time-consuming sample preparation protocols. Laser-induced breakdown spectroscopy (LIBS) was used in this study as an alternative instrumental methodology that uses direct introduction and analysis of chromium ions in wastewater. LIBS is a laser-based, non-destructive, rapid and sensitive optical diagnostic technique used to detect certain atomic species. A measuring device by LIBS has been established which included the Nd:YAG laser, Echelle spectrograph, intensified charge-coupled device (ICCD) and liquid jet.

17.1.4 Methods of Treating Wastewater

Several techniques for wastewater treatment have been used for the metal ion removal, including sedimentation, lime precipitation, floatation, coagulation, filtration, biological, membrane process, and ion exchange; some of these methods, however, have disadvantages that include the generation of toxic sludge, incomplete metal removal, continuous input of chemicals and high cost.

Adsorption is a widely used technique in the removal of pollutants because it is a simple and relatively economic method. Adsorption is a process that takes place when a gas or liquid solute accumulates on the surface of a solid or a liquid (adsorbent), forming a molecular or atomic film (the adsorbate). It is one of the more effective methods for removing heavy metals from industrial wastewaters, which is why it is used in this study. Depending on the type of pollutant, the adsorbents used may vary due to the change in adsorption conditions [3].

In the present chapter, the effectiveness of AC and TRH in removing chromium ions from synthetic wastewaters was studied using batch and fixed-bed column adsorption techniques. Batch experiments were conducted to know the influence of various parameters of adsorption on the removal of chromium metal ions by TRH, followed by a fixed-bed column adsorption technique to be more practical.

This study also aims at finding an alternative instrumental methodology that uses direct introduction and analysis of chromium ions in wastewater without undergoing time-consuming sample preparation steps. LIBS could be such technique.

In the batch adsorption technique, the adsorption of Cr (III) ions was found to be affected by solution pH, contact time, adsorbent dosage and initial metal ion concentration. It was found that the removal percentage of metal ions increased with increasing adsorbent dose and contact time. The adsorption efficiency increased with increasing initial metal concentration.

The equilibrium was analyzed using the Freundlich, Temkin and Dubinin–Radushkevich isotherm models. The data was found to have a closer correlation with the Freundlich isotherm, as evidenced by a higher correlation of determination (R^2). The kinetics data was also subjected to pseudo-first-order and the pseudo-second-order kinetic models using three different initial concentrations (50, 100, and 150 ppm). The data was better explained using the pseudo-second-order kinetic model.

The fixed-bed column experiments were carried out for different influents of heavy metal ion concentrations, bed depths and flow rates. In the fixed-bed adsorption it was found that the removal percentage of chromium ions increased with an increase in bed depth and decrease of flow rate. Although a slight increase in removal percentage was noticed when using charcoal, TRH was generally preferred owing to its low cost.

In the fixed-bed adsorption, Experiments were carried out at a temperature 25°C ± 2°C, three different flow rates with three different concentrations at three different heights were investigated (3, 4, 5 L/min; 30, 100, 150 ppm; 35, 58, 82 cm), respectively. It was found that the removal percentage of chromium ions increased with an increase in bed depth and decrease of flow rate. Although a slight increase in removal percentage was noticed when using charcoal, TRH was generally preferred owing to its low cost.

17.2 EXPERIMENTAL MATERIAL AND METHODS

17.2.1 Bio-sorbents

17.2.1.1 Rice Husk

Rice husk is a good adsorbent as it is insoluble in water, has high mechanical strength, has good chemical stability and possesses a granular structure. It consists mainly of crude protein (3%), ash (including 17% silica), lignin (20%), hemicellulose (25%) and cellulose (35%). Pretreatment of rice husks can remove lignin, hemicellulose, reduce cellulose crystallinity and increase the porosity or surface area [4].

Rice husk was supplied by the local rice mill in Sharkia, Egypt. Since the slots in the adsorption column used are less than 2 mm, the rice husk was sifted to avoid clogging of the slots. This was followed by washing the rice husk until water used in washing was almost clear; this process was repeated at least four times to remove particular material, then it was also washed in deionized water for about 15 min until all apparent excess material was removed from the rice husks. After that the clean rice husk was soaked in NaOH with a concentration 2N (8%) about 45 min. Bases increase adsorption efficiency due to negatively charged hydroxyl anions, allowing the adsorption of metal cations on the surface. The TRH was sun dried on a marble sheet at a temperature of about

30°C ± 2°C for about 30 h to ensure completely dryness. The cleaned TRH particles were stored in pre-cleaned polypropylene bottles and kept airtight to avoid absorption of moisture from the atmosphere. TRH prepared in this way were used for the experiments.

17.2.2 Instrumentation

These are the instruments that were used in this research:

17.2.2.1 Mechelle 7500 System

The high dispersion of the echelle grating produces a highly resolved spectrum in one direction, which consists, due to the limited free spectral range, of many overlapping orders. A proper order sorter modulus separates the different orders along the other direction, allowing the use of the entire surface of the detector for reconstructing a full broadband spectrum at high resolution, without the overlapping of the different diffraction orders. Echelle technology is not new; the first experiments were carried out with echelle spectrographs modified for combination with an ICCD camera, but it was not initially designed for this purpose.

Meanwhile, at least three systems are now commercially available with integrated ICCD cameras. The spectrometer used in this study is an echelle spectrometer (Mechelle 7500) with a focal length of 17 cm and f-number of 5.2. It provides a constant spectral resolution (CSR) of 7500, corresponding to 4 pixels FWHM, over a wavelength range of 200–900 nm displayable in a single spectrum. A gateable ICCD camera (DiCAM-Pro from PCO computer optics, equipped with a high-resolution CCD sensor of 1280 × 1024 pixels, pixel size 9 × 9 μm^2) was coupled to the spectrometer. The overall linear dispersion of the Mechelle spectrometer ranges from 0.0078 (at 200 nm) to 0.038 nm/pixel (at 1000 nm).

17.2.2.2 (DiCam Pro) ICCD

This is a high performance intensified CCD camera system with gating times down to 3 ns. With its 12-bit dynamic range and a high resolution CCD image sensor, it features an excellent signal-to-noise-ratio and single photon detection. This system is suited for applications in environments with high electromagnetic disturbances. A high-speed serial fiber-optic data link connects the system to the computer. The camera can be triggered by light or electrical input.

It has some advantages:

- Excellent sensitivity of the system allows single photon detection.
- Fast shutter down to 3 ns.
- High-resolution micro channel plate–image intensifier and CCD (1280 × 1024 pixel).
- 12-bit dynamic range.
- Exposure times from 3 ns to 1000 s.
- Spectral sensitivity from ultra violet to near infrared.

17.2.2.3 Quantel Ultra Compact Nd:YAG

The Ultra laser is a lamp-pumped 1064 nm Nd:YAG laser featuring a degree of ruggedization not found in exemplary scientific lasers. The Ultra design has been vibration tested and each laser is temperature cycled overnight, and tested again before shipping to ensure the laser arrives aligned and ready for use. Some advantages of the Quantel Ultra Compact Nd:YAG include: the alignment is guaranteed, the cables are quickly detachable, the resonators are Gaussian or multimode, and it is compact and portable.

17.2.2.4 PS 1000 R2 Balance

The PS.R2 balances (shown in Figure 17.1) represent a new standard of accuracy in balances. They have many advantages, including a readable LCD display that allows a clearer presentation of the weighing result. The display has a new text information line allowing the display of additional data and message.

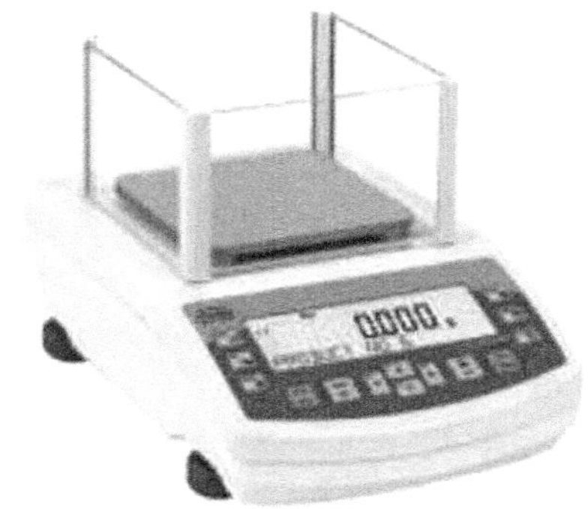

FIGURE 17.1 PS.R2 Balance.

The PS.R2 balances, like the previous PS series balances, have pans in two possible dimensions: 195 × 195 mm or 128 × 128 mm. Balances with a pan 128 ×128 mm have a draft shield. The balance measurement accuracy and precision is proven by automatic internal adjustment, which takes into consideration time flow and temperature change.

17.2.2.5 W5 Deep Bed Filter Column

The W5 Deep Bed Filter Column (shown in Figure 17.2) consists of a clear Perspex column (100 mm internal diameter × 1350 mm long) mounted in a floor standing framework approximately 2 m high; it includes a service system that consists of: two sump tanks (each 350 L capacity) one for influent solution and other for effluent solution, flow controller, rotameter (range: 0.5–5.0 L/min), control valves, tubing, sampling tubes and a bank of 41 differential manometers. The column can be operated as a pressure filter up to 1 bar g. The filtration medium is supported by a corrosion-resistant gauze mesh below, which is packed with 1 kg of 10 mm Ballotini to ensure good wash water distribution. Sampling and manometer probes are located at 20 mm depth intervals, but staggered in position, over 0.8 m depth.

FIGURE 17.2 W5 Deep bed filter column used in the experiments.

17.3 PREPARATION OF STANDARD SOLUTIONS FOR CALIBRATION CURVE

For the construction of the calibration curve a series of standard solutions of chromium (III) of known concentrations were prepared. Chromium solution samples of concentrations (2, 4, 6, 8, 10, 20, 30, and 40) mg/L were prepared by dissolving each of the calculated concentration amounts of chromium chloride ($CrC13.6H_2O$) into 1 L tap water; ten bottles were prepared, each labeled with its concentration. After that, 40 mL of each solution (with different concentrations) was taken and put in bottles, each with its label and sealed, and were thus ready for analysis.

17.3.1 Analysis of Chromium Ion in Standard Solutions for Calibration Curve

The analysis was carried out by LIBS. The samples were introduced to the LIBS experiment by adding 2 drops of the solution on an ashless filter paper (qualitative circles 125 mm diameter), which was left to dry at room temperature for 10 min before the LIBS experiments were performed; the filter paper was then placed in front of the laser pulse. The excitation laser used was an Nd:YAG laser of wavelength 1064 nm, laser pulse energy of 40 mJ and pulse duration 6 ns. The laser was focused on the sample using a 10 cm focal length quartz plano-concave lens. The light emitted by the plasma plume was collected using a quartz optical fiber with a 0.6 mm aperture. The fiber was mounted at 1.2 cm with an angle of 45° to ensure that the plasma plume was in its collection cone. The light was then introduced to an echelle spectrometer (Mechelle 7500) coupled with an ICCD camera (DiCam pro). In order to get the best signal to noise ratio, the camera was optimized to a 1500 ns delay time and a gate of 2500 ns.

To decrease the shot-to-shot fluctuations that may occur in a LIBS experiment, the spectrum was taken as the average of three spectra. Each spectrum is the accumulation of 10 shots. To be sure that the filter paper was not penetrated by the laser, each shot was taken on a fresh surface in addition to choosing a laser pulse energy of 40 mJ. The average spectrum of each sample was then normalized on the carbon spectral line at 247.8 nm.

The previous procedures were done to monitor the residual chromium ion in each sample in both batch and fixed bed experiments.

17.3.1.1 Preparation of Stock Solution

The synthetic wastewater solution for chromium ion was prepared from analytical grade chemicals. The stock solution of chromium (III) was prepared from $CrCl_3.6H_2O$ for the batch experiments. The stock solutions were prepared by dissolving the exact quantity of $CrCl_3.6H_2O$ in tap water to obtain the required concentration in ppm (mg/L) as described earlier.

The experiments were done by batch and fixed bed adsorption. Batch adsorption tests give information on adsorption equilibrium characteristics and adsorption kinetics. However, batch operations are not economical in practice, and the data from fixed-bed column operations were important for industrial adsorber design. Column experiments were conducted to understand the adsorption behavior in fixed-bed columns.

17.3.2 Batch Experiments

The adsorption experiments were carried out by a batch technique at ambient temperature (25°C ± 2°C). For each experimental run, a fixed amount of the TRH (0.1 g) was massed on the PS.R2 balance. The measured TRH was then transferred into 100 mL polypropylene bottles that were numbered and kept.

17.3.2.1 Batch Experiments to Determine the Optimum Time

Solutions of concentrations 50, 100, and 150 ppm (mg/L) were prepared by dissolving the calculated amounts of chromium chloride ($CrCl_3.6H_2O$) in tap water. Eight bottles from each concentration were prepared from 40 mL of each concentration and were each put in a bottle containing 0.1 g TRH.

Each bottle was left for a time that ranged from 0.25 min to 24 h before it was filtered. This process was done to determine the optimum time, that is, the maximum time after which 0.1 g of the adsorbent cannot remove more chromium ion. The solution was then filtered using filter paper and the filtrate was analyzed by LIBS to measure the residual concentration of the metal ion left in the solution after adsorption onto TRH.

17.3.2.2 Batch Experiments to Determine the Optimum Amount of Adsorbent

After determining the optimum pH and optimum time, it was essential to check the optimum amount of adsorbent. Five bottles were prepared, each containing 40 mL of the prepared solution at a concentration 100 ppm (mg/L). Different amounts of adsorbent were added to each bottle, ranging from 0.02 to 0.2 g. The pH was adjusted to the optimum pH (6) and the five bottles were left for the optimum time (10 h) and then were filtered to remove the adsorbent. The filtrate was analyzed by LIBS to measure the residual concentration of the metal ions left in the solution after adsorption onto the TRH.

17.3.2.3 Batch Experiments to Determine the Effect of Initial Concentration

To determine the effect of initial concentrations, seven bottles were prepared with different initial concentrations ranging from 25 to 700 ppm (mg/L). Each of the bottles contained 0.1 g adsorbent (TRH) at pH 6 and 40 mL of the prepared solution at a different initial concentration; the seven bottles were left for 10 h, and were then filtered to remove the adsorbent. The filtered solution was analyzed by LIBS to measure the residual concentration of metal ions left in the solution after adsorption onto the TRH.

- Calculation of the Percentage of Removal of Heavy Metal:

$$\text{Removal efficiency (\%)} = \frac{[(C_o - C_e) \times 100]}{C_o}$$

where:

C_o is the initial concentration of the metal ion in the solution (mg/L), calculated from the measured concentration of the stock solution.

C_e is the concentration of the metal ion (mg/L) left in the solution after adsorption equilibrium was reached.

- Calculation of the Amount of Metal Adsorbed on Adsorbent:

$$q_e = \frac{[(C_o - C_e) \times V]}{W}$$

where:

C_o is the initial concentration of metal ion (mg/L).

C_e is the concentration of the metal ion (mg/L) left in the solution after adsorption equilibrium was reached (mg/L).

V is the total volume (L) of the solution stock.

W is the mass of the adsorbent used (g).

17.3.3 Adsorption Isotherm and Models

Adsorption isotherm is an empirical relationship used to predict how much solute can be adsorbed by TRH. Adsorption isotherm is defined as a graphical representation showing the relationship between the amount adsorbed by a unit weight of adsorbent and the amount of adsorbate remaining in a test medium at equilibrium; it shows the distribution of absorbable solute between the liquid and solid phases at various equilibrium concentrations.

In the present study the adsorption from aqueous solution at equilibrium was described by Freundlich, Temkin and Dubinin–Radushkevich isotherms.

17.3.3.1 Freundlich Isotherm Model

The standard model of Freundlich equation used is represented as follows:

$$\ln q_e = \ln K_f + (1/n)\ln C_e$$

where:

K_f is the Freundlich characteristic constant [(mg·g^{-1}) (L·g^{-1})1/n], obtained from the intercept of ln (q_e) vs. ln C_e linear plot.

1/n is the heterogeneity factor of sorption, obtained from the slope of ln q_e vs. ln C_e linear plot.

q_e is the amount of metal adsorbed onto the adsorbent (mg/g) and is calculated from the following equation:

$$q_e = \frac{[(C_o - C_e)V]}{W}$$

The Freundlich isotherm was used for the concentration range from 50 to 900 ppm (mg/L), volume of solution used V is 40 mL and the weight of adsorbent W is 0.1 g.

17.3.3.2 Temkin Isotherm Model

The standard model of the Temkin equation used is represented as follows:

$$q_e = B\ln AT + B\ln C_e$$

$$B = \frac{RT}{bt}$$

where:

AT is the Temkin isotherm equilibrium binding constant (L/g).

bt is the Temkin isotherm constant.

R is the universal gas constant (8.314 J/mol/K).

T is the Temperature at 298 K.

B is the Constant related to heat of sorption (J/mol).

Then the quantity q_e against ln C_e was plotted and the constants were determined from the slope and intercept.

17.3.3.3 Dubinin–Radushkevich Isotherm Model

The standard model of Dubinin–Radushkevich equation used is represented as follows:

$$\ln q_e = \ln q_s - (K_{ad}\varepsilon^2)$$

$$E = \left[\frac{1}{(2\mathrm{BD})^{1/2}}\right]$$

$$\varepsilon = RT\ln\left[1 + \left(\frac{1}{C_e}\right)\right]$$

where:

q_e is the amount of adsorbate in the adsorbent at equilibrium (mg/g).

q_s is the theoretical isotherm saturation capacity (mg/g).

K_{ad} is the Dubinin–Radushkevich isotherm constant (mol²/KJ²).
ε is the Dubinin–Radushkevich isotherm constant.
R is the universal gas constant (8.314 J/mol K).
T is the absolute temperature 298 K.
C_e is the adsorbate equilibrium concentration (mg/L).

From the linear plot of the Dubinin–Radushkevich–Kaganer (DRK) model (ln q_e vs. ε^2), q_s was determined in mg/g, the mean free energy E in KJ/mol and R^2.

17.3.4 Kinetic Studies and Models

17.3.4.1 Pseudo-First-Order

The linearized pseudo-first-order equation used is represented by the following equation

$$\log(q_e - q_t) = \log q_e - \frac{K_{pf}}{2.303t}$$

where:

q_e is the amount of metal adsorbed at equilibrium (mg/g).
q_t is the amount of metal adsorbed at time t (mg/g).
K_{pf} is the adsorption rate constant (min^{-1}) [first order constant].

The time (t) in min was plotted versus log ($q_e - q_t$). K_{pf}, q_e, R^2 were determined from the plots and intercept of the curves.

The pseudo-first-order was calculated for the three concentrations 50, 100, and 150 ppm (mg/L), at times 15, 30, 60, and 240 min for each concentration.

$$q_t = \frac{[(C_o - C_e)V]}{W}$$

where:

C_e is the concentration at equilibrium after each certain time (mg/L).
C_o is the initial concentration (mg/L).
V is the volume of solution (L).
W is the weight of adsorbent (g).
q_t is the amount of adsorbed (mg/g) metal at time t.

17.3.4.2 Pseudo Second Order

The linearized Lagergren second-order kinetic equation used was represented by the following equation:

$$\frac{t}{q_t} = \frac{1}{K_{ps}q_e} + \frac{t}{q_e}$$

where:

q_e is the amount of adsorbed metal at equilibrium (mg/g).
q_t is the amount of adsorbed metal at time t (mg/g).
K_{ps} is the adsorption rate constant (g/(mg·min)) [second order constant].

We plotted time (t) in min versus t/q_t, K_{ps}, q_e, and R^2 were determined from the slope and intercept of the curve.

The pseudo-second-order was calculated for the three concentrations 50, 100, and 150 ppm (mg/L), at times 15, 30, 60, and 240 min for each concentration.

17.3.5 Fixed-Bed Adsorption Study

Fixed-bed column study was conducted using W5 Deep Bed Filter Column. The experiments were carried out at ambient temperature (25°C ± 2°C). At the top of the column the solution with initial chromium concentration was pumped into a column filled with adsorbent (TRH or AC). Three different flow rates (3, 4, and 5 L/min) with three different concentrations (30, 100, and 150 mg/L [ppm]) at three different heights (35, 58, and 82 cm) were investigated. A total of 54 samples were studied.

The first run was performed with an initial concentration of 30 mg/L (ppm) and height of 35 cm using the different flow rates 3, 4, and 5 L/min. The influent tank was filled with solute ion of an initial concentration of 30 mg/L chromium ion and the column filled with TRH as adsorbent.

The flow meter was initially adjusted at 3 L/min, the pump power to turn on and a tap at height 35 cm opened to take a 40 mL sample from the treated solution, in a polypropylene bottle which was sealed and thus ready for analysis. The previous procedure was repeated for flow rates of 4 and 5 L/min.

The second run was performed using the same initial concentration of 30 mg/L (ppm) but with a bed height of 58 cm and the three different flow rates 3, 4, and 5 L/min to check the effect of bed height on the percentage of removal. The influent tank was filled with chromium solution of concentration 30 mg/L. The flow meter was adjusted at 3 L/min. The tap of 58 cm height was opened to take a 40 mL sample of treated solution in a polypropylene bottle, which was sealed, thus ready for analysis. The previous steps were repeated using the flow rates of 4 and 5 L/min. The third run was performed using an initial concentration of 30 ppm, bed height of 82 cm and three different flow rates of 3, 4, and 5 L/min.

Prior to the fourth run, the residual solution was removed from the influent tank; the tank was carefully washed with tap water and then dried with a clean towel. The experiment was repeated using the different bed heights of 35, 58, and 82 cm and different flow rates.

17.4 RESULTS AND DISCUSSION

17.4.1 Calibration Curve

Exhibited calibration curves were constructed by running chromium standards at several wavelengths: 427.46, 428.96, 520.4, 520.6, and 520.84 nm. A linear regression approach was used to fit the appropriate trend lines to LIBS. All calibration curves demonstrated good linearity with correlation of determination R^2 = 0.9743, 0.8757, 0.8799, 0.874, and 0.9098, respectively. These chromium atomic lines were chosen because they are:

1. Not resonant
2. Isolated
3. Well-defined
4. Have no self-adsorption
5. Symmetric

Figure 17.3 represents the calibration curve with wavelength 427.46 nm and correlation of determination R^2 0.9743; this calibration curve was chosen because it has the best correlation of determination and all samples demonstrated good linearity with this wavelength.

The samples prepared for the calibration curve were of concentration ranging from 4 ppm to 40 ppm, as it was observed that there saturation was reached above 40 ppm, which could be attributed to self-absorption.

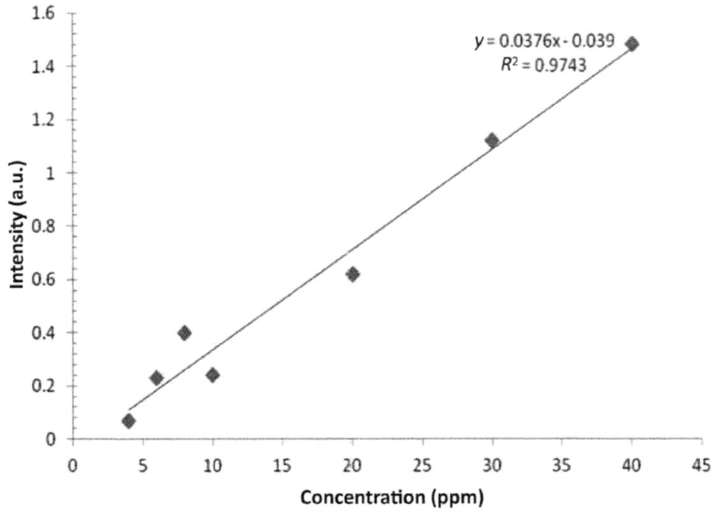

FIGURE 17.3 Calibration curve at wavelength 427.46 nm.

17.4.2 Batch Adsorption Studies

17.4.2.1 Effect of Contact Time with Different Concentrations

To establish an appropriate contact time between the TRH and metal ion solution, adsorption capacities were measured as function of time. Initial chromium concentrations of 50, 100, and 150 ppm were used with 0.1 g of TRH as adsorbent. Figure 17.4 shows that the percent of TRH removal is high at the beginning, which agrees with previous studies on other adsorbents and heavy metals [5–7]. For all initial concentrations used, greater 80% removal was achieved within the first 15 min. The initial rapid phase may be due to the large number of vacant sites available at the initial period of the sorption. As the surface adsorption sites become exhausted and the equilibrium between the heavy metal inside the adsorbent and in the solution occurs, the uptake rate is controlled by the rate at which the adsorbate is transported from the exterior to the interior sites of the adsorbent particles.

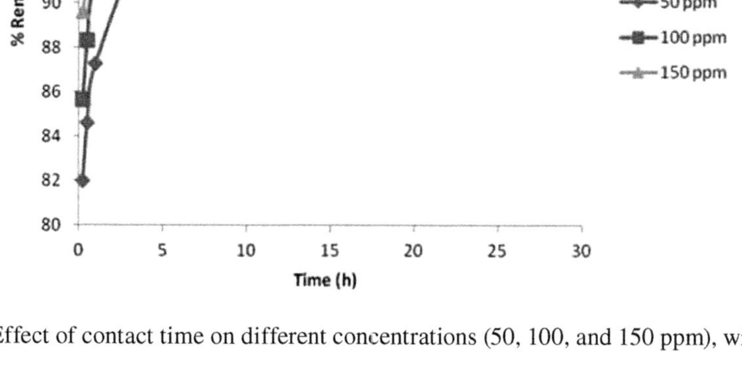

FIGURE 17.4 Effect of contact time on different concentrations (50, 100, and 150 ppm), with 0.1 g TRH.

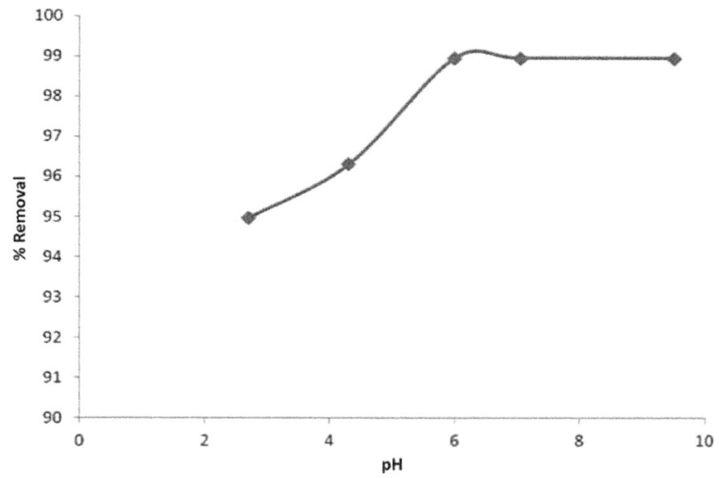

FIGURE 17.5 Removal efficiency of Cr(VI) ions by TRH versus pH (initial concentration 100 ppm, contact time 10 h, amount of adsorbent 0.1 g and 25°C ± 2°C).

It does not seem to be of much benefit to leave the TRH in the chromium ion solution longer than 10 h. The equilibrium time of 10 h was therefore selected for all further studies. Figure 17.4 is a collective plot of the removal percentage by time using different initial concentrations of 50, 100, and 150 ppm. It is clear that the percentage of removal reached 96.86%, 97.9%, and 98.2%, respectively.

17.4.2.2 Effect of pH

Figure 17.5 shows the pH adsorption edges on TRH of chromium ion concentration of 100 ppm. The lower the pH, the more H^+ ions competed with the metal ions for adsorption sites, thus reducing their adsorption. On the other hand, the higher the pH, the fewer the H^+ ions competing with metal ions for adsorption sites, thus increasing their adsorption, which explains the results presented in Figure 17.5. This result is in agreement with previous studies on other adsorbents [8–10]. The percent removal peaked at pH 6 with a removal efficiency of 98.97%, so pH 6 was selected for future tests. At a pH higher than 6 the amount adsorbed was found to be constant, as all the TRH pores become saturated.

17.4.2.3 Effect of Different Amount of Adsorbent

The accessibility and availability of adsorption sites are controlled by adsorbent dosage. The effect of mass of adsorbent loading on chromium ion removal using TRH was investigated by varying the adsorbent loading weight from 0.02 to 0.2 g per 40 mL of metal ion solution. Figure 17.6 represents the removal efficiency of chromium (III) ions vs. amount of TRH. It can easily be observed that the percentage removal of metal ions increased with the increasing weight of TRH.

This is due to the greater availability of exchangeable sites or surface area at a higher dose of the adsorbent. This explanation is in agreement with previous studies on many other adsorbents [9,11]. Increasing the amount of adsorbent increased the percentage of removal until it reached 98.96% at 0.1 g adsorbent. The percentage of removal was found to remain constant at 98.96% despite further increases to the amount of adsorbent.

17.4.2.4 Effect of Initial Chromium Concentrations

Figure 17.7 shows the effect of initial chromium ion concentration on the adsorption efficiency of TRH. The adsorption efficiency increased with an increasing initial chromium ion concentration. This result is in accordance with the work of Agbozu and Emoruwa [12].

It is generally expected that as the concentration of the adsorbate increased the amount of metal ions removed would also increase. It is believed that an increase in the concentration of the adsorbate

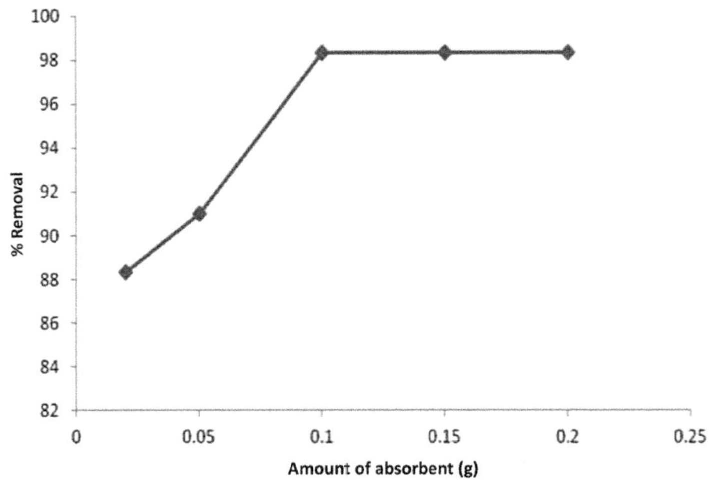

FIGURE 17.6 Removal efficiency of Cr (VI) ions versus amount of TRH (Initial concentration 100 ppm, pH 6, contact time 10 h and 25°C ± 2°C).

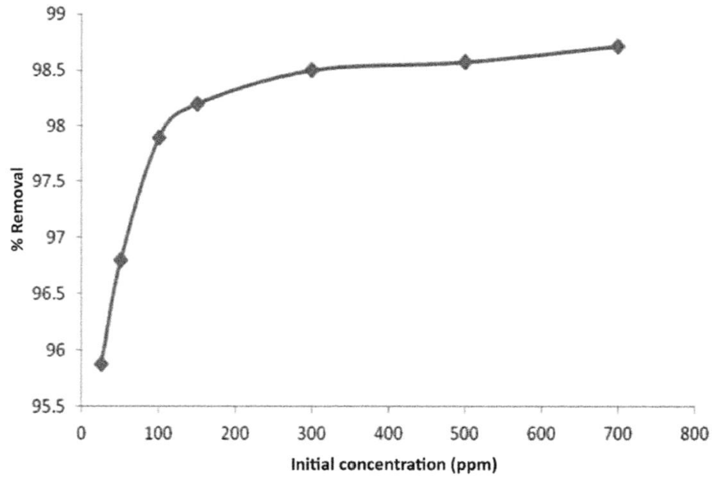

FIGURE 17.7 Removal efficiency of Cr (III) ions versus initial concentrations (pH 6, contact time 10 h and temperature 25°C ± 2°C).

brings about an increase in the completion of adsorbate molecules, for few available binding sites on the surface of the adsorbent increases the amount of metal ions removed. At low initial concentrations the percentage of removal is low and the concentration increased as the percentage of removal increased until it reached 98.71% at an initial chromium ion concentration of 700 ppm.

As the initial concentration increased the percent of removal increased due to the increase in the exchangeability between chromium ions and exchangeable sites (OH^-) until all sites are occupied with positive ions.

17.4.3 Adsorption Isotherms and Models

The linearized forms of Freundlich, Temkin and Dubinin–Radushkevich isotherms were compared. The correlation of determination demonstrates that the Freundlich and Dubinin–Radushkevich

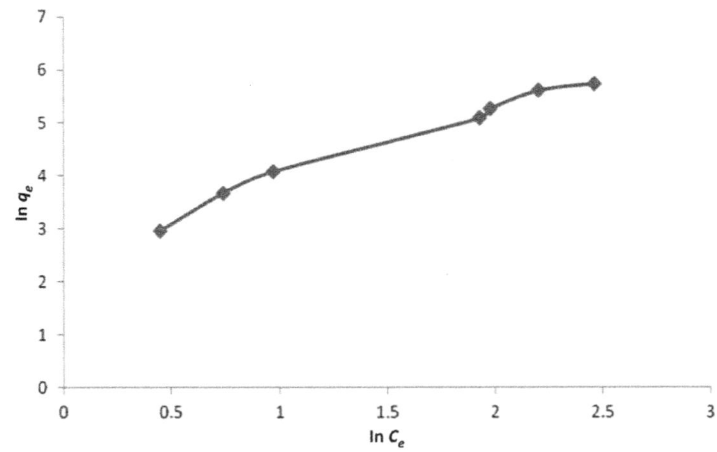

FIGURE 17.8 Freundlich isotherm for Cr (III) adsorption on TRH.

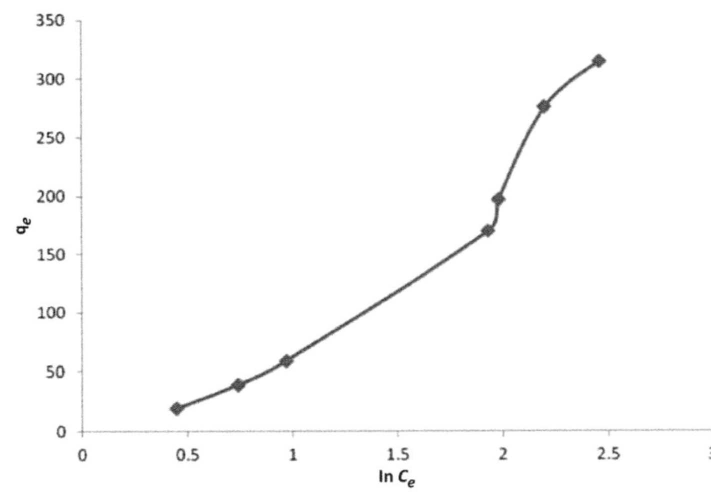

FIGURE 17.9 Temkin isotherm for Cr (III) adsorption on TRH.

models adequately fit the data for Cr^{+3} adsorption. However, the coefficient of determination (R^2) values is higher in the Freundlich model for chromium adsorption compared to other models. The experimental data for heavy metal ions fit well with the linearized Freundlich, Temkin and Dubinin–Radushkevich isotherms. R^2 values ranged from 0.9594 to 0.8706 for adsorption of Cr^{+3}. Figure 17.8 presents the plot of log q_e versus log C_e, which was found to be linear, indicating the applicability of the Freundlich model. The intercept of the line is roughly an indicator of the adsorption capacity, and the slope is an indication of the adsorption intensity.

Figure 17.9 presents the Temkin isotherm for Cr (III) adsorption on TRH and Figure 17.10 shows the Dubinin–Radushkevich Isotherm for Cr (III) adsorption on TRH.

17.4.4 Kinetic Study

Various adsorption kinetics models have been used to describe the adsorption of metal ions.

The first-order kinetics process has been used for reversible reaction with an equilibrium being established between liquid and solid phases. In most cases, the first-order equation of Lagergren did not apply well throughout the whole contact time.

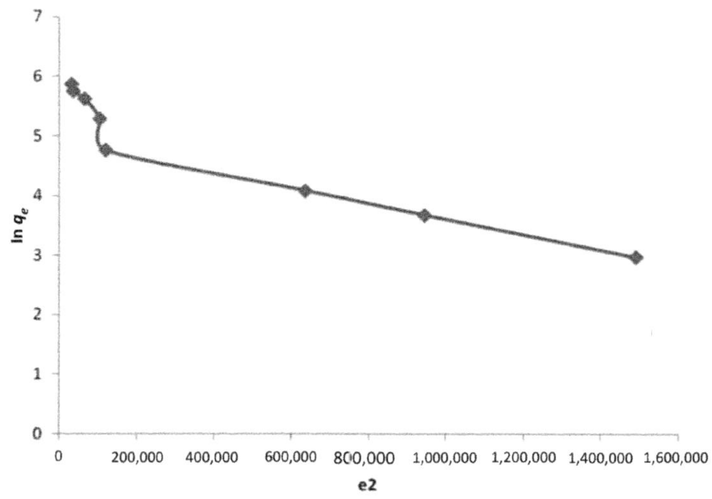

FIGURE 17.10 Dubinin–Radushkevich isotherm for Cr (III) adsorption on TRH.

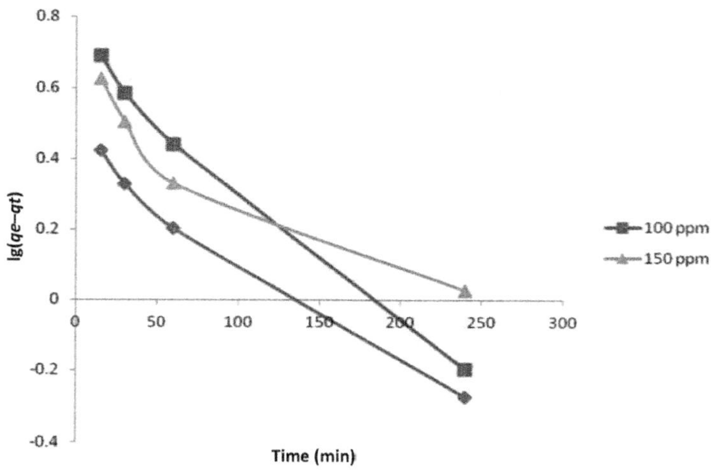

FIGURE 17.11 Pseudo-first-order adsorption kinetics of Cr (III) on TRH (initial concentrations 50, 100, and 150 ppm, weight of TRH = 0.1 g, and temperature = 25°C ± 2°C).

Figure 17.11 represents the plotting of log ($q_e - q_t$) versus time, which deviated considerably from the theoretical data after a short period. The plots and intercepts of curves were used to determine the first-order constant K_{pf}, capacity q_e and the corresponding linear correlation of determination R^2.

For the pseudo-second-order kinetic model, the values of K_{ps} and corresponding linear correlation of determination R_2^2 were determined from the slope of the plots.

Figure 17.12 shows the correlation of determination for the second-order kinetics model R_2^2 are greater, almost unity, indicating the applicability of this kinetics equation and the second-order nature of the adsorption process of chromium ion onto TRH. A similar phenomenon was observed in heavy metal ions (HMI) onto different adsorbents. So the pseudo-second-order model can be considered.

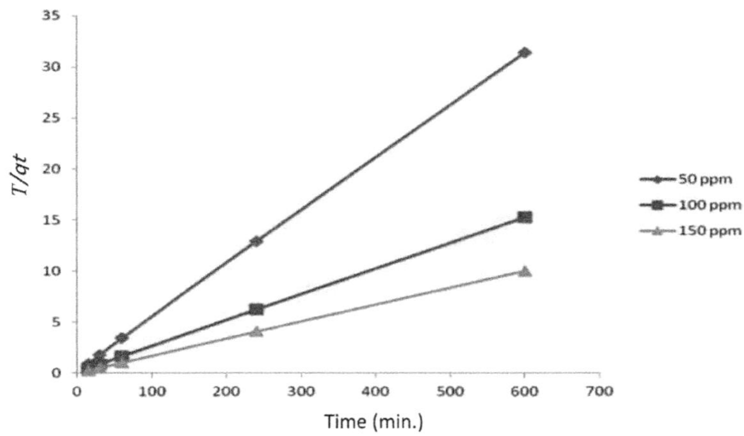

FIGURE 17.12 Pseudo-second-order adsorption kinetics of Cr (III) on TRH (initial concentrations 50, 100, and 150 ppm, weight of TRH = 0.1 g, and temperature = 25°C ± 2°C).

17.4.5 Column Adsorption Studies

17.4.5.1 TRH Results

17.4.5.1.1 Effect of Initial Concentration of Chromium Ions

Three different initial concentrations (30, 100, and 150 ppm) vs. residual concentration at different flow rates and different heights of adsorbent are presented in Figures 17.13 through 17.15. It was observed that adsorption capacity increased with the increase in initial inlet concentration. This is due to the high driving force for the adsorption process. At an initial concentration of 100 ppm at the flow rates 3 and 4 L/min, the percentage of removal is the same for the three different heights of adsorbent. This may be due to the greater availability of the exchangeable sites or surface area. This is in agreement with previous results [13]. At an initial concentration of 30 ppm, the ratio of surface active sites of the TRH to the total chromium ions in the solution is high, so all metal ions may interact with the adsorbent and be removed from the solution; this is shown in Figures 17.16 and 17.17.

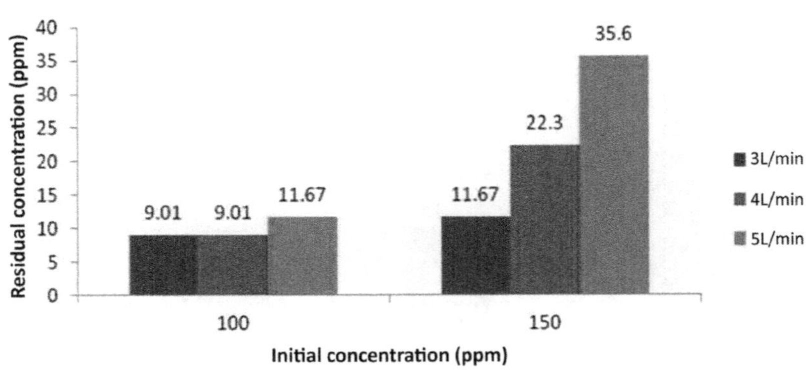

FIGURE 17.13 Effect of initial concentration on removal of Cr (III) by TRH at different flow rates (pH 6, column height 35 cm and temperature 25°C ± 2°C).

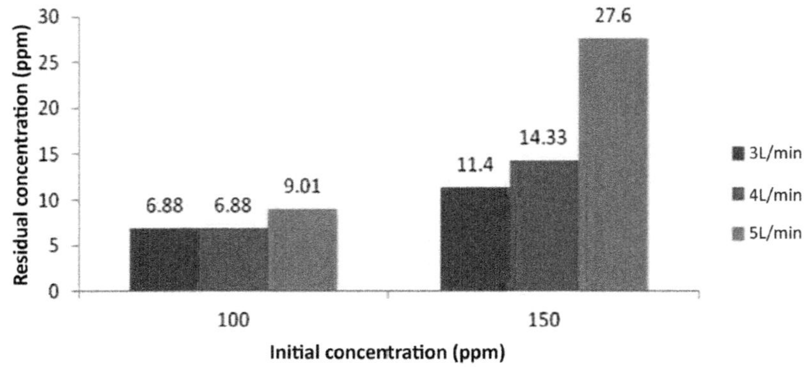

FIGURE 17.14 Effect of initial concentration on removal of Cr (III) by TRH at different flow rates (pH 6, column height 58 cm and temperature 25°C ± 2°C).

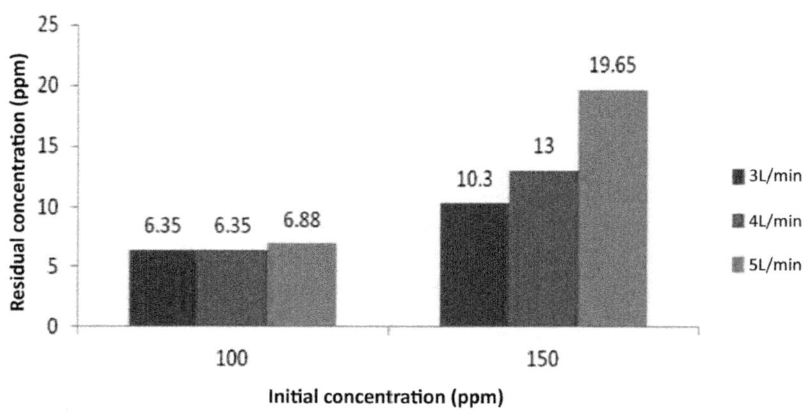

FIGURE 17.15 Effect of initial concentration on removal of Cr (III) by TRH at different flow rates (pH 6, column height 82 cm and temperature 25°C ± 2°C).

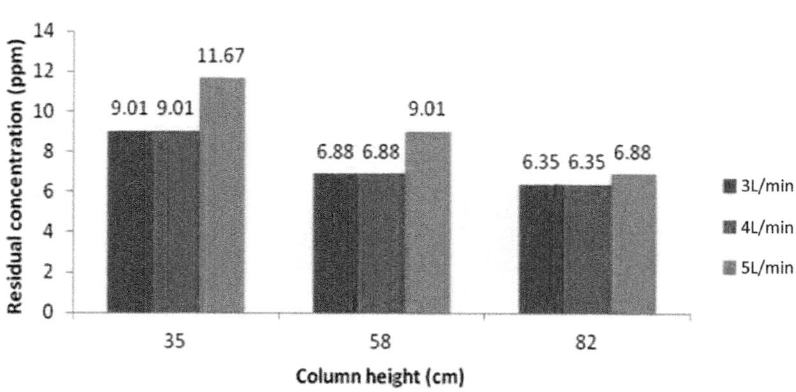

FIGURE 17.16 Effect of bed depth on removal of Cr (III) by TRH at different flow rates (pH 6, initial concentration 100 ppm and temperature 25°C ± 2°C).

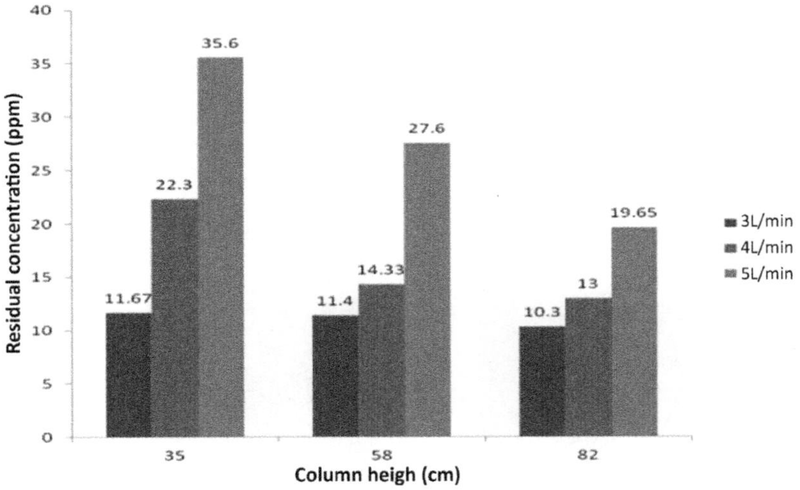

FIGURE 17.17 Effect of bed depth on removal of Cr(III) by TRH (pH 6, flow rates 3, 4, and 5 L/min, initial concentration 150 ppm and Temperature 25°C ± 2°C).

17.4.5.2 AC Results

17.4.5.2.1 Effect of Initial Concentration of Chromium Ions

Figures 17.18 through 17.20 are plots of three different initial concentrations vs. residual concentrations at different column heights and flow rates. It is clear from the results that when the initial concentration increases, the removal percentage of chromium ion increases. This is due to a high concentration difference between the bulk solution and the concentration of the solute on the solid phase. This will increase the rate of mass transfer of solute to attach to a free site(s) on the solid phase of the TRH. The driving force for adsorption is the concentration difference between the solute on the adsorbent and in the solution. If the initial concentration is high, the bed saturation is faster.

At an initial chromium ion concentration of 30 ppm the ratio of surface active sites on the AC to the total metal ions in the solution is high, so all metal ions may interact with the adsorbent and be removed from the solution.

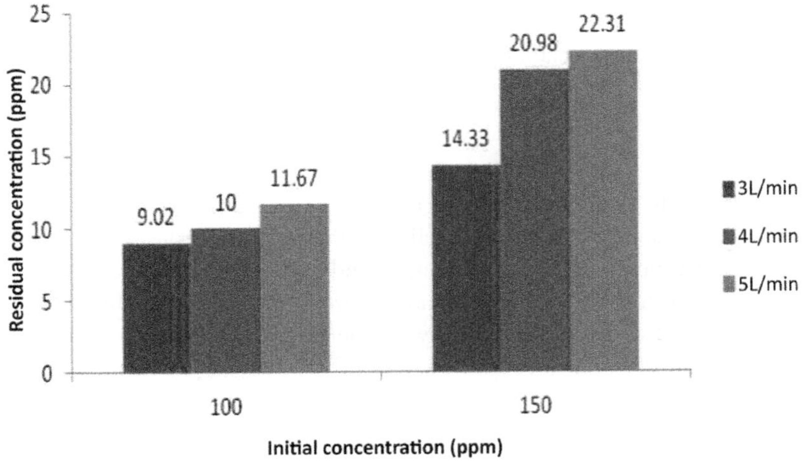

FIGURE 17.18 Effect of initial concentration on removal of Cr (III) by AC at different flow rates (pH 6, bed depth 35 cm and temperature 25°C ± 2°C).

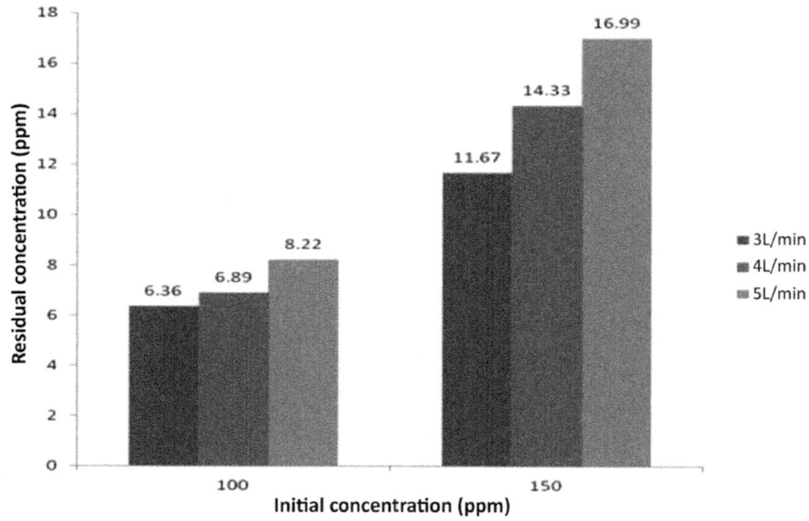

FIGURE 17.19 Effect of initial concentration on removal of Cr (III) by AC at different flow rates (pH 6, bed depth 58 cm and temperature 25°C ± 2°C).

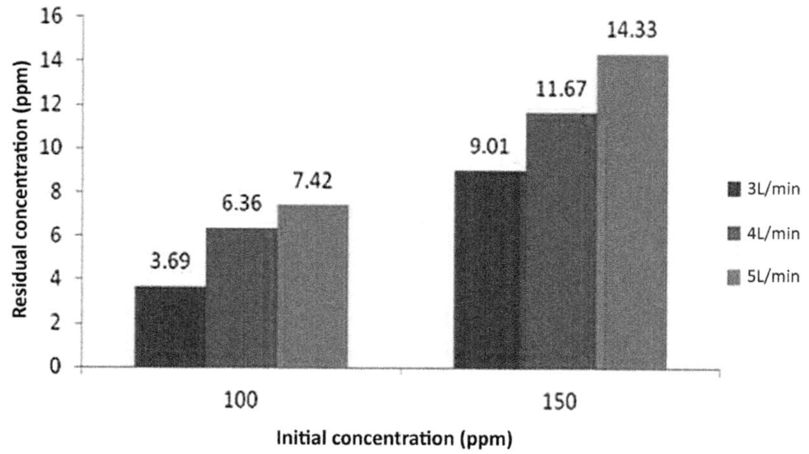

FIGURE 17.20 Effect of initial concentration on removal of Cr (III) by AC at different flow rates (pH 6, bed depth 82 cm and temperature 25°C ± 2°C).

17.5 CONCLUSION

Industrial wastewater is one of the main sources of pollution to the ecosystem. Heavy metals should be removed from wastewater to avoid water pollution that affects the health of living organisms. This study suggests that biosorption is an advanced and cost-effective process. TRH is indeed an efficient and promising adsorbent for the removal of chromium ion from polluted wastewater. The adsorption process was affected by various physicochemical parameters, including contact time, pH, initial concentration of metal ions and quantity of adsorbent. The removal percentage was found to be optimum when 0.1 g TRH was added to 100 ppm initial concentration of chromium ion at pH 6 with a contact time of 10 h. The kinetic study revealed that the adsorption data obeyed the pseudo-second-order model better than the pseudo-first-order model given the higher correlation of determination (R^2) = 1.

The results indicated that the Freundlich isotherm model fit the removal of chromium ions from aqueous solution better with $R^2 = 0.9594$.

For fixed-bed experiments with a concentration 30 ppm, all the results were beyond the limit of detection at any bed depth or flow rate for both THR and AC. Adsorption capacity was also increased with increases in the initial inlet concentration for both adsorbents. We noticed that as we increased the bed height the residual concentration decreased (an increase in the removal percentage) when other variables were constant for TRH and AC. As a result, we conclude that this method can effectively run in industrial units, because the percentage of removal is very high, reaching almost 92%, using a costless adsorbent. The filter column used (W5 Deep Bed Filter Column) is near to that which can be used in industries. The LIBS analytical limit of detection was 3 ppm (mg/L) and it can detect the concentration of metal ions in aqueous solution quickly and with no sample preparation, so it has the potential to be an ideal analytic tool for monitoring water quality, rather than the conventional methods currently being used.

TRH shows a higher removal percentage than AC for most of the experiments, although AC shows better adsorption than TRH for some experiments. It can be concluded that TRH is at least equally good as good as AC for the removal of chromium (III) from aqueous solutions, save for at a high flow rate. TRH still may be the best choice, cost-wise, as it is nearly costless.

REFERENCES

1. Hanchang, S. H. I., Industrial wastewater types, amounts and effects, *Point Sources of Pollution: Local Effects and It's Control*, Vol. I. Department of Environmental Science and Engineering, Tsinghua University, Beijing, China (EOLSS).
2. Agarwal, A., Kadu, M. S., Pandhurnekar, C. P. et al., Langmuir, Freundlich and BET adsorption isotherm studies for zinc ions onto coal fly ash, *International Journal of Application or Innovation in Engineering and Management (IJAIEM)*, 3(1):64–71, 2014.
3. Kumar, U. and Acharya, J., Fixed bed column study for the removal of lead from aquatic environment by NCRH, *Research Journal of Recent Sciences*, 2(1):9–12, 2013.
4. Mathew, F., Removal of heavy metals from electroplating wastewater using rice husk and coconut coir, A Thesis Presented to the Faculty of the Graduate School of the Missouri University of Science and Technology In Partial Fulfillment of the Requirements for the Degree Master of Science in Chemical Engineering, 2008.
5. Giwa, A. A., Bello, I. A., Oladipo, M. A. A., Adeoye, D. O., Removal of cadmium from wastewater by adsorption using the husk of melon (*Citrullus lanatus*) Seed, *International Journal of Basic and Applied Science*, 2(1):110–123, 2013.
6. Bulut, Y. and Baysal, Z., Removal of Pb(II) from wastewater using wheat bran, *Journal of Environmental Management*, 78(2):107–113, 2006.
7. Rajesh Kumar, P., Akila Swathanthra, P., Basava Rao, V. V., and Ram Mohan Rao, S., Adsorption of cadmium and zinc ions from aqueous solution using low cost adsorbents, *Journal of Applied Sciences*, 14:1372–1378, 2014.
8. Sivakumar, D., Hexavalent chromium removal in a tannery industry wastewater using rice husk silica, *Global Journal of Environmental Science and Management* 1(1):27–40, 2015.
9. Rahate, M. and Deshmukh, P. et al., Low cost efficient treatment for contaminated water, *13th Edition of the World Wide Workshop for Young Environmental Scientists (WWW-YES-2013)*, June 2013, Arcueil, France.
10. Al-Hashim, M. A. and Al-Safar, M. M., Removal of cadmium from polluted aqueous solutions using agriculture wastes, *Journal of Environmental Studies*, 6(2):131, 2012.
11. Thakur, L. S. and Semil, P., Adsorption of heavy metal (Cd^{2+}, Cr^{6+} and Pb^{2+}) from synthetic waste water by rice husk adsorbent, *International Journal of Chemical Studies*, 1(4):78–87, 2014.
12. Agbozu, I. E., Emoruwa, F. O., Batch adsorption of heavy metals (Cu, Pb, Fe, Cr and Cd) from aqueous solutions using coconut husk, *African Journal of Environmental Science and Technology*, 8(4):239–246, 2014.
13. Abbas, M. N. and Abbas, F. S., The feasibility of rice husk to remove minerals from water by adsorption and avail from wastes, 9(4): 2224–3496, 2013.

18 Low-Cost Adsorbent for Ammonia Nitrogen Removal

A Review

A. Y. Zahrim, L. N. S. Ricky, Y. Lija, and I. Azreen

CONTENTS

18.1 INTRODUCTION

Ammonia nitrogen (NH_3-N) is among the most common contaminants found in water bodies. The main sources of ammonia nitrogen pollution are agricultural runoff, untreated landfill leachate and sewage, urban activities, etc. Over-enrichment of ammonia nitrogen causes eutrophication, depletion of dissolved oxygen and brings toxicity to aquatic organisms (Wahab et al., 2010). There are several methods to remove ammonia nitrogen from water, including chemical precipitation, membrane, adsorption etc. Among these approaches, adsorption is the most convenient, as it is economical, environmental friendly and simple to operate in practice (Bhatnagar and Sillanpaa, 2010).

Recently, extensive studies have been conducted on using low-cost adsorbent as an alternative to conventional adsorbents. Low-cost adsorbents derived from agricultural waste could reduce solid waste disposal in landfills. This chapter aims to provide a review of low-cost adsorbents for removing ammonia nitrogen from wastewater. In addition, several types of adsorbent, including zeolite, activated carbon and hydrogel are also described. Finally, this chapter provides an insight overview of the uses for ammonium-loaded adsorbent.

18.2 NON-ADSORPTION TECHNOLOGIES FOR AMMONIA NITROGEN REMOVAL

Numerous approaches have been employed to remove ammonia nitrogen contaminants from waterways and this is generally accomplished by biological, physical or chemical processes, or by a combination of these methods. Several technologies used in this removal process include chemical precipitation, adsorption, membrane filtration, nitrification, air stripping and constructed wetlands.

Raw leachate with high ammonium nitrogen (NH_4^+-N) concentrations treated by using the chemical precipitation method has been studied by Li et al. (1999). This study reported on raw leachate collected from a local landfill in Hong Kong that was used as magnesium ammonium phosphate (MAP). The experiment was conducted by using three combinations of chemicals, $MgCl_2.6H_2O$ + $Na_2HPO_4.12H_2O$, MgO + 85% H_3PO_4 and $Ca(H_2PO_4)_2.H_2O$ + $MgSO_4.7H_2O$. The chemicals were used with different stoichiometric ratios to form the MAP precipitate effectively. The results showed a significant decreased when $MgCl_2.6H_2O$ and $Na_2HPO_4.12H_2O$ were used at $Mg^{2+}:NH_4^+:PO_4^{3-}$ mol ratio of 1:1:1. The initial NH_4^+-N concentration contained in the raw leachate decreased from 5618 to 112 mg/L within 15 min of the reaction time. Meanwhile, the other two combinations of chemicals were not able to remove NH_4^+-N effectively (Li et al., 1999). The chemical precipitation method was able to reduce the NH_4^+-N concentration, but there are several limitations associated with this removal method, such as an increase in dissolved solids (salinity) load and sludge production, as well as the need for pH and alkalinity correction (De Haas et al., 2000).

The removal of ammonia and nitrates using biological processes has been studied by Chen et al. (2013a). The bioretention system used in this study was characterized by low infiltration rates and long drainage times (five days). The system removed 33% of influent nitrate and 56% of total influent nitrogen. Effluent ammonia concentrations were substantially lower than those in the bioretention cell influent (<0.3 mg/L). Overall ammonia removal, however, was negligible or slightly negative, reflecting the low concentrations in the roadway runoff (Chen et al., 2013a).

Another study of nitrogen removal from landfill leachate using a batch reactor system was reported by Miao et al. (2014). A three-stage sequence batch reactor (SBR), comprising pre-treating SBR, nitritation SBR, and anaerobic ammonium oxidation (Anammox) SBR was constructed with a total volume capacity of 39 L for the nitrogen treatment process. The SBR system was fed with a synthetic solution consisting of different chemical solutions and the pH of the influent was adjusted and controlled at 7.5 ± 0.2. Then, the leachate sample was used as the influent after the first process of the acclimation period was complete. About 90% nitrogen removal efficiency was achieved in the SBR system, and the nitrogen load removal rates obtained were 0.81 and 0.76 kg N/(m^3 d), respectively. Despite the high removal efficiency of ammonia nitrogen, the biological method met with some difficulties, including the long reaction time and it was hindered by the presence of specific toxic substances and/or biorefractory organics such as humic substances (Wiszniowski et al., 2006).

The ability of photocatalytic ozonation to remove contaminants from wastewater was reported by Kern et al. (2013). The wastewater from a hospital laundry was treated by using a reactor column based on photocatalytic ozonation (O_3, UV, UV/O_3, UV/O_3/Fe^{2+} 50 mg/L and 150 mg/L). The reactions were carried out in acid media at adjusted pH value ranges from pH 3 to 3.5 with 180 min of contact time. Based on the results, the UV/O_3/Fe^{2+} 150 mg/L condition shows a good reduction value of total Kjeldahl nitrogen (TKN: 86.8%), chemical oxygen demand (COD: 59.1%) and 5-day biological oxygen demand (BOD_5: 50.3%) (Kern et al., 2013). However, several problems may be encountered during this process, including high energy consumption and formation of a deposit layer on the UV lamps (Wiszniowski et al., 2006).

Yuan et al. (2016) recently reported on the effectiveness of the air stripping method in removing ammonia from ammonia-rich stream. This study was performed in a continuous flow rotating packed bed (RPB) at temperature varying from 25°C to 40°C with an initial ammonia concentration of 1000 mg/L. Ammonia aqueous solution was pumped into the RPB at a controlled rate and the pH value adjusted to pH 11 by using sodium hydroxide solution. The RPB reached stripping efficiency of 69% and 81% at 30°C and 40°C respectively within 13.3 s (Yuan et al., 2016). The major drawbacks of this treatment method are its high operational cost due to high energy consumption and the need for additional chemicals in the treatment process. High pH values are also needed and the release of ammonia can cause air pollution if the ammonia contaminant cannot be absorbed properly (Wiszniowski et al., 2006, Renou et al., 2008, Yuan et al., 2016).

The application of mixed matrix membranes in ammonium ion recovery from aquaculture wastewaters was studied by Ahmadiannamini et al. (2017). A zeolite 13X molecular sieve was ground to an average particle size of 1.5 μm and the pore-filled membranes were prepared by using polyethersulfone (PES, MWCO = 30 kDa, EMD Millipore). The experiment was carried out by mixing a total of 0.5 g zeolite 13X powder in 50 mL of deionized water and sonicating for 10 min. The suspension was then filtrated through a 30 kD PES membrane. Overall, the performance of this mixed matrix membrane was good, as it recorded a high removal capacity of total ammonia nitrogen (TAN) of 19.8 mg/g (Ahmadiannamini et al., 2017). Nonetheless, the major problem of membrane technology is fouling or biofouling contributed by deposits of constituent inorganic, organic and microbiological substances deposited on the surface and inside the membrane pores (Wiszniowski et al., 2006, Renou et al., 2008).

A combination of biological and nanofiltration (NF) method has been studied by Bunani et al. (2012), where municipal wastewater was first treated by the biological method and then further treated by the NF method. Various types of NF membranes were used for this experimental study including CK, NF-90 and NF-270. Out of these membranes, NF-90 showed the best results and was able to produce higher quality water than the others. The ammonia nitrogen level in the wastewater decreased to an average value of 0.06 mg/L from 0.11 mg/L. This method was also able to treat other wastewater contaminants, conductivity, salinity and turbidity effectively (Bunani et al., 2012).

Nguyen et al. (2014) investigated the efficiency of a new technology system combining rotating hanging media bioreactor (RHMBR), submerged membrane bioreactor (SMBR) and electrocoagulation (EC) in removing pollutant nutrients from municipal wastewater. The hybrid pilot plant was operated repetitively up to 4 times with different conditions, which considered several factors such as flow rate, temperature, pH and specific aeration. To summarize, the treatment system exhibited excellent performance, as the initial ammonia nitrogen content ranging from 18.86 to 43.46 mg/L was completely removed to nil for all runs. Of the total nitrogen and phosphorus, 90.4% and 99.3%, respectively, were removed from the municipal wastewater (Nguyen et al., 2014).

Zhao et al. (2010a) studied the effect of the biological and photoelectrochemical (PEC) oxidation process in the landfill leachate treatment process. Leachate samples first underwent biological treatment and were then treated with PEC oxidation in a pilot-scale flow reactor, using DSA anode and UV light irradiation. The experiment was carried out in a continuous flow mode for a 270 min reaction time, and the leachate samples were periodically collected for analysis. The effect of applied current density on PEC oxidation of leachate was analyzed by varying the density (29.0, 41.2, 55.6, 67.1, and 78.5 mA/cm^2) and the results showed that the removal rates of ammonium ions increased with the current density. Almost all ammonia content in the samples was able to be removed by using the PEC oxidation method under appropriate conditions. This method also reduced a large amount of metal ion concentrations, COD and color from the leachate samples (Zhao et al., 2010a).

The performance of an air-cathode microbial fuel cell (MFC) coupled with granular activated carbon (GAC) adsorption for palm oil mill effluent treatment was reported by Tee et al. (2016). The GAC adsorption process with a large surface area was added to the hybrid system in order to enhance the performance of the system in removing wide ranges of contaminants compared to the standalone MFC system. Integration of MFC with adsorption has a better ability to generate electricity at low COD concentrations due to its adsorption capacity (Wu et al., 2015). Local earthen pots were used as material for the tubular hybrid system with thickness of 7 mm. The MFC adsorption system was operated under room temperature conditions with the anode connected to the cathode with a resistance of 50 Ω. The results show that the hybrid system was able to remove 89.75% and 84.3% of ammonia nitrogen (NH_3-N) and total nitrogen, respectively, from the substrate (Tee et al., 2016).

Various treatment methods of ammonia nitrogen and its efficiency have been reviewed, including biological, chemical precipitation, air stripping, and a combination of biological and physicochemical methods. The integrated system comprised of biological and physicochemical treatment methods had higher removal efficiency of ammonia nitrogen compared to standalone treatment processes with removal rates ranging from 89% to 99%. Despite the excellent performance, there are some major drawbacks associated with each treatment method reviewed. These include high production of sludge, long reaction times, complicated operation, high energy consumption and fouling problems. Continued research on the treatment methods that offer the most efficient and adaptable toxin removal that are also economically and environmental feasible is therefore needed.

18.3 ZEOLITES, ACTIVATED CARBON AND HYDROGEL ADSORBENT

The adsorption process is the most applied method for water and wastewater treatment as it offers efficiency and ease of operation and is both cost effective and environmental friendly (Liu et al., 2010a, Gupta and Verma, 2015). Several studies on ammonia nitrogen adsorption were conducted using different commercial adsorbents such as zeolites, hydrogel and activated carbon (Wang and Peng, 2010, Zheng et al., 2012, Boopathy et al., 2013).

Assessment of a solid-liquid adsorption system is usually carried out through equilibrium isotherm and kinetic modeling studies, which are also able to provide an effective and accurate design model for the removal of pollutants from aqueous solution. The adsorption isotherm model is a tool used to explain and predict the possible interactions between the adsorbent and adsorbate. Langmuir, Freundlich, Temkin and Dubinin–Radushkevich are some of widely known equilibrium models. Study of adsorption kinetics is important to analyze the effectiveness of the adsorption process and also to predict the optimum conditions needed, as kinetic modeling gives information on the mechanisms and possible rate-controlling steps. Several kinetic models, including pseudo-first order, pseudo-second order and intra-particle diffusion, are commonly used in kinetic studies. Table 18.1 tabulates the adsorption capacity, isotherm and kinetic studies of several research studies that have reported on the usage of these types of adsorbent.

Natural zeolites are abundant and mostly applied for very specific chosen pollutants (Erdem et al., 2004, Perić et al., 2004, Sarioglu, 2005). The application of natural zeolite as adsorbent for the separation and purification process has been investigated by several researchers (Karadag et al., 2006, Saltalı et al., 2007, Thornton et al., 2007, Wang and Peng, 2010, Alshameri et al., 2014, Martins et al., 2017). The experimental data summarized that the maximum adsorption capacity of natural Turkish zeolite was 5.95 mg/g (Karadag et al., 2006). Thornton et al. (2007) studied the ability of natural zeolite mesolite in removing ammonia nitrogen at different adsorbate concentrations and the adsorption capacities obtained were 55.0 and 49.0 at initial concentrations of 50 and 400 mg/L, respectively (Thornton et al., 2007).

Meanwhile, 30 g of natural clinoptilolite adsorbent was used to remove ammonium ions from landfill leachate with an initial concentration of 2292 mg/L at initial pH 7.0 for 6 h. The maximum adsorption capacity achieved was 10.80 ± 2.14 mg/g with regression of correlation coefficient best fitted to the Langmuir isotherm model; it was thus assumed that the ammonium removal occurred by monolayer adsorption on a homogeneous surface (Martins et al., 2017).

Since various other cations may be present in industrial wastewater, these ions would compete with the ammonium ion for the negatively charged surface adsorbent and thus affect adsorption capacity. Therefore, ammonium adsorption onto a ceramic adsorbent (20 g/L) in the presence of salt solutions of sodium, potassium, calcium and magnesium ions was investigated by Zhao et al. (2013). The sieve size (150 μm) of Kanuma mud and Akadama mud was first mixed with zeolite powder, soluble starch and Na_2SO_4 to homogeneity with optimum mass ratios (2:2:3:2:1). The adsorption process reached maximum capacity at 75.5 mg/g at an initial ammonium nitrogen concentration of 10,000 mg/L and a contact time of 480 min without addition of the salt solution. The adsorption capacities of ammonium nitrogen on the ceramic adsorbent were significantly reduced by the presence of competitive cations with a selectivity order of $Na^+ > K^+ > Ca^{2+} > Mg^{2+}$ with adsorption capacity of 13.6, 20.3, 24.1, and 27.9 mg/g, respectively (Zhao et al., 2013).

Modification of the adsorbents could increase their effectiveness in wastewater treatment applications and this has been shown by several research studies. Natural raw zeolite is modified by reacting the adsorbent in various modifying agents, such as NaCl, $FeCl_3$, and $Ca(OH)_2$ solutions. A maximum adsorption capacity of 9.07 mg/g was obtained when using zeolite adsorbent modified with NaCl solution in the removal of ammonium from landfill leachate. The experiment was conducted at a temperature of 25°C and agitated at 200 rpm with adsorbent dosage and particle size of 1100 g/L and –20 + 35 mesh respectively. Furthermore, the kinetic and isotherm equilibrium

TABLE 18.1
Ammonia Nitrogen Adsorption by Commercial Adsorbent

Adsorbent	Modifying Agent	Operating Condition	Adsorption Capacity (mg/g)	Isotherm Model	Kinetic Model	References
Natural Turkish zeolite (Clinoptilolite)	N/A	M: 0.5 g C_o: 150 mg/L S.R: 200 rpm pH: 8 Temperature: 25°C t_e: 40 min	5.95 mg/g	Langmuir	Pseudo-second order	Karadag et al. (2006)
Zeolite mesolite	N/A	M: 1 g, C_o: 50 mg/L t: 120 min t_e: 60 min Temperature: 25°C pH: 6–7	55.00	Langmuir	N/A	Thornton et al. (2007)
Zeolite mesolite	N/A	M: 1 g C_o: 400 mg/L t: 120 min t_e: 25 min Temperature: 25°C pH: 6–7	49.0	Langmuir	N/A	Thornton et al. (2007)
Rectorite hydrogel composite/ Chitosan-g-poly (acrylic acid)	Hydrogel composite chitosan grafted with poly(acrylic acid) (10 wt%) prepared from in-situ copolymerization	M: 0.05 g C_o: 100 mg/L S.R: 120 rpm t: 60 min, t_e: 30 min Temperature: 30°C pH: 6–7	61.95	Redlich–Peterson	Pseudo-first order	Zheng and Wang (2009)
NaA zeolite	2.6 g of NaOH powder	M: 0.5 g C_o: 100 mg/L S.R: 180 rpm t: 5–180 min	44.30	Langmuir	N/A	Zhao et al. (2010b)

(*Continued*)

TABLE 18.1 (*Continued*)
Ammonia Nitrogen Adsorption by Commercial Adsorbent

Adsorbent	Modifying Agent	Operating Condition	Adsorption Capacity (mg/g)	Isotherm Model	Kinetic Model	References
Clinoptilolite	2% sodium chloride (NaCl) and iron (III) chloride ($FeCl_3$)	M: 1.5 g C_o: 20 mg/L S.R: 200 rpm t: 5 h, t_e: 3 h Temperature: 30°C	39.38	Langmuir	Elovich	Huo et al. (2012)
Composite hydrogel	Chitosan grafted poly(acrylic acid)/unexpanded vermiculite (CTS-g-PAA/UVMT)	M: 0.05 g C_o: 100 mg/L S.R: 120 rpm t: 30 min, t_e: 5 min Temperature: 30°C pH: 6–7	21.70	Freundlich	N/A	Zheng et al. (2012)
Ceramic (Kanuma and Akadama clay)	Clay mixed with zeolite powder, soluble starch and Na_2SO_4 to homogeneity with optimum mass ratios (2:2:3:2:1)	M: 20 g/L C_o: 10,000 mg/L t: 480 min	75.5	Langmuir	Pseudo-second order	Zhao et al. (2013)
Yemeni natural zeolite	2.5M sodium chloride (NaCl)	M: 1.2 g C_o: 80 mg/L S.R: 260 rpm t: 250 min, t_e: 20 min Temperature: 25°C pH: 7	11.18	Langmuir	Pseudo-second order	Alshameri et al. (2014)
Yemeni natural zeolite	N/A	M: 1.2 g C_o: 80 mg/L S.R: 260 rpm t: 250 min, t_e: 100 min Temperature: 25°C pH: 7	8.29	Langmuir	Pseudo-second order	Alshameri et al. (2014)

(*Continued*)

TABLE 18.1 (*Continued*)
Ammonia Nitrogen Adsorption by Commercial Adsorbent

Adsorbent	Modifying Agent	Operating Condition	Adsorption Capacity (mg/g)	Isotherm Model	Kinetic Model	References
NaA zeolite	NaOH solution	*M*: 0.1 g C_o: 300 mg/L S.R: 120 rpm pH: 7, Temperature: 25°C t_e: 60 min	16.6	Freundlich	N/A	Jiang et al. (2016)
Natural Turkish zeolite	2 N of NaCl solution	*M*: 100 g/L C_o: 263.2 mg/L S.R: 200 rpm, pH: 7 Temperature: 25°C Particle size: –20 + 35 mesh	9.07	Langmuir and Temkin	Pseudo-second order	Aydin Temel and Kuleyin (2016)
Natural zeolite	1 M of NaOH solution and 0.4% of $LaCl_3$ solution	*M*: 2 g/L C_o: 20 mg/L *t*: 6 h, Temperature: 30°C	8.15	Freundlich	Pseudo-second order	He et al. (2016a)
Amphoteric hydrogel	Dimethyldiallylammonium chloride (DMDAAC), acrylic acid and NaOH	*M*: 20 mg C_o: 50 mg/L pH: 7, S.R: 140 rpm Temperature: 25°C t_e: 10 min	33.98	Freundlich	Pseudo-second order	Wei et al. (2016)
Clinoptilolite	N/A real case	*M*: 30 g/L C_o: 2292 mg/L pH: 7.0, S.R: 150 rpm *t*: 6 h Temperature: 25°C	10.80	Langmuir	N/A	Martins et al. (2017)

Note: N/A: not available, C_o: Initial ammonium concentration, S.R: stirring rate, *t*: contact time, t_e: equilibrium time.

studies of ammonium adsorption onto modified zeolite were well fitted to pseudo-second order and Langmuir–Temkin isotherm models, respectively (Aydın Temel and Kuleyin, 2016).

Another modification on the natural zeolite sample by using 2% of sodium chloride-iron (III) chloride solutions and sodium hydroxide-lanthanum (III) chloride were conducted by Huo et al. (2012) and He et al. (2016a), respectively. The modified clinoptilolite zeolite with salt (NaCl, $FeCl_3$) showed that the ammonia nitrogen removal rate reached up to 98.46% with the final content recorded at 0.80 mg/L with adsorption capacity of 39.38 mg/g (Huo et al., 2012).

Meanwhile, He et al. (2016b) studied the ability of alkaline-activated and lanthanum-impregnated zeolite (NLZ) to remove ammonium from synthetic solution. The batch adsorption experiments were conducted for 6 h at 303 K with 2 g/L of NLZ dose and initial concentration of 20 mg/L. The results indicated that the ammonium adsorption capacity increased steadily and reached maximum (8.15 mg/g) as the pH values increased from pH 3 to 7 but decreased gradually as the pH value approached alkaline (He et al., 2016b).

Jiang et al. (2016) reported on the potential of NaA zeolite prepared from coal fly ash (CFA) and modified with sodium hydroxide solution as adsorbent for ammonia nitrogen removal. The ammonium synthetic solution with various concentrations (5, 50, 100, 200, 300 and 400 mg/L) was agitated with 0.1 g of adsorbent at 120 rpm with pH values (4–10) for determination of the variables' effect on adsorption. Increasing the pH value from pH 4 to 7 resulted in increased removal efficiency up to 59.6%, but as the pH value further increased to pH 10, the removal of ammonium declined to 42.6%. The results obtained from the batch adsorption experiments revealed that the maximum adsorption capacity reached 16.6 mg/g and equilibrium time of 60 min (Jiang et al., 2016).

Modified zeolite showed better adsorption performance compared to natural, unmodified zeolite as reported by Alshameri et al. (2014). The modification of natural zeolite was done by using 0.5–2 mol/L sodium chloride solution, stirred at 120 rpm for 2 h at a temperature of 90°C. Both natural and modified zeolite weighing 1.2 g were then mixed in 80 mg/L of ammonium concentration solution with a stirring rate of 260 rpm at a temperature of 25°C. Based on the results, modified zeolite showed a higher ammonium uptake capacity (11.18 mg/g) than natural, unmodified zeolite (8.29 mg/g). However, the equilibrium time for the modified natural zeolite was quicker than the natural, unmodified zeolite, at 20 and 100 min, respectively (Alshameri et al., 2014).

Several researchers have discovered the ability of hydrogel as an adsorbent for the removal of ammonia nitrogen from aqueous solution. Hydrogel is a slightly cross-linked polymeric network that has the potential for the adsorption of many pollutants due to its high adsorption capacity and fast removal (Karadağ et al., 2002, Kaşgöz et al., 2008, Kaşgöz and Durmus, 2008). An amphoteric hydrogel prepared by cross-linking graft copolymerization of dimethyldiallylammonium chloride and acrylic acid onto soluble starch exhibited a fast adsorption rate and high adsorption capacity of 33.98 mg/g (Wei et al., 2016).

The quality of an adsorbent can be determined by its reusability in addition to its high adsorption capacity. The regeneration ability of an adsorbent might be crucial for minimizing the operational cost and for the purposes of possible multiple reuse. The used adsorbent can be reused and regenerated by using different chemical agents such as hydrochloric acid (HCL), sodium chloride (NaCl) and sodium hydroxide (NaOH). Few studies have reported on the adsorbent regeneration ability (Zheng and Wang, 2009, Zhao et al., 2010b, Zheng et al., 2012). Hydrogel composite chitosan grafted poly (acrylic acid)/rectorite prepared from *in situ* copolymerization was tested for ammonium removal and the maximum adsorption capacity obtained was 61.95 mg/g at a stirring rate of 120 rpm, adsorbent dosage of 0.05 g and equilibrium time of 30 min. The hydrogel composite adsorbent was able to be recovered and used again to adsorb ammonium ions, and the results showed that 0.1 mol/L of NaOH can act not only as desorbing agent but also as a regenerating agent (Zheng and Wang, 2009).

The feasibility of NaA zeolite was also tested for ammonium adsorption and regeneration (Zhao et al., 2010b). The adsorption capacity of NaA zeolite synthesized from natural halloysite mineral powder is 44.30 mg/g. The ammonium adsorption equilibrium is within 15 min. The zeolite also exhibited good regeneration ability, with only a slight decrease in adsorption capacity from 16.72 mg/g for the first cycle to 16.35 mg/g for the seventh cycle (Zhao et al., 2010b). Another study done by Zheng et al. (2012) investigated the use and reusability of composite hydrogel adsorbent. The adsorbent was synthesized from chitosan (CTS), acrylic acid (AA) and unexpanded vermiculite (UVMT) under the polymerization reaction and followed by neutralization using sodium hydroxide solution (Xie and Wang, 2009, Zheng et al., 2012). The adsorption process was found to be pH dependent, and the maximum adsorption (21.7 mg/g) was observed at pH 4 to 8 within 5 min of equilibrium contact time. The adsorbent showed no changes in adsorption capacity even after being used for five cycles of the adsorption-desorption process, thus confirming its good regeneration ability (Zheng et al., 2012).

Various other adsorbents, including local activated carbon (Okoniewska et al., 2007) and silty and sandy loam soil (Fernandoa et al., 2004), were used for ammonia nitrogen removal. Although these commercial adsorbents are efficient for ammonia nitrogen removal, their usage in real applications is sometimes restricted due to their higher cost (Bhatnagar et al., 2010). Overall, the sorption performance of chemically modified adsorbents was comparatively higher than unmodified adsorbents, which reflects that modification of adsorbent surface sites is able to enhance adsorption capacity. The highest adsorption capacity obtained was around 61.95 mg/g for modified rectorite hydrogel composite and chitosan-g-poly (acrylic acid) respectively (Table 18.1). Although the usage of conventional adsorbents has been established for years, great attempts have been made to identify new sustainable, effective and low-cost adsorbents as alternatives to replace conventional adsorbents. It is worth reviewing agricultural waste as a low-cost adsorbent for ammonia nitrogen removal since there is lack of other reviews discussing the matter.

18.4 LOW-COST AGRICULTURAL WASTES AS ADSORBENTS

Biosorption is an adsorption process using agricultural or lignocellulosic wastes for the removal of pollutants in wastewater, including ammonia nitrogen, by mechanisms such as electrostatic attraction, complexion, ion exchange, covalent binding, van der Waals' attraction, adsorption and/or micro-precipitation (Montazer-Rahmati et al., 2011, Witek-Krowiak, 2012, Abdolali et al., 2014). To date, there has been a remarkable growth in research related to the use of low-cost adsorbents derived from fruit and agricultural waste. The use of these waste products for developing low-cost adsorbents helps to minimize waste, recovery and reuse since such these agricultural waste products currently cause numerous disposal problems due to their bulk volume, toxicity and physical nature (Ali et al., 2012). In general, the basic components of these agricultural waste products mainly consist of lignin, cellulose and hemicellulose and include the functional groups of alcohols, ketones, carboxylic, phenolic, aldehydes and ether (Fengel and Wegener, 1983).

Previous studies suggested that low-cost adsorbents might be good materials to replace the available conventional sorbents. Usually, the cost estimation for a low-cost sorbent includes the cost of activating the materials, transportation, chemicals, and the electrical energy used. The total cost for the finished product of bagasse fly ash adsorbent was estimated to be Rs 400/ton, which is reasonable considering its efficiency in removing contaminants (Gupta et al., 1998). In another report, the cost estimated for the finished product of a low-cost adsorbent and a conventional adsorbent is US$ 12/ton and US$ 285/ton, respectively (Gupta et al., 2000, Ali et al., 2012).

Some studies correlating low-cost adsorbents for ammonia nitrogen removal from aqueous solution are reviewed in this section. Table 18.2 shows a summary of adsorption capacity, isotherm and kinetic studies from several adsorption studies using various agricultural-based adsorbents. Liu et al. (2010a) examined the suitability of leaves and stems from various agricultural by-products as inexpensive adsorbents for the removal of ammonia nitrogen from synthetic solution. The screening results showed that a few samples of agricultural waste products exhibited high removal capacities and the adsorption performances were comparable to those of mineral adsorbents. The adsorption capacities recorded were 3.29, 2.38, 2.77, 2.63, 2.35, and 2.37 for Boston ivy leaves and stems,

TABLE 18.2
Adsorption of Ammonia Nitrogen Using Low-Cost Adsorbent

Biosorbent	Modifying Agent	Operating Condition	Adsorption Capacity (mg/g)	Isotherm Model	Kinetic Model	References
Boston ivy (*Parthenocissus tricuspidata* Planch) leaves	N/A	*M*: 0.2 g C_o: 50 mg/L S.R: 250 rpm *t*: 24 h t_e: 18 h Temperature: 30°C	3.29	Langmuir	Logistic	Liu et al. (2010a)
Boston ivy (*Parthenocissus tricuspidata* Planch) stems			2.38	Langmuir		
Southern magnolia (*Magnolia grandiflora* L.) leaves			2.77	Langmuir		
Poplar (*Populus euramericana* cv. 'I-214') leaves			2.63	Langmuir		
Strawberry (*Fragaria ananassa* Duchesne) leaves			2.35	Freundlich		
Strawberry (*Fragaria ananassa* Duchesne) stems			2.37	Langmuir		
Sawdust (*Eucalyptus globulus* sawdust)	N/A	*M*: 5 g C_o: 50 mg/L S.R: 400 rpm *t*: 2 h t_e: 20 min Temperature: 20°C ± 1°C pH: 6	1.70	Langmuir	Pseudo-second order	Wahab et al. (2010)
Boston ivy (*Parthenocissus tricuspidata*) leaf powder (BPTL)	NA	*M*: 0.2 g C_o: 50 mg/L S.R: 4000 rpm *t*: 120 min t_e: 40 min Temperature: 35°C pH: 7	3.21	Langmuir	Logistic	Liu et al. (2010b)

(Continued)

TABLE 18.2 (*Continued*)
Adsorption of Ammonia Nitrogen Using Low-Cost Adsorbent

Biosorbent	Modifying Agent	Operating Condition	Adsorption Capacity (mg/g)	Isotherm Model	Kinetic Model	References
Posidonia oceanica (L.) fibers	N/A	*M*: 3 g C_o: 50 mg/L S.R: 400 rpm *t*: 120 min t_e: 40 min Temperature: 18°C ± 2°C pH: 6	1.73	Langmuir	Pseudo-second order	Jellali et al. (2011)
Cactus leaf fibers	N/A	*M*: 5 g C_o: 50 mg/L S.R: 400 rpm *t*: 2 h t_e: 40 min Temperature: 40°C pH: 6	1.80	Langmuir	Pseudo-second order	Wahab et al. (2012)
Coconut shell-activated carbon (CSAC)	85% of H_3PO_4 solution	*M*: 2 g C_o: 500 mg/L S.R: 100 rpm pH: 9 *t*: 120 min	2.48	Freundlich	Pseudo-second order	Boopathy et al. (2013)
Biochar (mixed hardwood shavings)	N/A	*M*: 2 g C_o: 1000 mg/L *t*: 1 week	5.29	Freundlich	N/A	Sarkhot et al. (2013)
Banana peels	N/A	*M*: 100 g C_o: 67–86 mg/L S.R: 150 rpm *t*: 40 min t_e: 25 min Temperature: 20°C ± 1°C pH: 7–12	5.37	N/A	N/A	Chen et al. (2013b)

(*Continued*)

TABLE 18.2 (*Continued*)
Adsorption of Ammonia Nitrogen Using Low-Cost Adsorbent

Biosorbent	Modifying Agent	Operating Condition	Adsorption Capacity (mg/g)	Isotherm Model	Kinetic Model	References
Brown seaweed *Cystoseira indica* (C. indica)	N/A	M: 3 g C_o: 40 mg/L t: 120 min t_e: 76 min Temperature: 25°C ± 1°C Initial pH: 7	15.21	N/A	Pseudo-second order	Mansuri et al. (2014)
Jatropha oil cake (JOC)	N/A	M: 3 g C_o: 40 mg/L t: 120 min t_e: 96 min Temperature: 25°C ⊥ 1°C pH: 3	13.59			
Pine cone powder	N/A	M: 0.3 g C_o: 50 mg/L S.R: 125 rpm t: 120 min t_e: 15 min Temperature: 22°C ± 0.5°C pH: 8	1.05	Langmuir	Pseudo-second order	Demırak et al. (2015)
Pine cone powder	Sodium hydroxide (0.05 M)		2.41			
Pine cone powder	Sodium hydroxide (0.10 M)		3.06			
Pine bark	N/A	M: 1 g C_o: 79 mg/L S.R: 30 rpm t_e: 24 h	0.47	N/A	N/A	Hina et al. (2015)
Biochar (pine wood chip)	N/A	M: 1 g C_o: 79 mg/L S.R: 30 rpm t_e: 24 h	0.52	N/A	N/A	Hina et al. (2015)

(*Continued*)

TABLE 18.2 (*Continued*)
Adsorption of Ammonia Nitrogen Using Low-Cost Adsorbent

Biosorbent	Modifying Agent	Operating Condition	Adsorption Capacity (mg/g)	Isotherm Model	Kinetic Model	References
Barbecue bamboo charcoal	N/A	M: 4 g C_o: 60 mg/L Temperature: 30°C pH: 7 t: 80 min	1.68	Freundlich	Pseudo-second order	Zhou et al. (2015)
Modified empty fruit bunch (EFB)	Sodium hydroxide	M: 7 g $C_o = 37.36$ mg/L t: 120 min t_e: 40 min	0.60	NA	NA	Ricky et al. (2016)
Watermelon rind	N/A Sodium hydroxide (NaOH), potassium hydroxide (KOH), sulfuric acid (H_2SO_4)	M: 20 g C_o: 500 g/L t_e: 40 min pH: 7.0	1.23 1.22 1.22 1.21	Langmuir	Pseudo-second order	Ibrahim et al. (2016), Zahrim et al. (2016)
Avocado seed	Methane-sulfonic acid	M: 3 g C_o: 150 mg/L Temperature: 25°C S.R: 125 rpm pH: 5 t: 6 h	5.4	Langmuir	Pseudo-second order	Zhu et al. (2016)
Jackfruit seed	N/A	M: 20 g C_o: 500 mg/L pH: 7 t: 40	3.37	N/A	N/A	Azreen et al. (2017)

Note: N/A: not available, C_o: Initial ammonium concentration, S.R: stirring rate, t: contact time, t_e: equilibrium time.

southern magnolia leaves, poplar leaves and strawberry leaves and stems, respectively. Moreover, the porous structure of biosorbent samples, which were mostly composed of holocellulose, lignin and protein, were suitable for adsorption of nitrogenous compounds. The experimental data of six screened samples were well described by both the Langmuir and Freundlich isotherm models, suggesting that the adsorption process might be due to physisorption. Ammonia nitrogen adsorption reached equilibrium at about 18 h, and the Logistic model fit well to the experimental data. The Logistic model is usually used to describe growth or diffusing with a limiting factor (Liu et al., 2010a).

The adsorption process onto porous adsorbents generally involves four main stages: bulk diffusion, boundary film diffusion, intra-particle diffusion and adsorption. Film and intra-particle diffusion are generally rate-controlling steps, whereas bulk diffusion and adsorption are usually assumed to be rapid and not rate determining. The adsorption mechanism of ammonium ions onto Boston ivy (*Parthenocissus tricuspidata*) leaf powder (BPTL) was reported by Liu et al. (2010b). The adsorption process reached equilibrium at 14 h with an optimum pH value range from 5 to 10, in which adsorption capacities were 3.02–3.21 mg/g. Based on the kinetic modeling results, three stages were involved in the adsorption of ammonium ion onto BPTL. Film diffusion was the rate-controlling step for the first stage of ammonium adsorption by BPTL, followed by intra-particle diffusion in the middle stage. The mechanism of adsorption involved in this study was compared to a previous study on ammonium adsorption by minerals, which was found to follow same three steps. Adsorption by minerals was, however, significantly faster than that by BPTL (Karadag et al., 2006; Lei et al., 2008) in the beginning stage of the adsorption process due to the regular surface structure of the mineral sorbent (Liu et al., 2010b).

Wahab et al. (2010) studied the effectiveness of sawdust for ammonium nitrogen removal from aqueous solution. The maximum sorption was found to be 1.70 mg/g for 5 g sawdust from 50 mg/L of ammonium nitrogen solution using a 2 h mixing time at pH 6 and 20°C, with an equilibrium time of 20 min. Fourier transform infrared spectroscopy (FTIR) analysis was conducted to determine the functional groups involved before and after the biosorption process as presented in Figure 18.1. Several functional groups, such as carboxyl and alcohol, undergo some changes in peaks due to the complexation reactions with ammonium cations (Wahab et al., 2010).

Aqueous ammonia nitrogen solutions with an initial concentration of 50 mg/L were treated using *Posidonia oceanica* (L.) fiber and cactus leaf fibers (CLF) (Jellali et al., 2011, Wahab et al., 2012). The adsorption capacity was reported to be 1.73 mg/g for ammonium removal with equilibrium

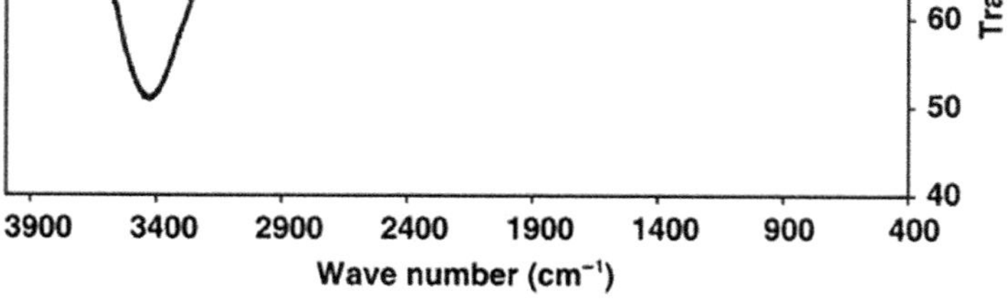

FIGURE 18.1 FTIR spectra of sawdust (a) before (b) after adsorption of ammonium ions. (From Wahab, M.A. et al., *Bioresour. Technol.*, 101, 5070–5075, 2010.)

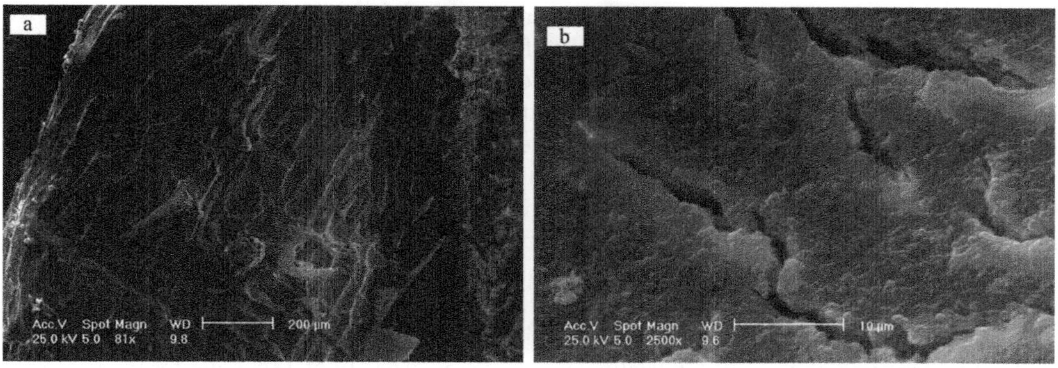

FIGURE 18.2 SEM micrograph of the surface CLF. (From Wahab, M.A. et al., *J. Hazardous Mater.*, 241, 101–109, 2012.)

established within 40 min at a temperature of 18°C and the optimum pH of 6 for ammonium removal (Jellali et al., 2011). Meanwhile, adsorption capacities of 1.40, 1.70, and 1.80 mg/g were recorded for the CLF adsorbent at 20°C, 30°C, and 40°C, respectively. Maximum adsorbent capacity was achieved at the optimum condition of pH 6 with 5 g of biosorbent at an equilibrium time of 40 min. The microscopic structure of CLF was evaluated using scanning electron microscopy (SEM) analysis as shown in Figure 18.2. The adsorbent surface appears to be very rough, as cracks and cavities are observed; this eventually has an important role in the biosorption process and intra-particle diffusion (Wahab et al., 2012). Both adsorption experimental data agreed well with the Langmuir isotherm and pseudo-second order kinetic model, thus indicating a monolayer uniform adsorption and chemisorption process (Jellali et al., 2011, Wahab et al., 2012).

The ability of coconut shell-activated carbon (CSAC) adsorbent prepared through pre-carbonization and chemical activation was investigated by Boopathy et al. (2013). The optimized conditions for effective ammonium ion adsorption were obtained at pH 9.0, temperature 283 K and contact time of 120 min (Boopathy et al., 2013). *Cystoseira indica* (brown seaweed) and Jatropha oil cake, which are abundantly available in India, were used to recover the excess ammonia nitrogen content in aqueous ammonium solution (Mansuri et al., 2014). As a result, *Cystoseira indica* and Jatropha oil cake were able to recovered 40 mg/L of ammonia nitrogen with maximum uptake of 15.21 mg/g and 13.59 respectively with equilibrium time of 76 min. The chemisorptive nature of the adsorption process was confirmed via the kinetic pseudo-second order model; the Weber–Morris diffusive model correlates well with experimental results, suggesting diffusion in the pores of the sorbent (Mansuri et al., 2014).

Adsorbent performance in real wastewater might deviate from its performance in synthetic aqueous solution due to the presence of co-contaminant. A few studies investigated agricultural waste adsorbents for removing ammonia nitrogen from real wastewater sources (Chen et al., 2013b, Sarkhot et al., 2013, Ricky et al., 2016). As excessive nutrients in dairy manure effluent could cause a number of environmental concerns for surface and groundwater resources, Chen et al. (2013b) studied the usage of dried banana (*Musa spp.*) peels treated with 30% sodium hydroxide (NaOH) solution for 8 h to remove ammonium from both landfill leachate and pure ammonium synthetic solution. The solution and biosorbent were mixed at 150 rpm for 30 min at room temperature (20°C ± 1°C) and the adsorption capacity obtained in pure solution and manure solution were 2.8 and 5.37 mg/g, respectively. This indicates that the sorption capacity is aided by other dissolved constituents in the manure slurry (Chen et al., 2013b).

Experimental studies performed by Sarkhot et al. (2013) tested the application of biochar prepared from mixed hardwood shavings for removing ammonium from dairy effluent. The biosorbent with a uniform particle size of 2 mm and weight of 2 g was added into the diary manure effluent. The amount of ammonium adsorbed increased as the initial concentration increased up to 1000 mg/L, with maximum ammonium uptake of 5.29 mg/g (Sarkhot et al., 2013).

Ricky et al. (2016) evaluated the performance of chemically modified fresh empty fruit bunches (EFB) with 10 mM of sodium hydroxide for ammonia nitrogen removal from urban drainage. Based on the results, the ammonia nitrogen levels in the wastewater were insignificantly reduced due to insufficient contact time. The maximum adsorption capacity obtained was in the range of 0.01–0.60 mg/g for 20 min of contact time. The adsorption capacity appeared to continue at a much slower rate after 20 min, until it reached equilibrium adsorption at 40 min. For urban drainage wastewater treatment, a minimum 80 min retention time was required for effective removal of ammonia nitrogen (Ricky et al., 2016).

A comparison study of natural and modified pine cone powder for ammonium removal was carried out by Demırak et al. (2015), revealing an adsorption capacity of 1.05 mg/g for the untreated pine cone powder in ammonium removal. The same authors (Demırak et al., 2015) also reported that the adsorption of ammonium onto modified pine cone powder with sodium hydroxide (NaOH) provided better ammonium removal compared to the untreated adsorbents in terms of their adsorption rate. The adsorption capacities for ammonium onto modified pine cone powder with 0.05 and 0.10 M of NaOH were 2.41 and 3.06 mg/g, respectively, after the equilibrium time of 15 min. The results showed that suitable conditions for the adsorption were: 0.3 g adsorbent; ammonium concentration, 50 mg/L; solution pH, 8; and temperature, 22°C (Demırak et al., 2015).

In another study, Hina et al. (2015) investigated pine bark and biochar as adsorbents for batch adsorption of ammonium nitrogen removal. Biochar showed a slightly better performance in ammonium removal compared to pine bark, with an adsorption capacity of 0.52 and 0.47 mg/g, respectively (Hina et al., 2015). Subsequent studies revealed that the ammonium ion uptakes by barbecue bamboo charcoal and avocado seed adsorbents were 1.68 and 5.4 mg/g, respectively, with both kinetic data fitting well to the pseudo-second-order model indicating a chemisorption process (Zhou et al., 2015, Zhu et al., 2016).

The use of watermelon rind (WR) as a low-cost adsorbent for ammonia nitrogen removal from aqueous solution was studied by Ibrahim et al. (2016). Four different types of watermelon rind sorbent were used, including: raw WR, and modified WR with sodium hydroxide (NaOH), potassium hydroxide (KOH) and sulfuric acid (H_2SO_4). The adsorption capacities obtained from the batch sorption experiment were in the range of 1.21 to 1.24 mg/g for all four types of sorbent, and the adsorption process was rapid, with an equilibrium state established within 40 min of contact time (Ibrahim et al., 2016). The same experimental data was then analyzed for isotherm and kinetic characteristics in another study (Zahrim et al., 2016). The results showed that the equilibrium data correlated well with the Langmuir isotherm model, with the highest correlation coefficient (R^2) value of 1.0, suggesting that the adsorption process involved monolayer adsorption. Meanwhile, the kinetic study was determined to follow the pseudo-second-order model, with the assumption that the adsorption was governed by sharing or exchange of cations (Zahrim et al., 2016).

The presence of hydroxyl group in starch-containing material (jackfruit seed) was reported to remove ammonium ions with adsorption capacity of 3.37 mg/g (Azreen et al., 2017). Jackfruit seed is high in starch, with approximately 93%–95%, and similar findings on the application of starch-containing materials in ammonium ion adsorption was reported by Wei et al. (2016).

Overall, the adsorption process of ammonia nitrogen tends to have a similar trend in that removal performance can be influenced by several factors such as dosage, pH value and the initial concentration of solution. The increase in dosage of biochar adsorbent has a significant effect on the adsorption process, as more active sites of adsorption are available, thus improving the adsorption rates. The initial pH value also played an important role in the adsorption process, as the pH can change the form of ammonia nitrogen in water (Liang et al., 2016).

Extensive studies on the feasibility of using low-cost adsorbents derived from agricultural waste and plant materials to remove ammonia nitrogen have been conducted. Based on the earlier review, the brown seaweed *Cystoseira indica* (*C. indica*) records the highest adsorption capacity with

15.21 mg/g at a low initial wastewater concentration (40 mg/L). The complexation of ammonium ions with the ionized O-H groups of "free" hydroxyl groups and bonded O-H bands of carboxylic acid in the inter- and intra-molecular hydrogen bonding suggest that the presence of both functional groups were the major contributors in the high adsorption capacity of the biosorbent (Wahab et al., 2010, Mansuri et al., 2014).

18.5 UTILIZATION OF SPENT ADSORBENT

The ammonium-loaded adsorbent should be properly handled and the best solution for handling this problem is through the production of soil conditioner or fertilizer, if the contaminants are not toxic. The market demand for slow-release fertilizers will continue to grow, as they offer an effective and efficient fertilizer alternative due to their sustainable characteristics (Vaneeckhaute et al., 2017). Adsorbent enriched with ammonia nitrogen offers a better alternative for agricultural applications, for example compost, as it can provide plants with required nutrients.

Zhao et al. (2013) has reported on the use of ammonium-loaded ceramic adsorbent as nitrogen fertilizer. The materials used for the ceramic adsorbent were Kanuma clay, Akadama clay and zeolite. Zeolite has been proved to be a good soil conditioner prior to its good adsorption qualities, and its ion exchange ability could improve soil performance and potentially increase crop yields (Malekian et al., 2011). The combination of three adsorbents (Kanuma clay, Akadama clay and zeolite) could improve soil quality, facilitating better plant growth. Ceramic adsorbents are also said to be more effective than conventional nitrogen fertilizer, as it has equal nitrogen content that is deposited at a deeper depth; the spent adsorbent also possesses nutrient-retention capacity. As a conclusion, nitrogen-rich adsorbent can act as an effective slow-release soil conditioner, owing to its strong affinity to ammonium ions and its high nutrient-retention capacity (Zhao et al., 2013).

Another report on the effectiveness of biochar as a soil fertilizer has been presented by Sarkhot et al. (2013). Previous studies have verified that biochar was able to remove contaminants from solution and recently, the potential of biochar as a soil fertilizer has been extensively studied. The ultimate goal of the study was to use biochar to adsorb nutrients from dairy manure and apply the nitrogen-rich adsorbent as a soil enhancer. The biochar used in this study was produced from a combination of mixed hardwood shavings, including maple, aspen, choke cherry and alder, which underwent slow pyrolysis at 300°C for 8–12 h of reaction time. The biochar was initially used to adsorb contaminant from dairy manure effluent collected from a sedimentation lagoon at the Vander Woude dairy farm in California before being assessed as soil fertilizer. The results show that the biochar was able to retain 78%–91% of the removed ammonium ions adsorbed from dairy manure solution. Biochar used for sorption of 50 g/L dairy effluent at the rate of 2 g per 40 mL, which equates to a requirement of 45 Mg ha^{-1} of biochar as nutrient-loaded soil fertilizer. However, further studies are required to establish the dosage of the fertilizer for farm-scale use, since the volumes and concentrations differ considerably with geography and season (Sarkhot et al., 2013).

Zeolite used in the ammonia nitrogen treatment process can be employed as a slow-release fertilizer due to its porous character, which allows the ammonium to leak out at a much slower rate than it was adsorbed. Ganrot (2012) examined the efficiency of a combination of struvite and ammonium-zeolite as a soil fertilizer for common wheat and barley growth. The experiments were short term and were performed for about 3–4 weeks under rigorously controlled climate chamber conditions. The performance of the zeolite-struvite mix (ZSM) fertilizer was then compared with two conventional, commercially available fertilizers. After a few periods, the soil with ZSM fertilizer added showed good results that were fully comparable to the commercial fertilizers, which confirmed the ability of nutrient-rich sorbent as a soil enhancer (Ganrot, 2012).

The growth of maize plants has been enhanced by the addition of ammonium-loaded sorbent as fertilizer to the soil. Perrin et al. (1998) successfully used ammonium-loaded clinoptilolite (A-Cp) as a nitrogen (N) fertilizer for sandy soil to decrease N leaching and increase N-use efficiency (Perrin et al., 1998). Moreover, Zahrim et al. (2015) examined the co-composting of empty fruit bunch (EFB) with nitrogen-rich palm oil mill effluent (POME) to reduce waste management problems and conserve plant nutrients. EFB fibers adsorb the nutrients from POME so that it becomes an ideal mixture for biodegradation. In conclusion, EFB composting with a suitable quantity of bulking agents was able to improve the quality of the composting process, both by increasing nutrient levels and accelerating the process (less than 60 days) (Zahrim et al., 2015).

Regeneration of adsorbent is another way to use the spent adsorbent, and a good adsorbent is measured both by its adsorption capacity and its good regeneration ability for multiples uses. Ji et al. (2007) carried out an investigation on the regeneration process of clinoptilolite sorbent used for adsorbing ammonium ions from wastewater. The clinoptilolite loaded with ammonium ions was regenerated by a 0.2 mol/L $Ca(OH)_2$ emulsion at 100°C–110°C with a by-product of ammonia (NH_3). The mass of the sorbent during regeneration and the adsorption capacity after regeneration were measured. Based on the results, the adsorbent can be regenerated once it reaches its saturation capacity, which makes the sorbent usable for a relatively long period of time (Ji et al., 2007).

The regeneration of low-cost sorbents enriched with ammonia nitrogen was also studied by Azhar and Aimi Shaza (2012). Chemical modified sand was initially used for ammonia nitrogen sorption in a fixed-bed column before being tested for regeneration process. The exhausted bed of sorbent was then treated with 1.0 M of sodium chloride at pH 12 for the regeneration process. The regeneration solution was pumped through the chemically modified adsorbent in the up-flow mode and the ammonia level was determined by using a breakthrough graph of solution. It was found that there was no significant difference in the percentage of ammonia removed during the initial sorption and regeneration studies, thus suggesting that the adsorbent has the ability to be reused (Azhar and Aimi Shaza, 2012).

Regeneration of modified EFB adsorbent was studied by Zahrim et al. (2017). The EFB fibers used in this study were initially modified by using 10 mM of sodium hydroxide (NaOH) for 12 h at room temperature. The EFB fibers showed maximum recovery of ammonia nitrogen at pH 2 with 0.0999 mg/g desorption capacity, which gradually decreased to 0.0271 mg/g at solution pH of 10 as displayed in Figure 18.3. At a lower pH, the high desorption ability of ammonia nitrogen may be influenced by the replacement of adsorbed ammonium ions by hydrogen (H^+) ions onto the carboxyl group (–RCOO–) of the fibers. This desorption study shows that the ammonium-rich modified EFB fibers can be regenerated for further use in the adsorption process (Zahrim et al., 2017).

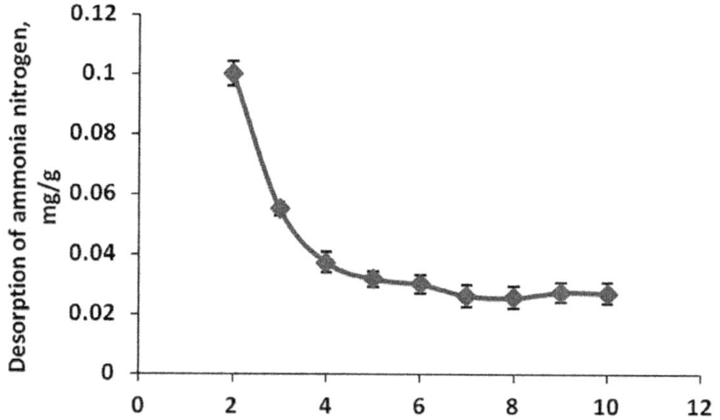

FIGURE 18.3 Effect of solution pH on the desorption of ammonia nitrogen from modified empty fruit bunch fibers (EFB). (From Zahrim, A. et al., *Indian J. Sci. Technol.*, 10, 1–4, 2017.)

18.6 CONCLUSION AND FUTURE REMARKS

The treatment method for ammonia nitrogen removal from water has been developed extensively over the few decades; some of the removal methods include the conventional biological method, nitrification and denitrification, chemical precipitation, as well as hybrid methods. Among these approaches, adsorption is considered to be the most efficient and effective treatment method, as it is low cost, environmental friendly, and easy to operate in practice. There are several types of adsorbent that have been discovered to effectively remove pollutants from wastewater, such as zeolite and hydrogel. Although these adsorbent are effective, the major drawback is their high production cost. One of the solutions provided is the utilization of waste as a low-cost adsorbent for pollutant removal, this is suitable because of its abundance, availability, inexpensiveness and ability to minimize solid waste generation. The feasibility of several agricultural residue–based sorbents was examined and, among these, the brown seaweed *Cystoseira indica* (*C. indica*) records the highest adsorption capacity with 15.21 mg/g.

Ammonium loaded adsorbents should be handled properly and the use of non-toxic spent adsorbent is recommended as a soil conditioner. The efficiency of nutrient-loaded adsorbent as a fertilizer for plant growth has been investigated by a few researchers. Further research is, however, still needed for in-depth assessment of the management of ammonium-loaded adsorbent as there information is still limited on the use of spent adsorbent as fertilizer.

REFERENCES

Abdolali, A., Guo, W. S., Ngo, H. H., Chen, S. S., Nguyen, N. C., and Tung, K. L. 2014. Typical lignocellulosic wastes and by-products for biosorption process in water and wastewater treatment: A critical review. *Bioresource Technology*, 160, 57–66.

Ahmadiannamini, P., Eswaranandam, S., Wickramasinghe, R., and Qian, X. 2017. Mixed-matrix membranes for efficient ammonium removal from wastewaters. *Journal of Membrane Science*, 526, 147–155.

Ali, I., Asim, M., and Khan, T. A. 2012. Low cost adsorbents for the removal of organic pollutants from wastewater. *Journal of Environmental Management*, 113, 170–183.

Alshameri, A., Ibrahim, A., Assabri, A. M., Lei, X., Wang, H., and Yan, C. 2014. The investigation into the ammonium removal performance of Yemeni natural zeolite: Modification, ion exchange mechanism, and thermodynamics. *Powder Technology*, 258, 20–31.

Andrew, G. F. 1991. Aerobic wastewater process models. In *Biotechnology*, 2nd ed., Rehm, H. J. and Reed, G. in cooperation with Pulher, A. and Stadler, P. Eds., pp. 408–439. VCH Publisher, New York.

Axelsson, A. and Persson, B. 1988. Determination of effective diffusion coefficients in calcium alginate gel plates with varying yeast cell content. *Applied Biochemistry and Biotechnology*, 18, 231.

Aydin Temel, F. and Kuleyin, A. 2016. Ammonium removal from landfill leachate using natural zeolite: Kinetic, equilibrium, and thermodynamic studies. *Desalination and Water Treatment*, 57, 23873–23892.

Azhar, A. and Aimi Shaza, C. 2012. Ammonia removal from an aqueous solution using chemical surface-modified sand. *Health Environment Journal*, 3, 17–24.

Azreen, I., Lija, Y., and Zahrim, A. Y. 2017. Ammonia nitrogen removal from aqueous solution by local agricultural wastes. *IOP Conference Series: Materials Science and Engineering*, 206, 012077.

Bakti, N. A. K. and Dick, R. I. 1992. A model for nitrifying suspended-growth reactor incorporating intraparticle diffusional limitation. *Water Research*, 26, 1681–1690.

Beccari, M., Pinto, A. C., Di Ramadori, R., and Tomei, M. C. 1992. Effects of dissolved oxygen and diffusion resistances on nitrification kinetics. *Water Research*, 26, 1099–1104.

Benefield, L. and Molz, F. 1983. A kinetic model for the activated sludge process which considers diffusion and reaction in the microbial floc. *Biotechnology and Bioengineering*, XXVI, 2591–2615.

Benefield, L. and Molz, F. 1984. A model for the activated sludge process which considers wastewater characteristics, floc behavior, and microbial population. *Biotechnology and Bioengineering*, 26(4), 352–361.

Benefield, L. and Molz, F. 1985. Mathematical simulation of a biofilm process. *Biotechnology and Bioengineering*, 27, 921–931.

Bhatnagar, A., Kumar, E., and Sillanpää, M. 2010. Nitrate removal from water by nano-alumina: Characterization and sorption studies. *Chemical Engineering Journal*, 163, 317–323.

Bhatnagar, A. and Sillanpaa, M. 2010. Utilization of agro-industrial and municipal waste materials as potential adsorbents for water treatment—A review. *Chemical Engineering Journal*, 157, 277–296.

Boopathy, R., Karthikeyan, S., Mandal, A., and Sekaran, G. 2013. Adsorption of ammonium ion by coconut shell-activated carbon from aqueous solution: Kinetic, isotherm, and thermodynamic studies. *Environmental Science and Pollution Research*, 20, 533–542.

Bunani, S., Yorukoglu, E., Sert, G., Yuksel, U., Yuksel, M., and Kabay, N. 2012. Application of nanofiltration for reuse of municipal wastewater and quality analysis of product water. *Desalination*, 315, 33–36.

Carl-Fredrik, L. 1997. *Control and Estimation Strategies Applied to the Activated Sludge Processes*. Uppsala University, Uppsala, Sweden.

Chen, X., Peltier, E., Sturm, B. S. M., and Young, C. B. 2013a. Nitrogen removal and nitrifying and denitrifying bacteria quantification in a stormwater bioretention system. *Water Research*, 47, 1691–1700.

Chen, Y.-N., Liu, C.-H., Nie, J.-X., Luo, X.-P., and Wang, D.-S. 2013b. Chemical precipitation and biosorption treating landfill leachate to remove ammonium-nitrogen. *Clean Technology Environmental Policy*, 15, 395–399.

De Haas, D., Wentzel, M., and Ekama, G. 2000. The use of simultaneous chemical precipitation in modified activated sludge systems exhibiting biological excess phosphate removal Part 1: Literature review. *Water Sa*, 26, 439–452.

Demirak, A., Keskin, F., Şahın, Y., and Kalemcı, V. 2015. Removal of ammonium from water by pine cone powder as biosorbent. *Mugla Journal of Science and Technology*, 1, 5–12.

Erdem, E., Karapinar, N., and Donat, R. 2004. The removal of heavy metal cations by natural zeolites. *Journal of Colloid and Interface Science*, 280, 309–314.

Fengel, D. and Wegener, G. 1983. *Wood: Chemistry, Ultrastructure, Reactions*. Berlin, Germany: Walter de Gruyter.

Fernandoa, W. A. R. N., Xia, K., and Rice, C. W. 2004. Sorption and desorption of ammonium from liquid swine waste in soils. *Soil Science Society of America Journal*, 69, 1057–1065.

Ganrot, Z. 2012. Use of zeolites for improved nutrient recovery from decentralized domestic wastewater. *Handbook of Natural Zeolites*. Bentham Science Publishers, Beijing, China.

Grady, N. F. 1990. Activated Sludge Theory and Practice. Oxford: Oxford University Press..

Grady, C. P. L. Jr. 1983. Modelling of biological fixed films: A state-of-the-art review, in *Proceedings of the 1st International Conference on Fixed-Film Biological Processes of Wastewater Treatment*, Wu, Y. C., Smith, E. D., Miller, R. D., and Patken, E. J. O. Eds., pp. 75–134. Purdue, Indiana.

Gujer, W., *Henze*, M., Mino, T., and *van Loosdrecht*, M. C. M. *1999*. Activated sludge model No. 3. *Water Science and Technology*, 39(1), 183–193.

Gujer, W. and Zehnder, A. J. B. 1983. Conversion processes in anaerobic digestion. *Water Science and Technology*, 15, 127–167.

Gupta, A. and Verma, J. P. 2015. Sustainable bio-ethanol production from agro-residues: A review. *Renewable and Sustainable Energy Reviews*, 41, 550–567.

Gupta, V. K., Mohan, D., Sharma, S., and Sharma, M. 2000. Removal of basic dyes (Rhodamine B and Methylene Blue) from aqueous solutions using bagasse fly ash. *Separation Science and Technology*, 35, 2097–2113.

Gupta, V. K., Sharma, S., Yadav, I. S., and Mohan, D. 1998. Utilization of bagasse fly ash generated in the sugar industry for the removal and recovery of phenol and p-nitrophenol from wastewater. *Journal of Chemical Technology and Biotechnology*, 71, 180–186.

Hannoun, B. J. M. and Stephanopoulos, G. 1986. Diffusion coefficients of glucose and ethanol in cell-free and cell-occupied calcium alginate membranes. *Biotechnology and Bioengineering*, 28, 829.

He, Y., Lin, H., Dong, Y., Liu, Q., and Wang, L. 2016a. Simultaneous removal of ammonium and phosphate by alkaline-activated and lanthanum-impregnated zeolite. *Chemosphere*, 164, 387–395.

He, Y., Lin, H., Dong, Y., Liu, Q., and Wang, L. 2016b. Simultaneous removal of phosphate and ammonium using salt–thermal-activated and lanthanum-doped zeolite: Fixed-bed column and mechanism study. *Desalination and Water Treatment*, 57, 27279–27293.

Hina, K., Hedley, M., Camps-Arbestain, M., and Hanly, J. 2015. Comparison of pine bark, biochar and zeolite as sorbents for NH_4^+-N removal from water. *CLEAN–Soil Air Water*, 43, 86–91.

Huo, H., Lin, H., Dong, Y., Cheng, H., Wang, H., and Cao, L. 2012. Ammonia-nitrogen and phosphates sorption from simulated reclaimed waters by modified clinoptilolite. *Journal of Hazardous Materials*, 229, 292–297.

Ibrahim, G. and Aba Saeed, A. E. 1995. Modeling of sequencing batch reactors. *Water Research*, 29(5), 1761–1766.

Ibrahim, H. M., Ibrahim, G., and El-Ahwany, A. H. 2002a. Processes Part I: Process modeling of activated sludge flocculation and sedimentation. *The Seventh International Conference for Chemical Engineering in Egyptian Engineering Society through November 1–3, 2002.*

Ibrahim, H. M., Ibrahim, G., and El-Ahwany, A. H. 2002b. Processes Part II: Process modeling of activated sludge flocculation and sedimentation. *The Seventh International Conference for Chemical Engineering in Egyptian Engineering Society through November 1–3, 2002.*

Ibrahim, A., Yusof, L., Beddu, N. S., Galasin, N., Lee, P. Y., Lee, R. N. S., and Zahrim, A. Y. 2016. Adsorption study of Ammonia Nitrogen by watermelon rind. *IOP Conference Series: Earth and Environmental Science*, 36, 012020.

Jellali, S., Wahab, M. A., Anane, M., Riahi, K., and Jedidi, N. 2011. Biosorption characteristics of ammonium from aqueous solutions onto Posidonia oceanica (L.) fibers. *Desalination*, 270, 40–49.

Ji, Z.-Y., Yuan, J.-S., and Li, X.-G. 2007. Removal of ammonium from wastewater using calcium form clinoptilolite. *Journal of Hazardous Materials*, 141, 483–488.

Jiang, Z., Yang, J., Ma, H., Ma, X., and Yuan, J. 2016. Synthesis of pure NaA zeolites from coal fly ashes for ammonium removal from aqueous solutions. *Clean Technologies and Environmental Policy*, 18, 629–637.

Karadag, D., Koc, Y., Turan, M., and Armagan, B. 2006. Removal of ammonium ion from aqueous solution using natural Turkish clinoptilolite. *Journal of Hazardous Materials*, 136, 604–609.

Karadağ, E., Üzüm, O. B., and Saraydin, D. 2002. Swelling equilibria and dye adsorption studies of chemically crosslinked superabsorbent acrylamide/maleic acid hydrogels. *European Polymer Journal*, 38, 2133–2141.

Kaşgöz, H. and Durmus, A. 2008. Dye removal by a novel hydrogel-clay nanocomposite with enhanced swelling properties. *Polymers for Advanced Technologies*, 19, 838–845.

Kaşgöz, H., Durmuş, A., and Kaşgöz, A. 2008. Enhanced swelling and adsorption properties of AAm-AMPSNa/clay hydrogel nanocomposites for heavy metal ion removal. *Polymers for Advanced Technologies*, 19, 213–220.

Kern, D. I., Schwaickhardt, R. D. O., Mohr, G., Lobo, E. A., Kist, L. T., and Machado, E. L. 2013. Toxicity and genotoxicity of hospital laundry wastewaters treated with photocatalytic ozonation. *Science of the Total Environment*, 443, 566–572.

Lei, L., Li, X., and Zhang, X. 2008. Ammonium removal from aqueous solutions using microwave-treated natural Chinese zeolite. *Separation and purification Technology*, 58, 359–366.

Li, X. Z., Zhao, Q. L., and Hao, X. D. 1999. Ammonium removal from landfill leachate by chemical precipitation. *Waste Management*, 19, 409–415.

Liang, P., Yu, H., Huang, J., Zhang, Y., and Cao, H. 2016. The review on adsorption and removing ammonia nitrogen with biochar on its mechanism. *MATEC Web of Conferences*, 67, 07006.

Licentiate Thesis, Department of Material Science, Uppsala University, Uppsala, Sweden.

Lindberg, C. F. 1997. Control and estimation strategies applied to the activated sludge process. Licentiate Thesis, Department of Material Science, Uppsala University, Uppsala, Sweden.

Liu, H., Dong, Y., Liu, Y., and Wang, H. 2010a. Screening of novel low-cost adsorbents from agricultural residues to remove ammonia nitrogen from aqueous solution. *Journal of Hazardous Materials*, 178, 1132–1136.

Liu, H., Dong, Y., Wang, H., and Liu, Y. 2010b. Adsorption behavior of ammonium by a bioadsorbent–Boston ivy leaf powder. *Journal of Environmental Sciences*, 22, 1513–1518.

Malekian, R., Abedi-Koupai, J., and Eslamian, S. S. 2011. Influences of clinoptilolite and surfactant-modified clinoptilolite zeolite on nitrate leaching and plant growth. *Journal of Hazardous Materials*, 185, 970–976.

Mansuri, N., Mody, K., and Basha, S. 2014. Biosorption of ammoniacal nitrogen from aqueous solutions with low-cost biomaterials: Kinetics and optimization of contact time. *International Journal of Environmental Science and Technology*, 11, 1711–1722.

Martins, T. H., Souza, T. S., and Foresti, E. 2017. Ammonium removal from landfill leachate by Clinoptilolite adsorption followed by bioregeneration. *Journal of Environmental Chemical Engineering*, 5, 63–68.

Matson, J. V. and Characklis, W. G. 1976. Difision into microbial aggregates. *Water Research*, 10: 877–885.

Metcalf, L. and Eddy, H. P. 1972. *Wastewater Engineering: Collection, Treatment, Disposal*. New York: McGraw-Hill.

Miao, L., Wang, K., Wang, S., Zhu, R., Li, B., Peng, Y., and Weng, D. 2014. Advanced nitrogen removal from landfill leachate using real-time controlled three-stage sequence batch reactor (SBR) system. *Bioresource Technology*, 159, 258–265.

Montazer-Rahmati, M. M., Rabbani, P., Abdolali, A., and Keshtkar, A. R. 2011. Kinetics and equilibrium studies on biosorption of cadmium, lead, and nickel ions from aqueous solutions by intact and chemically modified brown algae. *Journal of Hazardous Materials*, 185, 401–407.

Nguyen, D. D., Ngo, H. H., and Yoon, Y. S. 2014. A new hybrid treatment system of bioreactors and electrocoagulation for superior removal of organic and nutrient pollutants from municipal wastewater. *Bioresource Technology*, 153, 116–125.

Okoniewska, E., Lach, J., Kacprzak, M., and Neczaj, E. 2007. The removal of manganese, iron and ammonium nitrogen on impregnated activated carbon. *Desalination*, 206, 251–258.

Onuma, O. 1982. Mass-transfer characteristics within microbial systems. *Water Science and Technology*, 14(6–7), 553–568.

Perić, J., Trgo, M., and Vukojević Medvidović, N. 2004. Removal of zinc, copper and lead by natural zeolite—A comparison of adsorption isotherms. *Water Research*, 38, 1893–1899.

Perrin, T. S., Drost, D. T., Boettinger, J. L., and Norton, J. M. 1998. Ammonium-loaded clinoptilolite: A slow-release nitrogen fertilizer for sweet corn. *Journal of Plant Nutrition*, 21, 515–530.

Renou, S., Givaudan, J., Poulain, S., Dirassouyan, F., and Moulin, P. 2008. Landfill leachate treatment: Review and opportunity. *Journal of Hazardous Materials*, 150, 468–493.

Ricky, L. N. S., Shahril, Y., Nurmin, B., and Zahrim, A. Y. 2016. Ammonia-nitrogen removal from urban drainage using modified fresh empty fruit bunches: A case study in Kota Kinabalu, Sabah. *IOP Conference Series: Earth and Environmental Science*, 36, 012055.

Saltali, K., Sari, A., and Aydin, M. 2007. Removal of ammonium ion from aqueous solution by natural Turkish (Yıldızeli) zeolite for environmental quality. *Journal of Hazardous Materials*, 141, 258–263.

Sanin, D. F. and Vesilind, A. 1991. Synthetic sludge: A physical/ chemical model in understanding bioflocculation. *Water Environment Research*, 68(5), 927–933.

Sarioglu, M. 2005. Removal of ammonium from municipal wastewater using natural Turkish (Dogantepe) zeolite. *Separation and Purification Technology*, 41, 1–11.

Sarkhot, D., Ghezzehei, T., and Berhe, A. 2013. Effectiveness of biochar for sorption of ammonium and phosphate from dairy effluent. *Journal of Environmental Quality*, 42, 1545–1554.

Siegrist, H. and Gujer, W. 1985. Mass transfer mechanisms in a heterotrophic biofilm. *Water Science and Technology*, 19(11), 1369–1378.

Smith, P. G. 1984. A model of localised oxygen sinks around bacterial colonies within activated sludge. *Water Research*, 18(8): 1045–1051.

Smith, P. G. and Coakley, P. 1984. Diffusivity, tortuosity and pore structure of activated sludge. *Water Research*, 18, 117–122.

Tee, P.-F., Abdullah, M. O., Tan, I. A. W., Amin, M. A. M., Nolasco-Hipolito, C., and Bujang, K. 2016. Performance evaluation of a hybrid system for efficient palm oil mill effluent treatment via an air-cathode, tubular upflow microbial fuel cell coupled with a granular activated carbon adsorption. *Bioresource Technology*, 216, 478–485.

Thornton, A., Pearce, P., and Parsons, S. 2007a. Ammonium removal from solution using ion exchange on to MesoLite, an equilibrium study. *Journal of Hazardous Materials*, 147, 883–889.

Vaneeckhaute, C., Lebuf, V., Michels, E., Belia, E., Vanrolleghem, P. A., Tack, F. M. G., and Meers, E. 2017. Nutrient recovery from digestate: Systematic technology review and product classification. *Waste and Biomass Valorization*, 8, 21–40.

Van't Riet, K. 1979. Review of measuring methods and results in nonviscous gas-liquid mass transfer in stirred vessels. *Industrial and Engineering Chemistry Process Design and Development*, 18, 357–364.

Villadsen, J. and Nielsen, J. 1992. Modelling of microbial kinetics. *Chemical Engineering Science*, 47, 4225–4270.

Wahab, M. A., Boubakri, H., Jellali, S., and Jedidi, N. 2012. Characterization of ammonium retention processes onto Cactus leaves fibers using FTIR, EDX and SEM analysis. *Journal of Hazardous Materials*, 241, 101–109.

Wahab, M. A., Jellali, S., and Jedidi, N. 2010. Ammonium biosorption onto sawdust: FTIR analysis, kinetics and Adsorption isotherms modeling. *Bioresource Technology*, 101, 5070–5075.

Wang, S. and Peng, Y. 2010. Natural zeolites as effective adsorbents in water and wastewater treatment. *Chemical Engineering Journal*, 156, 11–24.

Wei, J., Liu, X., and Xu, S. 2016. Adsorption behaviors of ammonium nitrogen by an amphoteric hydrogel. *Desalination and Water Treatment*, 57, 5753–5759.

Wilkinson, J. F. 1959. The problem of energy-storage compounds in bacteria. *Experimental Cell Research*, 7, 111–130.

Williamson, K. J. and McCarty, P. L. 1976. Verification studies of the biofilm model for bacterial substrate utilization. *Journal of the Water Pollution Control Federation*, 48, 9–24.

Wiszniowski, J., Robert, D., Surmacz-Gorska, J., Miksch, K., and Weber, J. V. 2006. Landfill leachate treatment methods: A review. *Environmental Chemistry Letters*, 4, 51–61.

Witek-Krowiak, A. 2012. Analysis of temperature-dependent biosorption of Cu^{2+} ions on sunflower hulls: Kinetics, equilibrium and mechanism of the process. *Chemical Engineering Journal*, 192, 13–20.

Wu, S., Liang, P., Zhang, C., Li, H., Zuo, K., and Huang, X. 2015. Enhanced performance of microbial fuel cell at low substrate concentrations by adsorptive anode. *Electrochimica Acta*, 161, 245–251.

Xie, Y. and Wang, A. 2009. Study on superabsorbent composites XIX. Synthesis, characterization and performance of chitosan-g-poly (acrylic acid)/vermiculite superabsorbent composites. *Journal of Polymer Research*, 16, 143–150.

Xu, X., Gao, B.-Y., Yue, Q.-Y., and Zhong, Q.-Q. 2010. Preparation of agricultural by-product based anion exchanger and its utilization for nitrate and phosphate removal. *Bioresource Technology*, 101, 8558–8564.

Yuan, M.-H., Chen, Y.-H., Tsai, J.-Y., and Chang, C.-Y. 2016. Ammonia removal from ammonia-rich wastewater by air stripping using a rotating packed bed. *Process Safety and Environmental Protection*, 102, 777–785.

Zahrim, A., Asis, T., Hashim, M., Al-Mizi, T., and Ravindra, P. 2015. A review on the empty fruit bunch composting: Life cycle analysis and the effect of amendment (s). *Advances in Bioprocess Technology*. Cham, Switzerland: Springer.

Zahrim, A., Ricky, L., and Hilal, N. 2017. Ammonia-Nitrogen recovery from synthetic solution using agricultural waste fibers. *Indian Journal of Science and Technology*, 10, 1–4.

Zahrim, A. Y., Lija, Y., Ricky, L., and Azreen, I. 2016. Fruit waste adsorbent for ammonia nitrogen removal from synthetic solution: Isotherms and kinetics. *International Conference on Chemical and Bioprocess Engineering*, Sabah, Malaysia. IOP Publishing, pp. 012028.

Zhao, X., Qu, J., Liu, H., Wang, C., Xiao, S., Liu, R., Liu, P., Lan, H., and Hu, C. 2010a. Photoelectrochemical treatment of landfill leachate in a continuous flow reactor. *Bioresource Technology*, 101, 865–869.

Zhao, Y., Yang, Y., Yang, S., Wang, Q., Feng, C., and Zhang, Z. 2013. Adsorption of high ammonium nitrogen from wastewater using a novel ceramic adsorbent and the evaluation of the ammonium-adsorbed-ceramic as fertilizer. *Journal of Colloid and Interface Science*, 393, 264–270.

Zhao, Y., Zhang, B., Zhang, X., Wang, J., Liu, J., and Chen, R. 2010b. Preparation of highly ordered cubic NaA zeolite from halloysite mineral for adsorption of ammonium ions. *Journal of Hazardous Materials*, 178, 658–664.

Zheng, Y. and Wang, A. 2009. Evaluation of ammonium removal using a chitosan-g-poly (acrylic acid)/rectorite hydrogel composite. *Journal of Hazardous Materials*, 171, 671–677.

Zheng, Y., Xie, Y., and Wang, A. 2012. Rapid and wide pH-independent ammonium-nitrogen removal using a composite hydrogel with three-dimensional networks. *Chemical Engineering Journal*, 179, 90–98.

Zhou, Z., Yuan, J., and Hu, M. 2015. Adsorption of ammonium from aqueous solutions on environmentally friendly barbecue bamboo charcoal: Characteristics and kinetic and thermodynamic studies. *Environmental Progress & Sustainable Energy*, 34, 655–662.

Zhu, Y., Kolar, P., Shah, S. B., Cheng, J. J., and Lim, P. K. 2016. Avocado seed-derived activated carbon for mitigation of aqueous ammonium. *Industrial Crops and Products*, 92, 34–41.

Section VI

Wastewater Treatment

Biological Processes

19 Application of Natural Zeolite in Textile Wastewater Treatment

Integrated Photodegradation and Anaerobic Digestion System

Seth Apollo, Benton Otieno, and Aoyi Ochieng

CONTENTS

19.1 INTRODUCTION

Treatment of wastewater from industries using or generating organic dyes has been of great concern due to the negative effects of such wastes on the environment. These wastewater streams, if not appropriately handled, can percolate into ground water or contaminate surface water, causing adverse impact on life. Wastewater containing dyes from industries such as plastics, textile, pulp, paper and ink, is a major concern given the toxic and carcinogenic properties of the dyes (Rauf et al. 2010; Hammed et al. 2016). Moreover, the increase in global industrialisation, in addition to the copious amount of effluent generated from dye-based industries, has led to an increase in the number of industrial wastewaters discharged into the environment. Besides the high organic load, the effluents containing dyes such as methylene blue (MB) are characterised by intense color even at low dye concentration (Rauf et al. 2010). The intense color prevents light penetration in receiving streams, thereby hindering the photosynthetic activity of aquatic plants. This can lead to the death of the aquatic plants causing, with some other factors, conditions favourable for eutrophication (Otieno et al. 2016). To prevent eutrophication of water bodies, avoid human poisoning and eliminate the adverse effects of dye pollutants, Yan et al. (2016) suggested that wastewaters containing dyes should be properly treated before discharge into receiving streams.

Until now, various physical, biological or chemical methods such as coagulation, anaerobic digestion, oxidation, adsorption and membrane filtration have been applied in the treatment of the industrial wastewater containing dyes with varying degrees of success (Hammed et al. 2016). Anaerobic digestion (AD), a biological process for the treatment of wastewater with high organic content, is among the widely applied first-step treatment technologies in reducing the pollution load of these effluents (Apollo and Aoyi 2016). The major advantage of the AD technology is that it converts the organic pollutants into biogas, which is a renewable energy source (Apollo et al. 2016). Though a preferable technique in industrial wastewater treatment, the AD technology faces some challenges that can compromise its efficiency. Some of the major challenges include reactor instability, which can lead to digester failure; inability to degrade some industrial wastewaters that are toxic to microbes or recalcitrant in nature; and microbial washout during the degradation process (Abelleira-Pereira et al. 2015). Dye wastewater contains inhibitory recalcitrant and toxic organic substances such as phenolic compounds that can hinder vital steps during the AD, especially the rate-determining hydrolysis step (Chaiprapat and Laklam 2011). The recalcitrant organic pollutants are problematic because they may require very long retention time for total degradation, while the toxic pollutants may lead to either reactor failure during start-up or longer acclimatization periods since they are very harmful to the microbes (Chaiprapat and Laklam 2011; Brooms et al. 2016; Yan et al. 2016). Moreover, the aromatic structure of most dyes makes biological treatment ineffective (Banat et al. 2005).

To this end, it is necessary to devise ways of improving the efficiency of the anaerobic process to handle recalcitrant wastes and reduce microorganism washout. One possible way of achieving this is to employ a pretreatment technique that can improve the biodegradability of the recalcitrant wastes and introduce effective microbial support to minimize washout. Photocatalytic degradation, which is an advanced oxidation process (AOP), can be used to oxidize the recalcitrant compounds into more biodegradable intermediates prior to anaerobic digestion. Semiconductor photocatalysts, upon irradiation by light energy, can generate electrons (e^-) and holes (h^+), which react with water molecules leading to the generation of hydroxyl ($OH^\bullet$) radicals. The photogenerated hydroxyl radicals and holes can, in turn, react with the toxic recalcitrant compounds reducing them into less-toxic biodegradable products (Akpan and Hameed 2009; Otieno et al. 2016). Titanium dioxide (TiO_2) is a widely used semiconductor photocatalyst due to its desirable properties such as low cost, high activity and chemical stability (Asiri et al. 2011; Mabuza et al. 2017). However, the application of TiO_2 in powdered or nanosized form during photodegradation leads to particle aggregation and a high cost of catalyst separation after treatment (Asiltürk and Şener 2012). To overcome these challenges, TiO_2 can be supported onto a good support material to minimise aggregation, promote substrate adsorption to the surface of the catalyst and achieve easy catalyst separation.

Zeolite, being a good adsorbent, can be used in improving the performance of photodegradation and AD processes. In one way, it can be used in the bioreactor to limit microbial washout, thereby ensuring high microbial concentration during the anaerobic process. Also, due to its good adsorptive properties, it helps in concentrating the pollutants on its surface thus bringing them near the microorganism colonies. At the same time, zeolite can be used in the photoreactor as a support material for the TiO_2 powdered catalyst, thus ensuring integration of adsorption and photocatalysis as well as facilitating post-treatment catalyst recovery (Haque et al. 2005; Zainudin et al. 2010). This work, therefore, focused on integrating the photocatalytic degradation treatment technique with AD for remediating wastewater containing MB. Photodegradation was first employed as a pretreatment for improved substrate biodegradability followed by AD. In the two processes, zeolite was employed as a biomass and catalyst support. Of special interest was the comparison of methane yield and coefficient of the single AD and integrated photodegradation–AD systems.

19.2 METHODOLOGY

19.2.1 Materials and Equipment

Materials used in this study included zeolite (Pratley mining company, South Africa), Titanium dioxide (99% purity, Sigma Aldrich), prepared inoculum for bioreactor start-up and synthetic industrial wastewater obtained by dissolving commercial MB in distilled water. Photodegradation was carried out in a photocatalytic reactor (Figure 19.1a) with a capacity of 0.45 L and a 15 W UVC lamp. The photocatalytic reactor had evenly distributed holes at the bottom. The holes ensured uniform influent distribution during reactor feeding and recycling thus minimising channelling across the bed. The radial length of the annular space was minimised to ensure that the UV irradiation could reach every section of the reactor. A peristaltic pump was used to feed the reactor and to recirculate substrate during treatment. A laboratory scale fixed-bed bioreactor shown in Figure 19.1b was used for AD.

19.2.2 Photocatalytic Degradation

The zeolite was first crushed then sieved to obtain particles with sizes of around 2 mm before being washed with deionized water and finally dried at 105°C for 24 h. TiO_2 supported onto zeolite (TiO_2/Zeolite, 80% TiO_2 composition by weight) was prepared by thoroughly mixing TiO_2 and zeolite together using a mortar and an agate pestle, in the presence of ethanol. The resulting mixture was thereafter dried at 105°C to remove the solvent, obtaining TiO_2/Zeolite photocatalyst. Photodegradation was carried out in the photocatalytic reactor with a 15 W UVC lamp. The pH of the synthetic wastewater containing MB was adjusted to neutral conditions as found optimal from preliminary studies. An appropriate mass of the TiO_2/Zeolite catalyst was weighed, then mixed with the MB wastewater at 1 g/L catalyst loading, and stirred in the dark for one hour. After attaining adsorption-desorption equilibrium, the substrate was transferred into the photoreactor for irradiation. During photodegradation, samples were withdrawn and filtered (0.45 μm membrane filters) to remove all solid catalyst particles before being subjected to analyses. The photodegradation process was evaluated in terms of color and chemical oxygen demand (COD) reduction, and methane yield potential.

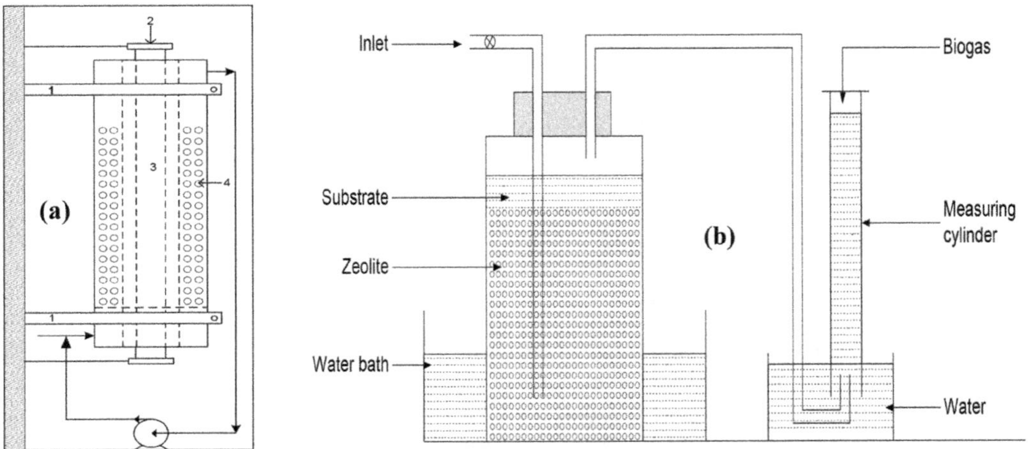

FIGURE 19.1 (a) Fluidised bed photoreactor (1) clamp, (2) lamp holder, (3) 15 W UVC lamp, (4) substrate and catalyst and (5) peristaltic pump, and (b) laboratory scale fixed bed anaerobic reactor.

19.2.3 Anaerobic Digestion

Seed inoculum was prepared by mixing fresh cow dung with two sets of sludge obtained from a secondary digester in a local wastewater treatment plant (Sebokeng Water Works, South Africa) and from an anaerobic digester treating distillery wastewater in a local distillery (Heineken distillery in Midvaal South Africa). The inoculum was prepared by gradually adding the synthetic industrial wastewater containing MB to the seed inoculum to acclimatize the microorganisms. Once the inoculum was ready, it was incubated at 37°C for 24 h to attain the mesophilic condition necessary for the anaerobic digestion. The wastewater containing MB was mixed with inoculum at a ratio of 7:3. It was then dosed with various nutrient supplements $KHPO_4$ (20 ppm), $MgSO_4.7H_2O$ (5 ppm), $NiSO_4$ (10 ppm), NH_4NO_3 (20 ppm) $FeSO_4$ (5 ppm) and Ca $(HCO_3)_2$ (20 ppm), and the pH adjusted to values between 6.8 and 7.20 before feeding it to the bioreactor. The annular space of the bioreactor was packed with zeolite for biomass support, then fed with the substrate and thereafter purged with nitrogen before incubation in a water bath at 37°C (Figure 19.1b). The Mariotte water displacement method was used for biogas collection and determination of the amount of gas produced within a specified period. The AD process was evaluated by biogas production and a reduction in color and COD.

19.2.4 Integrated Photodegradation and AD

To investigate the efficiency of the integrated photocatalytic degradation and anaerobic degradation, the photoreactor and bioreactor units were connected in series and operated batch-wise. A peristaltic pump was used to feed the reactors and to recirculate the wastewater during the photodegradation treatment process. The wastewater, after undergoing photodegradation, was subsequently fed to the bioreactor.

19.2.5 Chemical Analyses and Catalyst Characterization

The extent of organic load reduction during AD and photodegradation was determined from COD and biological oxygen demand (BOD) using standard analytical methods. For analysis of COD, the closed reflux method with dichromate oxidant, and Nanocolor colorimeter was used for analysis. Color analyses were done from a UV-Vis spectrophotometer (HACH, model DR 2800) at a maximum absorption wavelength of 664 nm (Eskizeybek et al. 2012). The volume of biogas produced was determined from the gas collected in the inverted measuring cylinder. Biogas methane composition was determined from a gas chromatograph (GC; SRI 8610C). The GC chromatograph had a thermal conductivity detector (TCD). The morphology of TiO_2/zeolite was studied from a scanning electron microscope (SEM, FEI NOVANANO 230). Nitrogen adsorption was used for the determination of the specific surface area of TiO_2 and zeolite based on the Brunauer–Emmett–Teller (BET) equation (ASAP 2020 V3.00 H).

19.3 RESULTS AND DISCUSSION

19.3.1 Catalyst Characterisation

The SEM analysis of TiO_2/Zeolite and the BET analyses of TiO_2 and zeolite are shown in Figure 19.2. From the SEM image of the TiO_2/zeolite material (Figure 19.2a), a uniform distribution of the TiO_2 particles on the surface of zeolite was observed. Figure 19.2b shows the adsorption isotherms of zeolite and TiO_2, with type II hysteresis loops indicating non-uniform pore sizes (Yang et al. 2010). Specific surface areas of 13.8481 and 21.3309 m^2/g were obtained for zeolite and TiO_2, respectively, from BET analysis. The zeolite particles, despite having relatively bigger particle sizes compared to TiO_2, still had a larger specific surface area, thus can be applied as good support for the catalyst.

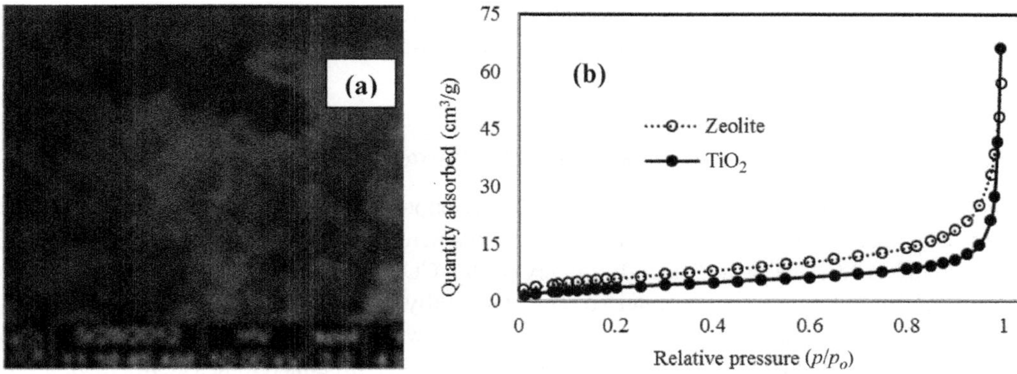

FIGURE 19.2 (a) SEM image of TiO_2/Zeolite and (b) zeolite and TiO_2 Nitrogen adsorption isotherms.

19.3.2 Photodegradation

During the photodegradation process, MB (initial concentration of 2000 ppm) was irradiated for 90 minutes to avoid over-oxidation, since the photoreactor effluent would be finally fed into the bioreactor. The performance of the photoreactor system when treating MB is shown in Figure 19.3a. After an irradiation time of 90 min, a higher color reduction of 71% was achieved and a high COD reduction of 54%. The observed color removal can be explained by the reaction mechanism involving photogeneration of hydroxyl molecules, which then react with the organic contaminants. Hydroxyl radicals react with the organic pollutants by addition to a double bond or abstraction of hydrogen (H) atoms from aliphatic organic compounds. This reaction leads to the degradation of chromophoric electron systems having conjugated double bonds such as –C=C– and –C=N– as contained in the MB structure shown in Figure 19.3b (Chandra et al. 2008; Mardani et al. 2015; Otieno et al. 2016).

Photodegradation led to the degradation of the chromophoric bonds as well as mineralization of some of the organic molecules. Also, there was a possibility that the organic molecules of the dye were degraded into simpler organic intermediates that required less oxygen for total oxidation, thus the higher reduction in COD recorded. A significant BOD increment of 48% was observed, signifying that photodegradation made the MB wastewater more inhabitable to the microorganisms

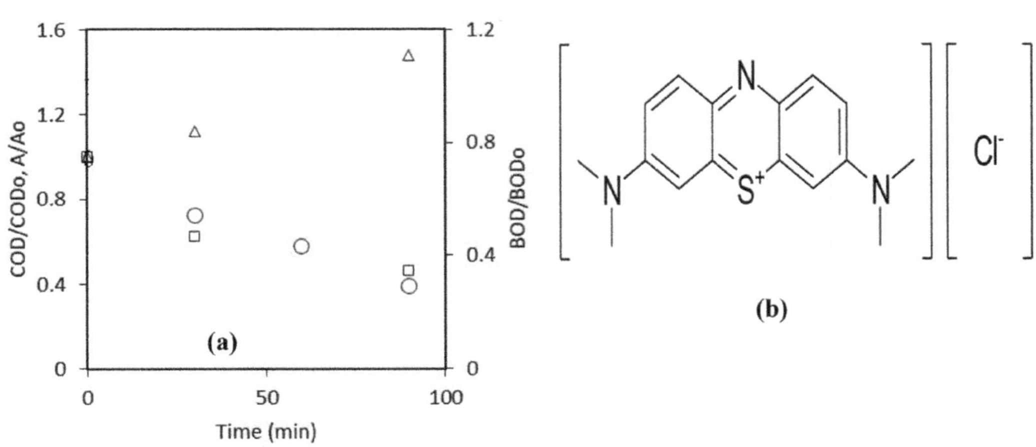

FIGURE 19.3 (a) Reduction in color (○), COD (□) and BOD (Δ) during photodegradation of MB dye, (b) basic structure of MB.

responsible for biodegradation. The degradation of the chromophoric bond opened the aromatic MB into an aliphatic compound thus making it easily biodegradable hence the increase in BOD (Banat et al. 2005).

19.3.3 Biodegradability Enhancement During Photodegradation

In Figure 19.4, the biodegradability index of MB before photodegradation was 0.33, which indicated that it was not readily biodegradable. Photodegradation increased the biodegradability index of MB dye from 0.33 to 1.0. The biodegradability index (BOD_5/COD ratio) has been used as a measure of biodegradability for various wastewater streams. Generally, a BOD_5/COD ratio higher than 0.4 is an indication that the wastewater is readily biodegradable, while a ratio below 0.4 is an indication of the presence of low-biodegradable contaminants in wastewater streams (Gomes et al. 2013).

The improved biodegradability index is an indication that photocatalysis degraded the MB molecules into lesser toxic organic intermediates. The huge decrease in color during the photodegradation of MB, as observed earlier (Figure 19.3a), implied that the original MB molecule was degraded or cleaved into simpler molecules. MB has aromatic rings in its structure and therefore its degradation was expected to produce aromatic amines, which are very toxic, as intermediate products. However, due to the observed increase in biodegradability after UV treatment, it can be concluded that this process did not produce such compounds. Similar observations were reported by Al-Momani et al. (2002) in a study involving the UV photodegradation of dyes that had aromatic rings in their structures. In the study, dye degradation by-products were tested for inhibition of AD of activated sludge, where no inhibition of oxygen consumption was reported. On the contrary, increased oxygen consumption due to the presence of more biodegradable compounds was observed.

19.3.4 Anaerobic Degradation of MB Waste Wastewater

During the anaerobic treatment of MB wastewater, a higher COD and BOD removal in the UV-treated stream in comparison to the non-UV treated stream was observed (Figure 19.5). The COD removals were 73% and 57% for the treated and non-treated streams, respectively, while the corresponding BOD removals were 85% and 58%, respectively. UV pretreatment enhanced biodegradation of MB because it improved its biodegradability (Figure 19.4).

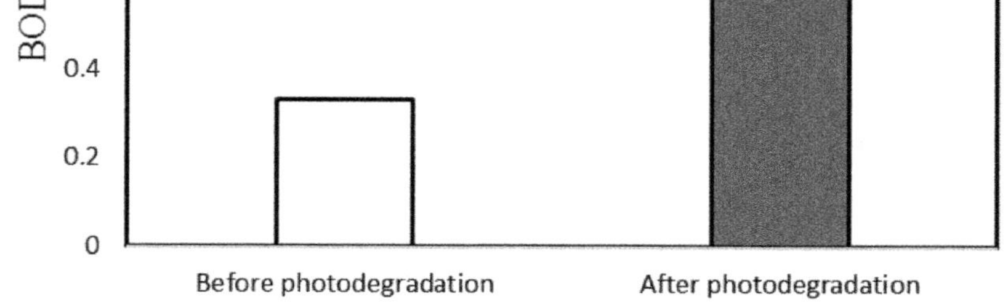

FIGURE 19.4 Effect of photodegradation on biodegradability of MB wastewater.

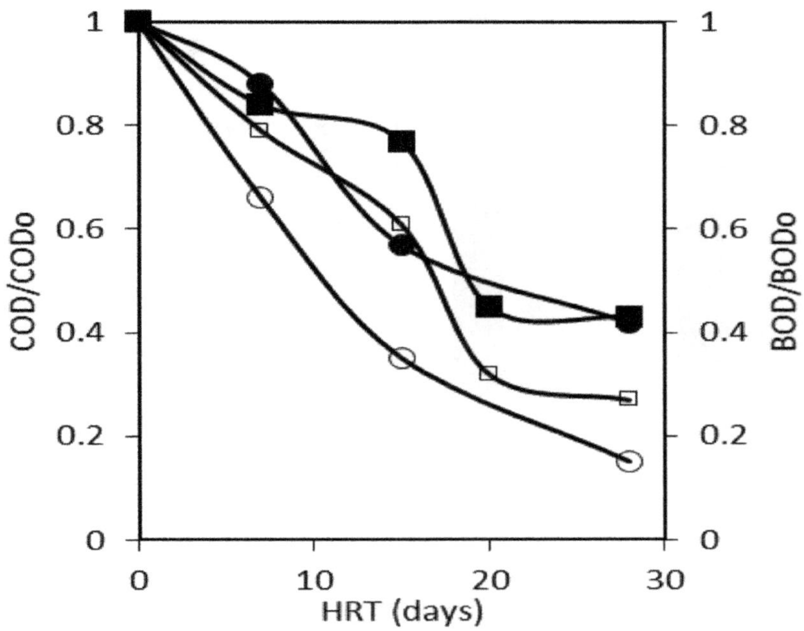

FIGURE 19.5 Anaerobic degradation of UV pretreated (BOD (○) and COD (□)), and non-UV treated (BOD (●) and COD (■)) MB wastewater streams.

The daily amount of biogas produced during anaerobic degradation of the UV-pretreated and non-UV-pretreated MB wastewater was recorded. The cumulative biogas amounts for the two wastewater streams are given in Figure 19.6. A shorter acclimatization period was needed for the UV-pretreated compared to the non-UV-treated MB wastewater streams. The longer acclimatization period of 7 days for the non-UV-treated wastewater is an indication that the MB dye was toxic to anaerobic microbes. The UV-pretreated MB solution had a better biogas production and a shorter acclimatization period because of enhanced biodegradability and reduced toxicity of the substrate before the AD.

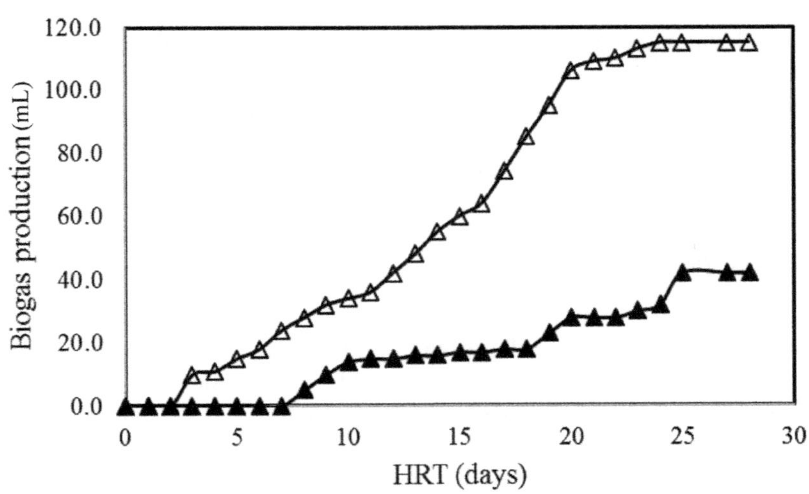

FIGURE 19.6 Biogas production during anaerobic digestion of UV pretreated MB (▲) and non-pretreated MB (Δ).

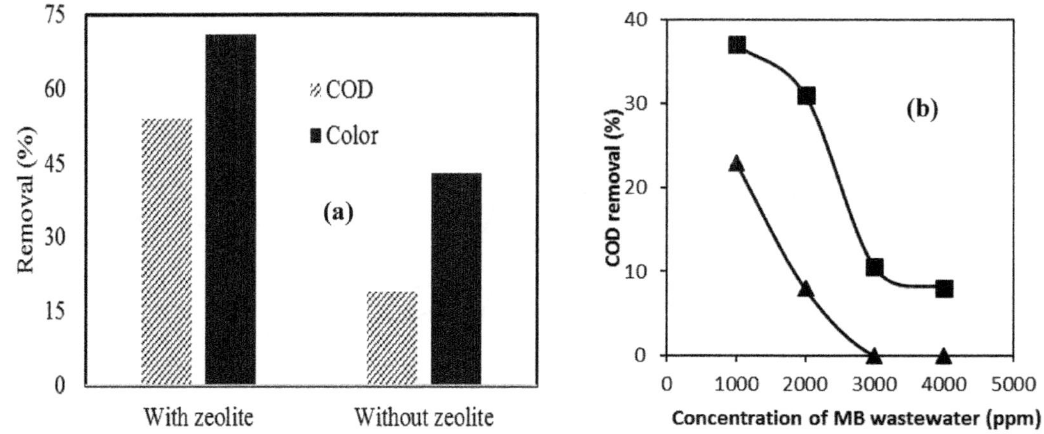

FIGURE 19.7 Effect of zeolite support on photodegradation (a) and anaerobic digestion (b) of MB with (■) and without (▲) zeolite wastewater.

19.3.5 Effect of Zeolite Support on Photodegradation and AD

The effect of zeolite as a catalyst and biomass support in the photodegradation and anaerobic processes, respectively, was investigated (Figure 19.7). In the AD of MB wastewater at different initial concentrations carried out for 42 days, zeolite was used to pack the annular space of the bioreactor, while in the control experiment no zeolite was used. For the photodegradation of MB wastewater (initial concentration 2000 ppm) process, zeolite was used to support TiO_2, while in the control TiO_2 was used on its own. As shown in Figure 19.7a, the TiO_2/Zeolite achieved a higher color and COD reduction than TiO_2. This could be attributed to a synergistic effect created between TiO_2 and the zeolite support material, where the zeolite support adsorbed organic compounds followed by their transportation to the surface of TiO_2 for immediate oxidation. Moreover, zeolite provided an increased surface area for organic substrate degradation by minimising the aggregation of the powdered TiO_2 catalysts (Zainudin et al. 2010; Otieno et al. 2016). The material consisting of TiO_2 and Zeolite integrated adsorption and catalytic activities resulting in an increased rate of degradation (Li et al. 2016; Otieno et al. 2017).

From Figure 19.7b, a higher COD removal was achieved by employing zeolite as biomass support. Various studies aimed at determining the effect of different types of microbial support material on the anaerobic degradation process have been documented, with high performance reported for zeolite due to its high capacity for the immobilization of microorganisms. Natural zeolite can also alter the concentration of ammonia/ammonium, thereby eliminating the inhibitory effect of free ammonia as well as reducing toxic ammonium ions (Milan et al. 2010).

19.3.6 Methane Yield and Coefficient of the Integrated Process

Methane yield (%) and coefficient (CH_4 L/g COD removed) for the integrated system (Phot-AD) and AD alone were determined, and the results are shown in Table 19.1. The integrated system had a higher CH_4 coefficient of 0.044 compared to AD alone, which had a coefficient of 0.012. Based on the methane yield and coefficient, the integrated system is suitable for MB wastewater treatment. Oller et al. (2011) conducted studies on the prospects and application of integrated advanced oxidation processes and biological methods on wastewater treatment. In the study, textile wastewater was among the wastewater streams identified as either having refractory components

TABLE 19.1
Methane Yield and Coefficients for AD and Integrated Photodegradation-AD

	Mean CH_4 Yield %	Biogas Produced After 25 Days (mL)	CH_4 Produced After 25 Days (mL)	CH_4 Coefficient (CH_4 L/g COD Removed)
Phot-AD	76	115	85	0.044
AD alone	73	42	31	0.012

or toxic compounds that would require a combination of different treatment techniques. Further, Oller et al. (2011) emphasized that if the wastewater to be treated contains a considerable number of non-biodegradable compounds and a high number of recalcitrant compounds, then it would be economical to employ a pretreatment that would improve biodegradability, followed by AD, as was the case in this study.

19.4 CONCLUSION

In the integrated process, UV photodegradation pretreatment improved the biodegradability of MB wastewater resulting in improved biogas yield and organic load reduction in the ensuing anaerobic process. Application of zeolite as catalyst and biomass support enhanced the two processes of AD and photodegradation. An integrated system of photodegradation and AD is thus suitable for the treatment of wastewater with high organic content and an appreciable amount of biorecalcitrant compounds. However, a kinetic model for the determination of the optimal conditions of operation for the integrated process should be developed.

ACKNOWLEDGMENT

This work was financially supported by the Water Research Commission (WRC, Project No. K5/2388), South Africa.

REFERENCES

Abelleira-Pereira, J. M., S. I. Pérez-Elvira, J. Sánchez-Oneto, R. de la Cruz, J. R. Portela, and E. Nebot. 2015. Enhancement of methane production in mesophilic anaerobic digestion of secondary sewage sludge by advanced thermal hydrolysis pretreatment. *Water Research* 71: 330–340. doi:10.1016/j.watres.2014.12.027.

Akpan, U. G., and B. H. Hameed. 2009. Parameters affecting the photocatalytic degradation of dyes using TiO_2-based photocatalysts: A review. *Journal of Hazardous Materials* 170 (2–3): 520–529. doi:10.1016/j.jhazmat.2009.05.039.

Al-Momani, F., E. Touraud, J. R. Degorce-Dumas, J. Roussy, and O. Thomas. 2002. Biodegradability enhancement of textile dyes and textile wastewater by VUV photolysis. *Journal of Photochemistry and Photobiology A: Chemistry* 153 (1–3): 191–197. doi:10.1016/S1010-6030(02)00298-8.

Apollo, S., and O. Aoyi. 2016. Combined anaerobic digestion and photocatalytic treatment of distillery effluent in fluidized bed reactors focusing on energy conservation. *Environmental Technology* 37 (17): 1–9. doi:10.1080/09593330.2016.1146342.

Apollo, S., M. S. Onyango, and A. Ochieng. 2016. Modelling energy efficiency of an integrated anaerobic digestion and photodegradation of distillery effluent using response surface methodology. *Environmental Technology* 3330: 1–12. doi:10.1080/09593330.2016.1151462.

Asiltürk, M., and Ş. Şener. 2012. TiO_2-activated carbon photocatalysts: Preparation, characterization and photocatalytic activities. *Chemical Engineering Journal* 180: 354–363. doi:10.1016/j.cej.2011.11.045.

Asiri, A. M., M. S. Al-Amoudi, T. A. Al-Talhi, and A. D. Al-Talhi. 2011. Photodegradation of Rhodamine 6G and phenol red by nanosized TiO_2 under solar irradiation. *Journal of Saudi Chemical Society* 15 (2): 121–128. doi:10.1016/j.jscs.2010.06.005.

Banat, F., S. Al-Asheh, M. Al-Rawashdeh, and M. Nusair. 2005. Photodegradation of methylene blue dye by the UV/H_2O_2 and UV/acetone oxidation processes. *Desalination* 181 (1–3): 225–232. doi:10.1016/j.desal.2005.04.005.

Brooms, T., B. O. Otieno, M. S. Onyango, and O. Aoyi. 2016. Photocatalytic degradation of aromatic compounds in abattoir wastewater. *International Journal of Environmental Sustainability* 12 (2): 1–16. doi:10.18848/2325-1077/CGP.

Chaiprapat, S., and T. Laklam. 2011. Enhancing digestion efficiency of POME in anaerobic sequencing batch reactor with ozonation pretreatment and cycle time reduction. *Bioresource Technology* 102 (5): 4061–4068. doi:10.1016/j.biortech.2010.12.033.

Chandra, R., R. N. Bharagava, and V. Rai. 2008. Melanoidins as major colourant in sugarcane molasses based distillery effluent and its degradation. *Bioresource Technology* 99 (11): 4648–4660. doi:10.1016/j.biortech.2007.09.057.

Eskizeybek, V., F. Sari, H. Gülce, A. Gülce, and A. Avci. 2012. Preparation of the new polyaniline/ZnO nanocomposite and its photocatalytic activity for degradation of methylene blue and malachite green dyes under UV and natural sun lights irradiations. *Applied Catalysis B: Environmental* 119–120: 197–206. doi:10.1016/j.apcatb.2012.02.034.

Gomes, A. C., L. Silva, R. Simões, N. Canto, and A. Albuquerque. 2013. Toxicity reduction and biodegradability enhancement of cork processing wastewaters by ozonation. *Water Science and Technology* 68 (10): 2214–2219. doi:10.2166/wst.2013.478.

Hammed, A. K., N. Dewayanto, D. Du, M. H. Ab Rahim, and M. R. Nordin. 2016. Novel modified ZSM-5 as an efficient adsorbent for methylene blue removal. *Journal of Environmental Chemical Engineering* 4 (3): 2607–2616. doi:10.1016/j.jece.2016.05.008.

Haque, F., E. Vaisman, C. H. Langford, and A. Kantzas. 2005. Preparation and performance of integrated photocatalyst adsorbent (IPCA) employed to degrade model organic compounds in synthetic wastewater. *Journal of Photochemistry and Photobiology A: Chemistry* 169 (1): 21–27. doi:10.1016/j.jphotochem.2004.05.019.

Li, H., J. Zhu, P. Xiao, Y. Zhan, K. Lv, L. Wu, and M. Li. 2016. On the mechanism of oxidative degradation of rhodamine B over $LaFeO_3$ catalysts supported on silica materials: Role of support. *Microporous and Mesoporous Materials* 221: 159–166. doi:10.1016/j.micromeso.2015.09.034.

Mabuza J., B. Otieno, S. Apollo, B. Matshediso, and A. Ochieng. 2017. Investigating the synergy of integrated anaerobic digestion and photodegradation using hybrid photocatalyst for molasses wastewater treatment. *Euro-Mediterranean Journal for Environmental Integration* 2 (1): 17. doi:10.1007/s41207-017-0029-6.

Mardani, H. R., M. Forouzani, M. Ziari, and P. Biparva. 2015. Visible light photo-degradation of methylene blue over Fe or Cu promoted ZnO nanoparticles. *Spectrochimica Acta Part A: Molecular and Biomolecular Spectroscopy* 141: 27–33. doi:10.1016/j.saa.2015.01.034.

Milan, Z., S. Montalvo, K. Ilangovan, O. Monroy, R. Chamy, P. Weiland, E. Sanchez, and R. Borja. 2010. The impact of ammonia nitrogen concentration and zeolite addition on the specific methanogenic activity of granular and flocculent anaerobic sludges. *Journal of Environmental Science and Health. Part A, Toxic/Hazardous Substances & Environmental Engineering* 45 (7): 883–889. doi:10.1080/10934521003709099.

Oller, I., S. Malato, and J. A. Sánchez-Pérez. 2011. Combination of advanced oxidation processes and biological treatments for wastewater decontamination—A review. *Science of the Total Environment* 409 (20): 4141–4166. doi:10.1016/j.scitotenv.2010.08.061.

Otieno, B., S. Apollo, B. Naidoo, and A. Ochieng. 2016. Photodegradation of molasses wastewater using TiO_2-ZnO nanohybrid photocatalyst supported on activated carbon. *Chemical Engineering Communications* 6445 (11): 1443–1454. doi:10.1080/00986445.2016.1201659.

Otieno, B. O., S. O. Apollo, B. E. Naidoo, and A. Ochieng. 2017. Photodecolorisation of melanoidins in vinasse with illuminated TiO_2-ZnO/activated carbon composite. *Journal of Environmental Science and Health, Part A* 52 (7): 1–8. doi:10.1080/10934529.2017.1294963.

Rauf, M. A., M. A. Meetani, A. Khaleel, and A. Ahmed. 2010. Photocatalytic degradation of methylene blue using a mixed catalyst and product analysis by LC/MS. *Chemical Engineering Journal* 157 (2–3): 373–378. doi:10.1016/j.cej.2009.11.017.

Yan, K., J. Mun, C. Chee, M. Nan, B. Jin, C. Saint, P. Eong, and R. Aryal. 2016. Evaluation of physicochemical methods in enhancing the adsorption performance of natural zeolite as low-cost adsorbent of methylene blue dye from wastewater. *Journal of Cleaner Production* 118: 1–13. doi:10.1016/j.jclepro.2016.01.056.

Yang, S., C. Sun, X. Li, Z. Gong, and X. Quan. 2010. Enhanced photocatalytic activity for titanium dioxide by co-modifying with silica and fluorine. *Journal of Hazardous Materials* 175 (1–3): 258–266. doi:10.1016/j.jhazmat.2009.09.158.

Zainudin, N. F., A. Z. Abdullah, and A. R. Mohamed. 2010. Characteristics of supported Nano-TiO_2/ZSM-5/silica gel (SNTZS): Photocatalytic degradation of phenol. *Journal of Hazardous Materials* 174 (1–3): 299–306. doi:10.1016/j.jhazmat.2009.09.051.

20 Mathematical Process Modeling and Biokinetics of Activated Sludge Processes

Ibrahim Hassan Mustafa, Asmaa Abdallah Awad, and Hamad Al-Turaif

CONTENTS

20.1 INTRODUCTION

This chapter shows the importance of using mathematical models for biological wastewater treatment systems. It also presents a classification of the models depending on the number of parameters in use at one time and dependent on the homogeneity of the activated sludge (AS) system at another time. We explain activated sludge models numbers 1, 2, and 3 (ASM1, ASM2, and ASM3)

TABLE 20.1
Definition of Kinetic Parameters

Symbol	Characterization	Units
K_h	Hydrolysis rate constant	$g\ X_S\ g^{-1}\ X_H d^{-1}$
K_X	Hydrolysis saturation constant	$g\ X_S\ g^{-1}\ X_H$
Heterotrophic Organisms, Denitrification, X_H		
k_{STO}	Storage rate constant	$g\ S_S\ g^{-1}\ X_H\ d^{-1}$
η_{NO}	Anoxic reduction factor	—
K_O	Saturation constant for S_o	$g\ O_2\ m^{-3}$
K_{NO}	Saturation constant for S_{NO}	$g\ NO_3$- N m^3
K_S	Saturation constant for substrate S_s	g COD m^3
K_{STO}	Saturation constant for X_{STO}	$g\ X_{STO}\ g^{-1}\ X_H$
μ_H	Heterotrophic max. growth rate	d^{-1}
K_{NH}	Saturation constant for ammonium S_{NH}	$g\ N\ m^{-3}$
K_{HCO}	Bicarbonate saturation constant of X_H	mole $HCO_3^-\ m^{-3}$
$b_{H,O2}$	Aerobic endogenous respiration rate of X_H	d^{-1}
$b_{H,NO}$	Anoxic endogenous respiration rate of X_H	d^{-1}
$b_{STO,O2}$	Aerobic respiration rate of X_{STO}	d^{-1}
$b_{STO,NO}$	Anoxic respiration rate of X_{STO}	d^{-1}
Autotrophic Organisms, Denitrification, X_A		
μ_A	Autotrophic max. growth rate	d^{-1}
$K_{A,NH}$	Ammonium substrate saturation for X_A	$g\ N\ m^{-3}$
$K_{A,O}$	Oxygen saturation for nitrifiers	$g\ O_2\ m^{-3}$
$K_{A,\ HCO}$	Bicarbonate saturation for nitrifiers	mole $HCO_3^-\ m^{-3}$
$b_{A,O2}$	Aerobic endogenous respiration rate of X_A	d^{-1}
$b_{A,NO}$	Anoxic endogenous respiration rate of X_A	d^{-1}

Source: Henze, M. et al., Table 9.2 (p. 118), from *Activated Sludge Models ASM1, ASM2, ASM2d and ASM3,* ©IWA Publishing 2000.

by International Association on Water Pollution Research and Control (IAWPRC), as well as the differences between them (Tables 20.1 and 20.2).

Mathematical models are powerful tools that the designers of biological wastewater treatment systems can use to investigate the performance of potential systems under a variety of conditions. They are particularly useful for those who are working with systems in which carbon oxidation, nitrification, and denitrification are accomplished with a single-sludge system, because the competing and parallel reactions in such systems are so complicated that it is difficult to intuitively estimate their response to changes in system configuration or load. Unfortunately, in spite of the benefits to be gained from the use of models, many engineers have not yet incorporated them into their routine practice.

Modeling and experimentation are interdependent, with each providing input to and taking information from the other. Consequently, as we have learned more about biofloc processes, we have been able to develop better models, which have helped us to see new applications and to develop better methods for design.

TABLE 20.2
Definition and Typical Values for the Kinetic Parameters of ASM2d

Temperature		**20°C**	**10°C**	**Units**
K_h	Hydrolysis rate constant	3	1	d^{-1}
η_{NO2}	Anoxic hydrolysis reduction factor	0.6	0.6	–
η_{am}	Anaerobic hydrolysis reduction factor	0.4	0.4	–
K_{O2}	Saturation coefficient for oxygen	0.2	0.2	$g\ O_2\ m^{-3}$
K_{NO2}	Saturation coefficient for nitrate	0.5	0.5	$g\ N\ m^{-3}$
K_X	Saturation coefficient for particulate COD	0.1	0.1	$g\ X_s/(g\ X_H\ d)$
Heterotrophic Organisms: X_H				
μ_H	Maximum growth rate on substrate	6	3	$g\ X_s/(g\ X_H\ d)$
q_{fe}	Maximum rate for fermentation	3	1.5	$g\ S_F/(g\ X_H\ d)$
η_{NO3}	Reduction factor for denitrification	0.8	0.8	–
b_H	Rate constant for lysis and decay	0.4	0.2	d^{-1}
K_{O2}	Saturation/inhibition coefficient for oxygen	0.2	0.2	$g\ O_2/m^3$
K_F	Saturation coefficient for growth on S_F	4	4	$g\ COD/m^3$
K_{fe}	Saturation coefficient for fermentation of S_F	4	4	$g\ COD/m^3$
K_A	Saturation coefficient for growth on acetate S_A	4	4	$g\ COD/m^3$
K_{NO3}	Saturation/inhibition coefficient for nitrate	0.5	0.5	$g\ N/m^3$
K_{NH4}	Saturation coefficient for ammonium (nutrient)	0.05	0.05	$g\ N/m^3$
K_p	Saturation coefficient for phosphate (nutrient)	0.01	0.01	$g\ P/m^3$
K_{ALK}	Saturation coefficient for alkalinity (HCO^-_3)	0.1	0.1	mole HCO_3^-/m^3
Phosphorus-Accumulating Organisms: X_{PNO}				
q_{PHA}	Rate constant for storage of X_{PHA} (base X_{PP})	3	2	$g\ X_{PHA}\ (gX_{PAO})^{-1}\ d^{-1}$
q_{PP}	Rate constant for storage of X_{PP}	1.5	1	$g\ X_{PP}\ (gX_{PAO})^{-1}\ d^{-1}$
q_{PAO}	Maximum growth rate of PAO	1	0.67	d^{-1}
μ_{NO2}	Reduction factor for anoxic activity	0.6	0.6	–
b_{PAO}	Rate for lysis of X_{PAO}	0.2	0.1	d^{-1}
b_{PP}	Rate for lysis of X_{PP}	0.2	0.1	d^{-1}
b_{PHA}	Rate for lysis of X_{PHA}	0.2	0.1	d^{-1}
K_{O2}	Saturation/inhibition coefficient for oxygen	0.2	0.2	$g\ O_2\ m^{-3}$
K_{NO2}	Saturation coefficient for nitrate, S_{NO3}	0.5	0.5	$g\ N\ m^{-3}$
K_A	Saturation coefficient for acetate, S_A	4	4	$g\ COD\ m^{-3}$
K_{NH4}	Saturation coefficient for ammonium (nutrient)	0.05	0.05	$g\ N\ m^{-3}$
K_{PS}	Saturation coefficient for phosphorous in storage of PP	0.2	0.2	$g\ P\ m^{-3}$
K_P	Saturation coefficient for phosphate (nutrient)	0.01	0.01	$g\ P\ m^{-3}$
K_{ALK}	Saturation coefficient for alkalinity (HCO_3^-)	0.1	0.1	mole $HCO_3^-\ m^{-3}$
K_{PP}	Saturation coefficient for poly-phosphate	0.01	0.01	$g\ X_{PP}\ (gX_{PAO})^{-1}$
K_{MAX}	Maximum ratio of X_{PP}/X_{PAO}	0.34	0.34	$g\ X_{PP}\ (gX_{PAO})^{-1}$
K_{IPP}	Inhibition coefficient for PP storage	0.02	0.02	$g\ X_{PP}\ (gX_{PAO})^{-1}$
K_{PHA}	Saturation coefficient for PHA	0.01	0.01	$g\ X_{PHA}\ (gX_{PAO})^{-1}$
Nitrifying Organisms (Autotrophic Organisms): X_{AUT}				
μ_{AUT}	Maximum growth rate of X_{AUT}	1	0.35	d^{-1}
b_{AUT}	Decay rate of X_{AUT}	0.15	0.05	d^{-1}
K_{O2}	Saturation coefficient for oxygen	0.5	0.5	$g\ O_2\ m^{-3}$
K_{NH4}	Saturation coefficient for ammonium (substrate)	1	1	$g\ N\ m^{-3}$
K_{ALK}	Saturation coefficient for alkalinity	0.5	0.5	mole HCO_3^-/m^3
K_P	Saturation coefficient for phosphorous (nutrient)	0.01	0.01	$g\ P\ m^{-3}$

(Continued)

TABLE 20.2 (*Continued*)
Definition and Typical Values for the Kinetic Parameters of ASM2d

Temperature		20°C	10°C	Units
Precipitation				
k_{PRE}	Rate constant for P precipitation	1	1	m^3 (g Fe $(OH)_3)^{-1}$ d^{-1}
k_{RED}	Rate constant for redissolution	0.6	0.6	d^{-1}
K_{ALK}	Saturation coefficient for alkalinity	0.5	0.5	mole HCO_3^- /m^3

Source: Henze, M. et al., Table 4.3 (p. 93), from *Activated Sludge Models ASM1, ASM2, ASM2d and ASM3,* ©IWA Publishing 2000.

20.2 CLASSIFICATION OF MODELS

Leslie Grady (1993) divided mathematical models into two categories: empirical and mechanistic. Empirical models simply relate operating input and output variables to each other and make little pretense of representing individual phenomena. Such "black box" descriptions are quite useful for design from pilot plant data and have found wide use in environmental engineering. Many of the models for biological film processes fall into this category.

Mechanistic models, on the other hand, express the influence and interrelationships of individual mechanistic phenomena in a manner that allows the investigator to discover how the system might respond to unexpected conditions. One might argue that the primary purpose of a mechanistic model is further understanding. This additional understanding will be of direct benefit to the practitioner, however, because it is the nature of practice to apply knowledge to areas in which no prior experience exists. Mechanistic models have broader utility than empirical ones. Consequently, this review will be limited to mechanistic models.

Mechanistic (physical) models of biochemical processes are generally developed by application of reactor engineering principles; that is, they combine expressions representing intrinsic kinetic and transport events with mass balance equations describing the characteristics of the particular physical system under consideration. Consequently, simulation with such models gives insight into the basic events (Grady 1993). In this section, models are classified according to a number of kinetic parameters as either simple or complex models, as well as whether they are homogeneous or heterogeneous.

20.2.1 Models Based on a Number of Parameters

Models can be classified into simple and complex depending on the number of parameters describing microbial growth processes. A simple kinetic model such as Monod kinetics is still used, and complex models such as ASM1, ASM2, and ASM3 have different degrees of complexity, as will be shown.

20.2.1.1 Simple Models

The simplest kinetic model has been, and is still being, used for the analysis of many microbial growth processes. These models are based on the assumption that the biomass concentration can be adequately described by a single parameter. Simple models are not interested in the internal structure of the cell nor the diversity of the cell forms. However, these models include the most fundamental observations concerning microbial growth processes: that the rate of cell-mass production is proportional to biomass concentration and that there is a decrease in cell mass also proportional to biomass concentration. The quality of model predictions increases when the substrate concentration in the reactor is high enough to permit equilibrium of the internal cell composition, the so-called "growth condition."

In batch fermentation, for example, the substrate concentration is usually high enough to assume an equal growth rate of all cell components, the so-called "balanced growth condition." Simple dynamic models can be used as a basis for control of industrial fermentation processes (Villadsen and Nielsen, 1992).

Simple kinetic models cannot realize the same success when applied to continuously-stirred tank bioreactors (CSTBR) (Mustafa et al., 2014). Because of the often low levels of the substrate in the chemostat, a transient behavior such as a sudden change in the volumetric flow rate or the recycle ratio can dramatically affect the cell environment, and the simple model may fail to predict system behavior (Nielsen, 1992). The Monod kinetics expression for biological synthesis is a clear example of a simple model. The Monod expression was used to describe the growth rate of both heterotrophic and autotrophic organisms in the International Association on Water Quality (IAWQ) model.

20.2.1.2 Complex Models

Complex models have different degrees of complexity. Their microbial kinetics are based on the knowledge accumulated in the fields of microbiology and biochemistry. Complex models such as ASM1, ASM2, and ASM3 vary in complexity. A good dynamic model should predict the dynamic behavior of the system experimentally and have a reasonable number of parameters to provide it with some level of flexibility (Sriyudthsak et al., 2016).

These models divide the biomass and wastewater into many constituents that give rise to many different processes, each with its own rate and yield equations. Sriyudthsak et al. (2016) have proposed a structured model that includes direct soluble substrate metabolism with concurrent storage of substrate.

20.2.1.2.1 Activated Sludge Model No. 1 (ASM1)

In 1983 IAWPRC formed a task group to facilitate the application of practical models to the design and operation of biological wastewater treatment systems. They presented the model development for a single-sludge system based on only two processes: carbon oxidation and nitrification. The first goal was to review existing models and the second goal was to reach a consensus concerning the simplest mathematical model that had the capability of realistically predicting the performance of single-sludge systems carrying out carbon oxidation, nitrification, and denitrification. The final result was presented in 1987 as the IAWQ Activated Sludge Model No. 1 (ASM1). The different processes incorporated into the IAWQ model are briefly described as follows.

- *Aerobic growth of heterotrophs*: A fraction of the readily biodegradable substrate is used for growth of heterotrophic biomass and the balance is oxidized for energy giving rise to associated oxygen demand. The growth is modeled using Monod kinetics. Ammonia is used as the nitrogen source for synthesis and incorporated into the cell mass. Both the concentrations of S_S and S_O may be rate limiting for the growth process. This process is the main contributor to the production of new biomass and removal of chemical oxygen demand (COD). It is also associated with an alkalinity change.
- *Anoxic growth of heterotrophs*: In the absence of oxygen, heterotrophic organisms are capable of using nitrate as the terminal electron acceptor with S_S as substrate. The process leads to the production of heterotrophic biomass and nitrogen gas (denitrification). The nitrogen gas is a result of the reduction of nitrate with an associated alkalinity change. The same Monod kinetics as used for aerobic growth is applied, except that the kinetic rate expression is multiplied by a factor η_g (< 1). This reduced rate could either be caused by a lower maximum growth rate under anoxic conditions or because only a fraction of the heterotrophic biomass can function with nitrate as an electron acceptor. Ammonia serves as the nitrogen source for cell synthesis, which in turn changes the alkalinity.
- *Aerobic growth of autotrophs*: Ammonia is oxidized to nitrate via a single-step process (nitrification) resulting in production of autotrophic biomass and giving rise to associated oxygen demand. Ammonia is also used as the nitrogen source for synthesis and incorporated into

the cell mass. The process has a marked effect on the alkalinity (both from the conversion of ammonia into biomass and by the oxidation of ammonia to nitrate) and the total oxygen demand. The effect on the amount of biomass formed is small, as the yield of the autotrophic nitrifiers is low. Once again the growth rate is modeled using Monod kinetics.

- *Decay of heterotrophs*: The process is modeled according to the death regeneration hypothesis. The organisms die at a certain rate and a portion of the material is considered to be non-biodegradable and adds to the X_P fraction. The remainder adds to the pool of slowly biodegradable substrate. The organic nitrogen associated with the X_S becomes available as particulate organic nitrogen. No loss of COD is involved and no electron acceptor is utilized. The process is assumed to continue at the same rate under aerobic, anoxic, and anaerobic conditions.
- *Decay of autotrophs*: The process is modeled in the same way as used to describe the decay of heterotrophs.
- *Ammonification of soluble organic nitrogen*: Biodegradable soluble organic nitrogen is converted to free and saline ammonia in a first-order process mediated by the active heterotrophs. Hydrogen ions consumed in the conversion process result in an alkalinity change.
- *Hydrolysis of entrapped organics*: Slowly biodegradable substrate enmeshed in the sludge mass is broken down extracellularly, producing readily biodegradable substrate available to the organisms for growth. The process is modeled on the basis of surface reaction kinetics and occurs only under aerobic and anoxic conditions. The rate of hydrolysis is reduced under anoxic conditions compared with aerobic conditions by a factor η_h (< 1). The rate is also first-order with respect to the heterotrophic biomass present, but saturates as the amount of entrapped substrate becomes large in proportion to the biomass.
- *Hydrolysis of entrapped organic nitrogen*: Biodegradable particulate organic nitrogen is broken down to soluble organic nitrogen at a rate defined by the hydrolysis reaction for entrapped organics described earlier.

With regard to denitrification, the task group separated the processes of hydrolysis and growth. Finally, the fate of the organic nitrogen and the source of organic nitrogen for synthesis were treated somewhat differently. The task group also introduced the concept of switching functions to gradually turn process rate equations on and off as the environmental conditions changed (mainly between aerobic and anoxic conditions). The switching functions are "Monod-like" expressions that are mathematically continuous and thereby reduce the problem of numerical instability during simulations. Furthermore, the work of the task group promoted the structural presentation of biokinetic models via a matrix format, which was easy to read and understand, and consolidated much of the existing knowledge on the AS process.

20.2.1.2.2 Activated Sludge Model No. 2 (ASM2 and ASM2d)

Activated Sludge Model No. 2 (ASM2) is considered to be a development of ASM1, but ASM2 is more elaborate and has many essential elements to elucidate both AS and wastewater, including many biological phenomena such as the process of biological phosphorus elimination and identification of the interior components of microorganism cells.

Activated Sludge Model No. 2d (ASM2d), was extended by Henze et al. (1999), as a development of ASM2. While ASM2 considers phosphorus-accumulating organisms (PAOs) to cultivate an aerobic environment, ASM2d considers denitrifying PAOs with the capability to deal with the organic cellular products of the denitrification process. Furthermore, ASM2 takes into consideration two further chemical processes that are helpful for the mathematical formulation of phosphorus precipitation. ASM2 covers the biological removal of polyphosphate in AS processes unable to labor with COD, which is one of the main outcomes dependent on ASM1. Total suspended solids (TSS), as well as mineral particulate solids, are therefore involved in ASM2.

ASM2 is a complex model that enables the characterization of many biological phenomena in wastewater plants and could be simplified to avoid elements with no significant effects on process kinetics. Furthermore, ASM2 is characterized by noncellular structuring, focusing on the cell as a whole entity, and it also differentiates among biomass composition. ASM2 depends on the average compositions of microbial cells.

20.2.1.2.2.1 Hydrolysis Processes In hydrolysis processes, microorganisms secrete enzymes to be able to catalyze the degrading organic substrates. Typically hydrolysis processes are considered surface reactions, which occur in close contact between the organisms that provide the hydrolytic enzymes and the slowly biodegradable substrates themselves. Furthermore, the hydrolysis process is accompanied by protozoan activity. Under anoxic and aerobic conditions, protozoa become idle and electron acceptors drive the hydrolysis processes.

In ASM2, it is not easy to accurately predict the constants of the hydrolysis rate within the various electron acceptor conditions. There are three different hydrolysis processes based on the available electron acceptors shown in Table 20.3:

1. Aerobic hydrolysis of slowly biodegradable substrate where hydrolysis is performed under aerobic conditions where ($S_{O2} > 0$). This process is faster than the following anoxic hydrolysis.
2. Anoxic hydrolysis of slowly biodegradable substrate where hydrolysis is performed under anoxic conditions ($S_{O2} \approx 0$) and $\left(S_{NO3} > 0\right)$.
3. Anaerobic hydrolysis of slowly biodegradable substrate where hydrolysis is performed under anaerobic conditions where $\left(S_{O2} \approx 0\right)$ and $\left(S_{NO3} \approx 0\right)$. This process is not well characterized and is probably slower than aerobic hydrolysis.

The hydrolysis process rates in ASM2 are similar to those of ASM1: the hyperbolic switching functions for S_{O2} and S_{NO3} consider the environmental conditions; and a limited surface reaction $\frac{X_S/X_H}{K_X+X_S/X_H}$ is assumed for the hydrolysis process itself. Hydrolysis is catalyzed only by heterotrophic organisms where hydrolysis in aerobic conditions is faster that in anoxic and anaerobic environments. Both the rates of anaerobic and anoxic hydrolysis are lowered by the factors η_{NO3} and η_{fe}, respectively.

Process 4 is characterized by aerobic growth of heterotrophic organisms on fermentable substrates S_F while process 5 is characterized by aerobic growth of heterotrophic organisms on fermentation products S_A. Both processes 4 and 5 are modeled as two-parallel processes, which consume the two degradable organic substrates S_F and S_A. For both processes, identical growth rates μ_m and yield coefficients Y_H are assumed. The rate equations are designed such that the maximum specific growth rate of the heterotrophic organisms does not increase above μ_m even if both substrates, S_F and S_A, are present in high concentrations. These processes require oxygen, S_{O2}, nutrients, S_{NH4} and S_{PO4}, and possibly alkalinity, S_{ALK}, and they produce suspended solids, X_{TSS}.

Process 6 is characterized by the anoxic growth of heterotrophic organisms on fermentable substrates, S_F, while process 7 is similar to process 6 but on fermentation products, S_A, denitrification. Both processes 6 and 7 are similar to the aerobic growth processes, but ASM2d requires nitrate, S_{NO3}, as the electron acceptor rather than oxygen; the stoichiometry for nitrate is computed based on the assumption that nitrate (S_{NO3}) is reduced to dinitrogen (S_{N2}) where denitrification releases alkalinity. Denitrification is assumed to be inhibited by oxygen S_{O2} and the maximum growth rate μ_m is reduced relative to its value under aerobic conditions, by the factor η_{NO3}. These account for the fact that not all heterotrophic organisms (X_H) may be capable of denitrification or that denitrification may only proceed at a reduced rate.

Process 8 is characterized by fermentation under anaerobic conditions where $\left(S_{O2} \approx 0,\ S_{NO3} \approx 0\right)$. In the latter process it is assumed that heterotrophic organisms have the capability to degrade readily biodegradable substrates (S_F) into fermentation products (S_A).

Although this process may cause the growth of heterotrophic organisms, it is introduced here as a simple transformation process. A growth process would require more complex kinetics, and more kinetic and stoichiometric parameters, which are difficult to obtain, and possibly different yield coefficients for S_f and S_A in processes 4–7. A fermentation release negatively charged fermentation product S_A and therefore requires alkalinity, S_{ALK}. This is predicted from charge conservation.

TABLE 20.3
Process Rate Equations (ASM2d)

j	Process	Process Rate Equations $\rho_j \geq 0$ ($M_1\ L^{-3}\ T^{-1}$)
Hydrolysis Processes		
1	Aerobic hydrolysis	$K_h \cdot \frac{S_{O2}}{K_{O2}+S_{O2}} \cdot \frac{X_S/X_H}{K_X+X_S/X_H} . X_H$
2	Anoxic hydrolysis	$K_h . \eta_{NO3} \cdot \frac{K_{O2}}{K_{O2}+S_{O2}} \cdot \frac{S_{NO2}}{K_{NO3}+S_{NO3}} \cdot \frac{X_S/X_H}{K_X+X_S/X_H} . X_H$
3	Anaerobic hydrolysis	$K_h . \eta_{fe} \cdot \frac{K_{O2}}{K_{O2}+S_{O2}} \cdot \frac{S_{NO2}}{K_{NO3}+S_{NO3}} \cdot \frac{X_S/X_H}{K_X+X_S/X_H} . X_H$
Heterotrophic Organisms: X_H		
4	Growth on fermentable substrate, S_F	$\mu_H \cdot \frac{S_{O2}}{K_{O2}+S_{O2}} \cdot \frac{S_F}{K_F+S_F} \cdot \frac{S_F}{K_F+S_A} \cdot \frac{S_{NH4}}{K_{NH4}+S_{NH4}} \cdot \frac{S_{PO4}}{K_P+S_{PO4}} \cdot \frac{S_{ALK}}{K_{ALK}+S_{ALK}} . X_H$
5	Growth on fermentable products, S_A	$\mu_H \cdot \frac{S_{O2}}{K_{O2}+S_{O2}} \cdot \frac{S_A}{K_A+S_A} \cdot \frac{S_A}{S_F+S_A} \cdot \frac{S_{NH4}}{K_{NH4}+S_{NH4}} \cdot \frac{S_{PO4}}{K_P+S_{PO4}} \cdot \frac{S_{ALK}}{K_{ALK}+S_{ALK}} . X_H$
6	Denitrification with fermentable substrates, S_F	$\mu_H . \eta_{NO3} \cdot \frac{K_{O2}}{K_{O2}+S_{O2}} \cdot \frac{K_{NO3}}{K_{NO3}+S_{NO3}} \cdot \frac{S_F}{K_F+S_F} \cdot \frac{S_F}{S_F+S_A} \cdot \frac{S_{NH4}}{K_{NH4}+S_{NH4}} \cdot \frac{S_{PO4}}{K_P+S_{PO4}} \cdot \frac{S_{ALK}}{K_{ALK}+S_{ALK}} X_H$
7	Denitrification with fermentable products, S_A	$\mu_H . \eta_{NO3} \cdot \frac{K_{O2}}{K_{O2}+S_{O2}} \cdot \frac{K_{NO3}}{K_{NO3}+S_{NO3}} \cdot \frac{S_A}{K_A+S_A} \cdot \frac{S_A}{S_F+S_A} \cdot \frac{S_{NH4}}{K_{NH4}+S_{NH4}} \cdot \frac{S_{PO4}}{K_P+S_{PO4}} \cdot \frac{S_{ALK}}{K_{ALK}+S_{ALK}} . X_H$
8	Fermentation	$q_{fe} \cdot \frac{K_{O2}}{K_{O2}+S_{O2}} \cdot \frac{K_{NO3}}{K_{NO3}+S_{NO3}} \cdot \frac{S_F}{K_F+S_F} \cdot \frac{S_{ALK}}{K_{ALK}+S_{ALK}} . X_H$
9	Lysis	$b_H \cdot X_H$
Phosphorous Accumulating Organisms (PAO): X_{PAO}		
10	Storage of X_{PHA}	$q_{PHA} \cdot \frac{S_A}{K_A+S_F} \cdot \frac{S_{ALK}}{K_{ALK}+S_{ALK}} \cdot \frac{X_{PP}/X_{PAO}}{K_{FP}+X_{PP}/X_{PAO}} . X_{PAO}$
11	Aerobic Storage of X_{PP}	$q_{PP} \cdot \frac{S_{O2}}{K_{O2}+S_{O2}} \cdot \frac{S_{PO4}}{K_{PS}+S_{PO4}} \cdot \frac{S_{ALK}}{K_{ALK}+S_{ALK}} \cdot \frac{X_{PHA}/X_{PAO}}{K_{PHA}+X_{PHA}/X_{PAO}} \cdot \frac{K_{MAX}-X_{PP}/X_{PAO}}{K_{PP}+K_{MAX}-X_{PP}/X_{PAO}} . X_{PAO}$
12	Anoxic Storage of X_{PP}	$\rho_{12} = \rho_{11} . \eta_{NO3} \cdot \frac{K_{O2}}{S_{O2}} \cdot \frac{S_{NO2}}{K_{NO3}+S_{NO3}}$
13	Aerobic growth of X_{PHA}	$\mu_{PAO} \cdot \frac{S_{O2}}{K_{O2}+S_{O2}} \cdot \frac{S_{NH4}}{K_{NH4}+S_{NH4}} \cdot \frac{S_{PO4}}{K_P+S_{PO4}} \cdot \frac{S_{ALK}}{K_{ALK}+S_{ALK}} \cdot \frac{X_{PHA}/X_{PAO}}{K_{PHA}+X_{PHA}/X_{PAO}} . X_{PAO}$
14	Aerobic growth of X_{PP}	$\rho_{14} = \rho_{13} . \eta_{NO3} \cdot \frac{K_{O2}}{S_{O2}} \cdot \frac{S_{NO3}}{K_{NO3}+S_{NO3}}$
15	Lysis of X_{PAO}	$b_{PAO} . X_{PAO} . S_{ALK}/(K_{ALK}+S_{ALK})$
16	Lysis of X_{PP}	$b_{PP} . X_{PP} . S_{ALK}/(K_{ALK}+S_{ALK})$
17	Lysis of X_{PHA}	$b_{PHA} . X_{PHA} . S_{ALK}/(K_{ALK}+S_{ALK})$
Nitrifying Organisms (Autotrophic Organisms): X_{AUT}		
18	Aerobic growth of X_{AUT}	$\mu_{AUT} \cdot \frac{S_{O2}}{K_{O2}+S_{O2}} \cdot \frac{S_{NH4}}{K_{NH4}+S_{NH4}} \cdot \frac{S_{PO4}}{K_P+S_{PO4}} \cdot \frac{S_{ALK}}{K_{ALK}+S_{ALK}} . X_{AUT}$
19	Lysis of X_{AUT}	$b_{AUT} . X_{AUT}$
Simultaneous Precipitation of Phosphorous with Ferric Hydroxide $Fe(OH)_2$		
20	Precipitation	$k_{PRE} . S_{PO2} . X_{MeOH}$
21	Redissolution	$k_{RED} . X_{MeP} . S_{ALK}/(K_{ALK}+S_{ALK})$

Fermentation needs more research in order to identify process kinetics and kinetic parameters and to obtain more reliable modeling experimental results. Reliable application of ASM2 requires that research is directed towards characterizing what is described here with the process of fermentation.

Process 9 is characterized by lysis of heterotrophic organisms and describes the phenomena of decay and loss of the heterotrophic organisms, such as endogenous respiration, lysis, and predation. This process is modeled in analogy to ASM1 where the effect of environmental conditions is ignored.

20.2.1.2.2.2 Process of Phosphorous-Accumulating Organisms PAOs are responsible for eliminating phosphorus in terms of polyphosphate. These organisms are included in the *Acinetobacter* genus and have the capability to denitrify. In the absence of nitrate, the rate of biological phosphorus removal becomes faster and incorporated in ASM2d.

In PAOs, glycogen is an internal cellular component; however, in order to simplify the model, it is not incorporated in ASM2. For designing wastewater treatment plants, the task group could successfully incorporate the process of phosphorous elimination by PAO through building simple mathematical models that neglect important phenomena such as denitrification in ASM2, although it is incorporated ASM2d. In ASM2d, it is assumed that PAOs can grow under both aerobic and anoxic conditions based on internal cellular organic components (X_{PHA}), resulting in the need to develop ASM2 to meet the require kinetics.

Process 10: Storage of cell internal organic storage material (X_{PHA}):

In this process, phosphate (S_{PO4}) is produced from the hydrolysis of poly-phosphate (X_{pp}) by PAOs using the released energy to accumulate the extracellular products (S_A) in terms of intracellular organic storage material (X_{PHA}). This process is performed under all conditions (anaerobic, aerobic, and anoxic). Therefore there are no inhibition terms in the kinetic rate equations. It is observed that the rate of phosphorus removal varies, while the rate of accumulating is constant. It is indicated that the phosphorus yield (Y_{PO4}) is based on pH.

Process 11 characterizes the aerobic storage of polyphosphate while process 12 focuses on the anoxic storage of polyphosphate. PAOs work to store phosphate (S_{PO4}) in the form of intracellular polyphosphate (X_{pp}), where the energy required for the growth of PAOS is released from the hydrolysis of X_{PHA}. Through the aerobic environment, the highest rate of storage of polyphosphate (q_{PP}) is lowered by the term η_{NO3}. In other words, this means that some PAOs are unable to perform the denitrification process and the rate of denitrification is low. It is observed that the process of anoxic storage of polyphosphate is incorporated only in ASM2d.

Process 13 characterizes aerobic growth of PAOs. Process 14 focuses on the anoxic growth of PAOs. The organisms in processes 13 and 14 are assumed to grow only at the expense of cell internal organic storage products X_{PHA}. PAOs grow on phosphorus released from the lysis of X_{pp} when soluble substrates (S_A) are depleted in both aerobic and anoxic conditions.

Processes 15, 16, and 17 focus on lysis of PAOs, and the storage products which are X_{pp} and X_{PHA}. Loss of PAOs exists because there are three significant processes (maintenance, death, and endogenous respiration) that affect the internal structure of biomass cells. The composition of the cells does not vary because of lysis if the rate constants of the three mentioned processes are the same. X_{PAO} and X_{PHA} could be lost slower than X_{pp}. Therefore, an increases rate for decay of polyphosphates could be chosen to express the increase in its lysis.

20.2.1.2.2.3 Nitrification Processes Ammonium (S_{NH4}) is hyrdolyzed into nitrate (S_{NO3}) in the nitrification process. Nitrite is not incorporated as a model element. The task group showed that nitrite could be consumed and produced through the denitrification process. However, considering nitrite in the process of nitrification in the modeling process and ignoring it in the denitrification may result in inaccuracy in modeling outcomes.

Process 18 considers the growth of nitrifying organisms under aerobic conditions. In this process ammonium is considered a substrate, and nitrate as a product. ASM1 takes into consideration the nitrification process but excludes the consumption of X_{pp}.

Process 19 indicates lysis of nitrifying organisms. In ASM2 and ASM2d, both X_s and S_f, which are the decay products of lysis, are considered as substrates for heterotrophs. Endogenous respiration of nitrifiers becomes obvious through the increased growth and oxygen consumption of heterotrophs. This is in analogy to ASM1.

20.2.1.2.2.4 Modeling of Phosphates Precipitation In the AS process, the high levels of soluble phosphate can combine with metals naturally existing in wastewater such as iron and calcium, leading to chemical precipitation of phosphorus. In addition, metal salts could be added as a common process for phosphorus precipitation, particularly when the phosphorus to carbon ratio high. The low levels of phosphate (S_{PO4}) in the exit stream in the AS process due to chemical precipitation could be modeled via a simple mathematical precipitation model built by the task group. ASM2 therefore incorporated two extra components (X_{MeOH} and X_{MeP}) and two processes (precipitation and redissolution).

Process 20 and process 21 focus on the precipitation and the redissolution of phosphate (S_{P04}), respectively. The two processes are mathematically formulated as reversible processes.

20.2.1.2.3 Activated Sludge Model No. 3 (ASM3)

ASM3 was presented in 1999 by the IAWQ task group as an extension for ASM1 (Gujer et al., 1999). The lysis (decay) process in ASM1 was replaced by endogenous respiration in ASM3, where the intracellular components are used for cellular survival when the external nutrients have been consumed. Stored polymers were therefore taken into consideration in ASM3 (Wilkinson, 1959; Van Loosdrecht and Henze 1999).

A major difference for the wastewater characterization between ASM1 and ASM3 is that soluble (S_s) and particulate (X_s) biodegradable components in ASM3 are supposed to be differentiated with filtration over 0.45 µm membrane filters, whereas a significant fraction of the slowly biodegradable organic substrates (X_s) in ASM1 would be contained in the filtrate of the influent wastewater (Gujer et al., 1999). The latter is most likely caused by the conversion of soluble biodegradable COD to storage polymers in the respiration tests. The kinetics of conversion of storage polymers more closely resembles the degradation rates of X_s than S_s in the model.

The aerobic storage process in ASM3 describes the storage of readily biodegradable substrate (S_s) in the form of cell internal storage products ($X_{storage}$). This process requires energy in the form of ATP, which is obtained from aerobic respiration. In ASM3 it is assumed that the readily biodegradable organic substrates are first taken up by the heterotrophic organisms and converted to stored material, which is subsequently assimilated to biomass.

Henze et al. (1999) showed that there are some defects that had become apparent with the application of ASM1:

- ASM1 does not include kinetic expressions that can deal the with nitrogen and alkalinity limitations of heterotrophic organisms. As a result, the computer code cannot be based on the original form of ASM1, which allows under some circumstances for negative concentrations of, for example, ammonium. These led to the creation of different versions of ASM1, which can hardly be differentiated any more.
- ASM1 includes biodegradable soluble and particulate organic nitrogen as model components. These cannot easily be measured and have, in the meantime, been eliminated in many versions of ASM1.
- The kinetics of ammonification in ASM1 cannot really be quantified. In many versions of ASM1 this process has been eliminated by assuming a constant composition of all organic components (constant N to COD ratio).
- ASM1 differentiates inert particulate organic material depending on origin: influent or biomass decay. It is impossible to differentiate these two fractions in reality.
- The process of hydrolysis has a dominating effect upon the prediction of oxygen consumption and denitrification by heterotrophic organisms. At the same time the quantification of the kinetic parameters for this process is difficult.

- Kinetic parameters are assessed inaccurately when the lysis process is accompanied by cell growth and hydrolysis. The lysis process includes many aspects of endogenous respiration such as cell death, storage of phosphorus, and cell trapping, leading to existing each viable, non-viable, and death cells together. Therefore, mathematical modeling should consider cell structuring, which, in turn, is more complicated than non-structuring. However, for simplicity, non-structuring mathematical modeling dominates neglecting other components of non-living and dead cells resulting in the predicted kinetic parameters become inaccurate.
- Storage of polyhydroxyalkanoates, and sometimes glycogen, is observed under aerobic and anoxic conditions in AS plants, provided that elevated concentrations of readily biodegradable organic substrates are available. This process is not included in ASM1.
- ASM1 does not include the possibility to differentiate the decay rates of nitrifiers under aerobic and anoxic conditions. At high solids retention times (STR) and high fractions of anoxic reactor volumes, this leads to problems with the prediction of maximum nitrification rates.
- ASM1 does not allow for the prediction of directly observable mixed liquor suspended solids.

Considering all these defects, the task group has decided to propose the Activated Sludge Model No. 3 (ASM3), which should correct all these defects and could become a standard. ASM3 relates to the same dominating phenomena as does ASM1: oxygen consumption, sludge production, nitrification, and denitrification in AS systems treating primarily domestic wastewater. Biological phosphorus removal is contained in ASM2 (Henze et al., 1995) and will not be considered in ASM3. Table 20.4 introduces the stoichiometric matrix v_{jI} of ASM3 together with the composition matrix $t_{jk,I}$ proposed by Gujer and Larsen (1995), while Table 20.5 shows the kinetic rate expressions (p_j) for ASM3 all $p_j > 0$.

20.2.2 Models Base on Homogeneity

This section shows the nature and limits of two main categories of models based on homogeneity: homogeneous and heterogeneous models.

20.2.2.1 Homogeneous Models

These models may be described by a set of first-order nonlinear differential equations. At steady state, it results in a set of non-linear algebraic equations that can be solved by traditional analytical or numerical methods according to the dimensionality of the system.

Numerous studies have been performed on this type of models, with significant success in designing and operating the process, as well as predicting the kinetic parameters. These models considered the AS system as one phase. They neglected the internal mass transfer inside the flocs, which makes the kinetic parameters rather inaccurate. In other words, these models assumed that the rate of the reaction at the center of the floc is the same as the outer surface, which implies that the whole floc is effective, there is no diffusional resistance, and the biochemical reaction is not limited by diffusion. It can be said that these models neglected the important role of mass transfer, as well as the importance of mechanical agitation, which helps to increase mass transfer rates. It is logical, when the rates of mass transfer of substrate, oxygen, and ammonia are increased, that the concentrations of these substances existing in the aerobic film of the AS floc will also increase, as will be shown later.

It is very important to consider the mass transfer operations in addition to the biochemical reactions to characterize these processes accurately and improve the efficiency of AS processes. The task group models can be considered good examples of homogeneous models.

20.2.2.2 Heterogeneous Models

These models consider the mass transfer operations in addition to the biochemical reaction processes. They consider the internal diffusion of the floc as well as the external mass transfer between gas, liquid, and solid phases through different processes occurring in the system. The resulting

TABLE 20.4
Stoichiometric Matrix n_j and Composition $t_{k,i}$ of ASM3. The Values of x_j, y_j, z_j, and t_j Can Be Obtained in the Sequence from Mass and Charge Conservation and Composition

	Components I >	1	2	3	4	5	6	7	8	9	10	11	12	13
J	Process	s_O	s_I	s_S	s_{NH}	s_{N2}	s_{NO}	s_{HC}	X_I	X_S	X_H	X_{STO}	X_A	X_{TS}
v	Expressed as >	O_2	COD	COD	N	N	N	Mole	COD	COD	COD	COD	COD	TSS
1	Hydrolysis		f_{SI}	X_I	y_I			z_I		−1				$-i_{XS}$
Heterotrophic Organisms, Denitrification														
2	Aerobic storage of COD	x_2		−1	y_2			Z_2				$Y_{STO,O2}$		$t2$
3	Anoxic storage of COD			−1	y_3	$-x_3$	x_3	Z_3				$Y_{STO,NO}$		$t3$
4	Aerobic growth	x_4			y_4			Z_4			1	$-1/Y_{H,O2}$		$t4$
5	Anoxic growth (denitrification)				y_5	$-x_5$	x_5	Z_5			−1	$-1/Y_{H,NO}$		$t5$
6	Aerobic endog. Respiration	x_6			y_6			Z_6	f_I		−1			$t6$
7	Anoxic endog. Respiration				y_7	$-x_7$	x_7	Z_7	f_I		−1			$t7$
8	Aerobic respiration of X_{STO}	x_8										−1		$t8$
9	Anoxic respiration of X_{STO}					$-x_9$	x_9	Z_9				−1		$t9$
Autotrophic Organism, Nitrification														
10	Nitrification	x_{10}			y_{10}			Z_{10}					1	$t10$
11	Aerobic endog. Respiration	x_{11}			y_{11}			Z_{11}	f_I				−1	$t11$
12	Anoxic endog. Respiration				y_{12}			Z_{12}	f_I				−1	$t12$
Composition Matrix $t_{k,i}$														
k	Conservatives													
1	COD g COD	−1	1	1		−1.71	−4.57		1	1	1	1	1	
2	Nitrogen g N		i_{NSI}	i_{NSS}	1	1	1		i_{NXI}	i_{NXS}	i_{NBM}		i_{NBM}	
3	Ionic charge Mole +				1/14		−1/14	−1						
	Observable													
4	TSS g TSS								i_{TSXI}	i_{TSXS}	i_{TSBM}	0.6	i_{TSBM}	

Source: Henze, M. et al., Table 5.1 (p. 110), from *Activated Sludge Models ASM1, ASM2, ASM2d and ASM3*, ©IWA Publishing, 2000.

kinetic parameters are more intrinsic than those resulting from homogenous models. In fact, there have been few studies concentrated on mass transfer within flocs of the AS processes, except by researchers such as Benefield and Molz, 1983, 1984; Mikesell, 1984; Andrews, 1991; Bakti and Dick, 1992 and Tyagi (1996). Benefield and Molz (1983, 1984) proposed a distributed parameter model including the material balance equations with Monod-type kinetics for the substrates inside the flocs and assumed an average floc size instead of considering the floc size distribution in the system in order to account for the effect of flocs on the dynamics of the system. Beccari et al. (1992) developed a simple floc model with emphasis on the nitrification process in suspended culture, taking into account the resistance related to oxygen diffusion inside the biofloc.

TABLE 20.5
Kinetic Rate Expressions p_j for ASM3 all $p_j < 0$

J	Process	Process rate equation p_j all $p_j < 0$
1	Hydrolysis	$K_H \frac{X_S/X_H}{K_X + X_S/X_H} X_H$
Heterotrophic Organisms, Denitrification		
2	Aerobic storage of COD	$k_{STO} \frac{S_O}{K_O + S_O} \frac{S_O}{K_S + S_{S\ g}} X_H$
3	Anoxic storage of COD	$k_{STO} . \eta_{NO} \frac{K_O}{K_O + S_O} \frac{S_{NO}}{K_{NO} + S_{NO}} \frac{S_S}{K_S + S_S} X_H$
4	Aerobic growth	$\mu_H . \frac{S_O}{K_O + S_O} \frac{S_{NH}}{K_{NH} + S_{NH}} \frac{S_{HCO}}{K_{HCO} + S_{HCO}} \frac{X_{STO} / X_H}{K_{STO} + X_{STO} / X_H} X_H$
5	Anoxic growth (denitrification)	$\mu_H . \eta_{NO} . \frac{K_O}{K_O + S_O} . \frac{S_{NO}}{K_{NO} + S_{NO}} \frac{S_{NH}}{K_{NH} + S_{NH}} \frac{S_{HCO}}{K_{HCO} + S_{HCO}} \frac{X_{STO} / X_H}{K_{STO} + X_{STO} / X_H} X_H$
6	Aerobic endog. respiration	$b_{H,O2} . \frac{S_O}{K_O + S_O} . X_H$
7	Anoxic endog. respiration	$b_{H,NO} . \frac{K_O}{K_O + S_O} . \frac{S_{NO}}{K_{NO} + S_{NO}} . X_H$
8	Aerobic respiration of X_{STO}	$b_{STO,O2} . \frac{S_O}{K_O + S_O} . X_{STO}$ $\quad b_{STO,O2} \geq b_{H,O2}$
9	Anoxic respiration of X_{STO}	$b_{STO,NO} . \frac{K_O}{K_O + S_O} . \frac{S_{NO}}{K_{NO} + S_{NO}} . X_{STO}$ $\quad b_{STO,NO} \geq b_{H,NO}$
Autotrophic Organism, Nitrification		
10	Nitrification	$\mu_A . \frac{S_O}{K_{A,O} + S_O} \frac{S_{NH}}{K_{A,NH} + S_{NH}} \frac{S_{HCO}}{K_{A,HCO} + S_{HCO}} X_A$
11	Aerobic endog. Respiration	$b_{A,O2} . \frac{S_O}{K_O + S_O} . X_A$
12	Anoxic endog. Respiration	$b_{A,NO} . \frac{K_O}{K_O + S_O} . \frac{S_{NO}}{K_{NO} + S_{NO}} . X_A$

Source: Henze, M. et al., Table 6.1 (p. 111), from *Activated Sludge Models ASM1, ASM2, ASM2d and ASM3*, ©IWA Publishing, 2000.

Tyagi et al. (1996) developed a simple floc model taking into account two growth processes, carbonaceous oxidation and nitrification, that were interacting through their competition for dissolved oxygen inside the floc. This study can be considered a good example for heterogeneous models. They did not consider two important points: (1) the anoxic decomposition of nitrate by denitrification was not incorporated into the floc model and they assumed the aerobic portion represents 100% of the total floc weight; and (2) they neglected the external mass transfer resistance due to boundary layer. The important role of the anoxic growth of heterotrophs was also neglected.

G. Ibrahim et al. (2002) developed an appropriate mathematical model for AS flocs based on the IAWPRC kinetic model, in order to study the biofloc characteristics from the kinetics-mass transfer interaction point of view. The model took into account three growth processes: carbon oxidation, nitrification, and denitrification in terms of four components (substrate, nitrate, ammonia, and oxygen). The effect of their bulk concentrations, diffusivity, and external mass transfer of substrates on the biofloc characteristics in terms of the aerobic portion weight to the total floc was studied. It can be said that the aerobic portion was found to be more sensitive to the change of the bulk concentrations of oxygen, substrate, and ammonia, in addition to the power input and substrate diffusivity. It was less sensitive to changes of nitrate bulk concentration. This model describes quantitatively the biofloc activity, as it may be totally active, which is totally aerobic or aerobic-anoxic, or it may be partially active.

20.3 AERATION AND MASS TRANSFER

Mass transfer is an important consideration in many wastewater treatment systems. In order to carry out chemical or biological reactions, it is necessary to transfer substances into or out of the wastewater as well as to move them adequately within the water to control concentration differences. The material transferred can be as diverse as gases, liquids, ions, charged colloids, or suspended solids. However, the rate at which these substances are transferred is the important consideration and is the primary concern of the field of mass transfer. The principles of mass transfer do not vary with each process.

Guellil et al. (2001) fractionated the organic matter of wastewater into settleable (i.e., particulate) and non-settleable (i.e., colloidal + soluble) fractions. Particulate, colloidal, and soluble proportions were found to be relatively constant (45%, 31%, and 24% of the total COD, respectively). Transfer of soluble fraction always occurred from the wastewater to the AS flocs, whereas bi-directional transfer occurred for the colloidal fraction. He showed that the transfer of soluble and colloidal matter reached a steady state after 40 min mixing and 20 min mixing, respectively.

The rate of oxygen transport is of great importance. We will consider the rate at which oxygen enters the water and the rate at which oxygen and other dissolved species are transferred to the biological floc.

20.3.1 Limiting Resistance for Mass Transfer

The study of Guellil et al. (2001) showed that a fraction of the organic matter is transferred between the aqueous phase and the AS flocs within a few minutes. On average, 45% of the non-settleable fraction of the wastewater from the Maxeville wastewater treatment plant (Nancy, France) was removed during this short contact time at an initial rate of about 14 mg COD g^{-1} TSS min^{-1}. Fractionation of the non- settleable matter into a colloidal and a soluble fraction revealed that steady state was obtained after 20 and 40 min, respectively. One can assume that the steady state obtained for soluble matter is delayed because of its diffusion into the floc matrix. This diffusion becomes a limiting step of soluble organic matter removal. Colloids do not penetrate into the matrix because of their size (Hamed et al., 2016), and may then be trapped very early in the outermost part of the floc.

When an oxygen molecule passes from the gas into the liquid phase and hence, to the biological floc, it must go through many separate resistances (Waite 1999). Before any species will move through the water, there must be some driving potential, usually concentration differences, which will make a molecule move from one region to another. The rate of flow of any substance is directly proportional to the driving potential and inversely proportional to the sum of the resistance between the two points of mass transfer. The important resistance occurs most often at the interface between two phases. Consider the transport of oxygen from the gas phase to a solid phase, for example, biological floc, as shown in Figure 20.1. The oxygen

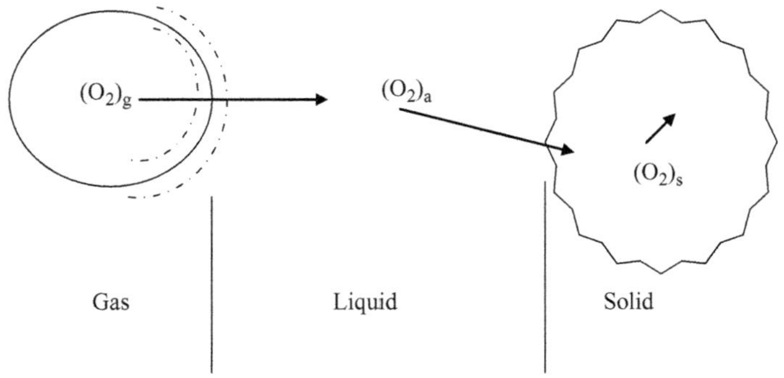

FIGURE 20.1 Resistance to oxygen transfer.

molecule in the gas phase must first overcome a resistance on the gas side within the bubble and then make the jump to the liquid side. where it again meets another resistance before reaching the bulk liquid. Once in the liquid, the oxygen molecule must now move through the fluid and eventually reach the solid phase and encounter another liquid–solid resistance, which it must overcome to become adsorbed onto the solid surface. Once the oxygen comes to the surface it must then diffuse into the pores or cells of the solids and reach the reaction site where it is consumed. Another reaction product, CO_2 for example, must go through the same resistances in the reverse order before they can reach the gas phase. Physically it can be said that these various resistances occur at points where the motion of the molecules is in the surrounding areas. The slower molecular motion may be caused by either decreased kinetic energy from lower temperature and forming chemical complexes, or by poor fluid mixing from increased shear stresses near a boundary.

In the aeration of water, the most important resistances are:

1. Liquid–film resistance between the gas–liquid interface and the bulk of the liquid
2. Bulk–liquid resistance caused by poor mixing in the liquid
3. Liquid–film resistance at the solid interface
4. Bulk–solid resistance caused by slow diffusion rates in the solid
5. Reaction resistance caused by slow chemical reaction rates at the site

20.3.2 Evaluation of Mass Transfer Coefficients

It has been found that at high agitation intensities, turbulence is expected to affect mass transfer rate at the biofloc surface. However, the actual floc velocity is unknown and conventional Reynolds numbers cannot be deduced. In this case, the concept of local isotropic turbulence may be applied (Moo-Young and Blanch, 1981). They proposed that an isotropic turbulence Re-no., Re_e, for the floc particle diameter d. This can be given as:

$$Re_e = \frac{d^{4/3}\rho^{2/3}(P/V)^{1/3}}{\mu} \tag{20.1}$$

They developed a correlation for rigid surface particle mass transfer in biochemical reactors in terms of the energy input to the system as follows:

$$Sh = 0.13 Re_e^{3/4} Sc^{1/4} \tag{20.2}$$

Where:

$$\text{Sh(Sherwood number)} = \frac{\text{total mass transfer}}{\text{diffusive mass transfer}} = \frac{K_l\ d}{D_l}$$

$$\text{Sc(Schmidit number)} = \frac{\text{momentum diffusivity}}{\text{diffusive mass transfer}} = \frac{\mu}{\rho D_l}$$

$$\text{Re(Reynolds number)} = \frac{\text{inertia forces}}{\text{vescous forces}} = \frac{\rho d v}{\mu}$$

The mass transfer coefficient (K_l) is seen to be dependent on $(P/V)^{1/4}$, which can be expressed by the effect of power input on interfacial area. Both of these relations by Moo-Young and Blanch (1981) are used to calculate the mass transfer coefficients of the four considered components as a function of the power input.

20.3.2.1 Estimation of K_la

The overall volumetric mass transfer coefficient or oxygen transfer function, K_la, is the common design parameter for specifying the rate of aeration of wastewater. When air is blown into the wastewater of an AS process, oxygen is transferred to the water.

The function that describes the oxygen transfer to the wastewater by the aeration system is called the K_la function. The most common way to describe the rate of change in dissolved oxygen (DO), due to the aeration of the water, is given by the following expression:

$$\frac{dC}{dt} = K_la\left(C^* - C(t)\right) \tag{20.3}$$

Integrating using the boundary conditions of $C = C^0$ at $t = 0$ gives,

$$\ln\left[\frac{C^* - C^0}{C^* - C}\right] = K_lat \tag{20.4}$$

Where $C(t)$ is the DO and C^* is the saturated DO.

In a real plant, K_la depends on several factors, including the type of diffusers, wastewater composition, temperature, design of aeration tank, and tank depth, but the main time-varying dependence is the airflow rate. K_la can be considered to consist of parts, K_l and a. K_l can be seen as a mass transfer coefficient and a as an area/volume ratio. Both K_l and a are, however, usually unknown so they are lumped into one parameter denoted K_la (Linberg, 1997).

Many different empirical correlations for volumetric mass transfer coefficients, K_la, have been presented in the literature (Moo-Young and Blanch, 1981). Most of these correlations can be written in the form:

$$K_la = K\ u_s^{\alpha}\left(\frac{P}{V}\right)^{\beta} \tag{20.5}$$

where:

u_s is the superficial gas velocity (the gas flow rate divided with the cross-sectional area of the tank. Unit; m/s)

K is constant

(P/V) is the power per unit volume (W/m^3).

Normally the correlation Equation (20.5) holds independently whether mixing is performed mechanically in a stirred tank reactor and or a bubble column. It is, however, possible to obtain much higher power input in stirred tank reactors than in bubble columns, and stirred tanks are therefore traditionally used in aerobic fermentation processes where there is a high oxygen demand (e.g., antibiotic fermentation). New fermenter designs based on cleverly designed static mixers or gas injection nozzles can, however, outperform the stirred tanks.

When the range of process variables for which the correlation Equation (20.5) holds is studied in more detail, it is observed that the mass transfer coefficient K_la for non-coalescing medium is greater by about a factor of 2 than for a coalescing medium at the same operating conditions. It is found that the influence of the power input is larger in the non-coalescing medium, whereas the influence of the superficial gas velocity is smaller compared with a coalescing medium (e.g., pure water). If one calculates the K_la value for a certain set of operating conditions, however, it is found that the variations using the different parameter values in Table 20.6 are relatively small.

TABLE 20.6
Parameter Values for the Empirical Correlation (20.5). The Parameter Values Are Specified with all Variables Being in SI-Unites, That Is the Power Input is in Units W/m³ and the Superficial Gas Flow Rate Is in Units m/s

Medium	K	α	β	Agitator	References
Coalescing	0.025	0.5	0.4	6-bladed Rushton turbines	Moo-Young and Blanch (1981)
	0.00495	0.4	0.953	6-bladed Rushton turbines	Linek et al. (1987)
	0.01	0.4	0.475	Different agitators	Moo-Young and Blanch (1981)
	0.026	0.5	0.4	Not specified	Van't Riet (1977)
Non-coalescing	0.0018	0.3	0.7	6-bladed Rushton turbines	Moo-Young and Blanch (1981)
	0.02	0.4	0.475	Different agitators	Moo-Young and Blanch (1981)
	0.002	0.2	0.7	Not specified	Van't Riet (1977)

It was found that the effect of (P/V) on the volumetric mass transfer coefficient, K_la, exhibits saturation-like behavior. This effect assumed a saturation-like correlation (20.6). This correlation is considered to be a possible alternative approach deduced to calculate K_la, as shown in correlation (20.5). The correlation (20.6) is a function of the power input per unit volume and the maximum overall volumetric mass transfer coefficients, K_la

$$K_la = \frac{K_la_{max.}\ (P/V)}{(P/V)+k} \tag{20.6}$$

where:

K_la_{max} is the maximum value of overall volumetric mass transfer coefficients
(P/V) is power input per unit volume in (W/m³)
k is constant.

20.4 DIFFUSION IN THE SLUDGE FLOC

This section explains flow-through AS flocs and explains how diffusion coefficients inside the flocs can be evaluated.

20.4.1 Flow through Activated Sludge Flocs

AS flocs are irregularly shaped, fragile almost transparent aggregates, with a high water content and spread over a wide size range (Li and Ganczarczyk, 1993; Biggs and Lant, 2000). Coinciding with the high water content, high porosity of AS flocs has been reported by many investigators (Smith and Coakley, 1984; Li and Ganczarczyk, 1993). However, diffusion models have been predominantly applied in the analysis of mass transfer within the AS floc (Hreiz et al., 2015), despite the fact that the flocs are highly porous and the outcomes from measurements of substrate uptake rates of the flocs are sometimes controversial. These diffusion models led to the conclusion that cells within a bacterial floc could never have a greater substrate uptake rate than dispersed cells (Hreiz et al., 2015). Based on the diffusion concept, many researchers hypothesized that an anoxic core existed in a bacterial floc of certain size as a result of oxygen transfer limitations (Mustafa et al., 2009a).

Hreiz et al., 2015 showed that for non-biological systems, some theoretical analyses described possible fluid flow through highly porous aggregates. The analyses showed that the hydrodynamic resistance experienced by an aggregate permeable to the liquid flow would be less than that by an impermeable aggregate. Correspondingly, the terminal settling velocity of a permeable aggregate would be higher than that of an impermeable aggregate of the same size and density.

Logan and Hunt (1988) indicated that the traditional diffusion models for substrate transport into microbial aggregates such as AS flocs were difficult to justify. From the genetic point of view, bacteria would not have expended energy to form flocs under nutrient conditions, if floc formation had resulted in reduced nutrient availability to the cells. They theorized that bioflocculation was an advantageous microbial response to substrate limitations. According to their prediction, the microbial aggregates were porous so that they might be permeable to fluid flow within flocs.

The hypothetical behavior of diffusion in the sludge floc was qualitatively described as follows: initially the sludge floc can be assumed to be fully penetrated with dissolved oxygen and lacking substrate. When substrate is added to the bulk liquid, the substrate diffuses into the AS floc, where aerobic microorganisms convert the substrate, consuming an amount of oxygen. This oxygen consumption induces a difference in oxygen concentration between the floc and the bulk, and oxygen diffuses from the bulk into the floc.

20.4.2 Estimation of Diffusion Coefficients

Although the aeration of the biological reactor may be excellent and sufficient agitation present to prevent local concentration gradients in the liquid, the rate of substrate removal may still be low because of either interfacial resistances around the biological floc, or poor diffusion of the oxygen and substrate into the interior of the floc. It can be said that transport within biological solids (films or flocs) is generally attributed to diffusion alone and assumed to adhere to Fick's law:

$$N_s = -D_s \frac{ds}{dr} \tag{20.7}$$

where:

N_s is the mass flux
D_s is the intra-film diffusivity
ds/dr is the spatial concentration gradient.

The internal mass transfer is credible to express the concentrations of solutes inside biological flocs or films. However, for simplification purposes, it is ignored and only external mass transfer from bulk fluid to the surface of biological solids is considered. Application of Fick's law requires knowledge of D_s. Investigators have reported widely varying ratios of intra-film to pure water diffusion coefficients (Grady, 1983).

Tanaka et al. (1984) found that the diffusion coefficient of glucose in calcium alginate is close to that of water, while Hannoun and Stephanonpoulus (1986) measured smaller values than for water. Merchant (1987), Itamunoala (1987), Axelsson and Persson (1988), and Scott et al. (1989) also observed that the effective diffusion coefficients of glucose and ethanol are smaller than that for water. Hence, in some cases diffusivities have been used as fitting parameters in substrate utilization models. Lack of certainty in the values for kinetic parameters adds to the uncertainty in estimates of the diffusion coefficient determined in this manner. Failure to consider external mass transfer resistance may also affect the resulting estimates of diffusivity.

Table 20.7 summarizes the published results of experiments that measured diffusivities through inactivated biomass (Kissel 1985). Substantial variability is seen even this narrowed sample of experiments. Possible explanations for this variability include differences in biomass growth conditions and film preparation and differences in hydraulic characteristics of the experimental systems. For example, Matson and Characklis (1976) and Onuma and Omura (1982) have reported variance in film diffusivities with carbon to nitrogen ratio in the media in which the biomass was grown interestingly, Smith and Coackley (1984) found essentially no dependence of intra-film oxygen diffusivity on biomass density. Siegrist and Gujer (1984) found reduced overall mass transfer resistance in their experimental apparatus with biofilms greater than a certain thickness. They attributed their results to penetration of the external laminar fluid layer by biological growth and subsequent induction of turbulent transport in the outer portion of the biofilm that effectively reduced the depth of biofilm through which solutes only diffused. The significance of this phenomenon in conventional treatment processes is unknown. Internal fluid flow probably occurs at least near the biofilm–liquid interface (which is often hard to define) in many fixed–film processes. However, fluid velocities relative to biomass in Siegrist and Gujer's reactor were almost certainly significantly greater than what would be found in treatment systems (their experiments were conducted at a Reynolds number greater than 4000 based on impeller speed and diameter).

Table 20.7 shows diffusion coefficients for various compounds through microbial aggregates that have been reported in the literature, mostly for floc particles. Matson and Characklis report variation in the diffusion coefficient for glucose and oxygen with respect to growth rate and the carbon–nitrogen ratio. In biofilms, the diffusion coefficient is most probably related to biofilm density.

It is noted that from Tables 20.6 and 20.8 there is uncertainty in the estimates of diffusion coefficients determined depending on measurement method.

There is a large range in the estimation of diffusivities of the substrate and other compounds and its evaluation with respect to the diffusivity of water (D_w). There seems to be confusion about

TABLE 20.7
Ratios of Experimentally Determined Diffusivities in Inactivated Biomass to Diffusivities in Water

D_f/D_w							Method of "Biofilm"	
Oxygen	Glucose	NH_4^+	NO_2^-	NO_3^-	Na^+	Br^-	Preparation	References
0.85	—	0.8–87	0.86	0.93–1.0	—	—	Filtered	Williamson and McCarty (1976)
0.2–1.0	0.3–0.5	—	—		—	—	Centrifuged, pressed into mold	Matson and Characklis (1976)
1.2	0.15–1.2	0.7	—	—	—	—	Settled/Filtered	Onuma and Omura (1982)
—	0.6	—	—	—	0.6	0.5	Grown in place	Siegrist and Gujer (1984)
0.3	—	—	—	—	—	—	Settled/Centrifuged	Smith and Coackley (1984)

TABLE 20.8
Experimental Diffusion Coefficient Measurements from the Literature

Reactant	Diffusivity 10^{-5} cm²/s	$F_{loc}/D_{H2O} \times 100\%$	Biomass Type	Growth System	Procedure
Oxygen	1.5	70	Bacterial slime	Rotating tube	Reaction products analysis
Oxygen	0.21	8	Fungi slime Zooglea ramigera	Fluidized reactor	Nonlinear curve fit
Glucose	0.048	8	Zooglea ramigera	Fluidized reactor	Nonlinear curve fit
Glucose	0.06–0.6	10–100	Nitrifier culture	Fluidized reactor	Two chamber
Oxygen	2.2	90	Mixed culture	Fluidized reactor	Two chamber
Ammonia	1.3	80			
Nitrate	1.4	90			
Oxygen	0.4–2.0	20–100	Mixed culture	Fluidized reactor	Two chamber
Glucose	0.06–0.21	10–30			

diffusivity estimation, which leads to confusion about the estimation of the mass transfer coefficient. To deal with this situation it has been assumed that the diffusivities in a sludge floc are 80% those in water Benefield and Randall (1985).

20.5 CONCLUSIONS

Mathematical process modeling and biokinetics of the AS process were reviewed considering different types of models. The task group models ASM1, ASM2, and ASM3, in the versions by Henze et al., were evaluated, considering the conditions and the different processes comprising each model (Henze et al., 1987a, and b; 1995, 1999, 2000). We showed that ASM1 contains some defects avoided in ASM3. Considering homogeneity, models can be classified into homogenous models characterized by taking the AS process as one phase. This type of model neglects the internal mass transfer inside the flocs, resulting in inaccuracy in the kinetic parameter produced. The other type of model is the heterogeneous model, which considers the mass transfer operations in addition to the biochemical reaction processes and produces kinetic parameters that can be considered more accurate than that of the homogenous model.

The mass transfer coefficients (K_l) of substances such as substrates, oxygen, nitrates, and ammonia were evaluated by Moo-Young and Blanch in relations as a function of power input. The overall volumetric mass transfer coefficient or oxygen transfer function, K_la, is the common design parameter for specifying the rate of aeration of wastewater. A large number of different empirical correlations for volumetric mass transfer coefficients, K_la, have been presented; however the most practical relations is that introduced by (Moo-Young and Blanch, 1981). These relations present K_la as a function of power input and superficial gas velocity. Confusion about diffusivity estimation remains, which leads to confusion about the estimation of mass transfer coefficient due to the uncertainty in the resulting estimates of diffusion coefficients depending on the method used for measuring. To deal with this situation, it has been assumed that the diffusivities in a sludge floc are 80% of those in water.

REFERENCES

Benefield, L. and Randall C. (1985) *Biological Process Design for Wastewater Treatment*. Charlottesville, VA: Prentice-Hall.

Biggs, C.A. (2000) Activated sludge flocculation: Investigating the effect of shear rate and cation concentration on flocculation dynamics. PhD thesis, University of Queensland, Australia.

Hamed, D., Mustafa, I.H., Ibrahim, G., Ahwany, A. and Elnashaie, S. (2016) Attached media performance in activated sludge wastewater treatment: Zenein pilot plant. *Int. J. Eng. Adv. Res. Technol.* 2(8): 4–10.

Gujer, W. and Larsen, T.A. (1995) The implementation of biokinetics and conservation principles in ASIM. *Wat. Sci. Technol.* 31 (2): 257–266.

Gujer, W., Henze, M., Mino, T. and van Loosdrecht, M.C.M. (1999) Activated sludge model no. 3. *Wat. Sci. Technol.* 39 (1): 183–193.

Henze, M., Grady, C.P.L. Jr., Gujer, W., Marais, G.V.R. and Matsuo, T. (1987a) Activated sludge model no. 1. (IAWPRC Scientific and Technical Report No. 1.) London, UK: IAWPRC.

Henze, M., Gujer, W., Mino, T., Matsuo, T., Wentzel, M.C. and Marais, G.V.R. (1995) Activated sludge model no. 2. (IAWQ Scientific and Technical Report No. 3.) London, UK: IAWQ.

Henze, M., Gujer, W., Mino, T., Matsuo, T., Wentzel, M.C., Marais, G.V.R. and van Loosdrecht, M.C.M. (1999) Activated sludge model no. 2d, ASM2d. *Wat. Sci. Technol.* 39 (1): 165–182.

Henze, M., Gujer, W., Mino, T. and van Loosedrecht, M. (2000) Table 5.1 (p. 110) from *Activated Sludge Models ASM1, ASM2, ASM2d and ASM3*. IWA Publishing.

Henze, M., Grady, C.P.L. Jr., Gujer W., Marais, G.V.R. and Matsuo, T. (1987b) A general model for single sludge wastewater treatment systems. *Wat. Res.* 21(5): 505–515.

Hreiz, R., Latifi, M.A. and Roche, N. (2015) Optimal design and operation of activated sludge processes: State-of-the-art. *Chem. Eng. J.* 281: 900–920.

Li, D.H. and Ganczarczyk, J.J. (1993) Factors affecting dispersion of activated sludge flocs. *Wat. Env. Res.* 65: 258–263.

Linberg, C.-F. (1997) Control and estimation strategies applied to the activated sludge process. PhD thesis, Uppsala University, Finland.

Logan, B. and Hunt, J.R. (1988) Bioflocculation as a microbial response to substrate limitations. *Biotechnol. Bioeng.* 31(2): 91–101.

Moo-Young, M. and Blanch, H.W. (1981) Advances in biochemical engineering: Reactors and reactions V. (19) Berlin, Germany, GDR.

Mustafa, I.H., Ibrahim, G., Elkamel, A. and Elahwany, A.H. (2009a) Modeling of activated sludge floc characteristics. *Am. J. Environ. Sci.* 5(1): 69–79.

Mustafa, I.H., Ibrahim, G., Elkamel, A. and Elahwany, A.H. (2009b) Heterogeneous modeling, identification and simulation of activated sludge processes. *Am. J. Environ. Sci.* 5(3): 352–363.

Mustafa, I.H., Elkamel, A., Lohi, A., Ibrahim, G. and Elnashaie, S. (2014) Structured mathematical modeling, bifurcation, and simulation for the bioethanol fermentation process using Zymomonas mobilis. *Eng. Indust. Chem. J.* 53: 5954–5972

Nielsen, J. and Villadsen, J. (1992) Modelling of microbial kinetics. *Chem. Eng. Sci.* 47: 4225–4270.

Tyagi, R.D., Du, Y.G. and Bhamidimarri, R. (1996) Dynamic behavior of the activated sludge under shock loading: Application of the floc model. *Wat. Res.* 30(7): 1605–616.

Grady, C.P., Cordone, L. and Cusack, L. (1993) Effects of media composition on substrate removal by pure and mixed bacterial cultures. *Biodegradation* 4: 23–38. doi:10.1007/BF00701452.

Waite, T.D. (1999) Measurement and implications of floc structure in water and wastewater treatment. *Colloids Surf A Physicochem. Eng. Asp.* 151: 27–41.

Sriyudthsak, K., Shiraishi, F. and Hirai, M.Y. (2016) Mathematical modeling and dynamic simulation of metabolic reaction systems using metabolome time series data. *Front. Mol. Biosci.* 3: 15. doi:10.3389/fmolb.2016.00015.

Tanaka, H., Matsumura, M. and Veliky, I.A. (1984) Diffusion characteristics of substrates in Ca-alginate gel beads. *Biotech. Bioeng.* 26: 53–8.

21 Hythane (H_2 and CH_4) Production from Petrochemical Wastewater via Anaerobic Digestion Process

Ahmed Tawfik and Ahmed Elreedy

CONTENTS

21.1 INTRODUCTION

Petrochemicals are mainly derived from petroleum or natural gas and represent the economic backbone of the developing and advanced countries [1]. The hydrocarbons used for petrochemical industries are olefins, aromatics, and paraffins (Table 21.1) [2]. The major by-products and intermediate chemicals from petrochemical industries are listed in Table 21.1. Petrochemical industries annually generate huge amounts of wastewater containing severe and harmful contaminants, which negatively affect human health and environment. One of the basic compounds for petrochemical industries is ethylene and its derivatives, 115 million tons of which is annually worldwide consumed. Among the ethylene derivatives, mono-ethylene glycol (MEG) has the highest demand, which expected to increase from 18.93 million tons in 2009 to 34.09 million tons in 2020 [3]. MEG is the main ingredient in the production of polyester fibers and film, polyethylene terephthalate (PET) resins, engine coolants, deicing fluids for airplanes, and runway deicers [4]. Furthermore, MEG is widely used as antifreeze, as well as a dewatering agent in the natural gas industry. The petrochemical industry consumes a lot of water and consequently huge amounts of MEG-containing wastewaters are generated, which undoubtedly contaminate water streams and have a negative effect on the environment and human health.

TABLE 21.1
Petroleum Raw Materials, Main Intermediate, Chemicals, and Products from Petrochemical Industries

Petroleum Raw Materials	First-Generation Petrochemicals		Main Intermediate Chemicals	Main Products	Main Industries
Crude petroleum Natural gas Natural gas liquids	Olefins	Ethylene	Mono-ethylene glycol, ethylene oxide and acetaldehyde	Acetic acid, mono-ethylene glycol and glycol ethers	Plastics resins Synthetic rubber Synthetic fiber Synthetic detergent Basic industrial chemicals Agricultural chemicals
		Propylene	Propylene oxide, phenol and acrolein	Isopropanolamines, polyamides and lubricating oil additives	
		Butylene	Butadiene and octyl phenols	Non-ionic detergents and sulpholase	
	Paraffins	Methane	Acetylene	Poly-vinyl chloride (PVC) and acetic acid	
		Ethane	Ethyl chloride	Tetraethyl lead	
		Propane	Formaldehyde and acetaldehyde	Formaldehyde, acetaldehyde and other oxygenated compounds	
	Aromatics	Benzene	Phenol, adipic acid and styrene	Epoxy resins, rubber anti-oxidant and benzene	
		Toluene	Toluene	Toluene	
		Xylene	Terephthalic acid, o-xylene and p-xylenes	Polyesters, isophthalic acid and phthalic anhydride	
		Complex Aromatics	Carbon black	Rubber reinforcing and aromatic extracts	

The annual production of PET in Egypt was around 420,000 tons in 2015, which is the third highest production among all the Egyptian petrochemical products with 191,600,000 USD worth of overseas sales [1,5]. The PET polyester, ethylene glycol/oxide industries, and the coolant liquid mainly discharge wastewater containing MEG with low quantity of aldehydes (<0.5%) [6–9]. Petrochemical industry wastewater containing MEG is mainly characterized as an odorless and colorless liquid with a chemical oxygen demand (COD) ranging from 500 to 30,000 mg/L, negligible amounts of suspended solids, and less than 1–2 mg/L of total Kjeldahl nitrogen (TKN), ammonium—nitrogen (NH_4-N) and phosphorous (PO_4-P). Dumping of untreated petrochemical wastewater rich in toxic pollutants will undoubtedly adversely affect the ecosystem and biodiversity. Proper treatment of petrochemical industry wastewater is necessary. Several treatment processes for petrochemical wastewater have previously been investigated, including the aerobic process [10], catalytic and electrochemical oxidations [11–13], nanofiltration and vacuum membrane distillation [6,14], and the chemical/coagulation/flotation process [13]. Although, these technologies are efficient for treatment of petrochemical industry wastewater, they are intensive in terms of energy, chemical, and equipment use. Meanwhile, because the challenge of treating petrochemical wastewater effluents will continue for the next few decades, searching for novel approaches and applications for biological degradative systems is an area of great interest among the scientific community. The application of anaerobic digestion (AD) process for energy production in terms of ethanol (Et-OH), hydrogen (H_2), and methane (CH_4) from petrochemical wastewater is a promising technology from an environmental and an economic point of view. This chapter aims to review the main factors affecting the AD for simultaneous treatment and bioenergy production from petrochemical wastewater containing MEG and phenol.

21.2 ANAEROBIC CONVERSION PATHWAYS OF PETROCHEMICAL WASTEWATER CONTAINING PHENOL AND MEG

The innovative anaerobic treatment technologies for energy production from industrial wastewater have globally received great attention due to the limitations of natural fossil fuels. Utilization of conventional fossil fuels greatly contributes to climate change due to emissions of greenhouse gases (e.g., CO_x, NO_x) [15–18]. Alternative and sustainable biofuels (H_2, CH_4, and Et-OH) production from industrial wastewater using AD is of great value for countries suffering from insufficient conventional fossil fuels and encourages the stakeholders to invest in this sector. AD technology produces little excess sludge and is preferable for the treatment of petrochemical wastewater due to its high resistance to shock organic loading rates and low operation and maintenance costs [19–21]. The mass heating values for H_2 and CH_4 gases have been reported to be 120 and 50 MJ/kg, respectively [22]. H_2 gas is an alternative energy source that provides an efficient and eco-friendly biofuel with only water as the end product of the combustion process [23–25]. The net heat energy generated from H_2 gas combustion is greater than that of the most other available fuels, such CH_4, ethane, and diesel [26]. Optimization of fermentative biofuel production from industrial wastewaters using AD processes has attracted much attention globally. However, a successful AD process needs additional effort to overcome the required energy for heating the digesters in cold-climate countries. The AD process proceeds via a series of metabolic reactions catalyzed by various microbial bacterium communities, namely acidogenesis and methanogenesis [27–28]. Three basic successive steps are occurred in AD process i.e., hydrolysis, acetogenesis and methanogenesis. AD pathways of phenol- and MEG-rich wastewater are illustrated in Figure 21.1. H_2, Et-OH, and CH_4 are mainly the biofuels produced from anaerobic degradation of petrochemical wastewater containing MEG and the degradation of phenol-rich wastewater generates only H_2 and CH_4. The anaerobic decomposition pathway of petrochemical wastewater containing MEG can be described by Equations 21.1 through 21.6 [29,30].

$$C_2H_6O_2\ (\text{MEG}) \rightarrow C_2H_4O\ (\text{Acetaldehyde}) + H_2O \tag{21.1}$$

$$C_2H_4O\ (\text{Acetaldehyde}) + H_2O \rightarrow CH_4COO^-\ (\text{Acetate}) + H^+ + H_2 \tag{21.2}$$

$$C_2H_4O\ (\text{Acetaldehyde}) + H_2 \rightarrow C_2H_6O\ (\text{Ethanol}) \tag{21.3}$$

$$C_2H_6O\ (\text{Ethanol}) + H_2O \rightarrow CH_3COO^-(\text{Acetate}) + H^+ + 2H_2 \tag{21.4}$$

$$4H_2 + CO_2 \rightarrow CH_4 + 2H_2O \tag{21.5}$$

$$C_2H_4O_2\ (\text{Acetate}) + 2H_2O \rightarrow CH_4 + CO_2 \tag{21.6}$$

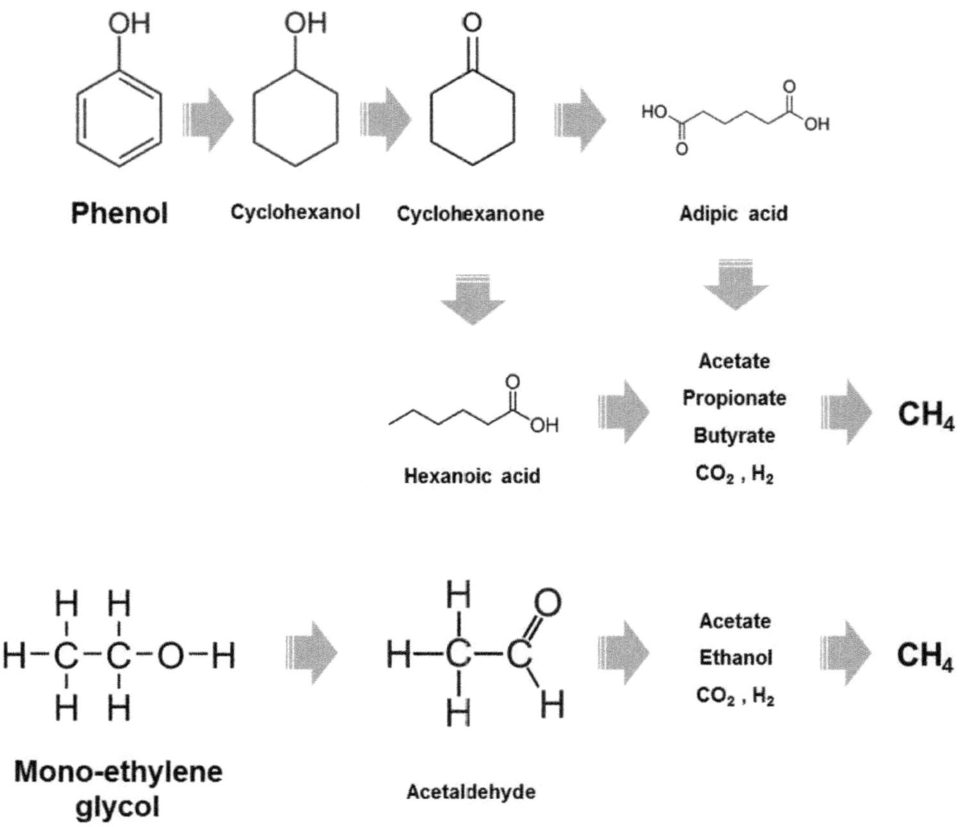

FIGURE 21.1 Anaerobic decomposition pathways of phenol and mono-ethylene glycol rich petrochemical wastewater.

21.3 FACTORS AFFECTING HYTHANE (H_2 AND CH_4) PRODUCTION FROM PETROCHEMICAL WASTEWATER CONTAINING MEG

21.3.1 Inoculum-to-Substrate Ratio (ISR)

The balance between inoculum (I) and substrate (S) ratio (ISR) is considered a crucial parameter affecting AD processes where acidogenesis and methanogenesis could occur at a minimum value of 1 gVS/gCOD [31]. The ISR can be calculated based on Equation 21.7,

$$\text{ISR} = \frac{\text{Weight of inoculum exist (gmVSS)}}{\text{Weight of substrate introduced (gm or gm/d)}} \tag{21.7}$$

A high ISR indicates insufficient substrate for microorganisms, which enhances the undesired growth rate of filament bacterium [32]. A low ISR enhances acidogenesis and would inhibit methanogenesis due to the acetate production associated with H_2 production [32]. Darlington and Kennedy [33] found that AD achieved 85% COD removal from MEG-containing wastewater at a constant ISR value of 3.7 gVSS/gCOD. Recently, Elreedy et al. [34] investigated the effect of ISRs on H_2 and CH_4 production from MEG-containing wastewater. They reported that a maximum methane yield (MY) and hydrogen yield (HY) of 151.86 ± 10.8 and 22.27 ± 1.1 mL/gCOD$_{\text{initial}}$ were registered at ISRs of 5.29 and 3.78 gVSS/gCOD, respectively. The H_2 and CH_4 contents were, respectively, 40.91% and

TABLE 21.2
Hythane (H_2 and CH_4) Yields from Different Substrate Composition Using AD Process

Substrate Composition	Temperature (°C)	$COD_{initial}$ (g/L)	ISR (gVSS/gCOD)	H_2/CH_4 Yields (L/gCOD)	References
Acetate	37	4.20	0.48	NA/0.126	[36]
Glucose	37	10.70	0.19	0.075/NA	[36]
Starch	37	5.00	0.49	NA/0.316	[37]
Cellulose	55	10.00	3.0	0.188/0.506	[38]
Protein	37	5.00	0.49	NA/0.246	[37]
PWW + AM	55	32.00	0.50	NA/0.017	[39]
Propylene glycol	35	2.00	NA	NA/1.910	[30]
Acrylic	35	30.33	NA	(0.041)	[4]
Ethylene oxide	35	30.33	NA	(0.062)	[4]
Mono-ethylene glycol	55	5.00	5.29	0.021/0.152	[34]

54.51% at ISR of 2.65 and 5.29 gVSS/gCOD. Partial substrate degradation and low HY occurred at low ISR as reported earlier by Sreela-Or et al. [35]. Table 21.2 summarizes the simultaneous H_2 and CH_4 production rates from various petrochemical wastewaters using AD process [4,34,36–39].

21.3.2 Nitrogen-to-Phosphorus (N/P) Ratio

Nitrogen (N) and phosphorous (P) are essential nutrients for bacterial growth and successful operation of the AD process [40]. These nutrients are mainly dependent on substrate compositions [41]. Speece [42] found that the required amount of phosphorus (P) is 15% of the nitrogen (N) demand resulting in an N/P ratio of 6.67. The minimum values of N and P are 45 and 8 mg/L, respectively, for treatment of petrochemical wastewater containing carboxylic acids [43]. Nevertheless, a wide range of COD/N/P (400/7/1, 500/5/1 and 200/4/1) ratios has been considered for AD of different substrates. Elreedy et al. [34] found that the biohydrogen production from MEG dropped from 58 ± 2 to 37 ± 2 mL at increasing N/P ratios from 4.6 to 8.5, respectively. This indicates that N/P balance is necessary for biohydrogen production. Moreover, it was found that the maximum total biogas production was recorded at N/P ratio of 5.5 [34]. The MY increased 1.6-fold by increasing N/P ratio from 4.6 to 5.5, resulting in a maximum MY of 290 ± 18 mL. Increase of the N/P ratio up to 8.5 slightly decreased the cumulative MY of 274 ± 21 mL. Apparently, N/P ratios of 4.6 and 5.5 are optimum nutrient sources for the sequential hydrogen and methane production, which is also consistent with the suitable N/P ratio of 5.0 for microbial bacterium growth [44]. The nitrogen content is the key factor for enhancement of methane production from industrial wastewater [45]. However, Demirer et al. [46] found that a high N/P ratios have no significant effect on the MY. Romero-Güiza et al. [41] stated that propionate and acetate utilization rates were relatively improved under the nutrient-amended conditions. The ammonification efficiency was also enhanced at an N/P ratio of 4.6 due to the release of NH_4-N, as well as acetate by acidogenesis consortium [34]. Increasing the initial ammonium nitrogen concentration would maintain a favorable environment for methanogenesis over acidogenesis, providing a higher acetate consumption pathway for methanogenesis [47].

21.3.3 Organic Loading Rate

Organic loading rate (OLR) is the daily influent organic content (expressed as gCOD/day) multiplied by flow rate (m^3/d) and divided by the reactor working volume (L). OLR mainly expresses the biological conversion capacity of the anaerobic digester and efficiently controls the preference of either the acidogenesis and methanogenesis processes. The OLR depends on the substrate composition, biodegradability, flow rate, reactor capacity, and hydraulic retention time (HRT). Elreedy et al. [48] found that the HY was substantially increased from 48.34 ± 4.52 to 359.01 ± 33.46 $mL/gCOD_{removed}$ when the OLR was increased from 0.33 to 1.67 gCOD/L/d., respectively. They also recorded that H_2 content and specific hydrogen production rate (SHPR) reached a maximum value of 49.84% ± 5.06% and 160.14 ± 12.30 mL/L/d, respectively, at an OLR of 1.67 gCOD/L/d. However, the maximum MY of 159.11 ± 14.72 $mL/gCOD_{removed}$ was achieved at an OLR of 0.67 gCOD/L/d, [48]. This indicates that high and low organic loading rates were preferable for acidogenesis and methanogenesis, respectively. This also emphasizes that methanogenesis inhibition takes place at increased OLR due to the toxicity of the accumulation of volatile fatty acids [49,50]. Total biogas production of 7.3 and 14.2 L/d was achieved from the anaerobic conversion of wastewater containing MEG at OLRs of 0.75 and 1.5 gCOD/L/d, respectively [51]. Elreedy et al. [48] also found that the major portion of metabolite production occurred at an OLR of 1.67 gCOD/L/d, resulting in 2159.95 ± 89.71 mg/L of total volatile fatty acids (TVFAs) in the treated effluent.

21.3.4 Effect of pH

The initial pH strongly affects the hydrolysis and acidification processes of wastewater in the AD process to maximize hydrogen production. A significant correlation coefficient ($P < 0.01$) has been reported between the hydrolysis rate and initial pH values [52]. The optimum initial pH value for H_2 production varied from 5.0 to 6.5, depending on the substrate composition [40,53]. The hydrogenase enzyme activity generated from acid-producing bacteria at pH value of 5.8 was approximately 2.2-fold higher than that from acidogenic bacteria at pH of 4.5 [54]. Elreedy et al. [34] found that a decrease in initial pH values from 7.0 to 5.0 significantly enhanced the cumulative H_2 production by 1.44-fold and the maximum value of 79 ± 6 mL was recorded at a pH value of 5.0. This corresponded to the maximum HY of 33.11 ± 2.1 and 47.55 ± 4.3 $mL/gCOD_{initial}$ (67.99 ± 5.6 to 97.65 ± 6.1 $mL/gMEG_{initial}$) at the pH values of 7.0 to 5.0, respectively. This is due to the positive effect of low pH values of 5.0–6.0 on hydrogenase enzyme activity [53]. Nonetheless, decreasing initial pH values (4–4.5) may inhibit the growth of acidogenic bacteria as well as their hydrogenase enzyme activities responsible for hydrogen production. The optimum pH value for hydrogen production is dependent on the wastewater composition, that is glucose (pH = 5.5), paperboard mill wastewater (pH = 5.0), palm oil mill wastewater (pH = 6.0), and rice slurry (pH = 4.5) [34]. Elreedy et al. [34] found that the COD removal efficiency peaked (33.08% ± 2.48%) at pH value of 5.0 for wastewater containing MEG. The residual ethanol (Et-OH) dropped significantly from 563.00 ± 40.81 at pH value of 7.0 to 39.84 ± 2.53 mg/L at pH 5.0. This can be attributed to the enhancement of the alcohol dehydrogenase enzyme (ADH) activity at low pH, which is responsible for the aforementioned reactions [55]. The pH value in the treated effluent dropped from 5.69 ± 0.51 to 3.98 ± 0.31 at an initial pH ranging from 7.0 to 5.0 due to acetate accumulation.

21.3.5 Temperature

Thermophilic conditions (45°C–70°C) are superior to mesophilic AD of wastewater (25°C–45°C) where higher reaction rates and stability under load-stress is attained. AD processes at ambient temperature in tropical and subtropical countries would substantially reduce the costs of heating where the temperature varied from 21°C to 40°C [56]. The anaerobic degradation of petrochemical wastewater containing MEG for hythane production at ambient temperature has been investigated [57,58]. The results were quite good at an ambient temperature of 22°C–34°C.

21.3.6 Inoculum Source and Pretreatment

The type of inoculum bacteria have an effect on AD efficiency due to the variations of the extracellular enzymatic activities, bacterial growth, nutrient demand, predominant species, and physiological characteristics [31]. A proper inoculum is selected based on the substrate decomposition pathway and associated enzymes. Forster-Carneiro et al. [59] found that pretreated sewage sludge is an efficient inoculum for H_2 production from organic fraction of municipal solid waste. Moreover, it is indispensable to properly choose the mixed culture's origin, which is capable of providing active species responsible for the MEG degradation pathway catalyzed by ADH. The excess sludge from baking yeast industry wastewater treatment has been recognized to have greater activity of ADH, which is responsible for MEG metabolism by baker's yeast, namely *Saccharomyces cerevisiae* [60]. Several studies indicated that thermal sludge pretreatment could effectively inhibit the hydrogen-consuming bacteria [61]. Mechanical, ultrasonic disintegration, alkali, heat, and thermochemical pretreatment have been efficiently used to inhibit methanogenesis [61]. Heat shock pretreatment of inoculum has been shown to suppress hydrogen consuming and harvesting spore-forming bacterium [32,62]. The heat-pretreated sludge was found to have higher a H_2 yield, compared to the acid and alkali pretreatment methods for conversion of glucose [61] and sucrose [63]. Elreedy et al. [34] found that the thermal pretreatment of the sludge increased the H_2 production from 48 ± 3 mL (untreated) to 55 ± 5 mL (thermal pretreated), and the HY was improved by 1.14-fold for AD of MEG-containing wastewater.

21.3.7 Supplementation of Nanomaterials

Supplementation with nanoparticles (NPs) would increase H_2 production from AD of wastewater. The NPs affect the activities of enzymes responsible for H_2 production by alternating the milieu around their active sites [64]. Appropriate doses of metal NPs can supply biologically available nutrients for bacteria via their dissolution in the culture bacterium medium, though excess metal NPs may inhibit microbial activity [65,66]. Previous studies have demonstrated the potential benefits of using Fe-based NPs for biohydrogen production from AD of glucose compared with supplementation with iron sulfate [66–68]. Henningsen et al. [69] and Catalanotti et al. [55] reported that the ADH and aldehyde dehydrogenase (ALDH) are responsible for MEG metabolism. When these enzymes were associated with NPs, they showed higher stability and performance compared with the free enzymes [64]. The release of hydrogen in this metabolic pathway is associated with electron transfer during a redox reaction, which is catalyzed by the hydrogenase enzyme activity [70]. Various types of metal-based NPs (Fe, Ni, Au, and Ag) have been used at different dosages to increase H_2 production from various substrates (Table 21.2). The optimum NP dosage strongly depends on the type of NP and the substrate composition. Nonetheless, most of previous studies have focused on improving H_2 production from glucose and sucrose. The impact of Ni NPs and nickel-graphene nanocomposite (Ni-Gr NC) on H_2 production from industrial wastewater containing MEG via the AD process was investigated by Elreedy et al. [71]. Batch anaerobic reactors were supplemented with different dosages of Ni NPs and Ni-Gr NC ranging from 0 to 100 mg/L. Maximum HYs of 24.73 ± 1.12 and 41.28 ± 1.69 mL/gCOD$_{initial}$ was achieved at a dosage of 60 mg/L for Ni NPs and Ni-Gr NC, respectively. Substantial improvements of 23% and 105% of H_2 production were registered at a dosage of 60 mg/L for Ni NPs and Ni-Gr NC, respectively, compared to the control batch reactors. However, increasing the dosage of Ni NPs and Ni-Gr NC up to 100 mg/L resulted in a significant decrease in HY of 20.80 ± 1.12 and 24.24 ± 1.13 mL/gCOD$_{initial}$, respectively.

21.4 CONCLUSION

Petrochemical industries annually generate huge amounts of wastewater containing severe and harmful contaminants, which negatively affect human health and the environment. The dumping of untreated petrochemical wastewater rich in toxic pollutants will undoubtedly affect the ecosystem and biodiversity. Treatment of petrochemical wastewater is necessary. The application of AD for simultaneous treatment and bio-hythane production from MEG-containing wastewater is recommended. HY from petrochemical wastewater was optimized at ISR, N/P ratio, and pH values of 3.78 gVSS/gCOD, 4.6, and 5.0, respectively. MY was maximized at ISR, N/P ratio, and pH value of 5.29 gVSS/gCOD, 5.5, and 7.0, respectively. Supplementation of Ni-Gr NC at 20 mg/L is recommended to maximize the biohydrogen production from wastewater containing MEG.

REFERENCES

1. Oxford Business Group. Healthy prospects for Egypt's chemicals and plastics sector 2015. http://www.oxfordbusinessgroup.com/news/healthy-prospects-egypt's-chemicals-and-plastics-sector (accessed January 1, 2017).
2. Gloyna EF, Ford DL. Petrochemical Effluents Treatment Practices. 1970. Washington, DC: EPA.
3. Demand for Ethylene Oxide and Ethylene Glycol to Rise 2010. http://www.chemicals-technology.com/news/news97708.html.
4. Schonberg JC, Bhattacharya SK, Madura RL, Mason SH, Conway RA. Evaluation of anaerobic treatment of selected petrochemical wastes. *J Hazard Mater* 1997;54:47–63.
5. Hussein S. Petrochemicals industry in Egypt. A look for the future. 3rd Annu. Downstr. summit 2015, 7–9th December, Algiers, Alger., 2015.
6. Mohammadi T, Akbarabadi M. Separation of ethylene glycol solution by vacuum membrane distillation (VMD). *Desalination* 2005;181:35–41. doi:10.1016/j.desal.2005.01.012.
7. Cocero MJ, Alonso E, Torı R, Vallelado D, Sanz T. Supercritical water oxidation (SCWO) for poly (ethylene terephthalate) (PET) industry effluents. *Ind Eng Chem Res* 2000;39:4652–7. doi:10.1021/ie000289t.
8. Fdz-Polanco F, Hidalgo MD, Fdz-Polanco M, García Encina PA. Anaerobic treatment of polyethylene terephthalate (PET) wastewater from lab to full scale. *Water Sci Technol* 1999;40:229–36. doi:10.1016/S0273-1223(99)00630-7.
9. Shakerkhatibi M, Monajemi P, Jafarzadeh MT, Mokhtari SA, Farshchian MR. Feasibility study on EO / EG wastewater treatment using pilot scale SBR. *Int J Environ Res* 2013;7:195–204.
10. Hassani AH, Borghei SM, Samadyar H, Ghanbari B. Utilization of moving bed biofilm reactor for industrial wastewater treatment containing ethylene glycol: Kinetic and performance study. *Environ Technol* 2013;35:499–507. doi:10.1080/09593330.2013.834947.
11. Zerva C, Peschos Z, Poulopoulos SG, Philippopoulos CJ. Treatment of industrial oily wastewaters by wet oxidation. *J Hazard Mater* 2003;97:257–65. doi:10.1016/S0304-3894(02)00265-0.
12. Kim KN, Hoffmann MR. Heterogeneous photocatalytic degradation of ethylene glycol and propylene glycol. *Korean J Chem Eng* 2008;25:89–94. doi:10.1007/s11814-008-0015-4.
13. Ishak S, Malakahmad A, Isa MH. Refinery wastewater biological treatment: A short review. *J Sci Ind Res* 2012;71:251–6.
14. Orecki A, Tomaszewska M, Karakulski K, Morawski A. Separation of ethylene glycol from model wastewater by nanofiltration. *Desalination* 2006;200(1–3):358–360. https://doi.org/10.1016/j.desal.2006.03.364
15. Das D. Hydrogen production by biological processes: a survey of literature. *Int J Hydrogen Energy* 2001;26:13–28. doi:10.1016/S0360-3199(00)00058-6.
16. Zou S, Wang H, Wang X, Zhou S, Li X, Feng Y. Application of experimental design techniques in the optimization of the ultrasonic pretreatment time and enhancement of methane production in anaerobic co-digestion. *Appl Energy* 2016;179:191–202. doi:10.1016/j.apenergy.2016.06.120.
17. Moraes BS, Junqueira TL, Pavanello LG, Cavalett O, Mantelatto PE, Bonomi A et al. Anaerobic digestion of vinasse from sugarcane biorefineries in Brazil from energy, environmental, and economic perspectives: Profit or expense? *Appl Energy* 2014;113:825–35. doi:10.1016/j.apenergy.2013.07.018.
18. Massé DI, Rajagopal R, Singh G. Technical and operational feasibility of psychrophilic anaerobic digestion biotechnology for processing ammonia-rich waste. *Appl Energy* 2014;120:49–55. doi:10.1016/j.apenergy.2014.01.034.
19. Yossan S, O-Thong S, Prasertsan P. Effect of initial pH, nutrients and temperature on hydrogen production from palm oil mill effluent using thermotolerant consortia and corresponding microbial communities. *Int J Hydrogen Energy* 2012;37:13806–13814. doi:10.1016/j.ijhydene.2012.03.151.

20. Ferguson LN. ANAEROBIC CO-DIGESTION OF AIRCRAFT DEICING FLUID AND MICROAEROBIC STUDIES. Marquette University, 1999.
21. Yu L, Ma J, Frear C, Zhao Q, Dillon R, Li X et al. Multiphase modeling of settling and suspension in anaerobic digester. *Appl Energy* 2013;111:28–39. doi:10.1016/j.apenergy.2013.04.073.
22. Elreedy A, Tawfik A. Effect of Hydraulic Retention Time on hydrogen production from the dark fermentation of petrochemical effluents contaminated with Ethylene Glycol. Energy Procedia 2015;74:1071–1078. doi:10.1016/j.egypro.2015.07.746.
23. Kamalaskar LB, Dhakephalkar PK, Meher KK, Ranade DR. High biohydrogen yielding Clostridium sp. DMHC-10 isolated from sludge of distillery waste treatment plant. *Int J Hydrogen Energy* 2010;35:10639–10644. doi:10.1016/j.ijhydene.2010.05.020.
24. Elsamadony M, Tawfik A, Suzuki M. Surfactant-enhanced biohydrogen production from organic fraction of municipal solid waste (OFMSW) via dry anaerobic digestion. *Appl Energy* 2015;149:272–282. doi:10.1016/j.apenergy.2015.03.127.
25. Yang Z, Guo R, Xu X, Fan X, Luo S. Fermentative hydrogen production from lipid-extracted microalgal biomass residues. *Appl Energy* 2011;88:3468–3472. doi:10.1016/j.apenergy.2010.09.009.
26. Hino R, Yan XL. *Nuclear hydrogen production handbook*. CRC Press; 2011.
27. Lee DJ, Jegatheesan V, Ngo HH, Hallenbeck PC. *Current developments in biotechnology and bioengineering: biological treatment of industrial effluents*. 2017.
28. Galway NUI. *Hydrolysis, methanogenesis and bioprocess performance during low-temperature anaerobic digestion of dilute wastewater*. 2015.
29. Dwyer DF, Tiedje JM. Degradation of Ethylene Glycol and Polyethylene Glycols by Methanogenic Consortia. *Appl Environ Microbiol* 1983;46:185–90.
30. Veltman S, Schoenberg T, Switzenbaum MS. Alcohol and acid formation during the anaerobic decomposition of propylene glycol under methanogenic conditions. *Biodegradation* 1998;9:113–8. doi:10.1023/A:1008352502493.
31. Ali Shah F, Qaisar M, Shah MM, Pervez A, Ahmed Asad S. Microbial Ecology of Anaerobic Digesters: The Key Players of Anaerobiosis. *Sci World J* 2014:1–21.
32. Farghaly A, Tawfik A, Danial A. Inoculation of paperboard sludge versus mixed culture bacteria for hydrogen production from paperboard mill wastewater. *Environ Sci Pollut Res* 2015. doi:10.1007/s11356-015-5652-7.
33. Darlington C, Kennedy KJ. Biodegradation of aircraft deicing fluid in an upflow anaerobic sludge blanket (UASB) reactor. *J Environ Sci Heal Part A* 1998;33:339–51. doi:10.1080/10934529809376735.
34. Elreedy A, Fujii M, Tawfik A. Factors affecting on hythane bio-generation via anaerobic digestion of mono-ethylene glycol contaminated wastewater: inoculum-to-substrate ratio, nitrogen-to-phosphorus ratio and pH. *Bioresour Technol* 2017;223:10–9. doi:10.1016/j.biortech.2016.10.026.
35. Sreela-Or C, Plangklang P, Imai T, Reungsang A. Co-digestion of food waste and sludge for hydrogen production by anaerobic mixed cultures: Statistical key factors optimization. *Int J Hydrogen Energy* 2011;36:14227–37. doi:10.1016/j.ijhydene.2011.05.145.
36. Elbeshbishy E, Hafez H, Nakhla G. Enhancement of biohydrogen producing using ultrasonication. *Int J Hydrogen Energy* 2010;35:6184–93. doi:10.1016/j.ijhydene.2010.03.119.
37. Elbeshbishy E, Nakhla G. Batch anaerobic co-digestion of proteins and carbohydrates. *Bioresour Technol* 2012;116:170–8. doi:10.1016/j.biortech.2012.04.052.
38. Lay CH, Chang FY, Chu CY, Chen CC, Chi YC, Hsieh TT et al. Enhancement of anaerobic blohydrogen/methane production from cellulose using heat-treated activated sludge. *Water Sci Technol* 2011;63:1849–54. doi:1021661wsl2011.390.
39. Siddique MNI, Munaim MSA, Zularisam AW. Effect of food to microbe ratio variation on anaerobic co-digestion of petrochemical wastewater with manure. *J Taiwan Inst Chem Eng* 2015;0:1–7. doi:10.1016/j.jtice.2015.06.038.
40. Mao C, Feng Y, Wang X, Ren G. Review on research achievements of biogas from anaerobic digestion. *Renew Sustain Energy Rev* 2015;45:540–55. doi:10.1016/j.rser.2015.02.032.
41. Romero-Güiza MS, Vila J, Mata-Alvarez J, Chimenos JM, Astals S. The role of additives on anaerobic digestion: A review. *Renew Sustain Energy Rev* 2016;58:1486–99. doi:10.1016/j.str.2014.12.012.
42. Speece RE. Anaerobic biotechnology for industrial wastewater treatment a description of several installations. *Environ Sci Technol* 1983;17:416A–427A. doi:10.1139/196-121.
43. Britz T, Noeth C, Lategan P. Nitrogen and phosphate requirements for the anaerobic digestion of a petrochemical effluent. *Water Res* 1988;22:163–9. doi:10.1016/0043-1354(88)90074-7.

44. Leite WRM, Gottardo M, Pavan P, Belli Filho P, Bolzonella D. Performance and energy aspects of single and two phase thermophilic anaerobic digestion of waste activated sludge. *Renew Energy* 2016;86:1324–31. doi:10.1016/j.renene.2015.09.069.
45. Lei Z, Chen J, Zhang Z, Sugiura N. Methane production from rice straw with acclimated anaerobic sludge: Effect of phosphate supplementation. *Bioresour Technol* 2010;101:4343–8. doi:10.1016/j.biortech.2010.01.083.
46. Demirer SU, Taskin B, Demirer GN, Duran M. The effect of managing nutrients in the performance of anaerobic digesters of municipal wastewater treatment plants. *Appl Microbiol Biotechnol* 2013;97:7899–907. doi:10.1007/s00253-012-4499-9.
47. Gallert C, Bauer S, Winter J. Effect of ammonia on the anaerobic degradation of protein by a mesophilic and thermophilic biowaste population. *Appl Microbiol Biotechnol* 1998;50:495–501. doi:10.1007/s002530051326.
48. Elreedy A, Tawfik A, Kubota K, Shimada Y, Harada H. Hythane (H2 + CH4) production from petrochemical wastewater containing mono-ethylene glycol via stepped anaerobic baffled reactor. *Int Biodeterior Biodegradation* 2015;105:252–61. doi:10.1016/j.ibiod.2015.09.015.
49. Fang HHP, Liu H. Effect of pH on hydrogen production from glucose by a mixed culture. *Bioresour Technol* 2002;82:87–93. doi:10.1016/S0960-8524(01)00110-9.
50. Sreethawong T, Niyamapa T, Neramitsuk H, Rangsunvigit P, Leethochawalit M, Chavadej S. Hydrogen production from glucose-containing wastewater using an anaerobic sequencing batch reactor: Effects of COD loading rate, nitrogen content, and organic acid composition. *Chem Eng J* 2010;160:322–32. doi:10.1016/j.cej.2010.03.037.
51. Marin J, Kennedy KJ, Eskicioglu C. Characterization of an anaerobic baffled reactor treating dilute aircraft de-icing fluid and long term effects of operation on granular biomass. *Bioresour Technol* 2010;101:2217–23. doi:10.1016/j.biortech.2009.11.055.
52. Zhang P, Chen Y, Zhou Q. Waste activated sludge hydrolysis and short-chain fatty acids accumulation under mesophilic and thermophilic conditions: Effect of pH. *Water Res* 2009;43:3735–42. doi:10.1016/j.watres.2009.05.036.
53. El-Bery H, Tawfik A, Kumari S, Bux F. Effect of thermal pre-treatment on inoculum sludge to enhance bio-hydrogen production from alkali hydrolysed rice straw in a mesophilic anaerobic baffled reactor. *Environ Technol* 2013;34:1965–72. doi:10.1080/09593330.2013.824013.
54. Elsamadony M. *Simultaneous treatment and hydrogen production from organic fraction of municipal solid waste using dry anaerobic digestion.* 2016.
55. Catalanotti C, Yang W, Posewitz MC, Grossman AR. Fermentation metabolism and its evolution in algae. *Front Plant Sci* 2013;4:150. doi:10.3389/fpls.2013.00150.
56. Elreedy A, Tawfik A, Enitan A, Kumari S, Bux F. Pathways of 3-biofules (hydrogen, ethanol and methane) production from petrochemical industry wastewater via anaerobic packed bed baffled reactor inoculated with mixed culture bacteria. *Energy Convers Manag* 2016;122:119–130. doi:10.1016/j.enconman.2016.05.067.
57. Bodík I, Herdová B, Drtil M. The use of upflow anaerobic filter and AnSBR for wastewater treatment at ambient temperature. *Water Res* 2002;36:1084–1088. doi:10.1016/S0043-1354(01)00308-6.
58. Wei S, Zhang H, Cai X, Xu J, Fang J, Liu H. Psychrophilic anaerobic co-digestion of highland barley straw with two animal manures at high altitude for enhancing biogas production. *Energy Convers Manag* 2014;88:40–48. doi:10.1016/j.enconman.2014.08.018.
59. Forster-Carneiro T, Perez M, Romero LI, Sales D. Dry-thermophilic anaerobic digestion of organic fraction of the municipal solid waste: Focusing on the inoculum sources. *Bioresour Technol* 2007;98:3195–3203. doi:10.1016/j.biortech.2006.07.008.
60. Leskovac V, Trivić S, Pericin D. The three zinc-containing alcohol dehydrogenases from baker's yeast, Saccharomyces cerevisiae. FEMS *Yeast Res* 2002;2:481–494. doi:10.1111/j.1567-1364.2002.tb00116.x.
61. Elbeshbishy E, Hafez H, Dhar BR, Nakhla G. Single and combined effect of various pretreatment methods for biohydrogen production from food waste. *Int J Hydrogen Energy* 2011;36:11379–11387. doi:10.1016/j.ijhydene.2011.02.067.
62. Elsamadony M, Tawfik A, William AD, Suzuki M. Optimization of hydrogen production from organic fraction of municipal solid waste (OFMSW) dry anaerobic digestion with analysis of microbial community. *Int J Energy Res* 2015;39(7):929–940. doi:10.1002/er.3297
63. Mu Y, Yu HQ, Wang G. Evaluation of three methods for enriching H2-producing cultures from anaerobic sludge. *Enzyme Microb Technol* 2007;40:947–953. doi:10.1016/j.enzmictec.2006.07.033.

64. Netto CGCM, Andrade LH, Toma HE. Association of Pseudomonas putida formaldehyde dehydrogenase with superparamagnetic nanoparticles: an effective way of improving the enzyme stability, performance and recycling. *New J Chem* 2015;39:2162–2167. doi:10.1039/C4NJ01716A.
65. Puntes VF. Using iron nanoparticles in anaerobic digestion. 2015. http://www.smartcbi.org/index.php/en/news-smart-quimic/561-using-iron-nanoparticles-in-anaerobic-digestion.
66. Han H, Cui M, Wei L, Yang H, Shen J. Enhancement effect of hematite nanoparticles on fermentative hydrogen production. *Bioresour Technol* 2011;102:7903–7909. doi:10.1016/j.biortech.2011.05.089.
67. Taherdanak M, Zilouei H, Karimi K. The effects of Fe0 and Ni0 nanoparticles versus Fe2+ and Ni2+ ions on dark hydrogen fermentation. *Int J Hydrogen Energy* 2016;41:167–173. doi:10.1016/j.ijhydene.2015.11.110.
68. Mohanraj S, Kodhaiyolii S, Rengasamy M, Pugalenthi V. Phytosynthesized iron oxide nanoparticles and ferrous iron on fermentative hydrogen production using Enterobacter cloacae: Evaluation and comparison of the effects. *Int J Hydrogen Energy* 2014;39:11920–11929. doi:10.1016/j.ijhydene.2014.06.027.
69. Henningsen BM, Hon S, Covalla SF, Sonu C, Aaron Argyros D, Barrett TF et al. Increasing anaerobic acetate consumption and ethanol yields in Saccharomyces cerevisiae with NADPH-specific alcohol dehydrogenase. *Appl Environ Microbiol* 2015;81:8108–8117. doi:10.1128/AEM.01689-15.
70. Xing D, Ren N, Rittmann BE. Genetic diversity of hydrogen-producing bacteria in an acidophilic ethanol-H2-coproducing system, analyzed using the [Fe]-hydrogenase gene. *Appl Environ Microbiol* 2008;74:1232–1239. doi:10.1128/AEM.01946-07.
71. Elreedy A, Ibrahim E, Hassan N, El-dissouky A, Fujii M, Yoshimura C et al. Nickel-graphene nanocomposite as a novel supplement for enhancement of biohydrogen production from industrial wastewater containing mono-ethylene glycol. *Energy Convers Manag* 2017;140:133–144. doi:10.1016/j.enconman.2017.02.080.

22 Anaerobic Degradation of Lipid-Rich Wastewater

Ahmed Tawfik and Mohamed Elsamadony

CONTENTS

22.1 INTRODUCTION

Lipids are composed of long-chain fatty acids (LCFA) normally bonded to glycerol, alcohols or other groups, such as ester. Fats and oils are a subgroup of lipids that contain the alcohol groups esterified with fatty acids to form triglycerides (glycerol with three LCFA) (Figure 22.1). Fats mainly contain saturated LCFA. Oils are composed of unsaturated fatty acids. Most agro-industrial effluent contains huge amounts of fats and oils from, i.e., dairy industries (Erdirencelebi, 2011), slaughterhouses (Rodríguez-Méndez et al., 2017), livestock farms (Ziels et al., 2017), wool scouring facilities (He et al., 2015) and edible oil–processing facilities (Kadhum and Shamma, 2015). The lipid content in the end-of-pipe effluent mainly depends on the raw materials and the type of industry; wool scouring and olive mills generate severe effluents with high lipid concentrations (5–25 g/L) (Ha et al., 2017). Low concentrations of lipids have been detected in sunflower oil mill wastewater, with LCFA concentrations (0.2–1.3 g/L) (Saatci et al., 2003). Dairy wastewater contains total lipid concentrations of 0.9–2.0 g/L (Kim et al., 2004). A total fat matter of 0.35–0.52 g/L exists in slaughterhouse effluents (Rodríguez-Méndez et al., 2017).

Dumping of lipid-rich wastewater will certainly cause accumulation inside the sewer networks, creating serious environmental problems. Deposition of fat, oil and grease (FOG) has resulted in blockage of 47% of sewer lines, due to narrowing of the pipes' cross sections (He et al., 2011; Elsheikh et al., 2013). Removal of FOG and/or lipids from wastewater prior to discharge into the sewer networks is therefore essential. The abatement device (Figure 22.2) is mainly used to remove FOG from industrial wastewater (Gallimore et al., 2011; Aziz et al., 2012; Long et al., 2012).

Anaerobic digestion (AD) of lipid-rich wastewater is a promising technique and produces energy in the form of methane (Long et al., 2012). Organic solid wastes (Elsamadony and Tawfik, 2015a, 2015b; Farghaly et al., 2017; Soltan et al., 2017), food wastes (Elsamadony et al., 2015; Mostafa et al., 2017; Elsamadony and Tawfik, 2018), cheese whey (Antonopoulou et al., 2008) and olive mill

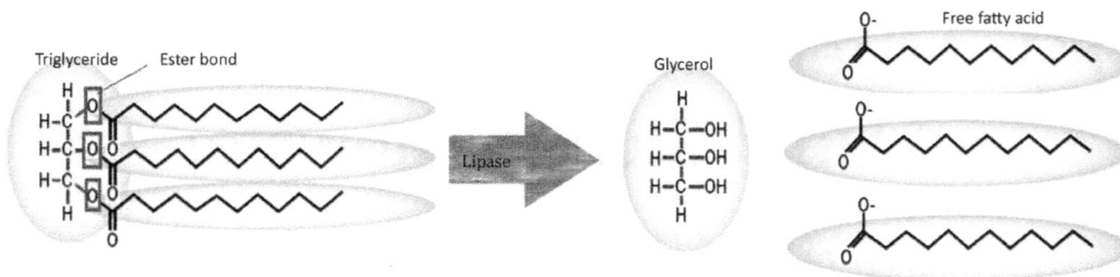

FIGURE 22.1 Lipolysis step of lipid-rich wastewater. (From van Lier, J.B. et al., *Environ. Sci. Technol.*, 20, 1200–1206, 2008.)

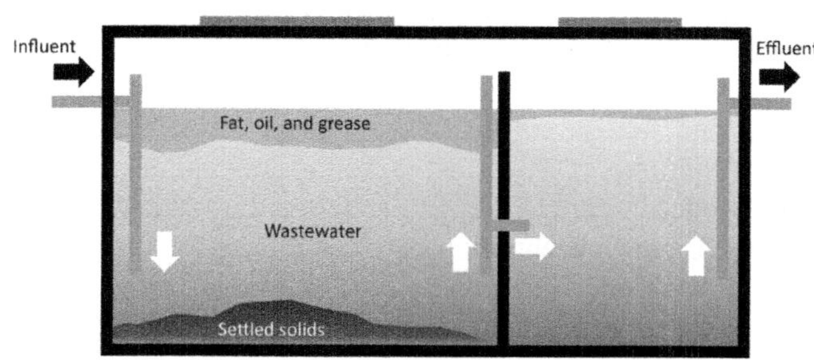

FIGURE 22.2 Fat, oil and grease (FOG) abatement device. (From Ragauskas, A.M.E. et al., *Energy Sci. Eng.*, 1, 42–52, 2013.)

wastewater (Koutrouli et al., 2009) has been efficiently used for H_2 and CH_4 production with the AD process. The AD of industrial wastewater has several advantages over conventional activated sludge treatment plants, because less energy is consumed, minimal sludge is produced and the AD process produces energy in the form of hydrogen and methane. Moreover, the methane production from AD of lipids has a great potential to be easily used for generating electricity (Equation 22.1) by burning it in a gas turbine or steam boiler (Linke, 2006). Each 1 m^3 of biogas produces 1.8 kWh (Ma, 1999).

$$CH_4(g) + 2O_2(g) \rightleftharpoons CO_2(g) + 2H_2O\ (L) + \text{electricity} \tag{22.1}$$

Methane could be also used for vehicle fuel and as a substitute for fossil fuels such as gasoline and diesel. Methane has been successfully converted into useful by-products such as methanol and syngas (the production of syngas is shown in Equations 22.2 and 22.3), where syngas consists of H_2 and carbon monoxide (CO).

$$CH_4 + H_2O \rightleftharpoons CO + 3H_2 \tag{22.2}$$

$$CO + H_2O \rightleftharpoons CO_2 + H_2 \tag{22.3}$$

Lipid-rich wastewater could be used also to generate organic acids via an anaerobic process. These organic acids could be used for a variety of industrial processes, that is, for polyhydroxy alkonates (PHA). *R. spheroids IFO 12203* efficiently converted 70% of the organics in palm oil mill effluent (POME) into organic acids, and 50% PHA yield was achieved from organic acids (Hassan et al., 2002). Itaconic acid was produced using *Aspergillus terreus IMI 282743* from filtered effluent (POME) (Jahim et al., 2006). POME and wheat flour was used by Aspergillus (A 103) to produce

TABLE 22.1
Methane Production from Organic Fractions of Wastewater

Organic Fractions	Methane Forming Reaction	Methane Production (L/g)
Carbohydrates	$C_6H_{10}O_5 + H_2O \rightarrow 3CH_4 + 3CO_2$	0.415
Proteins	$C_{16}H_{24}O_5N_4 + 14.5H_2O \rightarrow 8.25CH_4 + 3.75CO_2 + 4NH_4^+ + 4HCO_3^-$	0.634
Lipids	$C_{50}H_9O_6 + 24.5H_2O \rightarrow 34.75CH_4 + 15.25CO_2$	0.990

citric acid after incubation for two days (Jamal et al., 2005). Anaerobic digestion of co-substrate (glucose, wheat flour and a nitrogen source, ammonium nitrate) with POME produced citric acid at 5.2 g/L (Alam et al., 2008). However, hydrolysis and methanogenesis of lipids is the rate-limiting step. Since conventional anaerobic digesters require a high volumetric reactor and long hydraulic retention time, high-rate anaerobic reactors have therefore been extensively investigated for degrading lipids and producing methane from industrial wastewater, including studies on upflow anaerobic sludge blanket (UASB), upflow anaerobic filtration unit (Borja and Banks, 1994), anaerobic fluidized bed reactor (Borja et al., 1995) and upflow anaerobic sludge fixed-film reactor (Najafpour et al., 2006). Lipid-rich wastewaters are comprised of glycerol, long chain fatty acids (LCFAs), alcohols and esters. Anaerobic conversion of 1 g of lipid fractions could theoretically produce 0.990 L of methane. Only 0.415 L methane could be generated from the degradation of 1 g of glucose via the anaerobic digestion bioprocess presented in Table 22.1. Lipid-rich wastewaters are therefore considered an attractive source for methane production (Rasit et al., 2015). However, lipid hydrolysis is the rate-limiting step in the AD process and will be discussed in detail in this chapter. The mechanism of anaerobic degradation of lipid-rich wastewater, as well as the mitigation of LCFA accumulation onto the anaerobes during the AD process, will be assessed.

22.2 ANAEROBIC DEGRADATION PATHWAYS OF LIPIDS-RICH WASTEWATER

Lipids represent the major organic fraction in agro-industrial effluents. The hydrolytic anaerobes convert the lipids (tri-glyceride esters) into glycerol and long chain fatty acids (LCFAs) using extracellular lipases as shown in Figure 22.1 and as previously reported (Cuetos et al., 2008). The acidogenesis further degrades the glycerol into acetate, H_2 and CO_2 (Figure 22.3). However, Batstone et al. (2000) found that the glycerol is mainly converted into acetate, lactate and 1,3-propanediol. The LCFAs are degraded using syntrophic acetogenesis via the β-oxidation process into acetate, H_2 and CO_2 (Weng and Jeris, 1976) and finally the methanogens convert both acetate and H_2 into CH_4 (Table 22.2). LCFAs are initially stimulated by "coenzyme A" prior to entering the β-oxidation process, resulting in acetyl-CoA (Figure 22.4). Oxidation of acetyl-CoA occurs via the citric acid cycle and β-oxidation process, during which acetate and hydrogen are successively produced at the end of the β-oxidation pathway (Equation 22.4).

$$CH_3(CH_2)nCOOH + 2H_2O \rightarrow CH_3(CH_2)n\text{-}2COOH + CH_3COOH + 2H_2 \qquad (22.4)$$

The LCFAs are degraded into short chain fatty acids (acetate) via the β-oxidation pathway (Table 22.2). Syntrophism is defined as two different microorganisms that can degrade some substances together or separately. A syntrophic reaction is, for example, H_2 gas being produced by acidogenesis and consumed by methanogens. LCFAs are degraded using proton-reducing acetogens, acetoclastic and hydrogen-utilizing methanogens (Schink, 1997). Lipid hydrolysis is generally a rapid process in the AD process. However, the subsequent **β**-oxidation step is rather slow (Angelidaki and Ahring, 1992). This is mainly due to the presence of unsaturated LCFAs, which need further hydrogenation to be

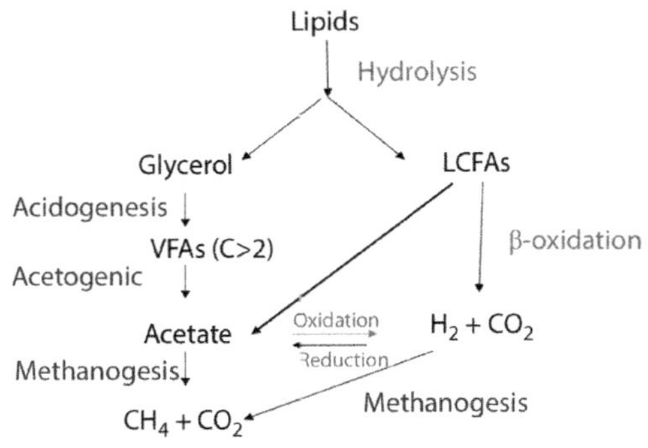

FIGURE 22.3 Lipid degradation in the anaerobic digestion (AD) process.

TABLE 22.2
Long Chain Fatty Acids Oxidation Reactions and Methane Formation Process

Bio-process	Reactions
β-oxidation process	
Linoleate (C 18:2)	$\text{Linoleate}^- + 16H_2O \rightarrow 9\ \text{acetate}^- + 14H_2 + 8H^+$
Oleate (C 18:1)	$\text{Oleate}^- + 16H_2O \rightarrow 9\ \text{acetate}^- + 15H_2 + 8H^+$
Stearate (C 18:0)	$\text{Stearate}^- + 16H_2O \rightarrow 9\ \text{acetate}^- + 16H_2 + 8H^+$
Palmitate (C 16:0)	$\text{Palmitate}^- + 14H_2O \rightarrow 8\ \text{acetate}^- + 14H_2 + 7H^+$
Methanogenic process	
Hydrogen	$4H_2 + HCO_3^- + H^+ \rightarrow CH_4 + 3H_2O$
Acetate	$\text{Acetate}^- + H_2O \rightarrow HCO_3^- + CH_4$

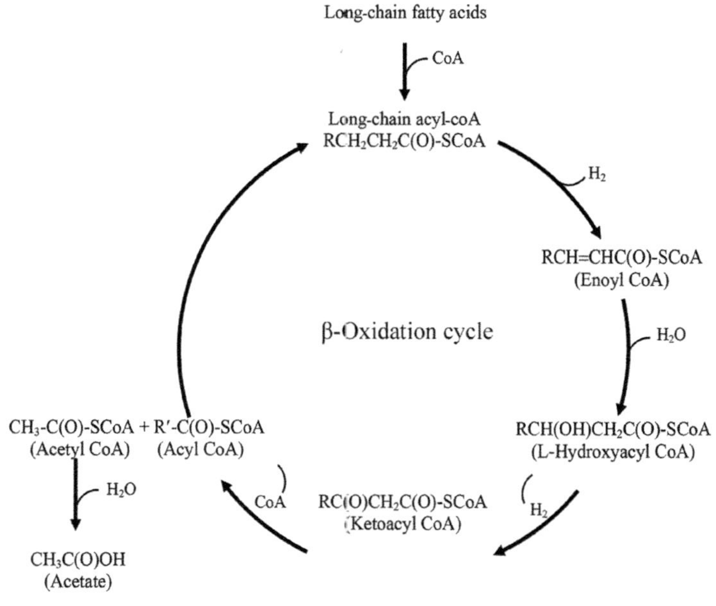

FIGURE 22.4 Anaerobic degradation pathway of long-chain fatty acids. (From Sousa, D.Z. et al., *FEMS Microbiol. Ecol.*, 68, 257–272, 2009.)

easily degradable. The oxidation of saturated LCFAs requires an external electron acceptor during the acetogenesis process. The hydrogenation of unsaturated LCFAs is necessary for further degradation.

22.3 LCFA ADSORPTION AND SLUDGE WASHOUT FROM ANAEROBIC DIGESTERS

Hydrolysis of fats and oils proceeds rapidly in the sewerage network, resulting in an increase of LCFAs in wastewater (Hanaki and Nagase, 1981), which adversely affects the subsequent AD process. Moreover, a high fraction of lipids are hydrolyzed in the anaerobic digester, which results in quite high concentrations of LCFAs that cover and inhibit the anaerobes, as presented in Figure 22.5. LCFAs are ionized at neutral pH, that is, as oleate and palmitate instead of oleic and palmitic acids. Palmitic and oleic acid are the most abundant saturated and unsaturated LCFAs, respectively. Application of anaerobic technology to the treatment of lipid/LCFA-rich wastewaters has drawn great attention. However, most anaerobic reactors treating lipids/LCFA-rich wastewaters have suffered from sludge flotation and biomass washout due to the adsorption of lipids/LCFAs onto the anaerobes, which inhibits the acetogenic bacteria and methanogenic archaea by accumulation of LCFAs in the anaerobic modules.

An industrial-scale UASB reactor treating milk fat failed due to sludge flotation (Samson et al., 1984). The high fat content provided poor biomass retention in four reactors treating ice-cream wastewater (Hawkes et al., 1995). Rinzema et al. (1993) observed severe washout caused by sludge flotation during treatment of LCFA-containing wastewaters in UASB reactors (Hwu and Lettinga 1997). Oleic acid degradation in a UASB reactor was investigated by Sam-soon et al. (1991) who found that the original inoculated granules suffered from disintegration and encapsulation by a gelatinous and whitish mass. Apparently at neutral pH, LCFAs act as surfactants, lowering the surface tension of the sludge. LCFAs have an amphiphilic structure and are composed of a hydrophobic aliphatic tail and a hydrophilic carboxylic head. The adhesion of the hydrophilic cells occurs at a low liquid surface tension. This was not the case for the adhesion of hydrophobic cells, which prefer a high surface tension (Thaveesri et al., 1995). Acetogens are mostly hydrophobic, and therefore the low-surface-tension environments created by LCFAs would cause a sloughing-off of granular sludge, which results in a washout of these anaerobes from the reactors (Daffonchio et al., 1995). Typical operational conditions of expanded granular sludge bed (EGSB) reactors (upflow velocity >4 m/h, hydraulic retention time <10 h) resulted in poor treatment performance of LCFA-containing wastewaters (Hwu et al., 1997a). However, the recirculation of the floated sludge enhanced reactor performance (Hwu et al., 1998). Complete sludge flotation of 0.203 $g_{COD}/g_{VSS}.d$ occurred at LCFs concentrations exceeding 263 mg LCFAs/L (Hwu et al., 1998), and the minimum inhibitory concentration for methanogenesis was 401 mg LCFAs/L. This is emphasized by the failure of the UASB reactor due to LCFA adsorption followed by the inhibition of the methanogenic archaea and resulting sludge washout from the reactor. Jeganathan et al. (2006) found that a UASB reactor treating complex oily wastewater from the food industry was efficient for converting approximately 75% of chemical oxygen demand (COD) into methane

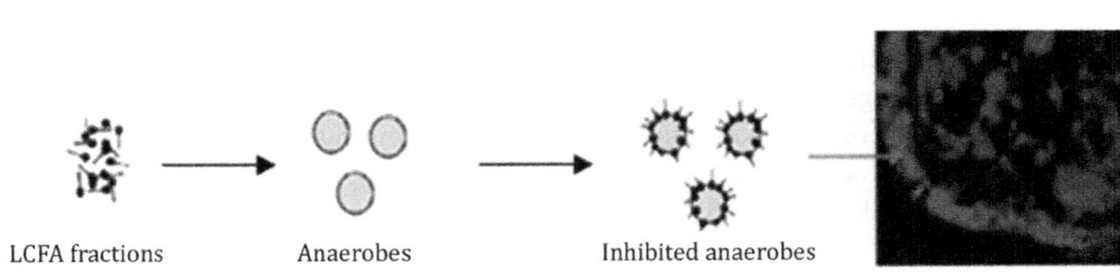

FIGURE 22.5 Anaerobe inhibition by long-chain fatty acids. (From Chen, J.L. et al., *Biotechnol. Adv.* 32, 1523–1534, 2014.)

at an organic loading rate (OLR) of 2.5 g COD/L.d. However reactor performance sharply declined at a higher loading rate of 5 g COD/L.d due to the accumulation of FOG in the sludge bed of the reactor. The sludge was washed out from the reactor and FOG replaced the anaerobes, creating a severe condition for biodegradation process.

22.4 MITIGATION OF LCFA INHIBITION

22.4.1 Co-digestion

The AD process is efficient for methane production from lipid-rich wastewater. However, the process is a failure, particularly at high lipid concentrations and in the presence of inhibitory compounds such polyphenols, lack of nutrients, etc. Dilution of wastewater, supplementation of nutrients (N&P) (Boari et al., 1984; Morelli et al., 1990), and buffering capacity ($NaHCO_3$, NaOH, or Ca $(OH)_2$ (Boari et al., 1984) have mainly been used to overcome sludge washout and minimize sludge flotation in anaerobic reactors. However, the dilution of wastewater provides unnecessarily large effluent volumes and the addition of chemicals is not economical or environmentally preferable. Anaerobic co-digestion of different substrates is a promising approach where the nutrients and buffering capacity could be sufficient to enhance the process and mitigate the inhibition effect of the LCFAs (Ahring et al., 1992). The cost of the anaerobic co-digestion process is minimal, when different types of wastewaters can be treated together and could produce a huge amount of bioenergy. Hydrogen sulfide was substantially reduced during co-digestion of pig manure with ferrous-containing waste (Davidsson et al., 2008). Moreover, detoxification of toxic elements could be overcome by the anaerobic co-metabolic process. The latter is defined as the microbial transformation capability of a compound by an organism that is unable to use it as an energy or carbon source. Dechlorination of pentachlorophenols was successfully achieved by co-digestion with glucose (Hendriksen et al., 1992). Angelidaki and Ahring (1997) found that olive mill effluent (OME) could be efficiently converted into biogas when co-digested with manure, household waste (HHW) or sewage sludge. An 87% reduction of lipid content occurred using co-digestion of OME with manure (50:50 and 75:25). Lipid reduction (73%) was achieved in co-digestion of OME with HHW (50:50 and 75:25). The high buffering capacity of manure, together with the supplementation of essential nutrients from other substrates, enhanced the degradation efficiency of OME without dilution, addition of external alkalinity or a nitrogen source. AD of meat processing wastewater containing protein and lipids suffered from accumulation of ammonia and LCFAs (Bustillo-lecompte and Mehrvar, 2016). However, co-digestion could be advantageous due to an improved C/N ratio and dilution of the inhibitory compounds (Soltan et al., 2017). A significant increase of the methane yield of cattle waste from 25 to 50 m^3 biogas/m^3 was achieved (Ahring, 2003) when fish oil (5%) was added to a manure digester. Co-digestion of sludge from grease traps and sewage sludge successfully increased the methane yield by 9%–27% when 10%–30% of the sludge from grease traps (on VS-basis) was added (Davidsson et al., 2008).

22.4.2 Calcium Addition

Calcium (Ca^{2+}) is an essential element for the growth of methanogens and the formation of microbial aggregates in anaerobic digesters (Elsamadony and Tawfik, 2015a). However, excessive amounts of Ca^{2+} leads to precipitation of carbonate and phosphate, resulting in scaling of reactors, reduction of the specific methanogenic activity, buffering capacity and essential nutrients for the anaerobic degradation process (Langerak et al., 1998). Nevertheless, the addition of calcium to the AD of lipid-rich wastewater could reduce LCFA inhibition due to the formation of insoluble salts. However, the addition of calcium cannot overcome the sludge flotation process (Roy et al., 1985). Further investigation is needed to confirm such phenomena. No inhibitory effect on anaerobic digestion was detected by Jackson-Moss and Duncan (1989) at Ca^{2+} concentrations of up to 7000 mg/L.

22.4.3 Feeding Mode

The pulse feeding sequence of anaerobic digesters represents an essential operation mode to overcome the accumulation of LCFAs. The latter are fed to the reactor several times to allow a sufficient reaction time between the lipids and anaerobes. This could minimize the shock and excessive load of LCFAs on the cell walls of the anaerobes (Cavaleiro et al., 2008). Nielsen and Ahring (2006) found that oleate pulses enhanced the degradation process and shortened the lag phase of the AD process. Anaerobes have the ability to tolerate LCFAs with a gradual, rather than a shock, feeding sequence (Angelidaki and Ahring, 1992).

22.4.4 Control the Lipid Concentration

Lipid concentration is the main parameter affecting hydrolysis and bio-methanation of wastewater. Increasing lipid concentration from 5% to 47% (w/w) negatively affected biogas production. The methane production rate was quite similar at lipid concentrations of 5%, 10%, and 18% (w/w, COD basis) (Cirne et al., 2007). However, methane production dropped at lipid concentrations of 31%, 40%, and 47% due to inhibition of LCFAs. The process was partially recovered with supplementation of lipase. Apparently, the enzyme enhanced the hydrolysis process with the excessive production of LCFAs, which caused inhibition of the later steps in the degradation process. Lipid concentrations of 1000 mg/L had an inhibitory effect on the AD process (Hanaki and Nagase, 1981). Lipid uptake up to 14,000 mg/L (3.5 g/L.d) removed 63%–89% of lipids removed in the UASB reactor, however, when the lipid concentration exceeded 9 g/L it caused a biomass flotation in the reactor (Erdirencelebi, 2011).

22.4.5 Adsorbent Addition

The combination of adsorption and the anaerobic digestion of lipid-rich wastewater has been reported earlier by Roy et al. (1985), who added calcium chloride to the media to mitigate LCFA toxicity and enhance its solubility. LCFA reacting with calcium chloride results in about 95% fatty acid calcium salts. The latter are precipitated, insoluble salts that are formed and simultaneously increase interfacial tension. Hanaki and Nagase (1981) found that a long lag-phase period for methane production could be overcome by the addition of calcium chloride ($CaCl_2$). Calcium chloride has a temporal ability to retard LCFA inhibition, which lessened for anaerobes after several hours. Other calcium salts such as calcium carbonate could not diminish the inhibition because of the insolubility of the produced fatty acid calcium salts. On the other hand, bentonite exhibited better performance than calcium salts for glyceride trioleate degradation. Bentonite is a type of clay mineral that has the same mechanism of LCFA precipitation (Angelidaki et al., 1990). As a result of its high porosity and surface area, bentonite provides superior adsorption capacity towards LCFAs (Palatsi et al., 2012). Furthermore, adding bentonite and fibers to LCFA-rich cow manure resulted in a reduction of the lag phase for methane production from 4–5 to 2–3 days (Palatsi et al., 2009).

22.5 CONCLUSIONS

AD of lipid-rich wastewater is a promising technique for wastewater treatment and produces energy in the form of methane. However, hydrolysis and methanogenesis of lipids is the rate-limiting step due to the adsorption of LCFAs onto the anaerobes during the AD process. Mitigating the accumulation of LCFAs is crucial for producing energy from lipid-rich wastewater. Anaerobic co-digestion of different substrates is a promising approach, where the nutrients and buffering capacity could be sufficient to enhance the process and mitigate the inhibition effect of the LCFAs. Addition of calcium to the AD of lipid-rich wastewater could reduce LCFA inhibition due to the formation of insoluble salts. Nevertheless, the addition of calcium cannot overcome the sludge flotation process. A pulse feeding sequence of the anaerobic digesters represents an essential operation mode to overcome the accumulation of LCFAs. The combination of adsorption and AD of lipid-rich wastewater is an excellent approach to mitigate LCFA toxicity, enhance its solubility and increase biogas production.

REFERENCES

Ahring, B.K., 2003. Perspectives for anaerobic digestion, in: *Biomethanation I*. Advances in Biochemical Engineering/Biotechnology. pp. 1–30.

Ahring, B.K., Angelidaki, I., Lohansen, K., 1992. Anaerobic treatment of manure together with industrial waste. *Wat. Sci. Tech.* 25, 311–318.

Alam, M.Z., Jamal, P., Nadzir, M.M., 2008. Bioconversion of palm oil mill effluent for citric acid production: Statistical optimization of fermentation media and time by central composite design. *World J. Microbiol. Biotechnol.* 24, 1177–1185. doi:10.1007/s11274-007-9590-5.

Angelidaki, I., Ahring, B.K., 1997. Codigestion of olive oil mill wastewaters with manure, household waste or sewage sludge. *Biodegradation* 8(4), 221–226.

Angelidaki, I., Ahring, B.K., 1992. Effects of free long-chain fatty acids on thermophilic anaerobic digestion. *Appl. Microbiol. Biotechnol.* 37, 808–812. doi:10.1007/BF00174850.

Angelidaki, I., Petersen, S.P., Ahring, B.K., 1990. Effects of lipids on thermophilic anaerobic digestion and reduction of lipid inhibition upon addition of bentonite. *Appl. Microbiol. Biotechnol.* 33, 469–472. doi:10.1007/BF00176668.

Antonopoulou, G., Stamatelatou, K., Venetsaneas, N., Kornaros, M., Lyberatos, G., 2008. Biohydrogen and methane production from cheese whey in a two-stage anaerobic process. *Ind. Eng. Chem. Res.* 47, 5227–5233.

Aziz, T.N., Holt, L.M., Keener, K.M., Groninger, J.W., Ducoste, J.J., 2012. Field characterization of external grease abatement devices. *Wat. Env. Res.* 84(3), 237–246. doi:10.2175/106143012X13347678384161.

Batstone, D.J., Keller, J., Newell, R.B., Newland, M., 2000. Modelling anaerobic degradation of complex wastewater. II: Parameter estimation and validation using slaughterhouse effluent. *Bioresour. Technol.* 75, 75–85.

Boari, G., Brunetti, A., Passinot, R., Rozzi, A., 1984. Anaerobic digestion of olive oil mill wastewaters. *Agric. Wastes* 10, 161–175.

Borja, R., Banks, C.J., 1994. Treatment of palm oil mill effluent by upflow anaerobic filtration. *J. Chem. Technol. Biotechnol.* 61, 103–109.

Borja, R., Banks, C.J., Wang, Z., 1995. Kinetic evaluation of an anaerobic fluidised-bed reactor treating slaughterhouse wastewater. *Bioresour. Technol.* 52, 163–167.

Bustillo-lecompte, C.F., Mehrvar, M., 2016. Treatment of actual slaughterhouse wastewater by combined anaerobic–aerobic processes for biogas generation and removal of organics and nutrients: An optimization study towards a cleaner production in the meat processing industry. *J. Clean. Prod.* 141, 278–289. doi:10.1016/j.jclepro.2016.09.060.

Cavaleiro, A.J., Pereira, M.A., Alves, M., 2008. Enhancement of methane production from long chain fatty acid based effluents. *Bioresour. Technol.* 99, 4086–4095. doi:10.1016/j.biortech.2007.09.005.

Chen, J.L., Ortiz, R., Steele, T.W.J., Stuckey, D.C., 2014. Toxicants inhibiting anaerobic digestion: A review. *Biotechnol. Adv.* 32, 1523–1534. doi:10.1016/j.biotechadv.2014.10.005.

Cirne, D.G., Paloumet, X., Björnsson, L., Alves, M.M., Mattiasson, B., 2007. Anaerobic digestion of lipid-rich waste—Effects of lipid concentration. *Renew. Energy* 32, 965–975. doi:10.1016/j.renene.2006.04.003.

Cuetos, M.J., Gómez, X., Otero, M., Morán, A., 2008. Anaerobic digestion of solid slaughterhouse waste (SHW) at laboratory scale: Influence of co-digestion with the organic fraction of municipal solid waste (OFMSW). *Biochem. Eng. J.* 40, 99–106. doi:10.1016/j.bej.2007.11.019.

Daffonchio, D., Thaveesri, J., Verstraete, W., 1995. Contact angle measurement and cell hydrophobicity of granular sludge from upflow anaerobic sludge bed reactors. *Appl. Environ. Microbiol.* 61, 3676–3680.

Davidsson, Å., Lövstedt, C., la Cour Jansen, J., Gruvberger, C., Aspegren, H., 2008. Co-digestion of grease trap sludge and sewage sludge. *Waste Manag.* 28, 986–992. doi:10.1016/j.wasman.2007.03.024.

Elsamadony, M., Tawfik, A., 2018. Maximization of hydrogen fermentative process from delignified water hyacinth using sodium chlorite. *Energy Convers. Manag.* 157, 257–265. doi:10.1016/j.enconman.2017.12.013.

Elsamadony, M., Tawfik, A., 2015a. Dry anaerobic co-digestion of organic fraction of municipal waste with paperboard mill sludge and gelatin solid waste for enhancement of hydrogen production. *Bioresour. Technol.* 191, 157–165. doi:10.1016/j.biortech.2015.05.017.

Elsamadony, M., Tawfik, A., 2015b. Potential of biohydrogen production from organic fraction of municipal solid waste (OFMSW) using pilot-scale dry anaerobic reactor. *Bioresour. Technol.* 196, 9–16. doi:10.1016/j.biortech.2015.07.048.

Elsamadony, M., Tawfik, A., Danial, A., Suzuki, M., 2015. Use of carica papaya enzymes for enhancement of H_2 production and degradation of glucose, protein, and lipids. *Energy Procedia.* 75, 975–980. doi:10.1016/j.egypro.2015.07.308.

Elsheikh, M.A., Saleh, H.I., Rashwan, I.M., El-samadoni, M.M., 2013. Hydraulic modelling of water supply distribution for improving its quantity and quality. *Sustain. Environ. Resour.* 23, 403–411.

Erdirencelebi, D., 2011. Treatment of high-fat-containing dairy wastewater in a sequential UASBR system: Influence of recycle. *J. Chem. Technol. Biotechnol.* 86, 525–533. doi:10.1002/jctb.2546.

Farghaly, A., Elsamadony, M., Ookawara, S., Tawfik, A., 2017. Bioethanol production from paperboard mill sludge using acid-catalyzed bio-derived choline acetate ionic liquid pretreatment followed by fermentation process. *Energy Convers. Manag.* 145, 255–264. doi:10.1016/j.enconman.2017.05.004.

Gallimore, E., Aziz, T.N., Movahed, Z., 2011. Assessment of internal and external grease interceptor performance for removal of food-based fats, oil, and grease from food service establishments. *Wat. Env. Res.* 83(9), 882–892. doi:10.2175/106143011X12989211840972.

Ha, M., Yasin, M., Mamat, R., Naja, G., Majeed, O., Fitri, A., Ha, M., 2017. Potentials of palm oil as new feedstock oil for a global alternative fuel: A review. *Renew. Sustain. Energy Rev.* 79, 1034–1049. doi:10.1016/j.rser.2017.05.186.

Hanaki, K., Nagase, M., 1981. Mechanism of inhibition caused by long chain fatty acids in anaerobic digestion process. *Biotechnol. Bioeng.* 23, 1591–1610. doi:10.1002/bit.260230717.

Hassan, M.A., Nawata, O., Shirai, Y., Abdul Rahman, N.A., Yee, P., Bin Ariff, A., Abdul Karim, M., 2002. A proposal for zero emission from palm oil industry incorporating the production of polyhydroxyalkanoates from palm oil mill effluent. *J. Chem. Eng. (Japan)* 35, 9–14. doi:10.1038/srep35194.

Hawkes, F.R., Donnelly, T., Anderson, G.K., 1995. Comparative performance of anaerobic digesters operating on ice-cream wastewater. *Water Res.* 29, 525–533.

He, X., Iasmin, M., Dean, L.O., Lappi, S.E., Ducoste, J.J., de los Reyes, F.L., 2011. Evidence for fat, oil, and grease (FOG) deposit formation mechanisms in sewer lines. *Environ. Sci. Technol.* 45, 4385–4391. doi:10.1021/es2001997.

He, X., Zhang, Q., Cooney, M.J., Yan, T., 2015. Biodegradation of fat, oil and grease (FOG) deposits under various redox conditions relevant to sewer environment. *Appl. Microbiol. Biotechnol.* 99, 6059–6068. doi:10.1007/s00253-015-6457-9.

Hendriksen, H.V., Larsen, S., Ahring, B.K., 1992. Influence of a supplemental carbon source on anaerobic dechlorination of pentachlorophenol in granular sludge. *Appl. Environ. Microbiol.* 58, 365–370.

Hwu, C.S., Lettinga, G., 1997. Acute toxicity of oleate to acetate-utilizing methanogens in mesophilic and thermophilic anaerobic sludges. *Enzyme Microb Technol.* 21(4), 297–301.

Hwu, C., Molenaar, G., Garthoff, J., Lier, J.B. Van, Lettinga, G., 1997. Thermophilic high-rate anaerobic treatment of wastewater containing long-chain fatty acids: Impact of reactor hydrodynamics. *Biotechnol. Lett.* 19, 447–451.

Hwu, C.S., Tseng, S.-K., Yuan, C.-Y. Kulik, Z., Lettinga, G., 1998. Biosoroption of long-chain fatty acids in UASB treatment process. *Water Res.* 32(5), 1571–1579.

Jackson-Moss, C.A., Duncan, J.R., 1989. The effect of calcium on anaerobic digestion. *Biotechnol. Lett.* 11, 219–224.

Jahim, J., Intan, N., Muhammad, S., Yeong, W.T., 2006. Factor analysis in itaconic acid fermentation using filtered POME by Aspergillus terreus IMI 282743. *J. Kejuruter.* 18, 39–48.

Jamal, P., Alam, Z., Ramlan, M., Salleh, M., Nadzir, M., 2005. Screening of Aspergillus for citric acid production from palm oil mill effluent. *Biotechnology* 4, 275–278.

Jeganathan, J., Nakhla, G., Bassi, A., 2006. Long-term performance of high-rate anaerobic reactors for the treatment of oily wastewater. *Environ. Sci. Technol.* 40, 6466–6472.

Kadhum, A.A.H., Shamma, M.N., 2015. Critical reviews in food science and nutrition edible lipids modification processes: A review. *Crit. Rev. Food Sci. Nutr.* 57, 48–58. doi:10.1080/10408398.2013.848834.

Kim, S., Han, S., Shin, H., 2004. Two-phase anaerobic treatment system for fat-containing wastewater. *J. Chem. Technol. Biotechnol.* 79, 63–71. doi:10.1002/jctb.939.

Koutrouli, E.C., Kalfas, H., Gavala, H.N., Skiadas, I.V., Stamatelatou, K., Lyberatos, G., 2009. Bioresource rechnology hydrogen and methane production through two-stage mesophilic anaerobic digestion of olive pulp. *Bioresour. Technol.* 100, 3718–3723. doi:10.1016/j.biortech.2009.01.037.

Langerak, E.P.A.V.A.N., Aelst, A.V.A.N., Hamelers, H.V.M., Lettinga, G., 1998. Effects of high calcium concentrations on the development of methanogenic sludge in upflow anaerobic sludge bed (UASB) reactors. *Water Res.* 32, 1255–1263.

Lier, J.B. Van, Mahmoud, N., Zeeman, G., 2008. Anaerobic wastewater treatment, biological wastewater treatment: Principles, modelling and design. *Environ. Sci. Technol.* 20, 1200–1206. doi:10.1021/es00154a002.

Linke, B., 2006. Kinetic study of thermophilic anaerobic digestion of solid wastes from potato processing. *Biomass Bioenergy* 30, 892–896. doi:10.1016/j.biombioe.2006.02.001.

Long, J.H., Aziz, T.N., Reyes, F.L.D.L., Ducoste, J.J., 2012. Anaerobic co-digestion of fat, oil, and grease (FOG): A review of gas production and process limitations. *Process Saf. Environ. Prot.* 90, 231–245. doi:10.1016/j.psep.2011.10.001

Ma, A., 1999. Innovations in management of palm oil mill effluent. *Plant.* (*Kuala Lumpur*) 75, 381–389. doi:10.1007/s11157-011-9253-8.

Morelli, A., Rindone, B., Andreoni, V., Villa, M., Sorlini, C., Balice, V., 1990. Fatty acids monitoring in the anaerobic depuration of olive oil mill wastewater. *Biol. Wastes* 32, 253–263.

Mostafa, A., Elsamadony, M., El-Dissouky, A., Elhusseiny, A., Tawfik, A., 2017. Biological H2 potential harvested from complex gelatinaceous wastewater via attached versus suspended growth culture anaerobes. *Bioresour. Technol.* 231, 9–18. doi:10.1016/j.biortech.2017.01.062.

Najafpour, G.D., Zinatizadeh, A.A.L., Mohamed, A.R., Isa, M.H., 2006. High-rate anaerobic digestion of palm oil mill effluent in an upflow anaerobic sludge-fixed film bioreactor. *Process Biochem.* 41, 370–379. doi:10.1016/j.procbio.2005.06.031

Nielsen, H.B., Ahring, K.A., 2006. Responses of the biogas process to pulses of oleate in reactors treating mixtures of cattle and pig manure. *Biotechnol. Bioeng.* 95, 96–105. doi:10.1002/bit.

Palatsi, J., Affes, R., Fernandez, B., Pereira, M.A., Alves, M.M., Flotats, X., 2012. Influence of adsorption and anaerobic granular sludge characteristics on long chain fatty acids inhibition process. *Water Res.* 46, 5268–5278. doi:10.1016/j.watres.2012.07.008.

Palatsi, J., Laureni, M., Andrés, M.V., Flotats, X., Nielsen, H.B., Angelidaki, I., 2009. Strategies for recovering inhibition caused by long chain fatty acids on anaerobic thermophilic biogas reactors. *Bioresour. Technol.* 100, 4588–4596. doi:10.1016/j.biortech.2009.04.046.

Ragauskas, A.M.E., Pu, Y., Ragauskas, A.J., 2013. Biodiesel from grease interceptor to gas tank. *Energy Sci. Eng.* 1, 42–52. doi:10.1002/ese3.4.

Rasit, N., Idris, A., Harun, R., Wan Ab Karim Ghani, W.A., 2015. Effects of lipid inhibition on biogas production of anaerobic digestion from oily effluents and sludges: An overview. *Renew. Sustain. Energy Rev.* 45, 351–358. doi:10.1016/j.rser.2015.01.066.

Rinzema, A., Alphenaar, A., Lettinga, G., 1993. Anaerobic digestion of long-chain fatty acids in UASB and expanded granular sludge bed reactors. *Process Biochem.* 28, 527–537.

Rodríguez-Méndez, R., Bihan, Y. Le, Béline, F., Lessard, P., 2017. Long chain fatty acids (LCFA) evolution for inhibition forecasting during anaerobic treatment of lipid-rich wastes: Case of milk-fed veal slaughterhouse waste. *Waste Manag.* 67, 51–58. doi:10.1016/j.wasman.2017.05.028.

Roy, F., Albagnac, G., Samain, E., 1985. Influence of calcium addition on growth of highly purified synthrophic cultures degrading long-chain fatty acids. *Appl. Environ. Microbiol.* 49, 702–705.

Saatci, Y., Arslan, E., Konar, V., 2003. Removal of total lipids and fatty acids from sunflower oil factory effluent by UASB reactor. *Bioresour. Technol.* 87, 269–272.

Sam-soon, P., Loewenthal, R.E., Wentzel, M.C., Marais, G., 1991. A long-chain fatty acid, oleate, as sole substrate in upflow anaerobic sludge bed (UASB) reactor systems. *Water SA*. 17, 7700.

Samson, O.A., Douglas, G.D., 1995. Calcium-induced destabilization of oil-in-water emulsions stabilized by Caseinate or by β-Lactoglobulin. Journal of food science, 60(2), 399–404.

Schink, B., 1997. Energetics of syntrophic cooperation in methanogenic degradation. *Microbiol. Mol. Biol. Rev.* 61, 262–280.

Soltan, M., Elsamadony, M., Tawfik, A., 2017. Biological hydrogen promotion via integrated fermentation of complex agro-industrial wastes. *Appl. Energy* 185, 929–938. doi:10.1016/j.apenergy.2016.10.002.

Sousa, D.Z., Smidt, H., Alves, M.M., Stams, A.J.M., 2009. Ecophysiology of syntrophic communities that degrade saturated and unsaturated long-chain fatty acids. *FEMS Microbiol. Ecol.* 68, 257–272. doi:10.1111/j.1574-6941.2009.00680.x.

Thaveesri, J., Daffonchio, D., Liessens, B., Vandermeren, P., Verstraete, W., 1995. Granulation and sludge bed stability in upflow anaerobic sludge bed reactors in relation to surface thermodynamics. *Appl. Environ. Microbiol.* 61, 3681–3686.

Weng, C., Jeris, J.S., 1976. Biochemical mechanisms in the methane fermentation of glutamic and oleic acids. *Water Res.* 10, 9–18.

Ziels, R.M., Beck, D.A.C., Stensel, H.D., 2017. Long-chain fatty acid feeding frequency in anaerobic codigestion impacts syntrophic community structure and biokinetics. *Water Res.* 117, 218–229. doi:10.1016/j.watres.2017.03.060.

Section VII

Water Networks

23 Regeneration-Recycling of Industrial Wastewater to Minimise Freshwater Usage with Water Cascade Analysis

Nicholas Nyamayedenga

CONTENTS

23.1 INTRODUCTION

Demand for fresh water has soared in the last two or more decades and negatively impacted the operational costs of both industrial and municipal processes. Mujtaba (2012) says that there is very little fresh water in the world that is pitted against an ever-growing population, water pollution, increasing standards of living and increasing demand for water. This is supported by Buros (2000) and Mujtaba et al. (2017), who state that the scarcity of fresh water is not temporary, but a long-term and substantial problem.

What to do about the scarcity of fresh water is no longer only a technological problem but a social one as well. Unavailability of fresh water has had negative outcomes all over the world where demand for fresh, potable water is known to have caused "water wars" in India, (Heimbuch, 2009) where it was reported that, "Water tankers, public taps are Madhya Pradesh's riot spots. Jeevan Malviaya, wife Sita Bai and son Raju were killed for drawing water from a supply line. The state is on a water-clash alert—50 violent incidents have already been reported this month." Also "Water conservation policy has only started becoming important, as a result of the increasing demand due to tourism, the rising number of immigrants, prolonged drought periods and occasionally bad management of the limited water resources. A recent example of the latter was the need to bring water to the island from

Greece in 2008, using water tankers." (IChemE Water Special Interest Group, 2014). Ferrell (2001) says, "A growing crisis in our world is a lack of freshwater. In fact, it is one of the greatest problems we will face in this new century. Yet five-sixths of the world is filled with water! The problem is how to inexpensively desalinise seawater. Researchers have worked on the problem for years, without success. Extracting salt from ocean water continues to be very expensive. Yet seabirds regularly do it, and without spending a penny. They drink seawater without any problems; for they have glands in their heads which discharge a highly concentrated salt solution into their nostrils, from where it drips back into the sea. With such a built-in desalination plant, seabirds never need to drink freshwater." Perhaps this is where the process of seawater desalination originates.

Seas hold most of the water that is accessible to us. Mujtaba (2012) says that 97% of the world's water is saline. This leaves only 3% as fresh water. Of this fresh water, 97% is held in glaciers Mujtaba (2012), making it difficult to use the water instantaneously. Desalination technology has become very important because it is now an alternative for the ever-increasing cost of conventional water supply technologies (Mujtaba et al., 2017). As naturally done by the seabirds, seawater desalination is the process of removal of the salt from the water by several processes such as (membrane) reverse osmosis and electrodialysis, multi-stage flash distillation, multiple effect distillation, vapour compression distillation and freeze desalination. Increasing use of water also results in the increase of wastewater. Wastewater is also becoming a supplementary source for producing high-quality recycled water. Wastewater is composed of organic and inorganic compounds and dissolved gases. A major problem in recent decades is groundwater contamination (including rivers) via organic chemical compounds resulting in a huge public health concern. These compounds constitute a very large group of toxic contaminants. A number of different techniques are being employed to treat wastewater, including membrane and oxidation processes. In recent years, reverse osmosis (RO) process has been successfully implemented to produce recycled water from wastewater containing toxic and harmful pollutants. Also, catalytic wet air oxidation (CWAO) using a trickle bed reactor (TBR) process has been considered a useful and powerful method for removing toxic compounds such as phenol from wastewater (Fujioka et al., 2014; Mohammed et al., 2016; Al-Obaidi et al., 2017).

Foo et al. (2005) say that the processing industry has had to find the means of reducing freshwater consumption and generation of wastewater because of the need for environmental sustainability and the spiralling costs of freshwater and treatment of effluent. Compounding the problem of water scarcity is contamination of ground water by industrial pollutants as described by Mohammed et al. (2016), who worked on the removal of phenol from wastewater by CWAO. Mohammed et al. (2016) also raise a point that industrial ground water contamination is a public concern showing that it is not only industry that is aware of the scarcity of water but also the community in which the industries operate.

These of course are alarming statistics, if a haphazard approach is employed in the use of water in industry, but if properly managed and integrated approaches are used, water minimisation could be a satisfying and rewarding experience. This is supported by Savelski and Bagajewicz (2000) who acknowledge that there has been a new approach to water usage because of water unavailability, spiralling energy costs and more stringent regulations on industrial discharge. Freshwater usage in industry can be minimised by several methodologies. The first method utilises wastewater in another operation unit where its level of contamination does not interfere with the operations of that unit, a process termed reuse. The second method partially treats wastewater to remove contaminants that could prevent its reuse, and the partially treated water is reused as a source for another unit; this is called regeneration and reuse. Finally, accumulated contaminants are removed by regeneration and the water is recycled through a process called regeneration recycling (Wang and Smith, 1994). This process is most favoured and easily employable to reduce both the flow of fresh water and the generation of wastewater in industry since all or a substantial fraction of the wastewater can be recycled back into the factory shop floor. Thus, process integration is a very important tool for water use minimisation as it can be applied to any or all of the points mentioned earlier. Water cascade analysis (WCA) is a very strong tool that can be used to study the operations of water using networks to minimise the flow of both fresh and wastewater. Recent research has concluded that not all units of a water-using network require fresh water. Take for example, a hotel: the kitchen would require water of the highest purity, but it may not be necessary to use fresh water in the toilets or for watering plants. The wastewater from the kitchen could be used to water plants or other purposes (Manan et al., 2006). Majozi et al. (2005) noted this and employed vertical mass transfer on batch processes of water using networks in which the water leaving a unit at a higher purity level or lower contaminant concentration than the next unit requires was used as a direct water source for that unit. This, however, was used on batch processes where various constraints were at play and is only referenced to demonstrate that water exiting one water-using unit at a low contaminant concentration can be used in another unit whose purity requirements are lower than the previous unit's.

This chapter focuses on wastewater treatment and water-use minimisation employing the WCA of Manan et al. (2006) on an industrial continuous process. Note that the same tool was successfully employed for a winery in Zimbabwe by Nyamayedenga and Mujtaba (2014) to minimise the use of fresh water and the generation of wastewater.

23.2 PROBLEM STATEMENT

In order to minimise the usage of fresh water and to reduce the generation of wastewater in a water-using network, given:

- A continuous system (That is, the flow of wastewater from the factory is continuous, contaminants are also added to the wastewater respectively and hence the treatment plant churns out water from the start of work in the morning until the end of operations in the evening. Contaminants must thus be removed continuously.)
- Existing boundary limits
- Required minimum and maximum contaminant loading of operations
- Partial recycling (If fully recycled, no fresh water would go into the factory at all. Processes that need freshwater would thus starve.)

Determine the minimum freshwater supply to the system in order to satisfy the requirements of the water-using network. It is assumed that the level of contaminants in the effluent from the waste treatment plant has been significantly reduced to the consent limits, so that recycling of the regenerated water is possible. Water coming out of the plant can thus be used elsewhere in the factory where the purity of fresh water is not required.

23.3 WATER CASCADE ANALYSIS

Water Cascade Analysis (WCA) is a tool developed by Manan et al. (2006) to analyse and minimise water usage in a mosque. Before commencing any work, the flow pattern of fresh water and wastewater generation must be determined. Rather than using individual contaminant concentrations of wastewater, biological oxygen demand (BOD) is used to determine the concentration levels, which are then ranked in ascending order. These concentration levels then form concentration intervals between one concentration level and the next. It is important to note that the lowest concentration level is the highest purity level, with the highest purity level having a BOD of zero (Manan et al. 2006 and Nyamayedenga and Mujtaba 2014); water with a BOD of zero would, therefore, be fresh water. It is remarkable to note that purity intervals are used to check if the water at that interval can be used in the next unit without affecting operations. Using the results of this analysis, a water cascade diagram that incorporates purity levels and corresponding flow rates is drawn. This then determines the interval freshwater demands showing how much water

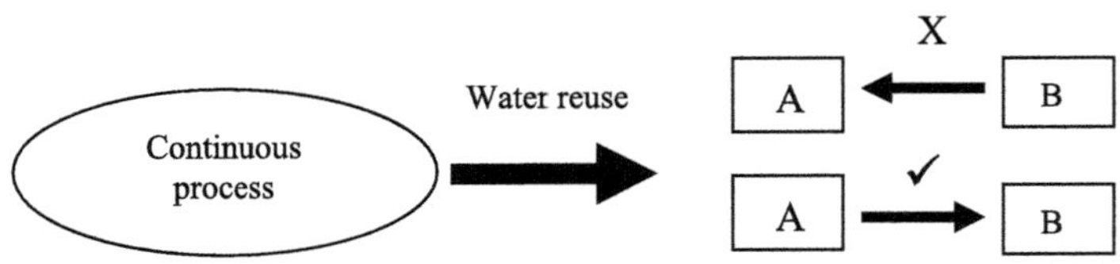

FIGURE 23.1 Direct water cascading on continuous process. (Adapted from Majozi, T. et al., *J. Environ. Manage.*, 78, 317–329, 2005. With permission.)

is required at each purity level. Figure 23.1 shows a simple example of direct water cascading for a continuous process; this assumes, however, that the purity level of A is higher than that of B. Hence water exiting unit A can safely be used as an inlet to unit B, but not the other way round.

Figure 23.2 is an existing network with three processes requiring water at purity levels of 0.999990 (30 ppm), 0.999950 (100 ppm) and 0.999900 (150 ppm), respectively. All three units are unnecessarily using fresh water, but only the first unit would require fresh water while the other two at 100 and 150 ppm could use the 50 ppm water leaving the first unit without affecting their processes. In Figure 23.2, 550 kg/min wastewater is generated, while only 300 kg/min wastewater is generated in Figure 23.3. This is a remarkable 45.5% reduction in wastewater generation, assuming that there are no leakages, no contamination from surrounding areas, and that, as in Figure 23.1, subsequent processes do not happen before that of the highest purity level.

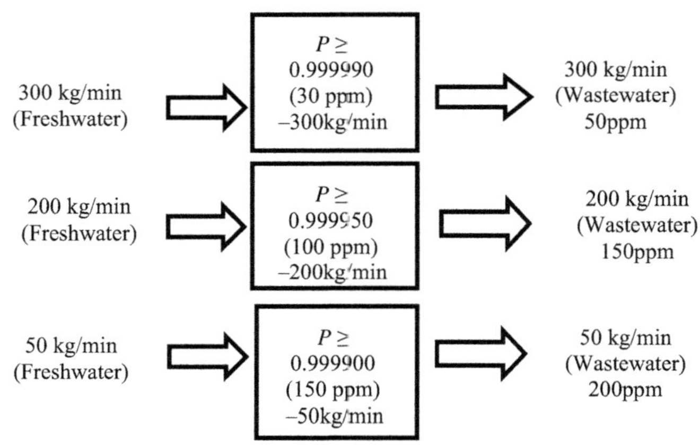

FIGURE 23.2 Water using network. (From Nyamayedenga, N. and Mujtaba, I.M., Minimisation of water usage in industry using water cascade analysis: A winery case study, *Proceedings of the International Conference on Chemical Engineering*, BUET, Dhaka, Bangladesh, pp. 218–222, 2014.)

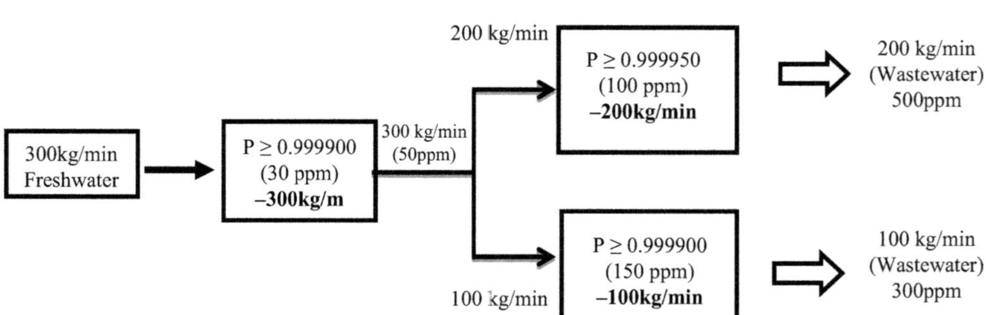

FIGURE 23.3 Direct water cascading. (From Nyamayedenga, N. and Mujtaba, I.M., Minimisation of water usage in industry using water cascade analysis: A winery case study, *Proceedings of the International Conference on Chemical Engineering*, BUET, Dhaka, Bangladesh, pp. 218–222, 2014.)

23.4 CALCULATIONS IN WCA

The following set of equations was used to draw an interval water balance table to determine the net water source or water demand at each purity level (Manan et al. 2006). Equations 23.1 through 21.3 can also be applied to the conditions in Figure 23.3.

$$P = \frac{1{,}000{,}000 - C}{1{,}000{,}000} \tag{23.1}$$

Assuming the highest possible pure water contaminant concentration of 1,000,000 ppm (Iancu, 2007; Manan et al., 2006) the purity (P) of a stream is thus described by Equation 23.1 where C is the concentration of water at that purity level (P_k)

$$n = N_D + N_S - N_{\text{DP}} \tag{23.2}$$

n is the number of purity levels in a unit, where in Equation 23.2 N_D and N_S are number of pure water demands and sources, respectively. In Figure 23.3 there are two pure water demands and one pure water source. This gives three purity levels. N_{DP} is the number of duplicate concentration levels, of which there are none in Figure 23.3.

$$\Delta P = P_n - P_{n+1} \tag{23.3}$$

ΔP becomes the difference in purity levels. Equation 23.4 applies to the water cascade diagram in Figure 23.3 to determine the flow of fresh water (F_{FW}) for a network.

$$F_{\text{FW}} = \frac{\text{cummulative pure water surplus/deficit}}{P_{\text{FW}} - P_k} \tag{23.4}$$

These equations are used as developed by Manan et al. (2006) without modification.

23.5 KARIMBA WINERY EXAMPLE

Using the data supplied, Nyamayedenga and Mujtaba (2014) studied the water distribution network of Karimba Winery (KW), an industrial establishment in Zimbabwe, using WCA on the principles developed by Manan et al. (2006) to minimise water usage in a plant. Before the study by Nyamayedenga and Mujtaba (2014), the winery was using a staggering 73.12 t/day of fresh water, with a wastewater generation of 31.32 t/day. Table 23.1 is a summary of the winery's water

TABLE 23.1
Summary of Water Demands for KW

Stream	Demands Description	Flow Rate (t/d)	BOD Concentration (ppm)
D1	WB	5.36	0
D2	WP	19.81	0
D3	CDP	31.25	0
D4	BW	9.75	10
D5	De-alkaliser	2	10
D6	Boiler	6.6	10
D7	Toilets	0.35	10

TABLE 23.2
Summary of Water Sources for KW

Stream	Source Description	Flow Rate (t/d)	BOD Concentration (ppm)
S1	WB	0.021	4
S2	Rainwater	12.9	10
S3	CDP	1.25	15
S4	Boiler	6.6	22
S5	BW and De-alkaliser	8.75	581
S6	WP	6.7	986

demands at the system audit. After the system audit, six possible water sources for the KW were established that would be used to reduce both the flow of fresh water and the generation of wastewater when redesigning the water-using network. BOD was used to compare water contaminant levels. Table 23.2 is a summary of the possible KW water sources.

Nyamayedenga and Mujtaba (2014) used a system model for the winery that determined a freshwater supply sufficiency of 53.35 t/day after the retrofit design. This was a good 27% reduction of the freshwater flow, while wastewater generation realised a 68.8% reduction from 31.32 t/day to only 9.75 t/day. A sensitivity test was also carried out to test the system response with an increase or decrease in contaminant level. The system called for more fresh water as the contaminant concentration rose, while the opposite happened when contaminant levels decreased (see Figure 23.6). However, it is important to note that the establishment did not generate any harmful or toxic waste, which reduced the operational costs in treating the wastewater. The bulk of the operational costs came from energy used to pump the water from a borehole about two miles from the plant.

23.6 SILCHROME PLATING, LTD, WATER DISTRIBUTION NETWORK

This work sought to improve and minimise freshwater usage at Silchrome Plating, LTD (SPL), by carefully and systematically analysing the whole water distribution system in the plant. Quite a large amount of water was expended through process rinses to the effluent plant via sumps 1 and 2 from the plant's surface finishing areas (Figure 23.4). Rinses are categorised into running and static, with running rinses having a holdup capacity of 33.5 t/time of running rinse water and 11.5 t/time of static rinse/dragout water. Approximately 40 t/day of wastewater was discharged and wasted into the sewer after treatment. This waste was 100% from the flowing rinses in the plant, so the flowing rinse water was immediately identified as a source for water reuse in the quest to minimise freshwater usage at SPL, because not all processes in the factory required the use of fresh water.

There was also the problem of lack of water in the factory when the effluent plant went into recirculation, due to high levels of metal concentration in the discharge from the plant or to the pH being outside of consent limits. Here, various limiting factors were at play. (1) The capacity of the plant was limited, so it could not properly settle the sludge in the settler tank before discharge to the sewer. This resulted in the wastewater from the two sumps being automatically shut off, which meant no water would go into the sumps lest they spilled contaminated water into the environment. To fix this problem, the capacity of the plant had to be expanded by increasing the residence time of the treated water in the settler tank. This can be summed up by the following equation, Time = Volume/Volumetric Flow OR $T = V/Q$.

There was in operation one 8,500 L settler tank and one 10,500 L sludge tank. With a wastewater flow of about 58 L/min, the residence time of the settler tank was about 2.5 h, which

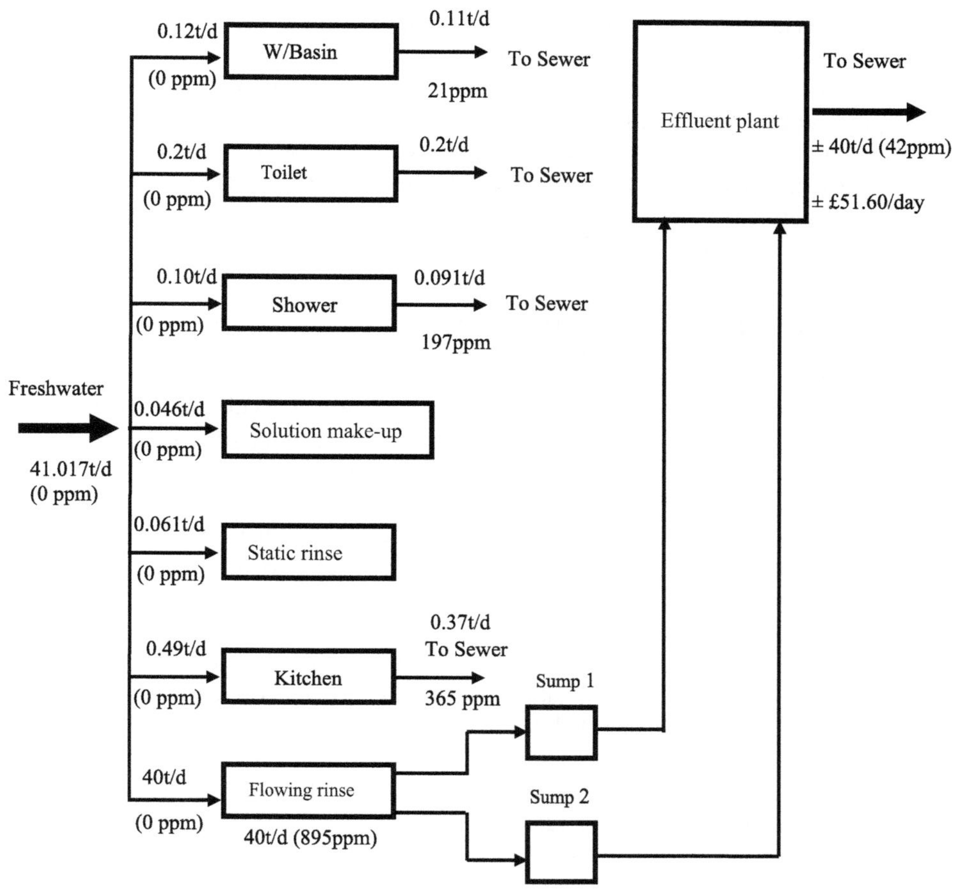

FIGURE 23.4 Existing water distribution network at Silchrome Plating, LTD.

resulted in a large quantity of suspended solids in the discharge wastewater. These suspended solids consequently held a significant concentration of metals intended for removal. This study proposed converting the sludge tank into a settler tank so that the capacity of the plant would increase to 19,500 L, thereby significantly increasing the residence time. After this conversion, the residence time increased to about 5.5 h, which saw the amount of suspended solids fall from 114 to below 20 ppm. Subsequently, the concentration of metal contaminants in the effluent decreased to below the consent limit. (2) The flocculation of metal contaminants is heavily dependent on the pH of the influent stream. Hexavalent chromium is carcinogenic, so it must be reduced to trivalent chromium before it can be discharged into the environment in whatever concentration permitted by the regulating authority. This reduction is done by sodium metabisulphite at a pH less than 3, whereas copper and nickel are precipitated out of solution at pHs of about 8.5 and 10, respectively; a balance therefore has to be reached where both can be precipitated without compromising the other, since they are treated in the same tank. Chromium is treated in a separate tank and only joins the high pH stream after reduction. Cyanide is destroyed in a separate tank by sodium hypochlorite at a pH higher than 10 before the treated wastewater can join the other streams. (3) In light of 1 and 2, the streams had, accordingly, to be separated from the factory before converging at the wastewater treatment plant; that is, alkaline-, acidic- and cyanide-bearing streams must be separated both for health and safety and for ease of treatment at the effluent treatment plant. We shall not dwell very much on the chemical reactions and manipulation of these factors at play, since this chapter focuses on wastewater reduction and minimisation of freshwater flow.

Figure 23.4 is a schematic representation of the water distribution network at SPL before this study. Mapping out the water distribution network as in Figure 23.4 clearly showed that 100% of all wastewater from the plant was going down the drain, yet three possible sources of reusable water were present: namely, the kitchen, the wash basin and the flowing rinse. There was also an alarming cost of approximately £51.60/day (about US$68/day) or £14,700/year (US$17,096/year), based on a 250-day working year at 8 h a day.

23.7 WCA ON SILCHROME PLATING, LTD

Not all processes at SPL required the use of fresh water. There are a total of seven processes or functions that require water at SPL, as shown in Table 23.3. We shall call these water demands. As shown in Table 23.3, the flowing rinse takes up quite a large share of the fresh water used at 40 t/day, whereas solutions can keep up to 5 years, as long as they are not contaminated. The static rinse/dragout tanks are emptied into intermediate bulk containers (IBCs) and disposed of to buyers about once every year. This yields about 950m^3 of flowing water every month, a total cost of approximately £1225.50/month. (The exchange rate at the time of writing and submission was GBP £1 = US$1.1629.) As no chemical reaction occurs at rinsing, there is no need for fresh water to be used for this purpose.

Rinsing water, together with toilet flushing, can be recycled water from the effluent plant. Looking at Table 23.3, the kitchen and solution make-up require the use of fresh water, leaving the shower, wash basin, flowing rinse, static rinse and toilet requiring recycled water. Table 23.4 displays the possible sources of recycled water at SPL. There are seven pure water demands in Tables 23.3 and 23.4, and pure water sources in Table 23.4. The data used in Table 23.4 for the acceptable pure water concentrations for various purposes is the same as used by Manan et al. (2006) and Nyamayedenga and Mujtaba (2014).

TABLE 23.3
Freshwater Demands at Silchrome Plating, LTD

Demand	Description	Water Usage (t/d)	Concentration (ppm)
D1	Kitchen	0.49	0
D2	Solution make-up	0.046	0
D3	Flowing rinse	40	10
D4	Static rinse	0.061	10
D5	Shower	0.10	10
D6	Wash basin	0.12	10
D7	Toilet	0.2	10

TABLE 23.4
Possible Water Sources at Silchrome Plating, LTD

Source	Description	Flow (t/d)	Concentration (ppm)
S1	Wash basin	0.12	21
S2	Showering	0.10	197
S3	Kitchen	0.49	365
S4	Flowing rinse	40	895

23.8 SETTING WATER TARGETS FOR SILCHROME PLATING, LTD

An excellent feature of WCA is its ability to predict and set water targets before redesign and implementation. These water targets are pure water demands, sources and corresponding contaminant concentrations. To determine these targets, an interval water balance table was constructed using data from Tables 23.3 and 23.4, applying the same technique used by Manan et al. (2006) and Nyamayedenga and Mujtaba (2014). This determines a balance between the water demands and sources within the water distribution system. This balance is struck when the pure water demand shifts from being a deficit to becoming a surplus or source. The point at which this happens is called the pinch point of the system (Manan et al. 2006), and usually occurs at a certain water purity level and concentration.

Table 23.5 is a combination of three tables namely: the interval water balance table (Table 23.5a), freshwater targets (Table 23.5b) and feasible water cascade table (Table 23.5c). These tables were combined to form, simulate and test sensitivity for the model of the WCA for SPL. The interval water balance table has already been explained, while Table 23.5b has an assumed injection of fresh water of 0 t/d to test the system's feasibility. Here the interval freshwater demand figures are all negative through to the flow of wastewater (F_{WW}). This only shows that the system is not feasible because of the 0 t/d injection of fresh water (Manan et al., 2006). However, injecting the system with the absolute value of the least negative interval freshwater demand from Table 23.5b, the system yielded a positive F_{WW} as shown in Table 23.5c. The system became feasible after the pinch point; thus the model for SPL determined a freshwater supply of 19.496 t/d to satisfy the requirements of the network.

23.9 REDESIGNING THE NETWORK

A proper analysis and layout of water usage at SPL before this study is shown in Figure 23.4, where all units were being supplied with fresh water. Figure 23.5 is the retrofitted design for SPL. Note that water lost through spillages and evaporation is not accounted for.

It is clear from Figure 23.5 that fresh water was solely directed to the kitchen, shower, wash basin and solution make-up, while the regenerated recycled water went to static rinses, flowing rinses and toilet flushing. After regeneration of the effluent, the discharge water before carbon filtration had a BOD concentration of approximately 42 ppm. It then came out of the carbon filter at a concentration of approximately 5 ppm, which by far surpassed the purity requirements of the toilet and static and flowing rinses. It was not possible to recycle all of the discharge water, because the plant is automated such that if the effluent plant and factory were self sufficient in water needs then other processes that require fresh water would be starved, so about half of the discharge was allowed to go to the drain as shown in Figure 23.5.

23.10 SENSITIVITY TEST OF THE SILCHROME PLATING, LTD, MODEL

As the system approaches the pinch point it slowly moves out of starvation into sufficient supply of water. For the SPL model, the pinch point occurred at purity interval P_4 ($P_k = 0.999803$; 197 ppm) as shown in Table 23.5. The pure water surplus/deficit just before the pinch point is −0.000082 t/d, while just after the pinch point it is 0.000000 t/d then 19.98144 t/d. Now, what happens if the system is disturbed by varying levels of contaminant loading? This question needs to be answered to validate the model, so a sensitivity test was conducted to investigate the different concentrations of contaminants at the pinch point. As production continues in the factory, so too are the flowing rinses contaminated with toxins. This also is apparent in the static rinses at SPL, where the levels of toxins rise as more rinses are done, such that when this water is finally released to the effluent plant the BOD shoots up and forces the plant to go into recirculation since the water cannot be released into the environment.

TABLE 23.5
Silchrome Plating WCA Model

Interval Water Balance Table (a)							
Column	1	2	3	4	5	6	7
Interval Water Balance (n)	Conc. C_n (ppm)	Purity P_n	ΔP	Σ_D (t/d)	Σ_S (t/d)	$\Sigma_D + \Sigma_S$ (t/d)	Net Water *s/d*
	0	1.000000		–0.076		–0.076	Demand
1			0.000010				
	10	0.999990		–40.131	20	–20.131	Demand
2			0.000011				
	21	0.999979			0.12	0.12	Source
3			0.000176				
	197	0.999803			0.1	0.1	Source
4			0.000168				
	365	0.999635			0.49	0.49	Source
5			0.000530				
	895	0.999105			20	20	Source
6			0.999105				
FWC	1000000	0.000000					

(*Continued*)

TABLE 23.5 (*Continued*)
Silchrome Plating WCA Model

Freshwater Targets (b)					Feasible Water Cascade (c)			
Water Cascade		Pure Water cascade			Water Cascade		Pure Water Cascade	
$F_{FW} = 0$ t/d	Cumu. Water s/d. F_C (t/d)	Pure water surp/def (t/d)	Cumu. Pure water surp/def (t/d)	Interval Fresh water demand F_{FW} (t/d)	$F_{FW} = 19.4963$ t/d	Cumu. Water s/d F_C (t/d)	Pure water surp/def (t/d)	Cumu. Pure water surplus (t/d)
↓					↓			
−0.076					−0.08	19.42	0.000194	
↓	−0.076	−0.000001			↓		↓	0.000194
−20.131		↓			−20.13	−0.71	−0.000008	
↓	−20.207	−0.000222	−0.000001	−0.0760	↓		↓	0.000186
0.12		↓			0.12	−0.59	−0.000104	
↓	−20.087	−0.003535	−0.000223	−10.6208	↓		↓	0.000082
0.1		↓			0.10	−0.49	−0.000082	
↓	−19.987	−0.003358	−0.003758	−19.0779	↓		**↓PINCH**	**0.00000**
0.49		↓			0.49	0.00	0.000000	
	−19.497	−0.010333	−0.007116	**−19.49634**	↓		↓	−3.485E-07
20		↓			20.00	20.00	19.98144	
↓								
F_{WW} = **0.503**			−0.017450	−0.01745	F_{WW} =**19.9993**			

Note: **Cumu** = cumulative; **sur** = surplus; **def** = deficit; **s** = source; **d** = demand.

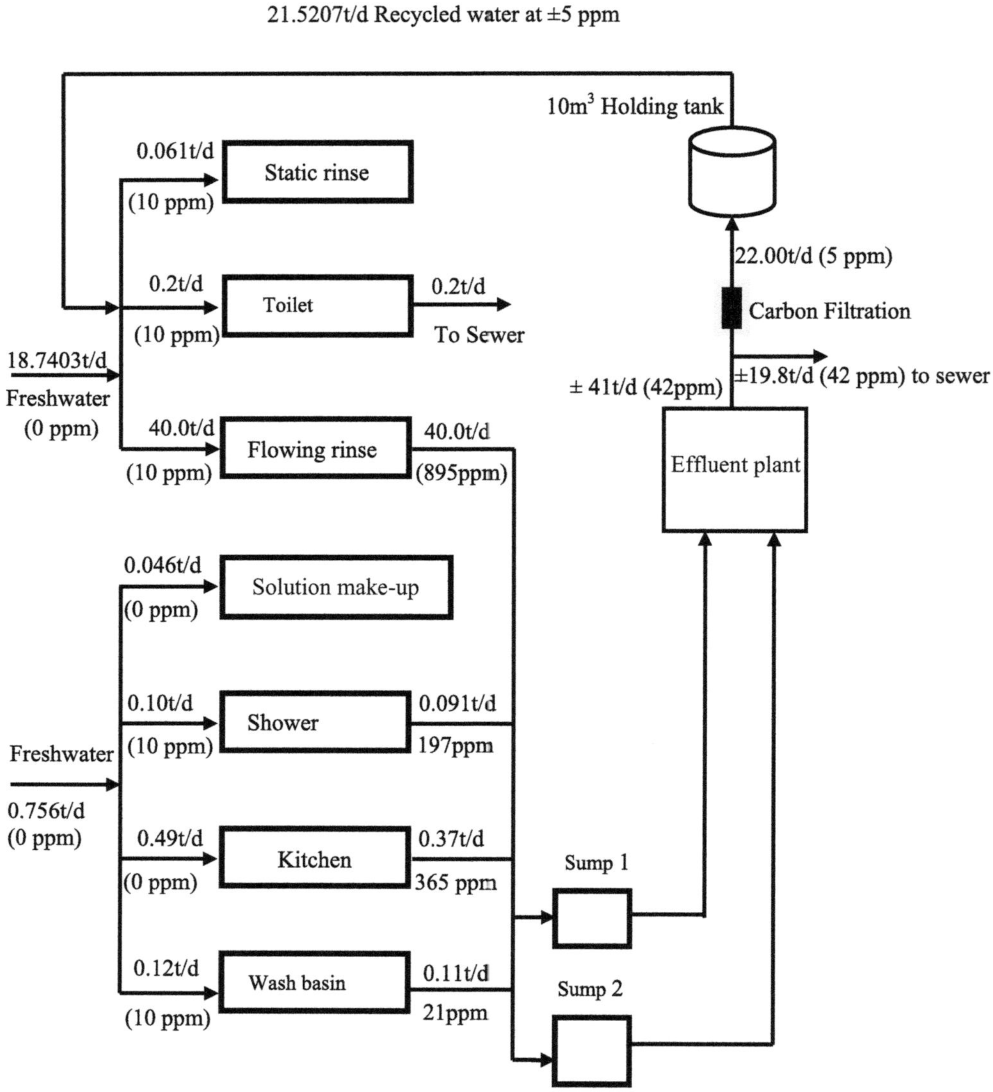

FIGURE 23.5 Retrofitted water using network design for Silchrome Plating, LTD.

Figure 23.6 is the KW model response to disturbances in contaminant concentration obtained at the pinch purity interval, while Figure 23.7 represents the SPL model response as the level of contamination is varied at the pinch point. Figure 23.6 is a negative curve because the values were taken as they are from the infeasible freshwater targets table. Using absolute values makes the curve positive as in Figure 23.7. Freshwater demand increased as contaminant concentration increased. The same pattern was observed when both the Mosque model by Manan et al. (2006) and the KW model by Nyamayedenga and Mujtaba (2014) were subjected to the same contaminant concentration disturbances. Since the system model responded to disturbances as expected, the model for SPL is valid and was therefore used to predict and determine the minimum freshwater requirement for the plant.

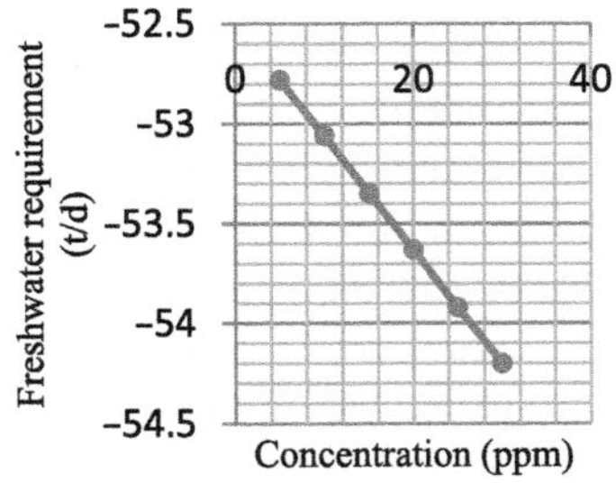

FIGURE 23.6 Freshwater demand at purity level 0.999985 (Karimba Winery).

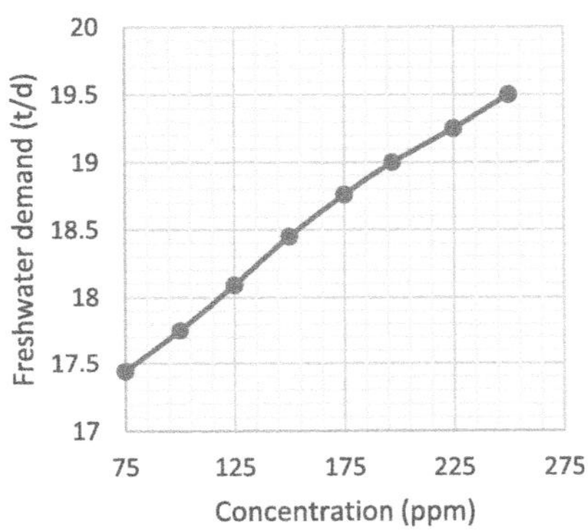

FIGURE 23.7 Freshwater demand at purity level P_4.

23.11 ECONOMIC CONSIDERATIONS

By reviewing the water distribution network of SPL, setting the minimum water targets through the WCA technique as developed by Manan et al. (2006) and calculating the new water network, it was possible to integrate and minimise water usage at the surface technology plant. The plant used approximately 41 t/d freshwater, with wastage of about the same magnitude every single working day before this study. WCA of the plant determined a freshwater requirement for the plant of 19.496 t/d. The fresh water was then distributed to processes that required freshwater, while those processes that did not necessarily require fresh water received the regenerated recycled water. This saw a reduction in the production of wastewater from about 40 to 19.8 t/d. This is a saving in freshwater usage of about 52.45% and a reduction in wastewater flow of 50.5%.

The effluent plant discharged about $40m^3$ of water to the sewer per day, amounting to about £51.60/day going down the drain before this study. The cost of freshwater supply from the mains is £1.29/m^3. The water authority also charges about £7,500/yr for the removal of effluent water to the sewer. Reducing the flow of wastewater by about 50.5% also cut the cost of effluent water removal by the same margin, further reducing the cost of this utility. This project was implemented in January 2015 and has since paid back through a period of 13 months.

23.12 CONCLUSION

Manan et al. (2006) developed and applied the technique of WCA on the activities of a mosque, and recommended it to most buildings and offices. Nyamayedenga and Mujtaba (2014) studied the activities of a winery, an industrial process, using the same technique of WCA and managed to reduce the flow of fresh water and the generation of wastewater by 27% and 68.87%, respectively. In this study, WCA has again been successfully implemented for the activities of a more sophisticated industrial water-using network and minimise both the usage of freshwater and the generation of wastewater. By regenerating and recycling water at SPL, the WCA technique reduced the flow of fresh water and the generation of wastewater by 52.45% and 50.5%, respectively. This makes WCA a very powerful and versatile tool for the minimisation of water usage in industry.

NOMENCLATURE

P = Purity level
ΔP = Difference in purity levels
n = Number of purity levels
N_S = Number of pure water sources
C = Concentration of pure water
Pn = Purity level n
F_{FW} = Flow of freshwater
F_{WW} = Flow of wastewater
P_{FW} = Purity of freshwater
FWC = Freshwater concentration
Σ_S = Sum of sources flow rate
Σ_D = Sum of demands flowrate
N_D = Number of pure water demands
N_{DP} = Number of duplicate concentration levels

REFERENCES

Al-Obaidi, M.A., Kara-Zaïtri, C. and Mujtaba, I.M. (2017) Wastewater treatment by spiral wound reverse osmosis: Development and validation of a two dimensional process model. *Journal of Cleaner Production* 140, 1429–1443.

Buros, O.K. (2000) *The ABCs of Desalting*. 2nd ed. International Desalination Association: Riyadh, Saudi Arabia.

Ferrell, V. (2001) *Science vs. Evolution*. Evolution Facts: Altamont, TN.

Foo, D.C.Y., Manan, Z.A. and Tan, Y.L. (2005) Synthesis of maximum water recovery network for batch process systems. *Journal of Cleaner Production* 13, 1381–1394.

Fujioka, T., Khan, S.J., Mcdonald, J.A. et al. (2014) Modelling the rejection of N-nitrosamines by a spiral-wound reverse osmosis system: Mathematical model development and validation. *Journal of Membrane Science* 454, 212–219.

Heimbuch, J. (2009) Violence over water already happening in India. Available at: http://www.treehugger.com/clean-water/violence-over-water-already-happening-in-india.html.

Iancu, P. (2007) Process Integration for water minimisation in oil processing and petrochemistry. University Politechnica of Bucharest, Bucharest.

IChemE Water Special Interest Group. (2014) *Wet News*. London, UK.

Majozi, T. (2005) Wastewater minimisation using central reusable water storage in batch plants. *Computers & Chemical Engineering* 29, 1631–1646.

Majozi, T., Brouckaert, C.J. and Buckley, C.A. (2005) A graphical technique for wastewater minimisation in batch processes. *Journal of Environmental Management* 78, 317–329.

Manan, Z.A., Wan Alwi, S.R. and Ujang, Z. (2006) Water pinch analysis for an urban system: A case study on the Sultan Ismail Mosque at the Universiti Teknologi Malaysia (UTM). *Desalination* 194, 52–68.

Mohammed, A.E., Jarullah, A.T., Gheni, S.A. and Mujtaba, I.M. (2016) Optimal design and operation of an industrial three phase reactor for the oxidation of phenol. *Computers and Chemical Engineering* 94, 257–271.

Mujtaba, I.M. (2012) The role of PSE community in meeting sustainable freshwater demand of tomorrow's world via desalination, In *Computer Aided Chemical Engineering*, Vol. 31, I.A. Karimi and R. Srinivasan (Eds.), pp. 91–98, Elsevier, Amsterdam, the Netherlands.

Mujtaba, I.M., Salih M. Alsadaie, Mudhar A. Al-Obaidi, Raj Patel, M. T. Sowgath, and Davide Manca (2017) *Model-Based Techniques in Desalination Processes: A Review.* Chapter 1 in The Water-Food-Energy Nexus, I.M. Mujtaba, R. Srinivasan, and N. Elbashir (Eds.), CRC Press: Boca Raton, FL.

Savelski, M.J. and Bagajewicz, M.J. (2000) On the optimality conditions of water utilization systems in process plants with single contaminants. *Chemical Engineering Science* 55, 5035–5048.

Nyamayedenga, N. and Mujtaba, I.M. (2014) Minimisation of water usage in industry using water cascade analysis: A winery case study. *Proceedings of the International Conference on Chemical Engineering*, BUET, Dhaka, Bangladesh, 218–222.

Wang, Y.P. and Smith, R. (1994) Wastewater minimisation. *Chemical Engineering Science* 49(7), 981–1006.

24 Total Site Water Integration Considering Multiple Water Reuse Headers

Ahmad Fikri Ahmad Fadzil, Sharifah Rafidah Wan Alwi, Zainuddin Abdul Manan, and Jiří Jaromír Klemeš

CONTENTS

24.1 INTRODUCTION

In total site water integration (TSWI) or interplant water integration, it is desirable to consider all integration possibilities of water exchange among sources and demands across industrial sites or regions on a wider scale. In practice, free or unrestrictive integration across the total site may be a major challenge to practically implement, due to the possibility of dealing with a highly complex water network across the total site that may not have considered the constraints of geography, layout, pressure and costs limitations and even the issues of policies, confidentiality and governance across companies. Consequently, the TSWI concept is introduced as one of the solutions to this problem. Total site centralised water integration (TS-CWI) involves the integration of water supply, demand and end-of-pipe treatment across the total site (TS). The key concept is the use of a centralised water reuse header that is managed and operated by a third party across the TS. A centralised water reuse header refers to water pipeline system that heads to the centralised wastewater treatment facilities as the end-of-pipe solution. The idea of using a centralised header has been applied to carbon total site planning by Mohd Nawi et al. (2016).

All the plants that are located across the TS could sell their wastewater to a specific header that accepts a certain range of concentrations and decides the amount of water sources to be extracted from the header to satisfy the water demands of their own plants. The accumulated water sources

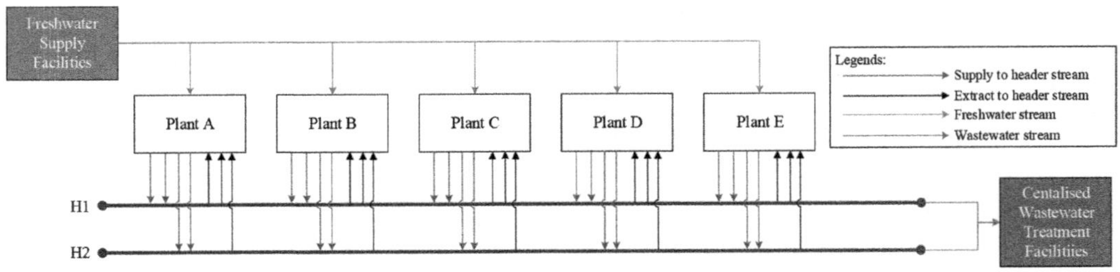

FIGURE 24.1 Illustration of TS-CWI network and arrangement of plants along the centralised water reuse header.

can only move in one direction along the header, aided by the water pump(s). The remaining unutilised water sources at the end of the header are sent to the centralised wastewater treatment facilities. Figure 24.1 illustrates a TS-CWI network and the arrangement of plants along the centralised water reuse header.

The significant aspect of TS-CWI is that it can protect the confidentiality between plants where they can opt to perform the methodology on their own. They only need to inform the centralised water reuse header system owner of the amount of wastewater they would like to sell and the amount of water sources they would like to extract from the centralised water reuse headers. The usage of centralised water reuse headers could minimise the number of water interconnections (piping) and pumping that make water networks very costly.

24.2 METHODOLOGY

The TS-CWI methodology for targeting the minimum fresh water required and wastewater generated is developed using a numerical method. The detailed methodology is explained as follows.

24.2.1 Step 1: Data Extraction

Water sources and the demands of the individual plant located along the centralised water reuse header are collected. Water sources refer to the outlet wastewater that is generated by a unit operation. Water demands refer to the inlet water that is required by a unit operation. The limiting water data consisting of stream flow rate and concentration of contaminants are extracted. The concentration of contaminants in water is an indicator of the quality of water. It is usually measured in ppm of total suspended solids (TSS), biological oxygen demand (BOD), chemical oxygen demand (COD) or others.

The water network can be modelled as having a single contaminant or multiple contaminants. Water network systems with multiple contaminants are more practical compared to single contaminant networks, but involve a complex mathematical optimisation technique. Therefore, by assigning boundaries and constraints, the water network system having multiple contaminants can be modelled as a single contaminant, that is, a pseudo-single contaminant system (Liu et al., 2004). The method is to select one limiting contaminant based on the contaminants existing in the extracted data. The other contaminants that are not considered in the water network system must be within the boundaries and constraints set earlier. If some contaminants are not within the limit, some pretreatment can be done on the water stream.

24.2.2 Step 2: Centralised Water Reuse Header Allocation

Based on the water quality that the water demands can accept, the number of centralised water reuse headers and their quality limit are set by the centralised system provider. For example, the first centralised water reuse header can be set for high purity water streams and be acceptable to be reused for toilet flushing purposes. The second centralised water reuse header is set for lower purity water streams for process cooling and gas scrubbing purposes. The purity range set should be ensured not to affect the equipment process operation, or cause fouling, bad odour and bacteria accumulation.

24.2.3 Step 3: Construction of Water Reuse Header Table

The water reuse header (WRH) table is constructed to determine the amount of water source received, extracted and unutilised by each plant. Water sources from each plant are distributed to the suitable centralised water reuse header. Water sources with a concentration higher than the upper limit concentration of the centralised water reuse headers are sent directly to the centralised wastewater treatment facilities. The calculation is conducted based on the number of centralised water reuse headers set out earlier following the arrangement of the plant across the TS.

The accumulated flow rate of the water source received by each plant is calculated using Equation 24.1.

$$F_i^{\text{received}} = \sum F_i^{\text{individual sources}} + F_{i-1}^{\text{unutilised}} \tag{24.1}$$

where F is the flow rate of the water stream in t/h, i is the number of plants.

The accumulated mass load of water source received by each plant is calculated using Equation 24.2.

$$m_i^{\text{received}} = \sum m_i^{\text{individual sources}} + m_{i-1}^{\text{unutilised}} \tag{24.2}$$

where m is the mass load of the water stream in t/h, i is the number of plants.

The accumulated concentration of the water source received by each plant is calculated using Equation 24.3.

$$C_i^{\text{received}} = \frac{\left(m_i^{\text{received}}\right)(1{,}000)}{\left(F_i^{\text{received}}\right)} \tag{24.3}$$

where C is the concentration of the water stream in ppm, i is the number of plants.

The amount of the water source extracted by each plant is taken from the individual total site centralised water cascade tables (TSC-WCT) conducted in Step 4. The flow rate is taken from Column 3, concentration from Column 2 and mass load is calculated using Equation 24.4.

$$m_i^{\text{extracted}} = \frac{\left(F_i^{\text{extracted}}\right)\left(C_i^{\text{extracted}}\right)}{(1{,}000)} \tag{24.4}$$

The flow rate of unutilised water source is calculated using Equation 24.5.

$$F_i^{\text{unutilised}} = F_i^{\text{received}} - F_i^{\text{extracted}} \tag{24.5}$$

The mass load of unutilised water source is calculated using Equation 24.6.

$$m_i^{\text{unutilised}} = m_i^{\text{received}} - m_i^{\text{extracted}} \tag{24.6}$$

The concentration of unutilised water source is the same as concentration of water source extracted earlier.

24.2.4 Step 4: Construction of TSC-WCT

The TSC-CWT is constructed to determine the amount of water exchanged across the TS, and to target the minimum fresh water required and wastewater generated. The TSC-CWT is constructed according to the arrangement of plants across the header. For example, a TSC-CWT for Plant A is constructed first and the remaining unutilised water sources from the centralised water reuse headers is sent to Plant B. The maximum water recovery targeting method by an individual plant is similar to the previous work by Foo (2007) where the lower quality sources are maximised first before the higher quality source is utilised. The main difference is that data used by each plant for TSC-WCT calculation is the cumulated flow rate and concentration of water sources received from the centralised water reuse headers, instead of using individual water sources.

The detailed construction of the TSC-WCT is shown per column as follows.

Column 1: The name of the streams.
Column 2: The concentration (C) is sorted in ascending order.
Column 3: The net flow rate (F_{net}) is calculated using Equation 24.7.

$$\sum F_i^{\text{net}} = \sum F_i^{\text{sources}} - \sum F_i^{\text{demands}} \tag{24.7}$$

Column 4: The cumulative flow rate ($F_{\text{cmltv net}}$) is calculated from top to bottom starting from zero using Equation 24.8.

$$F_i^{\text{cmltv net}} = F_{i-1}^{\text{cmltv net}} + F_i^{\text{net}} \tag{24.8}$$

Column 5: The net mass load (m_{net}) is calculated using Equation 24.9.

$$m_i^{\text{net}} = \frac{\left(F_i^{\text{net}}\right)\left(C_i\right)}{(1{,}000)} \tag{24.9}$$

Column 6: The cumulative mass load ($m_{\text{cmltv net}}$) is calculated from top to bottom starting from zero using Equation 24.10.

$$m_i^{\text{cmltv net}} = m_{i-1}^{\text{cmltv net}} + m_i^{\text{net}} \tag{24.10}$$

Column 7: Flow rate determination of H2 (F_{H2}) is targeted using Equation 24.11.

$$F_{\text{H2}} = \frac{\left(m_i^{\text{cmltv net}}\right)(1{,}000)}{\left(C_i - C_{\text{H2}}\right)} \tag{24.11}$$

Column 8: Flow rate determination of H1 (F_{H1}) is targeted using Equation 24.12.

$$F_{\text{H1}} = \frac{\left(m_i^{\text{cmltv net}}\right)(1{,}000)}{\left(C_i - C_{\text{H1}}\right)} \tag{24.12}$$

Column 9: Flow rate determination of fresh water (F_{FW}) is targeted using Equation 24.13.

$$F_{\text{FW}} = \frac{\left(m_i^{\text{cmltv net}}\right)(1{,}000)}{\left(C_i - C_{\text{FW}}\right)} \tag{24.13}$$

24.3 CASE STUDY

The developed methodology of the TS-CWI is illustrated with a case study that consists of five plants located along the centralised water reuse header. A single-contaminant (TSS) system is considered.

24.3.1 Step 1: Data Extraction

The limiting water data for Plant A is adapted from Chew et al. (2010a) as shown in Table 24.1. The initial fresh water required and initial wastewater generated for Plant A is 300.0 t/h.

The limiting water data for Plant B is adapted from Chew et al. (2010b) as shown in Table 24.2. The initial fresh water required and initial wastewater generated for Plant B is 490.0 t/h.

The limiting water data for Plant C is adapted from Bandyopadhyay and Ghanekar (2006) as shown in Table 24.3. The initial fresh water required and initial wastewater generated for Plant C is 190.0 t/h.

TABLE 24.1
Limiting Water Data for Plant A

1 Number of Streams	2 Name of Stream	3 Type of Stream	4 Flow Rate, *F* (t/h)	5 Concentration, *C* (ppm)	6 Mass Load, *m* (t/h)
1	S-A1	Source	50.0	50.0	2.5
2	S-A2	Source	100.0	100.0	10.0
3	S-A3	Source	80.0	150.0	12.0
4	S-A4	Source	70.0	250.0	17.5
1	D-A1	Demand	50.0	20.0	1.0
2	D-A2	Demand	100.0	50.0	5.0
3	D-A3	Demand	80.0	100.0	8.0
4	D-A4	Demand	70.0	200.0	14.0

Source: Chew, I.M.L. et al., *Ind. Eng. Chem. Res.* 49, 6439–6455, 2010a.

TABLE 24.2
Limiting Water Data for Plant B

1 Number of Streams	2 Name of Stream	3 Type of Stream	4 Flow Rate, *F* (t/h)	5 Concentration, *C* (ppm)	6 Mass Load, *m* (t/h)
5	S-B5	Source	80.0	50.0	4.0
6	S-B6	Source	20.0	100.0	2.0
7	S-B7	Source	50.0	125.0	6.3
8	S-B8	Source	300.0	150.0	45.0
9	S-B9	Source	40.0	800.0	32.0
5	D-B5	Demand	20.0	0.0	0.0
6	D-B6	Demand	80.0	25.0	2.0
7	D-B7	Demand	50.0	25.0	1.3
8	D-B8	Demand	40.0	50.0	2.0
9	D-B9	Demand	300.0	100.0	30.0

Source: Chew, I.M.L. et al., *Ind. Eng. Chem. Res.*, 49, 6456–6468, 2010b.

TABLE 24.3
Limiting Water Data for Plant C

1 Number of Streams	2 Name of Stream	3 Type of Stream	4 Flow Rate, F (t/h)	5 Concentration, C (ppm)	6 Mass Load, m (t/h)
10	S-C10	Source	50.0	200.0	10.0
11	S-C11	Source	50.0	300.0	15.0
12	S-C12	Source	30.0	500.0	15.0
13	S-C13	Source	60.0	500.0	30.0
10	D-C10	Demand	50.0	0.0	0.0
11	D-C11	Demand	50.0	100.0	5.0
12	D-C12	Demand	30.0	100.0	3.0
13	D-C13	Demand	60.0	300.0	18.0

Source: Bandyopadhyay, S. and Ghanekar, M.D., *Ind. Eng. Chem. Res.*, 45, 5287–5297, 2006.

TABLE 24.4
Limiting Water Data for Plant D

1 Number of Streams	2 Name of Stream	3 Type of Stream	4 Flow Rate, F (t/h)	5 Concentration, C (ppm)	6 Mass Load, m (t/h)
14	S-D14	Source	66.7	80.0	5.3
15	S-D15	Source	20.0	100.0	2.0
16	S-D16	Source	100.0	100.0	10.0
17	S-D17	Source	41.7	800.0	33.3
18	S-D18	Source	10.0	800.0	8.0
14	D-D14	Demand	20.0	0.0	0.0
15	D-D15	Demand	66.7	50.0	3.3
16	D-D16	Demand	100.0	50.0	5.0
17	D-D17	Demand	41.7	80.0	3.3
18	D-D18	Demand	10.0	400.0	4.0

Source: Chew, I.M.L. et al., *Ind. Eng. Chem. Res.*, 49, 6456–6468, 2010b.

The limiting water data for Plant D is adapted from Chew et al. (2010b) as shown in Table 24.4. The initial fresh water required and initial wastewater generated for Plant D is 238.4 t/h.

The limiting water data for Plant E is adapted from Chew et al. (2010b) as shown in Table 24.5. The initial fresh water required and initial wastewater generated for Plant E is 151.9 t/h.

From 5 plants, 23 water sources and 23 water demands were identified for integration in the TS-CWI. Without any integration, the initial fresh water required and initial wastewater generated is 1,370.3 t/h. The arrangement of the plants across the TS follows the arrangement as shown in Figure 24.1. The detailed explanation for the calculation of the first plant, Plant A, is shown as per the methodological steps.

24.3.2 Step 2: Centralised Water Reuse Header Allocation

In this case study, two headers were considered: Header 1 (H1) with a concentration in the range of 10 to 150 ppm and Header 2 (H2) with a concentration in the range of 150 to 400 ppm.

TABLE 24.5
Limiting Water Data for Plant E

1 Number of Streams	2 Name of Stream	3 Type of Stream	4 Flow Rate, *F* (t/h)	5 Concentration, *C* (ppm)	6 Mass Load, *m* (t/h)
19	S-E19	Source	66.7	80.0	5.3
20	S-E20	Source	20.0	100.0	2.0
21	S-E21	Source	15.6	400.0	6.3
22	S-E22	Source	42.9	800.0	34.3
23	S-E23	Source	6.7	1,000.0	6.7
19	D-E19	Demand	20.0	0.0	0.0
20	D-E20	Demand	66.7	50.0	3.3
21	D-E21	Demand	15.6	80.0	1.3
22	D-E22	Demand	42.9	100.0	4.3
23	D-E23	Demand	6.7	400.0	2.7

Source: Chew, I.M.L. et al., *Ind. Eng. Chem. Res.*, 49, 6456–6468, 2010b.

24.3.3 Step 3: Construction of the WRH Table

A WRH table is constructed to determine the amount of water source received, extracted and unutilised by Plant A from both H1 and H2. Since Plant A is the first plant located across TS, there is no unutilised water source from H1 and H2 received by Plant A.

The water sources from Plant A are added to H1 based on the concentration as shown in Step 2. S-A1, S-A2 and S-A3 are added to H1 (see Table 24.6, Row 2, 3 and 4). The accumulated flow rate of H1 received by Plant A is 230.0 t/h with a concentration of 106.5 ppm (see Table 24.6, Row 5). The amount of the flow rate of water source extracted by Plant A from H1 is taken from Step 4, which is 155.8 t/h with a concentration of 106.5 ppm (see S-AH1 in Table 24.13). The remaining water source of H1 that is not utilised by Plant A is 74.2 t/h with a concentration of 106.5 ppm (see Table 24.6, Row 7). H1 WRH table is shown in Table 24.6.

The water sources from Plant A are added to H2 based on the concentration as shown in Step 2. S-A4 is added to H2 (see Table 24.7, Row 2). The accumulated flow rate of H2 received by Plant A is 70.0 t/h with a concentration of 250.0 ppm (see Table 24.7, Row 3). The flow rate of water source extracted by Plant A from H2 is taken from Step 4, which is 45.6 t/h with a concentration of 250.0 ppm (see S-AH2 in Table 24.13). The remaining water source of H2 that is not utilised by Plant A is 24.4 t/h with a concentration of 250.0 ppm (see Table 24.7, Row 5). H2 WRH table is shown in Table 24.7.

TABLE 24.6
H1 WRH Table of Plant A

1 Name of Plant	2 Name of Stream	3 Type of Stream	4 Flow Rate, *F* (t/h)	5 Concentration, *C* (ppm)	6 Mass Load, *m* (t/h)
A	S-A1	Source	50.0	50.0	2.5
	S-A2	Source	100.0	100.0	10.0
	S-A3	Source	80.0	150.0	12.0
	S-AH1	Received	230.0	106.5	24.5
	S-AH1	Extracted	−155.8	106.5	−16.6
	S-AH1	Unutilised	74.2	106.5	7.9

TABLE 24.7
H2 WRH Table of Plant A

1 Name of Plant	2 Name of Stream	3 Type of Stream	4 Flow Rate, F (t/h)	5 Concentration, C (ppm)	6 Mass load, m (t/h)
A	S-A4	Source	70.0	250.0	17.5
	S-AH2	Received	70.0	250.0	17.5
	S-AH2	Extracted	–45.6	250.0	–11.4
	S-AH2	Unutilised	24.4	250.0	6.1

24.3.4 Step 4: Construction of TSC-WCT

The TSC-WCT is constructed to determine the amount of water exchange and to target the minimum fresh water required and wastewater generated by Plant A. Step by step calculation for the TSC-WCT for Plant A is shown as follows.

First, TSC-WCT is constructed with zero flow rates of S-AH2, S-AH1 and FW-A to obtain the infeasible TSC-WCT as shown in Table 24.8.

The flow rate of S-AH2 is targeted using the largest negative value of F_{H2}, which is 300.0 t/h, found in the concentration level of 1,000,000.0 ppm (see Table 24.8, Column 7). The value is then added to S-AH2 (see Table 24.9, Column 3). A negative value of $m_{cmltv\ net}$ (see Table 24.9, Column 6) indicates that Plant A needs a higher quality water source, and hence S-AH1 is needed.

The flow rate of S-AH1 is targeted using the largest negative value of F_{H1}, which is 342.3 t/h, found in the concentration level of 200.0 ppm (see Table 24.9, Column 8). The value is then added to S-AH1 (see Table 24.10, Column 3). A negative value of $m_{cmltv\ net}$ (see Table 24.10, Column 6) indicates that Plant A needs a higher quality water source, and hence FW-A is needed.

TABLE 24.8
TSC-CWT for Plant A (Infeasible)

1 Name of Streams	2 C (ppm)	3 F_{net} (t/h)	4 $F_{cmltv\ net}$ (t/h)	5 m_{net} (t/h)	6 $m_{cmltv\ net}$ (t/h)	7 F_{H2} (t/h)	8 F_{H1} (t/h)	9 F_{FW} (t/h)
			0.0					
FW-A	0.0	**0.0**			0.0			0.0
			0.0	0.0				
D-A1	20.0	–50.0			0.0			0.0
			–50.0	–1.5				
D-A2	50.0	–100.0			–1.5			–30.0
			–150.0	–7.5				
D-A3	100.0	–80.0			–9.0			–90.0
			–230.0	–1.5				
S-AH1	106.5	**0.0**			–10.5		0.0	–98.6
			–230.0	–21.5				
D-A4	200.0	–70.0			–32.0		–342.3	–160.0
			–300.0	–15.0				
S-AH2	250.0	**0.0**			–47.0	0.0	–327.6	–188.0
			–300.0	–299,925.0				
WW-A	1,000,000.0	0.0			–299,972.0	**–300.0**	–300.0	–300.0
			–300.0	300,000.0				

TABLE 24.9
TSC-CWT for Plant A (F_{H2} Targeting)

1 Name of Streams	2 C (ppm)	3 F_{net} (t/h)	4 $F_{cmltv\ net}$ (t/h)	5 m_{net} (t/h)	6 $m_{cmltv\ net}$ (t/h)	7 F_{H2} (t/h)	8 F_{H1} (t/h)	9 F_{FW} (t/h)
			0.0					
FW-A	0.0	**0.0**			0.0			0.0
			0.0	0.0				
D-A1	20.0	–50.0			0.0			0.0
			–50.0	–1.5				
D-A2	50.0	–100.0			–1.5			–30.0
			–150.0	–7.5				
D-A3	100.0	–80.0			–9.0			–90.0
			–230.0	–1.5				
S-AH1	106.5	**0.0**			–10.5		0.0	–98.6
			–230.0	–21.5				
D-A4	200.0	–70.0			–32.0		**–342.3**	–160.0
			–300.0	–15.0				
S-AH2	250.0	**300.0**			–47.0	0.0	–327.6	–188.0
			0.0	47.0				
WW-A	1,000,000.0	0.0			0.0	0.0	0.0	0.0
			0.0	–47.0				

TABLE 24.10
TSC-CWT for Plant A (F_{H1} Targeting)

1 Name of Streams	2 C (ppm)	3 F_{net} (t/h)	4 $F_{cmltv\ net}$ (t/h)	5 m_{net} (t/h)	6 $m_{cmltv\ net}$ (t/h)	7 F_{H2} (t/h)	8 F_{H1} (t/h)	9 F_{FW} (t/h)
			0.0					
FW-A	0.0	**0.0**			0.0			0.0
			0.0	0.0				
D-A1	20.0	–50.0			0.0			0.0
			–50.0	–1.5				
D-A2	50.0	–100.0			–1.5			–30.0
			–150.0	–7.5				
D-A3	100.0	–80.0			–9.0			–90.0
			–230.0	–1.5				
S-AH1	106.5	**342.3**			–10.5		0.0	**–98.6**
			112.3	10.5				
D-A4	200.0	–70.0			0.0		0.0	0.0
			42.3	2.1				
S-AH2	250.0	**300.0**			2.1	0.0	14.7	8.5
			342.4	342,287.0				
WW-A	1,000,000.0	0.0			342,289.1	342.4	342.3	342.3
			342.4	–342,372.6				

TABLE 24.11
TSC-CWT for Plant A (F_{FW} Targeting)

1 Name of Streams	2 C (ppm)	3 F_{net} (t/h)	4 $F_{cmltv\ net}$ (t/h)	5 m_{net} (t/h)	6 $m_{cmltv\ net}$ (t/h)	7 F_{H2} (t/h)	8 F_{H1} (t/h)	9 F_{FW} (t/h)
			0.0					
FW-A	0.0	**98.6**			0.0			0.0
			98.6	2.0				
D-A1	20.0	−50.0			2.0			98.6
			48.6	1.5				
D-A2	50.0	−100.0			3.4			68.6
			−51.4	−2.6				
D-A3	100.0	−80.0			0.9			8.6
			−131.4	−0.9				
S-AH1	106.5	**342.3**			0.0		0.0	0.0
			210.9	19.7				
D-A4	200.0	−70.0			19.7		210.9	98.6
			140.9	7.0				
S-AH2	250.0	**300.0**			26.8	0.0	**186.5**	107.0
			440.9	440,833.8				
WW-A	1,000,000.0	0.0			440,860.5	441.0	440.9	440.9
			440.9	−440,944.0				

The flow rate of FW-A is targeted using the largest negative value of F_{FW}, which is 98.6 t/h, found in the concentration level of 106.5 ppm (see Table 24.10, Column 9). The value is then added to FW-A (see Table 24.11, Column 3).

However, the pinch location of S-AH1 has disappeared. This is an indication that the flow rate of S-AH1 may be excessive (upon addition of FW-A). Consequently, the flow rate of S-AH1 needs to be adjusted.

The minimum value of F_{H1} that needs to be subtracted is 186.5 t/h (see Table 24.11, Column 8). The adjusted flow rate is 155.8 t/h and then added to S-AH1 (see Table 24.12, Column 3).

However, the pinch location of S-AH2 has disappeared. This is an indication that the flow rate of S-AH2 may be excessive (upon addition of S-AH1). Hence, the flow rate of S-AH2 needs to be adjusted.

The minimum value of F_{H2} that needs to be subtracted is 254.4 t/h (see Table 24.12, Column 7). The adjusted flow rate is 45.6 t/h and then added to S-AH2 (see Table 24.13, Column 3).

From Table 24.13, the minimum fresh water required is 98.6 t/h (see FW-A, Table 24.13) while utilizing 155.8 t/h of H1 (see S-AH1, Table 24.13) and 45.6 t/h of H2 (see S-AH2, Table 24.13). Plant A TS-CWI optimal TS water network is designed to achieve the minimum targeted fresh water required and wastewater generated as shown in Figure 24.2. Repeating the same calculation for all plants until the last plant, the minimum overall TS fresh water required is 549.4 t/h (59.9% reduction) and the minimum wastewater generated is 549.4 t/h (59.9% reduction). H1 generates 254.8 t/h of wastewater with a concentration of 102.2 ppm, while H2 generates 63.4 t/h of wastewater with a concentration 283.5 ppm to be sent to the centralised wastewater treatment facilities. Source S-B9, S-C12, S-C13, S-D17, S-D18, S-E22 and S-E23 are sent directly to the centralised wastewater treatment facilities because the concentration exceeds the maximum limit set for the headers. The total wastewater sent to the centralised wastewater treatment facilities is 549.4 t/h with a concentration of 370.0 ppm. The optimal TS Water Network for the case study is shown in Figure 24.3.

TABLE 24.12
TSC-CWT for Plant A (F_{H1} Adjustment)

1 Name of Streams	2 C (ppm)	3 F_{net} (t/h)	4 $F_{cmltv\ net}$ (t/h)	5 m_{net} (t/h)	6 $m_{cmltv\ net}$ (t/h)	7 F_{H2} (t/h)	8 F_{H1} (t/h)	9 F_{FW} (t/h)
			0.0					
FW-A	0.0	**98.6**			0.0			0.0
			98.6	2.0				
D-A1	20.0	−50.0			2.0			98.6
			48.6	1.5				
D-A2	50.0	−100.0			3.4			68.6
			−51.4	−2.6				
D-A3	100.0	−80.0			0.9			8.6
			−131.4	−0.9				
S-AH1	106.5	**155.8**			0.0		0.0	0.0
			24.4	2.3				
D-A4	200.0	−70.0			2.3		24.4	11.4
			−45.6	−2.3				
S-AH2	250.0	**300.0**			0.0	0.0	0.0	0.0
			254.4	254,377.3				
WW-A	1,000,000.0	0.0			254,377.3	**254.4**	254.4	254.4
			254.4	−254,441.0				

TABLE 24.13
TSC-CWT for Plant A (F_{H2} Adjustment)

1 Name of Streams	2 C (ppm)	3 F_{net} (t/h)	4 $F_{cmltv\ net}$ (t/h)	5 m_{net} (t/h)	6 $m_{cmltv\ net}$ (t/h)	7 F_{H2} (t/h)	8 F_{H1} (t/h)	9 F_{FW} (t/h)
			0.0					
FW-A	0.0	**98.6**			0.0			0.0
			98.6	2.0				
D-A1	20.0	−50.0			2.0			98.6
			48.6	1.5				
D-A2	50.0	−100.0			3.4			68.6
			−51.4	−2.6				
D-A3	100.0	−80.0			0.9			8.6
			−131.4	−0.9				
S-AH1	106.5	**155.8**			0.0		0.0	0.0
			24.4	2.3				
D-A4	200.0	−70.0			2.3		24.4	11.4
			−45.6	−2.3				
S-AH2	250.0	**45.6**			0.0	0.0	0.0	0.0
			0.0	0.0				
WW-A	1,000,000.0	0.0			0.0	0.0	0.0	0.0
			0.0	0.0				

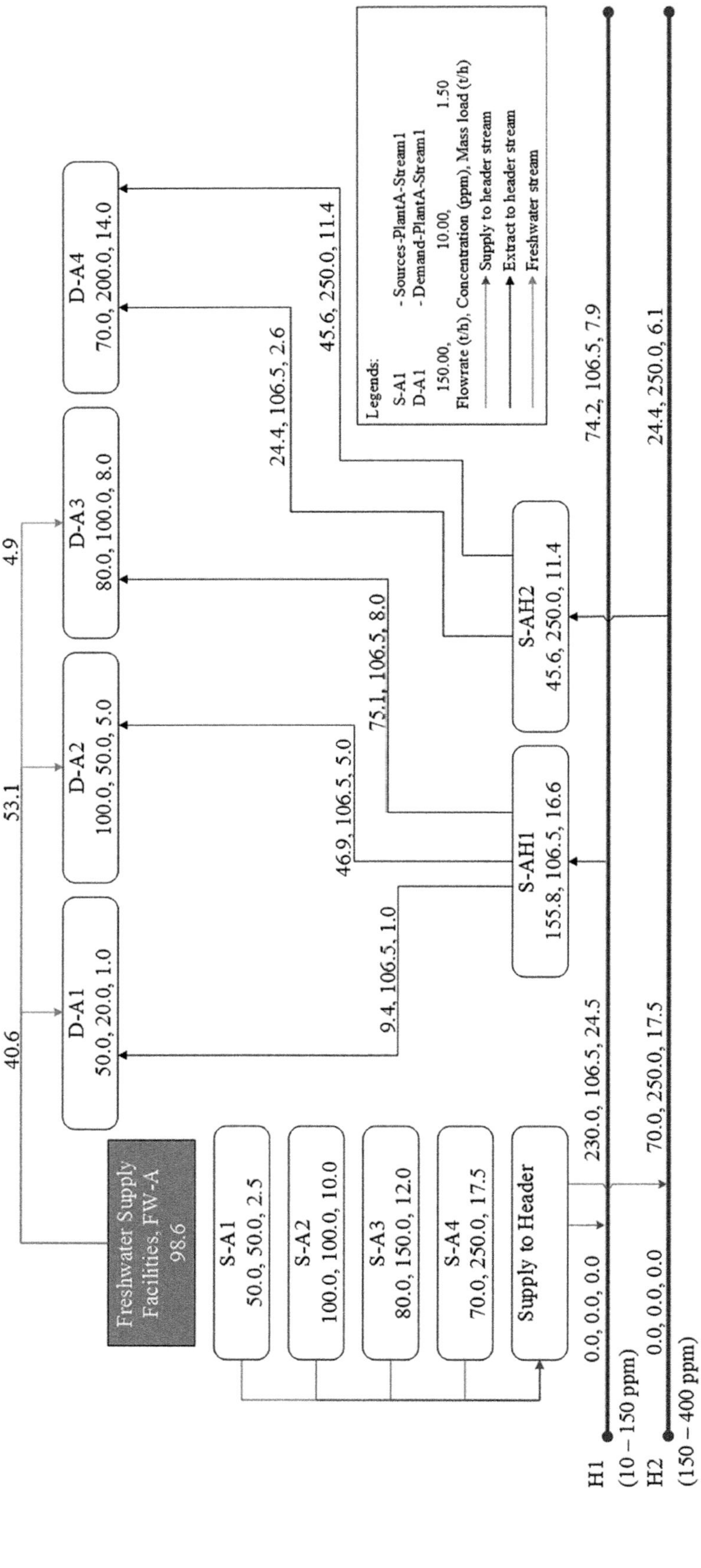

FIGURE 24.2 Plant A TS-CWI optimal total site water network.

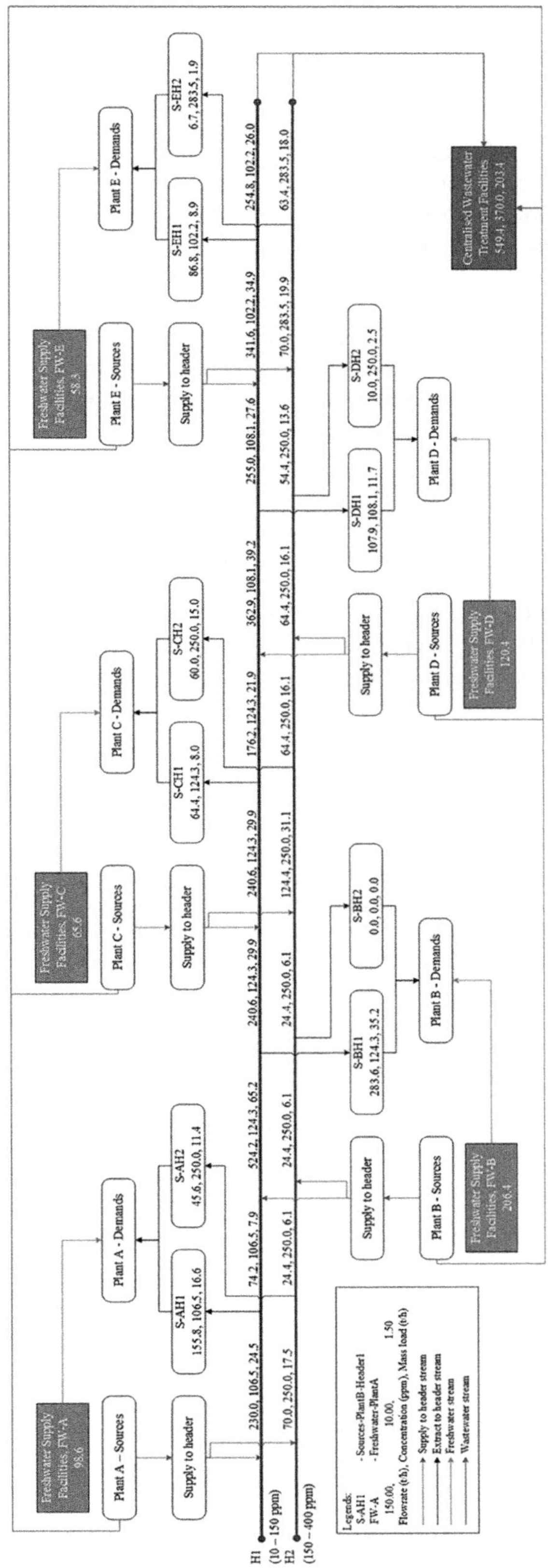

FIGURE 24.3 The case study TS-CWI optimal total site water network.

24.4 CONCLUSION

The methodology for TS-CWI has been developed and utilised to address the issue of water minimisation to target the minimum fresh water required and minimum wastewater generated for a TS that is arranged across the centralised water reuse header. This methodology has been applied to a case study with two centralised water reuse headers (high purity and low purity). The reduction of fresh water required and wastewater generated was 59.9%.

The optimal water network across TS of the case study was not complex and showed that the number of water interconnections (piping), including pumping, are at a considerable amount. This shows that the TS-CWI methodology is applicable in reality because it can reduce the excessive cost of interconnections (piping) and pumping. Using a centralised water reuse header can also protect the confidentiality of plants across the TS. Each plant could sell its wastewater to the centralised water reuse header where it can be utilised the plants located downstream. This could also reduce the cost of each plant treating its wastewater because all of the wastewater heads to the centralised wastewater treatment facilities as the end-of-pipe solution.

ACKNOWLEDGMENTS

Financial support from the Universiti Teknologi Malaysia Research University Grant Vote R.J130000.7809.4F918 and from the EC-funded project Sustainable Process Integration Laboratory–SPIL, funded as project No. CZ.02.1.01/0.0/0.0/15_003/0000456, by the Czech Republic Operational Programme Research and Development, Education, Priority 1: Strengthening capacity for quality research under the collaboration agreement with the UTM.

REFERENCES

Bandyopadhyay, S., and Ghanekar, M. D. 2006. Process water management. *Industrial and Engineering Chemical Research* 45, 5287–5297.

Chew, I. M. L., Foo, D. C. Y., and Tan, R. R. 2010a. Flowrate targeting algorithm for interplant resource conservation network. Part 1: Unassisted integration scheme. *Industrial and Engineering Chemical Research* 49(14), 6439–6455. doi:10.1021/ie901804z.

Chew, I. M. L., Foo, D. C. Y., and Tan, R. R. 2010b. Flowrate targeting algorithm for interplant resource conservation network. Part 2: Assisted integration scheme. *Industrial and Engineering Chemical Research* 49(14), 6456–6468. doi:10.1021/ie901804z.

Foo, D. C. Y. 2007. Water cascade analysis for single and multiple impure fresh water feed. *Chemical Engineering Research and Design* 85(8), 1169–1177. doi:10.1205/cherd06061.

Liu, Y. A., Lucas, B., and Mann, J. 2004. Up-to-date tools for water-system optimization. *Chemical Engineering Magazine* 111, 30–41.

Mohd Nawi, W. N. R., Wan Alwi, S. R., Manan, Z. A., and Klemeš, J. J. 2016. Pinch analysis targeting for CO_2 total site planning. *Clean Technologies and Environmental Policy* 18(7), 2227–2240. doi:10.1007/s10098-016-1154-7.

25 Exploring Water Reuse Opportunities in a Large-Scale Milk Processing Plant through Process Integration

Esther Buabeng-Baidoo, Nielsen Mafukidze, Sarojini Tiwari, Akash Kumar, Babji Srinivasan, Thokozani Majozi, and Rajagopalan Srinivasan

CONTENTS

25.1 INTRODUCTION

The process and manufacturing industry consumes substantial amounts of water and also generates considerable amounts of wastewater. The waste load mainly contains products lost during the processing operations. Sustainable water and wastewater management through improved operation and proper management practices may effectively reduce the freshwater consumed and wastewater generated in industry (Carawan et al., 1979). Since the amount of wastewater discharged is dependent on the amount of freshwater consumed, minimizing the freshwater consumption leads to minimizing the wastewater discharge (Halim et al., 2015). This applies to all industries including dairy processing, which is known to consume large amounts of water. As an example, in an effort to minimize freshwater consumption as well as wastewater generation, Singapore has succeeded in developing NEWater, which is the brand name given to reused (regenerated) wastewater that is now supplied to industries for non-potable use and is also discharged into the reservoirs for domestic processing (Evans, 2008).

The dairy sector in India, being the largest in the world, consumes approximately 62 billion m^3 of fresh water every year and, based on statistics, consumption is expected to rise above 400 billion m^3 by 2025 (Sustainability Outlook, 2014). According to the Central Pollution Control Board (CPCB) report the quality of surface water in India has deteriorated to an alarming level (CPCB India, 2011). World Bank reports that 60% of aquifers in India will be in critical condition in the next 20 years (WB, 2010). According to a report by Columbia Water Centre, over-exploitation of groundwater in northern Gujarat may exhaust this valuable water resource if left unchecked (Narula et al., 2011). In this regard, sustainable water consumption in India will be a primary concern in the near future (Amarasinghe et al. 2012). Amul Dairy is the biggest dairy cooperative in India. It processes milk, ghee (clarified butter), butter, flavored milk and milk powder. This plant consumes approximately 6 million m^3 of water annually. The dairy industry falls within the food-processing sector, and as such, there is a need to maintain high hygiene standards in the plants. In order to maintain these standards, dairy plants are usually equipped with cleaning-in-place (CIP) systems to ensure efficient and automated cleaning (Niamsuwan et al., 2011).

Water optimization is one of the prevailing methodologies being proposed in literature to minimize water usage, as it can be utilized to reduce freshwater consumption and wastewater generation through water reuse, recycling and regeneration (W3R; Khor et al. 2011). In the CIP, this can be achieved through the reuse or recycling of the first rinse water that, under normal circumstances, is sent directly to the effluent treatment plant (ETP) before being discharged to the ground. This wastewater can be treated by means of a reverse osmosis (RO) membrane/regenerator in order to maximize opportunities for W3R in the CIP process. RO membranes can separate a wastewater stream into a lean stream of low contaminant concentration, known as permeate, and a highly contaminated stream, known as the retentate stream, by means of a pressure-difference driving force. RO membranes are frequently used in the processing industries due to their relatively low energy consumption, ease of operation, high product recovery and quality (El-Halwagi, 1992; Al-Obaidi et al., 2017)

Two methods can be used in water use minimization, namely insight-based techniques and mathematical optimization methods. Insight-based techniques use a unified framework to match individual water demand with suitable supplies, depending on the quantity required and the quality offered. Minimum water targets are determined and alternative water network structures are derived, from which the best is selected (CanmetENERGY, 2003). On the other hand, mathematical programming techniques involve the building of models made up of mathematical equations that describe the system at hand and consider other design, physical and economic constraints. An objective for the problem is set and rigorous algorithms are used to solve the problem (Jezowski, 2010).

In this study, the mathematical optimization approach is used, as it allows the processing of complex systems with multiple contaminants and has the capacity to successfully integrate the water network (WN) model with the regeneration model. Mathematical optimization usually employs a superstructure that forms the basis of the mathematical model formulation that is ultimately solved to optimality. This optimization approach allows the designer to identify an optimal configuration for the process from a number of alternatives (Porn et al., 2008). In the context of this work, this is referred to as superstructure optimization.

The current work proposes a superstructure optimization approach for the synthesis of a detailed WN for minimization of fresh water use and wastewater generation in the raw milk receiving and processing department (RMRD) at Amul Diary. The rest of the chapter is organized as follows. Section 25.2 presents an overall picture of the Amul process, mainly the RMRD. It also describes the quality factors (COD, chemical oxygen demand; TSS, total suspended solids) that determine whether the processed water is still in its reusable form or not. The formulation of the problem is presented in Section 25.3. Section 25.4 provides a detailed explanation of the normal operation of the Amul process along with the two new scenarios that are proposed in the research work, including the corresponding mathematical models. The WN analysis shows that if water integration coupled with regeneration is implemented, it gives a better outcome than that given only by integration and much better than that given by the normal operation of Amul. These results are discussed in detail in Section 25.5, which leads to the conclusion in Section 25.6.

25.2 PROCESS DESCRIPTION

Amul is the largest food brand of India concentrating in the production of milk and milk products. It collects milk from 700 thousand villagers. This dairy processes 1800 m^3 of raw milk per day, for which it requires an average of 1600 m^3 of water per day. The CIP processes account for nearly 75% of the total water consumption; the rest is used in operations such as boiler feed and cooling tower makeup. The RMRD requires nearly 90% of the total water used for CIP. This section has three pasteurizers that process 1.2 million liters of milk daily. CIP is carried out for the pasteurizers, tanks, silos and other processing equipment at regular intervals to maintain cleanliness and hygiene standards.

The collected milk is first filtered to remove the unwanted solids. It is then chilled to 3°C–4°C and then sent to the buffer tank before being sent to the centrifugal clarifier for the removal of bacterial spores (Lelievre et al., 2002). The clarified milk is sent to the raw milk silos for storage whence it is sent to the pre-pasteurization regeneration unit. In this regeneration unit, the raw milk, which is initially entering at a temperature of 5°C–6°C, is heated up to a temperature of 40°C–50°C before being sent to the separator, where the cream is separated from the milk. The cream and the skimmed milk are stored in their respective holding tanks at a temperature of 62°C. From the holding tanks both the cream and the skimmed milk are sent to the blender where they are blended to achieve the fat content required for the final products. Blending is primarily done to have a uniform fat content in the finished dairy product. The blended product is then pasteurized in a second regeneration unit where hot water is used to heat the milk up from 62°C to 70°C. It is again heated by a heater to 78°C before being sent to the first regenerator to heat up the raw milk coming from the silos. The pasteurized milk is then transported to the respective processing units to manufacture the required products.

Sometimes a small amount of the pasteurized milk is homogenized and transported to the flavored milk manufacturing section. Homogenization is the process of breaking down the fat globules in milk so that they remain integrated rather than separated as cream (Tomasula et al., 2013). The cream is sent for cream pasteurization. Note that in the first regeneration unit, the milk that is recycled for pasteurization heats up the raw milk coming from the silos to raise temperature from 5°C–6°C to 40°C–50°C. The pasteurized milk cools from 78°C to 10°C before it is sent to the chiller and subsequently for homogenization (Figure 25.1).

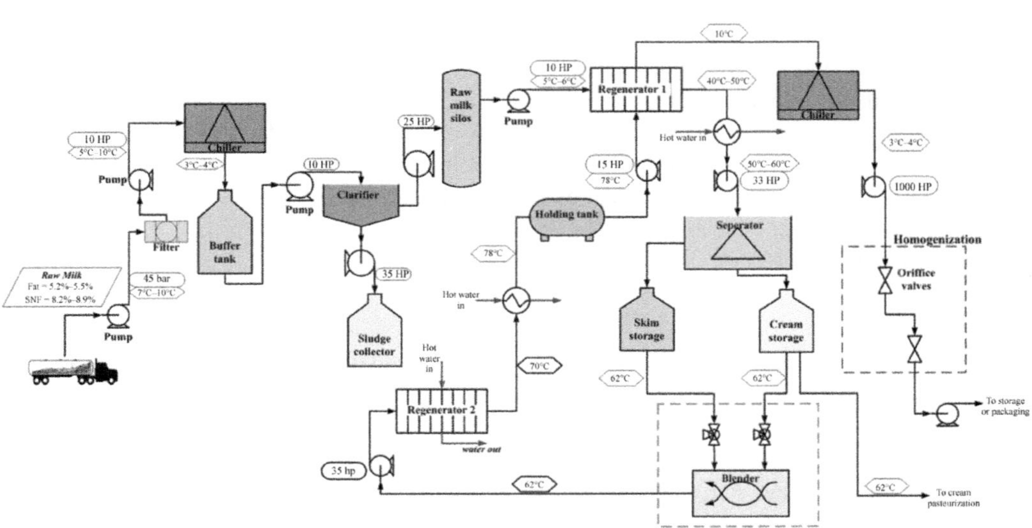

FIGURE 25.1 Process flow diagram of Amul dairy.

25.2.1 Quality Factors

The quality of water leaving the various process units is analyzed primarily on the basis of two contaminants: COD and TSS. A brief description on them are given as follows:

25.2.1.1 Chemical Oxygen Demand (COD)

Many organic substances, which are difficult to oxidize biologically by aerobic microbes or are toxic to microorganisms (such as lignin), can be oxidized chemically by using strong oxidizing agents like dichromate (Cr_2O_7) in acidic media. The COD is therefore, the measure of the oxygen equivalent of the organic material in wastewater that can be chemically oxidized to CO_2, ammonia and water in the presence of strong oxidizing agents using dichromate in an acid solution. A high COD value may occur because of the presence of "inorganic substances" in wastewater with which only the dichromate can react.

One limitation of COD is that it cannot distinguish between biologically active and biologically inactive organic substances; but the major advantage of COD is that it requires significantly less time compared to BOD. COD is measured in ppm or mg/L. It is very important to note that earthworms are versatile waste eaters and decomposers, and can ingest and remove several organic substances from the wastewater that otherwise cannot be oxidized by microbes, and thus bring down the COD values significantly.

25.2.1.2 Total Suspended Solids (TSS)

Solids in wastewater consist of organic and inorganic particles, and they can either be "suspended" or "dissolved." TSS is the dry weight of particles trapped by a filter. In wastewater of medium strength, about 75% of the suspended solids and 40% of the filterable solids are organic in nature. They provide adsorption sites for chemical and biological contaminants. As suspended solids degrade biologically, they can create toxic by-products. Suspended solids in wastewater directly affect the turbidity. TSS is measured in ppm or mg/L (Sinha et al., 2007).

25.3 PROBLEM STATEMENT

A fixed flow rate model that considers the concept of sources and sinks is adopted in this study (Khor et al., 2011). A water source is a unit that produces water and a water sink is a unit that requires water. The model takes into account streams with multiple contaminants. Figure 25.2 shows the general WN superstructure. CIP processes are batch operations however, so in this work a time-average approach to convert quantities to flow rates over a predefined time interval was adopted. This allows batch process to be treated as continuous. Worthy of mention, however, is the fact that this can result in oversimplification of batch-process behavior. Nonetheless, that is beyond the scope of this particular contribution.

The problem addressed in this work, in general, can be stated as follows:

Given:

1. A set of water sources, J, with known flow rates and known contaminant concentrations
2. A set of water sinks, I, with known flow rates and known maximum allowable contaminant concentrations
3. A set of regenerators, R, with known removal ratios and design parameters
4. A freshwater source, FW, with known contaminant concentration and variable flow rate
5. A wastewater sink, WW, with known maximum allowable contaminant concentration and variable flow rate

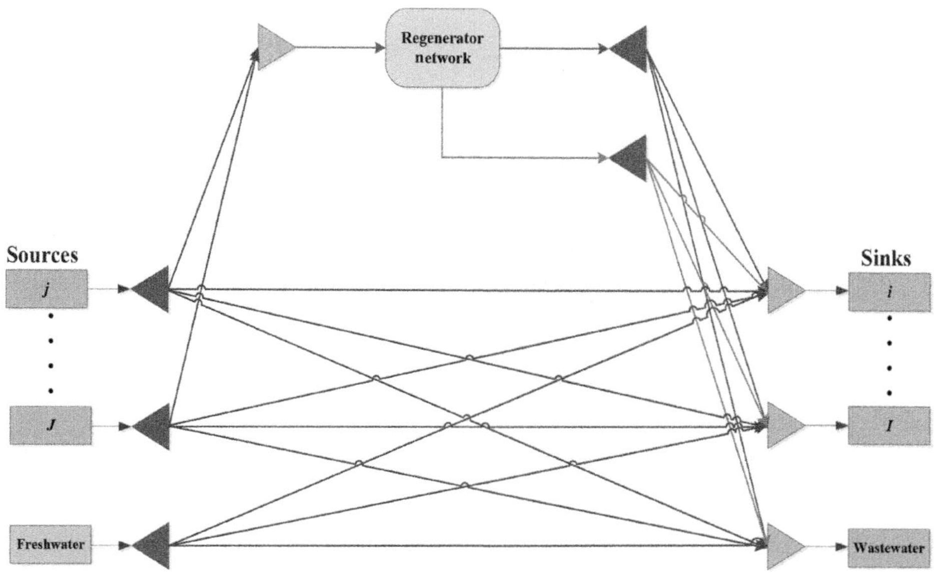

FIGURE 25.2 A general representation of the water network superstructure. (From Khor, C. et al., *Ind. Eng. Chem. Res.*, 50, 13444–13456, 2011.)

To be determined:

1. The minimum freshwater intake, wastewater generation and the total annualized cost (TAC)
2. The optimal configuration of the WN
3. The optimal operation and design conditions of the RO membrane, including feed pressure, number of hollow fiber modules per regenerator, stream distributions, and separation levels

25.4 MATHEMATICAL MODEL

The mathematical model entails the following constraints, which are based on the superstructure depicted in Figure 25.2. Constraint (25.1) describes how wastewater from a source can be distributed to the sinks and the regenerator as shown in Figure 25.3.

$$Q_j^{out} = \sum_{i=1}^{I} Q_{j,i}^a + \sum_{r=1}^{R} Q_{j,r}^d \qquad \forall j \in J \tag{25.1}$$

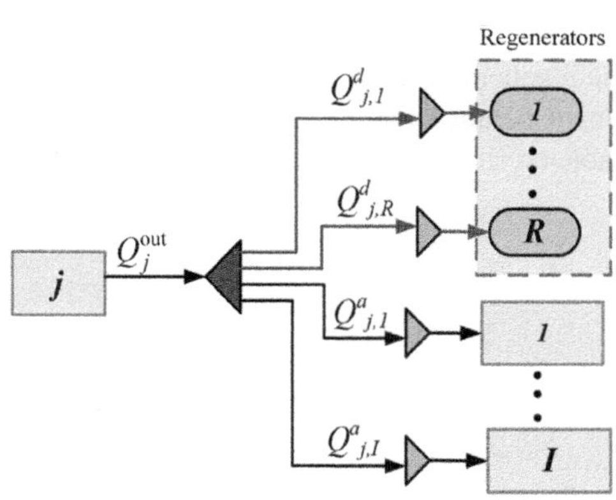

FIGURE 25.3 Schematic representation of a water source.

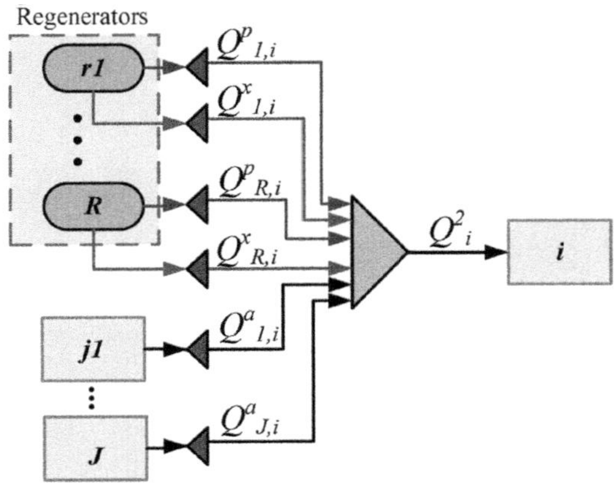

FIGURE 25.4 Schematic representation of a water sink.

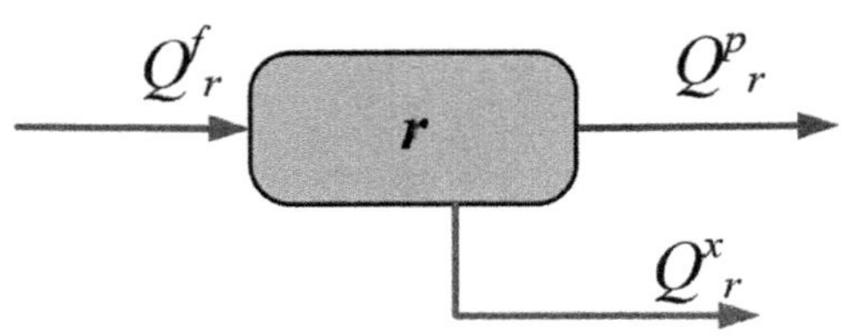

FIGURE 25.5 Schematic representation of a regenerator.

In a similar way, constraint (25.2) shows how water from the sources and regenerator can be distributed to satisfy the sinks as shown in Figure 25.4. The maximum allowable load into a particular sink is dependent on the sink operations and is described by the contaminant balance in constraint (25.3).

$$Q_i^{in} = \sum_{r=1}^{R} Q_{r,i}^{p} + \sum_{r=1}^{R} Q_{r,i}^{x} + \sum_{j=1}^{J} Q_{j,i}^{a} \qquad \forall i \in I \tag{25.2}$$

$$C_{i,m}^{U} \geq \frac{\sum_{r=1}^{R} Q_{r,i}^{p} C_{r,m}^{p} + \sum_{r=1}^{R} Q_{r,i}^{x} C_{r,m}^{x} + \sum_{j=1}^{J} Q_{j,i}^{a} C_{j,m}^{1}}{Q_i^2} \qquad \begin{matrix} \forall i \in I \\ \forall m \in M \end{matrix} \tag{25.3}$$

The water balance around the regenerator as shown in Figure 25.5 is presented in constraint (25.4). The corresponding contaminant balance for the regenerator is shown in constraint (25.5) where $C^U_{r,m}$ is the maximum allowable concentration into the regenerator.

$$Q_r^f = Q_r^p + Q_r^x \qquad \forall r \in R \tag{25.4}$$

$$C_{r,m}^{U} \geq \frac{Q_r^p C_{r,m}^{p} + Q_r^x C_{r,m}^{x}}{Q_r^f} \qquad \begin{matrix} \forall r \in R \\ \forall m \in M \end{matrix} \tag{25.5}$$

The performance of the regenerator is based on the removal ratio, $RR_{r,m}$, which is the fraction of mass load into the regenerator that exits in the retentate stream and defined by constraint (25.6).

$$RR_{r,m} = \frac{Q_r^x C_{r,m}^x}{Q_r^f C_{r,m}^f} \quad \begin{matrix} \forall r \in R \\ \forall m \in M \end{matrix} \tag{25.6}$$

The liquid recovery is another performance indicator of the regenerator that is defined by constraint (25.7).

$$LR_r = \frac{Q_r^p}{Q_r^f} \quad \forall r \in R \tag{25.7}$$

Detailed design equations for the RO regenerator are also included in the model. This includes the osmotic pressure, $\Delta\pi_q$, which is a function of the contaminant concentration on the feed side as shown in constraint (25.8)

$$\Delta_{\pi r} = OS\sum_{m=1}^{M} C_{r,m}^{av} \quad \begin{matrix} \forall r \in R \\ \forall m \in M \end{matrix} \tag{25.8}$$

where $C_{r,m}^{av}$ is the average concentration on the shell side and is given by constraint (25.9).

$$C_{r,m}^{av} = \frac{C_{r,m}^f + C_{r,m}^x}{2} \quad \begin{matrix} \forall r \in R \\ \forall m \in M \end{matrix} \tag{25.9}$$

The permeate contaminant concentration is a function of the pressure drop and the osmotic pressure. This is given in constraint (25.10).

$$C_{r,m}^p = \frac{k_m C_{r,m}^{av}}{A(\Delta P_r - \Delta\pi_r)\gamma} \quad \begin{matrix} \forall r \in R \\ \forall m \in M \end{matrix} \tag{25.10}$$

In constraint (25.10), γ is the design parameter that caters for deviation between theoretical and practical values. Its true representation can be found in Khor et al. (2011). The permeate flow rate per module is determined based on constraint (25.11), where N_r is the number of hollow fiber RO modules.

$$\frac{Q_r^p}{N_r} = AS(\Delta P_r - \Delta\pi_r) \tag{25.11}$$

Additionally, to forbid remixing of the regenerator product streams, some logical constraints are also introduced. One such constraint (25.12) prevents the mixing of permeate and retentate streams in the same sink.

$$y_{r,i}^{p} + y_{r,i}^{x} \le 1 \quad \begin{matrix} \forall i \in I \\ \forall r \in R \end{matrix} \tag{25.12}$$

To ensure that only practical flow rates are accommodated by the piping, appropriate bounds are allocated to all the flow rates such that any stream with a flow rate outside these limits is not allowed to exist. This is achieved by means of the big-M constraint adopted by Khor et al. (2011). In their formulation, M is a valid lower or upper bound denoted by U and L, respectively and this is shown in constraints (25.13 through 25.16).

$$M_{j,i}^L y_{j,i} \le Q_{i,j}^a \le M_{j,i}^U y_{j,i}, \quad \forall i \in I,\ j \in J \tag{25.13}$$

$$M_{r,i}^L y_{r,i}^p \le Q_{r,i}^a \le M_{r,i}^U y_{r,i}^p, \quad \forall r \in R,\ i \in I \tag{25.14}$$

$$M^L_{r,i} y^x_{r,i} \le Q^x_{r,i} \le M^U_{r,i} y^x_{r,i}, \qquad \forall r \in R,\ i \in I \tag{25.15}$$

$$M^L_{j,r} y^d_{j,r} \le Q^d_{j,r} \le M^U_{j,r} y^d_{j,r}, \qquad \forall r \in R,\ j \in J \tag{25.16}$$

The objective is set to simultaneously minimize the freshwater consumption, the wastewater generation, as well as the capital and operating cost of the piping on an annualized basis. The TAC of the RO membrane consists of the capital cost of hollcw fiber RO modules, energy recovery turbines and pumps; pretreatment of chemicals; as well as the cperating cost of pumps and turbines. The TAC also considers the operating revenue of the energy recovery turbines and is shown in constraint (25.17).

$$\begin{aligned}\mathrm{TAC}(r) = {} & C^{\mathrm{pump}}\left(\sum_{r=1}^{R} pwp_r\right)^{0.65} + C^{\mathrm{tur}}\left(\sum_{r=1}^{R} pwt_r\right)^{0.43} + C^{\mathrm{elec}}\,\mathrm{AOT}\left(\frac{\sum_{r=1}^{R} pwp_r}{\eta_{\mathrm{pump}}}\right) \\ & - C^{\mathrm{elec}}\,\mathrm{AOT}\left(\sum_{r=1}^{R} pwt_r\right)\eta_{\mathrm{turbine}} + C^{\mathrm{mod}}\sum_{r=1}^{R} N_r^{\,m} + C^{\mathrm{chem}}\,\mathrm{AOT}\sum_{j=1}^{J} Q^d{}_{j,r} \qquad \forall r \in R\end{aligned} \tag{25.17}$$

where Pwp_r and Pwt_r are the power of the pump and energy recovery turbines, respectively, and shown in constraints (25.18) and (25.19).

$$Pwp_r = Q^f_r\left(P^f_r - P^{\mathrm{atm}}\right) \qquad \forall r \in R \tag{25.18}$$

$$Pwt_r = Q^x_r\left(P^x_r - P^{\mathrm{atm}}\right) \qquad \forall r \in R \tag{25.19}$$

The piping cost of components was calculated by assuming a linear fixed-charge model. In this formulation, the cost of a pipe is incurred if the flow rate through the pipe falls below the threshold value. This is achieved by using 0–1 variables. It was assumed within the model that all the pipes share the same properties of p_c and q_c and a 1-norm distance, D.

The objective function of the water network is shown in constraint (25.20).

$$\min \left(\begin{aligned} & \sum_{r=1}^{R} \mathrm{TAC}_r + \mathrm{AOTC}^{\mathrm{water}}\mathrm{FW} + \mathrm{AOTC}^{\mathrm{waste}}\mathrm{WW} \\ & + \mathrm{AA}\left(\sum_{j=1}^{J}\sum_{i=1}^{I} D_{i,j}\left(\frac{p_c Q^a_{j,i}}{3600v} + q_c y_{j,i}\right)\right) \\ & + \mathrm{AA}\left(\sum_{r=1}^{R}\sum_{i=1}^{I} D^F_{r,i}\left(\frac{p_c Q^p_{r,i}}{3600v} + q_c y^p_{r,i}\right)\right) \\ & + \mathrm{AA}\left(\sum_{r=1}^{R}\sum_{i=1}^{I} D^x_{r,i}\left(\frac{p_c Q^x_{r,i}}{3600v} + q_c y^x_{r,i}\right)\right) \\ & + \mathrm{AA}\left(\sum_{j=1}^{J}\sum_{r=1}^{R} D^d_{j,r}\left(\frac{p_c Q^d_{j,r}}{3600v} + q_c y^d_{j,r})\right)\right) \end{aligned}\right) \tag{25.20}$$

where $AA = \left(\frac{m(1+m)^n}{(1+m)^n - 1^n} \right)$.

The overall model results in a nonconvex mixed-integer non-linear program (MINLP) due to the nonlinear terms as well as the integer variables in the constraints.

25.5 DETAILED STUDY OF THE WATER NETWORK

In this study scenarios are considered to demonstrate the impact of an integrated framework in freshwater and wastewater minimization.

- Base scenario: Normal operation of the plant
- Scenario 1: Water integration without regeneration
- Scenario 2: Water integration with regeneration

The base scenario refers to an informed estimate of the freshwater consumption and wastewater generation of the plant under its normal operation, without full exploration of reuse and recycling opportunities. To improve its efficiency with respect to minimizing freshwater intake and wastewater generation, two more scenarios are proposed that are discussed along with the base scenario in the following sub-sections.

25.5.1 Base Case Scenario

The base scenario describes the normal operation of the Amul plant. To understand this, we have to first understand the CIP mechanism deployed in the plant for efficient cleaning of the equipment. CIP is mostly a three-stage process, as described as follows.

Stage 1: Freshwater is circulated for 10–15 min to clean the equipment after the milk products are removed in what is called the pre-rinsing stage. This water is highly impure as it picks up the maximum amount of contaminants because it is the very first cleaning stage, and so it is sent directly for treatment. It might also contain some milk stains that remain inside the equipment after the removal of the products and that have to be recovered.

Stage 2: Hot lye solution, which is essentially NaOH solution, followed by clean water from the reverse osmosis (RO) membrane, is then circulated for about 10 more minutes to rinse the equipment. The lye solution prevents microbial growth. The water leaving this stage is sent for quality check where its conductivity levels are monitored. High conductivity means it is rich in negatively charged ions (OH^-, NO_3^-). If the conductivity exceeds the tolerance, set at 49 μS, it means it has a high alkaline content and is therefore stored in the alkali holding tank. Otherwise, it is sent to the rinse water tank for reuse. There is no processing required as the water is still in usable form.

Stage 3: Clean RO water is circulated just after the alkaline solution is used to clean up the tanks and wash away traces of the lye solution.

In some cases, some acid—preferably nitric acid—is also circulated for 5 min after the alkali solution (Sathit et al., 2011). In such cases, an intermediate rinsing with warm or cold water must be carried out between the cleaning steps to rinse out the alkali solution. The follow-up clean RO water to be sent after the acid is also sent for quality check after its exit and stored in the acid holding tanks for reuse if it is found to be highly conductive.

To ensure CIP, water from the freshwater supply line is brought to the RMRD of the plant where the equipment is cleaned. CIP is a batch operation. Some equipment may require cleaning twice daily, some after every 10 h. Accordingly the flow rate of freshwater is integrated with respect to the amount of time the water is circulated in each piece of equipment. The wastewater generated is sent to the effluent treatment plant for purification and discharge.

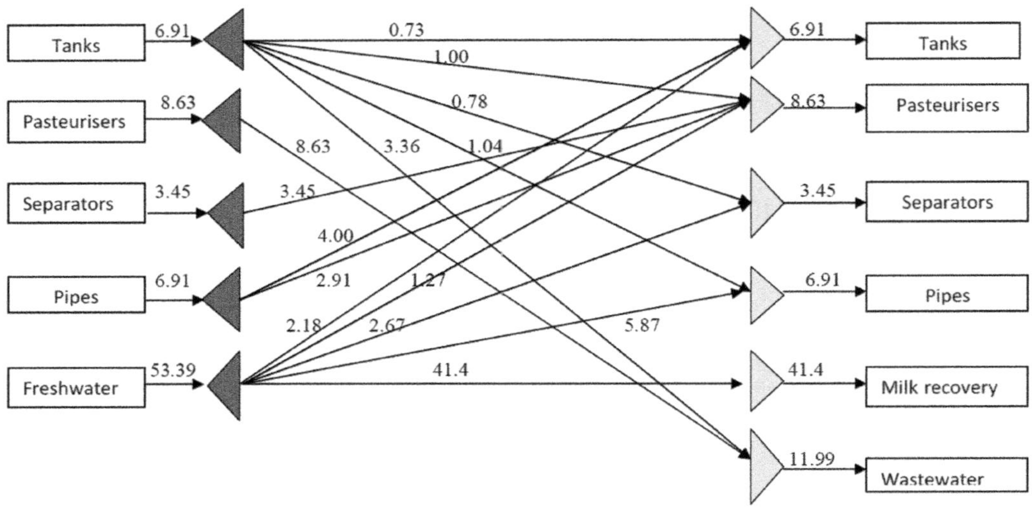

FIGURE 25.6 Water network structure for Scenario 1.

25.5.2 Scenario 1: Water Integration Without Regeneration

Similar to the base scenario, here water is also reused for further cleaning with the only difference being that the WN is made a lot more complicated by incorporating interconnections between the various process units. If there are n units, reusable water from the first unit can be sent for rinsing to all the remaining $n-1$ units, along with the first unit, and the same follows for all the other units. These interconnections increase the extent of water reuse, which in turn reduces the total intake of freshwater compared to the base scenario.

In the base scenario the amount of freshwater required is 67.28 kg/s and the wastewater generated is 25.88 kg/s whereas, in Scenario 1 the amount of freshwater intake and wastewater generated are 53.39 and 11.99 kg/s, respectively (refer to Table 25.4). This is achieved due to the presence of the numerous interconnections previously discussed. For instance, 0.73 kg/s of water coming out of the tanks is recycled back to itself for the next rinsing step, as shown in Figure 25.6. Along with that 1, 0.78, and 1.04 kg/s of the same processed water from the tanks is sent to the pasteurizers, separators and pipes, respectively, for rinsing, which proves to be beneficial instead of sending them directly for treatment without being reused, as happens in the normal operation of the plant. This saves a lot of fresh water and reduces the amount of wastewater generated.

25.5.3 Scenario 2: Water Integration with Regeneration

Here-along with the huge amount of interconnections between the different units—there is also a RO regeneration unit that is used to regenerate a portion of the contaminated water. Water is purified by RO and recycled back to the process units for further cleaning of the equipment. There are specific conditions for the different contaminants generated in each piece of equipment that determine whether the water is sent to the regeneration unit for recycling or not. Of the three scenarios, Scenario 2 gives the best result, since implementation of regeneration along with water integration minimizes the freshwater intake significantly, which ultimately reduces the wastewater generation.

Considering the same example of tanks, in Scenario 2, apart from being recycled, a portion of the processed water is sent to the RO regenerator for purification and later reused in the process units as depicted in Figure 25.7. Similarly, for the pasteurizers, all of the processed water is sent for regeneration. During recycling, the stream is divided into two parts, that is the permeate, which is the regenerated water, and the retentate, which is the wastewater; 3.64 kg/s of retentate is sent for treatment from the regenerator, while 2.58 and 5.93 kg/s of the permeate stream is sent to the

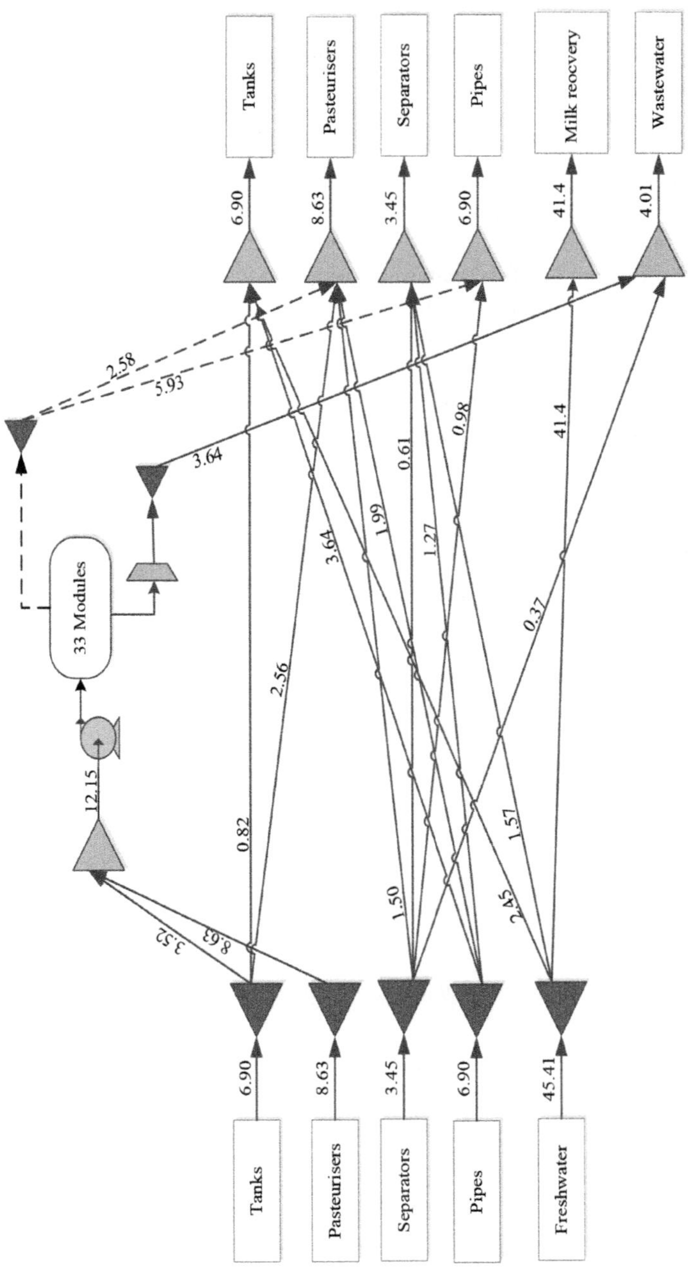

FIGURE 25.7 Water network structure for Scenario 2.

pasteurizers and pipes, respectively, for further rinsing. This added recycling mechanism reduces the amount of wastewater from 11.99 kg/s in Scenario 1 to 4.01 kg/s in Scenario 2. Consequently, the freshwater requirement is lowered from 53.39 kg/s in Scenario 1 to 45.41 kg/s in Scenario 2 due to the increased recycle/reuse potential.

25.6 RESULTS

A model used for a literature-based refinery case study based on the work by Buabeng-Baidoo and Majozi (2015) and implemented in GAMS 24.2 using the general purpose global optimization solver BARON has been applied to the CIP process of the RMRD at Amul Dairy in order to minimize the overall water usage. Table 25.1 presents the process and economic data for the detailed RO membrane, while the economic data and model parameters are given in Table 25.2. Table 25.3 shows the limiting water data that was used for the case study. The Manhattan distance between different units was assumed to be 100 m.

At first an informed estimate of freshwater consumption and wastewater generation in the current RMRD plant is reported, followed by the implementation of the proposed Scenarios 1 and 2. The results obtained from the three scenarios are shown in Table 25.4.

In Table 25.4, Scenario 1 shows that introducing water integration to the system has the potential of saving up to 20.6% of fresh water and reducing the wastewater by 53.7% in comparison to the base scenario. Scenario 2 shows that the introduction of a regenerator can further reduce these figures by

TABLE 25.1
Process and Economic Data for the Detailed RO Membrane

Parameter	Value
Pure water permeability, A	5.50×10^{-13} m/(s. Pa)
Shell side pressure drop per module per regenerator, P_m	4.05×10^4 Pa
Solute permeability coefficient, k_m	1.82×10^{-8} m/s
Fiber length, L	0.75 m
Seal length, L_s	0.075 m
Outside radius of fiber, r_o	42×10^{-6} m
Inner radius of fiber, r_i	21×10^{-6} m
Membrane area, S	180 m^2
Water viscosity, μ	0.001 kg/(m.s)
Dimensionless constant, Υ	0.69
Permeate pressure per regenerator, $P_p(q)$	101325 Pa
Pump efficiency, η_{pump}	0.7
Turbine efficiency, $\eta_{turbine}$	0.7
Liquid recovery for all regenerators, $\alpha(q)$	0.7
Removal ratio, RR	0.9
Osmotic constant, OS	4.14×10^{-7} Pa
Cost parameter for chemicals, C^{chem}	0.11 \$/kg
Cost of electricity, C^{elec}	0.06 \$/(kW.h)
Cost coefficient for pump, C^{pump}	6.5 \$/(yearW$^{0.65}$)
Cost coefficient for turbine, C^{tur}	18.4 \$/(yearW$^{0.43}$)
Cost per module of HFRO membrane, C^{mod}	2300 \$/(year.module)
Maximum flow rate per hollow fiber module, F^U	0.27 kg/s
Minimum flow rate per hollow fiber module, F^L	0.21 kg/s

Source: Khor, C. et al., *Ind. Eng. Chem. Res.*, 50, 13444–13456, 2011.

TABLE 25.2
Economic Data and the Model Parameters for the WNS

Parameter	Value
Annual operating time, AOT	8760 h
Unit cost of freshwater, C^{water}	1 $/t
Unit cost of wastewater, C^{waste}	1 $/t
Interest rate per year, *m*	5%
Number of years, *n*	5 year
Parameter *p* for carbon steel piping	7200
Parameter *q* for carbon steel piping	250
Velocity, *v*	1 m/s

Source: Khor, C. et al., *Ind. Eng. Chem. Res.*, 50, 13444–13456, 2011.

TABLE 25.3
Limiting Data for the Case Study

Sources				Sinks			
	Flow Rate (kg/s)	Conc. (mg/L) COD	Conc. (mg/L) TSS		Flow Rate (kg/s)	Conc. (mg/L) COD	Conc. (mg/L) TSS
Tanks	6.9	90	40	Tanks	6.9	50	10
Pasteurizers	8.625	120	90	Pasteurizers	8.625	70	20
Separators	3.45	80	30	Separators	3.45	40	9
Pipes	6.9	70	10	Pipes	6.9	20	6
Fresh water	—	0	0	Milk recovery	41.4	0	0
				Wastewater	—	760	600

TABLE 25.4
Summary of Results Obtained from the Three Scenarios Considered

	Base Scenario	Scenario 1	Scenario 2
Fresh water (kg/s)	67.28	53.39	45.41
Wastewater (kg/s)	25.88	11.99	4.01
TAC (million $/year)	2.94	2.15	1.96

11.9% and 30.8%, respectively. Scenario 2 is therefore the best and is capable of saving up to 33% of the TAC. This optimized water recovery using the regenerator requires a total of 33 hollow fiber RO modules, which we obtained after performing the optimization. Figures 25.6 and 25.7 show the optimal WN obtained for Scenario 1 and Scenario 2, respectively.

25.7 CONCLUSION

This work addresses water reuse/recycle opportunities in a large-scale milk processing plant through process integration based on comprehensive superstructure optimization. A comprehensive regenerator model embedded within a WN model was used. The model was applied to both a literature example and a case study from the Amul Dairy and was solved using GAMS/BARON in order to highlight its applicability. The results show that water integration coupled with a regeneration system can lead to a reduction in the total cost of the network, due to the significant reduction in freshwater consumption and wastewater generation. This shows that optimization of WNs for minimum water targets is a powerful and sustainable water management tool that can be valuable to the dairy industry. However, it is worth re-emphasizing that this work adopted a time-average model in order to suppress the time dimension that is inherent in batch plants. Future work would have to consider including time in the analysis.

ACKNOWLEDGMENTS

The authors gratefully thank the leadership and technical team at Amul Dairy, Anand, for their strong support throughout the course of this project and also the National Research Foundation (NRF) in South Africa for funding this work under the NRF/DST Chair in Sustainable Process Systems Engineering.

NOMENCLATURE

Sets

$J = \{j|j = \text{water sources}\}$
$I = \{i|i = \text{water sinks}\}$
$R = \{r|r = \text{regenerators}\}$
$M = \{m|m = \text{contaminants}\}$

Parameters

$RR_{r,m}$	Removal ratio
LR_r	Liquid Recovery
AOT	Annual operating time
C^{water}	Freshwater cost
C^{waste}	Wastewater treatment cost
C^{pump}	Cost coefficient for pumps
C^{tur}	Cost coefficient for turbines
C^{chem}	Cost parameter for chemicals
C^{mod}	Cost per module of HFRO membrane
C^{elec}	Cost of electricity
Pwp_r	Power consumed by the pump
Pwt_r	Power consumed by the energy recovery turbines
η_{pump}	Efficiency of pump
η_{turbine}	Efficiency of turbine
$C_{i,m}^{U}$	Maximum allowable concentration into the regenerator.
$C_{r,m}^{U}$	Maximum allowable concentration into the regenerator
$C_{i,m}^{\max}$	Maximum allowable concentration into sink, *i*, for contaminant *m*
$D_{j,i}$	Manhattan distance between source, *j*, and sink, *i*
$D_{r,i}^{p}$	Manhattan distance between regenerator, *r*, and sink, *i*
$D_{r,i}^{x}$	Manhattan distance between regenerator, *r*, and sink, *i*
$D_{j,r}^{d}$	Manhattan distance between source, *j*, and regenerator, *r*
A	Water permeability coefficient
S	Membrane area per module
N_r	Number of RO modules

ΔP_r Pressure drop across regenerator, r
γ Dimensionless constant
N_r Number of hollow fiber RO modules
AOT Annual operating time
v Velocity
m Interest per year
n Number of years
OS Osmotic constant
L Fiber length
L_s Seal length
r_o Outside radius of fiber
r_i Inner radius of fiber
k_m Solute permeability coefficient
P_m Shell side pressure drop per module per regenerator
P_p Permeate pressure per regenerator
F^U Maximum flow rate per HFRO module
F^L Minimum flow rate per HFRO module

Continuous variables

Q_j^{out} Flow rate of source, j
Q_i^{in} Flow rate of sink, i
$Q_{j,i}^{a}$ Flow rate between source, j and sink, i
$Q_{j,r}^{d}$ Flow rate between source and regenerator, r
Q_r^{f} Feed flow rate into regenerator, r
Q_r^{p} Permeate flow rate out of regenerator, r
Q_r^{x} Retentate flow rate out of regenerator, r
$Q_{r,i}^{p}$ Permeate flow rate between regenerator and sink, i
$Q_{r,i}^{x}$ Retentate flow rate between regenerator and sink, i
$C_{r,m}^{p}$ Concentration of contaminant, m, in the permeate
$C_{r,m}^{x}$ Concentration of contaminant, m, in the retentate
$C_{r,m}^{f}$ Concentration of the contaminant, m, in the feed
$C_{j,m}^{1}$ Concentration of contaminant, m, in the source
$C_{r,m}^{av}$ Average contaminant concentration on the feed side
$\Delta\pi_r$ Osmotic pressure
FW Freshwater flow rate
WW Wastewater flow rate

Binary variable

$$y_r^{ED} = \begin{cases} 1 \leftarrow \text{if regenerator } r \text{ exists} \\ 0 \leftarrow \text{otherwise} \end{cases}$$

$$y_{r,i}^{p} = \begin{cases} 1 \leftarrow \text{piping exists between reg. permeate and sink, } i \\ 0 \leftarrow \text{otherwise} \end{cases}$$

$$y_{r,i}^{x} = \begin{cases} 1 \leftarrow \text{piping exists between reg. retentate and sink, } i \\ 0 \leftarrow \text{otherwise} \end{cases}$$

$$y_{j,r}^{d} = \begin{cases} 1 \leftarrow \text{piping exists between source, } j \text{ and reg., } r \\ 0 \leftarrow \text{otherwise} \end{cases}$$

REFERENCES

Al-Obaidi, M.A., Li, J.P., Kara-Zaïtri, C., and Mujtaba, I.M. (2017). Optimisation of reverse osmosis based wastewater treatment system for the removal of chlorophenol using genetic algorithms. *Chemical Engineering Journal*, 316, 91–100.

Amarasinghe, U., Shah, T., and Smakhtin, V. (2012). Water–milk nexus in India: A path to a sustainable water future? *International Journal of Agricultural Sustainability*, 10(1), 93–108.

Buabeng-Baidoo, E., and Majozi, T. (2015). Effective synthesis and optimisation framework for intergrated water and membrane networks: A focus on reverse osmosis membranes. *Industrial & Engineering Chemistry Research*, 54, 9394–9406.

CanmetENERGY. (2003). *Pinch Analysis for the Efficient Use of Energy, Water and Hydrogen.* Retrieved July 20, 2015, from Natural Resources Canada: https://www.nrcan.gc.ca

Carawan, R., Chambers, J., and Zall, R. (1979). *Water and Wastewater Management in Food Processing.* North Carolina State University. Raleigh, NC: The North Carolina Agriculture Extension Service.

El-Halwagi, M. (1992). Synthesis of reverse osmosis networks for waste reduction. *American Institute of Chemical Engineers*, 38(8), 1185–1198.

Evans, G. (2008). *Singapore's Self-Sufficiency.* Retrieved January 2015, http://www.water-technology.net_

Halim, I., Adhitya, A., and Srinivasan, R. (2015). A novel application of genetic algorithm for synthesizing optimal water reuse network with multiple objectives. *Chemical Engineering Research and Design*, 100, 39–56.

Jezowski, J. (2010). Review of water network design methods with literature annotations. *Industrial & Engineering Chemistry Research*, 49, 4475–4516.

Khor, C., Foo, D., El-Halwagi, M., Tan, R., and Shah, N. (2011). A superstructure optimization approach for membrane separation-based water regeneration networks synthesis with detailed nonlinear mechanistic reverse osmosis model. *Industrial & Engineering Chemistry Research*, 50, 13444–13456.

Lelievre, C., Antonini, G., Faille, C., and Benezech, T. (2002). Cleaning-in-place modelling of cleaning kinetics of pipes soiled by Bacillus spores assuming a process combining removal and deposition. *Trans IChemE*, 80, 305–311.

Narula, K., Fishman, R., Modi, V., and Polycarpou, L. (2011). *Addressing the Water Crisis in Gujarat, India.* New York City: Columbia Water Center.

Niamsuwan, S., Kittisupakorn, P., and Mujtaba, I.M. (2011). Minimization of water and chemical usage in the cleaning-in-place process of a milk pasteurization plant. *Songklanakarin Journal of Science & Technology*, 33(4), 431–440.

Porn, R., Bjork, K.-M., and Westerlund, T. (2008). Global solution of optimization problems with signomial parts. *Discrete Optimization*, 5, 108–120.

Sinha, R.K., Bharambe, G., and Bapat, P. (2007). Removal of high BOD and COD loadings of primary liquid waste products from dairy industry by vermi-filtration technology using earthworms. *IJEP*, 27(6), 486–501.

Sustainability Outlook. (2014). I. Sustainability outlook, water: A business risk to the food and water: A business risk to the food and beverages industry? Quantifying lifecycle water risk and embodied value for, no. August. 2014. Retrieved September 10, 2015, from http://sustainabilityoutlook.in/content/page/water-business-risk-food-beverages-industries-quantifying-life-cycle-risk-embodied-valu

Tomasula, P.M., Yee, W.C.F., McAloon, A.J., Nutter, D.W., and Bonnaillie, L.M. (2013). Computer simulation of energy use, greenhouse gas emissions, and process economics of the fluid milk process. *Journal of Dairy Science*, 96(5), 3350–3368.

World Bank (WB). (2010). *Deep Wells and Prudence: Towards Pragmatic Action for Addressing Groundwater Overexploitation in India.* Washington, DC: World Bank.

Section VIII

Water Management

26 A Case of Wastewater Management Modeling in the Southern Singapore Sea
Application for Coral Reef Protection

Jaan H. Pu, Yakun Guo, Md. Arafatur Rahman, and Prashanth Reddy Hanmaiahgari

CONTENTS

26.1 INTRODUCTION: BACKGROUND AND RESEARCH MOTIVATION

Singapore, with a limited sea and marine territory of about 600 km^2, represents one of the busiest island countries connected to the world by its shipping and haulage activities. According to Chou (2006), Singapore supported over 133,000 vessels that called at its ports in 2004 alone, with a large proportion (35%) composed of regional shipping between Singapore and Indonesia. In order to accommodate the busy shipping loads, port territories (managed by Maritime and Port Authority of Singapore) cover over 80% of Singapore's sea and coastal areas. Due to the convenience of shipping routes, most of Singapore's ports are located in its southern sea areas, where various wharfs and harbors located on reclaimed coasts fill the majority of the southern and southwestern coastline.

The relatively small marine space of the Singapore coastal waters was found to support a huge variety of tropical marine habitats in its marine environment. The seabed and shores that accommodate this variety of marine species are rocky, sandy or muddy. Numerous patch and fringing-type coral reefs are found near the southern islands. In research works by Chou (2006) and Knoell (2008), the seafloor of Singapore's coastal waters was found to be dominantly formed by mud-bed due to sand deposition from reclamation activities. In addition to land reclamation, shipping activities, wastewater discharge and oil spills also contribute to increase suspended sediment and modify hydrodynamic wave and tidal conditions at different locations in Singapore coastal waters, which can further expose and affect reef flats.

Judging by these aforementioned reasons, more thorough hydrodynamic computational models for understanding wastewater transport in the sea regions of Singapore, including information gathering through a low-cost wireless network, are crucial to seek relief of the serious marine issues discussed earlier. As stated in various computational and modeling studies of Singapore's coastal waters (Chao et al., 1999; Hasan et al., 2011; Kurniawan et al., 2011; Pu, 2016), there are a number of difficulties in modeling this particular coastal region. These include the mix of diurnal and semi-diurnal tidal wave conditions surrounding Singapore and the neighboring Malay Peninsula and Indonesia; the large quantity of small islands in the Singapore Sea region; and the seasonal flow impacts from, e.g., the Monsoon Season. These modeling restrictions—coupled with representative computational resolution constraints in various parts of the Singapore coastal waters—give rise to challenging scenarios for accurate tidal modeling.

Various software modeling packages (i.e., Delft3D, MIKE, and FVCOM) have been studied for their usability in producing a hydrodynamic model for the Singapore coastal waters, which have included two-dimensional (Chen et al., 2005; Sun et al., 2009; Pu, 2016) and three-dimensional modeling approaches (Chao et al., 1999; Zhang and Gin, 2000). In these studies, both 2D and 3D models gave comparable accuracy to each other, while 2D models were more effective and consumed less computational time. In a study by Pu (2015), it was further suggested that for highly turbulent flow, more advance turbulence modeling, such as the proposed Kolmogorov k-ε model, is needed to gain representative and accurate simulated results in hydrodynamic modeling.

This works constructs a literature study of wastewater modeling and proposes a possible improvement to wastewater management in Singapore coastal waters. It starts by introducing the background of the studied problem and Singapore coastal waters. Then the use of a hypothetical hybrid-wireless network will be explored for its ability to manage wastewater in the southern Singapore Sea. After that, the coral reef studies surrounding the southern Singapore Sea region will be reviewed and analyzed. Finally, the hydrodynamic and thermal mapping approaches to study the shrinking and growing of coral reefs will also be proposed and discussed in terms of its links to wastewater management, before the concluding remarks being drawn for this study.

26.2 CORAL REEFS

It was estimated that around 60% of the coral reefs around Singapore's coastal waters have been destroyed in the past few decades due to reclamation activities, while the rest are subjected to sediment impact (Chou, 2006). To study changes in the coral map, the factors commonly considered include: (1) sedimentation and sediment turbidity (van Maren et al., 2010), (2) light-intensity penetration due to sediment concentration (Dikou and van Woesik 2006), or (3) anthropogenic activities such as industrial activities, land-reclamation, dredging and pollutant transport effects (Chou, 1988; Hilton and Chou, 1999; Doorn-Groen, 2007; Goh, 2008). As was found conclusively in a majority of these studies, even though the terrestrial sediment can increase the coral colony, its influences in the shallow Singapore coastal waters are not clear (Pu, 2016). Due to the complicated reasons that cause the reduction of coral mapping, deeper exploration of those various different reasons is needed to fully understand the survival criteria for coral reefs.

As Singapore is located in a tropical climate zone, its surrounding reefs exhibit characteristics of tropical species, which are fringing-type corals that concentrate on the coastline (Chia et al., 1988). These tropical fringing coral reefs can be threatened by coral bleaching due to: (1) thermal impacts (Fitt et al., 2001) and (2) the tidal waves of a shallow sea (Storlazzi et al., 2011). Since these characteristics describe the fringing coral in the Singapore coastal waters well, the investigations of the thermal and hydrodynamic tide impacts are crucial to study coral map progression, where wastewater has to be managed in conjunction with the wireless system to improve the management in the rapidly changing tidal regions of Singapore. Their literature studies will be introduced at the following sections.

26.3 A HYPOTHETICAL HYBRID-WIRELESS NETWORK FOR WASTEWATER MANAGEMENT IN THE SOUTHERN SINGAPORE SEA

This section suggests a hypothetical hybrid-wireless network to potentially study the wastewater management in the southern Singapore Sea. The hybrid-wireless network concept presented here is one that combines two different scales of networks. One of them is a short-scale network where the end devices communicate with an access point (AP) to transmit their acquired water level and flow information. These end devices will analyze the seawater to measure water sedimentation and temperature levels. After gathering all required information, the APs will forward these data and store them in a central server, exploiting the large-scale network. The short-scale network can be in either static or mobile form. If the end devices are placed in the seashore, this would result in a static scenario; otherwise, the APs could be tagged to boat or ship, which would result in the mobile scenario. The large-scale network would be static, since the APs will forward their information through an infrastructure-oriented network. There are several technologies available for establishing this wireless communication, such as Wi-Fi, ZigBee, XBee, WiMAX, LTE, and satellite communication, as shown in Table 26.1. For the hypothetical use in southern Singapore Sea wastewater management, we chose suitable technologies from among the existing technologies for maritime communication. There are several features that need to be taken into account when choosing a suitable technology, including data rate, coverage, low latency, reduced energy expenditure, operating band and diffusion (Rahman, 2014).

Based on Table 26.1, for small-scale network, XBee would be a suitable choice, since it is able to fulfill the required features in the southern Singapore Sea. On the other hand, satellite would be suitable for a large-scale network for southern Singapore Sea use due to its better maximum data rate compared with other services. However, the choice for selecting the technologies may vary due to the requirements of the network and the implementation budget. This introduced hybrid-wireless technology would work in conjunction with the computational approach discussed in the following section for wastewater management modeling and control in the southern Singapore Sea.

TABLE 26.1
Different Communication Technologies

Technology	Frequency	Maximum Data Rate	Coverage	Mobility	Low Latency
WiFi	2.4–2.4835 GHz	150 Mbps	250 m	Yes	Yes
ZigBee	2.4 GHz	250 kbps	10–150 m	Yes	No
XBee	2.4 GHz	250 kbps	1.6 km	Yes	No
WiMAX	10–66 GHz	32–134 Mbps	Up to 30 miles	Yes	Yes
LTE	1–3 GHz	Downlink 300 Mbps Uplink 75 Mbps	5 km	Yes	Yes
Satellite	4/6 GHz (C band) 19/29 GHz (Ka)	Downlink 1000 Gbps Uplink 1000 Mbps	100–6,000 km	Yes	Yes

26.4 HYDRODYNAMIC AND THERMAL CRITERIA FOR CORAL REEF SURVIVAL—A COMPUTATIONAL APPROACH

The Singapore Sea regions are influenced by complicated semi-diurnal and diurnal tides, and their hydrodynamic characteristics play an important role in determining coral mapping. In the computational study by Pu (2016), it was observed that the tides occurring in the sea regions of Singapore showed amplitude and phase characterized by K_1, O_1 and S_2 tides. According to Pu's study, the K_1 and O_1 tides evolved from the South China Sea, while the tidal contributor for S_2 tides was from Java Sea. This has also been observed by Kurniawan et al. (2011). According to the findings by Pu (2016) and Kurniawan et al. (2011), the tidal conditions are not too severe in the southern Singapore Sea due to the protection given by the island of Java Island from the wind and monsoon effects, in which it also serves as the gathering area for coral reefs.

Further study of the thermal mapping impact to coral reefs using a proposed statistical approach by Pu (2016) indicated that the thermal impact is crucial to influence the progression of coral mapping. This finding agreed with the findings of Fitt et al. (2001) and Brown (1997) on the negative impacts of thermal stress on tropical fringing coral reefs. Pu (2016) also showed that the coral reefs surrounding the southern Singapore Sea exhibit tropical coral behavior and are sensitive towards thermal sources. Apart from both hydrodynamic and thermal considerations, the flow turbulence (as discussed by the proposed model of Pu, 2015) is another factor affecting coral reefs and will need further investigation.

26.5 CONCLUDING REMARKS

This study provides a state-of-the-art review of current literature on wastewater modeling and environmental coral reef studies in the Singapore Sea region. A hypothetical hybrid-wireless network approach is suggested to manage wastewater effectively in the southern Singapore Sea, where its usage, together with the computational modeling approach, can provide effective monitoring of wastewater management. The literature findings also show that changes in the fringing coral reefs near Singapore are probably influenced by the hydrodynamic and thermal impacts from the sea tidal waves, which are closely related to wastewater management.

ACKNOWLEDGMENTS

The first and corresponding author acknowledges the support of the Major State Basic Research Development Grant No. 2013CB036402. The support from the Major State Basic Research Development Program (973 program) of China is also greatly appreciated.

REFERENCES

Brown BE (1997) Coral bleaching: Causes and consequences. *Coral Reefs* 16:S129–S138.

Chao X, Shankar NJ, Cheong HF (1999) A three-dimensional multi-level turbulence model for tidal motion. *Ocean Eng* 26:1023–1038.

Chen M, Murali K, Khoo BC, Lou J, Kumar K (2005) Circulation modelling in the Strait of Singapore. *J Coast Res* 21(5):960–972.

Chia LS, Khan H, Chou LM (1988) The coastal environmental profile of Singapore. Association of Southeast Asian Nations/United States Coastal Resources Management Project Technical Publication Series 3, *International Center for Living Aquatic Resources*, Manila, Philippines, p. 91.

Chou LM (1988) Community structure of sediment stressed reefs in Singapore. *Galaxea* 7(2):101–111.

Chou LM (2006) Marine habitats in one of the world's busiest harbours. In: Wolanski E (Ed.), *The Environment in Asia Pacific Harbours*. Springer, Dordrecht, the Netherlands, pp. 377–391.

Dikou A, van Woesik R (2006) Survival under chronic stress from sediment load: Spatial patterns of hard coral communities in southern islands of Singapore. *Mar Pollut Bull* 52:1340–1354.

Doorn-Groen SM (2007) Environmental monitoring and management of reclamations works close to sensitive habitats. *Terra et Aqua* 108:3–18.

Fitt WK, Brown BE, Warner ME, Dunne RP (2001) Coral bleaching: Interpretation of thermal tolerance limits and thermal thresholds in tropical corals. *Coral Reefs* 20:51–65.

Goh N (2008) Management and monitoring for coral reef conservation in the Port of Singapore. *Proceedings of the 11th International Coral Reef Symposium*, Ft. Lauderdale, FL, July 7–11, 2008, Session 23, pp. 1108–1111.

Hasan GMJ, van Maren DS, Cheong HF (2011) Improving hydrodynamic modelling of an estuary in a mixed tidal regime by grid refining and aligning. *Ocean Dyn* 62:395–409.

Hilton MJ, Chou LM (1999) Sediment facies of a low-energy, meso-tidal fringing reef, Singapore. *Singap J Trop Geogr* 20(2):111–130.

Knoell C (2008) Developing the concept of building a coral reef in Singapore for conservation, environmental education, and tourism. Master Thesis, Duke University, Durham, NC, pp. 1–33.

Kurniawan A, Ooi SK, Hummel S, Gerritsen H (2011) Sensitivity analysis of the tidal representation in Singapore Regional Waters in a data assimilation environment. *Ocean Dyn* 61:1121–1136.

Pu JH (2015) Turbulence modelling of shallow water flows using Kolmogorov approach. *Comput Fluids* 115:66–74.

Pu JH (2016) Conceptual hydrodynamic-thermal mapping modelling for coral reefs at south Singapore sea. *Appl Ocean Res* 55:59–65.

Rahman MA (2014) Enabling drone communications with WiMAX technology, *Fifth International Conference on Information, Intelligence, Systems and Applications* (IISA'2014), Greece.

Storlazzi CD, Elias E, Field ME, Presto MK (2011) Numerical modelling of the impact of sea-level rise on fringing coral reef hydrodynamics and sediment transport. *Coral Reefs* 30:83–96.

Sun Y, Sisomphon P, Babovic V, Chan ES (2009) Efficient data assimilation method based on chaos theory and Kalman filter with an application in Singapore Regional Model. *J Hydro Environ Res* 3:85–95.

van Maren DS, Liew SC, Hasan GMJ (2014) The role of terrestrial sediment on turbidity near Singapore's coral reefs. *Cont Shelf Res* 76:75–88.

Zhang QY, Gin KYH (2000) Three-dimensional numerical simulation for tide motion in Singapore's coastal waters. *Coast Eng* 39:71–92.

27 Catchments as Asset Systems

A Transdisciplinary Approach to Integrated Water Resources Management

Chrysoula Papacharalampou, Marcelle McManus, Linda B. Newnes, and Dan Green

CONTENTS

27.1 INTRODUCTION: BACKGROUND AND DRIVING FORCES

The World Forum for Natural Capital (i.e., the world's natural systems, such as aquatic systems, land and services derived from them) relates poor management of the natural environment to catastrophic consequences on ecosystem productivity, human well-being and financial resilience (Natural Capital Initiative 2015). On these grounds, the United Nations Natural Capital Declaration (NCD; UNEP 2012) demonstrated the commitment private and commodity financial institutions to integrate Earth's natural assets in their reporting, accounting and decision-making. A considerable number of business initiatives have emerged, aiming to integrate natural capital in financial decision-making with a special focus on awareness raising, business encouragement and publications (Maxwell et al. 2014).

The NCD requests companies to disclose the nature of their dependence and impact on natural capital through transparent qualitative and quantitative reporting. Several policy initiatives (e.g., the UN System for Environmental-Economic Accounting, SEEA) and programs (e.g., the World Bank Wealth Accounting and Valuation of Ecosystem Services, WAVES, https://www.wavespartnership.org/) provide a basis for resource accounting, using accounting techniques in environmental science and the management of natural flows. These mainly focus on the economic valuation of natural capital and its ecosystem services. Limited work has been undertaken to evaluate whole systems and to integration of accounting methods into systems modelling.

The creation of whole-systems modelling tools would allow for the reporting and analysis of mutual relationships among built, financial and natural assets. Such tools would also enable multi-viewpoint analysis and combined-systems analysis. This functionality would be of particular use for the commodity sector, for example the water industry, as the delivery of its services depends on the provision of both physical and natural assets. The UK water sector has been officially encouraged to become more resilient by adopting integrated an approach to its asset management to achieve balance between financial costs and environmental impact (DEFRA 2016, OFWAT 2015; UKWIR 2014). In the meanwhile, approaches to enable businesses to integrate natural capital in their planning and practice have been recognised as priority areas for future research (Natural Capital Initiative 2015).

The research described in this chapter was responsive to demands for approaches allowing for transparent reporting on the dependence of the water sector on natural assets. The research presented here introduced the concept of and modelling schema for catchment metabolism (CM), which is a structured, transdisciplinary approach for modelling catchment systems and gathering data for integrated asset-management purposes. A synthesis of well-established methods and tools available from other disciplines is used in synergy to shape the basis for integrating natural capital in the strategic planning schemes of the water industry. The whole-system approach developed is based on the principles of integrated CM (ICM), water accounting and environmental multi-regional input–output analysis (E-MRIO). It builds on a combination of concepts and methods that have been reviewed and approved for their ability to address sustainability issues (Little et al. 2016; Ma et al. 2015; Paterson et al. 2015; Rudell et al. 2014; Xue et al. 2015), and shape optimised planning strategies (Daniels et al. 2011; Ma et al. 2015; Rudell et al. 2014) for better resource efficiency. The CM schema offers an approach where researchers and end users can conceptualise catchment systems and their processes, which is essential for integrated water resources management (Macleod et al. 2007).

The remainder of the chapter outlines the rationale and methods underpinning the creation of the CM modelling schema. It discusses future steps for practical applications of the schema in the UK water sector.

27.2 SYSTEM AND RESEARCH BOUNDARIES

The catchment is selected as the unit of analysis as the most suitable scale to assess water sustainability (Hester and Little 2013; Nafi et al. 2014; Papacharalampou et al. 2015) and the interactions between different types of capital (Pérez-Maqueo et al. 2013). In the context of this study, catchments are defined as hybrid integrated systems, which include both natural elements (biosphere) and infrastructure (technosphere); thus, they are defined and conceptualised as complex asset systems. Following the principle of integrated water resources management and ecosystem services (Cook and Spray 2012), the ecosystem is considered to be a stakeholder who plays an active role within the boundaries of the catchment.

27.3 CREATING THE CM MODELLING SCHEMA

The CM schema is designed on a robust, transdisciplinary basis but is also practical, so that it can be easily used by water practitioners. Its feasibility to serve everyday practice has been validated in an industrial case study in collaboration with a regional water company. This section gives an overview of the rationale for the creation of the modelling schema and the concepts and tools underpinning it. An explanatory brainstorming diagram given in Figure 27.1 outlines the synthesis of the transdisciplinary methodology. The divergence of the work and the lack of previous relevant approaches in the field of asset management required an extensive literature review to be performed. This mainly focused on the identification and analysis of the tools for integrated environmental–economic accounting widely used in other fields and that have been applied at different scales (e.g., infrastructure asset systems, community, city).

Transdisciplinary approaches have emerged to address the complexity of systems; these require methods to be constructed around the research goal (Leavy 2011; Walter et al. 2007). The term 'transdisciplinary' is used to describe an approach involving collaboration between two or more disciplines with high levels of interaction, causing the development of new conceptual, theoretical and methodological frameworks, after Leavy (2011).

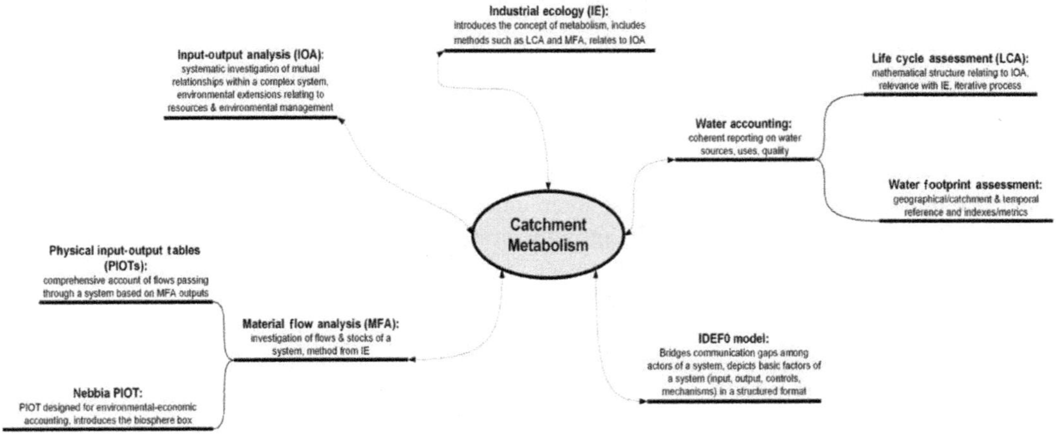

FIGURE 27.1 The formulation of catchment metabolism modelling schema based on a robust synthesis of methods available for systems engineering and environmental-economic flow accounting. (From Papacharalampou, C. et al., *J. Clean. Prod.*, 142, 1994–2005, 2016.)

To formulate CM, it was hypothesised that currently analysed tools could be used to create catchment-based approaches for asset-management purposes. For the hypotheses to be held true, the tools were required to account for both natural and built capital on a catchment basis.

The initial intention was to create an approach enabling the achievement of research goals through the application of life-cycle management and using the life-cycle assessment (LCA) tool at a catchment scale. To overcome the limitations of LCA in its spatial reference and applicability over delineated geographical areas (Baumann and Tillman 2004), a number of other tools were used. The study then explored how industrial ecology (IE)—which is the research field underpinning LCA—can be used to create CM schema. In order to do this, the development of the field of IE into other widely used concepts was explored, using a detailed literature search. Four main techniques were identified: water accounting, input–output analysis (IOA), material flow analysis (MFA) and the functional modelling language IDEF0. The structures and main knowledge blocks of a number of concepts and tools were analysed and then synthesised, based on their strength and contribution to specific objectives of the modelling schema. An overview of the concepts and techniques is presented in this section, along with the linkages among them.

IE outlines the analogy between the industrial system (technosphere) and the natural environment (biosphere) and embodies a framework oriented towards practical sustainability. It has been used in the optimisation of material cycles within industrial systems, as it serves for the development of symbiotic relationships among industries and treats the industrial system as a complex organism with unique metabolic rules (Suh and Kagawa 2005). The basic methodological concept of IE is that of 'industrial metabolism', which is a descriptive and analytical concept based on the principle of the conservation of mass applied for the understanding of the complex patterns and dynamics of flow and stocks of material and energy within the industrial system. Industrial metabolism has been widely applied in the urban context, as summarised by Clift et al. (2015) and involves a range of methods (e.g., LCA, MFA) that have served planning and development purposes, especially in the form of regional flow analysis (Brattebø 2003; Erkman 2003).

Despite the recognised value of the concept of IE in strategic sustainable development (Korhonen 2004), its application to water-related studies is rather limited (Núñez et al. 2010). Recent water-related IE applications have focused on the development of indicators for effective water management (Ziolkowska and Ziolkowski 2016; Farreny et al. 2013), the formulation of models for water demand and pricing (Dharmaratna and Harris 2012; Morales-Pinzón et al. 2012) and the environmental assessment of municipal and urban systems (Lemos et al. 2013; Oliver-Solà et al. 2013) and cultural services (Farreny et al. 2012).

LCA is technique of IE used to quantify the environmental impact associated with all the stages of a product, service or process from cradle to grave and has gained popularity as a sustainability assessment method (Guinée et al. 2011), as evidenced by the increasing number of publications and databases supporting its implementation. Until recently, water flows have been neglected in freshwater inventories and impact-assessment studies. The last few years, however, there has been a growing interest in the field of water accounting, followed by the development of metrics and indicators (Kounina et al. 2013) that can assist communication among water-related scientists, policy-makers and stakeholders. Water accounting is the systematic process of identifying, quantifying, reporting and publishing information about water as a resource (e.g., sources and uses of water). The information produced needs to be coherent and harmonised in order to prove useful to decision makers within the water sector.

There have been two main parallel developments in the water accounting community: the Water Footprint Network (WFN, Hoekstra 2011) and the LCA of water or the water-footprint (WF) standard (ISO/DIS 14046). As analysed in Boulay et al. (2013), both methodologies aim at helping practitioners to manage and sustain water resources. However, the quantitative indicators obtained from the LCA-type and WFA-type approaches are hardly comparable: LCA is largely focused on a product, whilst the crux of WFA is water management over a given geographic area. In recent years, a number of reviews of both methodologies have been conducted (Berger et al. 2010; Kounina et al. 2013) alongside critiques (Chenoweth et al. 2014; Tillotson et al. 2014; Wichelns 2015; Yang et al. 2013), mainly in regard to the limitations of these methodologies, in terms of their policy relevance, data accuracy, methodological approaches and conceptual consistency. Attempts to pursue methodological harmonisation between LCA and footprint research have been strongly encouraged in the literature.

Recent case studies (e.g., Hubacek et al. 2011; Yang et al. 2010; Yu et al. 2010; Zhi et al. 2014) have focused on the combined use of WFs with IOA as a means to inform regional or national decision-making.

IOA was introduced in the 1930s as an analytical framework to investigate the economic transactions between the various sectors of an economy. Since, it has evolved and been widely applied in a large number of studies and fields (Hubacek et al. 2011) as a method for systemically quantifying the mutual interrelationships within a complex economic system and has proven valuable in IE studies for the compilation of statistical data at a national or sectorial level (Suh and Kagawa 2005). Economic input–output modelling has also been used for environmental systems analysis. Environmental IOA (E-IO) and its multi-regional extensions (E-MRIO) have emerged as popular and promising frameworks for sustainability analysis (Hendrickson et al. 2007; Wiedmann et al. 2011). E-IO enables assessment of natural resources and pollutants embodied into goods and services and in their supply chains along the economy. Multi-regional input-output analysis enhances this capability by mapping the geography of resource use, emissions and other environmental effects, providing a spatially explicit framework that can assist in assessing environmental impacts. This ability of geo-position is vital for assessing sustainable scale and impacts for many environmental resources, especially for water, since its sustainability and management is considered at a local level (Daniels et al. 2011). Recent applications (Hubacek et al. 2011; Yang et al. 2010; Yu et al. 2010; Zhi et al. 2014) show progress in the integration of geographical information and process-based WFs in input–output models and accounting tables.

Physical input–output tables (PIOTs) are accounting tools that provide a comprehensive description of anthropogenic material flows (e.g., material and energy flows) passing through the economy of a country. For their construction, the mass–balance principle is utilised, and the economic system is depicted as being embedded in the larger natural system. An MFA study can form the basis for the quantitative information necessary to construct a PIOT. MFAs have been widely used to assess the material base and resource throughput of national economies (Brunner and Rechburger 2003; Giljum and Hubacek 2009), and its applications generally include the quantification of aggregated resource inputs and outputs of economic systems and are performed according to a methodological guidebook (Eurostat 2001).

The result of the transferal of MFA data to the PIOT is that the output produced by each production chain is split among various columns, where each column refers to a specific economic sector. A full PIOT can show material flows between sectors (industry by industry) or the materials required to transform other materials in the production process (material by material or commodity by commodity). In general, a PIOT is a tabular scheme in which a certain number of economic activities or sectors are represented by their material input and output. Nebbia (2000; 1975) outlines a type of PIOT aiming to capture the circularity of industrial metabolism in terms of a natural history of commodities—from the environment, and back to the environment. At the heart of Nebbia's PIOT is an economic–ecologic accounting, carried out by the principles of commodity science to determine intersectoral flows between and within the biosphere and the technosphere. The distinguishing feature of this approach is that the biosphere too, not just the sectors of economic system, can receive intersectoral flows. As analysed in De Marco et al. (2009), the general formation for the construction of a Nebbia's PIOT can be synthesised in a table, which is initially split in four different quadrants:

	Nature (*i*)	Technosphere (*j*)
Nature (*i*)	*aii*	*aij*
Technosphere (*j*)	*aji*	*ajj*

where *aii* represents flows within the biosphere, *aji* is resources "sold" from the biosphere to the technosphere (e.g., water used in production processes), *aji* is material flows from the technosphere to the biosphere (e.g., waste disposed or emissions) and *aij* is commodities exchanged between different technosphere sectors (e.g., electricity sold to production processes).

Using this PIOT, one can compute the "physical" mass of materials absorbed by final consumption, including exports and stocks, minus the imports. However, its application to date has excluded the mass of water that circulates through the natural and economic systems (e.g., embedded water in products). The major shortcoming of PIOTs is that all flows are accounted for in one single unit; thus, the consideration of the qualitative differences of materials flows in terms of different environmental impacts is very limited (Giljum and Hubacek 2009), and more research needs to be undertaken to overcome this issue.

Undertaking the steps to construct a PIOT that would represent outputs of the sectors within the complex catchment system, a tremendous amount of data is required, along with the contribution of multiple experts. To overcome this challenge, a IDEF0 was introduced in the schema. IDEF0 (a compound acronym deriving from Icam DEFinition for Function Modelling, National Institute for Standards and Technology, December 21, 1993) is a method designed to model the decisions, actions and activities of an organisation or a system. It has been applied, but is not limited, to topics such as strategic planning, hybrid systems design and business process reengineering (Feldmann, 1998) and has proven useful for handling complexity and bridging communications gaps between various actors involved in a system. Recent research (Settanni et al. 2015, 2014; Šerifi et al. 2009) highlights the applicability of the method across disciplines and sectors for the development of modelling approaches to product service systems to measure performance and outcomes of asset systems, as well as designing software packages.

An IDEF0 model (made of several IDEF0 diagrams) depicts constraint, not flow. The graphical elements of IDEF0 are very simple (Figure 27.2)—just boxes and arrows. The syntax and semantics for both IDEF0 diagrams and models are precisely defined in the FIPS for IDEF0 (FIPS PUB 1983). Each activity box on an IDEF0 diagram depicts the function described by the verb phrase written in the box. The arrows shown entering and leaving the boxes depict things that are needed or produced by the function. Unlike data flow diagrams, an IDEF0 model shows what controls each activity and who performs it, as well as the resources needed by each activity. The development of an IDEF0 model is a step-by-step procedure, beginning at the point where the author determines

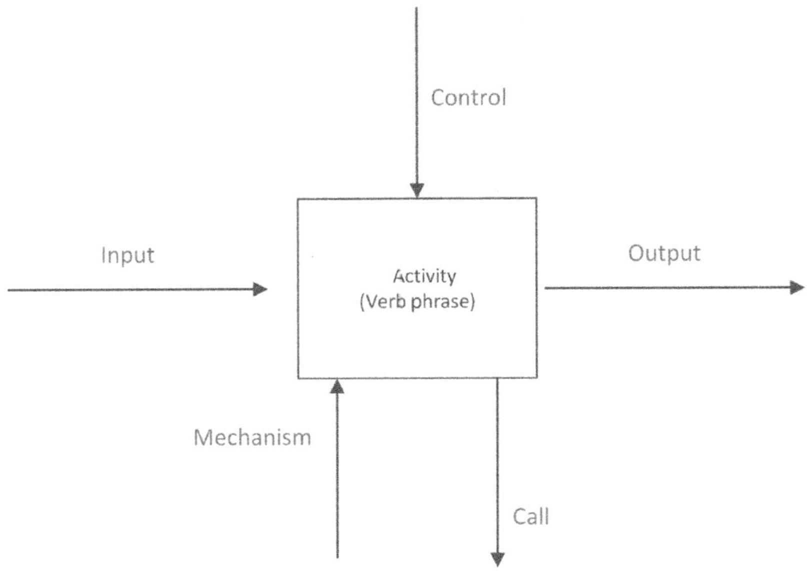

FIGURE 27.2 IDEF0 basic syntax. (From Papacharalampou, C. et al., *J. Clean. Prod.*, 142, 1994–2005, 2016.)

the basic parameters of the model: the purpose and the viewpoint. For the same system, different IDEF0 models can be created, based on the selected viewpoint.

Summing up, the creation of the CM modelling schema, a structured, creative and transdisciplinary approach was followed, and a number of concepts and techniques were synthesised. The concept of metabolism derives from the field of IE and was used as the conceptual basis for the modelling schema. MFA and its PIOTs were used to formulate the reasoning for flow accounting within the catchment systems and construct the format of the Catchment PIOT. IOA and its environmental extensions were used as tools to account for the multiple flows of the complex catchment system in a constructed approach. Water-accounting methods provide the metrics for water-flow accounting in multiple systems. The IDEF0 model was selected to serve as a method of collecting and depicting information for the subsystems of the catchment and to bridge communication gaps among the experts involved in the process of integrated catchment management.

27.4 STEPS FOR APPLYING THE CM MODELLING SCHEMA

A number of steps were undertaken in order to depict and map the metabolism of the selected system. These building blocks of the CM modelling schema synthesise a new approach to asset management in the water sector.

A catchment PIOT is the main output of the schema and is constructed through a sequel of interlinked stages which add value to the modelling schema (Figure 27.3). This catchment PIOT was developed as a structured way to map the metabolism a catchment, which essentially refers to the inter-industrial relationships taking place within the system's boundaries. The metabolic relationships of the catchment compartments were mapped over a period of a year. This time scale was chosen in order for practical and scientific purposes and also complies with the rules of the original PIOTs. In order to gain insight into the natural processes of the catchment system, the catchment PIOT included a breakdown of the biosphere into its sectors, such as the hydrosphere and atmosphere.

Following the example of the original PIOT, following the selection of the catchment system under study (step 1), an MFA of the catchment is performed (step 2). This focusses on water

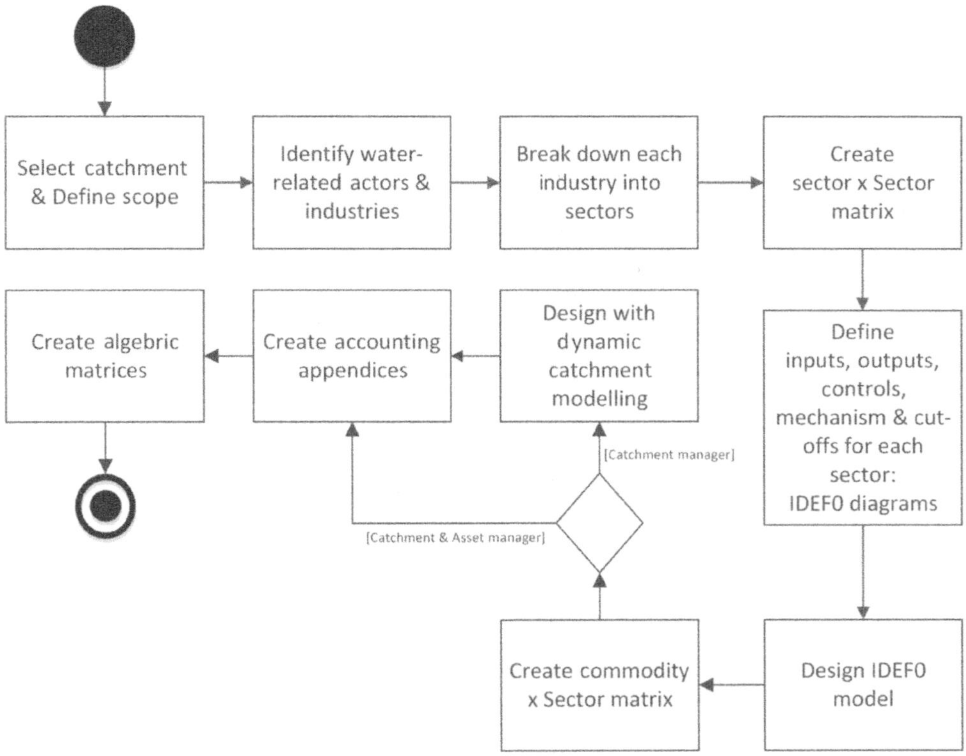

FIGURE 27.3 Steps undertaken to produce a catchment metabolism modelling schema for a selected catchment. (From Papacharalampou, C. et al., *J. Clean. Prod.*, 142, 1994–2005, 2016.)

circulation within the catchment's boundaries and can be depicted by a modified flowchart. After the key water actors and their roles are identified within the catchment boundaries (steps 3 & 4), the metabolism of the most critical subsystems needs to be studied. The criticality of the subsystems selected reflects both the scope of the work and the key issues in the designated catchment. IDEF0 diagrams were produced for each the identified industries or actors, analysing the inputs, outputs, controls and mechanisms of their subsystems (steps 5 & 6). At the next stage, the cells of the catchment PIOT were filled in using indexes from water accounting, where the output of each of the sectors (row) to the other sectors (column) are depicted (steps 7, 8 & 9). The later steps of the modelling schema require the input of multiple experts, as a combination of dynamic catchment and water accounting modelling is essential.

As a result of these processes, a catchment PIOT was constructed, with each column representing figures related to the inputs received by a single metabolic compartment of the system. Similarly, for the original PIOT, this procedure assists to the visualisation of the quantitative information relating to each component (sector) of the catchment in the form of inter-component exchanges. The catchment PIOT created was essentially a matrix of flows, both physical and economic, circulating within the catchment boundaries.

Applying the CM schema in practice requires input from a number of experts, because for the needs of this transdisciplinary methodology, a wide spectrum of expertise must be synthesised. Figure 27.4 demonstrates the types of experts and their individual contributions for the design and application of the CM schema.

The practical application of the schema is a comprehensive process, requiring collaborative action and input from multiple experts. Throughout the process, an asset manager and a catchment expert are heavily involved. These roles can be fulfilled by individuals or teams. Their common tasks include the definition of the scope of the application and the identification of the main water

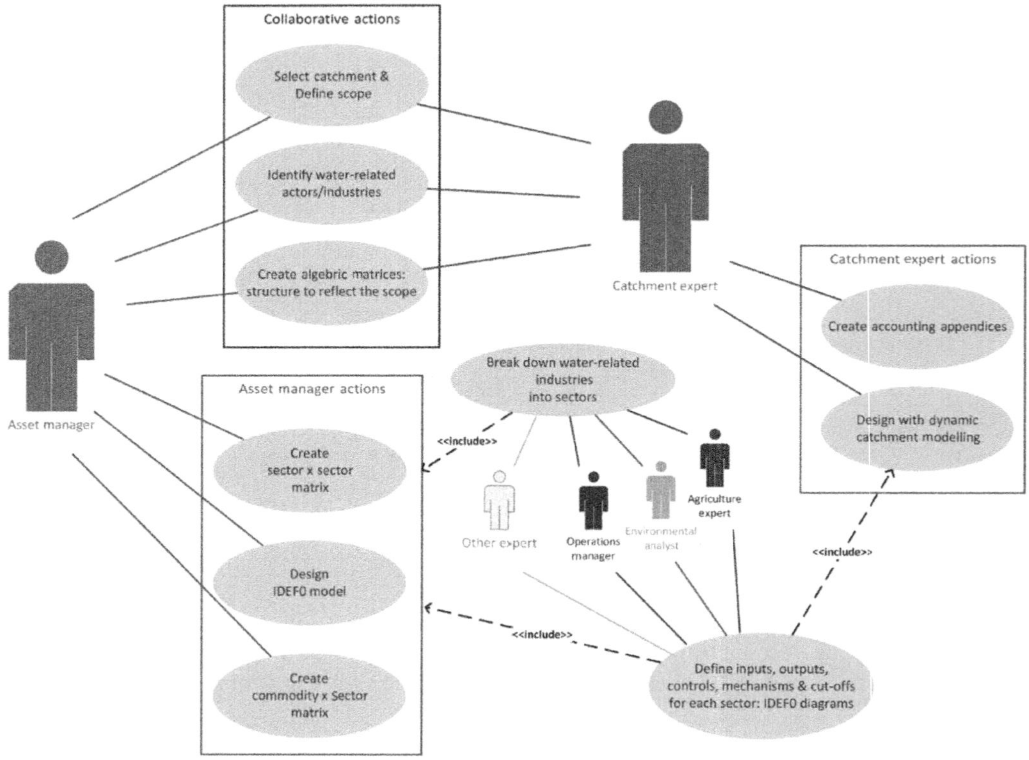

FIGURE 27.4 Expert input for the production and implementation of the catchment metabolism schema within a water company. (From Papacharalampou, C. et al., *J. Clean. Prod.*, 142, 1994–2005, 2016.)

actors in the catchment, that is, of the catchment metabolism. Their individual tasks reflect their particular skills and knowledge and are also aligned with input from other companies or external experts. In their individual tasks, the asset manager is responsible for the construction of the matrices representing the output of individual sectors or activities within catchment boundaries, while the catchment expert develops accounting mechanisms for computations of the output, making use of water-accounting techniques.

For the creation of the catchment PIOT and the IDEF0 model, a number of experts are required in order to perform the breakdown of water-related industries into sectors and to define their structural features (inputs, outputs, controls and mechanisms) respectively. For the case study presented in this work, the expertise of an environmental analyst, an operations manager and an agricultural expert were required to analyse the building blocks of the three main water actors identified within the given system. The further development of the CM schema and its application to diverse catchment typologies—in terms of their water sectors and activities—required involvement of and input from different experts. That serves to create knowledge blocks and ensures the quality of the data displayed and produced.

The data produced by the assembly of the separate IDEF0 diagrams constituted the heart of the entire schema, providing essential insights in the subsystems of the catchment under consideration. The asset manager then pulled the separate IDEF0 diagrams together in order to create the IDEF0 model and the input–output matrices for sectors and commodities. The data gathered for the development of the IDEF0 model served as the basis for the construction of a systems dynamic model by the catchment expert. The output of this type of model produced the information for the catchment PIOT.

27.5 CONCLUDING REMARKS

The work introduced a structured approach for conceptualising and modelling catchment as complex asset systems. The research outputs of this approach can enable water companies to meet UK national policy demands for integrated and resilient asset management. CM modelling schema responds to the need for evidenced-based approaches, which can be used in the practical application of sustainability and systems-thinking principles in the water industry. It is tailored to address current challenges in the water sector, and its design enables practitioners to apply research advancements. One of the advantages of the schema is that systems-thinking is required, hence, collaboration among experts within the water sector occurs. This reflects the transdisciplinary nature of the work.

Despite its structured and comprehensive design, it is a rather sophisticated and data-intensive methodology that requires collaboration among experts and the automatization of processes in a later stage. The application of the schema to diverse typologies of catchments is required to evaluate its flexibility and highlight areas for future improvement, and its use in multiple and diverse case studies may provide further practical insight and facilitate the integration of the approach in everyday practice.

ACKNOWLEDGMENTS

The research presented in this chapter is largely based on the publication: "Catchment Metabolism: Integrating Natural Capital in the Asset Management Portfolio of the Water Sector" (Papacharalampou et al. 2016). The author would like to thank the co-authors of the journal publication and the funders of the project: Wessex Water Services Ltd and the EPSRC Impact Acceleration Award (EP/K503897/1).

REFERENCES

Baumann, H., Tillman, A.M. (2004). *The Hitch Hiker's Guide to LCA: An Orientation in Life Cycle Assessment Methodology and Application*. Lund, Sweden: Studentlitteratur.

Berger, M., Finkbeiner, M. (2010). Water footprinting: How to address water use in life cycle assessment? *Sustainability*, 2, 919–944.

Boulay, A.M., Hoekstra, A.Y., Vionnet, S. (2013). Complementarities of water-focused life cycle assessment and water footprint assessment. *Environmental Science and Technology*, 2013(47), 11926–11927.

Brattebø, H. (2003). The impact of industrial ecology on university curricula. *Perspectives on Industrial Ecology*, Bourg, D., Erkman, S. (Eds.). Sheffield, UK: Greenleaf Publishing.

Brunner, P.H., Rechburger, H. (2003). *Practical Handbook of Material Flow Analysis*. Boca Raton, FL: CRC Press.

Cook, B.R., Spray, C.J. (2012). Ecosystem services and integrated water resources management: Different paths to the same end? *Journal of Environmental Management*, 109, 93–100.

Chenoweth, J., Hadjikakou, M., Zoumides, C. (2014). Quantifying the human impact on water resources: A critical review of the water footprint concept. *Hydrology and Earth System Sciences*, 18, 2325–2342. doi:10.5194/hess-18-2325-2014.

Clift, R., Druckmna, A., Christic, I., Kennedy, C., Keirstead, J. (2015). Urban metabolism: A review in the UK context. Government Office for Science, No GS/15/30, London, UK.

Daniels, P.L., Lenzen, M., Kenway, S.J. (2011). The ins and outs of water use- a review of multi-region input-output analysis and water footprints for regional sustainability analysis and policy. *Economic Systems Research*, 23(4), 353–370.

De Marco, O., Lagioia, G., Amicarelli, V., Sgaramella, A. (2009). Constructing physical input-output tables with material flow analysis (MFA) data: Bottom-up case studies. S. Suh (Ed.), *Handbook of Input-Output Economics in Industrial Ecology, Eco-Efficiency in Industry and Science*. Dordrecht, the Netherlands: Springer, p. 23.

Department for Environment Food & Rural Affairs (DEFRA). (2016). Creating a great place for living: Enabling resilience in the water sector. Report PB14418, London, UK.

Dharmaratna, D., Harris, E. (2012). Estimating residential water demand using the stone-geary functional form: The case of Sri Lanka. *Water Resources Management*, 26(8), 2283–2299. doi:10.1007/s22269-012-0017-1.

Erkman, S. (2003). *Perspectives on Industrial Ecology*. Bourg, D., Erkman, S. (Eds.). Sheffield, UK: Greenleaf Publishing Limited.

Eurostat. (2001). Economy-wide material flow accounts and derived indicators. A methodological guide. European Commission. Luxemburg. ISN 92-894-0459-0.

Farreny, R., Rieradevall, J., Barbassa, A.P., Teixeira, B., Gabarrell, X. (2013). Indicators for commercial urban water management: The case of retail parks in Spain and Brazil. *Urban Water Journal*, 10(4), 281–290.

Farreny, R., Oliver-Solà, J., Escuder-Bonilla, S., Roca-Marti, M., Sevigné, E., Gabarrell, X., Rieradevall, J. (2012). The metabolism of cultural services. Energy and flows in museums. *Energy and Buildings*, 47, 98–106.

Feldmann, C.G. (1998). *The Practical Guide to Business Process Reengineering Using IDEF0*. New York: Dorset House Publishing.

FIPS PUB. (1983). Integration Definition for Function Modelling (IDEF0), U.S. Department of Commerce, Technology, Administration, National Institute of Standards and Technology (Washington, DC: 1983).

Giljum, S., Hubacek, K. (2009). Conceptual foundations and applications of input-output tables. *Handbook of Input-Output Economics in Industrial Ecology*, S. Suh (Ed.). Dordrecht, the Netherlands: Springer.

Guinée, J.B., Heijungs, R., Huppes, G. (2011). Life cycle assessment: Past, present, future. *Environmental Science and Technology*, 45(1), 90–96. doi:10.1021/es101316v.

Hendrickson, C., Hawkins, T., Higgins, C., Matthews, S. (2007). A mixed input-output model for environmental life cycle assessment and material flow analysis. *Environmental Science and Technology*, 41, 1024–1031.

Hester, E.T., Little, J.C. (2013). Measuring environmental sustainability of water in watersheds. *Environmental Science and Technology*, 47, 8083–8090.

Hoekstra, A.Y., Chapagain, A.K., Aldaya, M.M., Mekonnen, M.M. (2011). *The Water Footprint Assessment Manual: Setting the Global Standard*. London, UK: Earthscan.

Hubacek, K., Feng, K., Siu, Y.L., Guan, D. (2011). Assessing regional virtual water flows and water footprints in the Yellow River Basin, China: A consumption based approach. *Applied Geography*, 32, 691–701.

Korhonen, J. (2004). Industrial ecology in the strategic sustainable development model: Strategic applications of industrial ecology. *Journal of Cleaner Production*, 12, 809–823.

Kounina, A., Margni, M., Bayart, J.B., Boulay, A.M., Berger, M., Bulle, C., Frischknecht, R. et al. (2013). Review of methods addressing freshwater use in life cycle inventory and impact assessment. *International Journal of Life Cycle Assessment*, 18, 707–721.

Leavy, P. (2011). *Essential for Transdisciplinary Research: Using Problem-Centred Methods*. Walnut Creek, CA: Left Coast Press.

Lemos, D., Dias, A.C., Gabarrell, X., Arroja, L. (2013). Environmental assessment of an urban water system. *Journal of Cleaner Production*, 54(1), 157–165.

Little, J.C., Hester, E.T., Carey, C.C. (2016). Assessing and enhancing environmental sustainability: A conceptual review. *Environmental Science and Technology*, 50, 6830–6845.

Ma, X., Xue, X., González-Mejía, A., Garland, J., Cashdollar, J. (2015). Sustainable water systems for the city of tomorrow—A conceptual framework. *Sustainability*, 7, 12071–12105. doi:10.3390/su70912071.

Macleod, C.J.A., Scholefield, D., Haygarth, P.M. (2007). Integration for catchment management. *Science of the Total Environment*, 373, 591–602.

Maxwell, D., McKenzie, E., Traldi, R. (2014). Natural Capital Coalition: Valuing Natural Capital in Business. ICAEW.

Morales-Pinzón, T., Rieradevall, J., Gasol, C.M., Gabarrell, X. (2012). Potential of rainwater resources based on ubran and social aspects in Colombia. *Water and Environmental Journal*, 26(4), 550–559.

Nafi, A., Bentarzi, Y., Granger, D., Cherqui, F. (2014). Eco-EAR: A method for the economic analysis of urban water systems providing services. *Urban Water Journal*, 11(6), 1–15.

Nebbia, G. (1975). The waste matrix. *Rassegna Economics*, 39(1), 37–62.

Nebbia, G. (2000). Monetary accounting and environmental accounting. *Economic Pubblica*, 30(6), 5–33.

Núñez, M., Oliver-Solà, J., Rieradevall, J., Gabarrell, X. (2010). Water management in integrated service systems: Accounting for water flows in urban areas. *Water Resources Management*, 24, 1583–1604.

Natural Capital Initiative. (2015). Valuing our life support systems 2014. Summit summary report.

OFWAT. (2015). Trust in water, Water 2020: Regulatory framework for wholesale markets and the 2019 price review. London, UK: OFWAT. Retrieved from https://www.ofwat.gov.uk/wp-content/uploads/2015/12/pap_con20150912 water2020.pdf.

Oliver-Solà, J., Armero, M., Martinez de Foix, B., Rieadevall, J. (2013). Energy and environmental evaluation of municipal facilities: Case study in the province of Barcelona. *Energy Policy*, 61, 920–930.

Papacharalampou, C., McManus, M., Newnwes, L.B., Green, D. (2016). Catchment metabolism: Integrating natural capital in the asset management portfolio of the water sector. *Journal of Cleaner Production*, 142(4), 1994–2005. doi:10.1016/j.jclepro.2016.11.084.

Papacharalampou, C., McManus, M., Newnes, L.B., Hayes, J., Wright, J. (2015). Thinking Catchments: a holistic approach to asset management in the water sector. *Proceedings of the IWA Cities of the Future: Transitions to the Urban Water Services of Tomorrow (TRUST) Conference*, Germany, April 2015. The Hague, the Netherlands: International Water Association (IWA), pp. 184–195.

Paterson, W., Rushforth, R., Rudell, B.L., Konar, M., Ahams, I.C., Gironás, J., Mijic, A., Mejia, A. (2015). Water footprint of cities: A review and suggestions for future research. *Sustainability*, 7, 8461–8490. doi:10.3390/su7078461.

Pérez-Maqueo, P., Martinez, M.L., Vázquez, G., Equihua, M. (2013). Using four capitals to assess watershed sustainability. *Journal of Environmental Management*, 51, 679–693.

Settanni, E., Newnes, L.B., Wright, J. (2015). Fifty shades of greywater: An outcome perspective on wastewater systems operations' performance. *Conference Proceedings of the 19th Cambridge International Manufacturing Symposium*, September 24–25, 2015.

Settanni, E., Newnes, L.B., Thenent, N.E., Parry, G., Goh, Y.M. (2014). A through-life costing methodology for use in product-service-systems. *International Journal of Production Economics*, 153, 161–177. doi:10.1016/j.ijpe.2014.02.016.

Šerifi, V., Dašic, P., Ječmenica, R., Labović, D. (2009). Functional and information modelling of production using IDEF methods. *Journal of Mechanical Engineering*, 55(2), 131–140.

Suh, S., Kagawa, S. (2005). Industrial ecology and input-output economics: An introduction. *Economic Systems Research*, 17(4), 349–364.

Tillotson, M.R., Liu, J., Guan, D., Wu, P., Zhao, X., Zhang, G., Pfister, S., Pahlow, M. (2014). Water footprint symposium: Where next for water footprint and water assessment methodology? *International Journal of Life Cycle Assessment*, 19, 1561–1565.

Rudell, B.L., Adams, E.A., Rushforth, R., Tidwell, V.C. (2014). Embedded resource accounting for coupled natural-human systems: An application to water resource impacts of the western U.S. electrical energy trade. *Water Resources Research*, 50, 7957–7972. doi:10.1002/2013WR014531.

UKWIR. (2014). Environmental impact assessment to compare the benefits of achieving tight BOD standards versus increase in whole lifecycle carbon. Report Ref. No. 14/WW/20/7.

United Nations Statistic Division, UNDS. (2012). System of Environmental-Economic Accounting for Water (SEEAW). United Nations, Department of Economic and Social Affairs, Statistics Division, New York.

Walter, A.I., Wiek, A., Scholz, R.W. (2007). Constructing Regional Development Strategies: A case study approach for integrated planning and synthesis. In: *Handbook of Transdisciplinary Research*, Hadorn, G.H. et al. (Eds.). Dordrecht, the Netherlands: Springer, pp. 223–243.

Wichelns, D. (2015). Virtual water and water footprints do not provide helpful insight regarding international trade or water scarcity. *Ecological Indicators*, 52, 277–283.

Wiedmann, T., Wilting, H., Lenzen, M., Lutter, S., Palm, V. (2011). Quo Vardis MRIO? Methodological, data and institutional requirements for multi-region input-output analysis. *Ecological Economics*, 70, 1937–1945.

Xue, X., Schoen, M.E., Ma, X., Hawkins, T.R., Ashbolt, N.J., Cashdollar, J., Garland, J. (2015). Critical insights for a sustainability framework to address integrate community water services: Technical metrics and approaches. *Water Research*, 77, 155–169.

Yang, H., Pfister, S., Bhaduri, A. (2013). Accounting for a scarce resource: Virtual water and water footprint in the global water system. *Current Opinion in Environmental Sustainability*, 5, 599–606.

Yang, H., Zhao, X., Yang, F., Chen, B., Qin, Y. (2010). Applying the input-output method to account for water footprint and virtual water trade in the Haihe River Basin in China. *Environmental Science and Technology*, 44, 9150–9156.

Yu, Y., Hubacek, K., Feng, K., Guan, D. (2010). Assessing regional and global water footprints for the UK. *Ecological Economics*, 69, 1140–1147.

Ziolkowska, J.R., Ziolkowski, B. (2016). Effectiveness of water management in Europe in the 21st century. *Water Resources Management*, 30(7), 2261–2274. doi:10.1007/s11269-016-1287-9.

Zhi, Y., Yang, F., Yin, X.A. (2014). Decomposition analysis of water footprint changes in a water-limited river basin: A case study of the Haihe River basin, China. *Hydrology and Earth System Science*, 18, 1549–1559.

28 Water Efficiency Lapses and Sustainable Solutions

Educational Buildings in Johannesburg, South Africa

Aghaegbuna O. U. Ozumba

CONTENTS

28.1 INTRODUCTION

The research interest for the study presented here is the apparent lack of sustainability, or the inefficiency, of water supply systems in buildings or facilities including the nature of existing problems and the sustainability of user-proffered solutions. In this case the concern is with educational buildings in institutions of higher learning. The chapter proceeds through an identification and analysis of the common lapses that occur within the plumbing systems of such facilities and considers suggestions for possible remedies. The consideration of user-suggested solutions is performed through the lens of selected sustainability principles for water efficiency in buildings. The study was carried out within the context of South Africa, on the basis of initial studies published in Aduda and Ozumba (2011), Silva-Afonso (2014), Silva-Afonso and Pimentel-Rodrigues (2014), and Ozumba and Aduda (2015).

The study presented here borders on the social, socio-technical, and socio-economic highlights of water conservation in the context of buildings. The particular context refers to the efficient provision of 'wet services', in buildings. Wet services refer to the system and components for water supply, distribution, and drainage in buildings. Water supply can be described at the regional, municipal, neighbourhood, and building levels. This chapter is, however, focused on water supply in buildings/facilities. For the purpose of the study, an existing perspective, the 5R Principle (Silva-Afonso and Pimentel-Rodrigues, 2014; Silva-Afonso, 2014) is adopted, through which water efficiency is explored in selected buildings. Through analysis of primary data and comparison with previous studies, a more accurate understanding of the problems, sustainability of the solutions, and applicability of the chosen sustainability perspectives are gained.

28.2 SUSTAINABILITY IN THE BUILT ENVIRONMENT, WATER EFFICIENCY AND THE 5R PRINCIPLE

Sustainability or sustainable development is defined as wholesome development that achieves balanced development in terms of addressing the three broad areas of environmental, economic, and social development (United Nations (UN), 1992).

The interpretation of sustainability in the built environment leads to sustainable construction. Essentially the application of sustainability principles in construction manifests as sustainable approaches to construction. The approach should create a built environment that is responsive to human, economic, and environmental factors (Kilbert, 2016:1). The sustainable built environment is expected to: enhance biodiversity, support communities, use resources efficiently, minimise pollution, and create healthy environments (Halliday, 2008:ix). Considering the assertions by the UN (1992) and Halliday (2008:ix), it is arguable that water is one of the critical areas for sustainability. As a natural resource, water fits into the environmental aspect of the sustainability concept. Following this argument, there is need to: conserve available water resources, use less water to achieve more, and efficiently use water to avoid wastage.

The UN defines water efficiency as being 'a multifaceted concept', which aims at achieving more value while lowering water use and its impact on the environment. It is described as involving every measure taken to safeguard and conserve the natural resource from excessive consumption and pollution, to achieve equity of allocation and higher 'socio-economic value', and maintain the natural flows and cycle of water (UNEP, 2014). The terms conservation and efficiency are used interchangeably in this chapter. Water conservation is a global need, which is emphasised by its inclusion in the Millennium Development Goals (MDGs; UN, 2016), and the UN Sustainable Development Goals (SDGs; UN Development Programme (UNDP), 2017). The SDGs highlight the importance of water and sanitation in facilities and their importance in emerging/developing communities around the world (Stockholm Environmental Institute, 2009; GRID-Ardenal, 2009). The same needs are stated in the 17th session report of the UN Division for Economic and Social Affairs (UN-DESA, 2009). Considering the MDGs and SDGs referenced here, it is equally arguable that the built environment has an appreciable role to play in achieving a substantial portion of the goals, especially air and water quality.

Sustainability applies to the built environment in all its ramifications. This is substantiated by the lifecycle sustainability perspective, through which the built asset has been considered in recent discourse on sustainable development such as Khasreen et al. (2009), Scheuer et al. (2003), and Zhang et al. (2006). Some of the contributions of the lifecycle sustainability discourse with regards to the built environment, manufacturing, and other sectors, are the general guidelines for resource conservation during the extraction and utilisation of natural resources. These guidelines particularly emphasise conservation of non-renewables resources. One of the guidelines is the 3R principle, which refers to 'Reduce, Re-use, and Recycle'. The application could be in physical development, manufacturing of products, or provision of services (Hill and Bowen, 1997). Another guideline is the 4R principle (Clark et al., 2009), which adds the element of reconsideration of the approach, with the idea of 'Redesign and Rethink' as core concepts of 'Design for Sustainability' (Clark et al., 2009; Charter and Tischner, 2001). In the case of waste management, it is the element of 'recovery of raw materials' (Charter and Tischner, 2001). The 6R principle refers to 'Reuse, Recover, Redesign, Remanufacture, Reduce, and Recycle' and is focused on sustainable manufacturing (Bi, 2011). There is also the 7R principle that adds 'Regulation, Renovation, Recovering, and Rethinking' to the 3R principle (Ceclan et al., 2011).

The lifecycle discourse on built environment sustainability looks at various stages of the built asset according to the principles of sustainability. There is an appreciable amount of discussion on construction extraction and processing activities and their impact on the environment. Building operation has however been noted as having appreciable impact on the environment, especially on water (Arpke and Hutzler, 2006; Balaras et al., 2005; Cheng, 2002; Coyle, 2014; Silva-Afonso, 2014). Some of the principles for sustainable building/construction include the conservation of natural resources used in construction activities (Hill and Bowen, 1997; Hill et al., 1994). Such resources would include water, which is used throughout the life cycle of buildings.

The 3R principle has been applied in the specific area of water efficiency. However a more robust guideline, the 5R principle, has been introduced by Silva-Afonso and Pimentel-Rodrigues (2014) and Silva-Afonso (2014). The 5R principle as articulated in Ozumba and Aduda (2015), stands for:

> Reduce consumption – Technical measures and non-technical measures (economics, general awareness); Reduce loss and waste – Control of losses (volume and heat); Re-use water – Rechanneling used water to another area of use; Recycle water – Recycling & re-introduction of water for another use; and Resort to alternative sources – Involving other sources (rainwater, groundwater, saltwater, etc.). Ozumba and Aduda (2015).

The 5R principle is a five-point sustainability principle for water efficiency in buildings. It is essentially a further development of the 3R and other such principles of sustainability. The 5R is focused on water conservation and water efficiency measures for buildings. It applies a range of water efficiency solutions, which could possibly occur in isolation or in any set of combinations within one building (Silva-Afonso and Pimentel-Rodrigues, 2014; Silva-Afonso, 2014). The 5R principle covers a wide area of considered approaches to achieving sustainability in building/facility water usage. The approach provides specific areas of measurement for upgrading the water efficiency performance of buildings. It also provides criteria for evaluating possible solutions to water efficiency in buildings. The five water efficiency measures are focused on the design and operational stages of buildings/facilities. Using such concepts, sustainability principles are critically applied to water supply systems of facilities. Due to the simplicity of the metrics, water efficiency can be evaluated in various scenarios, regardless of the level of development. This is particularly relevant for the application of sustainable water usage in the built environment.

Deductions from the discourse thus far reveal the global case for water conservation (especially in the built environment), the criticality of water in developing regions of the world, and the application of various guidelines such as the 5R principle. In view of the deductions and substantiations presented in the next section, it is arguable that South Africa is an emerging economy that faces common and peculiar challenges that have an impact on resource demand and depletion. As substantiated below, South Africa presents a water critical context; this chapter will therefore focus on South Africa, based on the theoretical background and findings from recent research.

28.3 THE SOUTH AFRICAN CONTEXT

South Africa faces a serious threat of acute shortage of potable water and the lingering threat of natural water degradation due to natural circumstances and the impact of human activities (Stephenson and Randell, 2009; du Plessis et al., 2003). The water situation in South Africa also has historical, industrial, and socio-economic dimensions. For example, its climatic zone, coupled with the impact of climate change, is partially responsible for demand increase on scarce water resources (Smakhtin et al., 2001; Naidoo and Constantinides, 2000). The history of racial segregation and oppression of the apartheid era left deep socio-economic realities, which currently challenge the equitable provision of water for the people as well as the preservation of available fresh water supplies (Méndez-Barrientos, 2016; du Plessis et al., 2003). South Africa has experienced steady population growth, and increased urbanisation has occurred since the inception of democracy in the country (Naidoo and Constantinides, 2000; du Plessis et al., 2003). The continuous urbanisation of South Africa increases the demand for housing and services such as electricity and water in buildings. There is also the continuous rural-to-urban migration of people seeking economic advantages, which increases the demands on building services such as water supply and therefore has an impact on scarce resources (du Plessis et al., 2003). The criticality of water resources is exemplified in the level of transactions in the nation's management of its water resources and the provision of the same to its peculiar demographic landscape, as reported in van Koppen and Schreiner (2014).

From the preceding discourse, there is obvious need for water conservation in places such as South Africa, especially in its buildings, as substantiated by Creemers and Pott (2002) and Otieno and Ochieng (2004). The necessity of conserving water in South Africa's buildings was explored in Aduda and Ozumba (2011). Specifically, the paper looked at the adoption of identified technologies, and clients, plumbing contractors, and suppliers were investigated. The technology dispersal pattern was found to be prescriptive, while clients' choice of technology was influenced more by cost rather than water conservation. Lack of awareness was indicated on the part of clients, while lack of capacity to recommend technologies was indicated for the retailers. The authors suggested a framework to facilitate the diffusion of water conservation technologies in the nation's built environment, using a combination of bottom-up and top-down approaches.

The need to apply efficiency measures to water supply and plumbing in South Africa's buildings was noted in Ozumba and Aduda (2015). Existing research was extended in the study by looking beyond technology adoption, to the adoption of a broader set of efficiency measures that could achieve water conservation in buildings. The study highlighted the relative success rates of identified water efficiency measures, as well as the challenges and some of the threats to further adoption. While previous research including Aduda and Ozumba (2011) proved that some degree of adoption exists for water efficiency technologies, it is arguably easier for new development than for existing stock. There was also little information on the dynamics of water-efficient technology adoption for existing building stock. Beyond the technologies, little is known about the adoption of other water-efficiency measures and the possible combination of such measures with available technologies. The densely built up Braamfontein area, part of the Johannesburg Central Business District (CBD), was chosen for Ozumba and Aduda (2015). To conceptualise the wider view on water-efficiency efforts in buildings, the 5R principle for water efficiency measures in buildings (Silva-Afonso and Pimentel-Rodrigues, 2014; Silva-Afonso, 2014), was identified and adopted, focusing on specific aspects of water conservation as described in the introduction to the 5R principle earlier in this chapter.

Different building and occupancy types were utilised, and observations were focused on identified lapses in the system and points of extreme consumption of water, as well as identifying water-efficiency measures adopted. Major lapses were found in the piping system and points of use such as the troughs and washbasins. Older bathroom systems also resulted in high water consumption. In terms of water efficiency measures, various technology-based solutions were identified. However, there was evidence of less technical measures and the combinations of non-technical measures with more technical measures that seemed to yield more promising results for facility/building managers (Ozumba and Aduda, 2015). Facility managers in the study were creatively adopting innovations by combining technical and non-technical measures. However, the study was limited in terms of generalizability of results to other building types and stakeholders, especially users. While various data sources were explored, there was still a lot of room for further study, especially about the application of the 5R principle in different scenarios. Thus far the studies focused on clients and their representatives, service providers, and suppliers, but left out users/occupants. Earlier studies did not look for users' views on solutions to water inefficiency. Furthermore, user suggestions on sustainable approaches to address shortcomings in the wet services had not been explored. Moreover, while it was not possible to explore all possible scenarios at once, it was necessary to continue extending the study series to other building and occupancy types. In view of the need to extend the study, educational buildings in institutions of higher learning were identified as a suitable subject for the current study.

Based on the foregoing, the current study uses the 5R principle as a lens to explore water efficiency lapses in the wet services of educational buildings and user's suggestions for sustainable solutions to identified problems. The objectives are outlined as follows:

- To determine the nature and occurrence of water efficiency lapses in educational buildings in higher education institutions
- To explore users' perspectives on sustainable solutions, which are applicable to the context, in view of environmental, social, and economic factors
- To build an understanding of the outcomes of the first two objectives based on the 5R principle for water efficiency in buildings

To achieve the stated objectives, fieldwork was introduced in the study, with the collection and analysis of primary data. The design for the fieldwork is discussed in Section 28.4.

28.4 METHODOLOGY FOR THE STUDY

The study presented here was carried out as part of building science class activities at the undergraduate level. An institution of higher learning in the Johannesburg area was chosen as the context for the current study to provide a common geographical scope with previous studies in Aduda and Ozumba (2011) and Ozumba and Aduda (2015). The study adopted an exploratory view, based on the analysis of qualitative data with inductive reasoning (Thorne, 2000), utilising a unique sample of buildings and users, and studying the same buildings over 2 years (2015–2016). An appreciably customised approach was adopted, whereby sets of students collected data on lapses in the wet services and generated data on user-suggested solutions for the 2 years of study.

28.4.1 The Units of Analysis

The fieldwork was treated as appreciably complex, being made up of two foci and essentially two stages of data collection and analysis, which complement each other. Data was primarily collected from two sources, namely the buildings and the users. As such, two units of analysis were determined: lapses in the wet services of buildings, and the sustainability of the solutions suggested by users. The uniqueness of the two units of analysis was adhered to in the sampling, approach to data collection, and analysis. For each stage of analysis, the following was determined: unit of study (where the research will be conducted); unit of observation (what will be observed when the study is taking place); and unit of analysis (what will be analysed from the observation). Table 28.1 presents the research stages with corresponding units of analyses and expected outcomes.

28.4.2 Sampling for the Study

Considering the data requirements, the determined units of analysis for the study, and the two sources of data, two stages of data collection were determined that needed two sub-groups in a hybrid sample. The sample population was made of two distinct sub-groups (buildings and users), from which the sample in each case was chosen. Each sub-group within the sample corresponded to one stage of data collection.

TABLE 28.1
Units in the Research According to Stages of Investigations (Data from Study)

Stage	Units of Observation	Unit of Study	Unit of Analysis
1	Educational buildings	Wet services of buildings	Lapses in the wet services of buildings.
2	Informed users	Informed user suggested solutions to lapses in the wet services of buildings	Sustainability of suggested solutions to lapses in the wet services of buildings.
3	Informed user suggested solutions to lapses in the wet services of buildings	Application of the 5R principle for water efficiency in buildings to suggested solutions	Relevance and applicability of the 5R principle to the study context. Nature and occurrence of suggestions of water efficiency measures, which lie outside the 5R principle.

The user sub-group within the study sample was classified as informed users. The classification of 'informed users' is based on the following:

- The participants were undergraduate students of building science.
- They were knowledgeable about wet services in buildings through building science discourse.
- They were knowledgeable about issues around sustainability, having gone through a semester of a highly relevant coursework that focused on sustainability in the built environment.
- They were knowledgeable about some data collection techniques, chiefly observation and documentary review.
- They were also investigating buildings that they use on daily basis on campus.

Furthermore, there was an opportunity to exploit the dual function of the informed user: a basic user and an informed solution provider. Additional benefits include accuracy of observation, cross-validation, and productive group discussions that yielded the suggested solutions. In the first stage of data collection, the undergraduate participants collected observational data on lapses in the wet services of buildings. In the second stage of data collection, they generated data by making suggestions about sustainable solutions. To achieve a richer, condensed, and more manageable data set, the sourcing of data focused on user groups. The user groups were purposively selected, being made up of students who enrolled for a building science course. The investigations were carried out between 2015 and 2016. The total number of undergraduate participants reported for the 2015 study was 138. They were organised into groups ranging from 5–8 members, which created 20 groups for the fieldwork. The total number of undergraduate participants reported for the 2016 study was 56. They were put into groups ranging from 4–6 members, which created 10 groups for the fieldwork. In total, 194 undergraduate participants worked on the study, constituting 30 groups of informed users.

The building sub-group within the study sample was made up of purposively selected buildings from two campuses of the same university. The sampling was based on achieving an appreciable degree of variability in the building and occupancy types on campus. There were five buildings in all that were investigated. In 2015, all five buildings were studied. In 2016, only four out of the original five buildings were studied, due to the lower number of groups available for the study.

28.4.3 The Research Strategy

To achieve the aim of the study, a bespoke and essentially pragmatic approach was utilised (Saunders et al. 2009:127). The research approach involved three levels of customisation, which was achieved through the following: the use of multiple units of analysis, multi-staged research design, and an integration of the features of various research strategies. The research approach for the study combined the strengths of case study (Yin, 1999), survey (Butts, 1983), focus group discussion (Calderón et al., 2000; Huer et al., 2003), and participant observer mode (Iacono et al., 2009; Kawulich, 2005).

The research instrument, as informed by the study objectives, provided lines of enquiry for:

- Identifying lapses in the system
- Suggesting sustainable solutions
- Achieving consensus on their suggestions (for cross-validation)
- Reporting observations (direct capture from field notes), and suggestions in written form

The groups also had to cross-validate individual observations on the spot. Captured information was submitted in the form of reports by each user group, which was analysed in this study as primary data. The reports were made up of text, photographs, and sketches of each group's observations. Another section of the report contained the group suggestions.

Analysis of data was performed through content analysis. Submitted reports were analysed thematically and categorised. The following software tools were utilised for data analysis: Microsoft Excel, Provalis Wordstat, and Wordclouds.com. Though the number of individual participants was nearly 200, there were only 30 participating groups. Through data analysis, categorisations were achieved, prevalence and commonality of lapses were explored, and solutions were analysed for sustainability in relation to relevant lapses. From the results of analyses, useful propositions regarding the context of the current study and results of previous studies were made. Furthermore, the relevance of the 5R principle to the identified range of efficiency measures in this context was evaluated; additional efficiency measures, if any, were identified and explored.

28.5 PRESENTATION OF RESULTS

The presentation of results follows the outlined objectives of the study. The lapses/shortcomings observed between the 2 years of the study are explored. The suggested sustainable improvements are also explored. Results between the 2 years are also compared for commonalities and prevalence of lapses. Outliers in relation to the 5R principles are highlighted and explored.

28.5.1 General Overview

Analysis of data from the fieldwork highlighted various types of identified shortcomings that cause more consumption of water and loss in terms of heat energy and volume of water. Data on the observed shortcomings/lapses were first collated according to each building and the year of observation. Data from the 2 years of study were merged and cluster analysis was performed to categorise the observations. The categories derived for observed lapses are shown in Table 28.2.

TABLE 28.2
Categories Derived for Analysis of Observed Lapses in the Plumbing Systems of Buildings (Data from Study)

Category	Description
Design	Shortcomings due to specification and provision for the system in the design of the facility.
Technology	Shortcomings due to the choice of technology installed/assembled.
Construction	Shortcomings due to poor workmanship or poor installation/assembly.
Functionality	Shortcomings due to inadequacies in the functioning of the system in place.
Usage	Shortcomings due to user interaction with the system at point of use.
Maintenance	Shortcomings due to poor maintenance or lack of maintenance.

28.5.2 Lapses/Shortcomings in the Wet Services of Observed Buildings

After determining the relevant categories, observations were then grouped into the categories for further analysis. Results of the analysis are presented in Table 28.3, according to categories and clusters of observed shortcomings.

As seen in Table 28.3, the least observed shortcomings are due to usage issues. The major problems emanate from inefficiencies in the design of technologies, inefficiencies in the functionality of units and their component fixtures, and poor maintenance of the plumbing system generally. While poor assembly/installation was highlighted, it was relatively less prevalent.

TABLE 28.3
Analysis of Observed Lapses in the Plumbing Systems of Buildings Studied (Data from Study)

Categories	Observations
Design	Poor access for maintenance.
	Exposed hot water storage.
	No rain harvesting.
	No storm water harvesting.
	Large bowl urinals that need more water.
	Front-mounted single toilet flush handles.
Technology	Inefficient basin fixtures.
	Inefficient taps.
	Single flush toilets.
	Large volume toilet cistern.
	Constant re-fill and re-heat boilers.
	Inefficient sprinkler system.
	Water flushed urinals.
	Energy consuming boilers.
	Inefficient automated timed flushing urinals.

(Continued)

TABLE 28.3 (*Continued*)
Analysis of Observed Lapses in the Plumbing Systems of Buildings Studied (Data from Study)

Categories	Observations
Construction	Poor workmanship in piping.
	Poor adjustment of flush pipes.
	Poor installation of piping.
	Lack of pipe and boiler insulation.
	Poor corrosion protection.
	Stiff wall-fitted toilet flush button.
Functionality	Long running washbasin.
	Long running push button taps.
	Long running urinals.
	Very short running self-stopping taps causing more consumption.
	Long running toilet flush valve.
	Long flushing toilet cistern.
	Non-calibrated dual flush toilet.
	Low pressure taps causing more consumption.
	Inconsistent flush volume of urinals.
	High flow rate per run-time of self-stopping taps.
	Inconsistent run-times for self-stopping taps.
Usage	Wrong use of dual flush toilets.
	Toilet flush handles not returned to position
Maintenance	Leakages at fixtures, units, joints, controls and along rusty pipes.
	Loose connections.
	Faulty/spoilt toilet flush handle.
	Malfunctioning fixtures and units.
	Corrosion linked to leakages.
	Malfunctioning toilets flushing when not in use.
	General poor maintenance.

For design deficiency, some of the poor installations were occasioned by poor design provisions that did not adequately consider constructability. Many of the observations highlighted the lack of provision for rain and storm water harvesting. It is also noteworthy that some of the design issues are linked to the choice of technology. In such cases it is not the age of the technology as such, but the design specifications such as the size that make them inefficient. For technology-related lapses, many observations were made of continued use of obsolete and inefficient technologies. Under construction, poor workmanship was the more general observation. A high degree of variation was observed in the functionality of similar and sometimes identical fixtures. For maintenance most observations were for breaches in the system resulting in leakages. Furthermore, some units were non-functional, in disrepair, or had malfunctioning parts. Identified user-related lapses referred more to awareness issues. Figures 28.1 and 28.2 show further exploration of the prevalence of observed lapses, using two iterations of text analysis that are presented as word-cloud generations.

Judging from the strength of the keywords in Figures 28.1 and 28.2, most of the problems emanate from the technology, functionality, design, and the installation or construction of components. While technology issues occurred appreciably in the study, the combination of functionality and maintenance-related issues were observed the most. It is arguable that some of the issues under

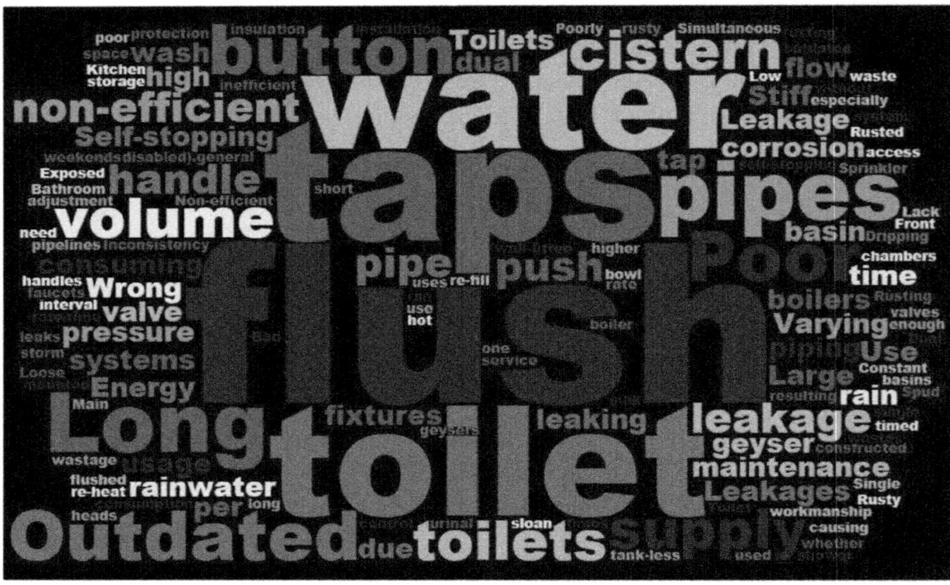

FIGURE 28.1 Analysis of the prevalence of observed lapses in buildings studied – Iteration 1. (Data from study.)

FIGURE 28.2 Analysis of the prevalence of observed lapses in buildings studied – Iteration 2. (Data from study.)

functionality could be traced partly to technology and assembly. However, commissioning, optimisation, continuous monitoring, and routine maintenance would play a big role in addressing most of the issues highlighted.

28.5.3 User-Suggested Sustainable Solutions

Data on suggested sustainable solutions were first collated according to the years of observation. Categorisation was applied using the categories in Table 28.2. The results are presented in Table 28.4. From the cluster analysis it was determined that no suggestions would sit under the functionality category.

TABLE 28.4
Analysis of Suggested Sustainable Solutions to Observed Lapses in the Plumbing Systems of Buildings Studied (Data from Study)

Categories	Observations
Design	Use of subterranean slow release watering pipes for lawns.
	Use smaller gauge pipes for urinal flush systems.
	Use smaller urinal bowls.
	Specifying PVC pipes.
	Specifying corrosion resistant titanium pipes.
	Insulate hot water pipes.
	Thermal cladding for exposed hot water storage.
	Reduce water volume by placing weighted solids in toilet cisterns.
	Add a roof storage tank to reduce demand on energy for pump.
	Combining a sink/wash basin with urinal bowl/toilets, to harness grey water.
	Rainwater harvesting for flushing toilets.
Technology	Change to waterless urinals.
	Smaller washbowl fitted with sensors that activate a spray of water, sanitizer and air.
	Use urinal with sensor controlled flush.
	Change trough urinals to pedestal urinals with manual flushing.
	Use dual flush toilet.
	Use sensor controlled taps.
	Use of solar powered boilers.
	Solar powered leak detectors.
	Use water efficient taps.
	Water re-use system to harness grey water from bath.
	Internal drainage system that harvests and recycles grey water for flushing toilets.
	Recycled grey water system for flushing toilets.
	Rainwater harvesting with purification system.
Construction	Use holding tanks for toilets and showers, instead of direct supply.
	Use automatic timed shower operation.
	Use pH level sensor controlled urinal flush.
Usage	Awareness and education of users.
	Use shower room speakers to stimulate user sensitivity.
	Signage for dual flush buttons.
	Use of harvested rainwater/purified harvested rainwater.
	Use of harvested and purified storm water for lawns and gardens.
Maintenance	Replace old and weak units and fixtures.
	Replace toilet flush handle.
	Repair/replace leakages, rusty components, and faulty components.

Most of the user-suggested solutions fall under the design and technology categories. Essentially, most of the suggestions focus on solutions that can be achieved through acquisition of new or unique technology, and solutions requiring design input. Nevertheless, there are suggestions that fit into the construction, usage, and maintenance categories. From the results, the design, technology, and construction categories would have a very close relationship for sustainability. The technologies suggested would need to be specified from the design level and factored in to the design provisions. In addition, the functionality, and appropriateness of the solution would depend largely on the construction. Therefore, regardless of the categorisation of the suggestions, their sustainability will depend on matching each suggested solution with the relevant components from other categories. Similarly, the usage and maintenance categories could stand alone. However, they are complimentary regarding water efficiency in buildings. The demand on any plumbing installation and the manner of usage will influence the need and frequency of maintenance activities. On the other hand, the maintenance regime and approach will influence functionality, user experience, and user response to the utilisation of the wet services in a building. The complementarities highlighted here will ultimately influence water efficiency in each case.

28.5.4 Application of the 5R Principles to Data on Suggested Solutions

The 5R principle was applied to the user-suggested solutions, as seen in Table 28.5. From the results, most of the suggested sustainable solutions, fall under the 'Reduce Consumption' principle. Most of the suggestions were about water-conserving technologies. The use of technologies was focused on addressing issues of functionality and size. Other suggestions addressed usage or user behavior control issues such as usage time control, usage volume control, user guidance, user education and awareness, and stimulating user sensitivity.

'Reduce Loss and Waste' was next in magnitude of suggestions, focusing on the need for repairs and reducing loss of heat and energy through exposed components and secondary systems such as pumps. Under 'Resort to Alternative Sources', there were many suggestions for design accommodation of rain and storm water harvesting systems to reduce the demand on the main water supply. The use of harvested water to flush toilets and urinals and water the lawns and gardens, was suggested.

Many of the suggested solutions did not fall under the 'Re-use Water' and 'Recycle Water' principles. However, the suggestions under these two categories all require appreciable degree of innovation. For example, combining a sink/wash basin with urinal bowl/toilets was suggested, to use

TABLE 28.5
Analysis of Suggested Sustainable Solutions to Observed Lapses in the Plumbing Systems of Buildings Studied (Data from Study)

Categories of Principles	Efficiency Measures
Reduce consumption	Use dual flush toilet.
	Use water efficient taps.
	Replace old and weak units and fixtures.
	Use of solar powered boilers.
	Solar powered leak detectors.
	Use of subterranean slow release watering pipes for lawns.
	Use sensor controlled taps.
	Signage for dual flush buttons.
	Change trough urinals to pedestal urinals with manual flushing.
	Use smaller gauge pipes for urinal flush systems.
	Use smaller urinal bowls.
	Use urinal with sensor controlled flush.
	Change to waterless urinals.
	Specifying corrosion resistant titanium pipes.
	Specifying PVC pipes.
	Use pH level sensor controlled urinal flush.
	Smaller washbowl fitted with sensors that activate a spray of water, sanitizer and air.
	Use automatic timed shower operation.
	Use holding tanks for toilets and showers, instead of direct supply.
	Awareness and education of users.
	Use shower room speakers to stimulate user sensitivity.

(Continued)

TABLE 28.5 (*Continued*)
Analysis of Suggested Sustainable Solutions to Observed Lapses in the Plumbing Systems of Buildings Studied (Data from Study)

Categories of Principles	Efficiency Measures
Reduce loss and waste	Replace toilet flush handle.
	Repair/replace leakages, rusty components, and faulty components.
	Insulate hot water pipes.
	Thermal cladding for exposed hot water storage.
	Reduce water volume by placing weighted solids in toilet cisterns.
	Add a roof storage tank to reduce demand on energy for pump.
Re-use water	Combining a sink/wash basin with urinal bowl/toilets, to harness grey water.
	Water re-use system to harness grey water from bath.
Recycle water	Internal drainage system that harvests and recycles grey water for flushing toilets.
	Recycled grey water system for flushing toilets.
Resort to alternative sources	Use of harvested rainwater/purified harvested rainwater.
	Use of harvested and purified storm water for lawns and gardens.
	Rainwater harvesting for flushing toilets.
	Rainwater harvesting with purification system.

the draining water to flush the system immediately after usage. While this has been explored in the case of toilets with existing products, it has not been tried with urinals. Bearing in mind the required volume of water for the respective flushing operations of toilets and urinals, this innovation would probably work better for urinal flushing. The configuration would still need to be properly conceptualised.

Furthermore an 'internal' drainage system to harvest and recycle grey water for flushing toilets was suggested. It is referred to as internal in the sense that it should be designed as part of the plumbing system; it would be in the service space and collect used water from wash basins and the bath into storage, from whence it would be piped to the toilets through the service ducts, thereby concealing the system. While it is innovative, the system would probably need some degree of water treatment. There might also be the need to utilise gravitational mechanics to assist the flow. Otherwise it might require an expensive and maintenance-intensive pump system.

Other innovative ideas include the use of subterranean slow-release watering pipes that use harvested and filtered storm water for the lawns the use of smaller wash bowls that are fitted with sensors that activate a spray of combined water, sanitizer and air; using signage to guide/influence usage of units; and the use of shower room speakers to stimulate user sensitivity to the need for water efficient practices by playing relevant messages.

28.5.5 Outlying Suggestions: From 5R to 6R

In analysing the solutions suggested by participants, room was made for suggestions that do not fall under the known 5R principles as described in Silva-Afonso and Pimentel-Rodrigues (2014); Silva-Afonso (2014). Suggestions that were deemed to be better understood outside the 5R principles of water efficiency measures are:

- Using access control to monitor and surcharge poor usage (water wastage) of toilets
- Meter-controlled monthly/weekly water supply to buildings
- Automated closure of water supply to unused bathroom on weekends
- Regular maintenance
- Rethink maintenance approach

These suggestions would not be adequately categorised under the existing 5R principle, which is focused specifically on systems, sub-systems, components, and users. They were therefore not discussed in Table 28.5. However, they form part of the possible solutions to inefficiencies in water supply that could be applied in combination with other categorised suggestions. Hence there is need to provide a categorisation for the aforementioned suggestions, thereby extending the 5R principle of Silva-Afonso and Pimentel-Rodrigues (2014). A sixth category is therefore proposed.

The outlying suggestions were grouped under the 'Re-think management' category. The 'Rethink' category label was derived by drawing on the 'Rethink/Re-design' principle of design for sustainability (Clark et al., 2009). In the original 5R principle (Silva-Afonso, 2014), the 'Rethink/Re-design' sustainability principle was applied as, 'Resort to Alternative Sources' (…of water). However, the new category emerging out of the current study focuses on 'Rethink (…the) management' (…of available water and its supply and distribution system in the building/facility). Following the analysis described here, the 'Rethink Management' category is proposed as an extension of the 5R into the 6R Principle for water efficiency measures in buildings.

The first two suggestions under the proposed category are noteworthy and relate to the control of water supply and introduction of individual user cost implications. The buildings studied are made up of offices, lecture rooms, and study and social activity spaces. The official occupants of each building are recognised as cost centres in terms of the University financial management structure. However, users of the ablution areas of the facilities could originate from any section of the university. Most members of the university community would have access to the ablution areas, at least up to the evening hours of each day. Hence the additional administrative demand needs to be considered when introducing access control because of the need to consider other users who are not directly linked to the building. Furthermore, there must be a balance between access control and the general rights of the members of the university community. Nevertheless, access control is an idea that should be explored beyond the current study in terms of determining how to best implement it.

28.6 CONCLUSION

The study presented in this chapter focused on observations of inefficiencies in water supply systems of buildings, otherwise described as wet services. The study also analysed the suggestions of informed user groups on possible solutions to the identified lapses/shortcomings. Selected educational buildings within a tertiary institution in the Johannesburg area of South Africa were investigated for the study. The current study is an extension of previous studies presented in Aduda and Ozumba (2011), Ozumba and Aduda (2015), Silva-Afonso and Pimentel-Rodrigues (2014), and Silva-Afonso (2014). The current study relates most to the last three studies.

Results of the current study are similar to Ozumba and Aduda (2015) in terms of identified breaches/lapses (faulty fixtures, fittings and piping). With regard to excessive consumption of water, there were similarities in the areas of toilets, urinals, bath and sprinkler systems. Regarding water efficiency measures adopted, there are similarities in the sense that most measures fall under the 'Reduce Consumption' category. Most of the measures are also of a technological/technical nature. Probably, the highest level of similarity in terms of efficiency measures between the two studies is in the dearth of suggestions for the 'Re-use Water', 'Recycle Water', and 'Resort to Alternative Sources' categories of the 5R principle.

The identified pattern of first response to the need for water efficiency is to adopt technologies. However, technology adoption will have financial, administrative, and user implications that could constitute challenges. On the other hand, the user-focused efficiency measure of 'Reduce Consumption' (which includes education, awareness, and sensitivity), usually costs less. Such approaches achieve appreciable results, especially in combination with other measures, as shown in Ozumba and Aduda (2015). Furthermore, as seen in Table 28.5, the categories of 'Re-use Water', 'Recycle Water', and 'Resort to Alternative Sources', present opportunities for original thought, creativity, and innovation.

Considering the current exploration of the 5R principles, there is full applicability and relevance for the approach by Silva-Afonso and Pimentel-Rodrigues (2014) and Silva-Afonso (2014) in the study context. The data on participants' 'plans for improvement' in Ozumba and Aduda (2015), which is the counterpart of the 'suggested solutions' in the current study, were not analysed with the 5R principle. As such, the possibility of outlying efficiency measures was not necessarily explored. However most of the suggested improvements could be classified under the proposed category of 'Rethink Management', thereby supporting the results of the current study, which proposed this new category. For example, the combination of water efficiency measures and user education produced positive outcomes. Furthermore, there was a suggested plan to use differential billing of water usage by metering each unit. Moreover, there were findings such as lack of 'programmed retrofitting' and lack of 'systematic approach to introducing water conservation measures' (Ozumba and Aduda, 2015). The highlighted examples from previous studies constitute efficiency measures that would be better understood under the proposed 'Rethink Management' category.

Judging from the discussion of the findings, it is probably necessary to prioritise the new category in the application of the proposed 6R principle, especially for existing stock. It would also be beneficial to develop the proposition into a criteria or framework for making decisions on water efficiency for existing stock and new facilities. It is important to have a system of factors linked to considerations and options of water efficiency measures for existing stock and building designs.

Considering the scope of the current study, appreciable extension of knowledge has been made with regards to water efficiency measures in buildings. Also the views of informed users have been explored with regard to water efficiency in buildings. The current study has also explored a slightly unique context, thereby enriching the body of knowledge constructed thus far by providing a basis for compounding of evidence. However, the unique focus does not allow for much generalisation.

Regardless of the limitations, highlights have been made and strong similarities have been drawn that can form the basis for future studies. Generally, the current study has contributed significantly through the bespoke methodology and the presentation of informed user group views, which extends Ozumba and Aduda (2015). The current study also introduced the sixth measure of water efficiency in buildings and proposed the 6R, which extends the work of Silva-Afonso and Pimentel-Rodrigues (2014) and Silva-Afonso (2014). Furthermore, the use of social science approaches to explore engineering principles in the study provides insight to the interaction between these two areas of knowledge in the context of water efficiency.

Future researchers along this path may choose to study a wider pool of buildings, for more representativeness, considering various types, designs, construction, occupancies, locations, age, or other factors. There is also the need to study a pool of participants who will be representative of the users, facility owners, facility managers, regulators, designers, builders, and other consultants. With such studies there will be a need to focus on the relative value placed on different efficiency measures, and the perceived challenges in each of the 5R or 6R categories. It would be interesting to cross-validate suggested solutions to water inefficiency in buildings from various stakeholder sub-groups. It may also be useful to explore the suggested solutions of other stakeholders. Finally, it would be necessary to explore the further development of water efficiency measures beyond the 5R principles and the proposed 6R. With such studies, it is possible to extrapolate this concept onto an nR framework for principles of water efficiency measures that can be expanded, revised and adapted to various uses. The development of such a framework would allow for the more accurate application of the principles explored in the study.

REFERENCES

Aduda, K. and Ozumba, A.O.U. (2011) Water conservation in South Africa's buildings. *Proceedings of ASOCSA2011, Association of Schools of Construction of Southern Africa conference 2011*, Johannesburg, South Africa, pp. 241–244.

Arpke, A. and Hutzler, N. (2006) Domestic water use in the United States: A life-cycle approach. *Journal of Industrial Ecology*, 10(1–2), 169–184. doi:10.1162/108819806775545312/full.

Balaras, C.A., Droutsa, K., Dascalaki, E. and Kontoyiannidis, S. (2005) Heating energy consumption and resulting environmental impact of European apartment buildings. *Energy and Buildings*, 37(5), 429–442.

Butts, D.P. (1983) The Survey—A research strategy rediscovered. *Journal of Research in Science Teaching*, 20(3), 187–193. doi:10.1002/tea.3660200302.

Bi, Z. (2011) Revisiting system paradigms from the viewpoint of manufacturing sustainability. *Sustainability*, 3(9), 1323–1340.

Calderón, J.L., Baker, R.S. and Wolf, K.E. (2000) Focus groups: A qualitative method complementing quantitative research for studying culturally diverse groups. *Education for Health*, 13(1), 91–5.

Ceclan, R.E., Ceclan, I. and Popa, I. (2011) Sustainable waste management in Europe. Statia ICPE, Agigea, July 28–29. http://www.inginerie-electrica.ro/acqu/2011/S2_1_Sustainable_Waste_Management_in_Europe.pdf.

Charter, M. and Tischner, U. (2001) Sustainable product design. In: M. Charter and E. Tischner (Eds.), *Sustainable Solutions: Developing Products and Services for the Future*, (Chapter 6, pp. 118–138) Sheffield, UK: Greenleaf.

Cheng, C. (2002) Study of the inter-relationship between water use and energy conservation for a building. *Energy and Buildings*, 34(3), 261–266.

Clark, G., Kosoris, J., Hong, L.N. and Marcel Crul, M. (2009) Design for sustainability: Current trends in sustainable product design and development. *Sustainability*, 1(3), 409–424.

Coyle, T. (2014) How does water efficiency impact a building? Green building 101, U.S. Green Building Council, May 21, 2014. http://www.usgbc.org/articles/green-building-101-how-does-water-efficiency-impact-building.

Creemers, G. and Pott, A.J. (2002) Development of a hydrological economic agricultural model on case studies in the upper mvoti catchment. Water Research Council Report No. 890/1/02.

du Plessis, C., Irurah, D.K. and Scholes, R.J. (2003) The built environment and climate change in South Africa. *Building Research & Information*, 31(3–4), 240–256.

Halliday, S. (2008) *Sustainable Construction*. Oxford, UK: Butterworth–Heinemann (Elsevier).

Hill, R.C. and Bowen, P.A. (1997) Sustainable construction: principles and a framework for attainment. *Construction Management & Economics*, 15(3), 223–239.

Hill, R.C., Bergmani, J.G. and Bowen, P.A. (1994) A framework for the attainment of sustainable construction, CIB TG 16 Sustainable Construction Conference, Tampa, FL, November 6–9. pp. 13–25. https://www.irbnet.de/daten/iconda/CIB_DC24776.pdf.

Huer, M.B. and Saenz, T.I. (2003) Challenges and strategies for conducting survey and focus group research with culturally diverse groups. *American Journal of Speech-Language Pathology*, 12(2), 209–220.

Kawulich, B.B. (2005) Participant observation as a data collection method. *Forum: Qualitative Social Research*, 6(2), 43.

Kilbert, C.J. (2016) *Sustainable Construction: Green Building Design and Delivery*, 4th ed. Hoboken, NJ: John Wiley & Sons.

Khasreen, M.M., Banfill, P.F.G. and Menzies, G.F. (2009) Life-Cycle Assessment and the Environmental Impact of Buildings: A Review. *Sustainability*, 1(3), 674–701.

Iacono, J., Brown, A. and Holtham, C. (2009) Research methods—A case example of participant observation. *The Electronic Journal of Business Research Methods*, 7(1), 39–46.

Méndez-Barrientos, L.E., Kemerink, J.S., Wester, P. and Molle, F. (2016) Commercial farmers' strategies to control water resources in South Africa: An empirical view of reform. *Water International*, 26(3), 314–334. doi:10.1080/02508060108686924.

Naidoo, D. and Constantinides, G. (2000) Integrated approaches to efficient water use in South Africa. *International Journal of Water Resources Development*, 16(1), 155–164.

Otieno, F.A.O. and Ochieng, G.M.M. (2004) Water management tools as means of averting a possible water scarcity in South Africa by the year 2025, *Water Institute of South Africa (WISA) Biennial Conference*, Cape Town, South Africa, May 2–6. www.wrc.org.za.

Ozumba, A.O.U. and Aduda, K. (2015) Adoption of water conservation measures in buildings: Case study. In R.H. Crawford and A. Stephan (Eds.), *Living and Learning: Research for a Better Built Environment: 49th International Conference of the Architectural Science Association 2015*, Melbourne, Australia. pp. 1–10.

Saunders, M., Lewis, P. and Thornhill, A. (2009) *Research Methods for Business Students*, 5th ed. London, UK: Pearson Education Limited.

Scheuer, C., Keoleian, G.A. and Reppe, P. (2003) Life cycle energy and environmental performance of a new university building: Modeling challenges and design implications. *Energy and Buildings*, 35(10), 1049–1064.

Smakhtin, V., Ashton, P., Batchelor, A., Meyer, R., Murray, E., Barta, B., Bauer, N. et al. (2009) Unconventional water supply options in South Africa. *Water International*, 26(3), 314–334.

Stephenson, D. and Randell, B. (2003) Water demand theory and projections in South Africa, *Water International*, 28(4), 512–518. doi:10.1080/02508060308691728.

Stockholm Environmental Institute. (2009) Comparing sanitation systems using sustainability criteria. EcoSanRes Series, 2009-1. http://www.ecosanres.org/pdf_files/ESR2009-1-ComparingSanitationSystems.pdf.

Thorne, S. (2000) Data analysis in qualitative research. *Evidence-Based Nursing*, 3(3), 68–70.

United Nations Environment Programme. (2014) Water and energy efficiency. Information brief. http://www.un.org/waterforlifedecade/pdf/01_2014_water_energy_efficiency.pdf.

United Nations. (1992) Sustainable Development, Agenda 21. *United Nations Conference on Environment & Development*, Rio. Rio de Janeiro, Brazil, June 3–14. United Nations. https://sustainabledevelopment.un.org/content/documents/Agenda21.pdf.

UN-DESA. (2009) E/CN.17/2009/19–Report on the 17th session of CSD. https://sustainabledevelopment.un.org/topics/ruraldevelopment/decisions.

United Nations Environment Programme. (2008) Inequity in access to clean water and sanitation. *Vital water graphics*, 2nd ed. http://www.unep.org/dewa/vitalwater/article63.html.

van Koppen, B. and Schreiner, B. (2014) Moving beyond integrated water resource management: Developmental water management in South Africa. *International Journal of Water Resources Development*, 30(3), 543–558. doi:10.1080/07900627.2014.912111.

Yin, R.K. (1999) Enhancing the quality of case studies in health services research. *Health Services Research*, 34(5), Part II, 1209–1224.

Zhang, Z., Wu, X., Yang, X. and Zhu, Y. (2006) BEPAS—A life cycle building environmental performance assessment model. *Building and Environment*, 41(5), 669–675.

Section IX

Water-Energy Nexus

29 Simultaneous Optimization of Water and Energy in Integrated Water and Membrane Networks

A Case for Water Energy Nexus

Esther Buabeng-Baidoo and Thokozani Majozi

CONTENTS

29.1 INTRODUCTION

The scarcity of water and strict environmental regulations have made sustainable engineering a prime concern in the processing and manufacturing industries.[1] Water minimization involves the reduction of fresh water use and effluent discharge in chemical plants. This is achieved through water reuse, water recycling and water regeneration. Water reuse involves the use of wastewater in other operations except the process where it was originally used. Water recycling, however, allows the effluent to be used in any process, including the process in which it was produced. In water regeneration, the effluent is partially treated before it is recycled or reused in other processes. Partial treatment can be achieved by using water purification units often classified as membrane and non-membrane processes, e.g., reverse osmosis (RO) membranes and steam stripping, respectively.[2]

The purification of water through membrane systems is an energy-intensive process. The minimization of energy within water networks is also needed for sustainable development. Energy usage within the water network is largely associated with the regeneration (membrane) units. In most published work, however, membrane systems have been represented using the "black-box" approach,[3–5] which uses a simplified linear model to represent the membrane systems. This approach does not give an accurate representation of the energy consumption and associated costs of membrane systems.[5] A more rigorous representation of the regeneration unit is therefore needed.[6]

The regenerator unit considered in this chapter is the RO membrane. RO membranes are highly favored over other separation units due to their low energy consumption, ease of operation, high product recovery and quality.[7] RO membranes separate a water stream into a lean stream of low contaminant concentration, known as the permeate stream, and a highly contaminated stream, known as the retentate stream. The process is achieved by applying external pressure to the feed solution in order to reverse the osmotic phenomenon. As a result of this process, retentate streams exit the membrane at a high pressure. Energy recovery turbines are often placed in the retentate streams to offset the energy requirements of the entire unit. Figure 29.1 depicts the principle of RO membranes.

The rigorous design of RO units has been extensively studied in literature. Following on Evangelista,[8] El-Halwagi[7] introduced the "state space approach" for the design of RO networks (RONs). A superstructure representation of the RON was introduced with the objective of synthesizing a network with the optimal number of RO units, booster pumps and energy recovery turbines that resulted in minimal freshwater consumption. Saif et al.[9] and Sassi and Mujtaba[10] have modified El-Halwagi's model[7] by including scheduling, adding further design rules, simplifying the modeling of the RON and including fouling effects in the modelling of the network. Saif et al.[11] extended the superstructure of Saif et al.[9] by applying an efficient branch-and-bound algorithm in order to obtain global optimality for the RON. This was achieved by the introduction of additional constraints to tighten the mathematical programming. Saif et al.[11] obtained a 14.8% lower total annualized cost (TAC) than that obtained by El-Halwagi.[7] There have, however, been few studies that consider a detailed RON superstructure within a water network synthesis.[6]

There are two major approaches adopted in addressing water network synthesis: insight-based techniques and mathematical model–based optimization methods. Insight-based techniques involve the water pinch analysis, which is a graphical method based on the concept of a limiting water profile that is the most contaminated water that can be fed into a particular operation. This method was first proposed by Wang and Smith.[12] Hallale[13] then proposed a graphical method based on non-mass transfer operations with single contaminants. Recent studies have extended water pinch analysis to

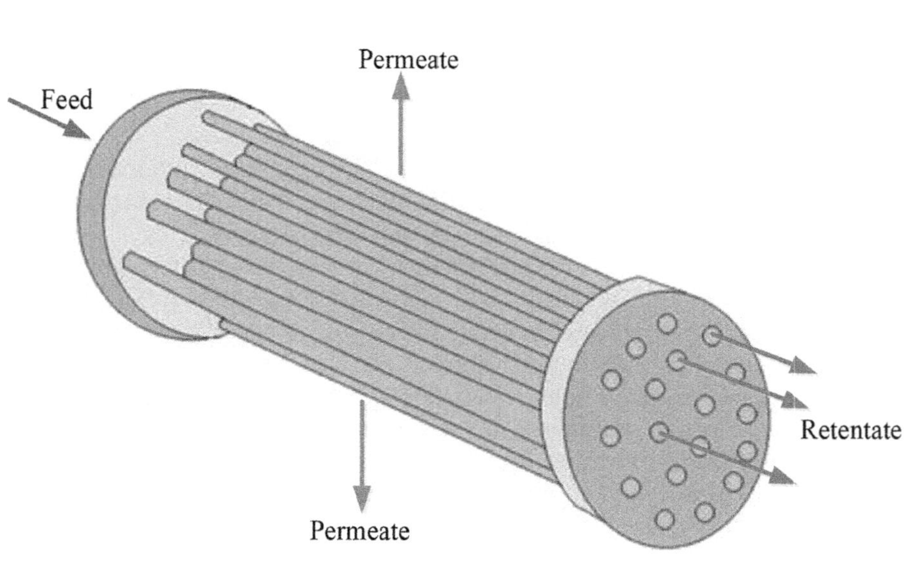

FIGURE 29.1 A schematic representation of a hollow fiber reverse osmosis membrane (HFRO).

algebraic methods, primarily water cascade analysis.[14,15] The water pinch method proves unsuccessful for complex problems involving multiple contaminants[16] and various topological constraints.[4] The computation burden of this method is, however, lower than that experienced by mathematical model–based optimization methods.

The mathematical optimization approach employs a superstructure identifying an optimal configuration for the process from a number of alternatives. This idea was first proposed in the work by Takama et al.[17] They proposed a nonlinear model that incorporates both water use and wastewater treating units for multiple contaminant systems. Significant developments in the area have been achieved, including the work of Galan and Grossmann,[18] Karuppiah and Grossmann,[19] and Tan et al.[5] who explored different techniques for modeling regenerators and developing strategies for complex mixed-integer non-linear programs (MINLPs). This method is, however, computationally expensive.

Khor et al.[2] presented a detailed membrane regenerator model incorporated into a water network superstructure (WNS). The model they proposed consisted of continuous variables for the contaminants and the flow rates as well as binary variables for the piping interconnections in conjunction with a nonlinear RON model. The resultant WNS was an MINLP in structure. The multi-contaminant model enabled direct water reuse/recycling, regeneration reuse or regeneration recycling. Khor et al.[2] assumed a single regenerator with a fixed design, which implies that the number of regenerators, pumps and energy recovery turbines were specified a priori. This limited the flexibility of the model, which could result in a suboptimal solution.

The current work proposes a superstructure optimization for the synthesis of a detailed RON within a WNS. The overall mathematical formulation is used to minimize both water and energy simultaneously. A rigorous nonlinear RON superstructure model, which is based on the state-space approach by El-Halwagi,[7] is included in the WNS to determine the optimum number of RO units, pumps and turbines required for an optimal WNS. A fixed–flow rate model that considers the concept of sources and sinks is adopted. The model takes into account streams with multiple contaminants. The idea of using a variable removal ratio to describe the performance of the regenerators is also explored in this work.

29.2 PROBLEM STATEMENT

The problem addressed in this work can be stated as follows:

Given:

1. A set of water sources, I, with known flow rates and known contaminant concentration
   A set of water sinks, J, with known flow rates and known maximum allowable contaminant concentration
2. A network of RO regenerators, Q, with known liquid recovery and design parameters
3. A freshwater source, FW, with known contaminant concentration and variable flow rate
4. A wastewater sink, WW, with known maximum allowable contaminant concentration and variable flow rate

Determine:

1. The minimal freshwater intake, wastewater generation and the energy consumed in the RON and the TAC
2. The optimal configuration of the water network
3. The optimal number of RO units, pumps and energy recovery turbines
4. The optimal operation and design conditions of the RON, including feed pressure, number of hollow fiber modules per regenerator, stream distributions, separation levels, etc.

29.3 MATHEMATICAL MODEL

The overall MINLP model is based on the superstructure represented in Figure 29.2, which is adapted from the work of El-Halwagi[7] and Khor et al.[2] In the state-space approach the RO networks are split into four distribution boxes as shown in Figure 29.2: a pressurization/depressurization stream distribution box (PDSDB), pressurization/depressurization matching box (PDMB), an RO stream-distribution box (ROSDB) and an RO matching box (ROMB). The purpose of the distribution boxes is to allow all possible combinations of stream mixing, splitting, recycling and bypass. In the PDSDB, the sources, permeate and retentate streams are fed into the box and are distributed to the final permeate and retentate stream or recycled to a regenerator. The streams then proceed to the PDMB where they can be sent to either a pump, energy recovery turbine or directly to a regenerator without undergoing any pressure change. Thereafter, streams are fed to the ROSDB where they are allocated to the appropriate unit in the ROMB. In the ROMB, the streams are fed to the RO membranes where they are separated into a permeate stream and a retentate stream. The permeate and retentate streams are then fed to the PDSDB and the cycle continues until an optimal network is obtained.

The RON superstructure proposed by El-Halwagi[7] must however be modified in order to incorporate it within the WNS. This is achieved by modifying the PDSDB and the PDMB section of the superstructure. The properties of the updated PDSBD and PDMB are detailed as follows.

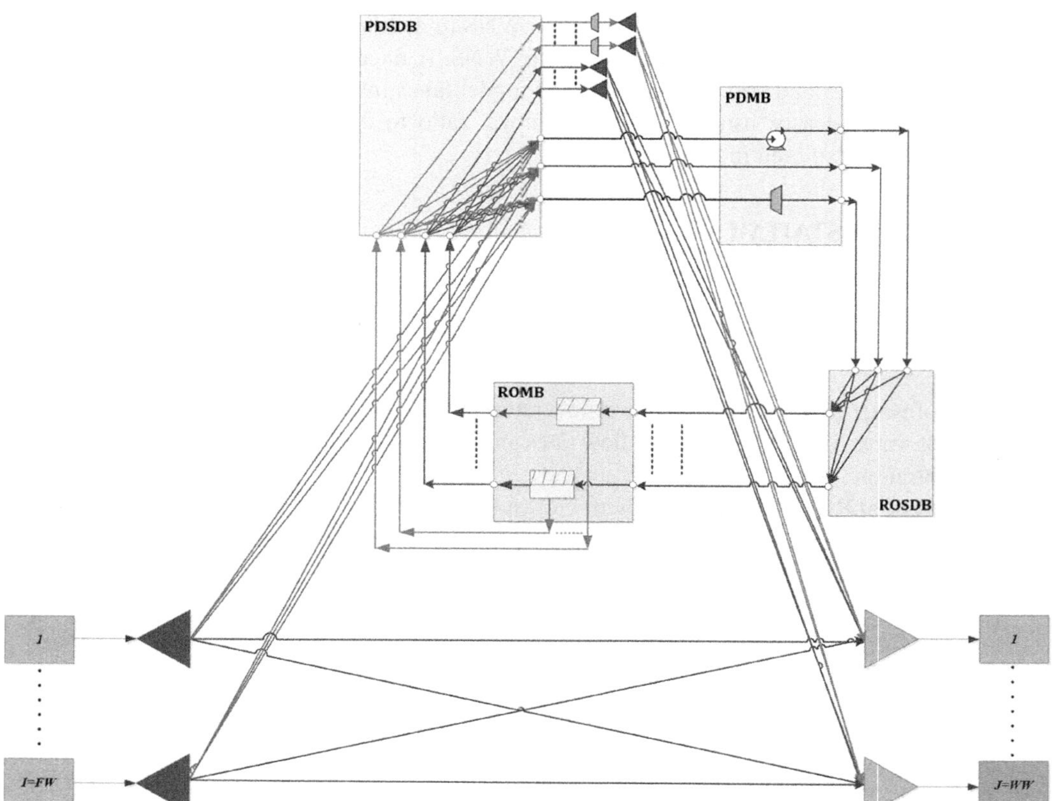

FIGURE 29.2 Superstructure representation of the RON superstructure within the WNS.

1. Water sources are fed directly to node n for regeneration and are not mixed with retentate or permeate streams. This was incorporated to ensure that each retentate and permeate stream leaves its respective regenerator, without further contamination.
2. Permeate and retentate streams are not allowed to mix in order for each stream to be fed directly from regenerator to the sinks. It is also assumed that each permeate stream will leave the regenerator at atmospheric pressure. Retentate streams, however, leave the RO at high pressure and are therefore passed through an energy recovery turbine for reduction in pressure to atmospheric pressure before distribution to the sinks.
3. Different retentate or permeate streams are also not allowed to mix in order to feed each stream directly to the water sinks. Mixing of the streams within the water sinks is decided by the water quality requirements of the sink.
4. Each retentate stream or permeate stream can, therefore, go directly to a retentate node or can be recycled back to node n for further cleaning by the regenerators.
5. A stream that does not require a pressure change can be fed directly to the ROSDB where it is then fed to the ROMB.
6. Inlet streamsto the PDMB caneither go to a pumportoan energy recovery turbine. The illustration of this idea is modified in order to clearly explain the original idea proposed by El-Halwagi[7].

These modifications are illustrated in Figure 29.3a and b. Figure 29.3a shows the original PDSDB proposed by El-Halwagi[7] and Figure 29.3b shows the modified PDSDB and PDMB that will be incorporated with the WNS.

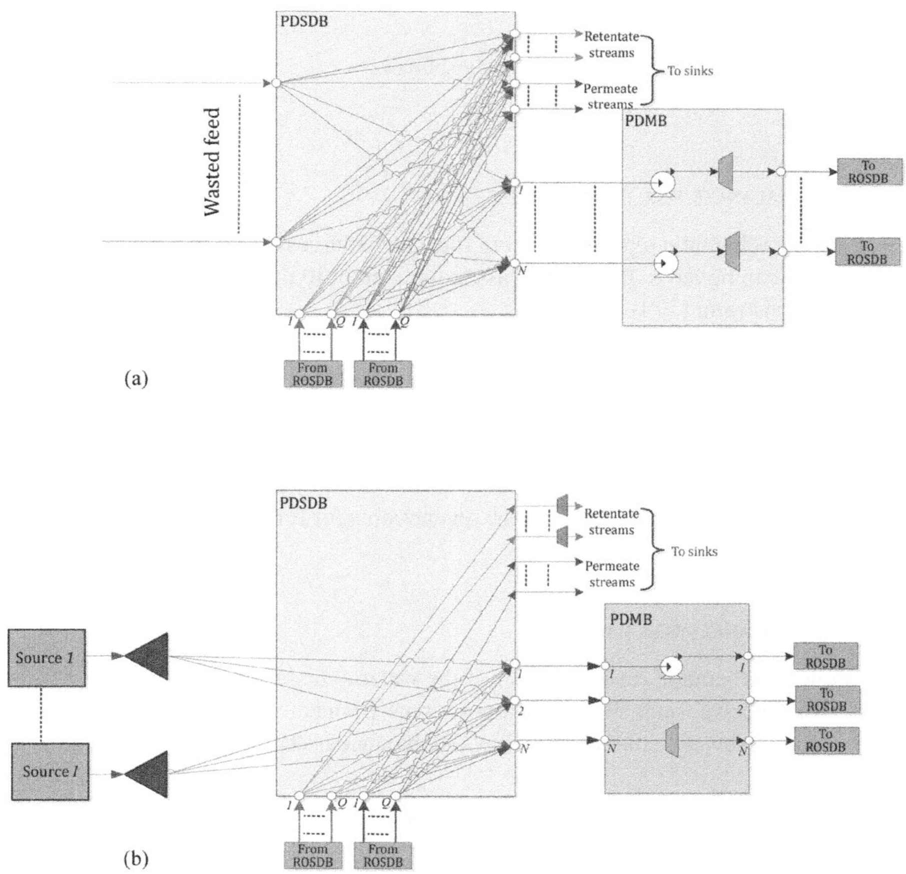

FIGURE 29.3 (a) Original PDSDB and PDMB by El-Halwagi.[7] (b) New modification to PDSDB and PDMB.

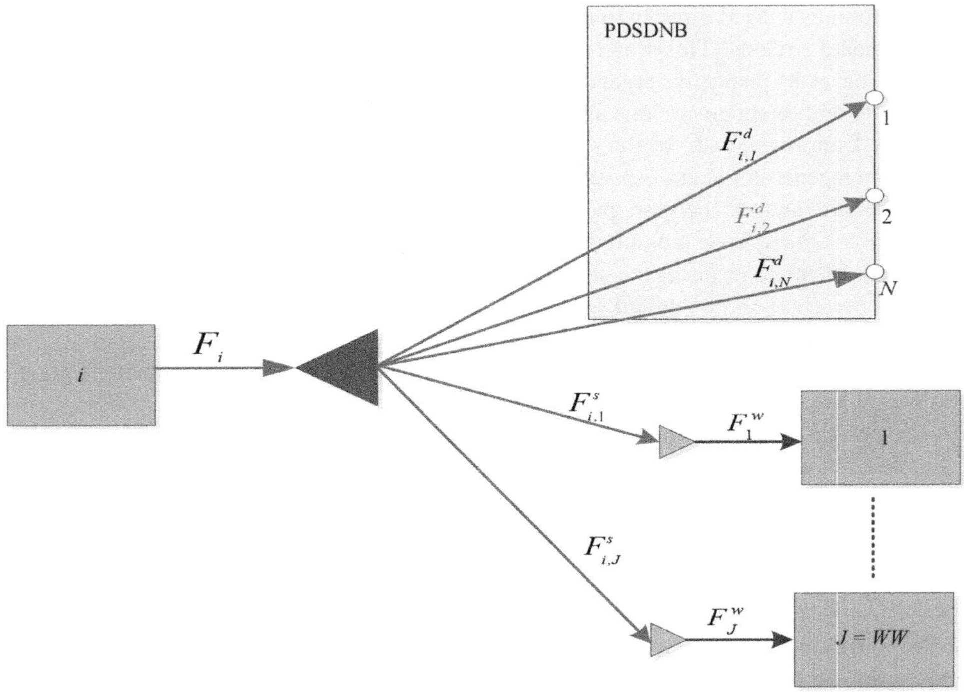

FIGURE 29.4 Schematic representation of the water sources.

29.3.1 Water Balances for the Sources

Figure 29.4 shows a schematic representation of the water sources. From the diagram it can be seen that a water source can be fed to the RON, wastewater sink or to the water sinks. The flow rate balance is shown in constraint (29.1).

$$F_i = \sum_{j=1}^{J} F^s_{i,j} + \sum_{n=1}^{N} F^d_{i,n} \qquad \forall i \in I \tag{29.1}$$

It should also be noted that the freshwater source is included in the model as the last source within the model formulation. It can also be sent to the regenerators for further cleaning, as its contaminant concentration is not zero.

29.3.2 Water Balances for the Sinks

Figure 29.5 shows a schematic representation of the water sinks. From the diagram it can be seen that the water sinks receive water from the water sources, permeate and retentate of the regeneration units as well as the freshwater source. This flow rate balance is shown in constraint (29.2).

$$F^w_j = \sum_{i=1}^{I} F^s_{i,j} + \sum_{q=1}^{Q} F^r_{q,j} + \sum_{q=1}^{Q} F^p_{q,j} \qquad \forall j \in J \tag{29.2}$$

Each sink can, however, handle a certain concentration limit. Constraint (29.3) implies that the load to each sink must not exceed the maximum allowable load to that particular sink.

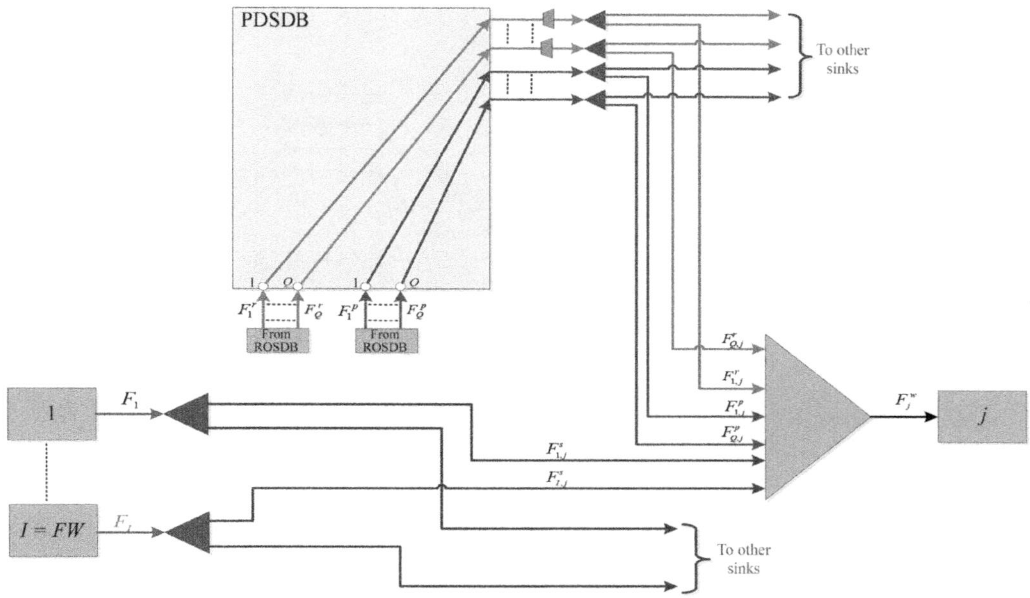

FIGURE 29.5 Schematic representation of the water sinks.

$$C^{U}_{j,m} \geq \frac{\sum_{i=1}^{I} F^{s}_{i,j}C^{s}_{i,m} + \sum_{q=1}^{Q} F^{r}_{q,j}C^{r}_{q,m} + \sum_{q=1}^{Q} F^{p}_{q,j}C^{p}_{q,m}}{F^{w}_{j}} \qquad \forall j \in J \quad \forall m \in M \tag{29.3}$$

It should be noted that the wastewater sink is considered the last sink. The maximum allowable load to this sink is also given in order to comply with the standard effluent discharge limits imposed by environmental regulations.

In order to forbid the mixing of permeate and retentate streams from one regenerator in the same sink, constraint (29.4) is added to the model as follows:

$$y^{p}_{q,j} + y^{r}_{q,j} \leq 1 \qquad \forall j \in J \quad \forall q \in Q \tag{29.4}$$

29.3.3 Regeneration Unit (RON Superstructure)

Figure 29.2 shows the schematic representation of the updated RON superstructure within the WNS superstructure. Figure 29.6 therefore shows the interaction of the PDSBD with the sources and sinks of the WN.

29.3.3.1 Performance Equations

The performances of the RO regenerators are represented by means of the liquid recovery α_q and removal ratio $RR_{q,m}$. The liquid recovery is the amount of the feed flow rate into the regenerator that exists in the permeate stream. The removal ratio $RR_{q,m}$ refers to the fraction of the inlet mass load that exists in the retentate stream of the regenerators.[2] Constraints (29.5) and (29.6) represent the α_q and $RR_{q,m}$ respectively.

$$\alpha_q = \frac{F^{p}_{q}}{F^{f}_{q}} \qquad \forall q \in Q \tag{29.5}$$

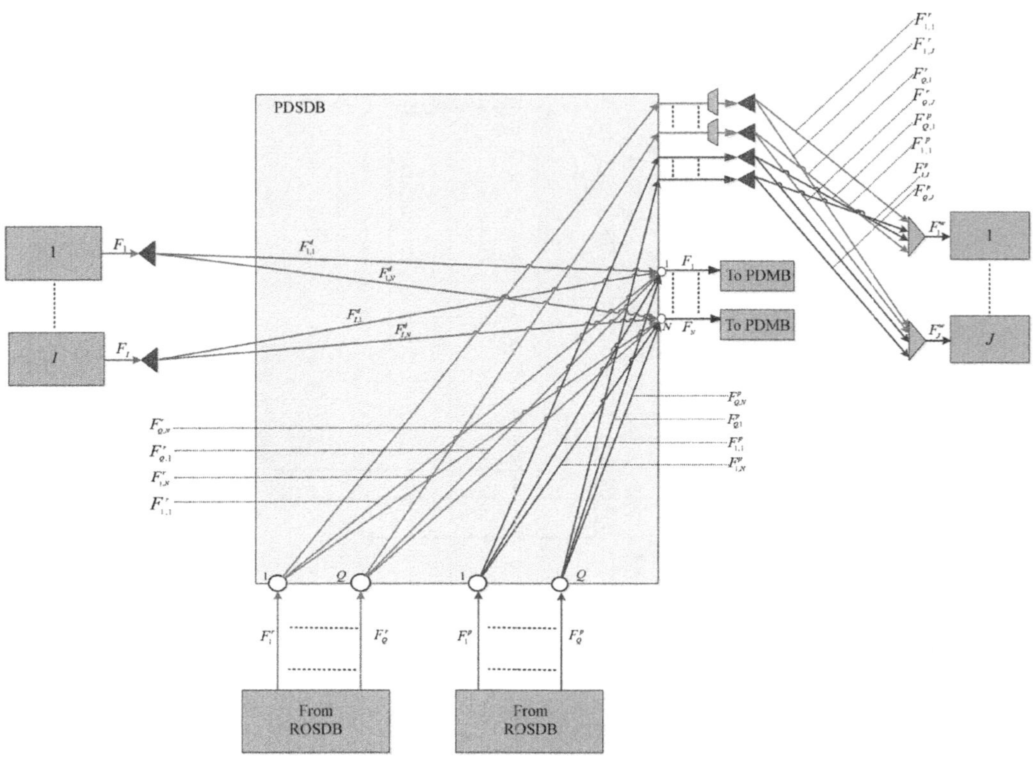

FIGURE 29.6 Schematic representation of the modified PDSDB.

$$RR_{q,m} = \frac{C^r_{q,m}F^r_q}{C^f_{q,m}F^f_q} \qquad \begin{array}{l}\forall q \in Q \\ \forall m \in M\end{array} \tag{29.6}$$

The recommended operating flow rate for RO modules is given in constraint (29.7) and is determined by the manufacturers. Constraint (29.8) gives the upper bound for the feed pressure into the RO membranes.

$$F^L \leq \frac{F^f_q}{N^s_q} \leq F^U \qquad \forall q \in Q \tag{29.7}$$

$$P^f_q \leq P_{\max} \qquad \forall q \in Q \tag{29.8}$$

29.3.3.2 RON Superstructure Equations

29.3.3.2.1 Constraints for Pressurization/Depressurization Stream Distribution Box

Constraint (29.9) shows the flow rate balance for the outlet junction of the PDSDB as can be seen in Figure 29.6. The node n represents a mixing junction at the outlet of the PDSBD.

$$F^a_n = \sum_{i=1}^{I} F^d_{i,n} + \sum_{q=1}^{Q} F^p_{q,n} + \sum_{q=1}^{Q} F^r_{q,n} \qquad \forall n \in N \tag{29.9}$$

Constraint (29.10) shows the corresponding concentration balance for the outlet junction of the PDSDB.

$$F_n^a C_{n,m}^a = \sum_{i=1}^{I} F_{i,n}^d C_{i,m}^d + \sum_{q=1}^{Q} F_{q,n}^p C_{q,m}^p + \sum_{q=1}^{Q} F_{q,n}^r C_{q,m}^r \qquad \begin{matrix} \forall n \in N \\ \forall m \in M \end{matrix} \tag{29.10}$$

The balance for the flow rate and concentration of the permeate stream entering the PDSDB to the sinks in shown in constraint (29.11) and (29.12), respectively.

$$F_q^p = \sum_{n=1}^{N} F_{q,n}^p + \sum_{j=1}^{J} F_{q,j}^p \qquad \forall q \in Q \tag{29.11}$$

$$F_q^p C_{q,m}^p = \sum_{n=1}^{N} F_{q,n}^p C_{q,m}^p + \sum_{j=1}^{J} F_{q,j}^p C_{q,m}^p \qquad \begin{matrix} \forall q \in Q \\ \forall m \in M \end{matrix} \tag{29.12}$$

The balance for the flow rate and concentration of the retentate stream entering the PDSDB to the sinks is shown in constraint (29.13) and (29.14), respectively.

$$F_q^r = \sum_{n=1}^{N} F_{q,n}^r + \sum_{j=1}^{J} F_{q,j}^r \qquad \forall q \in Q \tag{29.13}$$

$$F_q^r C_{q,m}^r = \sum_{n=1}^{N} F_{q,n}^r C_{q,m}^r + \sum_{j=1}^{J} F_{q,j}^r C_{q,m}^r \qquad \begin{matrix} \forall q \in Q \\ \forall m \in M \end{matrix} \tag{29.14}$$

Since the permeate and retentate streams from the regenerator are at different pressures, constraints have to be given in order to ensure that streams are at the same pressure before they mix. This is shown in constraint (29.15), (29.16), and (29.17) for the feed, permeate and retentate streams. Constraint (29.18) shows the isobaric mixing of streams within the ROSDB.

$$(P_n^a - P_i^w) F_{i,n}^d = 0 \qquad \begin{matrix} \forall n \in N \\ \forall i \in I \end{matrix} \tag{29.15}$$

$$(P_n^a - P_q^p) F_{q,n}^p = 0 \qquad \begin{matrix} \forall n \in N \\ \forall q \in Q \end{matrix} \tag{29.16}$$

$$(P_n^a - P_q^r) F_{q,n}^r = 0 \qquad \begin{matrix} \forall n \in N \\ \forall q \in Q \end{matrix} \tag{29.17}$$

$$(P_n^a - P_n^o) F_{n,q}^a = 0 \qquad \begin{matrix} \forall n \in N \\ \forall q \in Q \end{matrix} \tag{29.18}$$

29.3.3.2.2 Constraints for PDMB and ROSDB

In the PDMB, the turbine is used to reduce the pressure of a stream, while the pump is used to increase the pressure. Constraints (29.19) and (29.20) represent the principles of an energy recovery turbine and a pump, respectively. Figure 29.7 shows the schematic representation of the PDMB and RODB.

$$(P_n^i - P_n^a) \geq 0 \qquad \forall n \in N \tag{29.19}$$

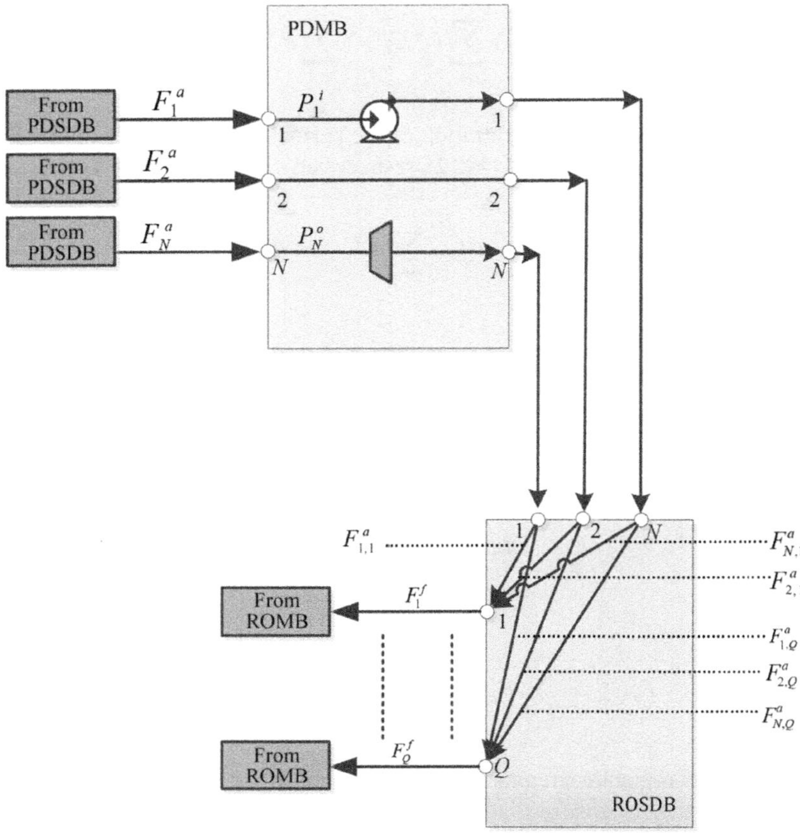

FIGURE 29.7 Schematic representation of the PDMB and ROSDB.

$$(P_n^i - P_n^o) \geq 0 \qquad \forall n \in N \tag{29.20}$$

The flow rate balance for the inlet of the ROSDB is given in constraint (29.21).

$$F_n^a = \sum_{q=1}^{Q} F_{n,q}^a \qquad \forall n \in N \tag{29.21}$$

The outlet flow rate and concentration balance for the ROSDB is given in constraints (29.22) and (29.23), respectively.

$$F_q^f = \sum_{n=1}^{N} F_{n,q}^a \qquad \forall q \in Q \tag{29.22}$$

$$F_q^f C_{q,m}^f = \sum_{q=1}^{Q} F_{n,q}^a C_{n,m}^a \qquad \begin{matrix} \forall q \in Q \\ \forall m \in M \end{matrix} \tag{29.23}$$

The maximum inlet concentration limit to the regenerators must also be specified, since not all of the waste streams can be fed to the RO membrane, and this is shown in constraint (29.24).

$$C^U_{q,m} \geq \frac{\sum_{q=1}^{Q} F^a_{n,q} C^a_{n,m}}{F^f_q} \qquad \forall q \in Q \quad \forall m \in M \tag{29.24}$$

29.3.3.2.3 Binary Variables for the Existence of Units

Constraint (29.25) shows that a booster pump exists in the RON if the P^i_n is larger than the pressure of the stream entering the PDMB and this forces the binary variable b_n to become one. A similar concept is used to represent the existence of an energy recovery turbine and is given in constraint (29.26). It is, however, illogical to pressurize and depressurize a stream simultaneously. Constraint (29.27) is therefore needed to prevent a turbine and pump from occurring in series.

$$P^L b_n \leq P^i_n - P^a_n \leq P^U b_n \qquad \forall n \in N \tag{29.25}$$

$$P^L t_n \leq P^i_n - P^o_n \leq P^U t_n \qquad \forall n \in N \tag{29.26}$$

$$b_n + t_n \leq 1 \qquad \forall n \in N \tag{29.27}$$

Constraint (29.28) indicates the existence of the RO unit, which is defined by the flow rate of the permeate stream from the regenerator q.

$$Fl^L r_q \leq F^p_q \leq Fl^U r_q \qquad \forall q \in Q \tag{29.28}$$

29.3.3.2.4 Constraints for the ROMB

The characteristic of the RO membrane needs to be described in order to relate flow rate to pressure. The pressure drop across the membrane ΔP_q is given in constraint (29.29).[2] The equation was simplified by assuming a linear–shell side concentration and pressure profile.[20] The schematic representation of the ROMB is given in Figure 29.8.

$$\Delta P_q = P^f_q - \left(\frac{\Delta P^m_q}{2} + P^p_q\right) \qquad \forall q \in Q \tag{29.29}$$

The osmotic pressure, $\Delta\pi_q$, is defined as a function of the contaminant concentration on the feed side[9] and is shown in constraint (29.30).

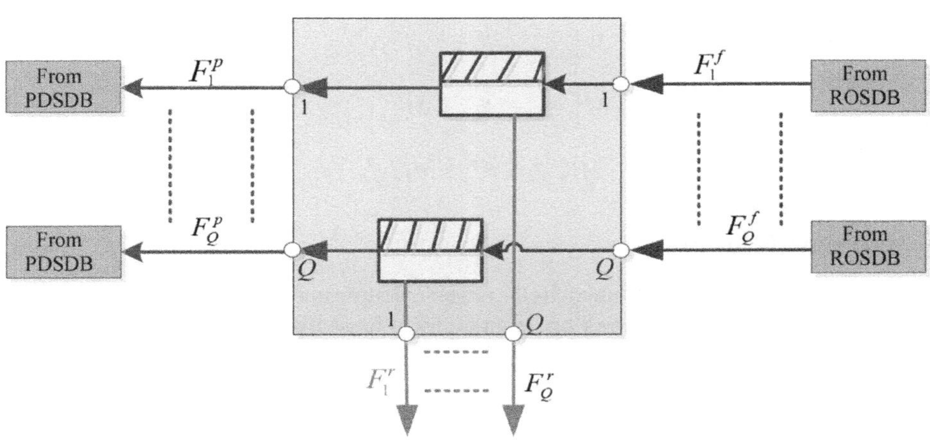

FIGURE 29.8 Schematic representation of the ROMB.

$$\Delta\pi_q = \text{OS}\sum_{m=1}^{M} C_{q,m}^{av} \qquad \forall q \in Q \tag{29.30}$$

The permeate flow rate per module is given in constraint (29.31).

$$\frac{F_q^p}{N_q^s} = \text{AS}_m(\Delta P_q - \Delta\pi_q) \qquad \forall q \in Q \tag{29.31}$$

The average concentration $C_{q,m}^{av}$ on the feed side is given by constraint (29.32).

$$C_{q,m}^{av} = \frac{C_{q,m}^f + C_{q,m}^r}{2} \qquad \begin{matrix}\forall q \in Q \\ \forall m \in M\end{matrix} \tag{29.32}$$

The concentration of contaminants on the feed side must also be described in terms of the pressure drop and the osmotic pressure. This is described in constraint (29.33).

$$C_{q,m}^p = \frac{k_m C_{q,m}^{av}}{A(\Delta P_q - \Delta\pi_q)\gamma} \qquad \begin{matrix}\forall q \in Q \\ \forall m \in M\end{matrix} \tag{29.33}$$

A mass and concentration balance around the regenerator is also needed and is described in constraint (29.34) and (29.35), respectively.

$$F_q^f = F_q^p + F_q^r \qquad \forall q \in Q \tag{29.34}$$

$$F_q^f C_{q,m}^f = F_q^p C_{q,m}^p + F_q^r C_{q,m}^r \qquad \begin{matrix}\forall q \in Q \\ \forall m \in M\end{matrix} \tag{29.35}$$

29.3.4 Big-M Constraints

In order to determine the existence of piping interconnections, logical constraints and discrete variables will be adopted. This formulation makes use of the big-M parameters adopted by Khor et al.[2] In the big-M parameters, M is a valid upper/lower bound denoted by U and L respectively. Constraints (29.36)–(29.39) represent the big-M parameters for the piping interconnections between the different units.

$$M_{i,j}^L y_{i,j} \le F_{i,j}^s \le M_{i,j}^U y_{i,j} \tag{29.36}$$

$$M_{q,j}^L y_{q,j}^p \le F_{q,j}^p \le M_{q,j}^U y_{q,j}^p \tag{29.37}$$

$$M_{q,j}^L y_{q,j}^r \le F_{q,j}^r \le M_{q,j}^U y_{q,j}^r \tag{29.38}$$

$$M_{i,n}^L y_{i,n}^d \le F_{i,n}^d \le M_{i,n}^U y_{i,n}^d \tag{29.39}$$

29.3.5 Objective Function

The objective function of the combined RON superstructure and WNS is used to minimize the overall cost of the regeneration network on an annualized basis that consists of:

1. TAC of the RON
2. cost of fresh water (FW)
3. treatment cost of wastewater (WW)
4. capital and operation costs of the piping interconnection

The TAC of the RON consists of the capital cost of the RO modules, pump and energy recovery turbines, operating cost of pumps and turbines, as well as pretreatment of chemicals. The operating revenue of the energy recovery turbine is also considered in the determination of the TAC and is shown in constraint (29.40).

$$\begin{aligned}\text{TAC}(q,n) = {} & C^{\text{pump}}\left(\sum_{n=1}^{N}(P_n^i - P_n^a)F_n^a\right)^{0.65} + C^{\text{tur}}\left(\sum_{n=1}^{N}(P_n^i - P_n^o)F_n^a\right)^{0.43} \qquad \forall q \in Q \\ & + C^{\text{tur}}\left(\sum_{j=1}^{J}(P_q^r - P_j^r)F_{q,j}^r\right)^{0.43} + C^{\text{elec}}\text{AOT}\left(\frac{\sum_{n=1}^{N}(P_n^i - P_n^a)F_n^a}{\eta_{\text{pump}}}\right) \\ & - C^{\text{elec}}\text{AOT}\left(\sum_{j=1}^{J}(P_q^r - P_j^r)F_{q,j}^r\right)\eta_{\text{turbine}} - C^{\text{elec}}\text{AOT}\left(\sum_{j=1}^{J}(P_n^i - P_n^o)F_n^a\right)\eta_{\text{turbine}} \\ & + C^{\text{mod}}\sum_{q=1}^{Q}N_q^s + C^{\text{chem}}\text{AOT}\sum_{i=1}^{I}F_{i,n}^d \end{aligned} \tag{29.40}$$

The piping cost of components will be formulated by assuming a linear fixed-charge model. In their formulation, a particular cost of a pipe is incurred if the particular flow rate through the pipe falls below the threshold value. This is achieved by using 0–1 variables. Constraint (29.41) represents the objective function of the total regeneration network. It is also assumed that all the pipes share the same properties of p_c and q_c and a 1-norm distance D.

$$\min\left(\begin{aligned} & \sum_{q=1}^{Q}\sum_{n=1}^{N}\text{TAC}_{q,n} + \text{AOTC}^{\text{water}}\text{FW} + \text{AOTC}^{\text{waste}}\text{WW} \\ & + \text{AA}\left(\sum_{i=1}^{I}\sum_{j=1}^{J}D_{i,j}\left(\frac{p_c F_{i,j}^s}{3600v} + q_c y_{i,j}\right)\right) \\ & + \text{AA}\left(\sum_{q=1}^{q}\sum_{j=1}^{J}D_{q,j}^p\left(\frac{p_c F_{q,j}^p}{3600v} + q_c y_{q,j}^p\right)\right) \\ & + \text{AA}\left(\sum_{q=1}^{q}\sum_{j=1}^{J}D_{q,j}^r\left(\frac{p_c F_{q,j}^r}{3600v} + q_c y_{q,j}^r\right)\right) \\ & + \text{AA}\left(\sum_{i=1}^{I}\sum_{n=1}^{N}D_{i,n}^d\left(\frac{p_c F_{i,n}^d}{3600v} + q_c y_{i,n}^d)\right)\right) \end{aligned}\right) \tag{29.41}$$

where $\text{AA} = \left(m(1+m)^n/(1+m)^n - 1^n\right)$

The overall model results in a nonconvex MINLP due to the bilinear terms, as well as the power function in the constraints.

29.4 ILLUSTRATIVE EXAMPLE

The earlier model was applied to a literature-based refinery case study.[2] The model was implemented in GAMS 24.2 using the general-purpose global optimization solver BARON, which obtains a solution by using a branch-and-reduce algorithm. The network consists of four sources and four sinks. The limiting water data for the sources and sinks is given in Table 29.1. Table 29.2 shows the Manhattan distances between different units. The distances between the regenerators and the sinks for both permeate and retentate streams are the same. Table 29.3 presents the process and economic data for the detailed RON. The economic data and the model parameters are given in Table 29.4.

Four scenarios will be compared in order to highlight the importance of incorporating a detailed RON superstructure within the water network.

1. First, the case in which no regeneration is considered within the water network is modeled in order to provide a basis (base case) for comparison (Case 1).
2. In the second case, a single regenerator is incorporated within the WNS with a fixed removal ratio (Case 2).
3. The third case looks at multiple regenerators within the WNS with a fixed removal ratio (Case 3).
4. The fourth case looks at multiple regenerators with a variable removal ratio (Case 4).

TABLE 29.1
Limiting Data for Water Network

Sources, *i*					Sinks, *j*				
i\	Unit	Flow Rate (kg/s)	Contaminant Concentration (kg/m³)		*J*	Unit	Flow Rate (kg/s)	Max Contaminant Concentration (kg/m³)	
			TDS	COD				TDS	COD
1	Amine sweeting	7.3	3.5	3.5	1	Caustic treating	0.83	2.5	2.5
2	Distillation	10.65	4	4	2	Menox-I sweeting	40	2	2
3	Hydrotreating	3.5	1	3	3	Desalting	5.56	2.5	2.5
4	Freshwater		2	1	4	Wastewater		25	25

TABLE 29.2
Manhattan Distance for the Case Study

	Sinks				Regenerator Unit	
Sources	**1**	**2**	**3**	**4**	**1**	**2**
1	50	50	50	60	50	50
2	60	50	60	70	40	40
3	50	50	50	60	65	50
4	60	50	60	70	100	50
Regenerator Unit						
1	80	70	60	70		
2	60	10	40	20		

TABLE 29.3
Process and Economic Data for the Detailed RON

Parameter	Value
Pure water permeability, A	5.50×10^{-13} m/(s.Pa)
Shell side pressure drop per module per regenerator, P_m	4.05×10^4 Pa
Solute permeability coefficient, k_m	1.82×10^{-8} m/s
Fiber length, L	0.75 m
Seal length, L_s	0.075 m
Outside radius of fiber, r_o	42×10^{-6} m
Inner radius of fiber, r_i	21×10^{-6} m
Membrane area, S_m	180 m
Water viscosity, μ	0.001 kg/(m.s)
Dimensionless constant, Υ	0.69
Permeate pressure per regenerator, $P_p(q)$	101325 Pa
Pump efficiency, η_{pump}	0.7
Turbine efficiency, η_{turbine}	0.7
Liquid recovery for all regenerators, $\alpha(q)$	0.7
Osmotic constant, OS	4.14×10^{-7} Pa
Cost parameter for chemicals, C_{chemical}	0.11 \$/kg
Cost of electricity, C_{elec}	0.06 \$/(kW.h)
Cost coefficient for pump, C_{pump}	6.5 \$/(yearW0.65)
Cost coefficient for pump, C_{tur}	18.4 \$/(yearW0.43)
Cost per module of HFRO membrane, C_{mod}	2300 \$/(year.module)
Maximum flow rate per hollow fiber module, F^U	0.27 kg/s
Minimum flow rate per hollow fiber module, F^L	0.21 kg/s

TABLE 29.4
Economic Data and the Model Parameters for the WNS

Parameter	Value
Annual operating time, AOT	8760 h
Unit cost of freshwater, C_{water}	1 \$/t
Unit cost of wastewater, C_{waste}	1 \$/t
Interest rate per year, m	5%
Number of years, n	5 year
Parameter p for carbon steel piping	7200
Parameter q for carbon steel piping	250
Velocity, v	1 m/s

The results obtained from the optimization are given Table 29.5 for Cases 1 to 3. In Cases 2 and 3, the regenerators had a fixed removal ratio of 0.95. In the first scenario, the water network with no regeneration had a higher total cost due to the high consumption of fresh water, which can be seen in Table 29.5; the network is shown in Figure 29.9. The second scenario where a single regenerator was used led to a 15.26% reduction in freshwater usage and a 43.36% in reduction in wastewater generation in comparison with the base case. The overall cost of the network was minimized by 17.6% due to the incorporation of the RO regenerator. The use of the energy recovery turbines in the RON led to a reduction in the regeneration cost of the network. Figure 29.10 shows the complete water network and RON obtained for Case 2. This diagram includes the ROSDBs as shown

TABLE 29.5
Summary of Results for Case 1 to 3

	No Regeneration (Case 1)	Single Regenerator Fixed RR (Case 2)	Two Regenerators Fixed RR (Case 3)
Freshwater flow rate (kg/s)	38.40	32.54	28.87
Wastewater flow rate (kg/s)	13.40	7.59	3.91
Cost of regeneration (million $/year)		0.068	0.23
Total cost (million $/year)	1.70	1.40	1.32
CPU time (h)	0	0.13	6

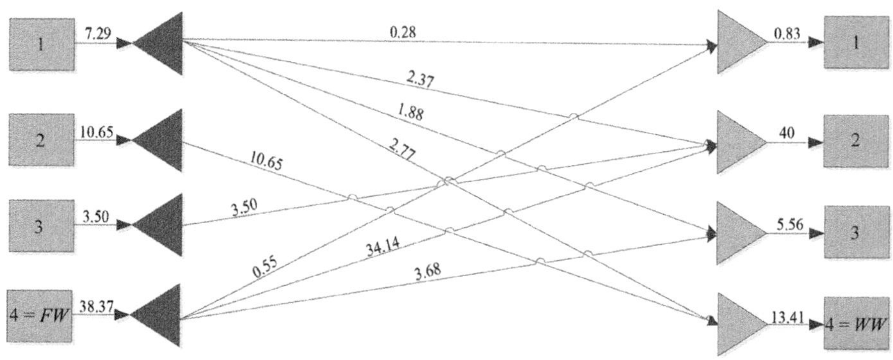

FIGURE 29.9 Network obtained for Case 1 (No regeneration).

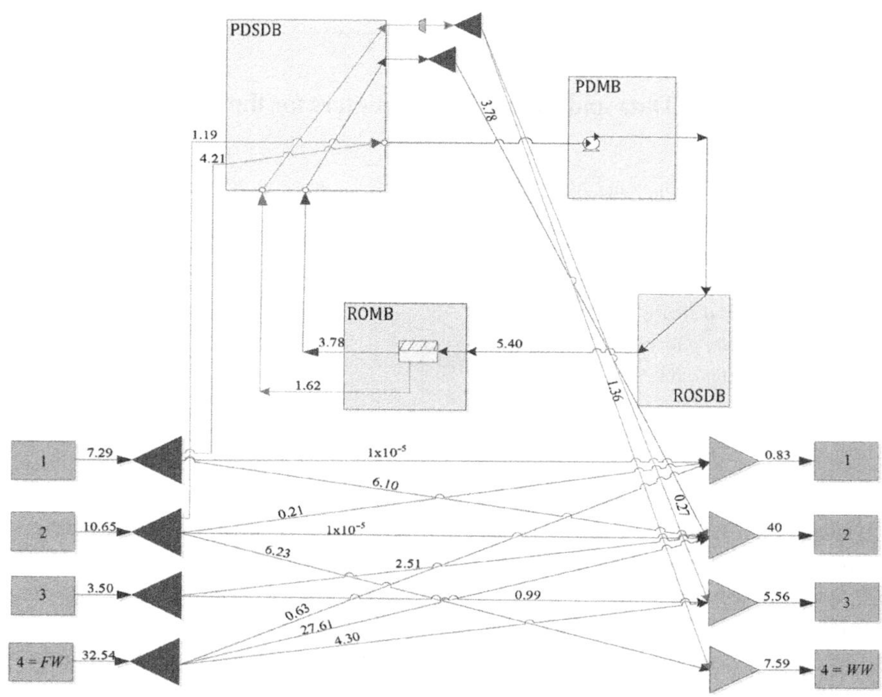

FIGURE 29.10 Network for Case 2 based on the distribution boxes.

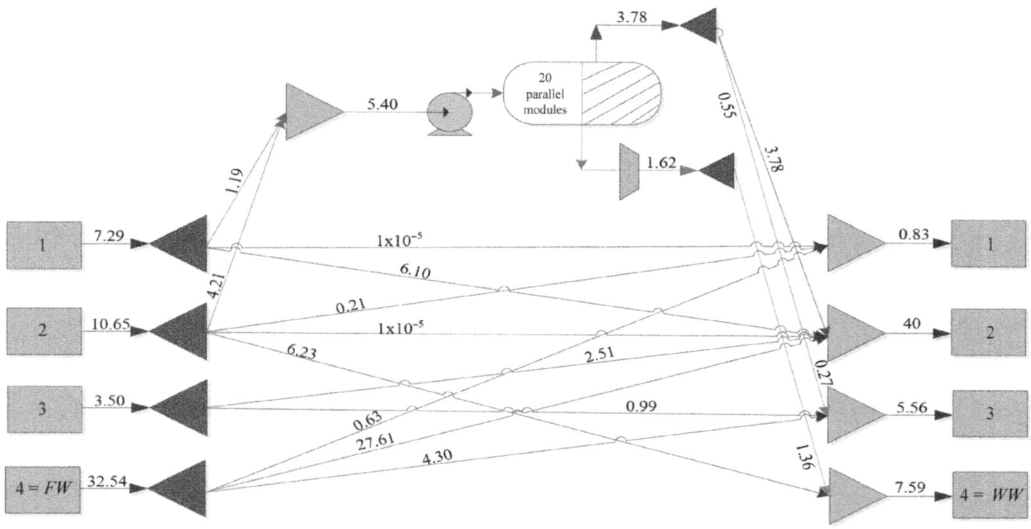

FIGURE 29.11 Network obtained for Case 2 (Single regenerator with fixed removal ratio).

in Figure 29.2. Figure 29.10 can be translated into a simplified schematic diagram showing only the relevant physical units, i.e., the RO membranes, pumps, turbines, mixes and splitters. Figure 29.10 shows the water network for Case 2. In Figure 29.11 it can be seen that one pump and turbine are needed for regeneration, as well as 20 hollow-fiber RO (HFRO) modules. For simplicity in Cases 3 and 4 only the simplified water network is presented.

Case 3 led to a 24.82% reduction in freshwater consumption and a 70.82% reduction in wastewater generation in comparison with Case 1. The total cost of the network was also reduced by 22.35%. The low cost of the water network is due to the low freshwater consumption and wastewater generation. The introduction of a second regenerator in Case 3 leads to further reduction in the total cost. This is due to the lower consumption of fresh water and wastewater generation. Figure 29.12 shows the water network for Case 3. In Figure 29.12 it can be seen that two pumps and turbines are needed for the regeneration, as well as 37 HFRO modules per regenerator. A parallel configuration was chosen by the model.

Table 29.6 shows the comparison between Cases 3 and 4. The removal ratio chosen by the model in Case 4 was 0.97 for all contaminants. Case 4 led to a 3.12% reduction in fresh water use and a 30.43% reduction in wastewater generation in comparison with Case 3. A 15.91% reduction in the total network cost was also achieved. The large decrease in the total cost of the network in Case 4 can be attributed to the high removal ratio, which was selected by the model at a higher value than that which was initially predicted. In comparison with the case where no regeneration was considered, Case 4 leads to a 28% reduction in freshwater consumption and an 80% reduction in wastewater generation. The modeling of Case 4 is, however, computationally expensive as can be seen in Table 29.6. The best case used 15 HFRO modules per regenerator. The model selected two regenerators, two pumps and two energy recovery turbines, as can be seen in Figure 29.13. It can also be seen that a parallel configuration of the network was chosen by the model. Flow rates obtained for the different streams are indicated on Figures 29.9 through 29.13.

The high computational time for solving the model in Case 3 was due to the complexity of the problem, as well as the large number of 0–1 variables. The model solves more quickly when tighter bounds are imposed on the feed and retentate pressure. The use of the energy recovery turbines in the RON led to a reduction in the regeneration cost of the network, and as a result, a reduction in energy usage by the system was achieved. The statistics of the model for all the four cases is shown in Table 29.7.

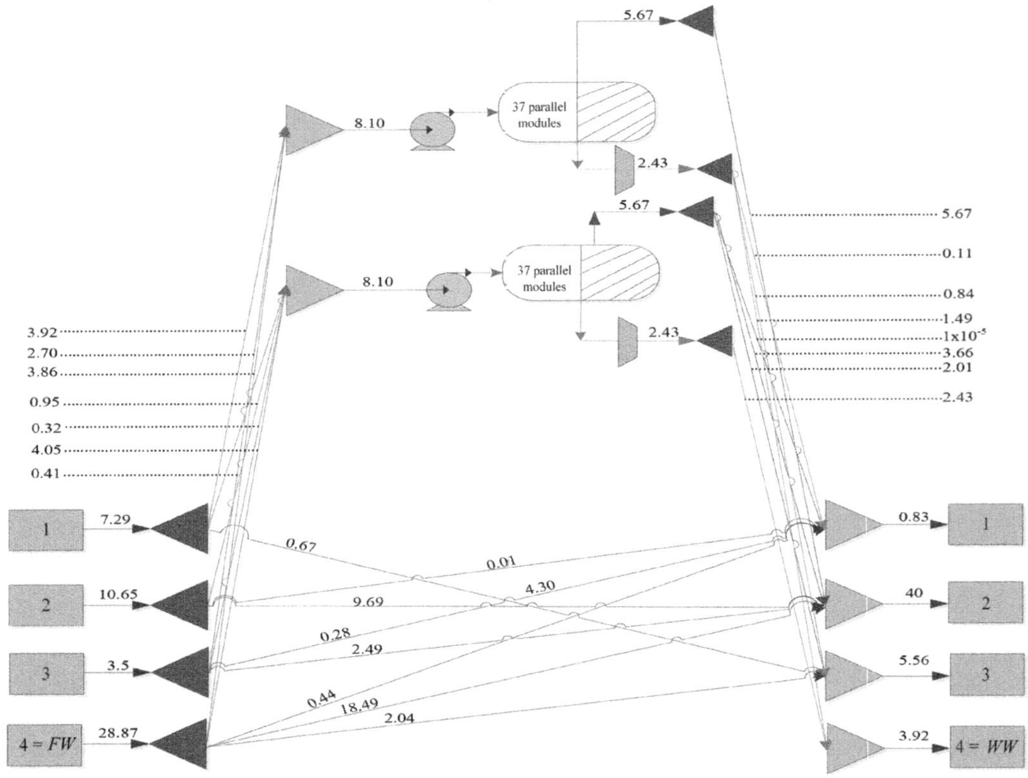

FIGURE 29.12 Network obtained for Case 3 (Single regenerator with variable removal ratio).

TABLE 29.6
Summary of Results for Case 3 and 4

	Multiple Regenerators	
	Fixed RR (Case 3)	**Variable RR (Case 4)**
Freshwater flow rate (kg/s)	28.87	27.68
Wastewater flow rate (kg/s)	3.91	2.72
Cost of regeneration (million $/year)	0.23	0.096
Total cost (million $/year)	1.32	1.11
Network configuration	Parallel	Parallel
Number of HFRO modules	37 for each regenerator	15 for each regenerator
CPU time (h)	6	54

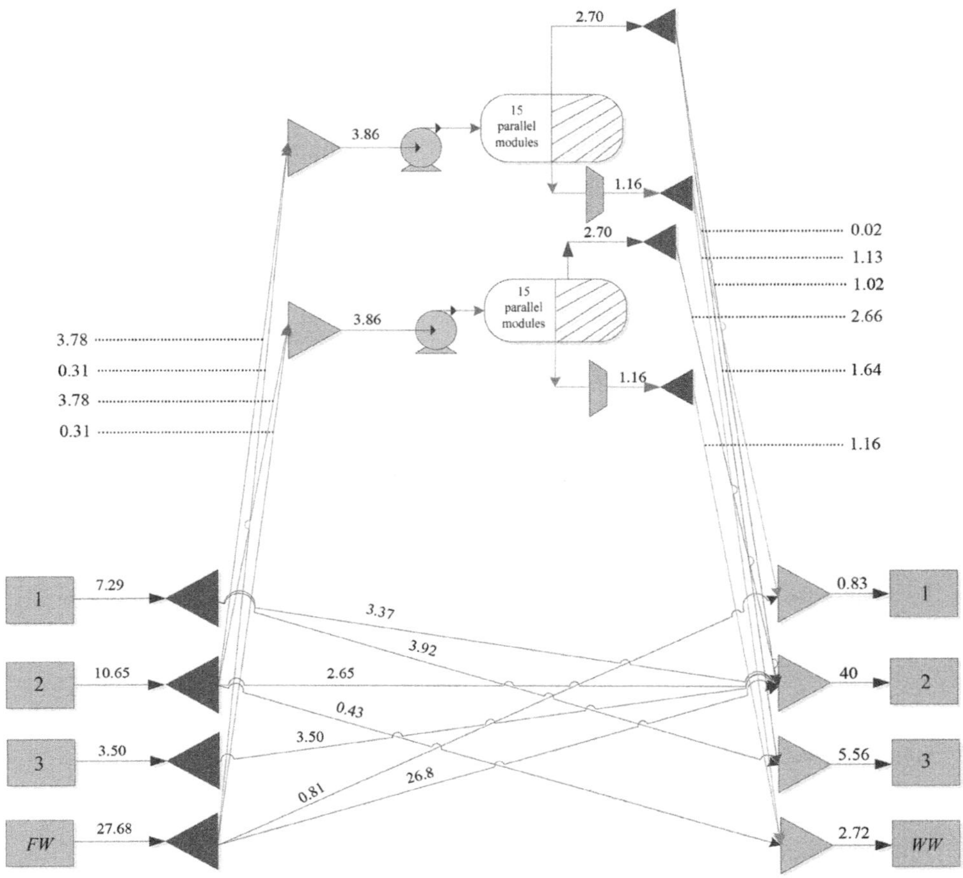

FIGURE 29.13 Network obtained for Case 4 (multiple regenerators with variable removal ratio).

TABLE 29.7
Model Statistics for Case 1 to 4

	No Regeneration	Single Regenerator	Multiple Regenerators	
	(Case 1)	Fixed RR (Case 2)	Fixed RR (Case 3)	Variable RR (Case 4)
Number of equations	60	168	282	282
Number of continuous variables	46	134	208	212
Number of discrete variables	16	32	48	48
Optimality gap	0.1	0.1	0.1	0.1

29.5 CONCLUSION

This chapter has addressed the synthesis of a water regeneration network that incorporates the detailed synthesis of a RON. The proposed model was applied to a literature case study and was then solved using GAMS/BARON in order to highlight its practicality. The results show that the use of multiple regenerators in the water network can lead to a reduction in the total cost of the network due to the significant reduction in freshwater consumption and wastewater generation. It can also be concluded that there is a significant benefit in allowing the removal ratio in the model to be a variable, as this has severe impact on the cost and structure of the network. The implications of this study show that detailed optimization of regenerators within water networks can significantly improve wastewater management within process plants. Large computational times were, however, incurred due to the complex nature and structure of the model. It is also noteworthy that the proposed model was limited to one membrane technology. Multiple membrane technologies such as ultrafiltration can, however, be incorporated in the membrane network and thus offering a scope for future work.

ACKNOWLEDGMENTS

The authors would like to thank the National Research Foundation (NRF) for funding this work under the NRF/DST Chair in Sustainable Process Engineering at the University of the Witwatersrand, Johannesburg, South Africa.

NOMENCLATURE

Sets

I = {i|i = water source}
J = {j|j = water sink}
M = {m|m = contaminants}
Q = {q|q = regeneration units}

Parameters

α_q	liquid recovery
$RR_{q,m}$	removal ratio
F^U	maximum flow rate per hollow fiber module
F^L	minimum flow rate per hollow fiber module
ΔP_q^m	shell side pressure drop per module
M^U	upper bound of big-M constant for interconnections between streams
M^L	lower bound of big-M constant for interconnections between streams
AOT	annual operating time
p_c	parameter for carbon steel piping based on Chemical Engineering Plant Cost Index (CEPCI) value of 318.3
q_c	parameter for carbon steel piping based on CEPCI value of 318.
v	velocity
A	water permeability coefficient
$P_{\max}$	maximum allowable pressure for the regenerators
k_m	solute permeability constant
L	fiber length
L_s	seal length
r_o	outside radius of fiber
r_i	inner radius of fiber
S_m	membrane area per module
P^U	an arbitrary big value for pressure
P^L	an arbitrary small value for pressure
Υ	a dimensionless constant
η_{pump}	pump efficiency
η_{turbine}	turbine efficiency
OS	proportionality constant between the osmotic pressure and average salt mass fraction on the feed side

$C^U_{j,m}$ maximum allowable contaminant concentration m in sink j
$C^U_{q,m}$ maximum allowable contaminant concentration m into a regenerator q
$D^a_{i,j}$ Manhattan distance between water source i and sink j
$D^p_{q,j}$ Manhattan distance between regenerator q and sink j
$D^r_{q,j}$ Manhattan distance between regenerator q and sink j
$D^d_{i,n}$ Manhattan distance between source i and node n
$C_{i,m}$ mass fraction of contaminant m within water source i
C^{chem} cost parameter for chemicals
C^{elec} cost of electricity
C^{mod} cost per module of HFRO membrane
C_{pump} cost coefficient for pump
C^{tur} cost coefficient for turbine
μ water viscosity
P^P_q pressure of a permeate stream from regenerator q
P^w_i pressure of source i
P^r_j pressure of the retentate stream in sink j
Fl^L lower bound on flow rate
Fl^U upper bound on flow rate
P^L lower bound on pressure
P^U upper bound on pressure

Continuous Variables

$F^s_{i,j}$ allocated flow rate between source i and sink j
$F^d_{i,n}$ allocated flow rate between source i and node n
F_i flow rate of sources i
$F^p_{q,j}$ flow rate of the permeate stream from regenerators q to sink j
$F^r_{q,j}$ flow rate of the retentate stream from regenerators q to sink j
$F^a_{n,q}$ flow rate of streams from node n to regenerator q
F^f_q flow rate leaving the outlet junction of ROSDB
F^p_q flow rate of permeate stream leaving the regenerator q
$F^p_{q,n}$ flow rate of permeate stream regenerator q to node n
F^r_q flow rate of retentate stream leaving the regenerator q
$F^r_{q,n}$ flow rate of retentate stream from regenerator q to node n
F^a_n flow rate of streams from node n
F^w_j flow rate of sink j
$C^a_{n,m}$ concentration of contaminant m in stream leaving node n
$C^f_{q,m}$ concentration of contaminant m in the feed to the regenerator q
$C^p_{q,m}$ concentration of contaminant m in permeate stream leaving regenerator q
$C^r_{q,m}$ concentration of contaminant m in retentate stream leaving regenerator q
$C^{av}_{q,m}$ average concentration of contaminant m in the high-pressure side of regenerator q
P^a_n pressure of streams leaving node n
P^i_n pressure of an inlet stream to an energy recovery turbine from node n
P^o_n pressure of an outlet stream from an energy recovery turbine from node n
ΔP_q pressure drop over regenerator q
P^f_q feed pressure into regenerator q
P^r_q pressure of a retentate stream from regenerator q
P^p_q pressure of a permeate stream from regenerator q
$\Delta\pi_q$ osmotic pressure on the retentate side of regenerator q
FW freshwater flow rate
WW wastewater flow rate

Binary Variables

$$b_n = \begin{cases} 1 \leftarrow \text{if a pump exists} \\ 0 \leftarrow \text{otherwise} \end{cases}$$

$$t_n = \begin{cases} 1 \leftarrow \text{if a turbine exits} \\ 0 \leftarrow \text{otherwise} \end{cases}$$

$$r_q = \begin{cases} 0 \leftarrow \text{if regenerator } q \text{ exits} \\ 1 \leftarrow \text{otherwise} \end{cases}$$

$$y^p_{q,j} = \begin{cases} 1 \leftarrow \text{if piping exits between the permeate streams and sink } j \\ 0 \leftarrow \text{otherwise} \end{cases}$$

$$y^p_{q,j} = \begin{cases} 1 \leftarrow \text{if piping exits between the retentate streams and sink } j \\ 0 \leftarrow \text{otherwise} \end{cases}$$

$$y_{i,j} = \begin{cases} 1 \leftarrow \text{if piping exits between source } i \text{ and sink } j \\ 0 \leftarrow \text{otherwise} \end{cases}$$

$$y^d_{i,n} = \begin{cases} 1 \leftarrow \text{if piping exits between source } i \text{ and node} n \\ 0 \leftarrow \text{otherwise} \end{cases}$$

Integer Variables

N^s_q the number of hollow fiber modules of regenerator q

REFERENCES

1. Bandyopadhyay, S., Cormos, C. Water management in process industries incorporating regeneration and recycle through a single treatment unit. *Industrial and Engineering Chemistry Research* **2008**, *47*, 1111–1119.
2. Khor, C., Foo, D., El-Halwagi, M., Tan, R., Shah, N. A superstructure optimization approach for membrane separation-based water regeneration networks synthesis with detailed nonlinear mechanistic reverse osmosis model. *Industrial and Engineering Chemistry Research* **2011**, *50*, 13444–13456.
3. Alva-Argáez, A., Kokossis, A., Smith, R. Wastewater minimisation of industrial systems using an integrated approach. *Computers and Chemical Engineering* **1998**, *22*, 741–744.
4. Khor, C., Chachuat, B., Shah, N. A superstructure optimization approach for water network synthesis with membrane separation-based regenerators. *Computers and Chemical Engineering* **2012**, *42*, 48–63.
5. Tan, R., Ng, D., Foo, D., Aviso, K. A superstructure model for the sybthesis of single-contaminant water network with partitioning regenerators. *Process Safety and Environmental Protection* **2009**, *87* (3), 197–205.
6. Khor, C., Chachuat, B., Shah, N. Optimization of water network synthesis for single-site and continous processes: Milstones, challenges, and future directions. *Industrial & Engineering Chemistry Research* **2014**, *53*, 10257–10275.
7. El-Halwagi, M. Synthesis of reverse osmosis networks for waste reduction. *AIChE Journal* **1992**, *38*, 1185–1198.
8. Evangelista, F. A short-cut method for the design of reverse-osmosis desalination plants. *Industrial & Engineering Chemistry Process Design and Development* **1985**, *24*, 211.
9. Saif, Y., Elkamel, A., Pritzker, M. Optimal design of reverse osmosis networks for wastewater treatment. *Chemical Engineering and Processing* **2008a**, *47*, 2163–2174.
10. Sassi, K., Mujtaba, I. Optimization of design and operation of reverse osmosis based desalination process using MINLP approach incorporating fouling effects. *21st European Symposium on Computer-Aided Process Engineering* **2011**, *21*, 206–210.
11. Saif, Y., Elkamel, A., Pritzker, M. Global optimization of reverse osmosis network for wastewater treatment and minimization. *Industrial and Engineering Chemistry Research* **2008b**, *47* (1), 3060–3070.

12. Wang, Y. P., Smith, R. Wastewater minimisation. *Chemical Engineering Science* **1994**, *49* (7), 981–1006.
13. Hallale, N. A new graphical targeting method for water minimisation. *Advances in Environmental Research* **2002**, *52*, 377–390.
14. Ng, D., Foo, D., Tan, R. Targeting for total water networks.1. waste stream identification. *Industrial and Engineering Chemistry Research* **2007**, *46*, 9107–9113.
15. Manan, Z., Tan, Y., Foo, D. Targeting the minimum water flow rate using water cascade analysis technique. *AIChE Journal* **2004**, *50* (12), 3169–3183.
16. Faria, D., Bagajewicz, M. On the appropriate modelling of process plant water systems. *American Institute of Chemical Engineering Journal* **2009**, *56*, 668–689.
17. Takama, N., Kuriyama, T., Shiroko, K., Umeda, T. Optimal water allocation in a petroleum refinery. *Computers and Chemical Engineering* **1980**, *4*, 251–258.
18. Galan, B., Grossmann, I. Optimal design of distributed wastewater treatment networks. *Industrial and Engineering Chemistry Research* **1998**, *37* (10), 4036–4048.
19. Karuppiah, R., Grossmann, I., Karuppiah, R., Grossmann, I. Global optimisation for the synthesis of integrated water systems in chemical processes. *Computers and Chemical Engineering* **2006**, *30* (4), 650–673.
20. El-Halwagi, M. *Pollution Prevention through Process Integration;* Academic Press: San Diego, CA, 1997.

30 Interaction of Energy Consumption, Energy Quality, and Freshwater Production in the Multiple Effect Desalination Process

Giacomo Filippini, Flavio Manenti, and Iqbal M. Mujtaba

CONTENTS

30.1 INTRODUCTION

Seawater desalination is a technology of paramount importance for providing fresh water in areas with few nature reserves and poor rainfall. We can classify desalination technologies into three main groups:

Thermal: heat-consuming processes in which seawater is desalted through evaporation. Evaporative processes include: multistage flash desalination (MSF), multiple-effect desalination (MED), single-effect evaporation (SEE), humidification–dehumidification, and solar stills. SEE and MED processes can be coupled with a unit for vapor compression to improve the thermal efficiency, including mechanical vapor compression (MVC), thermal vapor compression (TVC), chemical vapor compression, absorption vapor compression, and adsorption vapor compression. MSF is currently the most used in industrial reality, followed by MED-TVC [1].

Membrane: power-consuming processes in which seawater is desalted passing through a membrane. Membrane processes include: reverse osmosis (RO) and electrodialysis (ED). The RO process is widely used in the desalination industry, while ED sees little application [1].

Hybrid: processes that integrate thermal and membrane desalination to improve the overall performance and minimize the cost of fresh water. Examples include MSF+RO and MED-TVC/MVC+RO.

Much progress has been made in the desalination industry since the 1960s; however, research efforts are still required to reduce the cost of desalted water. In recent years, MED with TVC has attracted more attention than other thermal processes due to its straightforward operation and maintenance, high effectiveness and feasible economic characteristics. This is particularly true in the case of the low-temperature MED process, which shows high performance together with few fouling/scaling problems, negligible heat loss, and a reduced need for thermal insulation [2]. The MED process can be arranged with a forward, backward, parallel, or parallel/cross feed. In this study, the model developed is for the forward-feed MED, which was chosen because of the reduced fouling and corrosion problems associated with this configuration [2].

30.2 OVERVIEW OF MATHEMATICAL MODELS AVAILABLE FOR MED

One of the main research areas in desalination is the mathematical modeling and simulation of the process using advanced programming tools. Being able to correctly simulate the process is of paramount importance to identify trade-offs between the process variables, with the aim of determining feasible and optimal process design. There are several models in literature for the multiple-effect evaporation process. Here, five of the most cited ones are analyzed.

El-Sayed and Silver [3] developed a model for forward-feed MED, which can predict some performance parameters, but relies on thermodynamic simplifications, like constant fluid properties (specific heat, boiling point elevation, latent heat). An explicit formula for evaluating the performance ratio is derived. Despite its simplicity, the formula provides a very quick and quite reliable way to assess the performance of the process under known operating conditions; however it cannot be used for optimization or sensitivity analysis.

Darwish et al. [4] also developed a simplified model for MED without TVC and discussed the trade-off between heat transfer area and performance ratio. The model relies on thermodynamic simplifications like constant fluid properties and constant heat exchange coefficients. The temperature profile is assumed linear. With these assumptions, an explicit formula for performance ratio has been derived. In the present work, the original model by Darwish et al. has been modified to remove all the thermodynamic simplifications and the linearity of temperature profiles, by imposing an equalizing procedure for the areas of the effects. The TVC section has been modeled and coupled with the MED. The resulting model is still easy to implement but provides more realistic values of the performance parameters.

El-Dessouky and Ettouney [1] developed a very simplified model that does not require a numerical solver and it is able to predict some performance parameters; however their non-linear dependence on operative parameters is lost because of many simplifying assumptions, like constant fluid properties, constant thermal load in each effect, no flashed distillate, and no feed pre-heating (it is assumed that the feed enters the first effect at the first effect's saturation temperature, i.e., steam is used only to evaporate distillate, not for heating the feed).

El-Dessouky et al. did several studies modeling various configurations of the MED process, eventually coupling it with TCV or MVC [5–7]. El-Dessouky et al. [5] also presented a very detailed forward-feed MED model that considers the preheaters and flashing boxes. The model assumes that the heat transfer areas for both the evaporators and feed preheaters in all effects are equal, according to industrial reality. The impact of vapor leaks in the venting system and influence of incondensable gasses are also modeled. Several correlations are used to determine the properties of the fluids, their heat transfer coefficients, and their pressure losses. The authors developed correlations for the heat transfer coefficients in evaporators and pre-heaters, taking into account the dependence of temperature and fouling. Various trends of performance parameters are analyzed, with the result that the effect of temperature on the specific heat transfer area is more pronounced for a high number of effects.

Mistry et al. [8] modeled the MED process in a modular way, modeling each subcomponent individually and then linking them in the desired order to piece together the complete plant. Models of components are detailed and rely on few assumptions. Modular development has the advantage of being usable for studying various MED configurations with minimal code modifications.

Other literature models are also listed here for completeness. El-Allawy [9] performed a sensitivity analysis of performance with respect to steam temperature and number of effects, showing a significant increase of the gained output ratio for a high number of effects and, if TCV section is enabled, for low steam temperature. Aly and El-Figi [10] found that fresh water production is significantly dependent on the number of effects, while dependence on steam temperature is weak for MED without vapor compression. Al-Sahali and Ettouney [11] developed a MED-TVC model based on sequential solution instead of iterative solution, by assuming a linear temperature profile. Ameri et al. [12] studied the effect of seawater properties on process performance. Darwish and Alsairafi [13] compared the performance of MSF and MED, concluding that MED is preferable because of lower pumping power, less feed pre-treatment required and less heat losses. Choi et al. [14], Darwish and Al-Najem [15] conducted an exergetic analysis using the second law of thermodynamics, highlighting that the TVC section is responsible for a large part of exergy destruction. Hamed [16] conducted a sensitivity analysis of dependence of performance of the MED process with respect to many design and operative parameters.

30.3 DESCRIPTION OF THE FORWARD-FEED MED-TVC PROCESS

Figure 30.1 shows the diagram of forward-feed MED process with TVC. The process includes n effects, n–1 pre-heaters, n–1 flashing boxes (not represented), and 1 down condenser. In this study, the flashing boxes are not modeled for simplifying the flow rate patterns and write smoother balances; however, the quantity of flashed distillate is evaluated in the model with thermodynamic considerations. In the forward configuration, material and energetic streams flow from effect 1 to n. Each effect includes a horizontal falling film evaporator, some space for the vapor phase, a spray nozzle, a demister, and a feed pre-heater.

The intake seawater Mw at temperature Tw and salinity xf flows into the end condenser to absorb the latent heat associated with distillate from the last effect. Therefore, the distillate is condensed, and seawater temperature is increased. In other words, all the excess heat added in the first effect by the motive steam is absorbed by the seawater stream. A fraction of seawater

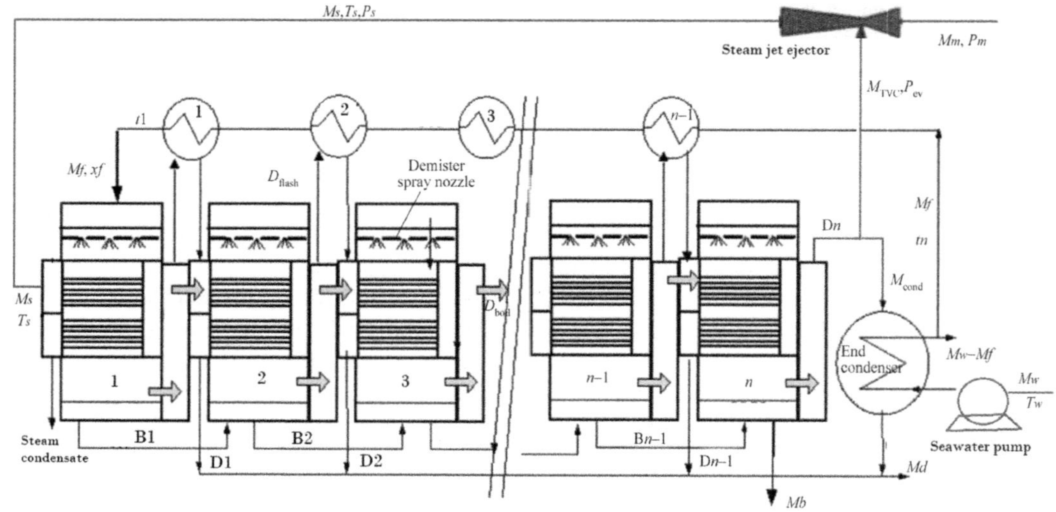

FIGURE 30.1 Schematic representation of forward-feed MED with TVC. Brine is always sprayed from the top and does not enter from the bottom of the effect. (Modified from Darwish, M., *Desalination*, 194, 22–39, 2006 [4].)

equal to *Mw*–*Mf* is rejected back to the sea, while the feed flow rate *Mf*, after chemical treatment and de-aeration, is pumped through a train of *n*–1 pre-heaters, increasing its temperature up to *t*1, which is the feed temperature in the first effect. Pre-heating is very important, since it allows the feed seawater in the first effect to be at a temperature near boiling point, so that the major part of the thermal load provided by the external steam will be converted into latent heat for evaporation instead of sensible heat for heating the feed. The heating utility in the pre-heaters is the flashed vapor, which is condensed in the process and sent back to the flashing-box.

The feed *Mf* is sprayed in the first effect, forming a thin film on the tubes of the horizontal heat exchanger. When the boiling temperature is reached, the vapor phase *D*1 is generated, at temperature *Tv*1. In each effect, the quantity of vaporized brine can be relatively small, so we need a relatively high number of effects to collect a good quantity of total distillate, *Md*. The quantity of vapor formed by boiling in an effect is slightly lower than the quantity produced in the previous one, because of the increase of latent heat of vaporization with the decrease of evaporation temperature. However, the difference is not relevant in the temperature window of a low-temperature MED process (Appendix Figure 30.3).

The boiled vapor *D*1 flows toward the second effect to provide the necessary thermal load. The remaining brine *B*1 also flows in the second effect and it is sprayed from the top spray nozzle, after passing through a lamination valve, which reduces the pressure and the boiling temperature. In that way, it is possible to guarantee a difference of temperature between the vapor at temperature Tv_{i-1} and the brine at temperature T_i. In the MED-TVC system, the vapor formed in the last effect is partially entrained by the steam jet ejector, where it is compressed by the external motive steam Mm to the desired temperature and pressure. The aim of the TVC is to increase the entrained steam pressure by converting the pressure energy of motive steam. When the motive steam expands, the conversion of its static pressure energy into kinetic energy takes place. The entrained vapor violently mixes with the motive steam, generating a stream at intermediate pressure *Ps*. In other words, a certain quantity of low-pressure steam can be reutilized in the process prior to recompression (upgrading); this increases the quantity of distillate produced per quantity of external steam provided. All the condensed distillate is collected together and stored, while the brine from the last effect is rejected into the sea. The salinity and temperature of this stream should not be too high for environmental reasons.

30.4 PROCESS MODELING

30.4.1 Model Hypotheses and Assumptions

1. Steady state process.
2. The vapor phase is salt free: demisters are capable of perfectly removing all the entrained droplets in the vapor phase, so that they do not contaminate the product.
3. Energy losses to the surroundings are negligible. Since the MED process operates at relatively low temperature, the heat exchange with the external environment can be neglected, providing that the effects and pipelines are well insulated.
4. All physical properties of water and steam are evaluated at the average temperature of inlet and outlet streams.
5. Equal transfer area in all the effects. This is in line with industrial reality, where evaporators are usually bought in stocks, also to take advantage of a possible discount.
6. Non-equilibrium allowance is neglected to lighten the model.
7. Pressure drops in pipelines, in lamination valves, and during condensation are neglected.
8. Boiling point elevation and specific heat are considered as a function of temperature and salinity.
9. Latent heat of evaporation and overall exchange coefficient are considered as a function of temperature. For heat exchange, experimental correlations that also consider fouling are implemented.
10. Steam from the external utility is provided saturated and leaves as saturated liquid.
11. Flashed distillate is evaluated, but the flashing box is not explicitly included in the process of Process Flow Diagram (PFD) to simplify the flow rate patterns.

30.4.2 Input and Output of the Model

The model can be operated in two modes (Table 30.1). In the first mode, fresh water demand and top brine temperature are fixed, while the steam flow rate and properties are evaluated. In the second mode, steam flow rate and temperature are assumed to be known from an upstream process (i.e., renewable energy process, co-generation plant), while fresh water production is evaluated. The output of the model is given in the form of intensive parameters. All those parameters are dimensionless, except for the specific area and specific heat consumption (Table 30.1).

$$\text{GOR} = \frac{Md}{Ms} \tag{30.1}$$

$$\text{GOR}_{\text{TVC}} = \frac{Md}{Mm} \tag{30.2}$$

The gained output ratio (GOR) is defined as the quantity of distilled freshwater produced by the process over the quantity of steam utilized as external utility. In the case of MED coupled with TVC, the external steam that must be provided to produce the same amount of fresh water is lower and indicated as motive steam, *Mm*.

$$\text{PR} = \text{GOR}\frac{2330\,\text{kJ/kg}}{\lambda(Ts)} \tag{30.3}$$

$$\text{PR}_{\text{TVC}} = \text{GOR}_{\text{TVC}}\frac{2330\ \text{kJ/kg}}{\lambda(Ts)} \tag{30.4}$$

PR is the amount of fresh water produced by condensing 1 kg of steam at an average latent heat of 2330 kJ/kg. The latent heat at steam temperature is $\lambda(Ts)$. The earlier formula is valid only under model

TABLE 30.1
Input of the Model

Design Parameters	Fixed Water Demand (Mode 1)	Fixed Steam Flow Rate (Mode 2)
Number of effects	Set to 12 or varied (4–20)	Set to 12 or varied (4–20)
External steam flow rate [kg/s]	Evaluated	Set to 100 or varied (1–600)
Freshwater demand [kg/s]	Fixed at 100 or varied (50–1000)	Evaluated
Operative Parameters		
Steam temperature [°C]	Evaluated	Set to 70 or varied (60–100)
Top brine temperature [°C]	Set to 65 or varied (50–80)	Evaluated
Rejected brine temperature [°C]	Set to 40	Set to 40
Intake Seawater temperature [°C]	Set to 25 or varied (15–30)	Set to 25 or varied (15–30)
Feed temperature after end condenser [°C]	Set to 35	Set to 35
Rejected brine salinity [w/w%]	Set to 7%	Set to 7%
Intake seawater salinity [w/w%]	Set to 4.2% or varied (3%–4.5%)	Set to 4.2% or varied (3%–4.5%)
Motive steam pressure [kPa]	Set to 500 or varied (100–3500)	Set to 500 or varied (100–3500)

assumption 10. It is more reasonable to use PR instead of GOR since the latter does not account for the enthalpy drop of the supplied steam; however, both parameters are considered here for completeness.

$$A_{tot_s} = \frac{A_{tot}}{Md} \tag{30.5}$$

$$Mw_s = \frac{Mw}{Md} \tag{30.6}$$

$$E_s = \frac{Ms \cdot \lambda(Ts)}{Md} \tag{30.7}$$

The total area is given by the sum of the areas of n evaporators, n–1 feed pre-heaters and final condenser. Dividing the total area by the quantity of produced fresh water, we obtain a specific parameter that does not depend on the plant capacity (Equation 30.5). The evaluation of this parameter is important to understand the dimension of the plant and eventually to estimate the capital investment required. The seawater intake in the end condenser can be also divided by the quantity of produced fresh water to obtain a specific parameter that does not depend on the plant capacity (Equation 30.6). The evaluation of this parameter is necessary to design the pumping system coupled with the process. The specific heat consumption is the energy consumed to produce 1 kg of fresh water (Equation 30.7).

Other interesting variables evaluated by the model are the following:

- Temperature trend in the plant: brine temperature, distillate temperature, feed temperature
- Salinity trend in the plant
- Use of thermal load: that is, the fraction of power provided to the first effect is used for evaporation with respect to the fraction used for feed heating
- Recovery ratio: the quantity of distillate obtained from 1 kg of seawater

30.4.3 Model Development

The model is made of a series of material and energetic balances together with thermodynamic correlations, which are provided in the appendix. The model for the MED before the converging procedure is adapted from Darwish et al. [4]. The equations are proposed in the logical order of implementation in an equation-oriented program (i.e., MATLAB®).

$$Md = Mf \cdot \frac{xb - xf}{xb} \tag{30.8}$$

$$Mb = Mf - Md \tag{30.9}$$

Equation 30.8 is a material balance on solute under the hypothesis 2, which implies that all the salt contained in intake seawater is also present in the rejected brine. Equation 30.9 is a global material balance. If the distillate flow rate is not a specification (mode 2), we can evaluate the feed flow rate with an energetic balance on the first effect.

$$Mf = \frac{Ms \cdot \lambda(Ts)}{Q_{\text{sensible}} + Q_{\text{latent}}} \tag{30.10}$$

$$Q_{\text{sensible}} = Mf \int_{t1}^{T1} cp(T1, x1)dT \tag{30.11}$$

$$Q_{\text{latent}} = D1 \cdot \lambda(Tv1) \tag{30.12}$$

Equation 30.11 evaluates the sensible power, which is required to heat the feed from temperature $t1$ to the boiling temperature in the first effect, $T1$. Equation 30.12 evaluates the latent power required for vaporizing a quantity of distillate D1. We can define linear temperature profiles as a *first attempt* by imposing an equal temperature drop among the effects and an equal temperature increase among the feed pre-heaters.

$$\Delta T = \frac{T1 - Tb}{n - 1} \text{ or } \Delta \text{T} = \frac{Ts - Tb}{n} \tag{30.13}$$

$$\Delta t = \Delta T \tag{30.14}$$

$$t1 = tn + (n-1)\Delta t \tag{30.15}$$

$$Tv = T - \text{BPE}(T, x) \tag{30.16}$$

Equation 30.15 evaluates the feed temperature in the first effect, after n–1 pre-heaters, starting from the temperature tn at the exit of the final condenser. Equation 30.16 describes the temperature of the vapor phase, which is lower than the brine temperature by the boiling point elevation.

A small fraction of brine rejected by each effect is flashed to a pre-heater for heating the feed stream. We can define α as the fraction of brine rejected by effect i–1 that is flashed in the associated pre-heater.

$$D_{\text{flash},i} = \alpha B_{i-1} \tag{30.17}$$

$$\alpha = \frac{cp(T_{\text{mean}}, x_{\text{mean}}) \cdot \Delta T}{\lambda(T_{\text{mean}})} \tag{30.18}$$

$$T_{\text{mean}} = \frac{T1 + Tb}{2} \tag{30.19}$$

$$x_{\text{mean}} = \frac{xf + xb}{2} \tag{30.20}$$

In Equation 30.18, it is an approximation to use the same value of temperature and salinity to evaluate the flashed distillate fraction among all the effects; however, since specific heat and latent heat are a *weak* function of those variables in their range for the MED process (Appendices Figures 30.2 and 30.3), this approximation is reasonable since it greatly simplifies the model and the convergence procedure. The fraction of total distillate produced by evaporation in each effect will be called β. This value can be evaluated as a function of known parameters (number of stages, initial salinity, final salinity, α) by rearranging the material balances.

$$D1 = D_{\text{flash1}} + D_{\text{boil1}} = \alpha Mf + \beta Md$$

$$B1 = Mf - D1 = (1-\alpha)Mf - \beta Md$$

$$D2 = D_{\text{flash2}} + D_{\text{boil2}} = \alpha B1 + \beta Md$$

$$B2 = B1 - D2 = (1-\alpha)B1 - \beta Md$$

$$B2 = (1-\alpha)[Mf(1-\alpha) - \beta Md] - \beta Md$$

$$B2 = (1-\alpha)^2 Mf - \frac{\beta Md}{\alpha}[1-(1-\alpha)^2]$$

Similarly, in the last effect:

$$Bn = Mb = (1-\alpha)^n Mf - \frac{\beta Md}{\alpha}[1-(1-\alpha)^n] \tag{30.21}$$

By substituting Equations 30.8 and 30.9 in Equation 30.21, we obtain:

$$\frac{xb - xf}{xb} - 1 = \frac{xb - xf}{xb}(1-\alpha)^n - \frac{\beta}{\alpha}[1-(1-\alpha)^n] \tag{30.22}$$

Equation 30.22 can be rearranged to explicit β:

$$\beta = \frac{\alpha[xb(1-\alpha)^n - xf]}{(xb - xf)[1-(1-\alpha)^n]} \tag{30.23}$$

The amount of distillate boiled in each effect, the total distillate, and the brine flow rates can now be evaluated.

$$D_{\text{boiled},i} = \beta Md \tag{30.24}$$

$$D_i = D_{\text{boilec},i} + D_{\text{flash},i} \tag{30.25}$$

$$B_i = B_{i-1} - D_i \tag{30.26}$$

$$x_i = \frac{x_{i-1}B_{i-1}}{B_i} \tag{30.27}$$

The distillate D_i at temperature Tv_i provides the thermal load for evaporation in the next effect. Finally, the salinity profile is calculated with Equation 30.27. The exchange areas of evaporators and pre-heaters can be calculated using simple energetic balances.

$$Q_i = U_{\text{ev},i} A_{\text{ex},i} \Delta T_{\text{ex},i} \tag{30.28}$$

$$Q_i = D_{\text{boiled},i-1}\lambda(Tv_{i-1}) \tag{30.29}$$

$$\Delta T_{\text{ex}} = Tv_{i-1} - T_i = T_{i-1} - \text{BPE}_{i-1} - T_i = \Delta T - \text{BPE}_{i-1} \tag{30.30}$$

In Equation 30.28, the overall heat exchange coefficient is a function of temperature (Appendix Figure 30.5). Since we are exchanging only latent heat between a vapor phase at Tv_{i-1} and a brine phase at T_i, temperatures are not changing during the process and there is no need to use a logarithmic mean temperature difference. On the other hand, in the first effect the thermal load is provided directly by the external steam.

$$Qs = Ms \cdot \lambda(Ts) = A_{ev,1}U_{ex,1}(Ts - T1) \tag{30.31}$$

When operating the model with known distillate production, Equation 30.31 must be used to compute the required steam flow rate. In the feed pre-heaters, heat exchange is between the flashed distillate at temperature Tv_i and liquid feed stream at temperature t_i. Since we are exchanging sensible heat and the feed temperature is increasing, we must evaluate a mean logarithmic temperature difference.

$$Mf \cdot \int_{t_{i+1}}^{t_i} cp(t, xf)dt = U_{ph,i}A_{ph,i}\Delta t_{\log,i} \tag{30.32}$$

$$\Delta t_{\log,i} = \frac{\Delta t}{\log\left(\dfrac{Tv_i - t_{i+1}}{Tv_i - t_i}\right)} \tag{30.33}$$

Since the exchange areas are evaluated using linear temperature profiles imposed by Equations 30.13 and 30.14, it is impossible to guarantee the fulfillment of hypothesis 5, that is, equal area in all the effects. Temperature profiles can be de-linearized according to the following procedure devised by the authors to achieve a fast equalization of exchange areas.

$$A_{ev,mean} = \frac{\sum_{i=1}^{n} A_{ev,i}}{n} \tag{30.34}$$

$$A_{ph,mean} = \frac{\sum_{i=1}^{n-1} A_{ph,i}}{n-1} \tag{30.35}$$

$$A_{ev,mean} - \frac{Q_i}{U_{ev,i}\Delta T_{ex,i}} = 0 \tag{30.36}$$

$$A_{ph,mean} - \frac{Mf \cdot \int_{t_{i+1}}^{t_i} cp(t, xf)dt}{U_{ph,i}\Delta t_{\log,i}} = 0 \tag{30.37}$$

$$\Delta T_i = \Delta T_{ex,i} + \mathrm{BPE}_i \tag{30.38}$$

$$\Delta t_{\log,i} - \frac{\Delta t_i}{\log\left[1 + (\Delta t_i / Tv_i - t_i)\right]} = 0 \tag{30.39}$$

Equations 30.36 and 30.37 must be zeroed by modifying the value of the vectors ΔT_{ex} and $\Delta t_{\log}$. Given the high non-linearity of those objective functions with respect to temperature, a numerical solver is required. Then, Equations 30.38 and 30.39 are solved to evaluate the vectors ΔT_i and Δt_i, which can be used to calculate the new non-linear temperatures profiles.

$$T_i = T_{i-1} - \Delta T_i \tag{30.40}$$

$$Tv_i = T_i - \mathrm{BPE}(T_i, x_i) \tag{30.41}$$

$$t_i = t_{i-1} - \Delta t_i \tag{30.42}$$

TABLE 30.2
Exchange Areas in Evaporators and Pre-Heaters. Subscript *Old* Means before Equalizing Procedure. Parameters for Simulation are Set According to Table 30.1

	$A_{ev,old}$ [m²]	$A_{ph,old}$ [m²]	A_{ev} [m²]	A_{ph} [m²]
Effect 1	4434.66	212.41	4495.78	219.96
Effect 2	4163.87	213.63	4509.15	218.56
Effect 3	4223.29	214.93	4505.87	219.33
Effect 4	4287.43	216.33	4502.36	218.35
Effect 5	4356.94	217.81	4498.60	218.92
Effect 6	4432.62	219.39	4500.64	220.54
Effect 7	4515.42	221.08	4496.31	220.97
Effect 8	4606.54	222.89	4494.57	218.46
Effect 9	4707.45	224.85	4505.05	219.97
Effect 10	4820.01	226.96	4499.44	220.54
Effect 11	4946.62	229.26	4494.17	218.98
Effect 12	5090.44	0[a]	4501.98	0[a]
Error %	**20.37%**	**7.66%**	**0.355%**	**1.39%**

[a] The pre-heater associated with the last effect is the final condenser, modeled later.

All the process variables are then re-evaluated considering the new temperature profiles. The equality of areas is checked according to Equations 30.43 and 30.44.

$$\Delta A_{ev}\% = \frac{\max(A_{ev}) - \min(A_{ev})}{A_{ev,mean}} \cdot 100\% \tag{30.43}$$

$$\Delta A_{ph}\% = \frac{\max(A_{ph}) - \min(A_{ph})}{A_{ph,mean}} \cdot 100\% \tag{30.44}$$

This procedure has proven effective in quickly equalizing the areas. As we can see in Table 30.2, after a single iteration the percentage error drops from 20% to 0.3% for the evaporator areas.

For the pre-heaters, the equalizing procedure is less effective, possibly due to the higher complexity of Equation 30.37, which involves a mean logarithmic temperature difference. However, since the pre-heater areas are in the order of 1/20 of the evaporator areas, some error is acceptable. For certain values of temperature, together with a high number of stages, the numerical solver can fail to solve Equation 30.37 and assign unfeasible values to some elements of vector $\Delta t_{\log}$. The problem can be bypassed by ignoring the equalizing procedure for pre-heaters and by imposing that $\Delta t_i = \Delta T_i$, for every i that causes a problem. After the equalizing procedure, it is possible to proceed with the TVC section modeling. The following model is adapted from Dessouky et al. [1]

$$\text{PCF} = 3e-7 \cdot Pm^2 - 0.0009 \cdot Pm + 1.6101 \tag{30.45}$$

$$\text{TCF} = 2e-8 \cdot Tv_n^2 - 0.0006 \cdot Tv_n + 1.0047 \tag{30.46}$$

Pm is the pressure of the motive steam, Tv_n is the temperature of the vapor phase that is separated from the distillate D_n and sent to the TVC section.

$$Pv = P_{crit} e^{\left(\frac{T_{crit}}{Tv_n} + 273.15\right)^{-1}} \cdot \sum_{j=1}^{8} f_j \tag{30.47}$$

TABLE 30.3
Values of Adaptive Coefficients

	f_1	f_2	f_3	f_4	f_5	f_6	f_7	f_8
Value	−7.4192	0.29721	−0.1155	0.00868	0.00109	−0.0043	0.00252	−0.00052

$$Ps = P_{\text{crit}} e^{\left(\frac{T_{\text{crit}}}{Ts}+273.15\right)-1} \cdot \sum_{j=1}^{8} f_j \tag{30.48}$$

Pv and Ps are the pressures of the saturated steam, respectively, at temperatures Tv_n and Ts. The critical temperature of water is 647.238 K, and the critical pressure is 22089 kPa. Values of adaptive coefficients f_j are reported in Table 30.3. The effective pressure of the steam that reaches the TVC section is lower than Pv, mainly because of pressure drops in the demister of the last effect. Those drops are usually not very relevant, but are evaluated here for completeness.

$$P_{\text{ev}} = Pv - \Delta P_{\text{demister}} \tag{30.49}$$

$$\Delta P_{\text{demister}} = \frac{L_{\text{demister}} 3.88178 \rho_{\text{demister}}^{0.375798} v_{\text{demister}}^{0.81317} D_{\text{wires}}^{-1.56114147}}{1000} \tag{30.50}$$

The demister's thickness is 0.1 m, density is 300 kg/m^3, flow velocity is 1.8 m/s, and wire diameter is 0.28 mm.

$$CR = \frac{P_{\text{ev}}}{Ps} \tag{30.51}$$

$$Ra = 0.296\left(\frac{P_s^{1.19}}{P_{\text{ev}}^{1.04}}\right)\left(\frac{P_m^{0.015}}{P_{\text{ev}}^{0.015}}\right)\left(\frac{\text{PCF}}{\text{TCF}}\right) \tag{30.52}$$

$$Mm = Ms \cdot \frac{Ra}{1+Ra} \tag{30.53}$$

Equation 30.51 defines the compression ratio in the thermos compressor, which cannot be lower than 1.81. Indeed, the steam ejector is critical when the compression ratio is more than or equal to the critical pressure ratio of the suction vapor, which is 1.81 for steam [1]. Equation 30.52 defines the entrainment ratio, which is the mass of motive steam per unit mass of entrained vapor. PCF and TCF are correction factors for steam pressure and temperature. Using Equation 30.53, it is possible to evaluate the mass of motive steam.

The last part of the process to be modeled is the final condenser, which receives a vapor flow rate to be condensed equal to the distillate from the last effect minus the vapor fraction entrained in the TVC section.

$$M_{\text{TVC}} = Ms - Mm \tag{30.54}$$

$$M_{\text{cond}} = D_n - M_{\text{TVC}} \tag{30.55}$$

In the final condenser we heat the seawater flow rate up to a fixed temperature tn, exchanging the latent heat provided by the condensation of M_{cond}. The unit can be modeled like a bigger pre-heater.

$$Q_{\text{cond}} = U_{\text{cond}} A_{\text{cond}} \Delta T_{\text{log,cond}} \tag{30.56}$$

$$Q_{\text{cond}} = M_{\text{cond}}\lambda(Tv_n) \tag{30.57}$$

$$\Delta T_{\text{log,cond}} = \frac{tn - Tw}{\log\left(\dfrac{Tv_n - Tw}{Tv_n - tn}\right)} \tag{30.58}$$

$$Q_{\text{cond}} = M_w \cdot \int_{Tw}^{tn} cp(T, xf)dT \tag{30.59}$$

Tw is the temperature of intake seawater. From Equation 30.56 it is possible to evaluate the area of the final condenser and from Equation 30.59 the total seawater flow rate in the plant.

30.4.4 Model Validation

Model validation has been carried out comparing the present model with other consolidated literature models in terms of performance parameters. Table 30.4 shows a numerical comparison between the present model and the five literature models described in the "state of the art" section.

The present model seems to show results similar to Darwish's model [4] for a low number of stages and low steam temperature, being an enhanced version of the latter. However, if the trend of performance parameters is indicated over a variable number of effects (Figures 30.2 and 30.3) or over variable steam temperatures (Figures 30.4 and 30.5), it is clear that the predictions are closer to those of complex models, that is, the ones that Mistry and Dessouky detailed, especially for high values of n and Ts. This comparison refers to an MED process without TVC, because the literature data about MED-TVC are difficult to find. Consequently, the TVC section has been deactivated when plotting the figures in the next page. However, a comparison between the present model and a simplified forward-feed MED-TCV developed by El-Dessouky [1] is shown in Table 30.5. Some differences are highlighted, especially in the value of the specific area, due to the greater complexity of the present model with respect to the simplified one by El-Dessouky.

30.4.5 Discussion of Model Output and Sensitivity Analysis

In the following paragraph, some important process variables and performance indicators are discussed through graphical representation. A sensitivity analysis is performed to study the functional dependence of plant performance with respect to design and operative parameters, together with the evaluation of energy-water production dependence. All the parameters for the simulations are set or varied according to Table 30.1.

TABLE 30.4
Comparison of PR and Specific Area, Values for Other Models

	Present Model	Dessouky et al. Simple	Dessouky et al. Detailed	Darwish et al.	El-Sayed et al.	Mistry et al.
PR [/]	6.7	7.8	5.8	6.8	6.1	6.2
$A_{\text{tot}_s}\left[\dfrac{m^2 \cdot s}{kg}\right]$	314	390	330	310	380	320

Source: Mistry, K.H. et al., *Desalin. Water Treat.*, 51, 807–821, 2013.
Note: Parameters for Simulation: $n = 8$, $Ts = 70°C$, $Tn = 40°C$, $Tw = 25°C$, $tn = 35°C$, $xf = 42{,}000$ ppm, $xf = 70{,}000$ ppm.

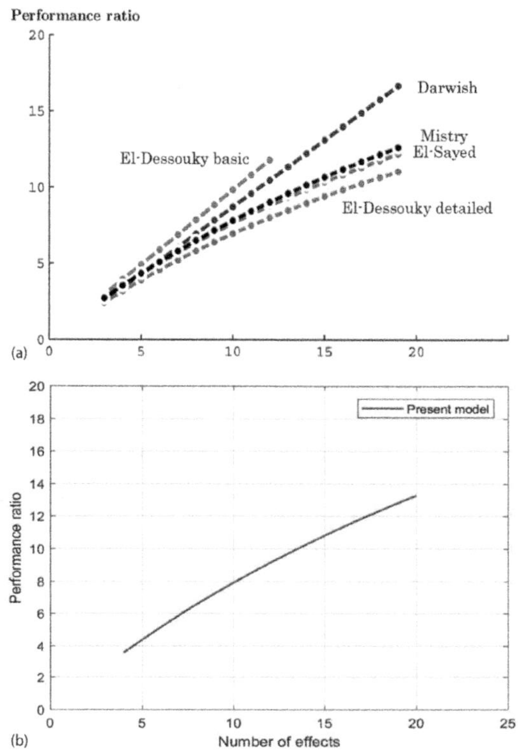

FIGURE 30.2 Comparison of performance ratio with variable n. ((a) Adapted from Mistry, K.H. et al., *Desalin. Water Treat.*, 51, 807–821, 2013. With permission (b) This work.)

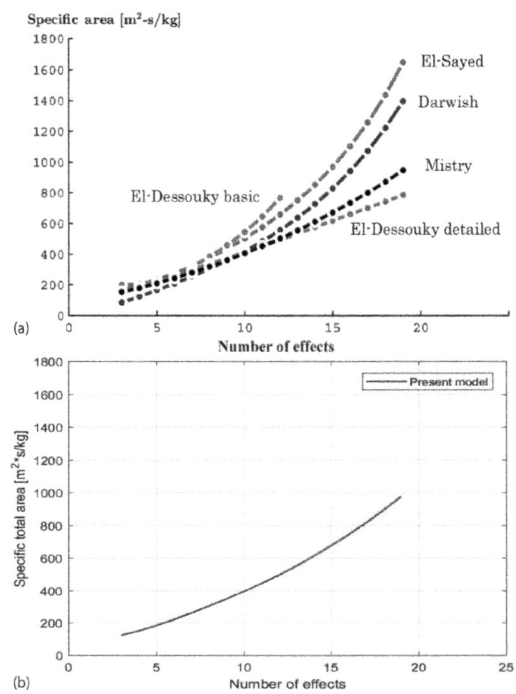

FIGURE 30.3 Comparison of specific total area with variable n. ((a) Adapted from Mistry, K.H. et al., *Desalin. Water Treat.*, 51, 807–821, 2013. With permission (b) This work.)

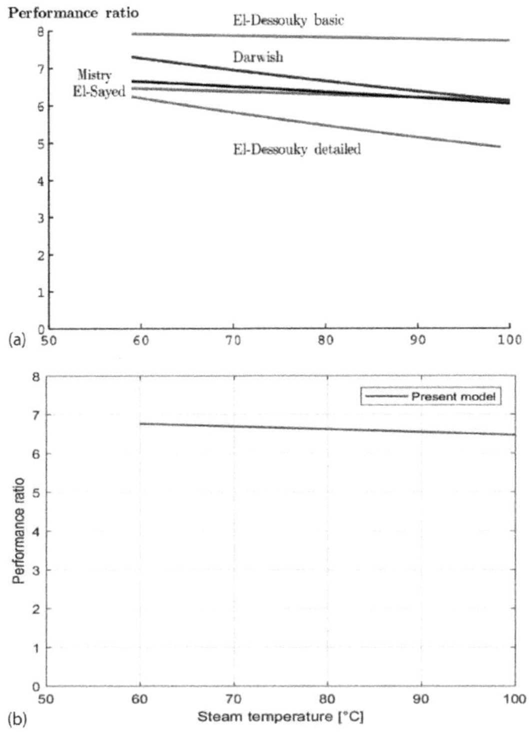

FIGURE 30.4 Comparison of performance ratio with variable *Ts*. ((a) Adapted from Mistry, K.H. et al., *Desalin. Water Treat.*, 51, 807–821, 2013. With permission (b) This work.)

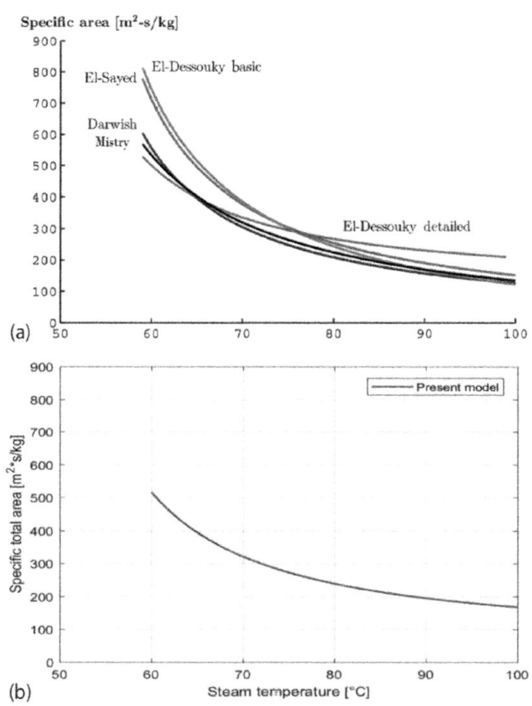

FIGURE 30.5 Comparison of specific total area with variable *Ts*. ((a) Adapted from Mistry, K.H. et al., *Desalin. Water Treat.*, 51, 807–821, 2013. With permission (b) This work.)

TABLE 30.5
MED-TVC, Comparison of Performance Parameters: Present Model vs. El-Dessouky

	PR [/]	$A_{tot_s}\left[\frac{m^2 \cdot s}{kg}\right]$	M_{w_s} [/]	$A_{cond_s}\left[\frac{m^2 \cdot s}{kg}\right]$
Present model	12.88	1545	2.8	13.1
Dessouky et al.	12.5	2119	2.2	15.8

Source: El-Dessouky, H.T. and Ettouney, H.M., *Fundamentals of Salt Water Desalination*, Elsevier, Amsterdam, the Netherlands, 2002.

Note: Parameters: $n = 12$, $Ts = 61°C$, $Tn = 40°C$, $Tw = 25°C$, $tn = 35°C$, $xf = 42000$ ppm, $xf = 70000$ ppm, $Pm = 500$ kPa.

30.4.5.1 Temperatures and Salinity Profiles

Figure 30.6 shows the non-linear profiles of brine temperature, distillate temperature, and feed temperature in the plant. Brine temperature is decreasing along the plant, as well as distillate temperature. The distance between the two trends is the boiling point elevation. Feed enters the train of pre-heaters at 35°C, after absorbing heat in the final condenser, and it is gradually heated until 60°C, which is its temperature when entering the first effect. The sensible heat transferred in the first effect is used to heat the feed from 60°C to the 65°C degrees, that is, the boiling temperature for brine. Figure 30.7 displays the process of brine concentration along the plant. Salinity is slowly but steadily increasing until reaching the fixed value for the discarded brine.

30.4.5.2 Use of External Steam Energy

The heat introduced in the first effect by the external steam is equal to the sum of latent heat used for brine evaporation and sensible heat used for feed heating. Equation 30.61 defines the ratio between latent and sensible heat, indicated as HR.

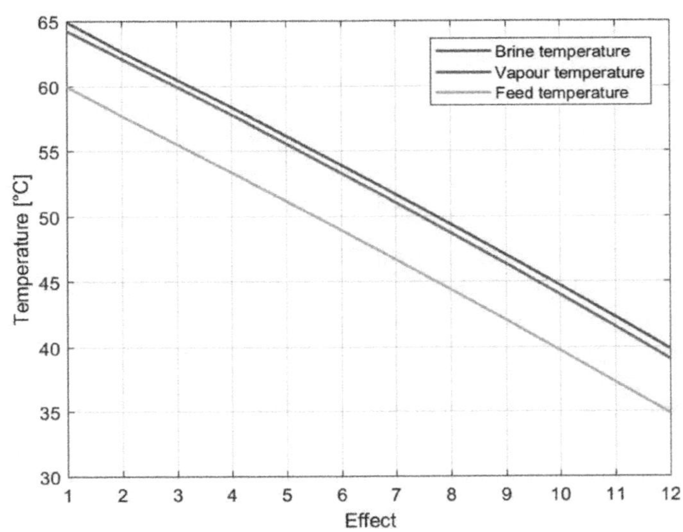

FIGURE 30.6 Temperature trend along the effects. Top = Brine. Bottom = Feed.

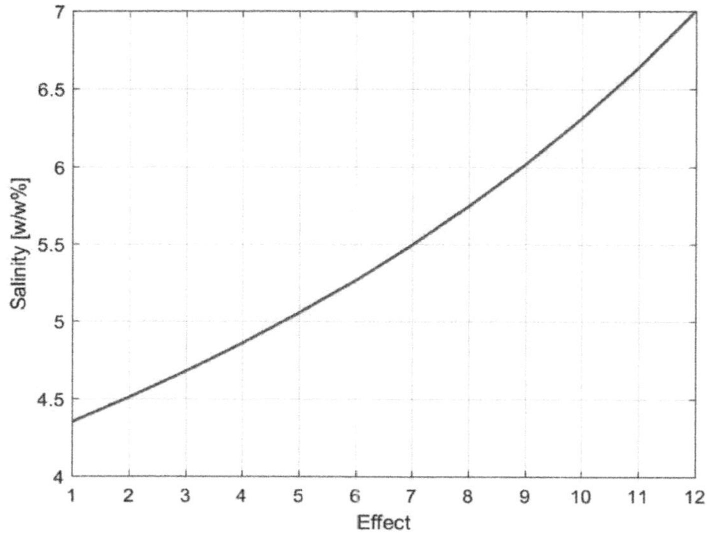

FIGURE 30.7 Salinity trends along the effects.

$$Qs = Ms \cdot \lambda(Ts) = Mf \int_{t1}^{T1} cp(T, xf) + D_1 \cdot \lambda(Tv_1)$$

$$Ms \cdot \lambda(Ts) = Mf \int_{t1}^{T1} cp(T, xf) + (\alpha Mf + \beta Md) \cdot \lambda(Tv_1)$$

$$Ms \cdot \lambda(Ts) = Mf \int_{t1}^{T1} cp(T, xf) + \left(\alpha Mf + \beta Mf \cdot \frac{xb - xf}{xb} \right) \lambda(Tv_1)$$

$$\frac{Ms \cdot \lambda(Ts)}{Mf} = \int_{t1}^{T1} cp(T, xf) + \left(\alpha + \beta \cdot \frac{xb - xf}{xb} \right) \lambda(Tv_1) \left[\frac{\text{kJ}}{\text{kg}} \right] \quad (30.60)$$

$$\text{HR} = \frac{\text{latent heat}}{\text{sensible heat}} = \frac{\left(\alpha + \beta \cdot \frac{xb - xf}{xb} \right) \lambda(Tv_1)}{\int_{t1}^{T1} cp(T, xf)} \quad (30.61)$$

The ratio HR is higher when the number of effects is low because, given a specification for the total fresh water production, more distillate is produced per effect. This means that higher latent heat is required. Similarly, lower feed salinity means more production of distillate per effect, and thus higher latent heat is required (Figure 30.8).

30.4.5.3 Influence of Motive Steam Pressure on Performance Parameters

In the MED-TVC process, the pressure of motive steam is an important operative parameter since it determines the effectiveness of the vapor re-compression. Figure 30.9 shows that there is an optimal value of motive pressure steam, around 1500 kPa, for which the quantity of motive steam required is minimized, and consequently the performance parameters are maximized. However, since the pressure of saturated motive steam is proportional to its temperature (Appendix Figure 30.5), it may not

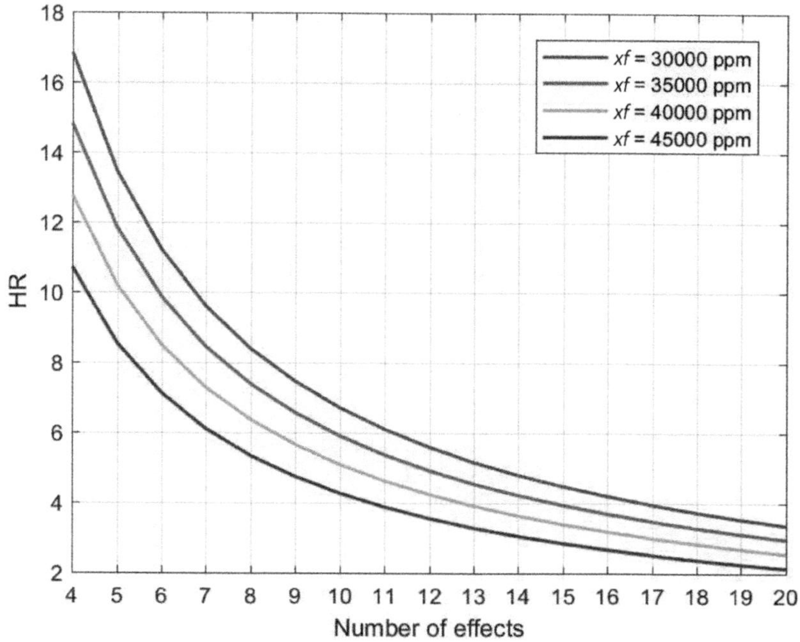

FIGURE 30.8 Latent heat over sensible heat in the first effect as a function of number of effects and feed salinity. Dependence on steam temperature is very weak, so it is not reported.

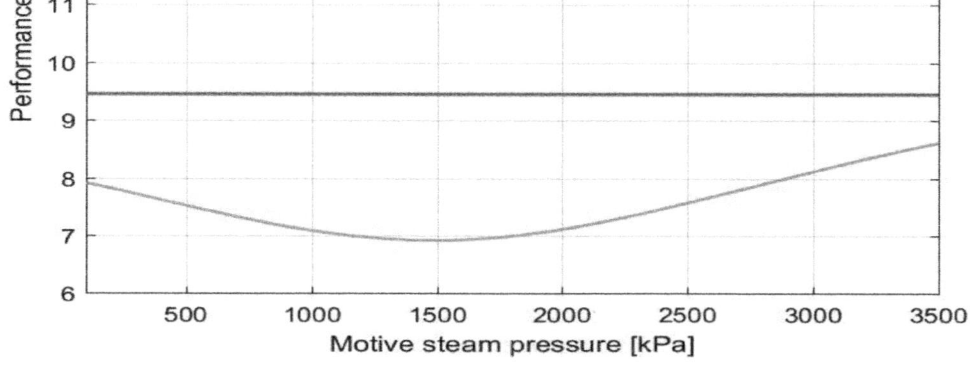

FIGURE 30.9 Performance Ratio as a function of pressure of the motive steam. Quantity of motive steam is minimized when $Pm = 1500$ kPa, consequently PR is maximized. Performance Ratio when the TCV section is deactivated is also plotted, it is significantly lower and obviously not dependent on Pm.

be possible in industrial reality to operate with such high-pressure steam, as it implies a very costly utility. Further research should be done to define the optimal value of pressure for the motive steam according also to energetic and economic considerations. For the next simulations, a more realistic value of 500 kPa is assumed for the motive steam pressure.

30.4.5.4 Influence of Number of Effects and Steam Temperature on Performance Parameters

The variation of PR and GOR with respect to number of effects and steam temperature when the TVC section is enabled are reported in Figures 30.10 and 30.11. Performance parameters

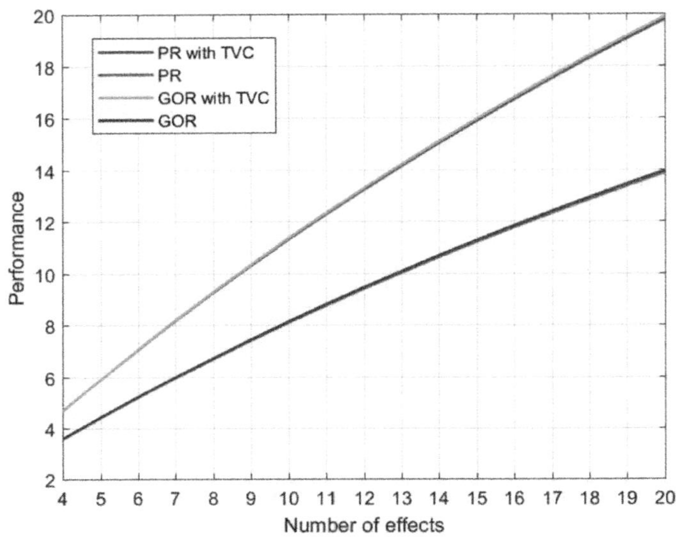

FIGURE 30.10 PR and GOR as a function of number of effects. The difference between MED and MED-TVC is highlighted.

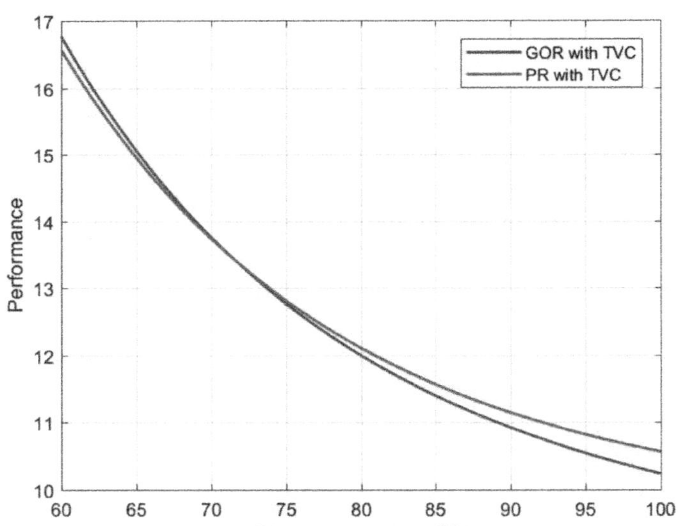

FIGURE 30.11 PR and GOR as a function of steam temperature. The difference between the two indicators is not negligible elevated temperatures.

have a strong dependence on the number of effects, since a high number of effects allows for increase in the amount of steam reuse and formation of additional distillate. It is interesting to notice that the introduction of the TVC section can vastly improve the performance parameters, obviously with the drawback of a more expensive plant, both to build and to operate. The dependence on steam temperature is also linked with the activation of the TVC section, because it determines the pressure of steam, Ps, and consequently the compression ratio. On the other hand, if the TCV is disabled, performance parameters weakly depended on steam temperature (Figure 30.4).

30.4.5.5 Influence of Number of Effects, Temperature and Seawater Properties on Specific Area

The evaluation of total specific area is important to assess the dimension of the plant and consequently the construction costs. Figure 30.12 shows how the specific area is strongly affected both by the number of effects (bigger plant) and by steam temperature. The reduction of specific area for high top temperature is explained by the fact that, when a bigger temperature window is available to operate the process, the temperature difference between the effects is higher, favoring heat exchange. On the other hand, the performance ratio is maximized when the number of effects is high and the steam temperature is low (Figures 30.10 and 30.11); it is not possible to maximize the performance ratio while minimizing the specific area. Other aspects to be considered in choosing the operative temperature of the plant are fouling and corrosion, both favored at high temperature and high salinity. In the forward-feed configuration those issues are minimized, because the highest temperature is in the first effect, where the brine has the lowest salinity. However, it should be considered that, even for this configuration, an elevated top brine temperature could cause significant fouling, consequently decreasing the heat exchange coefficient. In Figure 30.13, specific area is plotted as a function of seawater properties (temperature and salinity). It seems better to operate the plant with cold and low-salted water, since the final condenser will be smaller; however, those properties have a limited impact.

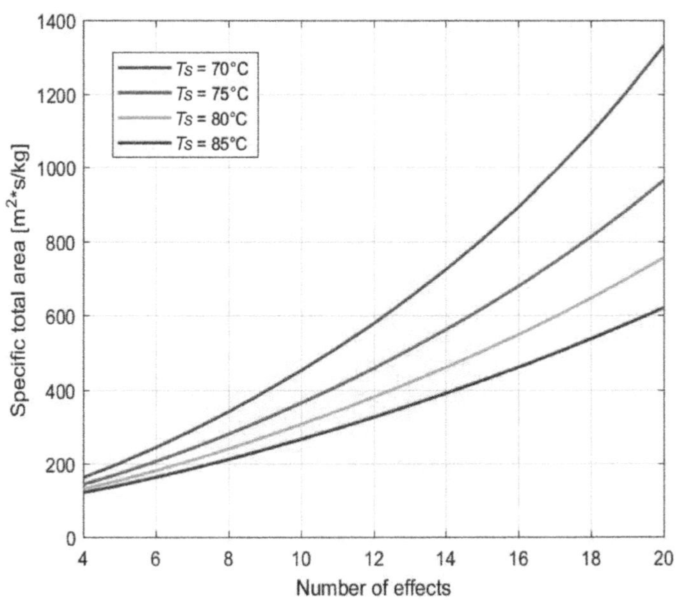

FIGURE 30.12 Specific area as a function of number of effects and steam temperature.

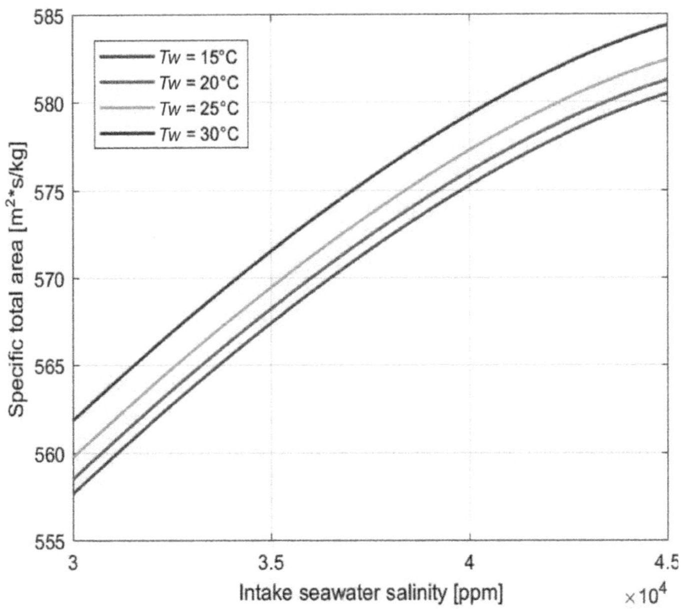

FIGURE 30.13 Specific area as a function of seawater temperature and salinity.

30.4.5.6 Influence of External Steam Flow Rate and Temperature on Fresh Water Production

As the motive steam flow rate increases the quantity of steam provided in the first effect linearly increases at $Ms = Mm + M_{TVC}$. Consequently, more distillate is produced and the capacity of the plant increases. From Figure 30.14 we can see that the increase of plant capacity is linear: if the distillate flow is divided by the motive steam flow, a constant equal to the GOR is obtained. Indeed, performance parameters will not be affected by variation of steam flow rate, because they are specific (intensive) quantities. Operating with lower steam temperatures reduces the quantity of steam required for a fixed plant capacity.

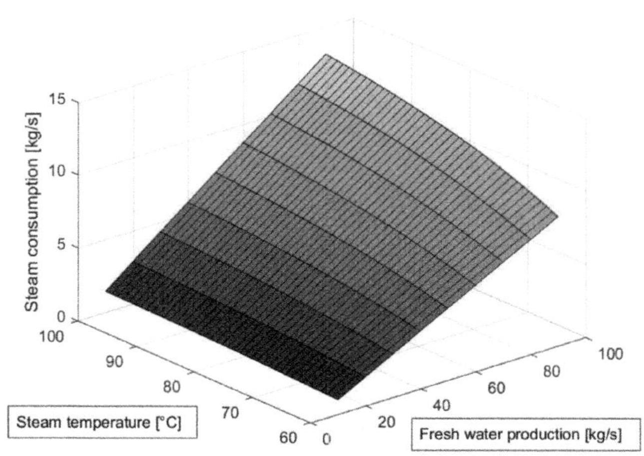

FIGURE 30.14 Plant capacity versus motive steam provided.

30.4.5.7 Influence of Number of Effects and Temperature on Motive Steam Consumption and Specific Heat Consumption

Figure 30.15 shows that, from a steam-economy perspective, it is better to design the plant with a high number of effects, because of the increase in the number of steam reuses. Once again, operating the MED plant at lower top brine temperatures looks favorable. Figure 30.16 shows the trend of specific heat consumption, together with the trend of total specific area, for a fixed plant capacity and variable steam temperature in the lower range, in order to obtain a good performance ratio and reduce fouling problems.

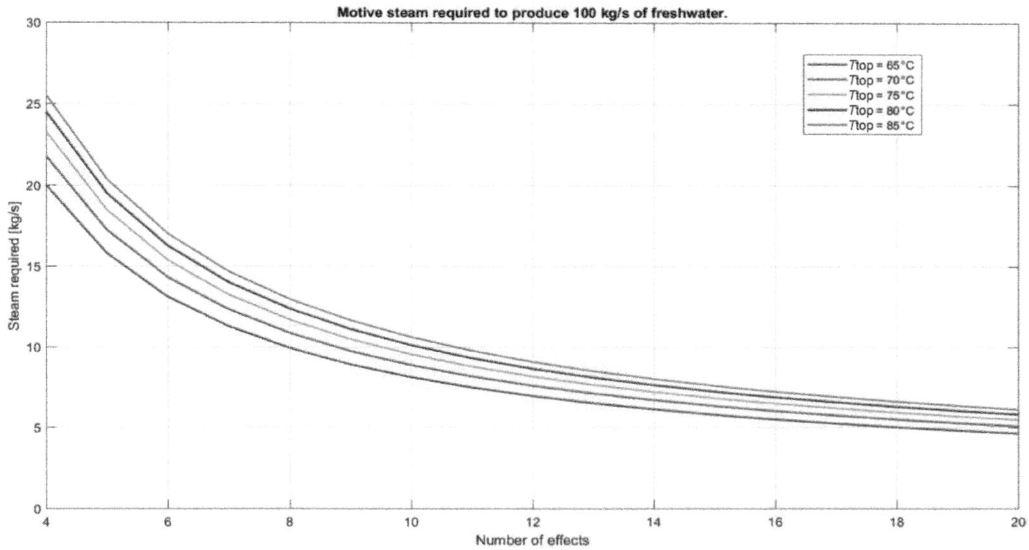

FIGURE 30.15 Steam consumption as a function of number of effects and top brine temperature.

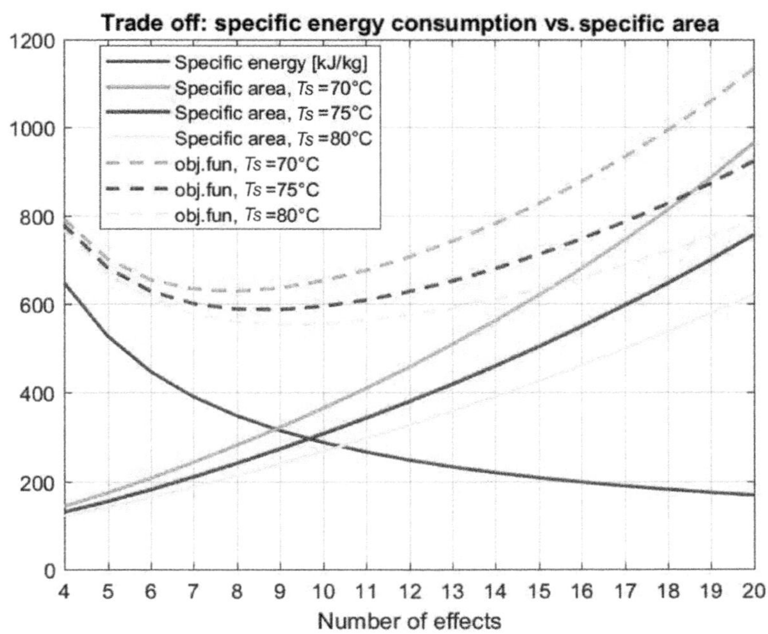

FIGURE 30.16 Specific heat consumption and specific total area as a function of number of effects. It is possible to identify the optimal number of stages that should minimize the cost of fresh water.

Since the specific heat consumption is an indicator of the operative expenses of the process, and the specific area is an indicator of the capital expenses, this figure helps us in understanding the trade-off between those two important indicators when deciding the optimal design of the plant. Summing the two indicators, we obtain the dotted lines, which can be considered as an indicator of total costs, and so they should be treated as objective functions to be minimized to have the most convenient fresh water production.

30.4.5.8 Influence of Number of Effects and Seawater Temperature on Pumping Power

The energy required to operate the seawater pump constitutes a significant portion of the plant's energetic consumption [15]. From Equation 30.59 we derive an expression to evaluate the seawater flow rate to be pumped in the final condenser. Looking at Equation 30.62, is it clear that operating with cooler water reduces the seawater intake by increasing the term at the denominator, with the obvious constrain that $Mw \geq Mf$. Also, operating with a high number of effects reduces the seawater intake since the condenser area and $\Delta T_{\text{log,cond}}$ are reduced. Functional dependence on steam temperature and seawater salinity is very weak and not considered for brevity.

$$M_w = \frac{U_{\text{cond}} A_{\text{cond}} \Delta T_{\text{log,cond}}}{\int_{Tw}^{tn} cp(T, xf)\, dT} \left[\frac{\text{kg}}{s}\right] \tag{30.62}$$

$$\frac{P}{\text{LU}} = \frac{Mw \cdot g}{1000 \cdot \eta} \left[\frac{\text{kg}}{s} \times \frac{m}{s^2}\right] = \left[\frac{kW}{\text{LU}}\right] \tag{30.63}$$

It is possible to evaluate the pumping power with Equation 30.63, where LU is a generic length unit and η is the pump efficiency, assumed to be equal to 0.8. If the pump head is known, with a multiplication we obtain the adsorbed power in kW. From that value, is it possible to evaluate the electrical consumption of the unit and the costs associated, considering the local prices for electricity. Figure 30.17 shows the functional dependence of pumping power on the number of

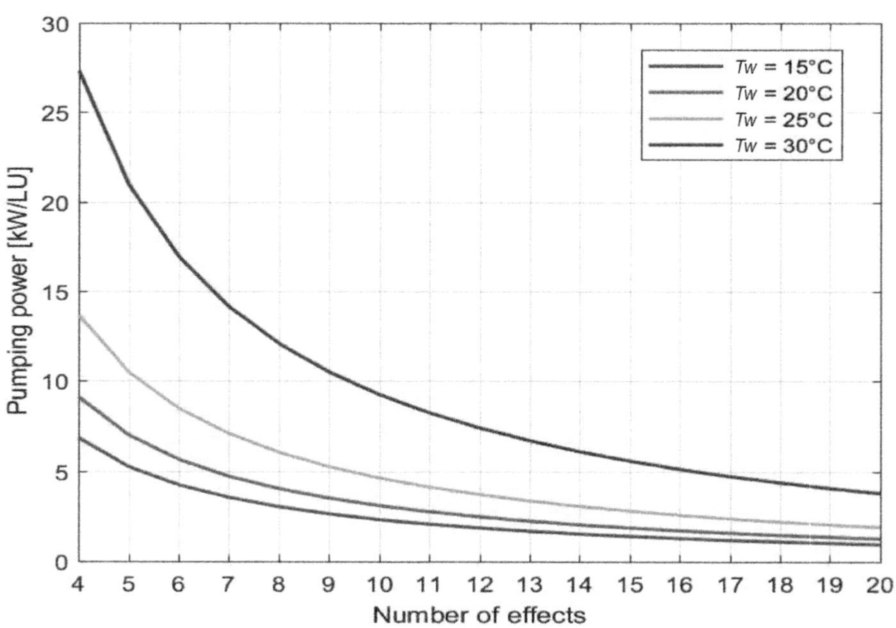

FIGURE 30.17 Functional dependence of pumping power on number of effects and seawater temperature.

effects and seawater temperature. Operating with a high number of effects can reduce the energetic demand of the seawater pumping system, especially when treating warm seawater.

30.5 CONCLUSION

This work analyzed the interaction between energy consumption and the amount of fresh water produced in a low-temperature MED plant. An improved version of an existing model from the literature has been presented and validated. The functional dependence of many process variables and performance parameters is investigated, confirming the literature findings. The authors described the effects of steam properties variation (flow rate and temperature) on plant capacity and, on the other hand, the effects of different design and operating conditions on energy consumption. The results highlight the greater energetic efficiency of a MED plant with a relatively high number of effects and operating at low top brine temperature. The trade-off between energy consumption and specific total area was then investigated, as an attempt to understand the optimal plant design not only from an energetic point of view, but also from an economic perspective. As a future development of this work, it would be appropriate to couple the energy assessment with a detailed economic evaluation of the process in order to assess the cost of fresh water in $/cubic meter, with the aim of determining the competitiveness of the process, also with respect to other desalination technologies.

APPENDIX

Boiling Point Elevation

Correlation valid in the range: $1\% < w < 16\%$, $10°C < T < 180°C$

$$w = x \cdot 10^{-5}\ [w/w\%]$$

$$\text{BPE}a = 8.325 \cdot 10^{-2} + 1.883 \cdot 10^{-4} \cdot T + 4.02 \cdot 10^{-6} \cdot T^2$$

$$\text{BPE}b = -7.625 \cdot 10^{-4} + 9.02 \cdot 10^{-5} \cdot T - 5.2 \cdot 10^{-7} \cdot T^2$$

$$\text{BPE}c = 1.522 \cdot 10^{-4} - 3 \cdot 10^{-6} \cdot T - 3 \cdot 10^{-8} \cdot T^2$$

$$\text{BPE} = \text{BPE}a \cdot w + \text{BPE}b \cdot w^2 + \text{BPE}c \cdot w^3\ [°C]$$

Specific Heat at Constant Pressure

Correlation valid in the range: 20,000 ppm $< x <$ 160,000 ppm, $20°C < T < 180°C$

$$s = x \cdot 10^{-3}\ [\text{gm/kg}]$$

$$\text{cp}a = 4206.8 - 6.6197 \cdot s + 1.2288 \cdot 10^{-2} \cdot s^2$$

$$\text{cp}b = -1.1262 + 5.4178 \cdot 10^{-2} \cdot s - 2.2719 \cdot 10^{-4} \cdot s^2$$

$$\text{cp}c = 1.2026 \cdot 10^{-2} - 5.3566 \cdot 10^{-4} \cdot s + 1.8906 \cdot 10^{-6} \cdot s^2$$

$$\text{cp}d = 6.8777 \cdot 10^{-7} + 1.517 \cdot 10^{-6} \cdot s - 4.4268 \cdot 10^{-9} \cdot s^2$$

$$\text{cp} = \frac{\text{cp}a + \text{cp}b \cdot T + \text{cp}c \cdot T^2 + \text{cp}d \cdot T^3}{1000} \quad \left[\frac{\text{kJ}}{\text{kg} \cdot °\text{C}}\right]$$

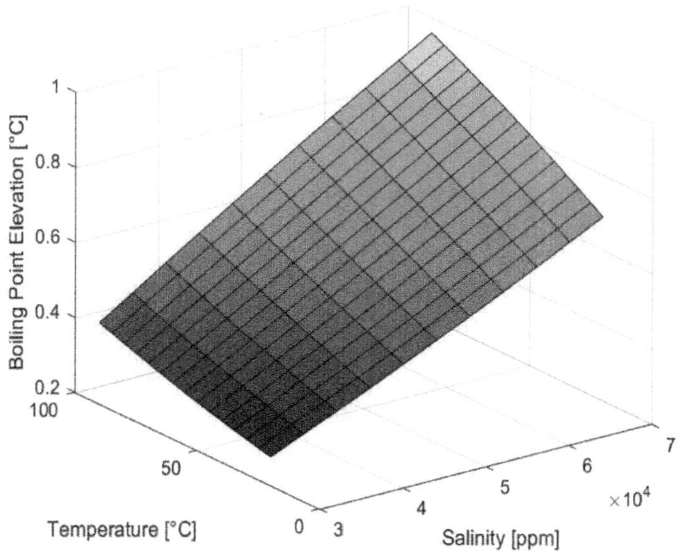

APPENDIX FIGURE 30.1 Variation of BPE over the temperature and salinity range interesting for our process.

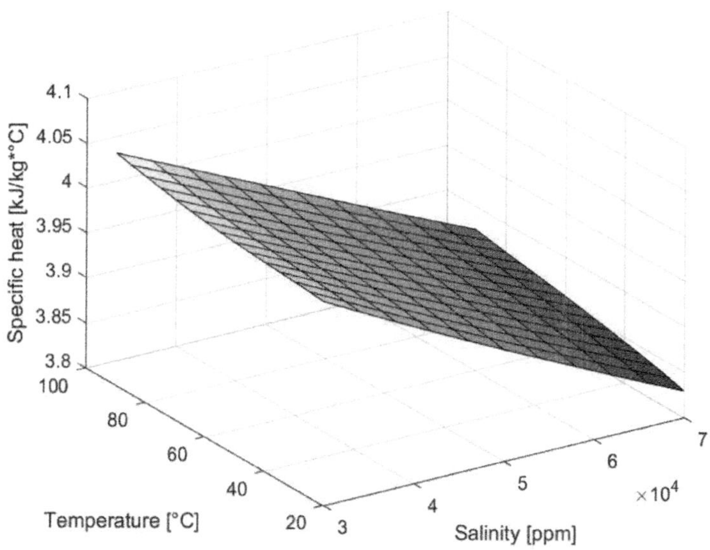

APPENDIX FIGURE 30.2 Variation of specific heat at constant pressure over the temperature and salinity range interesting for our process.

Latent Heat of Evaporation

$$\lambda = 2501.89715 - 2.40706 \cdot T + 1.19221 \cdot 10^{-3} \cdot T^2 - 1.5863 \cdot 10^{-5} \cdot T^3 \left[\frac{\text{kJ}}{\text{kg}} \right]$$

Global Heat Exchange Coefficients

$$U_{\text{ev}} = 1.9695 + 1.2057 \cdot 10^{-2} \cdot T - 8.5989 \cdot 10^{-5} \cdot T^2 + 2.5651 \cdot 10^{-7} \cdot T^3 \left[\frac{\text{kW}}{\text{m}^2 \cdot °\text{C}} \right]$$

$$U_{\text{cond}} = U_{\text{ph}} = 1.7194 + 3.2063 \cdot 10^{-3} \cdot T + 1.597 \cdot 10^{-5} \cdot T^2 - 1.9918 \cdot 10^{-7} \cdot T^3 \left[\frac{\text{kW}}{\text{m}^2 \cdot °\text{C}} \right]$$

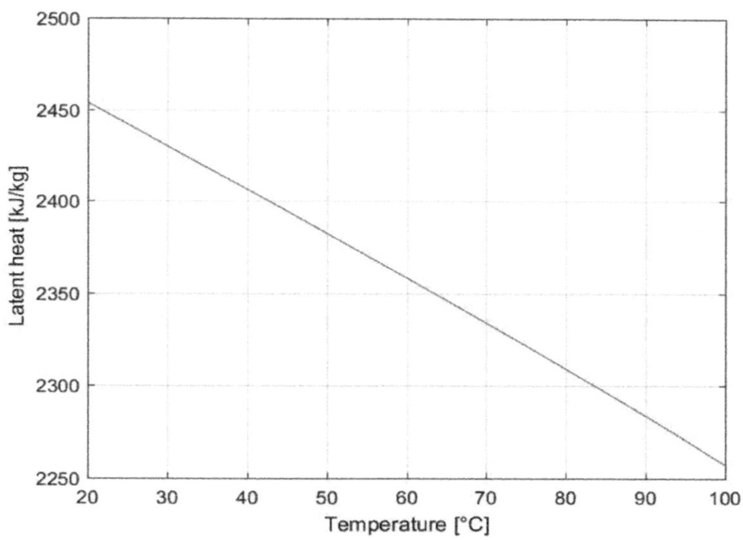

APPENDIX FIGURE 30.3 Variation of latent heat over the temperature range interesting for our process.

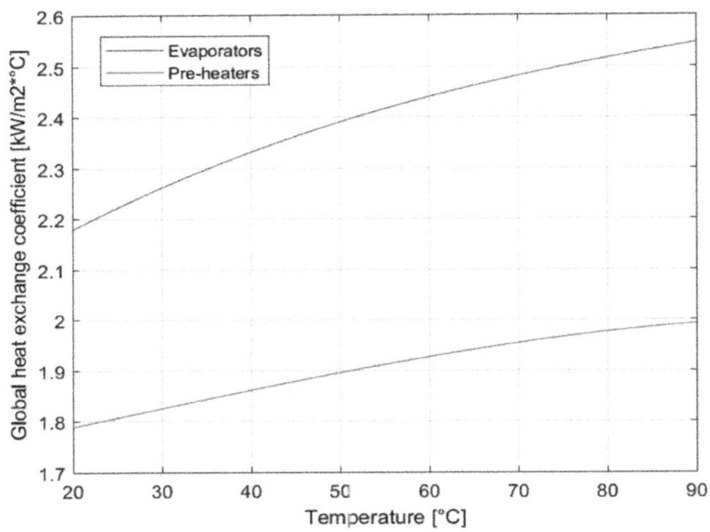

APPENDIX FIGURE 30.4 Variation of global heat exchange coefficients over the temperature range interesting for our process.

Motive Steam Temperature

Under the hypothesis of saturated motive steam, temperature can be evaluated as a function of pressure with the following correlation, valid in the range: 100 kPa < Pm < 3500 kPa.

$$Tm = 42.6676 - \frac{3892.7}{\log(Pm/1000) - 9.48654} - 273.15\,[°C]$$

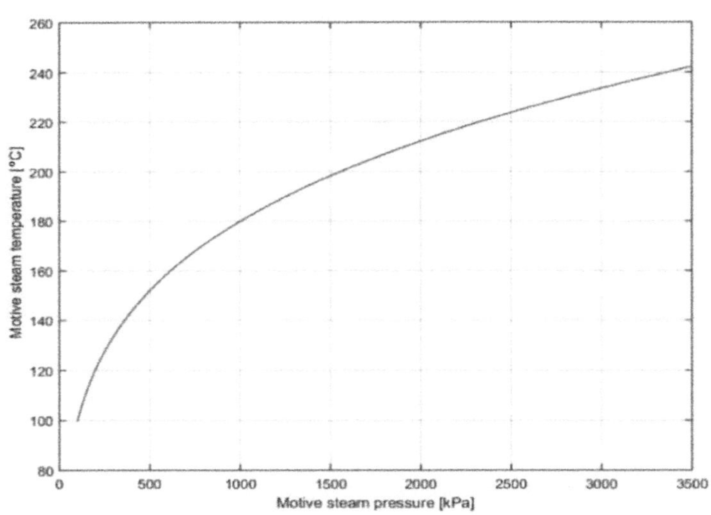

APPENDIX FIGURE 30.5 Saturated steam temperature as a function of pressure.

NOMENCLATURE

α	Fraction of rejected brine from previous effect flashed in the associated pre-heater [/]
β	Fraction of total distillate boiled in each evaporator [/]
λ	Latent heat [kJ/kg]
η	Seawater pump efficinecy [/]
$A_{ev,i}$	Exchange area of i-th evaporator [m^2]
$A_{ph,i}$	Exchange area of i-th pre-heater [m^2]
A_{cond}	Exchange area of final condenser [m^2]
$A_{ev,mean}$	Mean exchange area of evaporators [m^2]
$A_{ph,mean}$	Mean exchange area of pre-heaters [m^2]
ΔA_{ev} %	Percentage error on evaporators areas [%]
ΔA_{ph} %	Percentage error on pre-heaters areas [%]
A_{tot_s}	Specific total area [m^{2}s/kg]
A_{tot}	Total area [m^2]
B_i	Brine rejected by i-th effect [kg/s]
BPE	Boiling Point Elevation [°C]
CR	Compression Ratio in TVC section [/]
D_i	Total distillate produced in i-th effect [kg/s]
$D_{boil,i}$	Distillate produced by boiling in i-th evaporator [kg/s]
$D_{flash,i}$	Distillate produced by flashing in i-th flashing box [kg/s]
E_s	Specific heat consumption [kJ/kg]
f_i	Coefficient in equations (47), (48) [/]
g	Gravitational acceleration = 9.81 [m/s^2]
GOR	Gained Output Ratio [/]
Mb	Rejected brine flowrate [kg/s]
M_{cond}	Vapor flowrate entering the final condenser [kg/s]
Md	Total distillate flowrate (fresh water production) [kg/s]
Ms	Total steam flowrate [kg/s]
Mm	Motive steam flowrate [kg/s]
Mw	Intake seatware flowrate [kg/s]
Mw_s	Specific intake seawater flowrate [/]
M_{TVC}	Vapor flowrate entrained in TVC section [kg/s]
n	Number of effects [/]
PFC	Pressure Correction Factor [/]
Pv	Pressure of saturated steam at temperature Tv [kPa]
Ps	Pressure of saturated steam at temperature Ts [kPa]
Pm	Pressure of saturated steam at temperature Tm [kPa]
Pev	Pressure of saturated entrained steam [kPa]
P_{crit}	Critical pressure of water [kPa]
PR	Performance Ratio [/]
P/LU	Power consumed by seawater pump [kW/LU]
$Q_{sensible}$	Sensible heat consumed in first effect [kJ/kg]
Q_{latent}	Latent heat consumed in first effect [kJ/kg]
Q_i	Thermal load at i-th evaporator [kW]
Q_{cond}	Thermal load in the final condenser [kW]
Ra	Entrainment ratio [/]
$\Delta T_{ex,i}$	Driving force for heat exchange in i-th evaporator [°C]
$\Delta t_{log,i}$	Driving force for heat exchange in i-th pre-heater [°C]
$\Delta T_{log,cond}$	Driving force for heat exchange in final condenser [°C]
ΔT_i	Temperature drop between two effects [°C]

Δt_i	Temperature increase between two pre-heaters [°C]
t_i	Feed temperature after i-th pre-heater [°C]
tn	Feed temperature after final condenser [°C]
T_i	Brine temperature in i-th evaporator [°C]
$T1$	Top brine temperature [°C]
Tb	Temperature of rejected brine [°C]
Ts	Steam temperature when entering the first effect [°C]
Tv_i	Temperature of the vapor phase in i-th effect [°C]
Tw	Seawater temperature [°C]
T_{mean}	Mean temperature in the plant [°C]
T_{crit}	Critical temperature of water [°C]
$U_{\text{ev},i}$	Global heat exchange coefficient in i-th evaporator [kW/m^2 °C]
$U_{\text{ph},i}$	Global heat exchange coefficient in i-th pre-heater [kW/m^2 °C]
Uc	Global heat exchange coefficient in final condenser [kW/m^2 °C]
x_i	Salinity in i-th evaporator [ppm or w/w%]
xb	Rejected brine salinity [ppm or w/w%]
xf	Feed salinity [ppm or w/w%]
x_{mean}	Mean salinity in the plant [ppm or w/w%]

REFERENCES

1. El-Dessouky HT, Ettouney HM. *Fundamentals of Salt Water Desalination.* Amsterdam, the Netherlands: Elsevier, 2002.
2. Al-Shammiri M, Safar M. Multi-effect distillation plants: State of the art. *Desalination* 1999;126(1–3):45–59.
3. El-Sayed YM, Silver RS. Fundamentals of distillation, in: KS Spiegler, ADK Laird (Eds.), *Principles of Desalination*, 2nd ed., Part A. New York: Academic Press, 1980, pp. 55–109.
4. Darwish M, Al-Juwayhel F, Abdulraheim HK. Multi-effect boiling systems from an energy viewpoint. *Desalination* 2006;194(1–3):22–39.
5. El-Dessouky H, Alatiqi I, Bingulac S, Ettouney H. Steady-state analysis of the multiple effect evaporation desalination process. *Chemical Engineering Technology* 1998;21(5):437.
6. El-Dessouky HT, Ettouney H. Multiple-effect evaporation desalination systems. Thermal analysis. *Desalination* 1999;125(1–3):259–276.
7. El-Dessouky HT, Ettouney HM, Mandani F. Performance of parallel feed multiple effect evaporation system for seawater desalination. *Applied Thermal Engineering* 2000;20(17):1679–1706.
8. Mistry KH, Antar MA, Lienhard VJH. An improved model for multiple effect distillation. *Desalination and Water Treatment* 2013;51(4–6):807–821.
9. El-Allawy M. Predictive simulation of the performance of MED/TVC desalination distiller. *IDA Conference, International Desalination Association, Bahams* 2003.
10. Aly NH, El-Figi AK. Thermal performance of seawater desalination systems. *Desalination* 2003;158(1–3):127–142.
11. Al-Sahali M, Ettouney H. Developments in thermal desalination processes: Design, energy, and costing aspects. *Desalination* 2007;214(1–3):227–240.
12. Ameri M, Mohammadi SS, Hosseini M, Seifi M. Effect of design parameters on multi-effect desalination system specifications. *Desalination* 2009;245(1–3):266–283.
13. Darwish M, Alsairafi A. Technical comparison between TVC/MEB and MSF. *Desalination* 2004;170(3):223–239.
14. Choi H, Lee T, Kim Y, Song S. Performance improvement of multiple-effect distiller with thermal vapor compression system by exergy analysis. *Desalination* 2005;182(1–3):239–249.
15. Darwish MA, Al-Najem N. Energy consumptions and costs of different desalting systems. *Desalination* 1987;64:83–96.
16. Hamed O. Thermal assessment of a multiple effect boiling (MEB) desalination system. *Desalination* 1992;86(3):325–339.

Index

Note: Page numbers in italic and bold refer to figures and tables respectively.